Interdisciplinary Applied Mathematics

Volume 41

Problems in engineering, computational science, and the physical and biological sciences are using increasingly sophisticated mathematical techniques. Thus, the bridge between the mathematical sciences and other disciplines is heavily traveled. The correspondingly increased dialog between the disciplines has led to the establishment of the series: Interdisciplinary Applied Mathematics.

The purpose of this series is to meet the current and future needs for the interaction between various science and technology areas on the one hand and mathematics on the other. This is done, firstly, by encouraging the ways that mathematics may be applied in traditional areas, as well as point towards new and innovative areas of applications; and secondly, by encouraging other scientific disciplines to engage in a dialog with mathematicians outlining their problems to both access new methods as well as to suggest innovative developments within mathematics itself.

The series will consist of monographs and high-level texts from researchers working on the interplay between mathematics and other fields of science and technology.

More information about this series at https://link.springer.com/bookseries/1390

Paul C. Bressloff

Stochastic Processes in Cell Biology

Volume I

Second Edition

 Springer

Paul C. Bressloff
Department of Mathematics
University of Utah
Salt Lake City, UT, USA

ISSN 0939-6047   ISSN 2196-9973   (electronic)
Interdisciplinary Applied Mathematics
ISBN 978-3-030-72517-4   ISBN 978-3-030-72515-0   (eBook)
https://doi.org/10.1007/978-3-030-72515-0

Mathematics Subject Classification: 35K57, 35Q84, 35Q92, 35B25, 60G07, 60J10, 60J60, 60J65, 60K20, 60K37, 82B05, 82B31, 82C03, 82C31, 82C26, 82D60, 92C05, 92C10, 92C37, 92C40, 92C45, 92C15

This Springer imprint is published by the registered company Springer Nature Switzerland AG
The registered company address is: Gewerbestrasse 11, 6330 Cham, Switzerland

To Alessandra and Luca

Preface to 2nd edition

This is an extensively updated and expanded version of the first edition. I have continued with the joint pedagogical goals of (i) using cell biology as an illustrative framework for developing the theory of stochastic and nonequilibrium processes, and (ii) providing an introduction to theoretical cell biology. However, given the amount of additional material, the book has been divided into two volumes, with

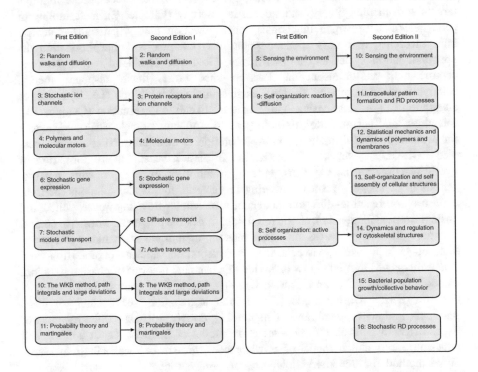

Mapping from the 1st to the 2nd edition

Volume I mainly covering molecular processes and Volume II focusing on cellular processes. The latter also includes significantly expanded material on nonequilibrium systems: intracellular pattern formation and reaction–diffusion processes, statistical physics, and the dynamics/self-organization of cellular structures. The mapping from the first to the second edition is shown in the diagram. In Volume I, the chapter on intracellular transport processes has been split into two chapters, covering diffusive and active processes, respectively. There are four completely new chapters in Volume II: statistical mechanics of polymers and membranes; self-organization and assembly of cellular structures; bacterial population growth and collective behavior; stochastic reaction–diffusion processes. The other three chapters have been significantly expanded.

Major new topics include the following: theory of continuous-time Markov chains (Chapter 3); first passage time problems with (nucleating) sticky boundaries (Chapter 4); genetic oscillators, the repressilator, the degrade-and-fire model, delay differential equations, theory of chemical reaction networks, promoter dynamics, transcriptional bursting and queuing theory, epigenetics, gene expression, and morphogen gradients (Chapter 5); molecular crowding and homogenization theory, percolation theory, narrow capture problems, extreme statistics, diffusion in randomly switching environments, and stochastically gated gap junctions (Chapter 6); reversible vesicular transport in axons, distribution of resources across multiple targets and queuing theory, and stochastic resetting (Chapter 7); metastability in gene networks, Brownian functionals, large deviation theory, generalized central limit theorems, and Levy stable distributions (Chapter 8); phosphorylation–dephosphorylation cycles and ultrasensitivity, Goldbeter–Koshland model, photoreceptors and phototransduction, Poisson shot noise, linear response theory, eukaryotic gradient sensing, the local excitation/global inhibition (LEGI) model of adaptation in gradient sensing, and maximum likelihood estimation (Chapter 10); robustness and accumulation times of protein gradients, non-classical mechanisms for protein gradient formation, pattern formation in mass conserving systems, coupled PDE-ODE systems, cell polarization in fission yeast, pattern formation in hybrid reaction-transport systems, pattern formation on growing domains, synaptogenesis in *C. elegans*, protein clustering in bacteria, multispike solutions far from pattern onset, reaction–diffusion models of intracellular traveling waves, pulled and pushed fronts (Chapter 11); elastic rod model of flexible polymers, worm-like chains, curvature and torsion, stress and strain tensors, membrane fluctuations and curvature, polymer networks, viscoelasticity and reptation, nuclear organization, and Rouse model of DNA dynamics (Chapter 12); classical theories of phase separation, spinodal decomposition and Ostwald ripening, phase separation of biological condensates, Becker–Döring model of molecular aggregation, self-assembly of phospholipids, and active membranes (Chapter 13); doubly stochastic Poisson model of flagellar length control, diffusion–secretion model of filament length control, cell adhesion, motor-clutch model of crawling cells, growth of focal adhesions, variational method for free energy minimization, and cytoneme-based morphogen gradients (Chapter 14); age-structured models of population growth and cell size regulation, bacterial persistence and phenotypic switching, stochastic models of population extinction, bacterial quorum sensing, synchronization of genetic oscillators, and biofilms (Chapter 15); stochastic reaction diffusion processes, stochastic Turing patterns, non-normality and noise-induced pattern amplification, statistical field theory, diagrammatic expansions and the renormalization group, and stochastic traveling waves (Chapter 16).

Meaning no disrespect to vegetarians, I do not explicitly cover plant cells. However, many of the mechanisms and concepts developed in this book would still apply. Chapter 15 on bacterial population growth suggests another natural extension of the current book, namely, stochastic and nonequilibrium processes at the multicellular and tissue levels, including biological neural networks, immunology, collective cell migration, cell development, wound healing, and cancer. This would involve additional topics such as cell-to-cell signaling, the propagation of intercellular signals, nonlocal differential and integral equations, physical properties of the extracellular matrix, and network theory. Clearly ripe themes for a possible third volume!

Acknowledgements

There are many applied mathematicians, physical scientists, and life scientists upon whose sturdy shoulders I have stood during the writing of this book, and whose work is featured extensively in the following pages. I apologize in advance if I have excluded anyone or didn't do proper justice to their contributions. It should also be noted that the relatively large number of self-citations is not a reflection of the significance of my own work in the field, but a consequence of the fact that I am most familiar with my own work! Finally, I would like to thank my wife Alessandra and son Luca (the Shmu) for their continuing love and support.

Salt Lake City, UT, USA   Paul C. Bressloff

Preface to 1st edition

In recent years there has been an explosion of interest in the effects of noise in cell biology. This has partly been driven by rapid advances in experimental techniques, including high-resolution imaging and molecular-level probes. However, it is also driven by fundamental questions raised by the ubiquity of noise. For example, how does noise at the molecular and cellular levels translate into reliable or robust behavior at the macroscopic level? How do microscopic organisms detect weak environmental signals in the presence of noise? Have single-cell and more complex organisms evolved to exploit noise to enhance performance? In light of the above, there is a growing need for mathematical biologists and other applied mathematicians interested in biological problems to have some background in applied probability theory and stochastic processes. Traditional mathematical courses and textbooks in cell biology and cell physiology tend to focus on deterministic models based on differential equations such as the Hodgkin-Huxley and FitzHugh-Nagumo equations, chemical kinetic equations, and reaction-diffusion equations. Although there are a number of well-known textbooks on applied stochastic processes, they are written primarily for physicists and chemists or for population biologists. There are also several excellent books on cell biology written from a biophysics perspective. However, these assume some background in statistical physics and a certain level of physical intuition. Therefore, I felt that it was timely to write a textbook for applied mathematicians interested in learning stochastic processes within the context of cell biology, which could also serve as an introduction to mathematical cell biology for statistical physicists and applied probabilists.

I started my interest in stochastic cell biology, as distinct from my work in mathematical neuroscience, around eight years ago when I volunteered to teach a course in biophysics for the mathematical biology graduate program at Utah. I was immediately fascinated by the molecular processes underlying the operation of a cell, particularly the mechanisms for transporting proteins and other macromolecules to the correct subcellular targets at the correct times. Such an issue is particularly acute for neurons, which are amongst the largest and most complex cells in biology. In healthy cells, the regulation of protein trafficking within a neuron provides an important mechanism for modifying the strength of synaptic connections between neurons, and synaptic plasticity is generally believed to be the cellular substrate of learning and memory. On the other hand, various types of dysfunction in protein trafficking appear to be a major contributory factor to a number of neurodegenerative diseases associated with memory loss including Alzheimer's disease.

In writing this book, I have gone back to my roots in theoretical physics, but refracted through the lens formed by many years working in applied mathematics. Hence, the book provides extensive coverage of analytical methods such as initial boundary value problems for partial differential equations, singular perturbation theory, slow/fast analysis and quasi-steady-state approximations, Green's functions, WKB methods and Hamilton-Jacobi equations, homogenization theory and multi-scale analysis, the method of characteristics and shocks, and reaction-diffusion equations. I have also endeavored to minimize the use of statistical mechanics, which is not usually part of a mathematician's tool-kit and requires a certain level of physical intuition. It is not possible to avoid this topic completely, since many experimental and theoretical papers in cell biology assume some familiarity with terms such as entropy, free energy and chemical potential. The reason is that microscopic systems often operate close to thermodynamic equilibrium or asymptotically approach thermodynamic equilibrium in the long-time limit. This then imposes constraints on any model of the underlying stochastic process. In most cases, one can understand these constraints by considering the Boltzmann-Gibbs distribution of a macromolecule in thermodynamic equilibrium, which is the approach I take in this book.

There are two complementary approaches to modeling biological systems. One involves a high level of biological detail and computational complexity, which means that it is usually less amenable to mathematical analysis than simpler reduced models. The focus tends to be on issues such as parameter searches and data fitting, sensitivity analysis, model reductions, numerical convergence, and computational efficiency. This is exemplified by the rapidly growing field of systems biology. The other approach is based on relatively simple conceptual or "toy" models, which are analytically tractable and, hopefully, capture essential features of the phenomena of interest. In this book I focus on the latter for pedagogical reasons and because of my own personal tastes. In the introductory chapter, I summarize some of the basic concepts in stochastic processes and non-equilibrium systems that are used throughout the book, describe various experimental methods for probing noise at the molecular and cellular levels, give a brief review of basic probability theory and statistical mechanics, and then highlight the structure of the book. In brief, the book is divided into two parts: Part I (Foundations) and Part II (Advanced Topics). Part I provides the basic foundations of both discrete and continuous stochastic processes in cell biology. It's five chapters deal with diffusion, random walks and the Fokker-Planck equation (chapter 2), discrete Markov processes and chemical reaction networks (chapter 3), polymers and molecular motors (chapter 4), gene expression and regulatory networks (chapter 5), and biochemical signaling and adaptation (chapter 6). Part II covers more advanced topics that build upon the ideas and techniques from part I. Topics include transport processes in cells (chapter 7), self-organization in reaction-diffusion models (chapter 8), self-organization of the cytoskeleton (chapter 9), WKB methods for escape problems (chapter 10), and some more advanced topics in probability theory (chapter 11). The chapters are supplemented by additional background material highlighted in gray boxes, and numerous exercises that reinforce the analytical methods and models introduced in the main body of the text. I have attempted to make the book as self-contained as possible. However, some introductory background in partial differential equations, integral transforms, and applied probability theory would be advantageous.

Finally, this book should come with a "government health warning." That is, throughout most of the book, I review the simplest mechanistic models that have been constructed in order to investigate a particular biological phenomenon or illustrate a particular mathematical method. Although I try to make clear the assumptions underlying each model, I do not carry out a comparative study of different models in terms of the degree of quantitative agreement with experimental data. Therefore, the reader should be cautioned that the models are far from the last word on a given phenomenon, and the real biological system is usually way more complicated than stated. However, it is hoped that the range of modeling and analytical techniques presented in this book, when combined with efficient numerical methods, provide the foundations for developing more realistic, quantitative models in stochastic cell biology.

Salt Lake City, UT, USA   Paul C. Bressloff

Organization of volumes I and II

Volume I: Molecular processes

The first volume begins with a short introduction to probability theory and statistical mechanics (Chapter 1). Chapter 2 presents two microscopic theories of diffusion in cells, one based on random walks and the other on overdamped Brownian motion. The latter leads to the theory of continuous Markov processes. Two complementary approaches to formulating continuous Markov process are developed, one in terms of the sample paths generated by a stochastic differential equation (SDE) or Langevin equation, and the other in terms of the Fokker–Planck (FP) equation describing the evolution of the probability density of possible paths. In the former case, a basic introduction to stochastic calculus is given, focusing on the rules for integrating an SDE in order to obtain an expression that can be used to generate moments of the stochastic process. The distinction between Ito and Stratonovich interpretations of multiplicative noise is explained in some detail. It is also shown how, in the case of linear SDEs, Fourier methods can be used to determine the power spectrum, which is important in quantifying the linear response properties of a noisy system. The FP equation, which is a deterministic partial differential equation (PDE) that generalizes the diffusion equation, is then analyzed using standard methods in the theory of linear PDEs: separation of variables, transform methods, Green's functions, and eigenfunction expansions. Many quantities measured by experimentalists can be interpreted mathematically in terms of the solution to a first passage time (FPT) problem. Using the fact that the distribution of first passage times satisfies a backward FP equation, the mean FPT is shown to satisfy a boundary value problem. This is then used to derive the classical Kramer's rate formula for escape across a potential barrier. Noise-induced changes in the effective potential (quasi-potential) in the presence of multiplicative noise are also discussed. Finally, some numerical methods for solving SDEs are reviewed.

Chapter 3 covers some of the main molecular players in cell signaling and transduction, namely, receptors and ion channels. After briefly summarizing the most common types of receptors, some simple kinetic models of cooperative binding are introduced, including the Monod-Wyman-Changeaux model and the Ising model. These provide one mechanism for a cell to amplify signals from the extracellular environment. Following a description of various single ion channel models, the stochastic dynamics of an ensemble of independent ion channels is formulated in terms of a birth–death process. The latter is an example of a discrete Markov process or Markov chain. It is shown how the probability distribution for the number of open ion channels evolves according to a corresponding birth–death master equation. Two models of stochastic ion channels are then explored, a conductance-based model of spontaneous action potential generation in a neuron, which is driven by the random opening and closing of voltage-gated ion channels, and the spontaneous release of calcium puffs and sparks by ligand-gated ion channels. In both cases, the occurrence of spontaneous events can be analyzed in terms of a FPT problem. Finally, the general theory of continuous-time Markov chains is reviewed, including a discussion of the Perron–Frobenius theorem and an introduction to Poisson processes. There are a number of systems considered in subsequent chapters where the signal received by a biochemical sensor involves a sequence of discrete events that can be modeled as a Poisson process. Examples include the arrival of photons at photoreceptors of the retina, and the arrival of action potentials (spikes) at the synapse of a neuron. Another type of event is the random arrival of customers at some service station, resulting in the formation of a queue. However, these processes are typically non-Markovian.

Chapter 4 describes how random walks and SDEs are used to model polymerization and molecular motor dynamics. Polymerization plays a major role in the self-organization of cytoskeletal structures, whereas molecular motors "walking" along polymer filaments constitute a major active component of intracellular transport. The analysis of polymerization focuses on the Dogterom–Leibler model of microtubule catastrophes, which takes the form of a two-state velocity-jump process for the length of a microtubule. The effects of nucleation and constrained growth are taken into account, and FPT problems with "sticky" boundaries are analyzed using the theory of conditional expectations, stopping times, and strong Markov processes. The FP equation for a Brownian particle moving in a periodic ratchet (asymmetric) potential is then analyzed. It is shown that the mean velocity of the Brownian particle is zero, which implies that the periodicity of the potential must be broken for a molecular motor to perform useful work against an applied load. One such mechanism is to rectify the motion, as exemplified by polymerization and translocation ratchets. A qualitative model of processive molecular motors is then introduced, based on a flashing Brownian ratchet. It is shown how useful work can be generated if the motor switches between different conformational states (and corresponding potentials) at rates that do not satisfy detailed balance; this is achieved via the hydrolysis of adenosine triphosphate (ATP). The theory of molecular motors is further developed by considering two examples of the collective motion of an ensemble of molecular motors: (i) the tug-of-war model of bidirectional vesicular transport by opposing groups of processive motors; (ii) a model of interacting motors attached to a rigid cytoskeletal backbone.

Chapter 5 covers the basics of stochastic gene expression and chemical reaction networks. First, various deterministic rate models of gene regulatory networks are described, including autoregulatory networks, the toggle switch, the *lac* operon, the repressilator, NK-βB oscillators, and the circadian clock. Brief reviews of linear stability analysis, Hopf bifurcation theory, and oscillations in delay differential equations are also given. The analysis of molecular noise associated with low copy numbers is then developed, based on the chemical master equation. Since chemical master equations are difficult to analyze directly, a system-size expansion is used to approximate the chemical master equation by an FP equation and its associated chemical Langevin equation. Gillespie's stochastic simulation algorithm for generating exact sample paths of a continuous-time Markov chain is also summarized. Various affects of molecular noise on gene expression are then explored, including translational bursting, noise-induced switching, and noise-induced oscillations. One of the assumptions of many stochastic models of gene networks is that the binding/unbinding of transcription factors at promoter sites is faster than the rates of synthesis and degradation. If this assumption is relaxed, then there exists another source of intrinsic noise known as promoter noise. The latter is modeled in terms of a stochastic hybrid system, also known as a piecewise deterministic Markov process. This involves the coupling between a continuous-time Markov chain and a continuous process that may be deterministic or stochastic. The evolution of the system is now described by a differential Chapman–Kolmogorov (CK) equation, which is a mixture of a master equation and an FP equation. In the limit of fast switching, a quasi-steady-state approximation is used to reduce the CK equation to an effective FP equation. This is analogous to the system-size expansion of chemical master equations. Various examples of networks with promoter noise are presented, including a stochastic version of the toggle switch. It is shown how one of the major effects of promoter noise, namely transcriptional bursting, can be analyzed using queuing theory. Some time-limiting steps in gene regulation are then described, including kinetic proofreading based on enzymatic reactions, and DNA transcription times. The penultimate section consists of a brief introduction to epigenetics. This concerns phenotypic states that are not encoded as genes, but as inherited patterns of gene expression originating from environmental factors, and maintained over multiple cell generations when the original environmental stimuli have been removed. A number of epigenetic mechanisms are discussed, including the infection of *E. coli* by the λ phage DNA virus, and local mechanisms such as DNA methylation and gene silencing by nucleosome modifications. Finally, the role of gene expression in interpreting morphogen gradients during early development is discussed.

Chapters 6 and 7 consider various aspects of intracellular transport, focusing on diffusive and active transport, respectively. Chapter 6 begins by describing the anomalous effects of molecular crowding and trapping, where the differences in diffusive behavior at multiple time scales are highlighted. The classical Smoluchowski theory of diffusion-limited reactions is then developed, with applications to chemoreception and to facilitated diffusion, which occurs when a protein searches for specific DNA binding sites. Extensions of the classical theory to stochastically gated diffusion-limited reactions and ligand rebinding in enzymatic reactions are also considered. Next it is shown how Green's functions and singular perturbation theory can be used to analyze narrow escape and narrow capture problems. The former concerns the escape of a particle from a bounded domain through small openings in the boundary of the domain, whereas the latter refers to a diffusion–trapping problem in which the interior traps are much smaller than the size of the domain. An alternative measure of the time scale for diffusive search processes is then introduced, based on the FPT of the fastest particle to find a target among a large population of independent Brownian particles, which is an example of an extreme statistic. This leads to the so-called "redundancy principle," which provides a possible explanation for the apparent redundancy in the number of molecules involved in various cellular processes, namely, that it accelerates search processes. In certain examples of diffusive search, regions of a boundary may randomly switch between open and closed states, which requires the analysis of PDEs in randomly switching environments. In particular, it is shown how a common switching environment can induce statistical correlations between noninteracting particles. The analysis of randomly switching environments is then extended to the case of molecular diffusion between cells that are coupled by stochastically gated gap junctions. Finally, diffusive transport through narrow membrane pores and channels is analyzed using the Fick–Jacobs equation and models of single-file diffusion. Applications to transport through the nuclear pore complex are considered.

Chapter 7 begins by considering population models of axonal transport in neurons. The stochastic dynamics of a single motor complex is then modeled in terms of a velocity-jump process, which focuses on the transitions between different types of motions (e.g., anterograde vs. retrograde active transport and diffusion vs. active transport) rather than the microscopic details of how a motor performs a single step. Transport on a 1D track and on higher-dimensional cytoskeletal networks are considered, including a model of virus trafficking. Next, the efficiency of transport processes in delivering vesicular cargo to a particular subcellular domain is analyzed in terms of the theory of random search-and-capture processes. The latter describe a particle that randomly switches between a slow search phase (e.g., diffusion) and a faster non-search phase (e.g., ballistic transport). In certain cases it can be shown that there exists an optimal search strategy, in the sense that the mean time to find a target can be minimized by varying the rates of switching between the different phases. The case of multiple search-and-capture events, whereby targets accumulate resources, is then analyzed using queuing theory. Another example of a random search process is then introduced, in which the position of a particle (searcher) is reset randomly in time at a constant rate. One finds that the MFPT to find a target is finite and has an optimal value as a function of the resetting rate. Stochastic resetting also arises in models of cell adhesion and morphogen gradient formation. Finally, it is shown how the effects of molecular crowding of motors on a filament track can be modeled in terms of asymmetric exclusion processes. In the mean-field limit, molecular crowding can be treated in terms of quasi-linear PDEs that support shock waves.

Chapters 8 and 9 cover more advanced topics. Chapter 8 focuses on methods for analyzing noise-induced transitions in multistable systems, such as Wentzel–Kramers–Brillouin (WKB) methods, path integrals, and large deviation theory. First, WKB theory and asymptotic methods are used to the analyze noise-induced escape in an SDE with weak noise. It is shown how the most likely paths of escape can be interpreted in terms of least action paths of a path integral representation of the SDE. An analogous set of analyses are also carried out for birth–death processes and stochastic hybrid systems, which are illustrated using the examples of an autoregulatory gene network and a conductance-based neuron model. The path integral representation of an SDE is then used to derive the Feynman–Kac formula for Brownian functionals. The latter are random variables defined by some integral measure of a Brownian path. Chapter 8 ends with a brief introduction to large deviation theory, as well as a discussion of generalized central limit theorems and Lévy stable distributions. Finally, Chapter 9 briefly reviews the theory of martingales and applications to branching processes and counting processes.

Volume II: Cellular processes

Chapter 10 explores the general problem of detecting weak signals in noisy environments. Illustrative examples include photoreceptors and shot noise, inner hair cells and active mechano-transduction, and cellular chemotaxis. Various mechanisms for signal amplification and adaption are described, such as phosphorylation–dephosphorylation cycles, ultrasensitivity, and receptor clustering. The basic principles of linear response theory are also introduced. The fundamental physical limits of cell signaling are developed in some detail, covering the classical Berg–Purcell analysis of temporal signal integration, and more recent developments based on linear response theory, and maximum likelihood estimation. One of the useful features of the latter approach is that it can be extended to take into account temporal concentration changes, such as those that arise during bacterial chemotaxis. Bacteria are too small to detect differences in concentrations across their cell bodies, so they proceed by measuring and comparing concentrations over time along their swimming trajectories. Some simple. PDE models of bacterial chemotaxis, based on velocity-jump processes, are also considered. In contrast to bacterial cells, eukaryotic cells such as the social amoeba *Dictyostelium discoideum* are sufficiently large so that they can measure the concentration differences across their cell bodies without temporal integration. Various models of spatial gradient sensing in eukaryotes are investigated, including the local excitation, global inhibition (LEGI) model, which takes into account the fact that cells adapt to background concentrations.

Chapter 11 explores intracellular pattern formation based on reaction–diffusion processes. First, various mechanisms for the formation of intracellular protein concentration gradients are considered, and the issue of robustness is discussed. Next, after reviewing the general theory of Turing pattern formation, two particular aspects are highlighted that are specific to intracellular pattern formation: (i) mass conservation and (ii) the dynamical exchange of proteins between the cytoplasm and plasma membrane. Various examples of mass-conserving reaction–diffusion models of cell polarization and division are then described, including Min protein oscillations in *E. coli*, cell polarization in budding and fission yeast, and cell polarization in motile eukaryotic cells. An alternative mechanism for intracellular pattern formation is then introduced, based on a hybrid transport model where one chemical species diffuses and the other undergoes active transport. Evolving the model on a slowly growing domain leads to a spatial pattern that is consistent with the distribution of synaptic puncta during the development of *C. elegans*. Next, asymptotic methods are used to study the existence and stability of multispike solutions far from pattern onset; the latter consist of strongly localized regions of high concentration of a slowly diffusing activator. The theory is also applied to a model of the self-positioning of structural maintenance of chromosome (SMC)-protein complexes in *E. coli*, which are required for correct chromosome condensation, organization, and segregation. Finally, various examples of intracellular traveling waves are analyzed, including polarization fronts in motile eukaryotic cells, mitotic waves, and CaMKII translocation waves in dendrites. An introduction to the theory of bistable and unstable waves is also given.

Chapter 12 presents an introduction to the statistical mechanics and dynamics of polymers, membranes, and polymer networks such as the cytoskeleton. First, the statistical mechanics of single polymers is considered, covering random walk models such as the freely jointed chain, and elastic rod models (worm-like chains). The latter type of model treats a polymer as a continuous curve, whose free energy contributions arise from the stretching, bending, and twisting of the polymer. The continuum mechanics of elastic rods is briefly reviewed in terms of curvature and torsion in the Frenet–Serret frame. A generalized worm-like chain model is used to account for experimentally obtained force–displacement curves for DNA. The statistical mechanics of membranes is then developed along analogous lines to flexible polymers, by treating membranes as thin elastic sheets. In order to construct the bending energy of the membrane, some basic results from membrane elasticity are reviewed, including stress and strain tensors, bending/compression moduli, and the theory of curved surfaces. The corresponding partition function is used to estimate the size of thermally driven membrane fluctuations. Since the membrane is modeled as an infinite-dimensional continuum, the partition function takes the form of a path integral whose associated free energy is a functional. The analysis of statistical properties thus requires the use of functional calculus. The next topic is the statistical dynamics of systems at or close to equilibrium. This is developed by generalizing the theory of Brownian motion to more complex structures with many internal degrees of freedom. Various results and concepts from classical nonequilibrium statistical physics are introduced, including Onsager's reciprocal relations, nonequilibrium forces, time correlations and susceptibilities, and a general version of the fluctuation–dissipation theorem. The theory is illustrated by deriving Langevin equations for fluctuating polymers and membranes. The chapter then turns to polymer network models, which are used extensively by biophysicists to understand the rheological properties of the cytoskeleton. Only the simplest classical models are considered: the rubber elasticity of a cross-linked polymer network, swelling of a polymer gel, and the macroscopic theory of viscoelasticity in uncross-linked polymer fluids. Reptation theory, which is used to model the dynamics of entangled polymers, is also briefly discussed. Finally, the dynamics of DNA within the nucleus is considered. After describing some of the key features of nuclear organization, a classical stochastic model of a Gaussian polymer chain (the Rouse model) is introduced. The latter is used to model the subdiffusive motion of chromosomal loci, and to explore mechanisms for spontaneous DNA loop formation. The mean time to form a loop requires solving an FP equation with nontrivial absorbing boundary condition. The Wilemski–Fixman theory of diffusion-controlled reactions is used to solve the problem by replacing the boundary condition with a sink term in the FP equation.

Chapter 13 considers the self-organization and assembly of a number of distinct cellular structures. First, there is a detailed discussion of the theory of liquid–liquid phase separation and the formation of biological condensates. This introduces various classical concepts in nonequilibrium systems, such as coexistence curves, spinodal decomposition, nucleation and coarsening, Ostwald ripening, and Onsager's principle. Recent developments that are specific to biological condensates are also described, including the effects of nonequilibrium chemical reactions and protein concentration gradients. The chapter then turns to the Becker and Döring model of molecular aggregation and fragmentation, which provides a framework for investigating the processes of nucleation and coarsening. An application of the model to the self-assembly of phospholipids in the plasma membrane is also included. Finally, a model for the cooperative transport of proteins between cellular organelles is introduced, which represents a self-organizing mechanism for organelles to maintain their distinct identities while constantly exchanging material.

Chapter 14 considers various models for the dynamics and regulation of the cytoskeleton. First, several mechanisms for filament length regulation are presented, including molecular motor-based control, protein concentration gradients, and diffusion-based secretion in bacterial flagella. The role of intraflagellar transport (IFT) in the length control of eukaryotic flagella is analyzed in terms of a doubly stochastic Poisson process. The dynamics of the mitotic spindle during various stages of cell mitosis is then described, including the search-and-capture model of microtubule–chromosome interactions and force-balance equations underlying chromosomal oscillations. Next, various models of biophysical mechanisms underlying cell motility are considered. These include the tethered ratchet model of cell protrusion and the motor-clutch mechanism for crawling cells. The latter describes the dynamical interplay between retrograde flow of the actin cytoskeleton and the assembly and disassembly of focal adhesions. The resulting dynamics exhibits a number of behaviors that are characteristic of physical systems involving friction at moving interfaces, including biphasic force–velocity curves and stick-slip motion. A mean-field analysis is used to show how these features can be captured by a relatively simple stochastic model of focal adhesions. In addition, a detailed model of the force-induced growth of focal adhesions is analyzed using a variational method for free energy minimization. Finally, a detailed account of cytoneme-based morphogensis is given. Cytonemes are thin, actin-rich filaments that can dynamically extend up to several hundred microns to form direct cell-to-cell contacts. There is increasing experimental evidence that these direct contacts allow the active transport of morphogen to embryonic cells during development. Two distinct models of active transport are considered. The first involves active motor-driven transport of morphogen along static cytonemes with fixed contacts between a source cell and a target cell. The second is based on nucleating cytonemes from a source cell that dynamically grow and shrink until making temporary contact with a target cell and delivering a burst of morphogen. The delivery of a single burst is modeled in terms of a FPT problem for a search process with stochastic resetting, while the accumulation of morphogen following multiple rounds of cytoneme search-and-capture and degradation is analyzed using queuing theory.

Chapter 15 presents various topics related to bacterial population growth and collective behavior. First, a continuum model of bacterial population growth is developed using an age-structured evolution equation. Such an equation supplements the continuously varying observational time by a second time variable that specifies the age of an individual cell since the last division. Whenever a cell divides, the age of the daughter cells is reset to zero. Although the total number of cells grows exponentially with time, the normalized age distribution approaches a steady state. The latter determines the effective population growth rate via a self-consistency condition. The age-structured model is then extended in order to keep track of both the age and volume distribution of cells. This is used to explore various forms of cell length regulation, including timer, sizer, and adder mechanisms. Further aspects of cell size regulation are analyzed in terms of a discrete-time stochastic map that tracks changes across cell generations. The chapter then turns to another important issue, namely, to what extent single-cell molecular variation plays a role in population-level function. This is explored within the context of phenotypic switching in switching environments, which is thought to be an important factor in the phenomenon of persistent bacterial infections following treatment with antibiotics. At the population level, phenotypic switching is modeled in terms of a stochastic hybrid system. The chapter then turns to a discussion of bacterial quorum sensing (QS). This is a form of collective cell behavior that is triggered by the population density reaching a critical threshold, which requires that individual cells sense their local environment. The next topic is an analysis of synchronization in a population of synthetic gene oscillators that are dynamically coupled to an external medium via a QS mechanism. In particular a continuity equation for the distribution of oscillator phases is constructed in the thermodynamic limit, and various methods of analysis are presented, including the Ott–Antonsen dimensional reduction ansatz. The chapter ends with a review of some mathematical models of bacterial biofilms.

Chapter 16 discusses various analytical methods for studying stochastic reaction–diffusion processes. First, the effects of intrinsic noise on intracellular pattern formation are investigated using the notion of a reaction–diffusion master equation. The latter is obtained by discretizing space and treating spatially discrete diffusion as a hopping reaction. Carrying out a linear noise approximation of the master equation leads to an effective Langevin equation, whose power spectrum provides a means of extending the definition of a Turing instability to stochastic systems, namely, in terms of the existence of a peak in the power spectrum at a nonzero spatial frequency. It is also shown how the interplay between intrinsic noise and transient growth of perturbations can amplify the weakly fluctuating patterns. The source of transient growth is the presence of a non-normal matrix in the linear evolution operator.

Next, using the canonical example of pair annihilation with diffusion, various well-known techniques from statistical field theory are used to capture the dimension-dependent asymptotic decay of the system. These include moment generating functionals, diagrammatic perturbation expansions (Feynman diagrams), and the renormalization group. Finally, a formal perturbation method is used to analyze bistable front solutions of a stochastic reaction–diffusion equation, which exploits a separation of time scales between fast fluctuations of the front profile and a slowly diffusing phase shift in the mean location of the front.

At the end of each chapter there is a set of exercises that further develop the mathematical models and analysis introduced within the body of the text. Additional comments and background material are scattered throughout the text in the form of framed boxes.

Chapters 1–5 can be used to teach a one-semester advanced undergraduate or graduate course on "Stochastic processes in cell biology." Subsequent chapters develop more advanced material on intracellular transport (Chaps. 6 and 7), noise-induced transitions (Chap. 8), chemical sensing (Chap. 10), cellular self-organization and pattern formation (Chaps. 11–14 and 16), and bacterial population dynamics (Chap. 15).

Volume II: Cellular

15. Bacterial population growth and collective behavior
• age-structured models • cell size control
• bacterial persistence/phenotypic switching
• extinction in bacterial populations
• bacterial quorum sensing • biofilms
• synchronization of genetic oscillators

16. Stochastic RD processes
• RD master equation
• stochastic Turing patterns
• path integral of RD master equation
• statistical field theory
• diagramatic expansions
• renormalization theory and scaling
• stochastic traveling waves

14. Regulation of the cytoskeleton
• filament length control
• intraflagellar transport
• doubly stochastic poisson processes
• cell mitosis • cell motility • cell adhesion
• cytoneme-based morphogenesis

12. Statistical mechanics/dynamics of polymers and membranes
• random walk models of polymers
• elastic rod models of polymers
• stress/strain tensors, membrane curvature
• elastic plate model of flluid membranes
• fluctuation-dissipation theorem for Brownian particles, Onsager relations, susceptibility
• polymer networks, viscoelsticity, reptation
• DNA dynamics in the nucleus

13. Self-organization and assembly of cellular structures
• phase separation/biological condensates
• active membranes
• nucleation and growth of molecular clusters
• self-assembly of micelles

11. Intracellular pattern formation and reaction-diffusion processes
• intracellular protein gradients
• Turing pattern formation
• mass-conserving systems
• coupled PDE/ODE systems
• cell polarization and division
• hybrid reaction-transport models
• intracellular traveling waves

10. Sensing the environment
• phosphorylation and ultrasensitivity
• photoreceptors and shot noise
• linear response theory
• hair cell mechanotransduction
• bacterial chemotaxis
• physical limits of chemical sensing
• spatial gradient sensing

Volume I: Molecular

9. Probability theory and martingales
• filtrations, martingales and stopping times
• branching processes
• counting process and biochemical networks

8. WKB method, path integrals,...
• WKB method for noise-induced escape
• path integral for SDEs
• Doi-Peliti path integral of a birth death process
• path integral for hybrid systems
• local and occupation times
• large deviation theory

5. Stochastic gene expression
• genetic switches and oscillators
• molecular noise and master equations
• system-size expansion • translational bursting
• noise-induced switching and oscillations
• promoter noise/stochastic hybrid systems
• transcriptional bursting and queuing theory
• time limiting steps in DNA transcription
• epigenetics
• morphogen gradients/gene expression

7. Active motor transport
• axonal transport
• PDE models of active transport
• transport on microtubular networks
• virus trafficking
• random intermittent search
• stochastic resetting
• exclusion processes

3. Protein receptors and ion channels
• receptor-ligand binding and cooperativity
• stochastic ion channels
• Markov chains and single channel kinetics
• birth-death process for channel ensembles
• voltage-gated ion channels
• calcium sparks in myocytes
• Poisson processes

6. Diffusive transport
• anomalous diffusion
• diffusion-limited reactions
• narrow capture and escape problems
• protein search for DNA target sites
• extreme statistics
• diffusion in switching environments
• gap junctions
• diffusion in channels and pores
• nuclear transport • diffusion on trees

2. Random walks and Brownian motion
• random walks and diffusion
• Wiener process and Ito stochastic calculus
• Langevin and Fokker-Planck equations
• first passage times
• Kramers escape rate
• multiplicative noise

4. Molecular motors
• polymerization
• microtubular catastrophes
• Brownian ratchets
• tethered ratchet and cell motility
• processive molecular motors
• collective motor transport

Contents

Boxes

Chapter 1
Introduction

One of the major challenges in modern biology is to understand how the molecular components of a living cell operate in a highly noisy environment. What are the specific sources of noise in a cell? How do cells attenuate the effects of noise in order to exhibit reliable behavior (robustness to noise)? In particular, how does a stochastic genotype result in a reliable phenotype through development? How does the noisy, crowded environment of a cell affect diffusive transport? How do molecular machines convert chemical energy to work? What are the physical limits of biochemical signaling, such as the sensitivity of biochemical sensors to environmental signals? Under what circumstances can a cell exploit noise to enhance its performance or the survival of its host organism? What is the role of self-organization in the formation and maintenance of subcellular structures such as the cytoskeleton, cell nucleus, and plasma membrane? The goal of this book is to use the theory of stochastic processes and nonequilibrium systems to investigate these types of biological questions at the cellular level. One can view the book either as an introduction to stochastic and nonequilibrium processes using cell biology as the motivating application or, conversely, as an introduction to mathematical cell biology with an emphasis on stochastic and nonequilibrium processes. Irrespective of the particular perspective, it is clear that there is a growing demand for mathematical biologists and other applied mathematicians to have some training in topics that have traditionally been the purview of physicists, chemists, and probabilists. This book provides the necessary background to tackle research problems in mathematical biology that involve stochastic and nonequilibrium processes.

For an excellent general introduction to molecular and cell biology, we refer the reader to the book *Molecular Biology of the Cell* by Alberts et al. [9]. For a more biophysics-oriented approach to cell biology, see *Biological Physics* by Nelson [604], *Physical Biology of the Cell* by Phillips et al. [658], and *Biophysics* by Bialek [94]. An extensive coverage of cell physiology with some examples of stochastic processes can be found in the first volume of *Mathematical Physiology* by Keener and Sneyd. Two standard references on the theory of stochastic processes, with an emphasis on physical and chemical processes, are *Handbook of Stochastic Methods*

© Springer Nature Switzerland AG 2021
P. C. Bressloff, *Stochastic Processes in Cell Biology*, Interdisciplinary Applied Mathematics 41, https://doi.org/10.1007/978-3-030-72515-0_1

by Gardiner [300] and *Stochastic Processes in Physics and Chemistry* by Van Kampen [822]. A more recent treatment can be found in the book *Thinking Probabilistically* by Amir [13]. For a more kinetic-based treatment of nonequilibrium processes, see *A Kinetic View of Statistical Physics* by Krapivsky, Redner, and Ben-Naim [468]. A book on stochastic processes oriented toward population biology is *An Introduction to Stochastic Processes with Applications to Biology* by Allen [10]. For a more mathematical formulation of discrete stochastic processes, see *Probability and Random Processes* by Grimmett and Strirzaker [343], and for a more rigorous treatment of stochastic differential equations, see *Stochastic Differential Equations* by Oksendal [624]. Finally, an analytical treatment of stochastic processes with a detailed description of asymptotic methods and large deviations can be found in the book *Theory and Applications of Stochastic Processes* by Schuss [739].

1.1 Stochastic processes in living cells

In this first section, we introduce some of the basic concepts that are useful in characterizing and analyzing noise in cells, starting at the level of individual macromolecules and building up to cellular structures.

Equilibrium and nonequilibrium systems. One of the fundamental features of a living cell is that it is an open system, that is, it interacts with the surrounding environment through the exchange of energy and matter, see Fig. 1.1. Moreover, a cell is maintained out of thermodynamic equilibrium, which means that there are nonzero fluxes of energy, matter, and charge flowing between the interior and exterior of the cell. If the dynamical variables are discrete (number of molecules in a biochemical reaction, conformational state of an ion channel, or molecular motor), then one needs to use the theory of discrete stochastic processes: Markov chains, birth–death processes, chemical master equations, Poisson processes, etc. On the other hand,

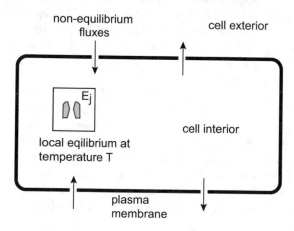

Fig. 1.1: The cell is an open, nonequilibrium system that exchanges energy and matter with the environment. However, a local subsystem such as an individual macromolecule may be in local thermodynamic equilibrium at a temperature T. The probability of being in intrinsic state j with energy E_j is then determined by the Boltzmann–Gibbs distribution.

if the dynamical variables are continuous (position of a molecule or a cell, protein concentration, and membrane voltage of a neuron), then one needs to use the theory of continuous stochastic processes: Brownian motion and diffusion, Langevin equation, Fokker–Planck equation, etc.

It is sometimes possible to approximate a subcellular system such as an individual macromolecule as being in thermodynamic equilibrium, provided that the rate of relaxation to local equilibrium is fast compared to other processes of interest. (Note, however, that a macromolecule such as a motor protein can only operate if it is maintained out of thermodynamic equilibrium.) In such cases, one can exploit the powerful machinery of equilibrium statistical mechanics [163, 429, 658]. In particular, one can make use of the Boltzmann–Gibbs distribution; see Sect. 1.3.

Internal and external states of a macromolecule. Consider a single macromolecule such as a motor protein, an enzyme, an ion channel, or a strand of DNA. Each macromolecule is subject to thermal fluctuations arising from the continual bombardment by molecules in the surrounding environment, which could be the interior aqueous solution of the cell (cytosol) or the surrounding plasma membrane. The size of molecular fluctuations is set by the basic unit of thermal energy $k_B T$, where T is the temperature (in degrees Kelvin K) and $k_B \approx 1.4 \times 10^{-23}$ J K^{-1} is the Boltzmann constant. A useful distinction at the molecular level is between internal conformational states of a macromolecule and external states, such as the position and momentum of the center of mass of the molecule. Often, the internal degrees of freedom are represented as a set of discrete states, and the stochastic dynamics within this state space is described in terms of a continuous-time Markov chain [300] (Chap. 3). That is, the state of the system takes values in some finite or countable set, and the time spent in each state has an exponential distribution. Moreover, the continuous-time stochastic process has the Markov property, which means that the future behavior of the system, both the remaining time in the current state and the identity of the next state, depends only on the current state and not on any prior history. A simple two-state continuous-time Markov process can be used to model the opening and closing of an ion channel, for example, or the binding and unbinding of a ligand molecule to a protein receptor, see Fig. 1.2(a). More generally, suppose that a macromolecule has m internal states labeled $j = 1, \ldots, m$. The probability $P_j(t)$ that the molecule is in state j at time t evolves according to the system of differential equations

$$\frac{dP_j}{dt} = \sum_{k=1}^{m} \left[W_{jk} P_k(t) - W_{kj} P_j(t) \right], \quad j = 1, \ldots, m, \quad (1.1.1)$$

where $W_{jk} \delta t$ is the probability that the molecule jumps to state j in an infinitesimal time interval δt, given that it is currently in state k; W_{jk} is called a state transition rate. Such a Markov process is said to satisfy detailed balance if there exists a stationary density Π_j such that for each pair of reversible transitions (jk)

$$W_{jk} \Pi_k = W_{kj} \Pi_j. \quad (1.1.2)$$

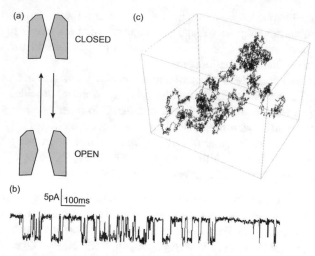

Fig. 1.2: Internal and external stochastic variables. (a) Internal open and closed states of an ion channel. (b) Patch-clamp recording of a glycine receptor showing stochastic variations in current due to the opening and closing of the ion channel. [Public domain figure downloaded from Wikimedia Commons.] (c) Sample 3D trajectory of a Brownian particle.

The detailed balance condition is stronger than that required merely for a stationary distribution—there are Markov processes with stationary distributions that do not have detailed balance. Detailed balance implies that, around any closed cycle of states, there is no net flow of probability.

In contrast to discrete internal states, the evolution of external variables such as the position of a macromolecule is modeled in terms of a stochastic differential equation (SDE) [300, 624], which is often called a Langevin equation in the physics literature [822] (Chap. 2). Mathematically speaking, an SDE is a differential equation in which one or more of the terms is a stochastic process, resulting in a solution that is itself a stochastic process. In the case of a macromolecule, the stochastic terms represent the effective forces due to collisions of the macromolecule with molecules in the surrounding medium. From this perspective, the external state of the macromolecule is described in terms of slow variables, whereas the degrees of freedom of the surrounding medium are treated as fast microscopic variables that are responsible for the stochastic nature of the SDE. A classical example of an SDE is Brownian motion, which refers to the random motion of a particle suspended in a fluid, see Fig. 1.2(c). This phenomenon is named after the botanist Robert Brown, who observed the erratic motion of pollen grains suspended in water through a microscope, but was unable to determine the mechanisms that caused this motion. Albert Einstein subsequently explained how the motion that Brown had observed was a result of the pollen being moved by individual water molecules, which served as a definitive confirmation that atoms and molecules actually exist.

An idealized mathematical representation of Brownian motion is the Wiener process, whose evolution can be described in terms of infinitesimal increments that are independent random variables generated from a zero mean Gaussian distribution whose variance scales as Δt, where Δt is the time step. One of the characteristic features of a Wiener process is that in the limit $\Delta t \to 0$, the time-dependent solution is continuous but its time derivative is everywhere infinite, reflecting the absence of a finite time scale. This means that the usual rules of calculus break down, and indicates that a Wiener process is an idealization of an actual random physical process, which always has a finite intrinsic time scale such as the time constant of second-order statistical correlations. Another important feature of a Wiener process is that it is the single-particle realization of diffusion. This can be seen from two perspectives. First, each realization of a Wiener process determines a sample trajectory through state space. The associated probability density of sample paths is the solution to a deterministic partial differential equation (PDE) known as a Fokker–Planck (FP) equation. In the case of idealized Brownian motion, the FP equation is formally identical to the classical diffusion equation. Indeed, if one were to consider a large number N of noninteracting Brownian particles then, in the large N limit, the concentration of particles evolves deterministically according to the diffusion equation. Second, if one were to discretize time and space, then a sample trajectory of a Wiener process reduces to an unbiased random walk, which is well known to be a discrete realization of a diffusing particle.

More generally, consider a molecule of mass m moving in one-dimension under the influence of an external force $F(x)$. In the absence of thermal fluctuations, the position of the molecule satisfies Newton's law of motion

$$m\frac{d^2x}{dt^2} + \gamma\frac{dx}{dt} = F(x),$$

where γ is a damping or drag coefficient. It turns out that at the microscopic length and velocity scales of molecular dynamics, the aqueous environment of a cell is highly viscous so that inertial terms can be ignored—the particle rapidly reaches the terminal velocity $F(x)/\gamma$. Under such circumstances, when the force due to thermal fluctuations is modeled as a Wiener process, the FP equation for the probability density $p(x,t)$ takes the form (Chap. 2)

$$\frac{\partial p}{\partial t} = -\frac{1}{\gamma}\frac{\partial}{\partial x}(F(x)p(x,t)) + D\frac{\partial^2 p(x,t)}{\partial x^2},$$

where D determines the level of noise. Moreover, since the molecules responsible for the fluctuating force are also responsible for the dissipation or damping, one finds that D and γ are related according to the so-called Einstein relation $D\gamma = k_BT$. This is a specific example of what is known as a fluctuation–dissipation theorem, since it relates the rate of relaxation to equilibrium to the size of thermal fluctuations (see also Chap. 10 and Chap. 12).

A number of important cellular processes at the macromolecular level involve a coupling between continuous external variables and discrete internal variables, with the latter given by a discrete Markov process. Such processes are modeled in terms of a stochastic hybrid system. Consider, for example, molecular motors, which are proteins that convert chemical energy into mechanical work [397] (Chap. 4). A motor protein undergoes a cyclic sequence of conformational changes after reacting with one or more molecules of a chemical such as adenosine triphosphate (ATP), resulting in the release of chemical energy. This allows the motor to perform work by exerting a force conjugate to a given external variable, e.g., pulling a load while moving along a protein filament—active intracellular transport. A second example of a stochastic hybrid system is a voltage-gated or ligand-gated ion channel, in which the opening and closing of the channel depend on an external variable such as membrane voltage or calcium concentration [431] (Chap. 3). Moreover, the dynamics of the given external variable itself depends on the internal state of the ion channel or, more precisely, a population of ion channels. Another important example occurs in gene regulatory networks, where a gene can switch between an inactive and active state, depending on whether or not a transcription factor is bound to the site of a promoter (Chap. 5).

Populations of molecules and chemical reactions. Another important issue is how noise at the level of an individual molecule scales up when one considers populations of molecules in different conformational states. Suppose that the number of molecules in each conformational state scales with some system size N. In the thermodynamic limit $N \to \infty$, the fraction of the molecules in a given state evolves according to a system of deterministic differential equations (kinetic equations). Moreover, for finite N, one can track the stochastic fraction of molecules in a given state using a so-called master equation. Let $\mathbf{n} = (n_1, n_2, \ldots, n_m)$ denote the number of molecules in each of m internal states with $\sum_{j=1}^{m} n_j = N$. The probability that the population is in the configuration state $\mathbf{n}$ at time t then evolves according to a master equation of the form

$$\frac{dP_{\mathbf{n}}}{dt} = \sum_{\mathbf{n}'} [W_{\mathbf{n}\mathbf{n}'} P_{\mathbf{n}'}(t) - W_{\mathbf{n}'\mathbf{n}} P_{\mathbf{n}}(t)].$$

Although it is generally difficult to analyze such a master equation, it is possible to carry out a perturbation expansion of the master equation in terms of the system size $1/N$ (system size expansion). For large but finite N, one can thus approximate the stochastic dynamics of the population using an FP equation or its equivalent Langevin equation [822]. However, now the stochastic variables are the fraction of molecules in a given intrinsic state rather than the position of a single macromolecule undergoing Brownian motion. The FP equation then provides an estimate for the size of fluctuations about the mean-field solutions.

In recent years, the system size expansion of master equations has become a major focus of work on genetic and other biochemical networks within a cell [425, 550, 645, 671, 720], where one now keeps track of changes in the numbers of different molecular species due to chemical reactions (Chap. 5). In the case of classical chemical reactions, the number of molecules involved is huge (comparable to Avagadro's number 6×10^{23}). In such cases, it is sufficient to model the chemical reactions in terms of deterministic kinetic equations based on the law of mass action: the rate of an elementary reaction (a reaction that proceeds through only one step) is proportional to the product of the concentrations of the participating molecules. In thermodynamic equilibrium, the rates of the forward and backward reactions must be equal, which allows one to express the ratio of the concentrations of reactants and products in terms of a constant known as the dissociation constant K_d. (An expression for K_d can be derived from first principles using the Boltzmann–Gibbs distribution; see Sect. 1.3.) However, the absolute values of the transition rates (rather than their ratios) cannot be determined from the theory of equilibrium systems. In contrast to classical chemical reactions, the number of molecular constituents involved in the transcription of DNA to produce a protein is small (tens or hundreds of molecules). The low copy numbers mean that fluctuations in the number of proteins produced are non-negligible and one has to deal with the corresponding master equation. An immediate issue that stems from this is how noisy processes at the level of DNA (genotypes) result in the robust development of single cells and multicellular organisms (phenotypes).

Once one considers noise at the level of more complex subcellular processes that involve populations of reacting macromolecules, it is useful to distinguish between intrinsic and extrinsic noise [247, 802]. Intrinsic noise refers to fluctuations due to the inherent stochasticity of the macromolecules, whereas extrinsic noise refers to fluctuations in the external environment (beyond the random molecular collisions that generate the intrinsic noise). For example, given a population of ion channels, intrinsic noise might correspond to fluctuations in the fraction of open ion channels whereas extrinsic noise could be due to random variations in the membrane voltage or calcium concentration. In the case of gene expression, intrinsic noise might refer to fluctuations in the number of bound protein promoters that repress or activate gene expression, whereas extrinsic noise could be due to fluctuations in the rates of binding and unbinding.

1.2 A brief introduction to probability theory

1.2.1 Random variables

We define the probability distribution (or mass function) of a discrete random variable X according to $P(x) = \mathbb{P}(X = x)$. The corresponding distribution function $F(x)$ is given by $F(x) = \sum_{x' \leq x} P(x')$. Given that X must take on one of its possible values, the total probability must be 1, that is, $\sum_x P(x) = 1$. Two important statistical quantities are the average, mean or expectation value of X, which is defined by

$$\mathbb{E}[X] \equiv \langle X \rangle = \sum_x x P(x),$$

and the variance of X,

$$\text{Var}[X] \equiv \langle X^2 \rangle - \langle X \rangle^2 = \sum_x x^2 P(x) - \left(\sum_x x P(x) \right)^2.$$

The standard deviation of X, denoted by σ_X, is defined to be $\sigma_X = \sqrt{\text{Var}[X]}$, and is a measure of how broad the probability distribution of X is. (In this book, we use both $\mathbb{E}[X]$ and $\langle X \rangle$ to denote expectation of X. Similarly, we use both $\text{Var}[X]$ and σ_X^2 to denote the variance of X.)

Two discrete random variables X and Y are said to be independent if the events $\{X = x\}$ and $\{Y = y\}$ are independent for all x and y. Introducing the joint probability distribution $P(x,y) = \mathbb{P}(X = x, Y = y)$, it follows that for independent variables $P(x,y) = P(x)P(y)$ and

$$\mathbb{E}[XY] = \sum_{x,y} xy P(x,y) = \sum_x x P(x) \sum_y y P(y) = \mathbb{E}[X]\mathbb{E}[Y].$$

It is also straightforward to show that $\text{Var}[X+Y] = \text{Var}[X] + \text{Var}[Y]$. A useful measure of the mutual dependence of two random variables is the correlation coefficient, which is defined by

$$\rho(X,Y) = \frac{\text{cov}[X,Y]}{\sqrt{\text{Var}[X]\text{Var}[Y]}}, \tag{1.2.1}$$

with $\text{cov}[X,Y]$ the covariance

$$\text{cov}[X,Y] = \mathbb{E}]XY] - \mathbb{E}[X]\mathbb{E}[Y]. \tag{1.2.2}$$

Note that the correlation coefficient vanishes if X and Y are independent. On the other hand, if X and Y are perfectly correlated ($X = cY, c > 0$) then $\text{cov}[X,Y] = 1$, and if they are perfectly anti-correlated ($X = cY, c < 0$) then $\text{cov}[X,Y] = -1$. Even in the case of dependent variables, one can still define a so-called marginal probability distribution for X, say, which applies when there is no information about the value of Y: $P(x) = \sum_y P(x,y)$.

Analogous results hold for a continuous random variable X. For concreteness, suppose that $X \in \mathbb{R}$. First, the expectation of X is now expressed as an integral

$$\mathbb{E}[X] = \int_{\mathbb{R}} x \, dF(x),$$

where $dF(x) = \text{Prob}(x \leq X \leq x + dx)$. The probability density function (if it exists) is then defined according to $dF(x) = p(x)dx$. (In the various applications to cell biology, we will usually assume that the probability density does exist.) In terms of $p(x)$, we have

$$\text{Prob}(a < x < b) = \int_a^b p(x)dx.$$

The total probability is again equal to 1, which can be expressed as the normalization condition $\int_{-\infty}^{\infty} p(x)dx = 1$. Similarly, the mean and variance are defined according to

$$\mathbb{E}[X] = \int_{-\infty}^{\infty} x p(x)dx,$$

and

$$\mathrm{Var}[X] = \int_{-\infty}^{\infty} x^2 p(x)dx - \mathbb{E}[X]^2.$$

If two continuous random variables X and Y are independent, then their joint probability density $p(x,y)$ can be written as the product of individual probability densities for each of the random variables, $p(x,y) = p(x)p(y)$, and

$$\mathbb{P}(a < X < b \text{ and } c < Y < d) = \int_a^b \int_c^d p(x,y)dydx$$

$$= \int_a^b \int_c^d p(x)p(y)dydx = \left(\int_a^b p(x)dx \right) \left(\int_c^d p(y)dy \right)$$

$$= \mathbb{P}(a < X < b) \cdot \mathbb{P}(c < Y < d).$$

It immediately follows that $\mathbb{E}[XY] = \mathbb{E}[X]\mathbb{E}[Y]$. In the case of dependent variables, the corresponding marginal density for X is

$$p(x) = \int_{-\infty}^{\infty} p(x,y)dy.$$

Some common probability distributions are as follows:

Poisson distribution. Let N be a discrete random variable that takes the values $N = 0, 1, 2, \ldots$. A Poisson distribution $P_n = \mathbb{P}[N = n]$ is defined according to

$$P_n = \frac{\lambda^n}{n!} e^{-\lambda}, \tag{1.2.3}$$

where the rate $\lambda > 0$. The mean and variance of the Poisson distribution are

$$\langle N \rangle = \lambda = \mathrm{Var}[N].$$

Binomial distribution. Suppose that a discrete random variable takes the values $N = 0, 1, \ldots N_T$ and introduces a parameter p_0 with $0 < p_0 < 1$. The binomial distribution takes the form

$$P_n = \frac{N_T!}{n!(N_T - n)!} p_0^n (1 - p_0)^{N_T - n}. \tag{1.2.4}$$

The mean and variance are

$$\langle N \rangle = N_T p_0, \quad \mathrm{Var}[N] = N_T p_0 (1 - p_0).$$

Gaussian distribution. Let X be a continuous random variable. A Gaussian distribution is one whose density takes the form

$$p(x) = \frac{1}{\sqrt{2\pi\sigma^2}} e^{-(x-x_0)^2/2\sigma^2}. \tag{1.2.5}$$

Using properties of Gaussian integration, one finds that

$$\langle X \rangle = x_0, \quad \text{Var}[X] = \sigma^2.$$

1.2.2 Conditional expectations

In the case of two discrete random variables, we define the conditional expectation of Y given X by

$$\mathbb{E}[Y|X] = \sum_y y \mathbb{P}(Y = y|X),$$

where $\mathbb{P}(Y = y|X)$ is the conditional probability distribution for Y given X. Note that the conditional expectation is still a random variable with respect to X. A very useful result is the law of total expectation, which is

$$\mathbb{E}[Y] = \mathbb{E}[\mathbb{E}[Y|X]]. \qquad (1.2.6)$$

That is, the expected value of the conditioned expectation of Y given X is the same as the expected value of Y. The proof is straightforward for discrete processes

$$\mathbb{E}[\mathbb{E}[Y|X]] = \mathbb{E}\left[\sum_y y \mathbb{P}(Y = y|X)\right] = \sum_x \left[\sum_y y \mathbb{P}(Y = y|X = x)\right] \mathbb{P}(X = x)$$

$$= \sum_y y \sum_x \mathbb{P}(Y = y|X = x)\mathbb{P}(X = x)$$

$$= \sum_{x,y} y \mathbb{P}(X = x, Y = y) = \sum_y y \mathbb{P}(Y = y) = \mathbb{E}[Y].$$

Using a similar argument, one can also derive the tower property

$$\mathbb{E}[\mathbb{E}[Y|X_1, X_2]|X_1] = \mathbb{E}[Y|X_1].$$

On the left-hand side, expectation is first taken with respect to X_2 and then X_1. These results carry over to continuous random variables with

$$\mathbb{E}[Y|X] = \int y p(y|X) dy,$$

where $p(y|X)$ is the   conditional probability density for Y given X.

Bayes Theorem. A basic result of conditional probabilities is

$$\mathbb{P}(X = x, Y = y) = \mathbb{P}(X = x|Y = y)\mathbb{P}(Y = y) = \mathbb{P}(Y = y|X = x)\mathbb{P}(X = x).$$

It follows that

$$\mathbb{P}(Y=y|X=x) = \frac{\mathbb{P}(X=x|Y=y)\mathbb{P}(Y=y)}{\sum_y \mathbb{P}(X=x|Y=y)\mathbb{P}(Y=y)}, \tag{1.2.7}$$

which is a statement of Bayes' theorem [537]. Although we will mainly use the frequency interpretation of probability in this book, we note that the Bayesian interpretation is often used in statistical and information theoretic approaches to systems biology. For example, suppose that X represents some observable data and Y represents a parameter of the system that produces the data. Then $\mathbb{P}(X=x|Y=y)$ is known as the likelihood function that a particular parameter value produces the observed data and $\mathbb{P}(Y=y)$ is the Bayesian prior, which expresses what was known (or thought to be known) about the parameter before the measurement. Finally, $\mathbb{P}(Y=y|X=x)$ is the updated posterior probability density for the parameter obtained by combining the prior information with the information gained from the measurement.

Law of total variance. The law of total variance, also known as Eve's law, states the following: if X and Y are random variables on the same probability space, then

$$\text{Var}[Y] = \mathbb{E}[\text{Var}[Y|X]] + \text{Var}[\mathbb{E}[Y|X]]. \tag{1.2.8}$$

The proof follows from the law of total expectation. First, from the definition of variance,

$$\text{Var}[Y] = \mathbb{E}[Y^2] - (\mathbb{E}[Y])^2.$$

Conditioning on the random variable X and applying the law of total expectation to each term gives

$$\text{Var}[Y] = \mathbb{E}\left[\mathbb{E}[Y^2|X]\right] - (\mathbb{E}[\mathbb{E}[Y|X]])^2.$$

Now we rewrite the conditional second moment of Y in terms of its variance and first moment:

$$\text{Var}[Y] = \mathbb{E}\left[\text{Var}[Y|X] + (\mathbb{E}[Y|X])^2\right] - (\mathbb{E}[\mathbb{E}[Y|X]])^2.$$

Since the expectation of a sum is the sum of expectations, the terms can now be regrouped as

$$\text{Var}[Y] = \mathbb{E}[\text{Var}[Y|X]] + \left\{\mathbb{E}[(\mathbb{E}[Y|X])^2] - (\mathbb{E}[\mathbb{E}[Y|X]])^2\right\}.$$

The result follows from observing that the terms in parentheses on the right-hand side give the variance of the conditional expectation $\mathbb{E}[Y|X]$.

1.2.3 Adding and transforming random variables

Suppose that N independent measurements of some quantity are made, which are denoted by $X_n, n = 1, \ldots, N$. The random variables are said to be independent, identically distributed (i.i.d.) random variables, and have the same mean μ and variance

σ^2. Averaging the results of these measurements generates a new random variable

$$X = \sum_{n=1}^{N} \frac{X_n}{N}.$$

It follows that

$$\langle X \rangle = \left\langle \sum_{n=1}^{N} \frac{X_n}{N} \right\rangle = \frac{1}{N} \sum_{n=1}^{N} \langle X_n \rangle = \mu,$$

and

$$\text{var}[X] = \left\langle \sum_{m=1}^{N} \frac{X_m}{N} \sum_{n=1}^{N} \frac{X_n}{N} \right\rangle - \mu^2 = \frac{1}{N^2} \sum_{m=1}^{N} \sum_{n=1}^{N} \langle X_m X_n \rangle - \mu^2$$

$$= \frac{1}{N^2} \sum_{m=1}^{N} \left[\langle X_m^2 \rangle + \sum_{n,n \neq m} \langle X_m X_n \rangle - N\mu^2 \right]$$

$$= \frac{1}{N^2} \sum_{m=1}^{N} \left[\sigma^2 + \sum_{n,n \neq m} \langle X_m \rangle \langle X_n \rangle - (N-1)\mu^2 \right] = \frac{\sigma^2}{N}.$$

We thus obtain the well-known result that the standard deviation of the average varies as $\sigma_X \sim N^{-1/2}$. This is an expression of the law of large numbers, which states that the average of the results obtained from a large number of trials should be close to the expected value, and will tend to become closer as more trials are performed.

A related result is the central limit theorem [343]. Suppose that $\{X_1, X_2, \ldots\}$ is a sequence of i.i.d. random variables with $\langle X_j \rangle = \mu$ and $\text{var}[X_j] = \sigma^2 < \infty$. Then as n approaches infinity, the random variables $\sqrt{n}(S_n - \mu)$ with $S_n = \sum_{j=1}^{n} X_n/n$ converge in distribution to a normal or Gaussian distribution $N(0, \sigma^2)$: The Gaussian distribution $N(\mu, \sigma^2)$ is defined according to

$$N(\mu, \sigma^2)(x) = p(x) \equiv \frac{1}{\sqrt{2\pi\sigma^2}} \exp\left(-\frac{(x-\mu)^2}{2\sigma^2} \right).$$

Convergence in distribution means that

$$\lim_{n \to \infty} \text{Prob}[\sqrt{n}(S_n - \mu) \leq z] = \frac{1}{\sqrt{2\pi\sigma^2}} \int_{-\infty}^{z} e^{-x^2/2\sigma^2} dx.$$

Suppose that we know the probability density $p_X(x)$ of a random variable X and we construct a new random variable $Y = g(X)$, where g is an invertible function. We would like to determine the probability density $p_Y(y)$. This can be achieved by considering the expectation value of a function $f(Y)$ in terms of $p_X(x)$ and performing a change of variables. That is,

$$\langle f(Y) \rangle = \int_{x=a}^{x=b} f(g(x)) p_X(x) dx = \int_{y=g(a)}^{y=g(b)} f(y) p_X(g^{-1}(y)) \frac{dx}{dy} dy$$

$$= \int_{y=g(a)}^{y=g(b)} f(y) \frac{p_X(g^{-1}(y))}{g'(x)} dy = \int_{y=g(a)}^{y=g(b)} f(y) \frac{p_X(g^{-1}(y))}{g'(g^{-1}(y))} dy.$$

It is possible that on transforming the limits of the integral, $g(b) < g(a)$, which means that the fraction is then negative. Therefore, the transformed density is

$$p_Y(y) = \frac{p_X(g^{-1}(y))}{|g'(g^{-1}(y))|}.$$

1.2.4 Moments and cumulants

A classical way to calculate moments of a discrete random variable $N \in \{0, \ldots, K\}$ is to use a moment generating function. This is defined as the discrete Laplace transform (or z-transform)

$$G(z) = \sum_{n=1}^{K} z^n P_n, \tag{1.2.9}$$

with $G(1) = 1$ (unit normalization of probability). Once $G(z)$ is known, the moments can be determined by successive derivatives:

$$\left. \frac{dG(z)}{dz} \right|_{z=1} = \sum_{n=1}^{K} n z^{n-1} P_n \bigg|_{z=1} = \langle N \rangle,$$

$$\left. \frac{d^2 G(z)}{dz^2} \right|_{z=1} = \sum_{n=2}^{K} n(n-1) z^{n-2} P_n \bigg|_{z=1} = \langle N^2 \rangle - \langle N \rangle.$$

In the case of a continuous random variable X with probability density $p(x)$, the expectation value $\langle X^n \rangle$ can be calculated using the so-called characteristic function of X, which is defined by

$$G(k) = \int_{-\infty}^{\infty} e^{ikx} p(x) dx.$$

It can be seen that $G(k)$ is the Fourier transform of $p(x)$. Taylor expanding $G(k)$, we have

$$G(k) = \sum_{n=0}^{\infty} \frac{G^{(n)}(0) k^n}{n!},$$

where $G^{(n)}(k)$ is the nth derivative of $G(k)$. An alternative series expansion of $G(k)$ is obtained by noting that

$$G(k) = \left\langle e^{ikx} \right\rangle = \left\langle \sum_{n=0}^{\infty} \frac{(ikX)^n}{n!} \right\rangle = \sum_{n=1}^{\infty} \frac{i^n \langle X^n \rangle k^n}{n!}.$$

Equating the two series representations of $G(k)$ shows that

$$\langle X^n \rangle = (-i)^n G^{(n)}(0).$$

A related quantity is the nth order cumulant of X, which we denote by κ_n. It is a polynomial in the first n moments, which is defined by

$$\kappa_n = (-i)^n \left. \frac{d^n}{dk^n} \ln G(k) \right|_{k=0}.$$

The first three cumulants are $\kappa_1 = \langle X \rangle$, $\kappa_2 = \mathrm{var}[X]$, and

$$\kappa_3 = \langle X^3 \rangle - 3 \langle X^2 \rangle \langle X \rangle + 2 \langle X \rangle^3.$$

One useful feature of cumulants is that $\kappa_n = 0$ for all $n \geq 3$ in the case of the Gaussian distribution. This implies that all higher moments of a Gaussian can be written in terms of the mean $\langle X \rangle$ and variance σ^2. A general formula for these moments can be derived using standard results from Gaussian integration, and one finds that

$$\langle (X - \langle X \rangle)^{2n} \rangle = \frac{1}{\sqrt{2\pi\sigma^2}} \int_{-\infty}^{\infty} y^{2n} e^{-y^2/2\sigma^2} \, dy = \frac{(2n-1)! \sigma^{2n}}{2^{n-1}(n-1)!},$$

and $\langle (X - \langle X \rangle)^{2n-1} \rangle = 0$ for $n \geq 1$. It also turns out that the characteristic function of a Gaussian is also a Gaussian, since

$$G(k) = \frac{1}{\sqrt{2\pi\sigma^2}} \int_{-\infty}^{\infty} e^{ikx} e^{-x^2/2\sigma^2} \, dx = \frac{e^{-\sigma^2 k^2/2}}{\sqrt{2\pi\sigma^2}} \int_{-\infty}^{\infty} e^{-(x-i\sigma^2 k)^2/2\sigma^2} \, dx = e^{-\sigma^2 k^2/2}.$$

(Technically speaking, the Gaussian integral is evaluated by completing the square and then using analytical continuation in the complex k-plane.)

Finally, note that this book is concerned with stochastic processes, which involve random variables evolving in time. Thus a random variable will have an additional time label: $X \to X_n, n \in \mathbf{Z}^+$ for discrete time processes and $X \to X(t), t \in \mathbb{R}^+$ for continuous-time processes. Roughly speaking, one can treat t (or n) as a parameter so that for fixed t, $X(t)$ is a random variable. However, various objects such as the probability density and characteristic function are now parameterized by t, and we write $p(x) \to p(x,t)$ and $G(k) \to G(k,t)$, etc.

1.3 Equilibrium systems

1.3.1 Boltzmann–Gibbs distribution

Suppose that a macromolecule such as DNA or a protein has a discrete set of intrinsic states labeled by j, and let E_j denote the energy of the molecule in the jth state. (These states could include different folded or twisted configurations of the underlying amino acid or nucleotide chain.) Furthermore, assume that the surrounding cellular environment maintains a constant temperature T. A basic result of equilibrium statistical mechanics is that the probability p_j that the molecule is in state j is given by the Boltzmann–Gibbs distribution [163, 429]

$$p_j = \frac{1}{Z} e^{-E_j/k_B T}, \quad Z = \sum_j e^{-E_j/k_B T}. \tag{1.3.1}$$

The partition function Z ensures that $\sum_j p_j = 1$. Statistical quantities of interest are often expressed in terms of derivatives of $\ln Z$ with respect to $\beta = 1/k_B T$. For example, the mean and variance of the energy of the molecule are as follows:

$$\langle E \rangle = \sum_j E_j p_j = -Z^{-1} \frac{\partial}{\partial \beta} \sum_j e^{-\beta E_j} = -\frac{\partial \ln Z}{\partial \beta}, \tag{1.3.2a}$$

$$\text{Var}[E] = \sum_j E_j^2 p_j - \langle E \rangle^2 = Z^{-1} \frac{\partial^2}{\partial \beta^2} \sum_j e^{-\beta E_j} - Z^{-2} \left(\frac{\partial}{\partial \beta} \sum_j e^{-\beta E_j} \right)^2$$

$$= Z^{-1} \frac{\partial^2 Z}{\partial \beta^2} - Z^{-2} \left(\frac{\partial Z}{\partial \beta} \right)^2 = \frac{\partial^2 \ln Z}{\partial \beta^2}. \tag{1.3.2b}$$

If we apply the Boltzmann–Gibbs distribution to the steady-state probability Π_j of detailed balance, equation (1.1.2), then we immediately see that the backward and forward transition rates satisfy the condition

$$\frac{W_{jk}}{W_{kj}} = \frac{\Pi_j}{\Pi_k} = \frac{p_j}{p_k} = e^{-(E_j - E_k)/k_B T}.$$

That is, the ratio of the forward and backward transition rates depends on the energy difference between the two states. As a simple example, consider a molecule that can exist in two states $j = C, O$ with energies E_C and E_O, respectively. The associated probabilities are

$$P_O = \frac{e^{-E_O/k_B T}}{e^{-E_O/k_B T} + e^{-E_C/k_B T}} = \frac{1}{1 + e^{\Delta E/k_B T}}, \quad P_C = 1 - P_O,$$

with $\Delta E = E_O - E_C$ the difference in energies between the two states. It follows that

$$\frac{P_O}{P_C} = e^{-\Delta E/k_B T}.$$

Examples of a two-state system include an ion channel that is either open or closed, a protein receptor that is either bound or unbound to a ligand (Chap. 3), and an active or inactive gene (Chap. 5).

Now consider a system of N identical molecules at temperature T, which may or may not be interacting. Suppose that each molecule can exist in some internal state $\sigma_i \in \Gamma = \{1, \dots, K\}$ for $i = 1, \dots, N$ and denote the microstate of the system by $\sigma = \{\sigma_1, \dots, \sigma_N\}$. Denoting the energy of the system by $E(\sigma)$, the probability of the state σ is

$$p(\sigma) = \frac{1}{Z} e^{-E(\sigma)/k_B T}, \tag{1.3.3}$$

with

$$Z = \sum_\sigma e^{-E(\sigma)/k_B T} = \sum_{\sigma_1 \in \Gamma} \cdots \sum_{\sigma_N \in \Gamma} e^{-E(\sigma)/k_B T}. \tag{1.3.4}$$

One of the typical features of such a system is that for large N, the mean and variance of the energy scale as

$$\langle E \rangle = N\varepsilon, \quad \mathrm{Var}[E] = N\sigma^2$$

for $\varepsilon, \sigma = O(1)$. That is, fluctuations vanish in the thermodynamic limit $N \to \infty$, and the total energy of the system can be approximated by $\langle E \rangle = -\partial \ln Z/\partial \beta$. Some examples of noninteracting particle systems are considered in Ex. 1.1 and Ex. 1.2.

1.3.2 Free energy and entropy

The Boltzmann–Gibbs distribution can also be applied to molecules with continuous states. For example, consider a molecule moving in $d = 1, 2, 3$ dimensions under the influence of a conservative force—a force that can be written as $\mathbf{f} = -\nabla U(\mathbf{x})$ where $U(\mathbf{x})$ is some potential energy function. Examples include gravitational and electrical potentials. Assuming that all other degrees of freedom are independent of $\mathbf{x}$ (momentum, internal microstates), the equilibrium probability density with respect to position $\mathbf{x}$ is given by

$$p(\mathbf{x}) = \frac{1}{Z} e^{-U(\mathbf{x})/k_B T}, \quad Z = \int e^{-U(\mathbf{x})/k_B T} d\mathbf{x}. \tag{1.3.5}$$

An example of molecules forming a vertical column under gravity is considered in Ex. 1.3.

Now suppose that the number of internal states of the molecule depends on position $\mathbf{x}$. That is, denoting the set of internal states at $\mathbf{x}$ by $I(\mathbf{x})$, we set $\sum_{j \in I(\mathbf{x})} = \Omega(\mathbf{x})$. If $p_j(\mathbf{x})$ is the joint probability density of being at position $\mathbf{x}$ and in internal state j, then the marginal probability density $p(\mathbf{x})$ is given by

$$p(\mathbf{x}) = \sum_{j \in I(\mathbf{x})} p_j(\mathbf{x}) = \frac{1}{Z} \sum_{j \in I(\mathbf{x})} e^{-U(\mathbf{x})/k_B T} = \frac{1}{Z} \Omega(\mathbf{x}) e^{-U(\mathbf{x})/k_B T},$$

with

$$Z = \int \sum_{j \in I(\mathbf{x})} e^{-U(\mathbf{x})/k_B T} d\mathbf{x} = \int \Omega(\mathbf{x}) e^{-U(\mathbf{x})/k_B T} d\mathbf{x}.$$

The probability density $p(\mathbf{x})$ can be reexpressed in terms of an effective Boltzmann–Gibbs distribution

$$p(\mathbf{x}) = \frac{1}{Z} e^{-F(\mathbf{x})/k_B T}, \quad Z = \int e^{-F(\mathbf{x})/k_B T} d\mathbf{x}, \tag{1.3.6}$$

where

$$F(\mathbf{x}) = U(\mathbf{x}) - k_B T \ln \Omega(\mathbf{x}). \tag{1.3.7}$$

One defines $F(\mathbf{x})$ to be the (Helmholtz) free energy of the molecule and the term

$$S = k_B \ln \Omega(\mathbf{x}) \tag{1.3.8}$$

to be the entropy. The total effective force on the molecule is

$$\mathbf{f}_{\text{tot}}(\mathbf{x}) = -\nabla F(\mathbf{x}) = -\nabla U(\mathbf{x}) + T \nabla S(\mathbf{x}) = f(\mathbf{x}) + T \nabla S(\mathbf{x}). \tag{1.3.9}$$

That is, there is an additional entropic force given by $T\nabla S(\mathbf{x})$. The entropic force has the following statistical mechanical interpretation: the total entropy of a closed system cannot decrease (second law of thermodynamics), so that if a change in position $\delta \mathbf{x}$ decreases the entropy of the molecule, $\nabla S(\mathbf{x}) \cdot \delta \mathbf{x} < 0$, then this results in the environment having to do work $\delta W = -T\nabla S(\mathbf{x}) \cdot \delta \mathbf{x} > 0$ to counteract the entropic force. Assuming the internal energy of the molecule does not change when it is displaced, the work done is dissipated as heat (conservation of energy), resulting in an increase in the environmental entropy, thus counterbalancing the decrease in entropy of the molecule. (For simplicity, we are treating the molecule and its surrounding environment as a closed system.) One important example of an entropic force arises in the uncoiling of a flexible polymer such as DNA (Chap. 12).

Let us now return to the example of a system of N particles with Boltzmann distribution (1.3.3), and suppose that there are $\Omega(E(\sigma))$ configurations with energy $E(\sigma)$. Then

$$Z = \sum_{\sigma} \Omega(E(\sigma)) e^{-E(\sigma)/k_B T} = \sum_{\sigma} e^{-F(\sigma)/k_B T},$$

where

$$F(\sigma) = E(\sigma) - k_B T \ln \Omega(E(\sigma)).$$

Exploiting the fact that both energy and entropy scale with the system size N, we can make the following approximation for large N:

$$Z \approx e^{-(\langle E \rangle - k_B T \ln \Omega(\langle E \rangle))/k_B T},$$

where $\langle E \rangle = -\partial \ln Z / \partial \beta$. This then leads to the thermodynamic definition of the free energy given by

$$F := -k_B T \ln Z = \langle E \rangle - k_B T \ln \Omega(\langle E \rangle). \tag{1.3.10}$$

1.3.3 Chemical potential.

Another important consequence of entropic effects arises from changes in the number of microstates when a solute molecule is removed from a dilute solution. Suppose that there are N solvent molecules and n solute molecules with $n \ll N$. For simplicity, we represent the solution in terms of $N + n$ boxes that can either be occupied by a solute molecule or a solvent molecule, see Fig. 1.3. The number of different configurations for given n, N is given by the combinatorial factor for distributing n items in $N + n$ boxes:

$$\Omega(n) = \frac{(N+n)!}{n!N!}.$$

Taking logs and using Stirling's formula

$$\ln N! \approx N \ln N - N + \frac{1}{2} \ln(2\pi N),$$

we have the entropy

$$S(n) = k_B [(N+n) \ln(N+n) - n \ln n - N \ln N]. \tag{1.3.11}$$

Thus there is an entropic contribution to the free energy of the solute of the form $-k_B T \ln \Omega(n)$. Now suppose that δn additional solute molecules are added to the solution. The change in free energy consists of two contributions: the change in energy $\varepsilon \delta n$ associated with the presence of the new solute molecules and the change in entropy of the solution due to $n \to n + \delta n$. Thus the total change in free energy δF (for $N \gg n$) is given by

$$\frac{\delta F}{\delta n} = \varepsilon - T \frac{dS(n)}{dn} \approx \varepsilon + k_B T \ln(n/N) = \varepsilon + k_B T \ln(c/c_{\text{sol}}),$$

where $c = n/V$ and $c_0 = N/V$ are the solute and solvent concentrations. This quantity leads to one definition of the chemical potential: The chemical potential μ of a solute with concentration c is

$$\mu = \mu_0 + k_B T \ln(c/c_0), \tag{1.3.12}$$

where μ_0 is the value at a reference concentration c_0 usually taken to be 1 molar. A molar is one mole of solute in one liter of solution, and one mole is equal to 6.022×10^{23} atoms or molecules of a given substance. A lattice model similar to the one shown in Fig. 1.3 will be used to explore the thermodynamics of phase sep-

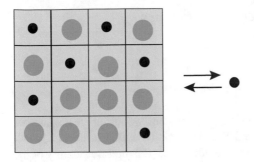

solvent molecule   solute molecule

Fig. 1.3: Lattice model of a mixture of solvent and solute molecules.

aration and biological condensates in Chap. 13. Some further examples of entropy calculations are given in Ex. 1.4 and Ex. 1.5.

Another interpretation of the chemical potential is as a Lagrange multiplier that is introduced to ensure particle conservation. For the sake of illustration, consider a large system of N particles with n_j particles in the state $j = 1, \ldots, K$. Conservation of particle number means that $\sum_{j=1}^{K} n_j = N$. Suppose that the total free energy of the system can be written in the form $F = Nf(\Phi)$, where $\mathbf{n} = (n_1, n_2, \ldots, n_K)$, $\Phi = (\phi_1, \phi_2, \ldots, \phi_K)$ with $\phi_j = n_j/N$, and $f = O(1)$. The total free energy F is said to be an extensive variable, whereas f is an intensive variable. The probability of the state Φ is given by the Boltzmann–Gibbs distribution

$$p(\Phi) = Z^{-1} e^{-Nf(\Phi)/k_B T}.$$

In the thermodynamic limit $N \to \infty$, the system has a probability approaching one that it exists in the macrostate Φ_{eq} that minimizes the free energy density (assuming the minimum is unique). However, such a minimization has to conserve particle number. This can be achieved by introducing a Lagrange multiplier and defining

$$\widehat{f}(\Phi) = f(\Phi) + \mu \left(1 - \sum_j \phi_j \right).$$

Minimization then yields the system of equations

$$\frac{\partial \widehat{f}}{\partial \phi_j} = \frac{\partial f}{\partial \phi_j} - \mu = 0. \tag{1.3.13}$$

1.3.4 Law of mass action and chemical kinetics

Consider the reversible chemical reaction

$$A + B \underset{k_-}{\overset{k_+}{\rightleftharpoons}} C.$$

In order to characterize the equilibrium state of the system, we introduce the number of molecules of each species, N_A, N_B, and N_C. In thermodynamic equilibrium, small changes in the number of each species should not change the total free energy of the system. This can be expressed by the condition

$$\Delta F = \mu_A dN_A + \mu_B dN_B + \mu_C dN_C = 0,$$

where μ_j is the chemical potential of species j. (In systems where the number of particles can change, F is often referred to as the Gibbs free energy and written using the symbol G.) When the numbers of molecules A and B decrease (increase) by 1, the number of C molecules increases (decreases) by 1. We can impose this condition by setting $dN_j = v_j dN$, with $v_A = v_B = -1$ and $v_C = 1$. The coefficients v_j are called stoichiometric coefficients. Similarly, for more general reversible chemical reactions, involving S species with associated integer stoichiometric coefficients, we have

$$0 = \sum_{j=1}^{S} \mu_j dN_j = \left(\sum_{j}^{S} \mu_j v_j\right) dN \implies \sum_{j=1}^{S} v_j \mu_j = 0.$$

Now applying the formula (1.3.12) to each species,

$$\mu_j = \mu_{0j} + k_B T \ln(c_j/c_{0j}),$$

gives

$$\Delta F_0 := \sum_{j=1}^{S} v_j \mu_{j0} = -k_B T \sum_{j=1}^{S} v_j \ln(c_j/c_{0j}) = -k_B T \ln\left[\prod_{j=1}^{S} \left(\frac{c_j}{c_{0j}}\right)^{v_j}\right].$$

Exponentiating both sides of this equation then yields the equilibrium law of mass action.

Equilibrium law of mass action. Given a set of S chemical species undergoing a reversible chemical reaction, the weighted product of their concentrations is equal to a constant known as the equilibrium constant K_{eq} according to

$$\prod_{j=1}^{S} \left(\frac{c_j}{c_{j0}}\right)^{v_j} = K_{\text{eq}}(T) := \exp\left(-\frac{\Delta G_0}{k_B T}\right). \tag{1.3.14}$$

Here c_j is the concentration of the jth species and v_j is the stoichiometric coefficient. In the case of the reversible reaction $A \rightleftharpoons B$, equation (1.3.14) with

$v_B = -v_A = 1$ implies that (for concentrations in molars so $c_{0j} = 1$)

$$\frac{c_B}{c_A} = e^{-(\varepsilon_B - \varepsilon_A)/k_B T}.$$ (1.3.15)

This is consistent with the Boltzmann–Gibbs distribution, which states that the probability distribution of A and B molecules is

$$p_A = \frac{e^{-\varepsilon_A/k_B T}}{e^{-\varepsilon_A/k_B T} + e^{-\varepsilon_B/k_B T}} = 1 - p_B.$$

When the concentrations of reacting molecules are not in thermodynamic equilibrium, one can write down differential equations for the evolution of the concentrations, so that for reversible reactions one approaches equilibrium in the large time limit. This is described by the kinetic version of the law of mass action, which will be used extensively in constructing deterministic models of gene networks (Chap. 5). For example, consider the single-step reaction $A + B \xrightarrow{k} C$, where k is the reaction rate. The rate of accumulation of the product C is given by the (kinetic) law of mass action

$$\frac{d[C]}{dt} = k[A][B].$$ (1.3.16)

The product of concentrations is motivated by the idea that in a well-mixed container there is a spatially uniform distribution of each type of molecule, and the probability of a collision depends on the probability that each of the reactants is in the same local region of space. Ignoring any statistical correlations, the latter is given by the product of the individual concentrations. The rate constant depends on the probability that a collision of the relevant molecules actually leads to a reaction, and will depend on the temperature, and shapes and sizes of the reactants. Similarly, for the reversible version of the above reaction with reaction rates $k_\pm$, we apply the law of mass action to the forward and reverse single-step reactions to obtain

$$\frac{d[A]}{dt} = k_-[C] - k_+[A][B].$$ (1.3.17)

At equilibrium,

$$\frac{[C]}{[A][B]} = \frac{k_+}{k_-} = K_{eq}(T).$$

Kinetic law of mass action. Consider a mixture of K chemical species X_j, $j = 1,\dots,K$. Let n_j be the number of molecules of X_j and set $\mathbf{n} = (n_1,\dots,n_K)$. A typical single-step chemical reaction takes the form

$$s_1 X_1 + s_2 X_2 + \dots \to r_1 X_1 + r_2 X_2 + \dots,$$

where s_j, r_j are the stoichiometric coefficients for the reactants and products. When one such reaction occurs, the state $\mathbf{n}$ is changed according to $n_i \to n_i + r_i - s_i$. Introducing the vector $\mathbf{S}$ with $S_i = v_i \equiv r_i - s_i$, we have $\mathbf{n} \to \mathbf{n} + \mathbf{S}$. The reverse reaction

$$\sum_j r_j X_j \rightarrow \sum_j s_j X_j$$

would then have $\mathbf{n} \rightarrow \mathbf{n} - \mathbf{S}$. The law of mass action for the concentrations x_i takes the general form

$$\frac{dx_i}{dt} = \kappa(r_i - s_i) \prod_{j=1}^{K} x_j^{s_j} \equiv S_i f(\mathbf{x}), \tag{1.3.18}$$

where $\mathbf{x} = (x_1, \ldots, x_K)$ where κ is the reaction rate. Finally, suppose that there are $a = 1, \ldots, R$ separate single-step reactions in a given biochemical or gene network. Then

$$\frac{dx_i}{dt} = \sum_{a=1}^{R} S_{ia} f_a(\mathbf{x}) \tag{1.3.19}$$

for $i = 1, \ldots, K$, where a labels a single-step reaction and $\mathbf{S}$ is the so-called $K \times R$ stoichiometric matrix for K molecular species and R reactions. Thus S_{ia} specifies the change in the number of molecules of species i in a given reaction a. The functions f_a are known as transition intensities or propensities. A few examples are considered in Ex. 1.6.

1.3.5 Michaelis–Menten kinetics

Many chemical processes in cell biology involve enzymatic reactions. Enzymes are generally protein catalysts that help convert other molecules called substrates into products, without themselves being changed by the reaction. In contrast to single-step reactions, the rate of reaction does not increase linearly with the concentration of substrate, since it saturates at high concentrations. A simple model to explain this behavior was first proposed by Michaelis and Menten. The basic reaction scheme involves an enzyme E converting a substrate S to a complex ES, which then breaks down to form the product P together with the original enzyme. This can be represented by the following two-step process:

$$S + E \underset{k_{-1}}{\overset{k_1}{\rightleftharpoons}} ES \overset{k_2}{\rightarrow} P + E. \tag{1.3.20}$$

Although all the reactions are reversible, reaction rates are typically measured under conditions in which the product P is continually removed from the system, which effectively prevents the final reverse reaction from occurring. Setting $s = [S], c = [ES], e = [E]$ and $p = [P]$, we have the system of kinetic equations

$$\frac{ds}{dt} = k_{-1}c - k_1 se, \tag{1.3.21a}$$

$$\frac{de}{dt} = (k_{-1} + k_2)c - k_1 se, \tag{1.3.21b}$$

$$\frac{dc}{dt} = -(k_{-1} + k_2)c + k_1 se, \tag{1.3.21c}$$

$$\frac{dp}{dt} = k_2 c. \tag{1.3.21d}$$

Note that the total concentration of enzyme is conserved, $e + c = E_0$ for some constant E_0.

If the total concentration of enzyme is small, then s changes relatively slowly, which suggests that the free enzyme and the enzyme–substrate complex ES reach a quasi-steady-state before s changes significantly. If we fix $[S]$, then the kinetics can be reduced to the irreversible (or corresponding reversible) catalytic cycle shown in Fig. 1.4 [57]:

$$E \underset{k_{-1}}{\overset{k_1[S]}{\rightleftharpoons}} ES \overset{k_2}{\rightarrow} E. \tag{1.3.22}$$

The corresponding pair of kinetic equations are

$$\frac{de}{dt} = (k_{-1} + k_2)c - k_1 se, \tag{1.3.23a}$$

$$\frac{dc}{dt} = -(k_{-1} + k_2)c + k_1 se, \tag{1.3.23b}$$

with s fixed. The corresponding steady-state concentrations are given by

$$e = \frac{k_{-1} + k_2}{k_{-1} + k_2 + k_1[S]} E_0, \quad c = \frac{k_1[S]}{k_{-1} + k_2 + k_1[S]} E_0. \tag{1.3.24}$$

The steady-state flux for the cycle $E \rightarrow ES \rightarrow E$ is equal to the rate of product formation, and is given by the so-called Michaelis–Menten form

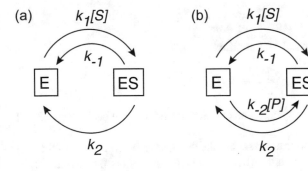

Fig. 1.4: The catalytic cycle for Michaelis–Menten kinetics. (a) Irreversible case. (b) Reversible case.

$$J[S] = \frac{k_1 k_2 [S] E_0}{k_{-1} + k_2 + k_1 [S]} = k_2 E_0 \frac{[S]}{[S] + K_M}, \tag{1.3.25}$$

where $K_M = (k_{-1} + k_2)/k_1$ is the Michaelis constant. Hence, the rate of production is linear in s when the substrate concentration is low but saturates when s is sufficiently large.

For a more extensive discussion of various schemes for enzymatic kinetics including Michaelis–Menten, see the books by Beard and Qian [57], Siegel and Edelstein-Keshet [744], and Keeener and Sneyd [431]. An important extension of Michaelis–Menten kinetics is the so-called Goldbeter–Koshland model of phosphorylation dephosphorylation cycles [316]. This will be analyzed in Chap. 10, and applied to certain examples of bacterial quorum sensing in Chap. 15. Here we briefly present a more detailed mathematical treatment of the quasi-steady-state approximation based on singular perturbation theory.

Asymptotic analysis of Michaelis–Menten kinetics. Take the initial concentrations at $t = 0$ to be $[S] = S_0$ and $[E] = E_0$, and introduce the dimensionless quantities $u = [S]/S_0$ and $v = [ES]/E_0$. Equations (1.3.21a, 1.3.21c) can then be rewritten as

$$\frac{du}{d\tau} = -u(1-v) + \frac{k_{-1}}{k_1 S_0} v, \tag{1.3.26a}$$

$$\varepsilon \frac{dv}{d\tau} = u(1-v) - \frac{K_M}{S_0} v, \tag{1.3.26b}$$

where $\tau = k_1 E_0 t$ and $\varepsilon = E_0/S_0$. If $\varepsilon \ll 1$ (fast catalytic reactions), then the second equation reaches a steady state much faster than the first equation. Hence, setting $\varepsilon = 0$ shows that $u(1-v) - (K_M/S_0)v = 0$, which implies

$$v = \frac{u S_0}{K_M + u S_0}. \tag{1.3.27}$$

Hence, as u slowly changes with time, v tracks these changes almost instantaneously so that the relationship (1.3.27) is maintained. Substituting into the equation for u then gives

$$\frac{du}{d\tau} = -\frac{k_2}{k_1} \frac{u}{K_M + u S_0},$$

which in terms of physical variables becomes

$$\frac{d[S]}{dt} = -J([S]).$$

Solving the ODE yields $u = u(\tau)$, and this can then be substituted back into (1.3.27) to give $v = v(\tau)$.

Note, however, that the above analysis breaks down at small times due to the fact that at $\tau = 0$ we have $u(0) = 1$ and $v(0) = 0$, which contradicts the result $v(0) = S_0/(K_M + S_0) \neq 0$ obtained from equation (1.3.27). This discrepancy arises from setting $\varepsilon = 0$ for all times $\tau > 0$, which consequently neglects the early kinetics where $dv/d\tau$ may be large compared to $du/d\tau$ so that $\varepsilon dv/d\tau$ is not small. The short-time behavior can be matched to the long-time behavior by carrying out the rescaling $\hat{\tau} = \tau/\varepsilon$. Equation (1.3.26) becomes

$$\frac{du}{d\hat{\tau}} = \varepsilon\left(-u(1-v) + \frac{k_{-1}}{k_1 S_0}v\right), \tag{1.3.28a}$$

$$\frac{dv}{d\hat{\tau}} = u(1-v) - \frac{K_M}{S_0}v. \tag{1.3.28b}$$

Now taking the limit $\varepsilon \to 0$, we see that on short time scales $du/d\hat{\tau} = 0$ and $u(\hat{\tau}) = 1$. On the other hand, $v(\hat{\tau})$ evolves according to the equation

$$\frac{dv}{d\hat{\tau}} = (1-v) - \frac{K_M}{S_0}v,$$

which has the solution

$$v(\hat{\tau}) = \frac{S_0}{K_M + S_0}\left(1 - e^{-(K_M+S_0)\hat{\tau}/S_0}\right).$$

Thus $v \to S_0/(K_M + S_0)$ as $\hat{\tau} \to \infty$ on the fast time scale, which matches the initial condition for v on the slow time scale ($\tau = 0$). Thus the two results match at $\hat{\tau} = \infty$ and $\tau = 0$. Within the context of singular perturbation analysis, this is called asymptotic matching. We will encounter several more complicated examples of singular perturbation methods in this book, including the narrow escape problem for diffusion (Chap. 6) and noise-induced escape in bistable systems (Chap. 8).

1.4 Exercises

Problem 1.1. Ensemble of identical, independent particles. Suppose that there are N independent, distinguishable molecules and each molecule can be in one of the set of energy states ε_n, $n = 0, 1, 2, \ldots$ The corresponding partition function is

$$Z_N = \sum_{n_1,\ldots n_N} e^{-(\varepsilon_{n_1} + \ldots \varepsilon_{n_N})/k_B T} = q^N,$$

where $q = \sum_n e^{-\varepsilon_n/k_B T}$.

(a) Show that the average total energy is given by

$$\langle E \rangle = N\langle \varepsilon \rangle = N\sum_n \varepsilon_n \eta_n, \quad \eta_n = \frac{e^{-\varepsilon_n/k_B T}}{q},$$

where η_n is the fraction of molecules in state ε_n. Calculate the corresponding variance $\mathrm{Var}[E]$.

(b) The specific heat is defined according to

$$C_V \equiv \frac{\partial \langle E \rangle}{\partial T}.$$

Using the fact that ε_n is independent of T, obtain the result

$$C_V = N \sum_n \varepsilon_n \frac{\partial \eta_n}{\partial T} = N \frac{\langle \varepsilon^2 \rangle - \langle \varepsilon \rangle^2}{k_B T^2}.$$

(c) Calculate q, E, C_V in the case where the energy states of a molecule are given by $\varepsilon_n = n\omega$ for $n = 0, 1, 2, \ldots$. Determine the behavior of E, C_V in the small T and high T limits.

Problem 1.2. Noninteracting spins in a magnetic field. Consider a system of N particles with so-called intrinsic spins $\sigma_i = \pm 1$, which are placed in a magnetic field H. A particle's spin is parallel (antiparallel) to the magnetic field if $\sigma_i = 1$ ($\sigma_i = -1$). Given a particular configuration $\sigma = \{\sigma_1, \ldots, \sigma_N\}$, the total energy of the system is

$$E(\sigma) = -\mu H \sum_{i=1}^{N} \sigma_i,$$

where μ is known as the size of the magnetic moment. The corresponding Boltzmann–Gibbs distribution is

$$p(\sigma) = \frac{1}{Z} e^{-E(\sigma)/k_B T}, \quad Z = \sum_{\sigma_1 = \pm 1} \cdots \sum_{\sigma_N = \pm 1} e^{-E(\sigma)/k_B T}.$$

(a) Show that the partition function is given by

$$Z = [2 \cosh(\beta \mu H)]^N, \quad \beta = \frac{1}{k_B T}.$$

(b) Define the magnetization M as

$$M(\sigma) = \sum_{i=1}^{N} \mu \sigma_i.$$

Show that the average magnetization, defined according to

$$\langle M \rangle = \sum_{\sigma_1 = \pm 1} \cdots \sum_{\sigma_N = \pm 1} M(\sigma) p(\sigma),$$

is given by

$$\langle M \rangle = N \mu \tanh(\beta \mu H).$$

(c) Show that

$$\text{Var}[M] = N\mu^2 \text{sech}^2(\beta\mu H) = \beta^{-1}\frac{\partial\langle M\rangle}{\partial H},$$

where $\partial\langle M\rangle/\partial H$ is known as the magnetic susceptibility.

(d) Determine the behavior of $\langle M\rangle$ and $\text{Var}[M]$ in the limit $T \to 0$.

> Remark 1.1. The magnetization statistics can be calculated by taking derivatives of $\ln Z$ with respect to βH. Analogs of magnetic spin systems arise within cell biology, including the cooperative binding of receptors (Chap. 3) and the configuration states of flexible polymers (Chap. 12). The relationship between $\text{Var}[M]$ and the magnetic susceptibility in (c) is an example of a fluctuation-dissipation theorem. That is, it expresses a relationship between the size of thermal fluctuations at equilibrium with the response of the system to an external perturbation (in this case changes in the external magnetic field.) Further examples will appear in Chap. 2, Chap. 6, Chap. 10 and Chap. 12.

Problem 1.3. Vertical column of gas. Consider a vertical column of N noninteracting gas particles in equilibrium at temperature T. Assume that the column has uniform cross-sectional area A. The potential energy of a particle at height z above the ground is mgz where m is the mass of each particle and g is the gravitational constant. Since the gas particles are noninteracting, the position of each particle is drawn independently from the single-particle Boltzmann distribution

$$p(z) = q^{-1}\exp(-mgz/k_BT),$$

where q is the single-molecule partition function

$$q = \int_0^\infty \exp(-mgz/k_BT)\,dz.$$

(a) Using the single-particle Boltzmann distribution, show that the mean particle density $n(z)$ varies with height above the ground according to

$$n(z) = \frac{mgN}{k_BTA}e^{-mgz/k_BT}.$$

(b) Let $P(z)$ be the gas pressure at a height z. By considering the balance between pressure and gravitational forces acting on a thin column of gas between z and $z+\delta z$, derive the equation

$$\frac{dP(z)}{dz} = -mgn(z).$$

Hence, deduce that the pressure at a given height satisfies $P(z) = k_BTn(z)$.

(c) What do cold mountain tops imply about the earth's atmosphere?

Problem 1.4. Excluded area. A two-dimensional box with sides of length L contains a gas of N hard disks of radius $r \ll L$. The area of the box is denoted by A. The gas is sufficiently dilute so that the summed area $N\pi r^2 \ll L^2$. Since each disk has a finite size, the closest the center of one disk can be to the center of another is

$2r$. Thus each disk has an excluded area of $\varepsilon = \pi(2r)^2$. In the following we assume that spatial units are non-dimensionalized.

(a) Suppose that the box is initially empty and that one particle at a time is placed into the box. Given that the number of states of the nth particle is equal to the available area of the box containing $n-1$ particles, and the order of placing particles does not matter, show that the total number of states is

$$\Omega(A) = \frac{1}{N!} \prod_{n=1}^{N} (A - (n-1)\varepsilon).$$

(b) Show that for large N, the entropy $S = k_B \ln \Omega(A)$ is approximately

$$S = Nk_B(1 + \ln(A/N - \varepsilon/2)).$$

Hint: Derive the formula

$$\sum_{n=1}^{N} \ln(A - (n-1)\varepsilon) = N\ln(A - (N-1)\varepsilon/2) + \mathcal{O}(\varepsilon^2),$$

and use Stirling's formula. Hence, calculate the entropic pressure of the gas, which is defined according to $p = T\partial S/\partial A$, and show that $p \to Nk_BT/A$ in the limit $\varepsilon \to 0$, which is a 2D version of the ideal gas law. Deduce that taking into account excluded volume increases the pressure compared to an ideal gas.

Problem 1.5. Particles in a box. Consider a box of volume V that is divided into K equal subvolumes V_0. Suppose that there are N particles in the box.

(a) Derive the following formula for the probability that there are n particles in a particular volume V_0:

$$p_n = \frac{1}{K^n} \left(1 - \frac{1}{K}\right)^{N-n} \frac{N!}{n!(N-n)!}.$$

(b) Show that in the limit $N, K \to \infty$ with $N_0 = N/K$ fixed and finite n, p_n reduces to the Poisson distribution

$$p_n = \frac{N_0^n e^{-N_0}}{n!}.$$

It follows that in this limit, the mean and variance of the number of particles in a volume V_0 are given by

$$\langle n \rangle = N_0, \quad \langle (n - N_0)^2 \rangle = N_0.$$

In particular, the mean distribution of particles is uniform. However, fluctuations are significant unless the mean number of particles in each subvolume, N_0, is also large.
(c) Now suppose that N_0 is large so that N is also large even for $K = 1$. Determine $\ln p_n$ from part (a) and apply Stirling's formula. Taylor expand the resulting expression for $\log p_n$ about $n = N_0$ up to second order in $n - N_0$. Hence show that

$$p_n \sim e^{-(n-N_0)^2/2\sigma_K^2}.$$

What is σ_K? Show that it is consistent with the result from part (b) in the limit $K \to \infty$.

Problem 1.6. Kinetic law of mass action.
(a) Consider the reaction network:

$$A \xrightarrow{k_1} X, \quad X \xrightarrow{k_2} Y, \quad X+Y \xrightarrow{k_3} B,$$

where the concentrations of A and B are buffered (i.e., [A] and [B] are fixed model parameters). Construct a differential equation model for the dynamics of [X] and [Y]. Determine the steady-state concentrations of X and Y as functions of [A] and the rate constants. Verify that the steady-state concentration of Y is independent of [A].

(b) Consider the reaction scheme

$$A+B \xrightarrow{k_1} C+D, \quad D \xrightarrow{k_2} B, \quad C \xrightarrow{k_3} E+F.$$

Write down the mass-action kinetic equations for $[A], [B], [C], [D]$ Using a conservation equation, determine the steady-state concentrations (some of which are zero).

(c) Repeat (b) when there is the additional reaction $\xrightarrow{k_0} A$, that is, A is produced at a rate k_0.

Chapter 2
Random walks and Brownian motion

When one first encounters the concept of diffusion, it is usually within the context of a conservation law describing the flux of many particles moving from regions of high concentration to regions of low concentration at a rate that depends on the local concentration gradient (Fick's law). However, there are some limitations of the standard macroscopic derivation of the diffusion equation. First, it does not take into account microscopic features of the environment within which the particles diffuse. This is crucial when considering diffusive processes within a cell, since the interior of the cell is highly heterogeneous. The same applies to surface diffusion within the plasma membrane. Second, with the use of advanced imaging techniques such as single-particle tracking, it is possible to observe the movement of individual molecules, which is highly stochastic, whereas the classical diffusion equation describes the collective motion of many particles and is deterministic.

In this chapter, we consider two different microscopic theories of diffusion: random walks and overdamped Brownian motion. Both approaches will be used to model diffusion within the complex cellular environment in Chap. 6. We begin in Sect. 2.1 by considering a discrete random walk on a 1D lattice, which is a simple example of a discrete Markov process. The probability distribution specifying the likelihood that the walker is at a particular lattice site after n time steps evolves according to a master equation. We show how the master equation can be solved using discrete Fourier and Laplace transforms, which in probability theory are known as characteristic functions and generating functions, respectively. The resulting solution is given by a Binomial distribution, which reduces to a Gaussian distribution in an appropriate continuum limit. Random walk models and various generalizations will later be used to model a variety of cellular processes, including molecular motors, polymerization of cytoskeletal filaments (Chap. 4), and anomalous diffusion (Chap. 6). We then show how the continuum limit of a random walk leads to the diffusion equation, which determines the probability density for the location of a single particle. The fundamental solution of the diffusion equation is a Gaussian, which naturally leads to the result that the mean-square displacement of the particle is linear in time.

© Springer Nature Switzerland AG 2021
P. C. Bressloff, *Stochastic Processes in Cell Biology*, Interdisciplinary Applied Mathematics 41, https://doi.org/10.1007/978-3-030-72515-0_2

In Sect. 2.2, we describe how individual trajectories of a diffusing particle correspond to realizations of a Wiener process (pure Brownian motion). It turns out that each sample path is continuous but non-differentiable everywhere, which means that we cannot write down a standard ordinary differential equation for the evolution of particle position. This leads to the theory of stochastic differential equations (SDEs) and Ito calculus. The SDE prescribes how the continuous variable $X(t)$ changes in each infinitesimal time step dt. Determining changes over finite times then requires evaluating an associated stochastic integral. We show how a particular type of SDE, known as the Langevin equation, can be used to model the motion of an overdamped Brownian particle moving in a fluid-like environment such as the cytoplasm of a cell. The Langevin equation describes the motion of the particle subject to a combination of frictional forces and a fluctuating diffusive force that are due to collisions with molecules in the surrounding fluid. Solutions of the Langevin equation represent random sample paths or trajectories of the particle. We also introduce the notion of the power spectrum of a continuous stochastic process, which is particularly useful when studying linear SDEs.

In Sect. 2.3, we consider the probability density on the space of sample paths of a Brownian particle, which evolves according to a Fokker–Planck (FP) equation. We establish the connection between SDEs and the FP equation, and illustrate the theory by considering the well-known Ornstein–Uhlenbeck (OU) process. We explicitly solve the corresponding FP equation using an important analytical tool in the theory of linear partial differential equations (PDEs), known as the method of characteristics. We then consider the case of multiplicative noise, in which the noise term explicitly depends on the state of the system. This could arise, for example, in the case of nonlinear Brownian motion where the diffusivity depends on position. We show how the interpretation of multiplicative noise is ambiguous due to the subtleties of stochastic integration, resulting in different versions of the FP equation in the presence of multiplicative noise, and the important distinction between Ito and Stratonovich stochastic calculus. In order to select the appropriate version, additional physical constraints are required. For example, consistency of nonlinear Brownian motion with equilibrium statistical physics yields the so-called kinetic interpretation (see also Chap. 12).

In Sect. 2.4, we introduce one of the most important characteristics of a diffusion process, namely the mean first passage time (MFPT) to reach a given target or boundary. We derive a differential equation for the MFPT by introducing the notion of a survival probability and showing that the latter evolves according to a backward FP equation. This is then used to estimate the mean time to escape a minimum of a double-well potential for a bistable system (Kramer's escape rate). We end the chapter with a summary of numerical methods for solving SDEs.

Remark 2.1. Note that for continuous stochastic processes, SDEs and FP equations will appear throughout the two volumes. In addition to describing the diffusive-like motion of microscopic particles in solution, they also frequently appear in diffusion approximations of discrete Markov processes, where the continuous variables now represent the concentrations of various chemical species. Examples include voltage-gated and ligand-gated ion channels (Chap. 3), gene networks (Chap. 5), and biochemical signaling networks (Chap. 10, Chap. 11 and Chap. 15). Moreover, many of these cellular processes exhibit bistability or limit cycle oscillations in the deterministic or thermodynamic limit, which will require estimating a noise-induced escape rate or analyzing stochastic phase dynamics.

2.1 Random walks and diffusion

2.1.1 Discrete-time random walks

Suppose that a particle hops at discrete times between neighboring sites on a lattice with unit spacing [403]. The examples of a one-dimensional (1D) lattice and a two-dimensional (2D) square lattice are shown in Fig. 2.1. For concreteness, let us focus on the 1D case. At each step, the random walker moves a unit distance to the right with probability p or to the left with probability $q = 1 - p$. Let $P_N(r)$ denote the probability that the particle is at site r at the Nth time step. The evolution of the probability distribution is described by the discrete-time master equation

$$P_N(r) = pP_{N-1}(r-1) + qP_{N-1}(r+1), \quad r \in \mathbb{Z}, \quad N \geq 1. \tag{2.1.1}$$

If $q = p = 1/2$ then the random walk is symmetric or unbiased, whereas for $p > q$ ($p < q$) it is biased to the right (left). We will analyze this equation using transform methods, since these can be generalized to more complex random walk models such as continuous-time random walks (see Sect. 6.1). An introduction to continuous and discrete transform methods can be found in Box 2A.

The characteristic function. Introduce the characteristic function (discrete Fourier transform) for fixed N

$$G_N(k) = \sum_{r=-\infty}^{\infty} e^{ikr} P_N(r), \quad k \in [-\pi, \pi]. \tag{2.1.2}$$

The characteristic function generates moments of the random displacement variable r according to

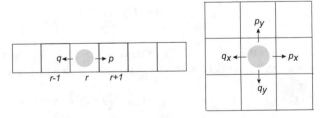

Fig. 2.1: A random walk on (a) a 1D lattice with $p + q = 1$ and (b) a 2D square lattice with $p_x + q_x + p_y + q_y = 1$.

$$\left(-i\frac{d}{dk}\right)^m G_N(k)\bigg|_{k=0} = \sum_{r=-\infty}^{\infty} r^m P_N(r) = \langle r^m \rangle, \tag{2.1.3}$$

where $\langle r^m \rangle$ is the mth order moment of r. Substituting for $P_N(r)$ using equation (2.1.1) gives

$$\begin{aligned}
G_N(k) &= \sum_{r=-\infty}^{\infty} e^{ikr}[pP_{N-1}(r-1) + qP_{N-1}(r+1)] \\
&= \sum_{r=-\infty}^{\infty} \left[pe^{ik}e^{ik(r-1)}P_{N-1}(r-1) + qe^{-ik}e^{ik(r+1)}P_{N-1}(r+1) \right] \\
&= u(k)G_{N-1}(k),
\end{aligned}$$

where $u(k) = pe^{ik} + qe^{-ik}$. Iterating this equation, we have

$$G_N(k) = u(k)G_{N-1}(k) = u(k)^2 G_{N-2}(k) = u(k)^N G_0(k).$$

Assuming that the particle starts at the origin, $P_0(r) = \delta_{r,0}$, so that $G_0(k) = 1$, it follows that $G_N(k) = u(k)^N$. Taking the inverse Fourier transform, one finds that (Ex. 2.1):

$$P_N(r) = \frac{1}{2\pi}\int_{-\pi}^{\pi} e^{-ikr}u(k)^N dk = \frac{N!}{\left(\dfrac{N+r}{2}\right)!\left(\dfrac{N-r}{2}\right)!} p^{(N+r)/2}q^{(N-r)/2} \tag{2.1.4}$$

when $N+r$ is an even integer and zero otherwise. The distribution (2.1.4) is known as the binomial distribution. In the unbiased case $p = q = 1/2$, it gives the probability of a total of r heads in tossing a fair coin N times and is known as the Bernoulli distribution.

Gaussian approximation of $P_N(r)$. Evaluating $\ln P_N(r)$ for large N using Stirling's approximation

$$\ln N! \approx N\ln N - N + \frac{1}{2}\ln(2\pi N), \tag{2.1.5}$$

and assuming $p, q = 1/2$, one finds that (see Ex. 2.1)

$$P_N(r) \sim \frac{1}{\sqrt{2\pi N}}e^{-[r-N(p-q)]^2/2N}. \tag{2.1.6}$$

Indeed, the Gaussian form of $P_N(r)$ in the large time limit arises universally whenever the mean and variance of the displacement $\Delta r = r - r'$ in a single step are finite, that is

$$\langle \Delta r \rangle = \sum_{\Delta r}\Delta r\, p(\Delta r) < \infty, \quad \langle \Delta r^2 \rangle = \sum_{\Delta r}(\Delta r)^2 p(\Delta r) < \infty,$$

where $p(\Delta r)$ is the probability of a step of length Δr. (In the standard 1D random walk, $\Delta r = \pm 1$ and $p(1) = p, p(-1) = q$.) One way to see this is to note that $u(k)$ has the small-k series expansion

$$u(k) = \sum_{\Delta r} e^{ik\Delta r} p(\Delta r) = 1 + ik\langle\Delta r\rangle - \frac{1}{2}k^2\langle\Delta r^2\rangle + \ldots \sim e^{ik\langle\Delta r\rangle - \frac{1}{2}k^2\langle[\Delta r - \langle\Delta r\rangle]^2\rangle}.$$

Substitute the above approximation into equation (2.1.4), after noting that the integral is dominated by the behavior in the region around $k = 0$ when N is large. The resulting Gaussian integral yields the approximation

$$P_N(r) \sim \frac{1}{\sqrt{2\pi N\sigma^2}} e^{-(r - N\langle\Delta r\rangle)^2/2N\sigma^2}, \tag{2.1.7}$$

with $\sigma^2 = \langle\Delta r^2\rangle - \langle\Delta r\rangle^2$. Such a result follows from the central limit theorem [343].

Transient and recurrent random walks. Another useful quantity when analyzing random walks is the generating function (discrete Laplace transform or one-sided z-transform)

$$\Gamma(r,z) = \sum_{N=0}^{\infty} z^N P_N(r). \tag{2.1.8}$$

It is often simpler to evaluate the generating function in Fourier space,

$$\widehat{\Gamma}(k,z) \equiv \sum_{r=-\infty}^{\infty} e^{ikr} \Gamma(r,z) = \sum_{N=0}^{\infty} z^N G_N(k)$$

assuming that we can reverse the order of summations. Since $G_N(k) = u(k)^N$, we can sum the resulting geometric series to obtain the result

$$\widehat{\Gamma}(k,z) = \frac{1}{1 - zu(k)}.$$

The generating function is thus given by the inverse Fourier transform

$$\Gamma(r,z) = \int_{-\pi}^{\pi} \frac{e^{-ikr}}{1 - zu(k)} \frac{dk}{2\pi}.$$

It can be shown (see Ex. 2.2) that for $r = 0$ and $p = q = 1/2$ (unbiased random walk),

$$\Gamma(0,z) = (1 - z^2)^{-1/2}.$$

One immediate consequence of this result is that an unbiased 1D random walk is recurrent, which means that the walker is certain to return to the origin; a random walk is said to be transient if the probability of returning to the origin is less than one. Recurrence follows from the observation that $\Gamma(0,1) = \sum_{N=0}^{\infty} P_N(0)$ is the mean number of times that the walker visits the origin, and

$$\lim_{z \to 1^-} \Gamma(0,z) = \infty$$

for the 1D random walk. Interestingly, although the 1D random walk is recurrent, the mean time to return to the origin for the first time is infinite. This result can also be established using transform methods and generating functions (see Ex. 2.3). An unbiased random walk in 2D is also recurrent but in 3D it is transient. Finally, note that discrete random walks have also been used to describe the coiling of flexible polymers [101, 409] (Chap. 12).

Box 2A. Transform methods

Throughout this book, we will make extensive use of transform methods, in particular, Laplace and Fourier integral transforms and their discrete analogs. Here we provide a basic introduction to such methods; see also [529].

Laplace transforms. Let $u(t)$ be a piecewise continuous function that is of exponential order, that is,

$$u(t) \leq ce^{at}, \text{ as } t \to \infty,$$

for constants $a, c > 0$. The Laplace transform of u is defined by

$$\mathscr{L}u(s) \equiv \widetilde{u}(s) = \int_0^\infty u(t)e^{-st}ds, \tag{2.1.9}$$

and one often writes $\mathscr{L}u = \widetilde{u}$. The Laplace transform operator $\mathscr{L}$ is linear, since

$$\mathscr{L}(c_1 u_1 + c_2 u_2) = c_1 \mathscr{L}u_1 + c_2 \mathscr{L}u_2$$

for constants c_1, c_2. One of the important features of the Laplace transform (and the Fourier transform) is that it converts differential operations in the time domain into multiplication operations in the transform domain. For example, setting $u' = du/dt$, etc.,

$$\mathscr{L}u'(s) = s\widetilde{u}(s) - u(0), \tag{2.1.10a}$$

$$\mathscr{L}u''(s) = s^2\widetilde{u}(s) - su(0) - u'(0), \tag{2.1.10b}$$

which can be proved using integration by parts. It follows that Laplace transforming an ordinary differential equation for $u(t)$ yields an algebraic equation for $\widetilde{u}(s)$. The most difficult step, once one has solved the algebraic equation, is to find the inverse Laplace transform to recover $u(t)$. The general formula for the inverse transform requires knowledge of contour integration, and takes the form

$$u(t) = \mathscr{L}^{-1}\widetilde{u}(t) = \frac{1}{2\pi i}\int_{a-i\infty}^{a+i\infty} \widetilde{u}(s)e^{st}ds. \tag{2.1.11}$$

The complex contour integral is taken over the infinite vertical line (the Bromwich path) in the complex plane from $a - i\infty$ to $a + i\infty$. The real number a is chosen so that the Bromwich path lies to the right of any singularities (poles, branch points and cuts, essential points) of the function $\widetilde{u}(s)$. The evaluation of the contour integral is often difficult. However, many of the Laplace transforms encountered in this book can be found in Table 2.1. One additional useful property of Laplace transforms is expressed by the convolution theorem:

Theorem 2.1. *Let u and v be piecewise continuous for $t \geq 0$ and of exponential order. Then*

$$\mathscr{L}(u*v)(s) = \widetilde{u}(s)\widetilde{v}(s), \qquad (2.1.12)$$

where

$$u*v(t) \equiv \int_0^t u(t-y)v(y)dy \qquad (2.1.13)$$

is the convolution of u and v. It immediately follows that $\mathscr{L}^{-1}(\widetilde{u}\widetilde{v}) = u*v$.

In the case of a discrete-time linear process, we can use a discrete version of the Laplace transform (also known as a one-sided z-transform)

$$\widetilde{u}(z) = \sum_{n=0}^{\infty} z^n u_n. \qquad (2.1.14)$$

Applying this to the first-order difference equation $u_n = au_{n-1}$ for $n \geq 1$ yields

$$\widetilde{u}(z) = az\widetilde{u}(z) + u_0 \implies \widetilde{u}(z) = \frac{u_0}{1-az} = u_0 \sum_{n=0}^{\infty} a^n z^n.$$

The series converges provided that $|az| < 1$, in which case we immediately see that $u_n = a^n u_0$. More generally, the inverse z-transform is given by the complex integral around a closed contour C around the origin in the z-plane that does not contain any singularities of $\widetilde{u}(z)$:

$$u_n = \oint_C \frac{\widetilde{u}(z)}{z^{n+1}} \frac{dz}{2\pi i}. \qquad (2.1.15)$$

However, one can often avoid using contour integration by simply Taylor expanding the z-transform in powers of z and reading off the coefficient of z^n, as in the above example.

Fourier transforms. The Fourier transform of a function of one variable $u(x)$, $x \in \mathbb{R}$, is defined by the equation

$$\mathscr{F}u(k) \equiv \widehat{u}(k) = \int_{-\infty}^{\infty} u(x)e^{ikx}dx. \qquad (2.1.16)$$

The corresponding inverse Fourier transform is

$$\mathscr{F}^{-1}\widehat{u}(x) = \frac{1}{2\pi} \int_{-\infty}^{\infty} \widehat{u}(k)e^{-ikx}dk. \qquad (2.1.17)$$

An important issue is to determine the set of functions for which the Fourier transform (and its inverse) are well-defined. For example, if u is integrable on $\mathbb{R}$ so that $\int_{-\infty}^{\infty} |u(x)| dx < \infty$, then

$$|\widehat{u}(k)| = \left| \int_{-\infty}^{\infty} u(x) e^{ikx} dx \right| \leq \int_{-\infty}^{\infty} |u(x)| dx < \infty,$$

and $\widehat{u}$ exists. However, the latter may itself not be integrable. Therefore, in the application of Fourier transforms, it is common to take u to belong to the space of square-integrable functions denoted by $L^2(\mathbb{R})$; see also Box 2D. A few important properties of the Fourier transform are as follows. First, it converts derivatives into algebraic expressions, that is,

$$\mathscr{F} u^{(n)}(k) = (-ik)^n \widehat{u}(k), \tag{2.1.18}$$

where $u^{(n)}$ denotes the nth derivative of u, and assuming that u and its derivatives are continuous and integrable. There also exists a convolution theorem:

Theorem 2.2. *If u and v are in $L^2(\mathbb{R})$, then*

$$\mathscr{F}(u * v)(k) = \widehat{u}(k)\widehat{v}(k), \tag{2.1.19}$$

where

$$(u * v)(x) \equiv \int_{-\infty}^{\infty} u(x - y)v(y) dy. \tag{2.1.20}$$

Proof. The theorem is established by interchanging the order of integration:

$$\mathscr{F}(u * v)(k) = \int_{-\infty}^{\infty} \left(\int_{-\infty}^{\infty} u(x - y)v(y) dy \right) e^{ikx} dx$$

$$= \int_{-\infty}^{\infty} \left(\int_{-\infty}^{\infty} u(x - y)v(y) e^{ikx} dx \right) dy$$

$$= \int_{-\infty}^{\infty} \left(\int_{-\infty}^{\infty} u(r)v(y) e^{ikr} e^{iky} dr \right) dy$$

$$= \int_{-\infty}^{\infty} u(r) e^{ikr} dr \int_{-\infty}^{\infty} v(y) e^{iky} dy = \widehat{u}(k)\widehat{v}(k).$$

Yet another useful property is Parseval's theorem

$$\int_{-\infty}^{\infty} |u(x)|^2 dx = \frac{1}{2\pi} \int_{-\infty}^{\infty} |\widehat{u}(k)|^2 dk. \tag{2.1.21}$$

Just as one can define a discrete Laplace transform for discrete-time processes, one can also introduce a discrete Fourier transform of spatial processes such as a random walk, which are defined on a discrete lattice. Therefore, suppose that u is a function on the space of integers $\mathbf{Z}$. The discrete Fourier transform of u is defined according to

$$(\mathscr{F}u)(k) \equiv \widehat{u}(k) = \sum_{r=-\infty}^{\infty} u(r)e^{ikr}, \qquad (2.1.22)$$

where k is now restricted to the finite domain $(-\pi, \pi)$. The intuition behind this is that for $|k| > \pi$, the spatial oscillations $\cos(kr)$ and $\sin(kr)$ probe the function on spatial scales smaller than a unit lattice spacing where there is no information, and are thus redundant. The inverse transform is

$$u(r) = \int_{-\pi}^{\pi} \widehat{u}(k)e^{-ikr}\frac{dk}{2\pi}. \qquad (2.1.23)$$

This is straightforward to prove using the identities

$$\int_{-\pi}^{\pi} e^{ik(r-s)}\frac{dk}{2\pi} = \frac{1}{2\pi i(r-s)}\left[e^{i\pi(r-s)} - e^{-i\pi(r-s)}\right] = 0 \text{ for } r \neq s,$$

and $\int_{-\pi}^{\pi} dk/2\pi = 1$. That is, substituting for $\widehat{u}(k)$ in the inverse transform and reversing the order of integration and summation,

$$\int_{-\pi}^{\pi} \widehat{u}(k)e^{-ikr}\frac{dk}{2\pi} = \int_{-\pi}^{\pi}\left(\sum_{s=-\infty}^{\infty} u(s)e^{iks}\right)e^{-ikr}\frac{dk}{2\pi}$$

$$= \sum_{s=-\infty}^{\infty} u(s)\int_{-\pi}^{\pi} e^{ik(s-r)}\frac{dk}{2\pi} = \sum_{s=-\infty}^{\infty} u(s)\delta_{s,r} = u(r).$$

Note that the discrete Fourier transform should be distinguished from a Fourier series, which is an expansion of a periodic function of x in terms of a countable set of Fourier components. In other words, in a Fourier series k is unbounded but takes discrete values. Finally, consider a higher-dimensional square lattice with points $\ell = n_1\ell_1 + n_2\ell_2$. The corresponding discrete Fourier transform (for $d = 2$) is

$$(\mathscr{F}u)(\mathbf{k}) \equiv \widehat{u}(\mathbf{k}) = \sum_{n_1=-\infty}^{\infty}\sum_{n_2=-\infty}^{\infty} u(\ell)e^{i\mathbf{k}\cdot\ell}, \qquad (2.1.24)$$

with $\mathbf{k}$ the dual vector

$$\mathbf{k} = k_1\hat{\ell}_1 + k_2\hat{\ell}_2, \quad k_1, k_2 \in (-\pi, \pi).$$

$u(t)$	$\tilde{u}(s)$	$u(t)$	$\tilde{u}(s)$
1	$s^{-1}, \quad s>0$	$f(t)e^{-at}$	$\tilde{f}(s+a)$
e^{at}	$\frac{1}{s-a}, \quad s>a$	$\delta(t-a)$	$\exp(-as)$
t^n	$\frac{n!}{s^{n+1}}, \quad s>0$	$H(t-a)f(t-a)$	$\tilde{f}(s)e^{-as}$
$\sin(at), \cos(at)$	$\frac{a}{s^2+a^2}, \frac{s}{s^2+a^2} \quad s>0$	$\mathrm{erf}(\sqrt{t})$	$s^{-1}(1+s)^{-1/2}, \quad s>0$
$\sinh(at), \cosh(at)$	$\frac{a}{s^2-a^2}, \frac{s}{s^2-a^2} \quad s>\lvert a\rvert$	$t^{-1/2}\exp(-a^2/4t)$	$\sqrt{\pi/s}\exp(-a\sqrt{s}), \quad s>0$
$e^{at}\sin(bt)$	$\frac{b}{(s-a)^2+b^2} \quad s>a$	$1-\mathrm{erf}(a/2\sqrt{t})$	$s^{-1}\exp(-a\sqrt{s}), \quad s>0$
$e^{at}\cos(bt)$	$\frac{s-a}{(s-a)^2+b^2}, \quad s>a$	$\frac{a}{2t^{3/2}}\exp(-a^2/4t)$	$\sqrt{\pi}\exp(-a\sqrt{s}), \quad s>0$

Table 2.1: Some common Laplace transforms.

2.1.2 Diffusion as the continuum limit of a random walk

Having analyzed the discrete random walk, it is now possible to take an appropriate continuum limit to obtain a diffusion equation in continuous space and time. First, introduce infinitesimal step lengths δx and δt for space and time and set $P_N(r) = p(x,t)\delta x$ with $x = r\delta x, t = N\delta t$. Substituting into the master equation (2.1.1) for $p = q = 1/2$ gives the following equation for the probability density $p(x,t)$:

$$p(x,t) = \frac{1}{2}p(x-\delta x,t-\delta t) + \frac{1}{2}p(x+\delta x,t-\delta t)$$

$$\approx \left[p(x,t) - \frac{\partial p}{\partial t}\delta t\right] + \frac{1}{2}\frac{\partial^2 p}{\partial x^2}\delta x^2,$$

where p has been Taylor expanded to first order in δt and to second order in δx. Dividing through by δt and taking the continuum limit $\delta x, \delta t \to 0$ such that the quantity D is finite, where

$$D = \lim_{\delta x, \delta t \to 0} \frac{\delta x^2}{2\delta t},$$

yields the diffusion equation with diffusivity D

$$\frac{\partial p(x,t)}{\partial t} = D\frac{\partial^2 p(x,t)}{\partial x^2}. \tag{2.1.25}$$

Note that the solution $p(x,t)$ to the diffusion equation describes the probability density function for the location of a single particle. In particular, $\int_\Omega p(x,t)dx$ can be identified as the probability that the particle is in region Ω at time t. A particle undergoing pure diffusion is said to exhibit pure Brownian motion.

Remark 2.2. Although we have derived the diffusion equation from an unbiased random walk, it is more typically interpreted in terms of an evolution equation for a conserved quantity such as particle number rather than a probability density for a single random walker. In order to link these two interpretations, consider N non-interacting, identical diffusing particles and let $u(x,t) = Np(x,t)$. For sufficiently large N, we can treat $u(x,t)dx$ as the deterministic number of particles in the infinitesimal interval $[x,x+dx]$ at time t, with $u(x,t)$ evolving according to the diffusion equation written in the conservation form

$$\frac{\partial u}{\partial t} = -\frac{\partial J}{\partial x}, \quad J(x,t) = -D\frac{\partial u}{\partial x},$$

where $J(x,t)$ is known as the Fickian flux. Integrating the diffusion equation over the interval $[x,x+dx]$ and reversing the order of integration and differentiation shows that

$$\frac{d}{dt}\int_x^{x+dx} u(y,t)dy = J(x,t) - J(x+dx,t),$$

which is an expression of particle conservation. That is, the rate of change of the number of particles in $[x,x+dx]$ is equal to the net flux crossing the endpoints of the interval.

One method for solving the diffusion equation, given the initial condition $p(x,0) = f(x)$, is to use Fourier transforms (see Box 2A). In particular, Fourier transforming equation (2.1.25) with respect to x gives

$$\frac{d\hat{p}(k,t)}{dt} = -k^2 D\hat{p}(k,t), \quad \hat{p}(k,t) = \int_{-\infty}^{\infty} p(x,t)e^{ikx}dx,$$

which is a differential equation in t with k treated as a parameter. Its solution is

$$\hat{p}(k,t) = c(k)e^{-k^2 Dt},$$

with the coefficient $c(k)$ determined by the initial data. That is, Fourier transforming the initial condition implies $\hat{p}(k,0) = \hat{f}(k)$ and, hence,

$$\hat{p}(k,t) = \hat{f}(k)e^{-k^2 Dt}.$$

Since $\hat{p}(k,t)$ is the product of two Fourier transforms, its inverse Fourier transform is given by a convolution:

$$p(x,t) = \int_{-\infty}^{\infty} K(x-y,t)f(y)dy,$$

where $K(x,t)$ is the inverse Fourier transform of $e^{-k^2 Dt}$:

$$K(x,t) = \frac{1}{2\pi}\int_{-\infty}^{\infty} e^{-ikx}e^{-k^2 Dt}dk = \frac{1}{\sqrt{4\pi Dt}}e^{-x^2/4Dt}. \tag{2.1.26}$$

The function $K(x,t)$ is known as the fundamental solution of the diffusion equation under the initial condition $f(x) = \delta(x)$, where $\delta(x)$ is the Dirac delta function (see Box 2B). (Strictly speaking, $p = K(x,t)$ is a weak solution of the underlying diffusion equation [717].) Also observe that the fundamental solution corresponds to the continuum limit of the Gaussian distribution (2.1.7) for an unbiased random walk. It is often useful to make explicit the initial position x_0 of a particle at some time t_0 by expressing the solution of the diffusion equation as the conditional probability density $p(x,t|x_0,t_0)$ with $p(x,t_0|x_0,t_0) = \delta(x-x_0)$. Solving the diffusion equation (in one spatial dimension) then gives

$$p(x,t|x_0,t_0) = \frac{1}{2\sqrt{\pi D(t-t_0)}} \exp\left(-\frac{(x-x_0)^2}{4D(t-t_0)}\right), \quad t > t_0. \tag{2.1.27}$$

It immediately follows from properties of Gaussians that the mean and variance, also known as the mean-square displacement (MSD), are (for $x_0 = t_0 = 0$)

$$\langle x \rangle = \int_{-\infty}^{\infty} x p(x,t|0,0)dx = 0, \quad \langle x^2 \rangle = \int_{-\infty}^{\infty} x^2 p(x,t|0,0)dx = 2Dt. \tag{2.1.28}$$

Similarly for normal diffusion in d spatial dimensions, one finds that $\langle \mathbf{x}^2 \rangle = 2dDt$.

Box 2B. Dirac delta function.

A heuristic definition of the Dirac delta function would be that it is a "function" with the following properties:

$$\delta(0) = \infty, \quad \delta(x) = 0 \text{ for all } x \neq 0, \quad \int_{\mathbb{R}} \delta(x)dx = 1.$$

However, this definition is not compatible with the classical concept of a function. A rigorous definition of the Dirac delta function requires the theory of generalized functions or distributions [717]. However, an operational definition of the Dirac delta function can be constructed in terms of the limit of a sequence of Heaviside functions. Let $H(x) = 1$ if $x \geq 0$ and $H(x) = 0$ if $x < 0$. It follows from this definition that

$$I_\varepsilon(x) \equiv \frac{H(x+\varepsilon) - H(x-\varepsilon)}{2\varepsilon} = \begin{cases} 1/2\varepsilon & \text{if } -\varepsilon \leq x < \varepsilon \\ 0 & \text{otherwise} \end{cases}.$$

It can be seen that $I_\varepsilon(x)$ has the following properties, see Fig. 2.2:
(i) For all $\varepsilon > 0$,

$$\int_{\mathbb{R}} I_\varepsilon(x)dx = \frac{1}{2\varepsilon} \times 2\varepsilon = 1.$$

(ii)

$$\lim_{\varepsilon \to 0} I_\varepsilon(x) = \begin{cases} 0 & \text{if } x \neq 0 \\ \infty & \text{if } x = 0 \end{cases}.$$

(iii) If $\varphi(x)$ is a smooth function that vanishes outside a bounded interval (a test function), then

$$\int_{\mathbb{R}} I_\varepsilon(x)\varphi(x)dx = \frac{1}{2\varepsilon}\int_{-\varepsilon}^{\varepsilon}\varphi(x)dx \xrightarrow[\varepsilon\to 0]{} \varphi(0).$$

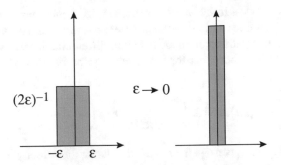

Fig. 2.2: One approximation of the Dirac delta function.

The third property suggests that we can define the Dirac delta function in terms of how it acts on test functions. Thus the Dirac delta function is defined as a distribution with the following properties:

$$\int_{\mathbb{R}} \delta(x)dx = 1, \qquad \int_{\mathbb{R}} \delta(x)\varphi(x)dx = \varphi(0).$$

One can also introduce a shifted Dirac delta function $\delta_y(x) \equiv \delta(x-y)$,

$$\int_{\mathbb{R}} \delta_y(x)dx = 1, \qquad \int_{\mathbb{R}} \delta_y(x)\varphi(x)dx = \int_{\mathbb{R}} \delta(x-y)\varphi(x)dx = \varphi(y).$$

The Heaviside construction also suggests that we can formally write $H'(x) = \delta(x)$, although again this only really makes sense in terms of test functions:

$$\int_{\mathbb{R}} H'(x)\varphi(x)dx = [H(x)\varphi(x)]_0^\infty - \int_{\mathbb{R}} H(x)\varphi'(x)dx = -\int_0^\infty \varphi'(x)dx = \varphi(0).$$

We have used integration by parts and the fact that $\varphi(x) = 0$ at $x = \infty$. Finally, note that alternative representations of the Dirac delta function include the Fourier integral,

$$\delta(x) = \frac{1}{2\pi}\int_{-\infty}^{\infty} e^{-ikx}dk,$$

and the $t \to 0$ limit of the fundamental solution (2.1.26),

$$\delta(x) = \lim_{t\to 0} \frac{1}{\sqrt{4\pi Dt}}e^{-x^2/4Dt}.$$

2.2 Stochastic differential equations and Ito calculus

The diffusion equation (2.1.25) determines the probability distribution that a single particle is at a particular place at some given time, but does not describe how that particle actually moves. If one were to run multiple simulations of the particle motion, then each time one would generate a different sample path or trajectory. It turns out that each sample path is continuous but non-differentiable everywhere. One way to see continuity is to note that for any $\varepsilon > 0$, the probability of escaping from a region of size ε in time Δt is

$$\int_{|x-z|>\varepsilon} p(x,t+\Delta t|z,t)dx = 2\int_{\varepsilon}^{\infty} \frac{1}{2\sqrt{\pi D\Delta t}} \exp\left(-\frac{x^2}{4D\Delta t}\right) dx = \int_{\frac{\varepsilon}{2\sqrt{D\Delta t}}}^{\infty} e^{-x^2} dx,$$

which approaches zero as $\Delta t \to 0$. On the other hand, the velocity of the particle can be arbitrarily large,

$$\mathbb{P}\left(\frac{X(t+\Delta t)-X(t)}{\Delta t} > k\right) = \int_{k\Delta t}^{\infty} \frac{1}{2\sqrt{\pi D\Delta t}} \exp\left(-\frac{x^2}{4D\Delta t}\right) dx$$

$$= \int_{\frac{k}{2}\sqrt{\frac{\Delta t}{\pi D}}}^{\infty} e^{-x^2} dx \to \frac{1}{2}$$

in the limit $\Delta t \to 0$. In other words, with unit probability, the absolute value of the velocity is larger than any number k, hence infinite. Since dX/dt does not exist, we cannot write down a standard ODE for the evolution of $X(t)$.

2.2.1 Wiener process

Recall from calculus that the ordinary differential equation $dx/dt = f(x)$ can be obtained from the equation

$$\Delta x(t) = x(t+\Delta t) - x(t) = f(x(t))\Delta t,$$

by dividing both sides by Δt and taking the limit $\Delta t \to 0$. On the other hand, for small but finite Δt, we can interpret the above equation as a difference equation in which the time interval $[0,t]$ is divided into N increments of size $\Delta t = t/N$. Setting $x_n = x(n\Delta t)$, we have

$$x_{n+1} = x_n + f(x_n)\Delta t,$$

which can be iterated to give

$$x_N = x_0 + \sum_{m=0}^{N-1} f(x_m)\Delta t.$$

Retaking the limit $\Delta t \to 0$ yields the integral equation

$$x(t) = x_0 + \int_0^t f(x(s))ds.$$

Analogously, we can write down a difference equation for the position of a diffusing particle according to

$$X_{n+1} - X_n = \sqrt{2D}\Delta W_n, \qquad (2.2.1)$$

where ΔW_n, $n = 0, \ldots, N-1$, are independent and identically distributed Gaussian variables with zero mean and variance $\sigma^2 = \Delta t$:

$$P(\Delta W) = \frac{1}{\sqrt{2\pi\Delta t}}e^{-(\Delta W)^2/2\Delta t}. \qquad (2.2.2)$$

(Note that a sequence of random variables is independent and identically distributed (i.i.d.) if each random variable has the same probability distribution as the others and all are mutually independent.) Iterating equation (2.2.1) gives (for $X_0 = 0$) $X_N = \sum_{m=0}^{N-1} \Delta W_m$. Using the fact that the sum of Gaussian random variables is also a Gaussian, it follows that the probability density for X_N is a Gaussian. Thus, we only need to determine its mean and variance. Since the ΔW_n are all independent, we have

$$\langle X_N \rangle = \sqrt{2D} \sum_{n=0}^{N-1} \langle \Delta W_n \rangle = 0, \quad \text{Var}(X_N) = 2D \sum_{n=0}^{N-1} \text{Var}(\Delta W_n) = 2DN\Delta t,$$

and

$$P(X_N) = \frac{1}{\sqrt{4\pi DN\Delta t}}e^{-X^2/(4DN\Delta t)}.$$

A corresponding continuous-time process can be constructed by taking the limit $N \to \infty$ such that $\Delta t \to 0$ with $N\Delta t = t$ fixed. In particular,

$$X(t) = \sqrt{2D} \lim_{N\to\infty} \sum_{n=0}^{N-1} \Delta W_n = \sqrt{2D} \int_0^t dW(s) \equiv \sqrt{2D}W(t).$$

(For the moment, we will not worry about the precise meaning of convergence and limits of stochastic variables—this will be addressed below.) $X(t)$ is still a Gaussian, whose mean and variance are obtained by taking the limit $N \to \infty$ of the results for X_N. We deduce that $X(t)$ has the Gaussian probability density

$$P(x,t) = \frac{1}{\sqrt{4\pi Dt}}e^{-x^2/4Dt},$$

which has zero mean and MSD that varies as $2Dt$. The corresponding function $W(s)$ is known as a Wiener process or pure Brownian motion—a Gaussian process with independent and stationary increments, zero mean, and variance $\sigma_W^2 = t$. That is,

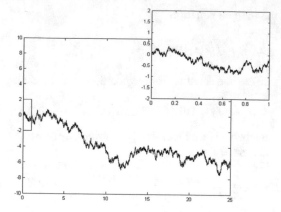

Fig. 2.3: Sample path of a Wiener process.

$$\mathbb{P}[w < W(t) < w + dw] = \frac{1}{\sqrt{2\pi t}} e^{-w^2/2t} dw,$$

and, for $s < t$,

$$\langle [W(t) - W(s)]W(s)\rangle = 0 = \langle W(t)W(s)\rangle - s.$$

Allowing for the possibility that $t < s$, we have

$$\langle W(t) = 0, \quad \langle W(t)W(s)\rangle = \min\{t,s\}. \tag{2.2.3}$$

An example of pure Brownian motion is shown in Fig. 2.3.

The above establishes that sample paths of a diffusing particle are given by a Wiener process, and the distribution of sample paths is the fundamental solution of the diffusion equation. Now consider a particle moving in an aqueous environment and subject to some external force F. Collisions with fluid molecules have two distinct effects: they induce the diffusive motion of the particle, and they generate an effective frictional force that opposes motion induced by the external force. In the case of microscopic particles, water acts as a highly viscous medium (low Reynolds number) so that any particle quickly approaches terminal velocity and inertial effects can be ignored. If we ignore the effects of diffusion, then we have the deterministic equation

$$\frac{dx}{dt} = \frac{F(x)}{\gamma} \equiv A(x), \tag{2.2.4}$$

where γ is known as the friction coefficient. The problem is that if we include the effects of diffusion, we again have to deal with non-differentiability. Mathematically speaking this issue is addressed using differentials. That is, $X(t)$ is taken to evolve according to an SDE given by the Langevin equation

$$dX(t) = A(X)dt + \sqrt{2D}dW(t). \tag{2.2.5}$$

Equation (2.2.5) is often referred to as an SDE for overdamped Brownian motion due to the absence of inertial terms. The non-differentiability of $W(t)$ means that one has to modify the usual rules of calculus, which requires the theory of Ito calculus. We will present an informal introduction to SDEs and Ito calculus, following along the lines of Jacobs [412]. A more detailed treatment can be found in Gardiner [300] and Duan [229], and a rigorous mathematical account can be found in [624]. We begin by introducing the basic notion of a continuous Markov process.

Remark 2.3. In the physics literature one often finds SDEs expressed in terms of differential equations and white noise processes. For example, equation (2.2.5) may be written as

$$\frac{dx}{dt} = A(x) + \sqrt{2D}\xi(t),$$

with $\xi(t)$ identified as Gaussian white noise:

$$\langle \xi(t) \rangle = 0, \quad \langle \xi(t)\xi(s) \rangle = \delta(t-s). \tag{2.2.6}$$

One heuristic way to understand the origin of the Dirac delta function is to note from equation (2.2.3) that

$$\langle W(t)W(s) \rangle = tH(s-t) + sH(t-s),$$

and differentiating both sides with respect to t and s yields the second equation in (2.2.6) under the identification $\xi(t) = dW(t)/dt$. However, one should not literally interpret $\xi(t)$ as $dW(t)/dt$, since $W(t)$ is not differentiable. Instead, one should view $\xi(t)dt$ as a convenient representation of $dW(t)$ that provides an efficient way of performing certain calculations.

2.2.2 Continuous Markov processes

Suppose that an experiment is carried out over a time interval of length T and has a given set of possible outcomes Ω. In the case of tracking a single molecule diffusing in the cell membrane, Ω could be the set of all possible trajectories. Each time the experiment is run, one obtains a particular realization of a continuous-time stochastic process (or random function) $X(\omega,t) \in \Sigma \subset \mathbb{R}^d$ with $\omega \in \Omega$. For fixed ω, $X(\omega,t) = X_\omega(t)$ is a function of time corresponding to a particular trajectory in state space, which is specified by the parameter ω. On the other hand, fixing time t yields a family of continuous random variables $X(\omega,t) = X_t(\omega)$ that are parameterized by t. In physical and biological applications, the explicit dependence on the events ω and the nature of the underlying probability space are ignored, and one simply writes $X = X(t)$. The cumulative distribution function of the stochastic process is defined according to

$$P(x,t) = \mathbb{P}[X(t) \leq x]. \tag{2.2.7}$$

Using the frequency interpretation of probability, this represents the fraction of trials for which the trajectory through state space does not exceed the value x at time t. One can then define the corresponding probability density (assuming it exists) according to

$$p(x,t) = \frac{\partial P(x,t)}{\partial x}. \tag{2.2.8}$$

Moreover, we can introduce joint cumulative distributions and densities

$$P(x_1,\ldots,x_n;t_1,\ldots,t_n) = \mathbb{P}[X(t_1) \leq x_1,\ldots,X(t_n) \leq x_n],$$

and

$$p(x_1,\ldots,x_n;t_1,\ldots,t_n) = \frac{\partial^n P(x_1,\ldots,x_n;t_1,\ldots,t_n)}{\partial x_1 \ldots \partial x_n}.$$

A further extension is the conditional probability density, which takes the form (for $n=2$)

$$p(x_2,t_2|x_1,t_1) = \frac{p(x_1,x_2;t_1,t_2)}{p(x_1,t_1)}. \tag{2.2.9}$$

Thus, $p(x_2,t_2|x_1,t_1)$ is the probability density for $X(t_2)$ conditioned on $X(t_1) = x_1$.

Given a probability density p, one can define various moments of the stochastic process. Some important examples are the mean

$$x(t) = \langle X(t) \rangle = \int_{-\infty}^{\infty} xp(x,t)dx, \tag{2.2.10}$$

and the two-point autocorrelation function

$$\langle X(t_1)X(t_2) \rangle = \int_{-\infty}^{\infty} \int_{-\infty}^{\infty} x_1 x_2 p(x_1,x_2;t_1,t_2)dx_1 dx_2. \tag{2.2.11}$$

The corresponding covariance is given by

$$C(t_1,t_2) = \langle (X(t_1)-x(t_1))(X(t_2)-x(t_2)) \rangle = \langle\langle X(t_1)X(t_2) \rangle\rangle. \tag{2.2.12}$$

The equal-time covariance $C(t,t)$ is the corresponding variance. Double brackets are often used to denote cumulants of the stochastic process. The latter are defined using a generating function:

$$\left\langle \exp\left(-i\int_0^t X(t')dt'\right)\right\rangle \tag{2.2.13}$$

$$= \exp\left[\sum_{n=1}^{\infty} \frac{(-i)^m}{m!} \int_0^t \cdots \int_0^t \langle\langle X(t_1)X(t_2)\ldots X(t_n)\rangle\rangle dt_1 dt_2 \ldots dt_n\right].$$

An important concept in stochastic processes is stationarity: a stochastic process $X(t)$ is stationary if every joint probability distribution for finite n is invariant under a global time shift:

$$P(x_1,x_2,\ldots,x_n;t_1+\tau,t_2+\tau,\ldots,t_n+\tau) = P(x_1,x_2,\ldots,x_n;t_1,t_2,\ldots,t_n)$$

for arbitrary τ. It follows that $P(x,t)$ is time-independent and the covariance $C(t.t') = C(t-t')$.

The continuous stochastic process also satisfies the Markov property if

$$p(x_n,t_n|x_1,\ldots,x_{n-1};t_1,\ldots,t_{n-1}) = p(x_n,t_n|x_{n-1},t_{n-1}). \tag{2.2.14}$$

In other words, given (x_{n-1},t_{n-1}), the process has no "memory" of values at earlier times. It follows that a Markov process is fully determined by the two functions $p(x_1,t_1)$ and $p(x_2,t_2|x_1,t_1)$. For example,

$$\begin{aligned}
p(x_1,x_2,x_3;t_1,t_2,t_3) &= p(x_3,t_3|x_1,x_2;t_1,t_2)p(x_1,x_2;t_1,t_2) \\
&= p(x_3,t_3|x_1,x_2;t_1,t_2)p(x_2,t_2|x_1,t_1)p(x_1,t_1) \\
&= p(x_3,t_3|x_2,t_2)p(x_2,t_2|x_1,t_1)p(x_1,t_1).
\end{aligned} \tag{2.2.15}$$

However, the functions $p(x_1,t_1)$ and $p(x_2,t_2|x_1,t_1)$ cannot be chosen arbitrarily, since they must obey two important identities. The first is obtained by integrating

$$p(x_1,x_2;t_1,t_2) = p(x_2,t_2|x_1,t_1)p(x_1,t_1)$$

with respect to x_1:

$$p(x_2,t_2) = \int_\Sigma p(x_2,t_2|x_1,t_1)p(x_1,t_1)dx_1. \tag{2.2.16}$$

The other is obtained by integrating (2.2.15) with respect to x_2, assuming that $t_1 < t_2 < t_3$:

$$p(x_1,x_3;t_1,t_3) = p(x_1,t_1)\int_\Sigma p(x_3,t_3|x_2,t_2)p(x_2,t_2|x_1,t_1)dx_2.$$

Since $p(x_1,x_3;t_1,t_3) = p(x_3,t_3|x_1,t_1)p(x_1,t_1)$, this reduces to the Chapman–Kolmogorov equation

$$p(x_3,t_3|x_1,t_1) = \int_\Sigma p(x_3,t_3|x_2,t_2)p(x_2,t_2|x_1,t_1)dx_2. \tag{2.2.17}$$

Note that the Chapman–Kolmogorov equation is automatically satisfied by the fundamental solution (2.1.27) of the diffusion equation, reflecting the fact that the position $X(t)$ of a diffusing particle evolves according to a continuous-time Markov process.

2.2.3 Stochastic integrals and Ito calculus

One way to make sense of an SDE is to construct the corresponding stochastic integral. Consider, for example, the general one-dimensional SDE

$$dX = A(X)dt + B(X)dW(t). \tag{2.2.18}$$

Comparing with the Langevin equation (2.2.5) for overdamped Brownian motion, we see that the term $A(X)$ could represent some external force applied to a Brownian particle, whereas $B(X)$ could arise from heterogeneities in the surrounding medium.

The integrated version of equation (2.2.18) is given by

$$X(T) = X_0 + \int_0^T A(X(t))dt + \int_0^T B(X(t))dW(t), \qquad (2.2.19)$$

with the final term defined according to the so-called Ito stochastic integral

$$\int_0^T B(X(t))dW(t) = \lim_{N \to \infty} \sum_{n=0}^{N-1} B(X_n)\Delta W_n. \qquad (2.2.20)$$

The integral equation is not very useful for generating an explicit solution for $X(t)$. However, from the definition of the Ito stochastic integral, we have

$$\left\langle \int_0^T B(X(t))dW(t) \right\rangle = 0, \qquad (2.2.21)$$

since X_n is a function of previous Wiener increments $\Delta W_{n-1}, \ldots, \Delta W_0$ so it is un-correlated with ΔW_n. One major difference from standard Riemann integrals is that the interpretation of the integral (2.2.20) is sensitive to the choice of discretization scheme used to evaluate the integral; see Sect. 2.3.5. We now consider some examples that highlight some important properties of stochastic integrals and Wiener processes.

Example 2.1. Consider the SDE

$$dX(t) = f(t)dW(t),$$

where $f(t)$ is a deterministic function of time. We will evaluate the stochastic integral appearing in the equation

$$X(t) = X_0 + \int_0^t f(s)dW(s).$$

After discretizing time, we find that $X(t_N)$ is a Gaussian random variable, with

$$\langle X(t_N) \rangle = \sum_{n=0}^{N-1} \langle f(t_n)\Delta W_n \rangle = 0, \quad \mathrm{Var}(X(t_N)) = \sum_{n=0}^{N-1} \mathrm{Var}(f(t_n)\Delta W_n) = \sum_{n=0}^{N-1} f(t_n)^2 \Delta t.$$

Taking the continuum limit along identical lines to the previous case yields the continuous time Gaussian variable

$$X(t) = \lim_{N \to \infty} \sum_{j=0}^{N-1} f(t_j)\Delta W_j \equiv \int_0^t f(s)dW(s),$$

with zero mean and variance

$$\mathrm{Var}(X(T)) = \int_0^T f(s)^2 ds.$$

Example 2.2. The increment dW^2. An important object in stochastic calculus is $(dW)^2$, which plays a crucial role in deriving Ito's lemma and the FP equation for an SDE. In terms of Wiener increments,

$$\int_0^T (dW(t))^2 = \lim_{N \to \infty} \sum_{n=0}^{N-1} (\Delta W_n)^2.$$

Taking the expectation of both sides and using the fact that each ΔW_n is i.i.d., gives

$$\left\langle \int_0^T (dW(t))^2 \right\rangle = \int_0^T \langle (dW(t))^2 \rangle = \int_0^T dt = T.$$

What about the variance? Using the Gaussian probability density (2.2.2), it is simple to show that $\mathrm{Var}[(\Delta W)^2] = 2(\Delta t)^2 = 2T^2/N^2$. Hence,

$$\mathrm{Var}\left[\int_0^T (dW(t))^2\right] = \lim_{N\to\infty} \mathrm{Var}\left[\sum_{n=0}^{N-1}(\Delta W_n)^2\right] = \lim_{N\to\infty}\sum_{n=0}^{N-1}\mathrm{Var}\left[(\Delta W_n)^2\right]$$

$$= \lim_{N\to\infty}\frac{2T^2}{N} = 0.$$

We thus obtain the surprising result that the integral of $(dW)^2$ is deterministic, and thus equal to its mean:

$$\int_0^T (dW(t))^2 = T = \int_0^T dt. \tag{2.2.22}$$

In other words, we can set $(dW)^2 = dt$ inside integrals. Using the higher moments of Gaussians, it can also be shown that $dW^m = 0$ for $m > 2$.

Example 2.3. We now obtain a result from stochastic calculus, which will be useful when discussing numerical simulations, in particular Milstein's method, see Sect. 2.5. Consider the discrete sum

$$S_n = \sum_{n=0}^{N-1} W_n \Delta W_n = \frac{1}{2}\sum_{n=0}^{N-1}\left[(W_n + \Delta W_n)^2 - W_n^2 - (\Delta W_n)^2\right]$$

$$= \frac{1}{2}[W(t)^2 - W(t_0)^2] - \frac{1}{2}\sum_{n=0}^{N-1}(\Delta W_n)^2,$$

where $W_0 = W(t_0)$ and $W_{N-1} = W(t)$. We now calculate the mean and variance of the last term. First,

$$\left\langle \sum_{n=0}^{N-1}(\Delta W_n)^2 \right\rangle = \sum_{n=0}^{N-1}\langle (\Delta W_n)^2 \rangle = \sum_{n=0}^{N-1}(t_{n+1} - t_n) = t - t_0.$$

Second,

$$\left\langle \left[\sum_{n=0}^{N-1}(\Delta W_n)^2 - (t - t_0)\right]^2 \right\rangle$$

$$= \left\langle \left[\sum_{n=0}^{N-1}(\Delta W_n)^4 + 2\sum_{n<m}(\Delta W_n)^2(\Delta W_m)^2 - 2(t-t_0)\sum_{n=0}^{N-1}(\Delta W_n)^2 + (t-t_0)^2\right]\right\rangle.$$

Since ΔW_n and ΔW_m are independent Gaussian random variables for $n \neq m$, we have

$$\langle (\Delta W_n)^2(\Delta W_m)^2 \rangle = (t_{n+1} - t_n)(t_{m+1} - t_m),$$

and the fourth moment of a Gaussian is given by

$$\langle (\Delta W_n)^4 \rangle = 3\langle (\Delta W_n)^2 \rangle^2 = 3(t_{n+1} - t_n)^2.$$

Hence,

$$\mathrm{Var}\left[\sum_{n=0}^{N-1}(\Delta W_n)^2\right] = 2\sum_{n=0}^{N-1}(t_{n+1}-t_n)^2$$

$$+\sum_{n,m}[(t_{n+1}-t_n)-(t-t_0)][(t_{m+1}-t_m)-(t-t_0)]$$

$$= 2\sum_{n=0}^{N-1}(t_{n+1}-t_n)^2 \to 0 \text{ as } N \to \infty.$$

We deduce that

$$\int_{t_0}^{t} W(t')dW(t') = \frac{1}{2}[W(t)^2 - W(t_0)^2 - (t-t_0)]. \tag{2.2.23}$$

Ito's lemma. The result $dW(t)^2 = dt$ has important implications for how one carries out a change of variables in stochastic calculus. This is most directly established by considering the SDE for an arbitrary function $f(X(t))$ with $X(t)$ evolving according to equation (2.2.18):

$$df(X) = f(X+dX) - f(X)$$

$$= f'(X)dX + \frac{1}{2}f''(X)dX^2 + \dots$$

$$= f'(X)[A(X)dt + B(X)dW(t)] + \frac{1}{2}f''(X)B(X)^2dW(t)^2,$$

where all terms of higher order than dt have been dropped. Now using $dW(t)^2 = dt$, we obtain the following SDE for f, which is known as Ito's lemma:

$$df(X) = \left[A(X)f'(X) + \frac{1}{2}B(X)^2f''(X)\right]dt + B(X)f'(X)dW(t). \tag{2.2.24}$$

Hence, changing variables in Ito calculus is not given by ordinary calculus unless f is a constant or a linear function.

Example 2.4. Consider the SDE with linear multiplicative noise interpreted in the sense of Ito:

$$dX = Xa(t)dt + Xb(t)dW(t). \tag{2.2.25}$$

One way to solve this equation is to eliminate the multiplicative factor by performing the change of variables $Y(t) = \ln X(t)$. However, care must be taken when calculating infinitesimals, since the normal rules of calculus no longer apply for Ito stochastic variables. In particular,

$$dY(t) = \ln(X(t+dt)) - \ln X(t) = \ln(X(t)+dX(t)) - \ln X(t)$$

$$= \ln(1+dX(t)/X(t)) = \frac{dX(t)}{X(t)} - \frac{dX(t)^2}{2X(t)^2} + o(dt)$$

$$= a(t)dt + b(t)dW(t) - \frac{1}{2}[a(t)dt + b(t)dW(t)]^2 + o(dt)$$

$$= a(t)dt + b(t)dW(t) - \frac{b(t)^2}{2}dt + o(dt).$$

Integrating this equation gives

$$Y(t) = Y_0 + \int_0^t \left[a(s) - \frac{1}{2} b(s)^2 \right] ds + \int_0^t b(s) dW(s),$$

and exponentiating

$$X(t) = X_0 \exp \left(\int_0^t \left[a(s) - \frac{1}{2} b(s)^2 \right] ds + \int_0^t b(s) dW(s) \right).$$

Correlations and the power spectrum. A very useful quantity is the power spectrum of a stationary stochastic process $X(t)$, which is defined as the Fourier transform of the autocorrelation function $C_X(\tau)$,

$$S_X(\omega) = \int_{-\infty}^{\infty} e^{i\omega\tau} C_X(\tau) d\tau, \quad C_X(\tau) = \langle X(t) X(t+\tau) \rangle. \tag{2.2.26}$$

Consider the covariance of two frequency components of $X(t)$:

$$\langle \widehat{X}(\omega) \widehat{X}(\omega') \rangle = \left\langle \int_{-\infty}^{\infty} e^{i\omega t} X(t) dt \int_{-\infty}^{\infty} e^{i\omega' t'} X(t') dt' \right\rangle$$
$$= \int_{-\infty}^{\infty} e^{i\omega t} \int_{-\infty}^{\infty} e^{i\omega' t'} \langle X(t) X(t') \rangle dt' dt.$$

Using the definition of the power spectrum, we have

$$= \int_{-\infty}^{\infty} e^{i\omega t} \int_{-\infty}^{\infty} e^{i\omega' t'} \left[\int_{-\infty}^{\infty} e^{-i\Omega(t-t')} S_X(\Omega) \frac{d\Omega}{2\pi} \right] dt' dt$$
$$= \int_{-\infty}^{\infty} S_X(\Omega) \left[\int_{-\infty}^{\infty} e^{i(\omega-\Omega)t} dt \right] \left[\int_{-\infty}^{\infty} e^{i(\omega'+\Omega)t'} dt' \right] \frac{d\Omega}{2\pi},$$

assuming that it is possible to rearrange the order of integration. Using the Fourier representation of the Dirac delta function, $\int_{-\infty}^{\infty} e^{i\omega t} dt = 2\pi\delta(\omega)$, we have

$$\langle \widehat{X}(\omega) \widehat{X}(\omega') \rangle = \int_{-\infty}^{\infty} S_X(\Omega) \cdot 2\pi\delta(\omega - \Omega) \cdot 2\pi\delta(\omega' + \Omega) \frac{d\Omega}{2\pi},$$

which establishes the Wiener–Khinchin theorem:

$$\langle \widehat{X}(\omega) \widehat{X}(\omega') \rangle = 2\pi S_X(\omega) \delta(\omega + \omega'). \tag{2.2.27}$$

The Fourier transform of a real-valued variable satisfies $\widehat{X}(-\omega) = \widehat{X}^*(\omega)$ so

$$\langle \widehat{X}(\omega) \widehat{X}^*(\omega') \rangle = 2\pi S_X(\omega) \delta(\omega - \omega'). \tag{2.2.28}$$

In the case of linear SDEs, it is possible to calculate the spectrum explicitly by formally setting $dW(t) = \xi(t) dt$ with $\xi(t)$ a white noise process; see equation (2.2.6). The solution can be expressed formally in terms of the integral solution

$$X(t) = \sqrt{2D} \int_{-\infty}^{\infty} G(\tau) \xi(t - \tau) d\tau, \tag{2.2.29}$$

where $G(\tau)$ is known as the causal Green's function or linear response function with the important property that $G(\tau) = 0$ for $\tau < 0$. The main point to emphasize is that although $\xi(t)$ is not a mathematically well-defined object, one still obtains correct answers when taking expectations. For example, it is clear that in the stationary state $\langle X(t) \rangle = 0$ and (for $s > 0$)

$$\langle X(t)X(t+s) \rangle = 2D \int_{-\infty}^{\infty} \int_{-\infty}^{\infty} G(\tau)G(\tau')\langle \xi(t-\tau)\xi(t+s-\tau') \rangle d\tau d\tau'$$

$$= 2D \int_{-\infty}^{\infty} \int_{-\infty}^{\infty} G(\tau)G(\tau')\delta(s+\tau-\tau') d\tau d\tau'$$

$$= 2D \int_{-\infty}^{\infty} G(\tau)G(\tau+s) d\tau. \tag{2.2.30}$$

One of the useful features of formally expressing a solution to a linear SDE in the form (2.2.29) is that one can view the dynamical system as acting as a filter of the white noise process. Applying the Wiener–Khinchin theorem to the white noise autocorrelation function, we see that the spectrum is given by the Fourier transform of a Dirac delta function, which is unity. However, once the noise has been passed through a filter with linear response function $G(t)$, the spectrum is no longer flat. This follows from applying the convolution theorem of Fourier transforms to equation (2.2.29):

$$\widehat{X}(\omega) = \sqrt{2D}\widehat{G}(\omega)\widehat{\xi}(\omega),$$

so

$$2\pi S_X(\omega)\delta(\omega-\omega') = 2D\widehat{G}(\omega)\widehat{G}^*(\omega')\langle \widehat{\xi}(\omega)\widehat{\xi}^*(\omega') \rangle,$$

and

$$\langle \widehat{\xi}(\omega)\widehat{\xi}^*(\omega') \rangle = \int_{-\infty}^{\infty} e^{i\omega t} \int_{-\infty}^{\infty} e^{-i\omega't'} \langle \xi(t)\xi(t') \rangle dt' dt = 2\pi \cdot \delta(\omega-\omega').$$

That is,

$$S_X(\omega) = 2D|\widehat{G}(\omega)|^2. \tag{2.2.31}$$

2.2.4 Ornstein–Uhlenbeck process (OU)

One of the simplest, nontrivial examples of a continuous stochastic process is the OU process. This evolves according to the SDE

$$dX(t) = -kX(t)dt + \sqrt{2D}dW(t). \tag{2.2.32}$$

There are two ways to obtain moments of the OU process. The first is to derive moment equations directly from the SDE. For example, taking expectations of both sides using $\langle dW(t) \rangle = 0$ and, we have

$$\langle dX(t) \rangle = -k\langle X(t) \rangle dt.$$

This yields the deterministic first-order moment equation

$$\frac{d\langle X \rangle}{dt} = -k\langle X \rangle,$$

which has the solution $\langle X \rangle = X_0 e^{-kt}$. Similarly, using $\langle dX(t)dW(t) \rangle = 0$ and $dW(t)^2 = dt$, we have

$$\langle X(t+dt)X(t+dt) \rangle = (1-kdt)^2 \langle X(t)X(t) \rangle + 2Ddt.$$

Subtracting $\langle X(t)X(t) \rangle$ from both sides, dividing through by dt, and taking the limit $dt \to 0$ lead to the second-order moment equation

$$\frac{d\langle X^2 \rangle}{dt} = -2k\langle X^2 \rangle + 2D,$$

which has the solution

$$\langle X^2 \rangle = e^{-2kt}X_0^2 + \frac{D}{k}\left(1 - e^{-2kt}\right).$$

It immediately follows that

$$\mathrm{Var}[X] = \frac{D}{k}\left(1 - e^{-2kt}\right). \tag{2.2.33}$$

An alternative approach is to integrate the SDE (2.2.32) by multiplying both sides by the integrating factor e^{kt}, and noting that

$$d[X(t)e^{kt}] = e^{kt}dX(t) + X(t)ke^{kt}dt\sqrt{2D}e^{kt}dW(t).$$

It follows that the integrated version of the OU equation is

$$e^{kt}X(t) - X(0) = \sqrt{2D}\int_0^t e^{ks}dW(s).$$

Taking expectations with respect to the Gaussian process immediately shows that

$$\langle X(t) \rangle = X(0)e^{-kt} + \sqrt{2D}\int_0^t e^{-k(t-s)}\langle dW(s) \rangle = X_0 e^{-kt}. \tag{2.2.34}$$

Similarly,

$$\langle [X(t) - \langle X(t) \rangle]^2 \rangle = 2D\int_0^t e^{-k(t-s)}\int_0^t e^{-k(t-s')}\langle dW(s)dW(s') \rangle$$

$$= 2D\int_0^t e^{-k(t-s)}\int_0^t e^{-k(t-s')}\delta(s-s')dtdt' = 2D\int_0^t e^{-2k(t-s)}ds = \frac{D}{k}(1-e^{-2kt}).$$

Note that in the limit $k \to 0$ for fixed t, we recover the mean-square displacement of 1D Brownian motion, since $1 - e^{-2kt} \approx 1 - [1 - 2kt + (2kt)^2/2\ldots] \approx 2kt$.

Multivariate OU process. The multivariate version of the OU process (2.2.32) takes the form

$$dX_i = \sum_{j=1}^{d} A_{ij} X_j dt + \sum_{j=1}^{d} B_{ij} dW_j(t).$$
(2.2.35)

It can be shown that the covariance matrix $\Sigma(t)$ with components

$$\Sigma_{ij}(t) = \langle [X_i(t) - \langle X_i(t) \rangle][X_j(t) - \langle X_j(t) \rangle] \rangle$$

satisfies the matrix equation (see Ex. 2.4 and 2.5)

$$\frac{d\Sigma(t)}{dt} = \mathbf{A}\Sigma(t) + \Sigma(t)\mathbf{A}^T + \mathbf{BB}^T.$$

It follows that if $\mathbf{A}$ has distinct eigenvalues with negative real part, then $\Sigma(t) \to \Sigma_0$ where Σ_0 is the stationary covariance matrix satisfying the Ricatti equation

$$\mathbf{A}\Sigma_0 + \Sigma_0\mathbf{A}^T + \mathbf{BB}^T = 0.$$
(2.2.36)

The multivariate OU process will play an important role in the analysis of gene networks (Chap. 5).

Power spectrum of the OU process. In the case of the OU process, the casual Green's function is $G(\tau) = e^{-\tau k} H(\tau)$, where $H(t)$ is the Heaviside function. It follows from (2.2.30) that

$$\langle X(t)X(t+s) \rangle = 2D \int_0^{\infty} e^{-k(2\tau+s)} d\tau = \frac{D}{k} e^{-ks}.$$

This is the expected result for the autocorrelation function of the OU process. Moreover,

$$\widehat{G}(\omega) = \int_{-\infty}^{\infty} e^{i\omega t} G(t) dt = \int_0^{\infty} e^{i\omega t} e^{-kt} dt = \frac{1}{k - i\omega},$$

and thus

$$S_X(\omega) = \frac{2D}{k^2 + \omega^2}.$$
(2.2.37)

The spectrum can be used to recover the variance by noting that

$$\langle X(t)^2 \rangle = \int_{-\infty}^{\infty} S_X(\omega) \frac{d\omega}{2\pi}.$$

Substituting for $S_X(\omega)$ and using the identity

$$\int_{-\infty}^{\infty} \frac{d\omega}{\omega^2 + k^2} = \frac{\pi}{k},$$

we see that $\langle X(t)^2 \rangle = D/k$. (A similar calculation is performed for the stochastic damped harmonic oscillator in Ex. 2.6.) The above analysis can also be extended to multivariate linear SDEs. For example, formally setting $dW_j(t) = \xi_j(t)dt$ in the multivariate OU equation (2.2.35), with

$$\langle \xi_i(t) \rangle = 0, \quad \langle \xi_i(t)\xi_j(t') \rangle = \delta_{i,j}\delta(t-t'),$$

one can show that the spectrum of $X_i(t)$ is (see Ex. 2.5)

$$S_i(\omega) = \sum_{j=1}^{d} \sum_{j'=1}^{d} \Phi_{ij}^{-1}(\omega)\Phi_{ij'}^{-1}(-\omega)D_{jj'},$$

where $D_{ij} = \sum_{k=1}^{d} B_{ik}B_{jk}$ and $\Phi_{ij}(\omega) = -i\omega\delta_{i,j} - A_{ij}$ is known as the transfer matrix.

2.3 The Fokker–Planck equation

2.3.1 1D Fokker–Planck equation

Just as pure Brownian motion generates sample paths that are distributed according to the solution to the diffusion equation (2.1.25), the Langevin equation (2.2.5) generates sample paths distributed according to a Fokker–Planck equation. That is, the probability density $p(x,t) = p(x,t|x_0,0)$ for the position of the particle satisfies the forward Fokker–Planck (FP) equation

$$\frac{\partial p(x,t)}{\partial t} = -\frac{\partial [A(x)p(x,t)]}{\partial x} + D\frac{\partial^2 p(x,t)}{\partial x^2}, \tag{2.3.1}$$

with $A(x) = F(x)/\gamma$ and the initial condition $p(x,0) = \delta(x-x_0)$. The 1D FP equation (2.3.1) can also be rewritten as a probability conservation law according to

$$\frac{\partial p(x,t)}{\partial t} = -\frac{\partial J(x,t)}{\partial x}, \tag{2.3.2}$$

where

$$J(x,t) = A(x)p(x,t) - D\frac{\partial p(x,t)}{\partial x} \tag{2.3.3}$$

is the probability flux. One direct way to derive the FP equation is to apply Ito's lemma (2.2.24) to an arbitrary smooth function $f(X(t))$, with $X(t)$ evolving according to equation (2.2.5):

$$\frac{\langle df(X(t))\rangle}{dt} = \langle A(X(t))f'(X(t)) + Df''(X(t))\rangle$$

$$= \int [A(x)f'(x) + Df''(x)]\, p(x,t)dx,$$

$$= \int f(x)\left[-\frac{\partial}{\partial x}(A(x)p(x,t)) + D\frac{\partial^2}{\partial x^2}(p(x,t))\right] dx, \qquad (2.3.4)$$

after integration by parts, where $p(x,t)$ is the probability density of the stochastic process $X(t)$ under the initial condition $X(0) = x_0$. However, we also have

$$\frac{\langle df(X(t))\rangle}{dt} = \left\langle \frac{df(X(t))}{dt} \right\rangle = \frac{d}{dt}\langle f(X(t))\rangle = \int f(x)\frac{\partial}{\partial t}p(x,t)dx. \qquad (2.3.5)$$

Comparing equations (2.3.4) and (2.3.5) and using the fact that $f(x)$ is arbitrary, we obtain the FP equation (2.3.1). An equilibrium steady-state solution of (2.3.1) corresponds to the conditions $\partial p/\partial t = 0$ and $J \equiv 0$. This leads to the first-order ODE for the equilibrium density $P(x)$:

$$DP'(x) - \gamma^{-1}F(x)P(x) = 0,$$

which has the solution

$$P(x) = \mathcal{N}e^{-\Phi(x)/\gamma D}.$$

Here $\Phi(x) = -\int^x F(y)dy$ is a potential energy function and $\mathcal{N}$ is a normalization factor (assuming that it exists).

Remark 2.4. *Einstein relation.* Comparison of the equilibrium distribution with the Boltzmann-Gibbs distribution (1.3.5) yields the Einstein relation

$$D\gamma = k_B T, \qquad (2.3.6)$$

where T is the temperature (in degrees Kelvin) and $k_B \approx 1.4 \times 10^{-23}$ J K^{-1} is the Boltzmann constant. This formula relates the variance of environmental fluctuations to the strength of dissipative forces, and is thus an example of a fluctuation-dissipation theorem (see also Chap. 10 and Chap. 12). In the case of a sphere of radius R moving in a fluid of viscosity η, Stoke's formula can be used, that is, $\gamma = 6\pi\eta R$. For water at room temperature, $\eta \sim 10^{-3}$ kg m^{-1} s^{-1} so that a particle of radius $R = 10^{-9}$ m has a diffusion coefficient $D \sim 100\,\mu\text{m}^2/\text{s}$. However, this overestimates the diffusion coefficient of molecules moving in the cytoplasm, since the latter is a complex heterogeneous environment, see Chap. 6.

Remark 2.5. In the limit $D \to 0$, the FP equation reduces to the so-called Liouville equation. The latter has a general solution of the form

$$p(x,t) = \int_{\mathbb{R}} \delta(x - \phi(x_0,t))\rho(x_0)dx_0,$$

where $\phi(x_0,t)$ is the solution to the deterministic equation $\dot{x} = F(x)/\gamma$ with initial condition $x(0) = x_0$ and $\rho(x_0)$ is a probability density over initial conditions. Thus $p(x,t)$ represents a distribution of deterministic trajectories with $p(x,0) = \rho(x)$.

FP equation for an OU process. The FP equation for the OU process (2.2.32) is

$$\frac{\partial p(x,t)}{\partial t} = \frac{\partial[kxp(x,t)]}{\partial x} + D\frac{\partial^2 p(x,t)}{\partial x^2}. \tag{2.3.7}$$

As we show below, given the initial condition $p(x,0) = \delta(x-x_0)$, the solution to the FP equation is a Gaussian with time-dependent mean and variance:

$$p(x,t) = \frac{1}{\sqrt{2\pi D[1-e^{-2kt}]/k}} \exp\left(-\frac{(x-x_0e^{-kt})^2}{2D[1-e^{-2kt}]/k}\right). \tag{2.3.8}$$

The mean and variance of the OU process are therefore given by

$$\langle X(t)\rangle = x_0 e^{-kt}, \quad \langle [X(t)-\langle X(t)\rangle]^2\rangle = \frac{D}{k}(1-e^{-2kt}), \tag{2.3.9}$$

which agrees with equation (2.2.33). In the large time limit, we obtain a stationary Gaussian process with zero mean and time-independent variance

$$\langle X(t)\rangle = 0, \quad \langle [X(t)-\langle X(t)\rangle]^2\rangle = \frac{D}{k}. \tag{2.3.10}$$

The Gaussian solution (2.3.8) can be derived in terms of the characteristic function (Fourier transform)

$$\Gamma(z,t) = \int_{-\infty}^{\infty} e^{izx} p(x,t)dx.$$

Fourier transforming the FP equation shows that Γ satisfies the PDE:

$$\frac{\partial\Gamma}{\partial t} + kz\frac{\partial\Gamma}{\partial z} = -Dz^2\Gamma. \tag{2.3.11}$$

This can be solved using separation of variables (see Ex. 2.7) or the method of characteristics [717]; see Box 2C. We find that

$$\Gamma(z,t) = \exp\left[-\frac{Dz^2}{4k}(1-e^{-2kt}) + izx_0e^{-kt}\right].$$

Applying the inverse Fourier transform, we obtain the probability density (2.3.8).

Box 2C. Method of characteristics.

Equation (2.3.11) for the characteristic function of the OU process is an example of a general class of so-called quasi-linear PDEs of the form [717]

$$a(x,t,u)\frac{\partial u(x,t)}{\partial t} + b(x,t,u)\frac{\partial u(x,t)}{\partial t} = c(x,t,u). \tag{2.3.12}$$

Quasi-linearity refers to the fact that all terms are linear with respect to the first derivatives, but can depend nonlinearly on u. In addition to equations for generating functions, quasi-linear PDEs crop up in various models of transport processes, and support nontrivial solutions such as shock waves and rarefaction waves [717]. An application to the analysis of exclusion effects in molecular motors will be developed in Sect. 7.6.

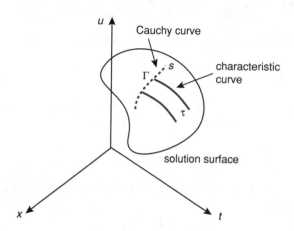

Fig. 2.4: Schematic illustration of the method of characteristics.

A standard method for solving this class of equations is the method of characteristics. The basic idea is that the solution surface $u = u(x,t)$ can be represented as the union of characteristic (or integral) curves that lie in the solution surface, see Fig. 2.4. Let τ be the parameter of one of these characteristic curves. In terms of the three-dimensional space with coordinates (x,t,u), such a curve is given by the points $(x(\tau),t(\tau),u(\tau))$ evolving according to the following characteristic equations:

$$\frac{dt}{d\tau} = a(x,t,u), \quad \frac{dx}{d\tau} = b(x,t,u), \quad \frac{du}{d\tau} = c(x,t,u). \tag{2.3.13}$$

We can check that these are consistent with the original PDE by setting $u(x(\tau),t(\tau)) = u(\tau)$ and noting that along the characteristic curve

$$c(x,t,u) = \frac{du}{d\tau} = \frac{\partial u}{\partial t}\frac{dt}{d\tau} + \frac{\partial u}{\partial x}\frac{dx}{d\tau}$$

$$= a(x,t,u)\frac{\partial u}{\partial t} + b(x,t,u)\frac{\partial u}{\partial x}.$$

The full solution surface is then given by the union of characteristic curves generated by the initial conditions. The latter are specified along a so-called Cauchy curve, which is parameterized by $s \in I \subset \mathbb{R}$. That is, for each $s \in \Gamma$ we specify the characteristic solution as $x = x(s,\tau), t = t(s,\tau), u = u(s,\tau)$, with the corresponding initial conditions

$$x(s,0) = f(s), \quad t(s,0) = g(s), \quad u(s,0) = h(s).$$

Finally, assuming that we can invert the pair of equations $x = x(s, \tau), t = t(s, \tau)$ such that $s = s(x,t), \tau = \tau(x,t)$, then we can express the solution in terms of x and t:

$$u(x,t) = u(s(x,t), \tau(x,t)).$$

Under certain smoothness conditions on the functions a, b, c, f, g, h, one can prove that the method of characteristics generates a smooth solution surface, at least within some local neighborhood of the Cauchy curve [717].

OU process. Let us now apply the above construction to the linear PDE for the generating function of the Ornstein–Uhlenbeck process with $(x,t,u) \to (z,t,\Gamma)$:

$$\frac{\partial \Gamma}{\partial t} + kz \frac{\partial \Gamma}{\partial z} = -\frac{D}{2} z^2 \Gamma.$$

In this case $a = 1$, $b = kz$, and $c = -Dz^2/2$. The characteristic equations are

$$\frac{dt}{d\tau} = 1, \quad \frac{dz}{d\tau} = kz, \quad \frac{d\Gamma}{d\tau} = -\frac{D}{2} z^2 \Gamma.$$

We can identify τ with t and solve for z as $z(t) = z_0 e^{kt}$. Take z_0 to parameterize the initial data for the characteristics and solve the equation

$$\frac{d\Gamma}{dt} = -\frac{D}{2} z_0^2 \Gamma e^{2kt}.$$

The latter has a solution of the form

$$\ln \Gamma(t) = \ln \Gamma_0 - \frac{D}{4k} z_0^2 e^{2kt}.$$

Using the fact that the integration constant Γ_0 can be a function of z_0, and making the replacement $z_0 = ze^{-kt}$, we finally obtain

$$\Gamma(z,t) = \Gamma_0(ze^{-kt}) e^{-Dz^2/4k}.$$

The initial condition $p(x,0) = \delta(x - x_0)$ implies that $\Gamma(z,0) = e^{izx_0}$, which then determines the function Γ_0:

$$\Gamma_0(z) = e^{izx_0} e^{Dz^2/4k}.$$

Hence,

$$\Gamma(z,t) = \exp\left[-\frac{Dz^2}{4k}(1 - e^{-2kt}) + izx_0 e^{-kt} \right].$$

Alternative derivation of FP equation. Here we consider an alternative derivation of the FP equation based on the Chapman–Kolmogorov equation (2.2.17) [300]. Consider an infinitesimal version of this equation by taking $t \to t + \Delta t, t' \to t$ and setting $w(x,t;u,\Delta t) = p(x+u,t+\Delta t|x,t)$:

$$p(x,t+\Delta t) = \int_{-\infty}^{\infty} w(x-u,t;u,\Delta t)p(x-u,t)du,$$

where the initial argument (x_0,t_0) has been suppressed. Now suppose that over a sufficiently small time window Δt, large jumps u in position are highly unlikely, so that u can be treated as a small variable. (It is possible to relax this requirement— one then obtains integral terms in the evolution equation for $p(x,t)$ that represent finite jumps between states.) Taylor expanding with respect to u gives

$$p(x,t+\Delta t) = \alpha_0(x,t)p(x,t) - \partial_x[\alpha_1(x,t)p(x,t)] + \frac{1}{2}\partial_{xx}^2[\alpha_2(x,t)p(x,t)] + \dots$$

$$(2.3.14)$$

where

$$\alpha_n(x,t) = \int_{-\infty}^{\infty} w(x,t;u,\Delta t)u^n du.$$

The SDE can be used to calculate the coefficients α_n. First, rewrite equation (2.2.5) in the infinitesimal form

$$X(t+\Delta t) = x + A(x)\Delta t + \sqrt{2D}\Delta W(t),$$

given that $X(t) = x$. This implies that the transition probability w can be written as

$$\begin{aligned}
w(x,t;u,\Delta t) &= \langle \delta(x+u-X(t+\Delta t)) \rangle \\
&= \langle \delta(u-A(x)\Delta t - \sqrt{2D}\Delta W(t)) \rangle, \\
&= \int_{-\infty}^{\infty} \delta(u-A(x)\Delta t - \sqrt{2D}\Delta W(t))p(\Delta W(t)),
\end{aligned}$$

where p is the probability density of $\Delta W(t)$. Since

$$\Delta W(t) = \int_{t}^{t+\Delta t} dW(s),$$

it follows that $\Delta W(t)$ is a Gaussian random variable with zero mean and variance Δt; the corresponding probability density is

$$p(\Delta W) = \sqrt{\frac{1}{2\pi\Delta t}}e^{-\Delta W^2/2\Delta t}.$$

Hence, averaging with respect to $\Delta W(t)$,

$$w(x,t;u,\Delta t) = \sqrt{\frac{1}{4\pi D\Delta t}} e^{-(u-A(x)\Delta t)^2/4D\Delta t}.$$

It follows that

$$\alpha_0 = 1, \quad \alpha_1 = A(x)\Delta t, \quad \alpha_2 = 2D\Delta t + \alpha_1^2,$$

and $\alpha_m = O(\Delta t^2)$ for $m > 2$. Substituting these results into equation (2.3.14), rearranging and taking the limit $\Delta t \to 0$ finally lead to the FP equation (2.3.1).

Boundary conditions. Now suppose that a particle is confined to an interval $\Omega = [0, L]$. It is then necessary to supplement the FP equation (2.3.1) by boundary conditions at the ends $x = 0, L$. The two most common choices are reflecting (no-flux) and absorbing boundary conditions. A reflecting or Neumann boundary condition at $x = 0$ is

$$J(0,t) = A(0)p(0,t) - D\frac{\partial p(0,t)}{\partial x} = 0, \quad t > 0, \tag{2.3.15}$$

which means that the particle cannot leave the interval Ω. On the other hand, an absorbing or Dirichlet boundary condition at $x = 0$ is given by

$$p(0,t) = 0, \quad t > 0, \tag{2.3.16}$$

which means that as soon as the particle reaches $x = 0$ it is removed from the system. (Either type of boundary condition can also be applied at $x = L$.) Integrating the FP equation, written in the conservation form (2.3.2), shows that the total probability

$$P(t) = \int_\Omega p(x,t)dt$$

evolves as

$$\frac{dP}{dt} = \int_\Omega \frac{\partial p(x,t)}{\partial t}dx = -\int_\Omega \frac{\partial J(x,t)}{\partial x}dx = J(0,t) - J(L,t).$$

Hence, total probability is conserved if there are reflecting boundary conditions at both ends. One the other hand, if $J(0,t) = 0$ and $p(L,t) = 0$ then

$$\frac{dP}{dt} = D\frac{\partial p(x,t)}{\partial x}\bigg|_{x=L} \leq 0,$$

since $p(x,t)$ is a positive quantity and $P(L,t) = 0$. In the limit $t \to \infty$, we have $P(t) \to 0$.

Backward FP equation. An important property of the solution $p(x,t|x_0,0)$ to the forward FP equation (2.3.1) on some domain Ω is that it satisfies the Chapman–Kolmogorov equation (2.2.17), which we rewrite as

$$p(x,t|x_0,0) = \int_\Omega p(x,t|y,\tau)p(y,\tau|x_0,0)dy. \tag{2.3.17}$$

This essentially states that the probability of x given x_0 is the sum of the probabilities of each possible path from x_0 to x. We will assume time translation invariance so that $p(x,t|y,\tau) = p(x,t-\tau|y,0)$. The Chapman–Kolmogorov equation can be used to derive a corresponding backward FP equation for $q(y,t) = p(x,t|y,0)$. We will illustrate this by taking $\Omega = [0,L]$. First, differentiating both sides of equation (2.3.17) with respect to the intermediate time τ gives

$$0 = \int_0^L \partial_\tau p(x,t|y,\tau)p(y,\tau|x_0,0)dy + \int_0^L p(x,t|y,\tau)\partial_\tau p(y,\tau|x_0,0)dy.$$

Using the fact that $p(y,\tau|x_0,0)$ satisfies the forward FP equation, $\partial_\tau[p(y,\tau|x_0,0)]$ can be replaced by terms involving derivatives with respect to y. Integrating by parts with respect to y then leads to the result

$$0 = \int_0^L \left[\partial_\tau q(y,t-\tau) + A(y)\partial_y q(y,t-\tau) + D\partial_{yy}^2 q(y,t-\tau)\right] p(y,\tau)dy$$
$$+ \left[-A(y)q(y,t-\tau)p(y,\tau) + Dq(y,t-\tau)\partial_y p(y,\tau) - D\partial_y q(y,t-\tau)p(y,\tau)\right]\Big|_{y=0}^{y=L},$$

where $p(y,\tau) = p(y,\tau|x_0,0)$. We choose adjoint boundary conditions so that the boundary contributions vanish (see also Box 2D). First, if $p = 0$ on one of the boundaries then we require $q = 0$ on the same boundary. On the other hand, if $-Ap + \partial_y p = 0$ then we require $q' = 0$. Finally, setting $\tau = 0$ with $p(y,0|x_0,0) = \delta(y-x_0)$ and noting that $\partial_\tau q(y,t-\tau) = -\partial_t q(y,t-\tau)$, we obtain the backward FP equation

$$\frac{\partial q}{\partial t} = A(y)\frac{\partial q}{\partial y} + D\frac{\partial^2 q}{\partial y^2}. \tag{2.3.18}$$

Assuming a reflecting boundary at $x = 0$ and an absorbing boundary at $x = L$, we have the boundary conditions

$$\partial_y q(y,t)|_{y=0} = 0, \quad q(L,t) = 0.$$

2.3.2 Higher dimensions

It is straightforward to generalize the Langevin (2.2.5) and FP equation (2.3.1) to higher dimensions ($d > 1$). Assuming, for simplicity, isotropic diffusion and friction, equation (2.2.5) becomes

$$dX_i = \frac{F_i(\mathbf{X})}{\gamma}dt + \sqrt{2D}dW_i(t), \tag{2.3.19}$$

with $i = 1,\ldots,d$, $\mathbf{X} = (X_1,X_2,\ldots,X_d)$, and

$$\langle W_i(t) \rangle = 0, \quad \langle W_i(t)W_j(t') \rangle = \delta_{i,j}\min\{t,t'\}. \tag{2.3.20}$$

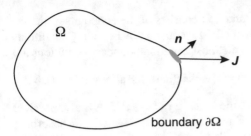

Fig. 2.5: Higher-dimensional bounded domain Ω with boundary $\partial\Omega$ and unit outward normal vector $\mathbf{n}$.

The corresponding FP equation is

$$\frac{\partial p(\mathbf{x},t)}{\partial t} = -\frac{1}{\gamma}\nabla\cdot[\mathbf{F}(\mathbf{x})p(\mathbf{x},t)]+D\nabla^2 p(\mathbf{x},t), \tag{2.3.21}$$

and the probability flux is given by the vector field

$$\mathbf{J}(\mathbf{x},t) = \frac{\mathbf{F}(\mathbf{x})}{\gamma}p(\mathbf{x},t)-D\nabla p(\mathbf{x},t). \tag{2.3.22}$$

Here ∇ denotes the gradient operator, which in Cartesian coordinates $\mathbf{x} = (x,y,z)$ (for $d = 3$) takes the form

$$\nabla = \mathbf{e}_x\frac{\partial}{\partial x}+\mathbf{e}_y\frac{\partial}{\partial y}+\mathbf{e}_z\frac{\partial}{\partial z},$$

with $\mathbf{e}_x$ the unit vector in the x-direction, etc. Similarly, ∇^2 is the Laplacian operator

$$\nabla^2 = \nabla\cdot\nabla = \frac{\partial^2}{\partial x^2}+\frac{\partial^2}{\partial y^2}+\frac{\partial^2}{\partial z^2}.$$

One simple example is the FP equation for the multivariate OU process; see equation (2.2.35), which takes the form

$$\frac{\partial p}{\partial t} = -\sum_{i,j=1}^{d}A_{ij}\frac{\partial x_j p(\mathbf{x},t)}{\partial x_i}+\frac{1}{2}\sum_{i,j=1}^{d}D_{ij}\frac{\partial^2 p(\mathbf{x},t)}{\partial x_i\partial x_j}, \tag{2.3.23}$$

with $D_{ij} = \sum_k B_{ik}B_{jk}$. The corresponding stationary density is

$$p_s(\mathbf{x}) = \frac{1}{\sqrt{\det(2\pi\Sigma_0)}}\exp\left(-\frac{1}{2}\mathbf{x}^\top\Sigma_0^{-1}\mathbf{x}\right), \tag{2.3.24}$$

where the stationary covariance matrix Σ_0 satisfies equation (2.2.36). The stationary density only exists if all of the eigenvalues of the matrix $\mathbf{A}$ have negative real part.

Boundary conditions. Suppose that the Brownian particle is confined to a bounded domain $\Omega\subset\mathbb{R}^d$ and denote the boundary by $\partial\Omega$, see Fig. 2.5. (In the case of 1D motion, Ω is typically a finite interval and $\partial\Omega$ consists of the endpoints of the interval.) Again it is necessary to supplement the FP equation by boundary conditions on $\partial\Omega$. As in the one-dimensional case, the most common boundary conditions are

$$p(\mathbf{x},t) = 0 \text{ (absorbing) or } \mathbf{J}(\mathbf{x},t)\cdot\mathbf{n}(\mathbf{x}) = 0 \text{ (reflecting) for all } \mathbf{x}\in\partial\Omega, \quad (2.3.25)$$

where $\mathbf{n}(\mathbf{x})$ is the unit outward normal to the boundary at $\mathbf{x}\in\partial\Omega$. It is also possible to have mixed boundary conditions, in which $\partial\Omega = \partial\Omega_D\cup\partial\Omega_N$ with an absorbing boundary $\partial\Omega_D$ and a reflecting boundary $\partial\Omega_N$. Alternatively, a boundary may be partially absorbing, in which case we have the so-called Robin boundary condition

$$p(\mathbf{x},t) + \alpha\mathbf{J}(\mathbf{x},t)\cdot\mathbf{n}(\mathbf{x}) = 0, \quad \alpha > 0.$$

Consider the particular case of a reflecting boundary condition. Integrating the FP equation (2.3.21) over the domain Ω, and reversing the order of integration and time differentiation, yields

$$\frac{d}{dt}\int_\Omega p(\mathbf{x},t)d\mathbf{x} = -\int_\Omega \nabla\cdot\mathbf{J}(\mathbf{x},t)d\mathbf{x} = -\int_{\partial\Omega}\mathbf{J}(\mathbf{x},t)\cdot\mathbf{n}(\mathbf{x})d\mathbf{x} = 0, \quad (2.3.26)$$

where we have used the divergence theorem [529] and imposed the boundary condition. Hence, in the case of a Fokker–Planck equation with reflecting boundaries, the total probability $P = \int_\Omega p(\mathbf{x},t)d\mathbf{x}$ is conserved, that is, $dP/dt = 0$, and the system typically converges to a nontrivial stationary state. On the other hand, the total probability is not conserved in the case of an absorbing boundary, which arises in first passage time problems (Sect. 2.4).

Remark 2.6. The Langevin equation (2.3.19) represents diffusive-like motion from the probabilistic perspective of a single microscopic particle moving in a fluid medium. However, as in the case of pure diffusion, it is possible to reinterpret the FP equation (2.3.21) as a deterministic advection–diffusion equation for the concentration $u(\mathbf{x},t)$ of many particles. That is, ignoring any interactions or correlations between the particles, set $u(\mathbf{x},t) = Np(\mathbf{x},t)$ where N is the total number of particles (assumed large). Multiplying both sides of equation (2.3.21) by N then leads to the corresponding Smoluchowski equation for $u(\mathbf{x},t)$ with $N\mathbf{J}(\mathbf{x},t)$ interpreted as the particle flux arising from a combination of advection and Fickian diffusion. One example is the well-known Nernst-Planck equation for electrodiffusion, see Ex. 3.9. However, the relationship between macroscopic and microscopic formulations is more complicated when chemical reactions are included. Macroscopically, reactions are described in terms of the deterministic law of mass action, whereas microscopically they are modeled stochastically using a chemical master equation. Differences between the two levels of modeling become significant when the number of interacting molecules becomes small [822]. From the macroscopic picture of Fickian diffusion, the conservation equation $\partial_t u = -\nabla\cdot\mathbf{J}$ can lead to two different forms of the diffusion equation, depending on whether $\mathbf{J}(\mathbf{x},t) = -\nabla[D(\mathbf{x})u(\mathbf{x},t)]$ or $\mathbf{J}(\mathbf{x},t) = -D(\mathbf{x})\nabla u(\mathbf{x},t)$. (These are equivalent when D is a constant.) In order to distinguish between the two cases, it is necessary to incorporate details regarding the microscopic dynamics using, for example, kinetic theory [140]. The situation is even more complicated in anisotropic heterogeneous media, where it is no longer possible to characterize the rate of diffusion in terms of a single coefficient. One now needs to consider a diffusion tensor, see the example of active transport on microtubular networks in Sect. 7.2.3. Finally, note that a generalized version of Fickian flux will be introduced in Sect. 13.1, which is defined in terms of the gradient of a chemical potential.

Polar coordinates. It is often useful to determine the form of higher-dimensional FP and Langevin equations in non-Cartesian coordinates, particularly in cells where there is an underlying radial symmetry. In the two-dimensional case $(x,y) \in \mathbb{R}^2$, introduce the polar coordinates (r, ϕ)

$$x = r\cos\phi, \quad y = r\sin\phi,$$

with $0 < r < \infty$ and $0 \le \phi < 2\pi$. Define the unit vectors

$$\mathbf{e}_r = \cos\phi\,\mathbf{e}_x + \sin\phi\,\mathbf{e}_y, \quad \mathbf{e}_\phi = -\sin\phi\,\mathbf{e}_x + \cos\phi\,\mathbf{e}_y,$$

where $\mathbf{e}_x$ and $\mathbf{e}_y$ are the orthogonal unit vectors in the Cartesian coordinate system. Using standard properties of the gradient operator ∇ in polar coordinates, and setting

$$\frac{\mathbf{F}}{\gamma} = A_r\mathbf{e}_r + A_\phi\mathbf{e}_\phi,$$

the FP equation (2.3.21) becomes

$$\frac{\partial p}{\partial t} = -\frac{1}{r}\frac{\partial}{\partial r}(rA_r p) - \frac{1}{r}\frac{\partial}{\partial \phi}(A_\phi p) + D\left(\frac{\partial^2 p}{\partial r^2} + \frac{1}{r}\frac{\partial p}{\partial r} + \frac{1}{r^2}\frac{\partial^2 p}{\partial \phi^2}\right). \quad (2.3.27)$$

The corresponding SDE in polar coordinates $(R(t), \Phi(t))$ is thus

$$dR = \left[A_r + \frac{D}{R}\right]dt + \sqrt{2D}\,dW_r \qquad (2.3.28a)$$

$$d\Phi = \frac{1}{R}A_\phi dt + \frac{\sqrt{2D}}{R}dW_\phi, \qquad (2.3.28b)$$

where $W_r(t)$ and $W_\phi(t)$ are independent Wiener processes with unit variance. In order to show that these are the appropriate equations, we rewrite the FP equation as a PDE for the probability density $\pi(r,\phi,t) = rp(r,\phi,t)$. Note that (2.3.28) could also be derived by performing a change of variables in the SDE (2.3.19) and using Ito's lemma.

Spherical polar coordinates. In the three-dimensional case $(x,y,z) \in \mathbb{R}^3$, introduce the spherical polar coordinates (r, θ, ϕ)

$$x = r\sin\theta\cos\phi, \quad y = r\sin\theta\sin\phi, \quad z = r\cos\theta,$$

with $0 < r < \infty$, $0 \le \theta \le \pi$, and $0 \le \phi < 2\pi$. Define the unit vectors

$$\mathbf{e}_r = \sin\theta\cos\phi\,\mathbf{e}_x + \sin\theta\sin\phi\,\mathbf{e}_y + \cos\theta\,\mathbf{e}_z, \quad \mathbf{e}_\phi = -\sin\phi\,\mathbf{e}_x + \cos\phi\,\mathbf{e}_y,$$

and

$$\mathbf{e}_\theta = \mathbf{e}_\phi \times \mathbf{e}_r = \cos\theta\cos\phi\,\mathbf{e}_x + \cos\theta\sin\phi\,\mathbf{e}_y - \sin\theta\,\mathbf{e}_z.$$

Using standard properties of the gradient operator ∇ in spherical polar coordinates, and setting

$$\frac{\mathbf{F}}{\gamma} = A_r \mathbf{e}_r + A_\theta \mathbf{e}_\theta + A_\phi \mathbf{e}_\phi,$$

the FP equation (2.3.21) becomes

$$\frac{\partial p}{\partial t} = -\frac{1}{r^2}\frac{\partial}{\partial r}\left(r^2 A_r p\right) - \frac{1}{r\sin\theta}\frac{\partial}{\partial \theta}\left(\sin(\theta)A_\theta p\right) - \frac{1}{r\sin\theta}\frac{\partial}{\partial \phi}\left(A_\phi p\right)$$
$$+ D\left(\frac{\partial^2 p}{\partial r^2} + \frac{2}{r}\frac{\partial p}{\partial r} + \frac{1}{r^2\sin\theta}\frac{\partial}{\partial \theta}\left(\sin\theta\frac{\partial p}{\partial \theta}\right) + \frac{1}{r^2\sin^2\theta}\frac{\partial^2 p}{\partial \phi^2}\right). \quad (2.3.29)$$

The corresponding SDE in spherical polar coordinates $(R(t), \Theta(t), \Phi(t))$ is thus

$$dR = \left[A_r + \frac{2D}{R}\right]dt + \sqrt{2D}dW_r, \quad (2.3.30a)$$

$$d\Theta = \left[\frac{A_\theta}{R} + \frac{D}{R^2}\cot(\Theta)\right]dt + \frac{\sqrt{2D}}{R}dW_\theta, \quad (2.3.30b)$$

$$d\Phi = \frac{A_\phi}{R\sin\Theta}dt + \frac{\sqrt{2D}}{R\sin\Theta}dW_\phi, \quad (2.3.30c)$$

where $W_r(t)$, $W_\Theta(t)$ and $W_\Phi(t)$ are independent Wiener processes with unit variance. These equations can be obtained by rewriting the FP equation in terms of the probability density $\pi(r, \theta, \phi) = r^2\sin\theta\, p(r, \theta, \phi, t)$ or applying Ito's lemma to the SDE (2.3.19).

Finally, suppose that we consider pure Brownian motion on the surface of a sphere of fixed radius r^* by setting $\mathbf{A} \equiv 0$ and $R(t) = r^*$. The resulting SDE for the angular coordinates $\Theta(t), \Phi(t)$ is

$$d\Theta = \frac{D}{r^{*2}}\cot(\Theta)dt + \frac{\sqrt{2D}}{r^*}dW_\theta, \quad d\Phi = \frac{\sqrt{2D}}{r^*\sin\Phi}dW_\phi. \quad (2.3.31)$$

The drift term in the first equation is a consequence of the fact that the sphere is a curved manifold [141, 172, 398].

2.3.3 Boundary value problems

As we have already highlighted, diffusion within the cell is bounded and often restricted to some subcellular compartment with complex geometry. This means that one has to solve an initial boundary value problem for the Fokker–Planck equation on a bounded domain $\Omega \subset \mathbb{R}^d$ with $d = 1, 2, 3$. There are a number of methods for solving the FP equation in a bounded domain, including separation of variables, transform methods, and eigenfunction expansions. Here we illustrate these methods in the specific case of the diffusion equation; see also Refs. [529, 717]. The same methods can be applied to the FP equation with drift, but the analysis is more involved [300, 699]; see Sect. 2.3.4.

Let $p = p(\mathbf{x}, t)$ satisfy the initial boundary value problem

$$\frac{\partial p}{\partial t} = D\nabla^2 p, \quad \mathbf{x} \in \Omega, t > 0, \tag{2.3.32}$$

$$p(\mathbf{x}, t) = 0, \quad \mathbf{x} \in \partial\Omega, t > 0, \quad p(\mathbf{x}, 0) = f(\mathbf{x}), \quad \mathbf{x} \in \Omega, \tag{2.3.33}$$

where $\partial\Omega$ denotes the boundary of Ω. For the sake of illustration, we consider the Dirichlet boundary condition $p(\mathbf{x}, t) = 0, \mathbf{x} \in \partial\Omega$. However, the same methods can be applied to the Neumann or no-flux boundary condition

$$J(\mathbf{x}, t) \equiv -D\mathbf{n_x} \cdot \nabla p(\mathbf{x}, t) = 0, \quad \mathbf{x} \in \partial\Omega, \tag{2.3.34}$$

where $\mathbf{n_x}$ is the unit normal to the boundary at $\mathbf{x} \in \partial\Omega$ (with the convention that it points outward from the domain Ω). A standard method for solving this initial boundary value problem is separation of variables. The first step is to substitute the solution $p(\mathbf{x}, t) = U(\mathbf{x})T(t)$ into the diffusion equation to give

$$U(\mathbf{x})T'(t) = DT(t)\nabla^2 U(\mathbf{x}),$$

which we rewrite as

$$\frac{T'(t)}{DT(t)} = \frac{\nabla^2 U(\mathbf{x})}{U(\mathbf{x})} = -\lambda.$$

The essential idea of the method is that λ is a constant, since it cannot be both a function of only t and only $\mathbf{x}$. It follows that we can separate the PDE into a spatial part and a temporal part according to

$$T'(t) = -\lambda DT(t), \tag{2.3.35a}$$

$$-\nabla^2 U(\mathbf{x}) = \lambda U(\mathbf{x}), \quad \mathbf{x} \in \Omega, \quad U(\mathbf{x}) = 0, \mathbf{x} \in \partial\Omega, \tag{2.3.35b}$$

where the Dirichlet boundary condition has been imposed on the spatial part.

Equation (2.3.35b) is an example of a boundary value problem for the negative Laplacian $-\nabla^2$. For each value of λ for which (2.3.35b) has a nontrivial solution $U(\mathbf{x})$, λ is called an eigenvalue and $U(\mathbf{x})$ is the corresponding eigenfunction (defined up to an arbitrary, nonzero, scalar multiplication). More precisely, λ is an element of the discrete spectrum of the linear differential operator $\mathbb{L} = -\nabla^2$ acting on the Hilbert space $L^2(\Omega)$. A brief introduction to linear differential operators and function spaces is given in Box 2D. (In the case of differential operators acting on an unbounded domain such as $\mathbb{R}^n$, the spectrum is more complicated; see Box 11G. Note, in particular, that $\mathbb{L}$ is self-adjoint with respect to the $L^2(\Omega)$ norm. Hence, its spectrum has the following properties [717]:

1. The eigenvalues are real.

2. There are infinitely many eigenvalues that can be ordered as $0 < \lambda_1 \leq \lambda_2 \leq \lambda_2 \leq \dots$ with $\lambda_n \to \infty$ as $n \to \infty$.

3. Eigenfunctions corresponding to distinct eigenvalues are orthogonal with respect to the standard inner product on Ω, that is,

$$\langle \phi_n | \phi_m \rangle \equiv \int_\Omega \phi_n(\mathbf{x}) \phi_m(\mathbf{x}) d\mathbf{x} = 0,$$

when $\lambda_n \neq \lambda_m$. The number of linearly independent eigenfunctions associated with a degenerate eigenvalue is finite, so that a Schmidt orthogonalization procedure can be used to make them orthogonal to each other, which we assume below.

4. The set of eigenfunctions $\phi_n(\mathbf{x})$ is complete in the sense that any square-integrable function $F \in L^2(\Omega)$ can be uniquely represented by a generalized Fourier series

$$F(\mathbf{x}) = \sum_{n=1}^{\infty} c_n \phi_n(\mathbf{x}), \quad c_n = \frac{\langle F | \phi_n \rangle}{\|\phi_n\|^2},$$

where c_n are the generalized Fourier coefficients and the norm is $\|\phi_n\| = \sqrt{\langle \phi_n | \phi_n \rangle}$. This means that the truncated Fourier series converges in the $L^2(\Omega)$ sense,

$$\int_\Omega \left(f(\mathbf{x}) - \sum_{n=1}^{N} c_n \phi_n(\mathbf{x}) \right)^2 d\mathbf{x} \to 0 \text{ as } N \to \infty.$$

Note that the same properties hold when the Dirichlet boundary condition is replaced by the Neumann boundary condition, except that there now exists a zero eigenvalue $\lambda_0 = 0$ whose eigenfunction $\phi_0(\mathbf{x}) = \text{constant}$. (This reflects the fact that the diffusion equation has a nontrivial steady state in the case of a no-flux boundary condition.)

Returning to equation (2.3.35), we immediately see that we can identify the constant λ with one of the eigenvalues λ_n of $-\nabla^2$. Solving the equation for T then shows that we have an infinite set of solutions of the form $p_n(\mathbf{x},t) = \phi_n(\mathbf{x}) e^{-D\lambda_n t}$. Since the diffusion equation is linear, we can apply the principle of superposition to write down the general solution

$$p(\mathbf{x},t) = \sum_{n=1}^{\infty} c_n \phi_n(\mathbf{x}) e^{-\lambda_n D t}. \tag{2.3.36}$$

Finally, imposing the initial condition requires that

$$f(\mathbf{x}) = \sum_{n=1}^{\infty} c_n \phi_n(\mathbf{x}),$$

so that we can identify the c_n as the generalized Fourier coefficients of f. That is,

$$c_n = \frac{1}{\|\phi_n\|^2} \int_\Omega f(\mathbf{x}) \phi_n(\mathbf{x}) d\mathbf{x}. \tag{2.3.37}$$

Substituting for c_n into the general solution and taking the eigenfunctions to have unit normalization ($\|\phi_n\|^2 = 1$) yields

$$p(\mathbf{x},t) = \sum_{n=1}^{\infty} \left(\int_0^L f(\mathbf{y})\phi_n(\mathbf{y})d\mathbf{y} \right) e^{-\lambda_n Dt} \phi_n(\mathbf{x}).$$

Formally switching the order of summation and integration, we have

$$p(\mathbf{x},t) = \int_0^L K(\mathbf{x},\mathbf{y},t)f(\mathbf{y})d\mathbf{y}, \qquad (2.3.38)$$

where

$$K(\mathbf{x},\mathbf{y},t) = \sum_{n=1}^{\infty} e^{-\lambda_n Dt} \phi_n(\mathbf{x})\phi_n(\mathbf{y}). \qquad (2.3.39)$$

Finally, taking the limit $t \to 0$, we deduce the completeness relation

$$\sum_{n=1}^{\infty} \phi_n(\mathbf{x})\phi_n(\mathbf{y}) = \delta(\mathbf{x}-\mathbf{y}). \qquad (2.3.40)$$

Some example boundary value problems are considered in Ex. 2.8–2.10.

Example 2.5. Consider the following initial boundary value problem for the 1D diffusion equation:

$$\frac{\partial p}{\partial t} = D\frac{\partial^2 p}{\partial x^2}, \quad 0 < x < L, t > 0,$$
$$p(0,t) = 0 = p(L,t), \quad t > 0,$$
$$p(x,0) = f(x), \quad 0 < x < L.$$

After performing separation of variables, we obtain the eigenvalue problem

$$-U''(x) = \lambda U(x), \quad 0 < x < L, \quad U(0) = U(L) = 0.$$

The eigenvalues and eigenfunctions are thus

$$\lambda_n = \frac{n^2 \pi^2}{L^2}, \quad \phi_n(x) = \sin\frac{n\pi x}{L}, \quad n = 1,2,\ldots$$

It follows that the general solution is given by

$$p(x,t) = \sum_{n=1}^{\infty} c_n e^{-n^2\pi^2 Dt/L^2} \sin\frac{n\pi x}{L}.$$

Comparison with the initial data shows that the c_n are the Fourier coefficients in the series expansion of $f(x)$,

$$f(x) = \sum_{n=1}^{\infty} c_n \sin\frac{n\pi x}{L},$$

and thus

$$c_n = \int_0^L f(\xi)\sin\frac{n\pi\xi}{L}d\xi \Big/ \int_0^L \sin^2\frac{n\pi x}{L}dx.$$

Evaluating the denominator and substituting for c_n into the general solution yields

$$p(x,t) = \frac{2}{L} \sum_{n=1}^{\infty} \left(\int_0^L f(\xi) \sin \frac{n\pi\xi}{L} d\xi \right) e^{-n^2\pi^2 Dt/L^2} \sin \frac{n\pi x}{L}.$$

Formally switching the order of summation and integration (which is valid provided that the functions are sufficiently well-behaved), the solution can be re-expressed in the compact form

$$p(x,t) = \int_0^L K(x,\xi,t) f(\xi) d\xi, \qquad (2.3.41)$$

where

$$K(x,\xi,t) = \frac{2}{L} \sum_{n=1}^{\infty} e^{-n^2\pi^2 Dt/L^2} \sin \frac{n\pi\xi}{L} \sin \frac{n\pi x}{L}. \qquad (2.3.42)$$

Note that equation (2.3.42) represents an eigenvalue expansion of the associated 1D Green's function (see Box 6C). An alternative series expansion of the Green's function can be obtained using the Poisson summation formula. For appropriate functions, the Poisson summation formula states that

$$\sum_{n=-\infty}^{\infty} f(n) = \sum_{k=-\infty}^{\infty} \widehat{f}(2\pi k), \qquad (2.3.43)$$

where $\widehat{f}$ is the Fourier transform of f. We now note that equation (2.3.42) can be rewritten as

$$K(x,\xi,t) = \frac{1}{2L} \sum_{n=-\infty}^{\infty} e^{-n^2\pi^2 Dt/L^2} \left[\cos \frac{n\pi(x-\xi)}{L} - \cos \frac{n\pi(x+\xi)}{L} \right].$$

Setting

$$f(n) = \frac{1}{2L} e^{-n^2\pi^2 Dt/L^2} \cos \frac{n\pi x}{L}$$

for fixed x, we have

$$\widehat{f}(2\pi k) = \text{Re} \int_{-\infty}^{\infty} e^{i2\pi kn} e^{-n^2\pi^2 Dt/L^2} e^{in\pi x/L} dn$$

$$= \frac{1}{\sqrt{4\pi Dt}} \exp \left\{ -\frac{(x+2kL)^2}{4Dt} \right\}.$$

An application of the Poisson summation formula then yields the following alternative series representation of the Green's function:

$$K(x,\xi,t) = \frac{1}{\sqrt{4\pi Dt}} \sum_{k=-\infty}^{\infty} \left[\exp \left\{ -\frac{(x-\xi+2kL)^2}{4Dt} \right\} - \exp \left\{ -\frac{(x+\xi+2kL)^2}{4Dt} \right\} \right].$$

$$(2.3.44)$$

The latter can also be derived using the method of images.

Box 2D. Linear differential operators.

Here we summarize some of the basic results regarding linear differential operators acting on a function space. For simplicity, we will restrict ourselves to real-valued functions $f : \mathbb{R} \to \mathbb{R}$, although it is straightforward to generalize the results to complex-valued functions.

Function spaces. Consider the set of all real functions $f(x)$ on the interval $[a,b]$. This is a vector space over the set of real numbers: given two functions $f_1(x), f_2(x)$ and two real numbers α_1, α_2, we can form the sum $f(x) = \alpha_1 f_1(x) + \alpha_2 f_2(x)$ such that $f(x)$ is also a function on $[a,b]$. Either on physical grounds or for mathematical convenience, we usually restrict ourselves to a subspace of functions that are differentiable to some given order. For example, the space of functions on $[a,b]$ with n continuous derivatives is denoted by $C^n[a,b]$, and the space of analytic functions (those whose Taylor expansion converges to the given function) is denoted by $C^\omega[a,b]$. In order to describe the convergence of a sequence of functions $f_n, n = 1, 2, \ldots$ to a limit function f, we need to introduce the concept of a norm, which is a generalization of the usual measure of the length of a finite-dimensional vector. The norm $\|f\|$ of a function f is a real number with the following properties:

(i)   positivity: $\|f\| \geq 0$, and $\|f\| = 0$ if and only if $f = 0$;
(ii) the triangle inequality: $\|f + g\| \leq \|f\| + \|g\|$;
(iii) linearity: $\|\lambda f\| = |\lambda| \|f\|$ for $\lambda \in \mathbb{R}$.

Common examples of norms are the "sup" norm

$$\|f\|_\infty = \sup_{x \in [a,b]} |f(x)|,$$

and the L^p norm

$$\|f\|_p = \left(\int_a^b |f(x)|^p dx \right)^{1/p}.$$

Given the L^p norm, we can introduce another important function space $L^p[a,b]$, which is the space of real-valued functions on $[a,b]$ for which $\|f\|_p < \infty$. However, there is one subtlety here, namely that it is possible for $\|f\| = 0$ without f being identically zero. For example, f may vanish at all but a finite set of points (set of measure zero). This violates the positivity property of a norm. Therefore, one should really treat elements of $L^p[a,b]$ as equivalence classes of functions, where functions differing on a set of measure zero are identified.

Cauchy sequences and completeness. Given a normed function space, convergence of a sequence $f_n \to f$ can be expressed as

$$\lim_{n \to \infty} \|f_n - f\| = 0.$$

In the case of the "sup" norm, f_n is said to converge uniformly to f, whereas for the L^1 norm, it is said to converge in the mean. An important property of a function space is that of being complete. First, consider the following definition of a Cauchy sequence: A sequence f_n in a normed vector space is Cauchy if for any $\varepsilon > 0$ we can find an integer N such that $n, m > N$ implies that $\|f_m - f_n\| < \varepsilon$. In other words, elements of the sequence become arbitrarily close together as $n \to \infty$. A normed vector space is then complete with respect to its norm if every Cauchy sequence converges to some element in the space. A complete normed vector space is called a Banach space $\mathscr{B}$. In many applications, the norm of the function space is taken to be the so-called natural norm obtained from an underlying inner product. For example, if we define an inner product for $L^2[a,b]$ according to

$$\langle f,g \rangle = \int_a^b f(x)g(x)dx,$$

then the $L^2[a,b]$ norm can be written as

$$\|f\|_2 = \sqrt{\langle f,f \rangle}.$$

A Banach space with an inner product is called a Hilbert space $\mathscr{H}$.

Linear differential operators. Suppose that $\mathbb{L}$ is a linear differential operator acting on a subspace of $L^2(\mathbb{R})$, which we denote by the domain $\mathscr{D}(\mathbb{L})$. Linearity of the operator means that for $f_1, f_2 \in \mathscr{D}(\mathbb{L})$ and $a_1, a_2 \in \mathbb{R}$,

$$\mathbb{L}(a_1 f_1 + a_2 f_2) = a_1 \mathbb{L}f_1 + a_2 \mathbb{L}f_2.$$

Given the standard inner product on $L^2(\mathbb{R})$, we define the adjoint linear operator $\mathbb{L}^\dagger$ according to

$$\langle f, \mathbb{L}g \rangle = \langle \mathbb{L}^\dagger f, g \rangle, \quad f, g \in \mathscr{D}(L).$$

The operator is said to be self-adjoint if $\mathbb{L}^\dagger = \mathbb{L}$. Note that, in practice, one determines $\mathbb{L}^\dagger$ using integration by parts. For functions defined on finite intervals, this generates boundary terms that only vanish if appropriate boundary conditions are imposed. In general, this can result in different domains for $\mathbb{L}$ and $\mathbb{L}^\dagger$. Therefore, the condition for self-adjointness becomes $\mathbb{L} = \mathbb{L}^\dagger$ and $\mathscr{D}(\mathbb{L}) = \mathscr{D}(\mathbb{L}^\dagger)$. Given these definitions, we can write the FP equation on a domain $[0,L]$ in the more compact form

$$\frac{\partial p}{\partial t} = \mathbb{L}p, \quad \mathbb{L} = -\frac{\partial}{\partial x}A(x) + D\frac{\partial^2}{\partial x^2},$$

where $\mathbb{L}$ is a linear operator acting on the Hilbert space $L^2[0,L]$ with inner product

$$\langle f,g \rangle = \int_0^L f(x)g(x)dx.$$

Introducing the adjoint operator

$$\mathbb{L}^\dagger = A(x)\frac{\partial}{\partial x} + D\frac{\partial^2}{\partial x^2},$$

one can then write the backward FP equation as $\partial_t q = \mathbb{L}^\dagger q$.

2.3.4 FP equation and symmetrization

One of the simplifications in the analysis of boundary value problems for the diffusion equation is that the associated Laplacian is self-adjoint. This is no longer true in the case of the FP equation with drift; see equation (2.3.1), defined on a bounded interval Ω. The differential operator takes the form

$$\mathbb{L} = -\frac{\partial}{\partial x}A(x) + D\frac{\partial^2}{\partial x^2}, \qquad (2.3.45)$$

whose formal adjoint (after integration by parts) is

$$\mathbb{L}^\dagger = A(x)\frac{\partial}{\partial x} + D\frac{\partial^2}{\partial x^2}. \qquad (2.3.46)$$

Since $\mathbb{L}^\dagger \neq \mathbb{L}$, we see that $\mathbb{L}$ is not self-adjoint with respect to the $L^2(\Omega)$ norm. However, one can carry out a transformation that yields a self-adjoint version of the FP equation [699], which is often referred to as a form of symmetrization. The basic idea is to introduce the new function $q(x,t)$ according to

$$p(x,t) = e^{-U(x)/2D}q(x,t), \quad U'(x) = -A(x).$$

Substituting $q(x,t) = q(x)e^{-\lambda t}$ into the corresponding FP equation gives

$$-\lambda q(x) = e^{U(x)/2D}\left[-\frac{d}{dx}A(x) + D\frac{d^2}{dx^2}\right]e^{-U(x)/2D}q$$

$$= \left(-A(x)\frac{dq(x)}{dx} - A'(x)q(x) + A(x)q(x)\frac{U'(x)}{2D}\right)$$

$$+ De^{U(x)/2D}\frac{d}{dx}\left(\frac{dq(x)}{dx} - q(x)\frac{U'(x)}{2D}\right)e^{-U(x)/2D}.$$

It can be checked that the first-order derivatives cancel so that we obtain the ODE

$$\lambda q = -D\frac{d^2q}{dx^2} + V(x)q, \qquad (2.3.47)$$

with

$$V(x) = \frac{A^2(x)}{4D} + \frac{A'(x)}{2}. \qquad (2.3.48)$$

The differential operator is now self-adjoint. Mathematically speaking, this ODE is identical to the time-independent Schrodinger equation for a particle of mass $m = 1$ and energy λ moving in a potential well $V(x)$. (Under this equivalence, Planck's constant becomes $\hbar = \sqrt{2D}$).

For certain choices of the drift $A(x)$, one can exploit knowledge of the energy eigenvalues of a quantum particle to determine the spectrum of the FP operator. For example, suppose that we have an OU process with $A(x) = -kx$. In that case, $V(x) =$

$k^2/4D - k/2$ and the quantum-mechanical version of equation (2.3.47) describes a harmonic oscillator of mass $m = 1$ and natural frequency $\omega = k/\sqrt{2D}$. Even though the domain $\mathbb{R}$ is unbounded, the spectrum is real and discrete because $V(x)$ is a confining potential, that is, $V(x) \to \infty$ as $x \to \pm\infty$. The eigenvalues are

$$\lambda_n = \left(n + \frac{1}{2}\right)\hbar\omega - \frac{k}{2} = nk \tag{2.3.49}$$

for integers $n \geq 0$, and the normalized eigenfunctions are

$$\phi_n(x) = \frac{1}{\sqrt{2^n n! \sqrt{\pi}}} \left(\sqrt{\frac{\omega^2}{2D}}\right)^{1/4} H_n\left((\omega^2/2D)^{1/4}x\right) e^{-\sqrt{\omega^2/2D}x^2/2},$$

with $H_n(x)$ the Hermite polynomial of integer order n.

2.3.5 Multiplicative noise: Ito versus Stratonovich

So far we have assumed that the diffusion coefficient in the Langevin equation (2.2.5) is position-independent, that is, the noise term is additive. The situation is considerably more involved when the term multiplying $dW(t)$ depends on $X(t)$, that is, when the noise term is multiplicative. In particular, consider an SDE of the form

$$dX(t) = A(X)dt + B(X)dW(t). \tag{2.3.50}$$

(Here we do not necessarily interpret $X(t)$ as the position of Brownian particle.) The difficulty arises since, in order to construct a solution of the SDE, we have to deal with stochastic integrals of the form

$$\mathscr{I}(t) = \int_0^t B(X(\tau))dW(\tau). \tag{2.3.51}$$

Suppose for the moment that $X(t)$ and $W(t)$ are deterministic functions of time, and we can apply the theory of Riemann integration. That is, we partition the time interval $[0, t]$ into N equal intervals of size Δt with $N\Delta t = t$ and identify the value of the integral with the unique limit (assuming it exists)

$$\mathscr{I}(t) = \lim_{N \to \infty} \sum_{n=0}^{N-1} B([1-\alpha]X_n + \alpha X_{n+1})\Delta W_n \tag{2.3.52}$$

for $0 \leq \alpha < 1$, where $\Delta W_n = W((n+1)\Delta t) - W(n\Delta t)$ and $X_n = X(n\Delta t)$. In the deterministic case, the integral is independent of α. Unfortunately, this is no longer true when we have a stochastic integral. One way to see this is to note that the ΔW_n are independent random variables. Hence, the function B is only statistically independent of ΔW_n when $\alpha = 0$, which is the Ito definition of stochastic integration (see Sect. 2.2). On the other hand, when $\alpha = 1/2$ we have the Stratonovich version. It turns out that the form of the corresponding FP equation also depends on α, as we now show.

Let us Taylor expand the nth term in the sum defining the integral $\mathscr{I}(t)$ about the point X_n and set $B_n = B(X_n)$:

$$B((1-\alpha)X_{n+1} + \alpha X_n) = B_n + \alpha\Delta X_n \frac{\partial B_n}{\partial x} + \frac{1}{2}(\alpha\Delta X_n)^2 \frac{\partial^2 B_n}{\partial x^2} + \cdots,$$

with

$$\Delta X_n = X_{n+1} - X_n = A_n\Delta t + (B_n + O(\Delta X_n))\Delta W_n.$$

Substituting for ΔX_n and dropping terms that are higher order than Δt shows that

$$B((1-\alpha)X_{n+1} + \alpha X_n) = B_n + \left(\alpha A_n \frac{\partial B_n}{\partial x} + \frac{\alpha^2 B_n^2}{2}\frac{\partial^2 B_n}{\partial x^2}\right)\Delta t + \left(\alpha B_n \frac{\partial B_n}{\partial x}\right)\Delta W_n.$$

Applying this result to the sum appearing in the definition of the integral, equation (2.3.51), and again dropping higher-order terms in Δt yields the result

$$\sum_{n=0}^{N-1} B((1-\alpha)X_{n+1} + \alpha X_n)\Delta W_n = \sum_{n=0}^{N-1} B_n\Delta W_n + \alpha\sum_{n=0}^{N-1} B_n \frac{\partial B_n}{\partial x}(\Delta W_n)^2.$$

Finally, taking the continuum limit with $dW(t)^2 = dt$, we have

$$\mathscr{I}(t) = \int_0^t B(X(s))dW(s) + \alpha \int_0^t \frac{\partial B(X(s))}{\partial x} B(X(s))ds. \qquad (2.3.53)$$

We can now rewrite the solution in terms of an Ito integral according to

$$X(t) = X_0 + \int_0^t \left[A(X(s)) + \alpha\frac{\partial B(X(s))}{\partial x}B(X(s))\right]ds + \int_0^t B(X(s))dW(s). \quad (2.3.54)$$

The latter is the solution to an equivalent Ito SDE of the form

$$dX = \left[A(X) + \alpha B(X)\frac{\partial B(X)}{\partial x}\right]dt + B(X)dW(t). \qquad (2.3.55)$$

Finally, given that we know the FP equation corresponding to an Ito SDE, we can immediately write down the FP equation corresponding to the modified SDE (2.3.50):

$$\frac{\partial}{\partial t}p(x,t) = -\frac{\partial}{\partial x}([A(x) + \alpha B(x)B'(x)]p(x,t)) + \frac{1}{2}\frac{\partial^2}{\partial x}\left(B(x)^2 p(x,t)\right). \qquad (2.3.56)$$

In summary, the mathematical interpretation of multiplicative noise is ambiguous due to the subtleties of stochastic integration, resulting in different versions of the FP equation.

(a) Ito multiplicative noise arises when carrying out a system-size expansion of a birth–death or chemical master equation (intrinsic noise); see Sect. 5.2. The usual rules of calculus no longer hold, and

$$\langle B(X)dW(t)\rangle = \langle B(X)\rangle\langle dW(t)\rangle = 0.$$

Mathematicians use Ito because they can prove theorems! The Ito FP equation is

$$\frac{\partial p(x,t)}{\partial t} = -\frac{\partial [A(x)p(x,t)]}{\partial x} + \frac{1}{2}\frac{\partial^2}{\partial x^2}B(x)^2 p(x,t). \qquad (2.3.57)$$

An example of a 1D stochastic process with Ito multiplicative noise is considered in Ex. 2.11.

(b) Stratonovich multiplicative noise tends to be used when a system is driven by fluctuations in the environment (extrinsic noise). In this case

$$\langle B(X)dW(t)\rangle \neq \langle B(X)\rangle\langle dW(t)\rangle.$$

Besides having the advantage that the usual rules of calculus hold, it also has a deeper physical origin. Since noise terms represent at some coarse-grained level the effects of microscopic degrees of freedom that have finite (but short) correlation times, any multiplicative noise terms should be physically interpreted as the limit in which these correlation times approach zero. This limit produces white noise terms that must be interpreted in the sense of Stratonovich. This is shown explicitly below using a backward FP equation. The Stratonovich FP equation is

$$\frac{\partial p(x,t)}{\partial t} = -\frac{\partial [A(x)p(x,t)]}{\partial x} + \frac{1}{2}\frac{\partial}{\partial x}B(x)\frac{\partial}{\partial x}B(x)p(x,t). \qquad (2.3.58)$$

(c) Yet another interpretation of multiplicative noise arises for a system in contact with a thermal bath at constant temperature T and with a space-dependent friction coefficient. The Einstein relation (2.3.6) then implies that the diffusion coefficient (noise strength) is also space-dependent, resulting in a multiplicative noise term in the corresponding SDE. The requirement that the system should approach the equilibrium Boltzmann–Gibbs distribution in the large time limit leads to the so-called kinetic interpretation of multiplicative noise [449, 491] for which $\alpha = 1$ and the FP equation takes the form

$$\frac{\partial p(x,t)}{\partial t} = -\frac{\partial [A(x)p(x,t)]}{\partial x} + \frac{1}{2}\frac{\partial}{\partial x}B(x)^2\frac{\partial}{\partial x}p(x,t). \qquad (2.3.59)$$

Consider for example an overdamped Brownian particle with a position-dependent friction coefficient $\Gamma(x)$ such that

$$dX(t) = \Gamma(X)F(X)dt + B(X)dW(t), \qquad (2.3.60)$$

with $F(X) = -U'(X)$ and the Einstein relation $b(X) = \sqrt{2k_B T \Gamma(X)}$. It is straight-forward to show that the steady-state solution of equation (2.3.59) with $A(x) = -\Gamma(x)U'(x)$ converges to the Boltzmann–Gibbs distribution, that is,

$$\lim_{t \to \infty} p(x,t) \sim e^{-U(x)/k_B T}.$$

This result would not hold for the other interpretations of multiplicative noise. The thermodynamic interpretation of multiplicative noise will be used in the analysis of Brownian potentials; see Sect. 12.3.

Derivation of the Stratonovich FP equation using a backward equation. Let $X(t)$ denote the position of the particle at time t, which is taken to evolve according to the SDE

$$dX(t) = [A(X) + \frac{1}{\kappa}B(X)Y(t)]dt, \tag{2.3.61}$$

where $Y(t)$ is a stochastic external input evolving according to the OU process

$$dY(t) = -\frac{1}{\kappa^2}Y(t)dt + \frac{1}{\kappa}dW(t). \tag{2.3.62}$$

and $W(t)$ is a Wiener process. Heuristically speaking, in the limit $\kappa \to 0$ we can set $Y(t)dt = \kappa dW(t)$ such that we obtain the scalar SDE (2.3.50) However, since we have multiplicative noise, there is an ambiguity with regards the interpretation of the noise term from the perspective of stochastic calculus, that is, whether one should choose the Ito or Stratonovich versions. This means that the form of the corresponding FP equation is also ambiguous.

One way to resolve the above issue is to start with the full 2D Fokker-Planck equation and to reduce it to a scalar FP equation in the limit $\varepsilon \to \infty$ using an adiabatic reduction [300]. Here we follow an alternative method whose starting point is the backward FP equation, and which makes use of the so-called Fredholm alternative theorem (see Box 2E) [647]. This takes the form

$$\frac{\partial q(x,y,t)}{\partial t} = \left(\frac{1}{\kappa^2}\mathbb{L}_1^\dagger + \frac{1}{\kappa}\mathbb{L}_2^\dagger + \mathbb{L}_3^\dagger \right) q(x,y,t), \tag{2.3.63}$$

where

$$\mathbb{L}_1^\dagger = -y\frac{\partial}{\partial y} + \frac{1}{2}\frac{\partial^2}{\partial y^2}, \quad \mathbb{L}_2^\dagger = B(x)y\frac{\partial}{\partial x}, \quad \mathbb{L}_3^\dagger = A(x)\frac{\partial}{\partial x}. \tag{2.3.64a}$$

Substituting the following power series expansion,

$$q = q^{(0)} + \kappa q^{(1)} + \kappa^2 q^{(2)} + \cdots$$

into (2.3.63) yields the following hierarchy of equations:

$$\mathbb{L}_1^\dagger q^{(0)} = 0, \tag{2.3.65a}$$

$$\mathbb{L}_1^\dagger q^{(1)} = -\mathbb{L}_2^* q^{(0)} \equiv h^{(1)}, \tag{2.3.65b}$$

$$\mathbb{L}_1^\dagger q^{(2)} = -\mathbb{L}_2^* q^{(1)} - \mathbb{L}_3^* q^{(0)} + \frac{\partial}{\partial t}q^{(0)} \equiv h^{(2)}. \tag{2.3.65c}$$

It immediately follows that $q^{(0)}(x,y,t) = q(x,t)$ for some function $q(x,t)$.

Now observe that the nullspace of $\mathbb{L}_1$ is spanned by the stationary density $p_s(y)$ of $Y(t)$, which is given by

$$p_s(y) = \sqrt{\frac{1}{\pi}} e^{-y^2}. \tag{2.3.66}$$

Therefore, the righthand side of (2.3.65b) is orthogonal to the nullspace of $\mathbb{L}_1$ since

$$\int_{-\infty}^{\infty} \int_{-\infty}^{\infty} h^{(1)}(x,y) p_s(y) \, dx \, dy = -B(x) \frac{\partial}{\partial x} q(x,t) \int_{-\infty}^{\infty} y p_s(y) \, dy = 0.$$

Hence, the Fredholm alternative ensures that (2.3.65b) is solvable. Indeed, it is straightforward to check that

$$q^{(1)}(x,y,t) = B(x) y \frac{\partial}{\partial x} q(x,t) \tag{2.3.67}$$

solves equation (2.3.65b). Again appealing to the Fredholm alternative, in order for equation (2.3.65c) to be solvable, we require

$$\int_{-\infty}^{\infty} p_s(y) \left\{ \mathbb{L}_2^{\dagger} q^{(1)} + \mathbb{L}_3^{\dagger} q - \frac{\partial q}{\partial t} \right\} dy = 0.$$

Now, it is immediate that

$$\int_{-\infty}^{\infty} p_s(y) \frac{\partial q}{\partial t} \, dy = \frac{\partial q}{\partial t}, \quad \int_{-\infty}^{\infty} p_s(y) \mathbb{L}_3^{\dagger} q \, dy = A(x) \frac{\partial}{\partial x} q,$$

since $\int_{-\infty}^{\infty} p_s(y) \, dy = 1$ and q is independent of y. Furthermore, using (2.3.67) we have that

$$\int_{-\infty}^{\infty} p_s(y) \mathbb{L}_2^{\dagger} q^{(1)} \, dy = \left(\int_{-\infty}^{\infty} y^2 p_s(y) \, dy \right) B(x) \frac{\partial}{\partial x} \left[B(x) \frac{\partial}{\partial x} q(x,t) \right]$$

$$= \frac{1}{2} B(x) \frac{\partial}{\partial x} \left[B(x) \frac{\partial}{\partial x} q(x,t) \right],$$

since $\int_{-\infty}^{\infty} y^2 p_s(y) \, dy = 1/2$. Putting this together yields the limiting backward FP equation,

$$\frac{\partial q}{\partial t} = A(x) \frac{\partial}{\partial x} q + \frac{1}{2} B(x) \frac{\partial}{\partial x} \left[B(x) \frac{\partial}{\partial x} q(x,t) \right].$$

The corresponding forward FP equation is given by the Stratonovich version (2.3.58).

Box 2E. Fredholm alternative theorem.

Consider an M-dimensional linear inhomogeneous equation $\mathbf{A}\mathbf{z} = \mathbf{b}$ with $\mathbf{z}, \mathbf{b} \in \mathbb{R}^M$. Suppose that the $M \times M$ matrix $\mathbf{A}$ has a nontrivial nullspace and let $\mathbf{u}$ be a null vector of the adjoint matrix $\mathbf{A}^{\dagger}$, that is, $\mathbf{A}^{\dagger}\mathbf{u} = 0$. The Fredholm alternative theorem for finite-dimensional vector space states that the inhomogeneous equation has a (non-unique) solution for $\mathbf{z}$ if and only if $\mathbf{u} \cdot \mathbf{b} = 0$ for all null vectors $\mathbf{u}$.

The Fredholm alternative theorem for matrices can be extended to the case of linear operators acting on infinite-dimensional function spaces [717]. Recall from Box 2D the definition of the Hilbert space $L^2([a,b])$ with inner product for any two functions $f,g \in L^2([a,b])$ given by

$$\langle f,g \rangle = \int_a^b f(x)g(x)dx.$$

Suppose that $\mathbb{L}$ is a linear differential operator acting on a subspace of $L^2([a,b])$ consisting of functions that are differentiable to the appropriate order, which we denote by the domain $\mathcal{D}(\mathbb{L})$. Linearity of the operator means that for $f_1, f_2 \in \mathcal{D}(\mathbb{L})$ and $a_1, a_2 \in \mathbb{R}$, Given the standard inner product on $L^2(\mathbb{R})$, we define the adjoint linear operator $\mathbb{L}^\dagger$ according to

$$\langle f, \mathbb{L}g \rangle = \langle \mathbb{L}^\dagger f, g \rangle, \quad f, g \in \mathcal{D}(L).$$

(The operator is said to be self-adjoint if $\mathbb{L}^\dagger = \mathbb{L}$ and $\mathcal{D}(\mathbb{L}) = \mathcal{D}(\mathbb{L}^\dagger)$; see Box 2D.) Given a differential operator $\mathbb{L}$ on $L^2([a,b])$, we can now state an infinite-dimensional version of the Fredholm alternative theorem: The inhomogeneous equation $\mathbb{L}f = h$ has a solution if and only if

$$\langle h, u \rangle = 0 \quad \text{for all u satisfying} \quad \mathbb{L}^\dagger u = 0.$$

2.4 First passage time problems

One of the most important characteristics of a diffusion process is the mean first passage time (MFPT) for a particle to reach a given target within or on the boundary of a given domain. We begin by considering the simple case of a particle diffusing in a bounded interval, with an absorbing boundary at one end. In particular, we derive a differential equation for the MFPT by introducing the notion of a survival probability and showing that the latter evolves according to a backward FP equation. In the case of an absorbing boundary at both ends, one can condition on the particular end from which the particle escapes, which leads to the notion of a splitting probability. We then turn to the classical problem of noise-induced escape from a double-well potential, and derive Kramer's formula for the MFPT. We end by briefly discussing extensions to the case of multiplicative noise, and introducing the notion of a quasi-potential. The theory of noise-induced escape is developed in much more detail in Chap. 8.

2.4.1 Mean first passage time on an interval

Consider a particle whose position evolves according to the 1D Langevin equation (2.2.5) with motion restricted to the bounded domain $x \in [0,L]$. Suppose that the corresponding FP equation (2.3.1) has a reflecting boundary condition at $x = 0$ and an absorbing boundary condition at $x = L$ (see Fig. 2.6):

$$J(0,t) = 0, \quad p(L,t) = 0. \tag{2.4.1}$$

We would like to determine the stochastic time $T(y)$ for the particle to exit the right-hand boundary given that it starts at location $y \in [0,L]$ at time $t = 0$. As a first step, we introduce the survival probability $Q(y,t)$ that the particle has not yet exited the interval at time t:

$$Q(y,t) = \int_0^L p(x,t|y,0)dx. \tag{2.4.2}$$

It follows that

$$\mathbb{P}[T(y) \leq t] = \text{probability that particle exits before time } t = 1 - Q(y,t),$$

and we can define the first passage time (FPT) density according to

$$f(y,t)\Delta t = \mathbb{P}[t < T(y) < t + \Delta t] = \mathbb{P}[T(y) \leq t + \Delta t]] - \mathbb{P}[T(y) \leq t]$$
$$= 1 - Q(y,t + \Delta t) - (1 - Q(y,t)) = Q(y,t) - Q(y,t + \Delta t).$$

Dividing both sides by Δt and taking the limit $\Delta t \to 0$ gives

$$f(y,t) = -\frac{\partial Q(y,t)}{\partial t}. \tag{2.4.3}$$

Using the definition of the survival probability, we obtain an alternative expression for f as follows:

$$f(y,t) = -\int_0^L \frac{\partial}{\partial t} p(x,t|y,0)dx = \int_0^L \frac{\partial J(x,t|y,0)}{\partial x}dx = J(L,t|y,0), \tag{2.4.4}$$

where we have used the FP equation written in conservation form (2.3.2) and the reflecting boundary condition $J(0,t|y,0) = 0$. Hence, the FPT density is equal to the flux through the absorbing boundary at $x = L$. In certain simple cases, the flux

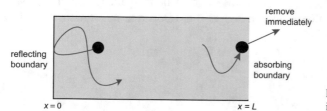

reflecting boundary

absorbing boundary

remove immediately

$x = 0$

$x = L$

Fig. 2.6: Brownian motion in a 1D bounded domain.

can be calculated explicitly. However, for more general cases, it is useful to derive explicit differential equations for moments of the FPT density, in particular, the first moment or MFPT. The MFPT $\tau(y)$ is defined according to

$$\tau(y) := \mathbb{E}[T(y)] = \int_0^\infty f(y,t)t\,dt. \tag{2.4.5}$$

Using equation (2.4.3) and integrating by parts yields

$$\tau(y) = -\int_0^\infty t\frac{\partial Q(y,t)}{\partial t}\,dt = \int_0^\infty Q(y,t)\,dt. \tag{2.4.6}$$

Laplace transforming equations (2.4.3) and (2.4.4) show that

$$\tilde{J}(y,s) = 1 - s\tilde{Q}(y,s), \tag{2.4.7}$$

where $\tilde{Q}(y,s)$ is the Laplace transform of $Q(y,t)$ and $\tilde{J}(y,s)$ is the Laplace transform of $J(L,t|y,0)$. Equation (2.4.6) can thus be written in the alternative form

$$\tau(y) = \tilde{Q}(y,0) = -\left.\frac{\partial}{\partial s}\tilde{J}(y,s)\right|_{s=0}. \tag{2.4.8}$$

Similarly, higher-order moments of the FPT density are obtained by differentiation:

$$\tau_n(y) := \int_0^\infty t^n f(y,t)\,dt = n\int_0^\infty t^{n-1}Q(y,t)\,dt = n\left(-\frac{\partial}{\partial s}\right)^{n-1}\left.\tilde{Q}(y,s)\right|_{s=0}. \tag{2.4.9}$$

Equations (2.4.6) and (2.4.8) suggest that there are two alternative methods for determining the MFPT, both of which will appear in subsequent chapters. The first is to solve the FP equation (2.3.1) in Laplace space and use (2.4.8). The second is to exploit the fact that the survival probability satisfies a backward FP equation and use equation (2.4.6). We will consider the latter formulation here. Integrating the backward FP equation (2.3.18) for $q(y,t) = p(x,t|y,0)$ with respect to x implies that $Q(y,t)$ also satisfies a backward FP equation:

$$\frac{\partial Q(y,t)}{\partial t} = A(y)\frac{\partial Q(y,t)}{\partial y} + D\frac{\partial^2 Q(y,t)}{\partial y^2}, \tag{2.4.10}$$

with boundary conditions

$$\partial_y Q(y,t)|_{y=0} = 0, \quad Q(L,t) = 0.$$

Laplace transforming equation (2.4.10) and using the initial condition $Q(y,0) = 1$ for $0 < y < L$, we have

$$A(y)\frac{\partial\tilde{Q}}{\partial y} + D\frac{\partial^2\tilde{Q}}{\partial y^2} - sS = -1, \quad x \in (0,L), \tag{2.4.11}$$

with the same boundary conditions as $Q(y,t)$. Now taking the limit $s \to 0$ and using equation (2.4.6) generates a boundary value problem for the MFPT $\tau(y)$:

$$A(y)\frac{d\tau(y)}{dy} + D\frac{d^2\tau(y)}{dy^2} = -1, \quad \tau'(0) = 0, \quad \tau(L) = 0. \tag{2.4.12}$$

Note that equation (2.4.12) could also be obtained by integrating both sides of equation (2.4.10) with respect to time t and using equation (2.4.6). One of the useful features of working in Laplace space is that one can easily generate corresponding ODEs for higher-order moments of the FPT density. For example, differentiating equation (2.4.11) with respect to s, taking the limit $s \to 0$ and using (2.4.9) yields an equation for the second-order moment:

$$A(y)\frac{d\tau_2(y)}{dy} + D\frac{d^2\tau_2(y)}{dy^2} = -2\tau(y). \tag{2.4.13}$$

It is straightforward to solve equation (2.4.12) by direct integration [300]. First, introduce the integration factor

$$\psi(y) = \exp\left(\frac{1}{D}\int_0^y A(y')dy'\right) = \exp\left(-U(y)/k_BT\right),$$

where we have set $F(y) = -U'(y)$ with $U(y)$ a potential energy, used the Einstein relation (2.3.6), and integrated with respect to y. Equation (2.4.12) becomes

$$\frac{d}{dy}\left[\psi(y)\tau'(y)\right] = -\frac{\psi(y)}{D},$$

so that

$$\psi(y)\tau'(y) = -\frac{1}{D}\int_0^y \psi(y')dy',$$

where the boundary condition $\tau'(0) = 0$ has been used. Integrating once more with respect to y and using $\tau(L) = 0$ then gives

$$\tau(y) = \int_y^L \frac{dy'}{\psi(y')}\int_0^{y'} \frac{\psi(y'')}{D}dy''. \tag{2.4.14}$$

This formula will be the starting point for analyzing escape problems in Sect. 2.4.2. In the case of pure diffusion ($A(x) = 0$), we have $\psi(y) = 1$ and

$$\tau(y) = \frac{L^2 - y^2}{2D}. \tag{2.4.15}$$

In particular, suppose the particle starts at the left-hand boundary. The corresponding MFPT is then $\tau(0) = L^2/2D$. (Further examples of FPT calculations can be found in Ex. 2.12–2.15.)

Remark 2.7. Diffusion in the cytosol. Within the cytosol of cells, macromolecules such as proteins tend to have diffusivities $D < 1\mu m^2 \; s^{-1}$, which is due to effects such as molecular crowding. This implies that the mean time for a diffusing particle to travel a distance $100\mu m$ is at least 10^4s (a few hours), whereas to travel a distance 1mm is at least 10^6s (10 days). Since neurons, for example, which are the largest cells in humans, have axonal and dendritic protrusions that can extend from 1mm up to 1m, the mean travel time due to passive diffusion becomes prohibitively large, and an active form of transport becomes essential (Chap. 7).

Splitting probabilities. It is also possible to extend the above 1D analysis to the case where the particle can exit from either end [300, 674]. If we don't care about which end the particle escapes from, then the MFPT satisfies equation (2.4.12) with the modified boundary conditions $\tau(0) = \tau(L) = 0$. On the other hand, if we wish to keep track of which end the particle exits, then it is necessary to introduce the concept of a splitting probability. Equations (2.4.3) and (2.4.4) show that the survival probability can be written as

$$Q(y,t) = \int_t^\infty J(L,t'|y,0)dt' - \int_t^\infty J(0,t'|y,0)dt' \equiv \Pi_L(y,t) + \Pi_0(y,t), \quad (2.4.16)$$

where $\Pi_0(y,t)$ ($\Pi_L(y,t)$) denotes the probability that the particle exits at $x = 0$ ($x = L$) after time t, having started at the point y. Differentiating with respect to t and using the backward FP equation (2.3.18), we see that

$$\frac{\partial \Pi_0(y,t)}{\partial t} = J(0,t|y,0) = -\int_t^\infty \frac{\partial J(0,t'|y,0)}{\partial t'}dt'$$

$$= A(y)\frac{\partial \Pi_0(y,t)}{\partial y} + D\frac{\partial^2 \Pi_0(y,t)}{\partial y^2}. \quad (2.4.17)$$

The hitting or splitting probability that the particle exits at $x = 0$ (rather than $x = L$) is $\pi_0(y) = \Pi_0(y,0)$. Taking the limit $t \to 0$ in equation (2.4.17) and noting that $J(0,0|y,0) = 0$ for $y \neq 0$,

$$A(y)\frac{\partial \pi_0(y)}{\partial y} + D\frac{\partial^2 \pi_0(y)}{\partial y^2} = 0, \quad (2.4.18)$$

with boundary conditions $\pi_0(0) = 1, \pi_0(L) = 0$. A similar analysis can be carried out for exit through the other end $x = L$ such that $\pi_0(y) + \pi_L(y) = 1$.

Let $T_0(y)$ be the FPT that the particle exits at $x = 0$ having started at y. Since there is a nonzero probability that the particle never exits at $x = 0$, it follows that the unconditional MFPT $\mathbb{E}[T_0(y)] = \infty$. This motivates the introduction of the conditional MFPT

$$\tau_0(y) = \mathbb{E}[T_0(y)|T_0(y) < \infty] = -\int_0^\infty t\frac{\partial \text{Prob}(T_0(y) > t)}{\partial t}dt = \int_0^\infty \frac{\Pi_0(y,t)}{\pi_0(y)}dt. \quad (2.4.19)$$

Integrating equation (2.4.17) with respect to t then gives

$$A(y)\frac{\partial \pi_0(y)\tau_0(y)}{\partial y} + D\frac{\partial^2 \pi_0(y)\tau_0(y)}{\partial y^2} = -\pi_0(y), \qquad (2.4.20)$$

with boundary conditions $\pi_0(0)\tau_0(0) = \pi_0(L)\tau_0(L) = 0$. Equations of the form (2.4.18) and (2.4.20) will occur throughout subsequent chapters, whenever considering a stochastic process with more than one absorbing boundary.

Higher spatial dimensions. The construction of the FPT density can also be extended to higher spatial dimensions. Suppose that a particle evolves according to the Langevin equation (2.3.19) in a compact domain Ω with boundary $\partial\Omega$. Suppose that at time $t = 0$ the particle is at the point $\mathbf{y} \in \Omega$ and let $T(\mathbf{y})$ denote the first passage time to reach any point on the boundary $\partial\Omega$. The survival probability that the particle has not yet reached the boundary at time t is then

$$Q(\mathbf{y},t) = \int_\Omega p(\mathbf{x},t|\mathbf{y},0)d\mathbf{x},$$

where $p(\mathbf{x},t|\mathbf{y},0)$ is the solution to the multivariate FP equation (2.3.21) with an absorbing boundary condition on $\partial\Omega$. The FPT density is again $f(\mathbf{y},t) = -\partial Q(\mathbf{y},t)/\partial t$ which, on using equation (2.3.21) and the divergence theorem, can be expressed as

$$f(\mathbf{y},t) = -\int_{\partial\Omega}[-\mathbf{A}(\mathbf{x})p(\mathbf{x},t|\mathbf{y},0) + D\nabla p(\mathbf{x},t|\mathbf{y},0)] \cdot d\sigma,$$

with $\mathbf{A} = \mathbf{F}/\gamma$. Similarly, by constructing the corresponding backward FP equation, it can be shown that the MFPT satisfies the equation

$$\mathbf{A}(\mathbf{y}) \cdot \nabla \tau(\mathbf{y}) + D\nabla^2 \tau(\mathbf{y}) = -1, \qquad (2.4.21)$$

with $\tau(\mathbf{y}) = 0$ for $\mathbf{y} \in \partial\Omega$. Finally, note that an analogous formulation of first passage times can be developed for discrete Markov processes and stochastic hybrid systems, which are introduced in Chap. 3 and Chap. 5, respectively.

2.4.2 Noise-induced transitions in bistable systems

Suppose that the deterministic version of equation (2.2.5) is written as

$$\frac{dx}{dt} = -\gamma^{-1}\frac{dU}{dx}, \qquad (2.4.22)$$

where $U(x)$ is a deterministic potential. The minima and maxima of the potential $U(x)$ correspond to stable and unstable fixed points of the deterministic dynamics, respectively; see also Box 5A. Suppose that there are two stable fixed points $x_\pm$ separated by an unstable fixed point x_0, see Fig. 2.7(a). In the absence of noise, one can represent the dynamics in terms of a "ball rolling down the potential hill," and the final state will depend on the initial conditions. When noise is included, it is possible for a stochastic trajectory to push the system beyond the top of the hill at x_0 leading to a noise-induced transition to the other minimum.

We wish to calculate the mean time to escape from x_- to x_+; an almost identical calculation holds for the transition $x_+ \to x_-$. Since the system will rapidly approach the state x_+ once it has passed the maximum at x_0, the major contribution to the escape time will be due to the fluctuation-driven transition from x_- to x_0. We can model this process by supplementing the FP equation (2.3.1) with an absorbing boundary condition at a point x beyond x_0, $p(x,t) = 0$, and using equation (2.4.14) for the MFPT starting from x_-:

$$\tau_- = \frac{1}{D} \int_{x_-}^{x} e^{U(x')/k_B T} dx' \int_0^{x'} e^{-U(x'')/k_B T} dx''.$$

We have used the Einstein relation (2.3.6). If the central peak of $U(x)$ around x_0 is large and $k_B T$ is small, then $e^{U(x')/k_B T}$ is sharply peaked around x_0. On the other hand, $e^{-U(x'')/k_B T}$ is very small near x_0 so $\int_0^{x'} e^{-U(x'')/k_B T} dx''$ is slowly varying around $x' = x_0$. Hence, $\int_0^{x'} e^{-U(x'')/k_B T} dx''$ is approximately constant for values of x' such that $e^{U(x')/k_B T}$ is well above zero. We can thus approximate the MFPT by a product of two independent integrals

$$\tau_- = \frac{1}{D} \left[\int_0^{x_0} e^{-U(x'')/k_B T} dx'' \right] \left[\int_{x_-}^{x} e^{U(x')/k_B T} dx' \right].$$

The first integral is dominated by a small region around $x = x_-$, whereas the second is dominated by a small region around $x = x_0$. Thus, the integrals are insensitive to the values of the limits and we can take

$$\tau_- = \frac{1}{D} \left[\int_{-\infty}^{\infty} e^{-U(x'')/k_B T} dx'' \right] \left[\int_{-\infty}^{\infty} e^{U(x')/k_B T} dx' \right].$$

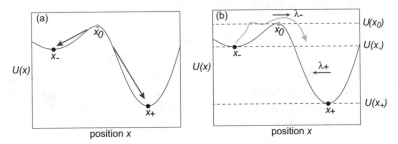

Fig. 2.7: (a) Deterministic double-well potential $U(x)$. Two stable fixed points $x_\pm$ are separated by an unstable fixed point x_0. (b) Effective potential (quasi-potential) of a stochastic model with additive noise, where noise induces transitions between the two metastable states.

Taylor expanding $U(x'')$ to second order about $x = x_-$ with $U'(x_-) = 0$, we obtain the Gaussian integral

$$\int_{-\infty}^{\infty} e^{-U(x'')/k_B T} dx'' \approx \int_{-\infty}^{\infty} e^{-[U(x_-) + U''(x_-)(x'' - x_-)^2/2]/k_B T} dx'' = \sqrt{\frac{2\pi\gamma k_B T}{U''(x_-)}} e^{-U(x_-)/k_B T}.$$

Similarly, expanding $U(x')$ to second order about $x = x_0$,

$$\int_{-\infty}^{\infty} e^{U(x')/k_B T} dx' \approx \int_{-\infty}^{\infty} e^{[U(x_0) - |U''(x_0)|(x' - x_0)^2/2]/k_B T} dx' = \sqrt{\frac{2\pi k_B T}{|U''(x_0)|}} e^{U(x_0)/k_B T}.$$

Combining these results, we finally arrive at one version of Kramer's formula for the mean time to escape from a minimum of a double-well potential:

$$\tau_- \sim \frac{1}{D} \frac{2\pi k_B T}{\sqrt{|U''(x_0)| U''(x_-)}} e^{[U(x_0) - U(x_-)]/k_B T}. \tag{2.4.23}$$

Note that the transition rate $\lambda_- = 1/\tau_-$ (inverse MFPT) from x_- to x_+ varies exponentially with the barrier height $U(x_0) - U(x_-)$, that is, $\lambda_- \sim e^{-[U(x_0) - U(x_-)]/k_B T}$. This is known as the Arrhenius formula. Similarly, the transition rate from the active state x_+ to x_- satisfies

$$\lambda_+ \sim e^{-[U(x_0) - U(x_+)]/k_B T}.$$

It follows that the ratio of the forward and backward transition rates is

$$\frac{\lambda_-}{\lambda_+} = \kappa e^{[U(x_-) - U(x_+)]/k_B T}, \tag{2.4.24}$$

where κ is a constant that is independent of $U(x_0)$. The exponential dependence on the energy difference $\Delta E = U(x_-) - U(x_+)$ is consistent with equilibrium thermodynamics. For the sake of illustration, suppose that we idealize the above stochastic process as a two-state Markov process involving transitions between two discrete states $O_\pm$ (corresponding to $x = x_\pm$) with rates

$$O_- \underset{\lambda_+}{\overset{\lambda_-}{\rightleftharpoons}} O_+.$$

At equilibrium we have $\lambda_- P_- = \lambda_+ P_+$, where $P_\pm$ is the steady-state probability of being in state $O_\pm$. Equilibrium thermodynamics requires that $P_\pm$ are given by a Boltzmann–Gibbs distribution (see Sect. 1.4) so that $P_\pm = Z^{-1} e^{-U_\pm/k_B T}$ where Z is a normalization factor. We deduce that

$$\frac{\lambda_-}{\lambda_+} = \frac{P_+}{P_-} = e^{[U_- - U_+]/k_B T}.$$

Note that this simplified model cannot account for the prefactor κ, and its precise form is still a matter of debate.

2.4.3 *Multiplicative noise and the quasi-potential*

Kramer's formula (2.4.23) can be extended to the case of multiplicative noise. Consider the Ito FP equation (2.3.57) written as

$$\frac{\partial p}{\partial t} = -\frac{\partial A(x)p(x,t)}{\partial x} + \frac{\partial^2 D(x)p(x,t)}{\partial x^2}, \qquad (2.4.25)$$

where we have set $B(x)^2 = 2D(x)$. We will assume that $D(x) > 0$ for all x. An equilibrium steady-state solution of the FP equation corresponds to the condition $\partial p/\partial t = 0$, which leads to a first-order ODE for the equilibrium density $P(x)$:

$$\frac{dD(x)P(x)}{dx} - A(x)P(x) = 0.$$

Rewriting this equation as

$$\frac{dQ(x)}{dx} - \frac{A(x)}{D(x)}Q(x) = 0, \quad Q(x) = D(x)P(x),$$

and using an integrating factor, we obtain the steady-state density

$$P(x) = \frac{\mathcal{N}}{D(x)}e^{-\Phi(x)}, \quad \Phi(x) = -\int^x \frac{A(y)}{D(y)}dy, \qquad (2.4.26)$$

where $\mathcal{N}$ is a normalization factor (assuming that it exists).

The function $\Phi(x)$ is known as a quasi-potential or stochastic potential, and has the following properties:

$$\Phi'(x) = -\frac{A(x)}{D(x)}, \quad \Phi''(x) = -\frac{A'(x)}{D(x)} + \frac{A(x)}{D(x)^2}D'(x).$$

Hence, it has the same stationary points x^* as the deterministic potential $U(x)$, that is, $A(x^*) = 0$. In addition, at the stationary points we have $\Phi''(x^*) = -A'(x^*)/D(x^*)$,

Fig. 2.8: Double-well potential $U(x)$ (light curve) for a Brownian particle exhibiting bistability in the deterministic limit. Two stable fixed points $x_\pm$ are separated by an unstable fixed point x_0. Noise-induced transitions from x_- to x_+ have to surmount the potential barrier ΔE. In the case of multiplicative noise, $U(x)$ is replaced by the quasi-potential $\Phi(x)$, which shares the same minima and maxima as $U(x)$.

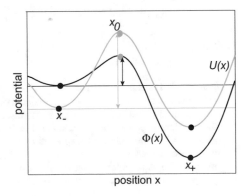

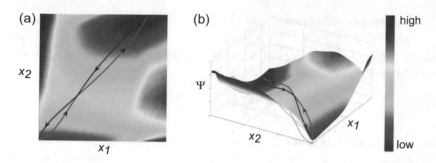

Fig. 2.9: Schematic illustration of stochastic trajectories moving along an energy landscape generated by the quasi-potential of a two-dimensional bistable system. (a) Contour plot. (b) 3D potential plot.

which means that $\Phi''(x^*)$ has the same sign as $-A'(x^*)$ and hence $U''(x^*)$. Thus the locations of the maxima and minima of $\Phi(x)$ and $U(x)$ coincide, see Fig. 2.8. (It is quite common to include the factor $D(x)^{-1}$ appearing in equation (2.4.26) into the definition of the stochastic potential, that is, $\Phi(x) \rightarrow \Phi(x) + \ln D(x)$. However, in many of the escape problems arising in cell biology, the quasi-potential $\Phi(x)$ is scaled by a large constant such as the system size, so that the term $\ln D(x)$ is relatively small.)

The effects of noise can now by visualized as stochastic trajectories moving along the quasi-potential $\Phi(x)$. Moreover, it is straightforward to extend the analysis of additive noise to obtain the following result concerning the rates of noise-induced switching: The mean rate of escape λ_- from the metastable state x_- to the metastable state x_+ in the case of multiplicative noise is

$$\lambda_- = \frac{D(x_-)}{2\pi} \sqrt{|\Phi''(x_0)| \Phi''(x_-)} e^{-[\Phi(x_0) - \Phi(x_-)]}. \tag{2.4.27a}$$

Similarly, the mean rate of escape λ_+ from the metastable state x_+ to the metastable state x_- is

$$\lambda_+ = \frac{D(x_+)}{2\pi} \sqrt{|\Phi''(x_0)| \Phi''(x_+)} e^{-[\Phi(x_0) - \Phi(x_+)]}. \tag{2.4.27b}$$

If D is a constant and $A(x) = F(x)/\gamma$, then $\Phi(x) = U(x)/k_B T$ and we recover the classical Kramer's formula.

Remark 2.8. Bistable or multistable dynamical systems arise in several areas of cell biology, most notably biochemical or gene networks and ion channel models. The analysis of noise-induced transitions and metastability in such systems often involves approximating a discrete Markov process by a continuous Markov process, resulting in an SDE with multiplicative noise. The latter could represent the stochastic dynamics of protein concentrations in the case of a gene network (Chap. 5), or the membrane voltage in the case of a neuron (Chap. 3). A major challenge in analyzing such systems is determining the underlying quasipotential, and calculating the associated transition rates. This is particularly difficult in higher-dimensional systems, where stochastic trajectories move along an energy land-scape determined by Φ, see Fig. 2.9. However, one can often exploit the fact that the noise is weak, so that the large fluctuations required for noise-induced transitions become rare events, and one can use various mathematical methods such as large deviation theory and asymptotic expansions. These will be developed in Chap. 8.

2.5 Simulation of stochastic differential equations

Consider the scalar SDE

$$dX = A(X)dt + B(X)dW(t), \qquad (2.5.1)$$

where $W(t)$ is a Wiener process. As with ordinary differential equations, the simplest numerical scheme is to use a direct Euler method. That is, given the solution $X(t)$ at time t, the solution at time $t + \Delta t$ is given by $X(t + \Delta) = X(t) + \Delta X$, where ΔX is determined explicitly by the equation

$$\Delta X = A(X(t))\Delta t + B(X(t))\Delta W \qquad (2.5.2)$$

for a Gaussian random variable ΔW with zero mean and variance equal to Δt. Iterating this process using a random number generator to choose a new value of ΔW at each time step Δt results in an approximation of a sample path of the stochastic process $X(t)$. Repeating the simulation over many independent trials up to a time T then generates a histogram of values of $X(T)$, which can be used to determine an approximation of the probability density for $X(T)$, and to estimate the mean and variance. The direct Euler method is easily extended to multivariate SDEs and those with non-autonomous coefficients A, B.

The accuracy of Euler's method increases with decreasing step size Δt, and the approximate sample path converges in mean-square to the true sample path in the limit $\Delta t \to 0$. For a rigorous discussion of estimating the accuracy of a stochastic numerical algorithm, see the book by Kloeden and Platen [450]. Here we give a heuristic definition of the numerical error; see also [300, 412]. Suppose that the time interval $(0, T)$ is divided into N infinitesimal subintervals of size $\tau = T/N$, so that the stochastic process $X(t)$ is sampled at the times τ_n with $n = 0, \ldots, \tau_N$. Let $x_n = X(\tau_n)$ be the exact solution on a given sample path and y_n the corresponding numerical approximation of the solution on the same sample path. At the nth time step, let $e_n = x_n - y_n$ and define the error at time T to be the root mean-square value $E(T) = \sqrt{\langle e_N^2 \rangle}$. In the case of the direct Euler method, $E(T) \sim \tau^{1/2}$, and the Euler method is said to be accurate to $\tau^{1/2}$ or that the order of convergence is $\tau^{1/2}$. (In general, the order of convergence will depend on how we define the numerical error, that is, the particular measure of convergence. We will restrict ourselves to mean-square convergence.) One practical method for checking the accuracy of a numerical simulation of a given sample path is to repeat the simulation after halving the time step Δt. Suppose that $T = N\Delta t$ and the sample path is generated by the N random increments ΔW_n, $n = 0,\ldots,N-1$. If we then halve the time step then in order to generate an approximation to the same sample path, it is necessary to produce a set of $2N$ Gaussian random numbers $\Delta \widehat{W}_m$, $m = 1,\ldots 2N$, such that

$$\Delta \widehat{W}_{2n} + \Delta \widehat{W}_{2n+1} = \Delta W_n, \quad n = 0,\ldots,N-1.$$

Given the values ΔW_n, this can be realized by generating N random variables r_n with zero mean and variance $\Delta t/2$, and setting

$$\Delta \widehat{W}_{2n} = r_n, \quad \Delta \widehat{W}_{2n+1} = \Delta W_n - r_n.$$

One can thus successively halve the time step until errors are within acceptable bounds for the given application. The method can also be used to estimate the rate of convergence.

Milstein's method. The direct Euler method is a low-order numerical method for SDEs due to the fact that in approximating an SDE one needs to take into account the fact that $dW(t)$ is of order $\sqrt{dt}$. Suppose that we rewrite the scalar SDE (2.5.1) as the integral equation

$$X(t) = X(t_0) + \int_{t_0}^{t} A(X(s))ds + \int_{t_0}^{t} B(X(s))dW(s). \tag{2.5.3}$$

We recover Euler's method by taking $t = t_0 + \Delta t$, with $X(t_0)$ known, and approximating the functions A, B in the interval $s \in (t_0, t_0 + \Delta t)$ according to

$$A(X(s)) \approx A(X(t_0)), \quad B(X(s)) \approx B(X(t_0)).$$

In order to obtain a more accurate approximation, we apply Ito's formula (2.2.24) to the functions A and B. For example,

$$B(X(s)) = B(X(t_0)) + \int_{t_0}^{s} \left[A(X(s'))B'(X(s')) + \frac{1}{2}B(X(s'))^2 B''(X(s')) \right] ds'$$
$$+ \int_{t_0}^{s} B(X(s'))B'(X(s'))dW(s'), \tag{2.5.4}$$

and similarly for $A(X(s))$. Iterating these equations by successively applying Ito's formula to $A(X(s'))$ and so forth generates an approximation of $A(X(s))$ and $B(X(s))$ in terms of $A(X(t_0)), B(X(t_0))$, higher-order derivatives of $A(x), B(x)$ evaluated at x_0, and a remainder. Substituting such an expansion of $A(X(s))$ and $B(X(s))$ into (2.5.3) generates a higher-order numerical scheme. The Milstein method is the next higher- order approximation to the stochastic integral (2.5.3) after Euler's method. It is obtained by substituting

$$A(X(s)) \approx A(X(t_0)), \quad B(X(s)) \approx B(X(t_0)) + B(X(t_0))B'(X(t_0)) \int_{t_0}^{s} dW(s')$$

into equation (2.5.3) for $t = t_0 + \Delta t$ and $s \in (t_0, t_0 + \Delta t)$. This yields the following equation for ΔX:

$$\Delta X = A(X(t_0)) \int_{t_0}^{t_0 + \Delta t} ds + B(X(t_0)) \int_{t_0}^{t_0 + \Delta t} dW(s)$$

$$+ \frac{1}{2} B(X(t_0)) B'(X(t_0)) \int_{t_0}^{t_0 + \Delta t} \int_{t_0}^{s} dW(s) dW(s').$$

The double integral can be evaluated using equation (2.2.23). That is,

$$\int_{t_0}^{t_0 + \Delta t} \int_{t_0}^{s} dW(s) dW(s') = \int_{t_0}^{t_0 + \Delta t} [W(s) - W(t_0)] dW(s)$$

$$= \int_{t_0}^{t_0 + \Delta t} W(s) dW(s) - W(t_0) W(t_0 + \Delta t)$$

$$= \frac{1}{2} [W(t_0 + \Delta t)^2 - W(t_0)^2 - \Delta t] - W(t_0) W(t_0 + \Delta t) = \frac{1}{2} [(\Delta W)^2 - \Delta t].$$

Hence, we arrive at the Milstein algorithm

$$\Delta X = \left[A(X(t_0)) - \frac{1}{2} B(X(t_0)) B'(X(t_0)) \right] \Delta t + B(X(t_0)) \Delta W$$

$$+ \frac{1}{2} B(X(t_0)) B'(X(t_0)) \Delta W^2. \tag{2.5.5}$$

It turns out that this algorithm has order Δt accuracy, which improves upon the $\sqrt{\Delta t}$ accuracy of Euler's method. The complexity of Milstein's method increases when there are multiple noise sources. Consider the multivariate SDE

$$dX_i = A_i(\mathbf{X}(t), t) dt + \sum_{j=1}^{M} B_{ij}(\mathbf{X}(t), t) dW_j(t),$$

where $W_i(t)$ are independent Wiener processes. The Milstein approximation of this equation takes the form

$$\Delta X_i = A_i \Delta t + \sum_{j=1}^{M} B_{ij} \Delta W_j + \sum_{j,k=1}^{M} \left[\sum_{m=1}^{M} B_{mj} \frac{\partial B_{ik}}{\partial X_m} \right] \int_{t_0}^{t_0 + \Delta t} \int_{t_0}^{s} dW_j(t') dW_k(s).$$

It can be shown that the symmetrized double integral is reducible according to

$$\int_{t_0}^{t_0 + \Delta t} \int_{t_0}^{s} [dW_i(s) dW_j(s') + dW_j(s) dW_i(s')] = \Delta W_i \Delta W_j - \delta_{i,j}(t - t_0).$$

It follows that when the matrix $\mathbf{B}$ satisfies the set of relations (commutative noise)

$$\sum_{m=1}^{M} B_{mj} \frac{\partial B_{ik}}{\partial x_m} = \sum_{m=1}^{M} B_{mk} \frac{\partial B_{ij}}{\partial x_m} \tag{2.5.6}$$

for all i, j, k, the double integral can be symmetrized and Milstein's algorithm becomes

$$\Delta X_i = A_i \Delta t + \sum_{j=1}^{M} B_{ij} \Delta W_j + \frac{1}{2} \sum_{j,k=1}^{M} \left[\sum_{m=1}^{M} B_{mj} \frac{\partial B_{ik}}{\partial X_m} \right] \Delta W_j \Delta W_k$$

$$- \frac{1}{2} \sum_{j,k=1}^{M} \left[\sum_{m=1}^{M} B_{mj} \frac{\partial B_{ij}}{\partial X_m} \right] \Delta t.$$

Runge–Kutte and implicit methods. One limitation of the Milstein method is that it requires an evaluation of the first derivative of the function $B(X)$ or its higher-dimensional matrix version. In a similar fashion to deterministic equations, one can use a Runge–Kutta method to eliminate the need to evaluate any derivatives. A first-order method that builds upon the Milstein algorithm has been developed by Platen [450]. It is based on using the approximation

$$B(X)B'(X) \approx \frac{1}{\sqrt{\Delta t}} [B(\hat{X}) - B(X)],$$

where

$$\hat{X} = X + A\Delta t + A\sqrt{\Delta t}.$$

Substituting into equation (2.5.5) yields the Milstein–Platen method

$$\Delta X = a\Delta t + B\Delta W + \frac{1}{2\sqrt{\Delta t}} [B(\hat{X}) - B(X)][(\Delta W)^2 - \Delta t]. \tag{2.5.7}$$

Similarly, for a multivariate process, one substitutes into the Milstein method the approximation

$$\sum_{m=1}^{M} B_{mj}(\mathbf{X}, t) \frac{\partial B_{ik}(\mathbf{X}, t)}{\partial X_m} \approx \frac{1}{\sqrt{\Delta t}} [B_{ij}(\hat{\mathbf{X}}^{(k)}) - B_{ij}(\mathbf{X})],$$

with

$$\hat{X}_i^{(k)} = X_i + a_i \Delta t + B_{ik}\sqrt{\Delta t}.$$

Another issue that numerical methods for solving stochastic differential equations share with their deterministic counterparts is instability. This refers to a rapid, exponential increase in numerical error, which can occur spontaneously even though the algorithm appears to be converging to a numerically accurate solution prior to the instability. This feature is a particular problem for "stiff" differential equations, that is, those that have two or more disparate time scales. Often an instability can be fixed by using an implicit rather than an explicit method. For example, consider a simple Euler scheme for a single variable,

$$X(t+\Delta t) = X(t) + \Delta X(t) = X(t) + A(X(t),t)\Delta t + B(X(t),t)\Delta W(t).$$

The implicit version is obtained by replacing $X(t)$ with $X(t+\Delta t)$ in the functions A, B:

$$X(t+\Delta t) = X(t) + \Delta X(t) = X(t) + A(X(t+\Delta t),t)\Delta t + B(X(t+\Delta),t)\Delta W(t).$$

This is clearly an implicit equation for $X(t+\Delta t)$, which can be solved numerically using the Newton–Raphson method.

Finally, note that we have focused on the speed and accuracy of numerical methods for generating sample paths of a stochastic differential equation. Convergence to a sample path is known as strong convergence. If one is only interested in properties of the corresponding probability density such as the mean and variance, then these properties are determined by averaging over many sample paths. For a given numerical method, the rate of convergence to the mean or variance tends to differ from the rate of strong convergence, and is thus referred to as weak convergence.

Ito integration and convergence. So far we have not been specific about the form of convergence used to take the continuum limit of a discrete sum of random variables in order to define a stochastic integral. Following Gardiner [300], we now revisit some results on Ito calculus using the notion of convergence in the mean-square (see also Chap. 9). That is, we define a random variable X to be the limit of a sequence of random variables $\{X_1, X_2, \ldots, X_n\}$ if

$$\lim_{n\to\infty} \left\langle |X - X_n|^2 \right\rangle = 0, \tag{2.5.8}$$

that is, for any $\varepsilon > 0$, there exists an integer $N = N(\varepsilon)$ such that for all $n > N$, $\left\langle |X - X_n|^2 \right\rangle < \varepsilon$. Given this definition of convergence, a stochastic process $X(t)$ is said to be mean-square integrable on the interval $(0,t)$ if there exists a random process $Z(t)$ such that the following limit exists:

$$\lim_{n\to\infty} \left\langle (Z_n - Z(t))^2 \right\rangle = 0, \tag{2.5.9}$$

where

$$Z_n = \Delta t \sum_{j=0}^{n} X(j\Delta t), \quad n\Delta t = t.$$

We then formally write $Z(t) = \int_0^t X(s)ds$. Suppose that $G(t)$ is a non-anticipating function, that is, $G(t)$ is statistically independent of $W(s) - W(t)$ for all $s > t$, where $W(t)$ is a Wiener process. We will show that

$$\int_0^t G(t')[dW(t')]^2 = \int_0^t G(t')dt',$$

in the mean-square sense, that is, equation (2.5.9) holds with

$$Z_n = \sum_{j=0}^{n} G_j [\Delta W_j]^2, \quad Z(t) = \int_0^t G(t')dt',$$

where $G_j = G(j\Delta t)$ and $\Delta W_j = W((j+1)\Delta t) - W(j\Delta t)$. Consider

$$I = \lim_{n \to \infty} \left\langle \left[\sum_j G_j (\Delta W_j^2 - \Delta t) \right]^2 \right\rangle$$

$$= \lim_{n \to \infty} \left\langle \sum_j G_j^2 (\Delta W_j^2 - \Delta t)^2 + 2 \sum_{i>j} G_i G_j (\Delta W_j^2 - \Delta t)(\Delta W_i^2 - \Delta t) \right\rangle.$$

Note that G_j^2 is statistically independent of $(\Delta W_j^2 - \Delta t)^2$ and $G_i G_j (\Delta W_j^2 - \Delta t)$ is statistically independent of $(\Delta W_i^2 - \Delta t)$ for $j < i$. Using the Gaussian nature of ΔW_i, we have

$$\langle \Delta W_i^2 \rangle = \Delta t, \quad \langle (\Delta W_i^2 - \Delta t)^2 \rangle = 2\Delta t^2.$$

Thus we find that

$$I = 2 \lim_{n \to \infty} \sum_j G_j^2 \Delta t^2 = 0,$$

assuming that $G(t)$ is bounded. Thus, for Ito integrals $dW(t)^2$ acts like dt.

2.6 Exercises

Problem 2.1. 1D random walk.

(a) Using the Binomial theorem to expand $(pe^{ik} + qe^{-ik})^N$ show that

$$P_N(r) = \frac{1}{2\pi} \int_{-\pi}^{\pi} e^{-ikr} u(k)^N dk = \frac{N!}{\left(\dfrac{N+r}{2}\right)! \left(\dfrac{N-r}{2}\right)!} p^{(N+r)/2} q^{(N-r)/2}$$

when $N+r$ is an even integer and zero otherwise. You will need the identity

$$\int_{-\pi}^{\pi} e^{-ikn} \frac{dk}{2\pi} = \frac{1}{2i\pi n} (e^{in\pi} - e^{-in\pi}) = 0 \text{ for } n \neq 0.$$

(b) Consider the probability distribution for a 1D unbiased random walk

$$P_N(r) = \frac{1}{2^N} \frac{N!}{\left(\dfrac{N+r}{2}\right)! \left(\dfrac{N-r}{2}\right)!}.$$

Using Stirling's formula

$$\ln N! \approx N \ln N - N + \frac{1}{2} \ln(2\pi N),$$

derive the Gaussian approximation

$$P_N(r) \sim \frac{1}{\sqrt{2\pi N}} e^{-r^2/2N}.$$

This result includes a factor of $1/2$ in order to take into account the fact that r is even (odd) when N is even (odd).

Problem 2.2. Random walk on a lattice. Consider a random walker on a 1D lattice with sites ℓ and displacement distribution $p(\ell)$. The probability $P_n(\ell)$ that the walker is at site $\ell + \ell_0$ after n steps, starting at ℓ_0, satisfies the recurrence relation

$$P_n(\ell) = \sum_{\ell'} p(\ell - \ell') P_{n-1}(\ell').$$

(For a homogeneous random walk, ℓ_0 is arbitrary so we can set $\ell_0 = 0$.) Define the generating function $\Gamma(\ell, z)$ according to

$$\Gamma(\ell, z) = \sum_{n \geq 0} z^n P_n(\ell).$$

(a) Show that the generating function satisfies the equation

$$\Gamma(\ell, z) = \delta_{\ell,0} + z \sum_{\ell'} p(\ell - \ell') \Gamma(\ell', z).$$

(b) Introduce the discrete Fourier transform

$$\widehat{\Gamma}(k, z) = \sum_{\ell} e^{ik\ell} \Gamma(\ell, z),$$

and define the structure function of the walk to be

$$\lambda(k) = \sum_{\ell} e^{ik\ell} p(\ell).$$

From part (a), show that

$$\widehat{\Gamma}(k, z) = 1 + z\lambda(k) \widehat{\Gamma}(k, z),$$

so that

$$\widehat{\Gamma}(k, z) = \frac{1}{1 - z\lambda(k)}.$$

(c) For a standard RW with $p(\ell) = (\delta_{\ell,1} + \delta_{\ell,-1})/2$, we have $\lambda(k) = \cos(k)$. Using the inverse transform

$$\Gamma(\ell, z) = \frac{1}{2\pi} \int_{-\pi}^{\pi} e^{-ik\ell} \widehat{\Gamma}(k, z) dk,$$

and the result of part (b), evaluate the integral to show that

$$\Gamma(0, z) = (1 - z^2)^{-1/2}.$$

Hint: make the change of variables $t = \tan(k/2)$.

(d) For a general structure function

$$\Gamma(\ell, z) = \frac{1}{2\pi} \int_{-\pi}^{\pi} \frac{e^{-ik\ell}}{1 - z\lambda(k)} dk,$$

divergence of the integral as $z \to 1^-$ is only possible if $\lambda(k_0) = 1$ for some $k = k_0$. If this holds then the integral will be dominated by the region around k_0. Show that $e^{ik_0\ell} = 1$ for all ℓ where $p(\ell) > 0$ and hence $\lambda(k) = \lambda(k - k_0)$. It follows that the local behavior of $1 - \lambda(k)$ near k_0 is the same as the local behavior of $1 - \lambda(k)$ around the origin. Use this to show that for small k and an unbiased RW, $\sum_\ell \ell p(\ell) = 0$, then we have

$$\lambda(k) \approx 1 - \frac{\sigma^2 k^2}{2}, \quad \sigma^2 = \sum_\ell \ell^2 p(\ell).$$

Hence, deduce that an unbiased 1D RW is recurrent when the mean-square displacement per step is finite.

Problem 2.3. First passage time for random walks on a lattice. Consider a random walker on a 1D lattice with sites ℓ and displacement distribution $p(\ell)$. The probability $P_n(\ell)$ that the walker is at site ℓ after n steps starting at $\ell_0 = 0$ satisfies the recurrence relation (see Ex. 2.2)

$$P_n(\ell) = \sum_{\ell'} p(\ell - \ell') P_{n-1}(\ell').$$

Let $F_n(\ell)$ denote the probability of arriving at site ℓ for the first time on the nth step, given that the walker started at $\ell_0 = 0$.

(a) $P_n(\ell)$ and $F_n(\ell)$ are related according to the recurrence relation

$$P_n(\ell) = \delta_{\ell,0}\delta_{n,0} + \sum_{m=1}^{n} F_m(\ell) P_{n-m}(0), \quad n \geq 0.$$

Explain what this relation means physically.

(b) Show that the corresponding generating functions are related according to

$$\Gamma_F(\ell, z) = \frac{\Gamma_P(\ell, z) - \delta_{\ell,0}}{\Gamma_P(0, z)},$$

where

$$\Gamma_P(\ell,z) = \sum_{n\geq 0} z^n P_n(\ell), \quad \Gamma_F(\ell,z) = \sum_{n\geq 0} z^n F_n(\ell).$$

Hence, use Ex. 2.2c to show that for a standard, unbiased RW

$$\Gamma_F(0,z) = 1 - \sqrt{1-z^2}.$$

(c) Let $R(\ell)$ denote the probability that site ℓ is ever reached by a walker starting at $\ell_0 = 0$:

$$R(\ell) = \sum_{n=1}^{\infty} F_n(\ell) \leq 1.$$

Use part (b) to show that $R(0) = 1$ for an unbiased RW (recurrent rather than transient) while the MFPT $\tau(0)$ to return to the origin is infinite, where

$$\tau(\ell) = \sum_{n=1}^{\infty} n F_n(\ell).$$

Problem 2.4. Multivariate Ornstein–Uhlenbeck process I. Consider the Langevin equation

$$dX_i(t) = \sum_{j=1}^{d} A_{ij} X_j(t) dt + \sum_{j=1}^{d} B_{ij} dW_j(t),$$

with $W_i(t)$ an independent Wiener process,

$$\langle dW_j(t) \rangle = 0, \quad \langle dW_j(t) dW_{j'}(t) \rangle = \delta_{j,j'} dt.$$

(a) Taking expectations of both sides of the Langevin equation, obtain the first moment equation

$$\frac{d\overline{X}_i}{dt} = \sum_{j=1}^{d} A_{ij} \overline{X}_j(t),$$

where $\overline{X}_i(t) = \langle X_i(t) \rangle$.

(b) Determine $\langle X_i(t+dt) X_{i'}(t+dt) \rangle$ in powers of dt by carrying out the following steps: (i) use the Langevin equation for $\langle X_i(t+dt) \rangle$ and $X_{i'}(t+dt) \rangle$; (ii) multiply out all terms; (iii) take averages using the mean and variance of the Wiener process together with the conditions $\langle X_i(t) dW_j(t) \rangle = 0$ for all i, j. Finally, divide through by dt and take the limit $dt \to 0$ to obtain the equation

$$\frac{d\Sigma_{ii'}(t)}{dt} = \sum_{j=1}^{d} A_{ij} \Sigma_{ji'} + \sum_{j=1}^{d} A_{i'j} \Sigma_{ij} + \sum_{j=1}^{d} B_{ij} B_{i'j},$$

where

$$\Sigma_{ij} = \langle X_i(t) X_j(t) \rangle - \langle X_i(t) \rangle \langle X_j(t) \rangle.$$

Problem 2.5. Multivariate Ornstein–Uhlenbeck process II. Consider the OU Langevin equation

$$dX_i = \sum_{j=1}^{d} A_{ij} X_j(t) dt + \sum_{j=1}^{d} B_{ij} dW_j(t),$$

with

$$\langle dW_j(t) \rangle = 0, \quad \langle dW_j(t) dW_{j'}(t') \rangle = \delta_{j,j'} \delta(t - t') dt.$$

(a) Show that the solution in vector form is given by

$$\mathbf{X}(t) = e^{-\mathbf{A}t} \bar{\mathbf{x}} + \int_0^t e^{-\mathbf{A}(t-t')} \mathbf{B} d\mathbf{W}(t').$$

(b) Introduce the correlation function $\mathbf{C}(t,s) = \langle \mathbf{X}(t), \mathbf{X}^T(s) \rangle$ with components

$$C_{ij}(t,s) = \langle X_i(t), X_j(s) \rangle = \langle [X_i(t) - \langle X_i(t) \rangle][X_j(s) - \langle X_j(s) \rangle] \rangle.$$

Using part (a), show that

$$\mathbf{C}(t,s) = \int_0^{\min(t,s)} e^{\mathbf{A}(t-t')} \mathbf{B} \mathbf{B}^T e^{\mathbf{A}^T(s-t')} dt'.$$

(c) Introduce the covariance matrix $\Sigma(t) = \mathbf{C}(t,t)$ with components

$$\Sigma_{ij}(t) = \langle [X_i(t) - \langle X_i(t) \rangle][X_j(t) - \langle X_j(t) \rangle] \rangle.$$

Derive the matrix equation

$$\frac{d\Sigma(t)}{dt} = \mathbf{A}\Sigma(t) + \Sigma(t)\mathbf{A}^T + \mathbf{B}\mathbf{B}^T.$$

Hence, show that if $\mathbf{A}$ has distinct eigenvalues with positive real part, then $\Sigma(t) \to \Sigma_0$ where Σ_0 is the stationary covariance matrix satisfying

$$\mathbf{A}\Sigma_0 + \Sigma_0 \mathbf{A}^T + \mathbf{B}\mathbf{B}^T = 0.$$

(d) By formally setting $dW_i(t) = \eta_i(t) dt$ with

$$\langle \xi_i(t) \rangle = 0, \quad \langle \xi_i(t)\xi_j(t') \rangle = \delta_{i,j}\delta(t - t'),$$

show that the spectrum of the ith component $X_i(t)$ is

$$S_i(\omega) = \sum_{j=1}^{d} \sum_{j'=1}^{d} \Phi_{ij}^{-1}(\omega) \Phi_{ij'}^{-1}(-\omega) D_{jj'},$$

where $D_{ij} = \sum_k B_{ik} B_{kj}$ and

$$\Phi_{ij}(\omega) = -i\omega \delta_{i,j} - A_{ij}.$$

Problem 2.6. Power spectrum of a damped stochastic oscillator. Consider the Langevin equation for a noise-driven, damped harmonic oscillator:

$$dX(t) = V(t)dt, \quad mdV = [-\gamma V(t)dt - kX(t)]dt + \sqrt{2D}dW(t),$$

where $W(t)$ is a Wiener process.

(a) By formally setting $dW(t) = \xi(t)dt$, with

$$\langle \xi(t) \rangle = 0, \quad \langle \xi(t)\xi(t') \rangle = \delta(t - t'),$$

calculate the spectrum $S_X(\omega)$ for $X(t)$.

(b) Plot the spectrum $S_X(\omega)$ as a function of the angular frequency ω for $\omega_0 \equiv \sqrt{k/m} = 1$, $2D/m^2 = 1$, and various values of $\beta = \gamma/m$. What happens in the limit $\beta \to 0$? What is the significance of ω_0?

Problem 2.7. FP equation for an OU process. The FP equation for the OU process is

$$\frac{\partial p(x,t)}{\partial t} = \frac{\partial [kxp(x,t)]}{\partial x} + \frac{D}{2}\frac{\partial^2 p(x,t)}{\partial x^2}.$$

Taking the fixed (deterministic) initial condition $X(0) = x_0$, the initial condition of the FP equation is

$$p(x,0) = \delta(x - x_0).$$

(a) Introducing the characteristic function (Fourier transform)

$$\Gamma(z,t) = \int_{-\infty}^{\infty} e^{izx} p(x,t)dx,$$

show that

$$\frac{\partial \Gamma}{\partial t} + kz\frac{\partial \Gamma}{\partial z} = -\frac{D}{2}z^2\Gamma.$$

Use separation of variables to obtain a solution of the form

$$\Gamma(z,t) = \Gamma_0(ze^{-kt})e^{-Dz^2/4k},$$

with Γ_0 determined by the initial condition for p. Hence, obtain the result

$$\Gamma(z,t) = \exp\left[-\frac{Dz^2}{4k}(1 - e^{-2kt}) + izx_0e^{-kt}\right].$$

[Hint: Set $\Gamma(z,t) = T(t)Z(z)$ to obtain the ODEs

$$\frac{T'}{T} = -kz\frac{Z'}{Z} - \frac{D}{2}z^2 = -\lambda$$

for some constant λ. Solve these equations for $\lambda = nk$ and positive integers n, $n \geq 0$, and write down the general solution using the principle of linear superposition.

Determine the coefficients of the general solution by matching with the initial condition for $\Gamma(z,0)$.]

(b) The probability density $p(x,t)$ can be obtained from $\Gamma(z,t)$ using the inverse Fourier transform

$$p(x,t) = \frac{1}{2\pi} \int_{-\infty}^{\infty} e^{-izx} \Gamma(z,t) dz.$$

Substituting for Γ using part (a), show that $p(x,t)$ is a Gaussian with mean and variance

$$\langle X(t) \rangle = x_0 e^{-kt}, \quad \mathrm{Var}[X(t)] = \frac{D}{2k}[1 - e^{-2kt}].$$

(c) Show that the solution to the steady-state FP equation is

$$p_s(x) = (2\pi D/k)^{-1/2} e^{-kx^2/2D},$$

and that this is consistent with the time-dependent solution in the limit $t \to \infty$.

Problem 2.8. Rotational diffusion. Consider a Brownian particle undergoing diffusion on the circle $\theta \in [-\pi, \pi]$. The corresponding FP equation for $p(\theta,t)$

$$\frac{\partial p}{\partial t} = D \frac{\partial^2 p}{\partial \theta^2}, \quad -\pi < \theta < \pi, \quad p(-\pi,t) = p(\pi,t), \, p'(-\pi,t) = p'(\pi,t),$$

where D is the rotational diffusion coefficient.
(a) Using separation of variables and the initial condition $p(\theta,0) = \delta(\theta)$, show that the solution of the FP equation is

$$p(\theta,t) = \frac{1}{2\pi} \sum_{n=-\infty}^{\infty} e^{in\theta} e^{-Dn^2 t}.$$

(b) If t is sufficiently small then $p(\theta,t)$ is strongly localized around the origin $\theta = 0$. This means that the periodic boundary conditions can be ignored and we can effectively take the range of θ to be $-\infty < \theta < \infty$. That is, performing the rescalings $x = \theta/\varepsilon$ and $\tau = \varepsilon^2 t$, show that $p(\theta,t)$ can be approximated by a Gaussian $p(x,t)$ and deduce the small time approximation

$$\langle \theta^2 \rangle = 2Dt, \quad t \ll \pi^2/D.$$

(c) What happens in the limit $t \to \infty$?

Problem 2.9. Circular statistics and the von Mises distribution. Consider the phase SDE

$$d\Theta(t) = -A \sin \Theta(t) dt + \sqrt{2D} dW(t).$$

The stochastic phase $\Theta(t)$ is known as a von Mises process, which can be regarded as a circular analog of the Ornstein–Uhlenbeck process on a line. It generates distributions that frequently arise in circular or directional statistics [558], and has also been used to model the directionally biased trajectories of swimming microorganisms [186, 372, 376]; see also Chap. 10. The corresponding FP equation is

$$\frac{\partial p(\theta,t)}{\partial t} = \frac{\partial}{\partial \theta}[A\sin(\theta)p(\theta,t)] + D\frac{\partial^2 p(\theta,t)}{\partial \theta^2}$$

for $\theta \in [-\pi, \pi]$ with periodic boundary conditions $p(-\pi,t) = p(\pi,t)$.

(a) Show that the steady-state solution of the FP equation is the von Mises distribution

$$p(\theta) = \frac{1}{2\pi I_0(\kappa)}\exp(\kappa\cos\theta), \quad \kappa = \frac{A}{D}.$$

Here $I_0(\kappa)$ is the modified Bessel function of the first kind and zeroth order ($n = 0$), where

$$I_n(\kappa) = \frac{1}{2\pi}\int_{-\pi}^{\pi}\exp(\kappa\cos\theta)\cos(n\theta)d\theta.$$

Plot $p(\theta)$ for $\kappa = 0.1, 1, 10$. Show that $p(\theta) \to 1/2\pi$ as $\kappa \to 0$ and

$$p(\theta) \approx \frac{1}{\sqrt{2\pi\kappa^{-1}}}e^{-\kappa\beta^2/2}$$

for large κ.

(b) Moments of the von Mises distribution are usually calculated in terms of the circular moments of the complex exponential $x = e^{i\beta} = \cos\beta + i\sin\beta$. The nth circular moment is given by

$$\mu_n := \langle z^n \rangle = \int_{-\pi}^{\pi} z^n p(\theta)d\theta = \frac{I_n(\kappa)}{I_0(\kappa)}.$$

Use the circular moments to show that

$$\text{Var}[\cos(\Theta + \beta)] = \frac{1}{2}\left\{1 - \left(\frac{I_1(\kappa)}{I_0(\kappa)}\right)^2 - \left[\left(\frac{I_1(\kappa)}{I_0(\kappa)}\right)^2 - \frac{I_2(\kappa)}{I_0(\kappa)}\right]\cos 2\beta\right\}.$$

Problem 2.10. Diffusion in a sphere. Consider the diffusion equation in a spherical cell of radius R:

$$\frac{\partial u(\mathbf{x},t)}{\partial t} = D\nabla^2 u(\mathbf{x},t), \quad 0 < |\mathbf{x}| < R,$$

with boundary condition $u(|\mathbf{x}| = R,t) = u_1$ and initial condition $u(\mathbf{x},0) = u_0$ with u_0, u_1 constants.

(a) Assume a radially symmetric solution $v(r,t) = u(r,t) - u_1$ so that

$$\frac{\partial v(r,t)}{\partial t} = D\frac{\partial^2 v}{\partial r^2} + \frac{2}{r}D\frac{\partial v}{\partial r}, \quad 0 < r < R,$$

with $v(R,t) = 0$ and $v(r,0) = u_0 - u_1$. Use separation of variables $v(r,t) = V(r)T(t)$ to derive the general solution

$$u(r,t) = \sum_{n=1}^{\infty} c_n e^{-tDn^2\pi^2/R^2} \frac{1}{r} \sin(n\pi r/R) + u_1.$$

Hint: in order to solve the boundary value problem for $V(r)$, perform the change of variables $\hat{V}(r) = rV(r)$.

(b) Setting $t = 0$ in the general solution and using $v(r,0) = u_0 - u_1$, determine the coefficients c_n. Hint: you will need to use the identity

$$\int_0^R \sin(n\pi r/R)\sin(m\pi r/R)dr = \frac{R}{2}\delta_{n,m}.$$

(c) Determine an approximation for the concentration $u(0,t)$ at the center of the sphere by taking the limit $r \to 0$, with $r^{-1}\sin(\theta r) \to \theta$. Keeping only the leading order exponential term ($n = 1$), show that the time τ for the center to reach a concentration u^*, $u_1 < u^* < u_0$, is approximately

$$\tau = \frac{R^2}{D\pi^2}\ln\frac{2(u_0 - u_1)}{u^* - u_1}.$$

Problem 2.11. 1D Fokker–Planck equation with multiplicative noise. Consider the Ito Fokker–Planck equation

$$\frac{\partial P}{\partial t} = -a\frac{\partial P}{\partial x} + \frac{1}{2}\frac{\partial^2}{\partial x^2}[D(x)P],$$

with a a constant.

(a) Calculate the steady-state probability density when $a = 0$ and $D(x) = k(b+|x|)$ for constants $k > 0, b > 1$. Take $x \in [-1,1]$ with reflecting boundary conditions. What happens when $b \to \infty$?

(b) Calculate the steady-state density (up to a normalization factor) when $a \neq 0$ and $D(x) = bx$ for constant b. Take $x \in [0,1]$ with reflecting boundary conditions.

(c) Using part (b), calculate the steady-state density for $y = 1/x$ and determine the normalization factor—use the change of random variables formula from Sect. 1.2. Hence, evaluate $\langle 1/x \rangle$ as a function of a and b.

Problem 2.12. FPT for a Brownian particle in a semi-infinite domain. Consider a Brownian particle restricted to a semi-infinite domain $x \in [0, \infty)$ with an absorbing boundary condition at $x = 0$. The FP equation is given by

$$\frac{\partial p}{\partial t} = D\frac{\partial^2 p}{\partial x^2}, \quad 0 < x < \infty,$$

with $p(0,t) = 0$.

(a) Check that the solution of the FP equation for the initial condition $x(0) = x_0$ is

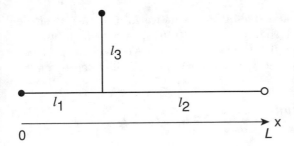

Fig. 2.10: Single side-branch network consisting of two backbone segments of lengths ℓ_1, ℓ_2 and a side branch of length ℓ_3. There is an absorbing boundary at $x = L$ and reflecting boundary conditions at the other two terminals.

$$p(x,t) = \frac{1}{\sqrt{4\pi Dt}}e^{-(x-x_0)^2/4Dt} - \frac{1}{\sqrt{4\pi Dt}}e^{-(x+x_0)^2/4Dt}.$$

(Such a solution can be derived using the method of images, in which one imagines initially placing a fictitious Brownian particle at the image point $x = -x_0$.)

(b) Show that for large times where $\sqrt{Dt} \gg x_0$, the probability density can be approximated by

$$p(x,t) \approx \frac{1}{\sqrt{4\pi Dt}} \frac{xx_0}{Dt} e^{-(x^2+x_0^2)/4Dt}.$$

(c) Calculate the FPT density $f(x_0,t)$ to reach the origin starting from x_0 by calculating the flux through the origin using part (a):

$$f(x_0,t) = D\frac{\partial p(x,t|x_0,0)}{\partial x}\bigg|_{x=0}.$$

Hence show that when $\sqrt{Dt} \gg x_0$, we have the asymptotic behavior

$$f(x_0,t) \sim \frac{x_0}{t^{3/2}}.$$

Deduce that the MFPT to reach the origin is infinite. Also show that the survival probability is given by an error function:

$$Q(t) = \text{erf}(x_0/\sqrt{4Dt}),$$

so that $Q(t) \to 0$ as $t \to \infty$. Interpret this result.

Problem 2.13. First passage time problem in a piecewise linear potential. Find the mean first exit time from the piecewise linear potential

$$U(x) = \begin{cases} -\frac{E_0 x}{L} & \text{for } -L < x < 0 \\ \frac{E_0 x}{L} & \text{for } 0 < x < L, \end{cases}$$

with a reflecting boundary at $x = -L$ and an absorbing boundary at $x = L$. What is the corresponding MFPT to cross $x = 0$ starting at $x = L$?

Problem 2.14. FPT problem on a side-branched network. Consider a particle diffusing on the simple network shown in Fig. 2.10 [674]. A particle diffuses on two backbone segments $x \in I_1 = (0, \ell_1)$ and $x \in I_2 = (\ell_1, \ell_1 + \ell_2)$ with a side branch of length ℓ_3 attached at $x = \ell_1$. Let $p_j(x,t)$ denote the probability density on the segment of length ℓ_j, with

$$\frac{\partial p_j}{\partial t} = D\frac{\partial^2 p_j}{\partial x^2} = -\frac{\partial J_j}{\partial x}, \quad x \in I_j, \quad j = 1, 2,$$

and

$$\frac{\partial p_3}{\partial t} = D\frac{\partial^2 p_3}{\partial y^2} = -\frac{\partial J_3}{\partial y}, \quad y \in (0, \ell_3).$$

These are supplemented by the boundary conditions

$$\frac{\partial p_1}{\partial x}(0,t) = 0, \quad p_2(L,t) = 0, \quad \frac{\partial p_3}{\partial y}(\ell_3, 0) = 0,$$

where $L = \ell_2 + \ell_2$. That is, the terminals at $x = 0$ and $y = \ell_3$ are reflecting, while the terminal at $x = L$ is absorbing. Continuity of probability and conservation of probability flux at the junction are ensured by imposing the additional conditions

$$p_1(\ell_1, t) = p_2(\ell_1, t) = p_3(0, t),$$
$$J_1(\ell_1, t) = J_2(\ell_1, t) + J_3(0, t).$$

(a) Suppose that a searcher starts at $x = 0$ and consider the FPT to reach the target at $x = L$. The FPT density $f(t)$ is identical to the flux $J_2(L,t)$ through $x = L$. Laplace transforming the above equations, derive the expression

$$\widetilde{f}(s) = \left(\cosh(\sqrt{s/D}L) + \cosh(\sqrt{s/D}\ell_1)\sinh(\sqrt{s/D}\ell_2)\tanh(\sqrt{s/D}\ell_3) \right)^{-1}.$$

(b) Taylor expanding $\widetilde{f}(s)$ to first order in s, show that the MFPT is

$$T = \frac{1}{D}(L^2/2 + \ell_2\ell_3). \tag{2.6.10}$$

Note that in the case of an infinite-length side branch, $T \to \infty$.

(c) Extend the analysis to the case of biased diffusion on the two backbone segments and pure diffusion on the side branch. Now

$$\frac{\partial p_j}{\partial t} = -v\frac{\partial p_j}{\partial x} + D\frac{\partial^2 p_j}{\partial x^2} \quad x \in I_j, \quad j = 1, 2,$$

with $-vp_1(0,t) + Dp_1'(0,t) = 0$ and $p_2(L,t) = 0$. In particular, obtain the expression

$$\widetilde{f}(s) = -\frac{(v^2/D + 4s)\,e^{v\ell_1/D}}{su(\ell_1)u(-\ell_2) + \sqrt{sD}\tanh(\sqrt{s/D}\ell_3)u'(\ell_1)u(-\ell_2) + Du'(\ell_1)v(-\ell_2)},$$

with

$$\eta_\pm = \frac{v \pm \sqrt{v^2 + 4Ds}}{2D}, \quad u(\ell) = e^{\eta_+\ell} - e^{\eta_-\ell}, \quad v(\ell) = \eta_- e^{\eta_+\ell} - \eta_+ e^{\eta_-\ell}.$$

Show that the result of part (a) is recovered in the limit $v \to 0$.

(d) Using $T = -\widetilde{f}'(0)$, numerically determine T and plot it as a function of v for $\ell_1 = 1/3$, $\ell_2 = 2/3$, $D = 1$, and $\ell_3 = 1, 2, 5$. Show that in each case $T \to \sum_{i=1,2,3} \ell_i/v$ as $v \to \infty$.

Problem 2.15. FPT problem for an integrate-and-fire model. Consider the piecewise linear SDE

$$dX = [-X(t) + I_0]dt + \sigma dW(t),$$

with $W(t)$ a Wiener process, supplemented by the following threshold-reset condition: whenever $X(t)$ reaches a threshold κ, it is immediately reset to zero. That is,

$$X(t) = \kappa \implies X(t_+) = 0.$$

This is a version of the well-known integrate-and-fire model of a spiking neuron [250].

(a) Let T_n denote the nth time that the system reaches threshold or fired. Show that in the absence of noise ($\sigma = 0$) and for $I_0 > \kappa$, the equation has periodic solutions with period

$$T_{n+1} - T_n = \Delta_0 = \ln\frac{I_0}{I_0 - \kappa}.$$

What happens when $I_0 < \kappa$?

(b) Suppose that $\sigma > 0$ and the system last fired at time $t = 0$. The distribution of next firing times T is determined by an OU process on $(-\infty, \kappa)$ with $X(0) = 0$ and an absorbing boundary condition at $X = \kappa$. Hence, show that the mean next spike time is given by

$$\langle T \rangle = \frac{2}{\sigma^2} \int_0^\kappa \frac{dy}{\psi(y)} \int_{-\infty}^y \psi(z)dz,$$

with

$$\psi(x) = \exp\left(-\frac{2}{\sigma}(u^2/2 - I_0 u)\right).$$

Substituting for ψ obtain the result

$$\langle T \rangle = \frac{2}{\sigma^2} \int_{-I_0/\sigma}^{(\kappa - I_0)/\sigma} (1 + \mathrm{erf}(y))dy,$$

where erf denotes the error function. Take $\kappa = 1$ and plot $\langle T \rangle$ as a function of I_0. How does it compare to the deterministic period $\Delta_0 = \Delta_0(I_0)$ of part (a)?

Problem 2.16. A planar FP equation. Consider the planar dynamical system

$$\frac{dx}{dt} = A_1(x,y) = \mu x + y - x(x^2 + y^2), \quad \frac{dy}{dt} = A_2(x,y) = -x + \mu y - y(x^2 + y^2).$$

(a) Transform to polar coordinates (r, θ) by setting $x = r\cos\theta, y = r\sin\theta$ and using

$$\dot{r} = \frac{x\dot{x} + y\dot{y}}{r}, \quad \dot{\theta} = \frac{x\dot{y} - y\dot{x}}{r^2}.$$

Show that the resulting equations become

$$\frac{dr}{dt} = r(\mu - r^2), \quad \frac{d\theta}{dt} = -1.$$

Hence show that the system undergoes a Hopf bifurcation with respect to the parameter μ. (See Box 5A for a discussion of Hopf bifurcations.)

(b) Now consider a stochastic version of the model given by the 2D Langevin equation

$$dX(t) = A_1(X,Y)dt + \sqrt{2D}dW_1(t), \quad dY(t) = A_2(X,Y)dt + \sqrt{2D}dW_2(t),$$

where

$$\langle dW_j(t) \rangle, \quad \langle dW_j(t)dW_{j'}(t') \rangle = \delta_{j,j'}\delta(t - t')dt\, dt'.$$

The corresponding FP equation for $p(x,y,t)$ is

$$\frac{\partial p}{\partial t} = -\frac{\partial}{\partial x}A_1(x,y)p - \frac{\partial}{\partial y}A_2(x,y)p + D\left(\frac{\partial^2 p}{\partial x^2} + \frac{\partial^2 p}{\partial x^2}\right).$$

This can be rewritten in polar coordinates as

$$\frac{\partial p}{\partial t} = -\frac{\partial}{\partial r}[\mu(1 - r^2)r^2 p] - \frac{\partial}{\partial \theta}p + D\left(\frac{1}{r}\frac{\partial}{\partial r}\left(r\frac{\partial p}{\partial r}\right) + \frac{1}{r^2}\frac{\partial^2 p}{\partial \theta^2}\right).$$

At large times we expect the phase around the limit cycle to be uniformly distributed. Therefore, look for a stationary solution $p^*(r)$ by setting all time and θ derivatives to zero, and solving the resulting ODE for $p^*(r)$. Show that the solution takes the form

$$p^*(r) = B\exp(ar^2 - br^4),$$

and determine the coefficients a, b. How would one determine the constant B?
(c) Setting $r = 1 + \rho$ with $\rho \ll 1$, show that the behavior of $p^*(r)$ near the deterministic limit cycle is a Gaussian centered at $r = 1$ and has width (standard deviation) $\sigma = \sqrt{\mu/2D}$.

Problem 2.17. Computer simulations: Langevin equation. Use the algorithms of Sect. 2.5 to solve the following problems.

(a) Consider the Ornstein–Uhlenbeck process

$$dX(t) = -\lambda X(t)dt + dW(t), \quad X(0) = x_0,$$

where $W(t)$ is a Wiener process. Use direct Euler to simulate 1000 trajectories on the time interval $[0,1]$ for $\lambda = 1/2$, $\Delta t = 0.01$, and $x_0 = 1$. Compare the mean and covariance of the trajectories with the theoretical values of the OU process.

(b) Use Milstein's method to simulate the following SDE on the time interval $[0,1]$

$$dX(t) = -\lambda X(t)dt + \mu X(t)dW(t), \quad X(0) = x_0$$

for $\lambda = 0.1, \mu = 0.1$ and $x_0 = 1$. Compare the cases $\Delta t = 0.1$, $\Delta t = 0.001$, and $\Delta t = 10^{-5}$. Check that the histogram of values at $t = 1$ is similar to the histogram obtained by simulating the exact solution

$$X(t) = x_0 \exp\left[(-\lambda - \mu^2/2)t + \mu W(t)\right].$$

Chapter 3
Protein receptors and ion channels

One essential requirement of sensory eukaryotic cells and single-cell organisms such as bacteria is to monitor the environment and to respond appropriately to external stimuli. For many cells, this could include communication with neighboring cells. Cell signaling or signal transduction consists of several stages, see Fig. 3.1:

1. Detection of the stimulus, which typically occurs via the binding of extracellular ligands to cell surface receptors.

2. Transfer of the signal to the cytoplasmic side. This often occurs via the opening of an ion channel.

3. Transmission of the signal to effector molecules via a signaling transduction pathway, where proteins at each stage undergo a change in conformational state, most commonly via the enzymatic action of kinases (phosphorylation) or phosphatases (dephosphorylation).

4. Triggering of a cell's response such as the activation of gene transcription.

In this chapter, we consider two of the major molecular components of the initial stages of cell signaling, namely, protein receptors and ion channels. After briefly summarizing the most common types of receptors in Sect. 3.1, we introduce some simple kinetic models of cooperative binding, including the Monod-Wyman-Changeaux (MWC) model and the Ising model (Sect. 3.2). These provide one mechanism for a cell to amplify signals from the extracellular environment. (In Chap. 10 we will consider an alternative, non-equilibrium mechanism for signal amplification, based on so-called temporal cooperativity in phosphorylation-dephosphorylation cycles (PdPCs). This is a common type of signaling pathway, occurring during the formation of intracellular protein concentration gradients (Chap. 11), for example, and in certain forms of bacterial quorum sensing (Chap. 15).) In Sect. 3.3, we consider some stochastic models of ligand-gated ion channels. Following a description of various single ion channel models, we construct the birth–death master equation for an ensemble of independent ion channels, and show how it can

© Springer Nature Switzerland AG 2021
P. C. Bressloff, *Stochastic Processes in Cell Biology*, Interdisciplinary
Applied Mathematics 41, https://doi.org/10.1007/978-3-030-72515-0_3

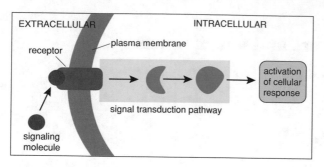

Fig. 3.1: Main stages of cell signaling: receptor-ligand binding, (ii) membrane transport, (iii) signal transduction, (iv) cellular response.

be reduced to a Fokker–Planck (FP) equation by carrying out a system size expansion with respect to the number of ion channels. (An analogous diffusion approximation will be carried out for more general chemical master equations in Chap. 5.) In Sect. 3.4, we analyze a simple stochastic conductance-based model for membrane voltage fluctuations in a neuron, driven by the random opening and closing of a finite number of voltage-gated ion channels. It is usually assumed that the number of ion channels is sufficiently large so that one can use the law of large numbers to obtain a deterministic conductance-based model of the Hodgkin–Huxley form. However, the resulting deterministic equations cannot account for spontaneous events driven by ion channel fluctuations. We illustrate this by showing how initiation of a spontaneous action potential can be formulated in terms of a first passage time problem in a bistable potential. In Sect. 3.5, we consider examples of the spontaneous opening and closing of ligand-gated ion channels, namely, the spontaneous release of calcium from ooycytes (eggs) and cardiac myocytes. The latter is thought to be related to delayed after-depolarizations, which are, in turn, believed to initiate fatal cardiac arrhythmias [536, 557]. (Yet another example is the stochastic opening and closing of high-conductance K^+-Ca^{2+} channels, which is thought to be responsible for the highly stochastic bursting patterns of isolated pancreatic β cells [750].) Finally, in Sect. 3.6 we consider the theory of continuous-time Markov processes, which not only include the various stochastic ion channel models of this chapter, but also chemical reaction networks (Chap. 5) and Poisson processes. The latter are often used to model the sequence of events describing signals received by biochemical sensors.

3.1 Types of receptors

A biochemical receptor is a protein molecule that receives a chemical signal in the form of a ligand. The ligand binds to the receptor and changes its conformational state, initiating some form of cellular response. Hence, receptor-ligand interactions play a major role in environmental sensing, signal transduction, and cell-to-cell signaling. Although receptors and ligands have many different forms, they come in closely matched pairs, with a receptor recognizing just one (or a few) specific ligands, and a ligand binding to just one (or a few) target receptors. Receptors can be divided into two broad categories: cell surface receptors, which are located in the

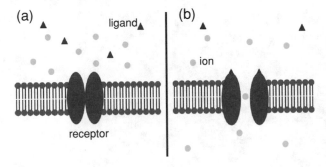

Fig. 3.2: Ligand-gated ion channel. (a) If the receptor is not bound by ligand, then the channel is closed. (b) Channel opens when ligand binds to the receptor, allowing ions to freely flow through the membrane.

plasma membrane, and intracellular receptors, which are either located in the cytoplasm and nucleus, or in the membrane of subcellular compartments such as the endoplasmic reticulum. We will focus on cell surface receptors. For a much more detailed review of receptors and ligands, see the book by Lauffenburger and Linderman [492].

A typical cell surface receptor can be divided into three domains: an extracellular ligand binding domain, a hydrophobic domain extending through the membrane, and an intracellular domain, which transmits any signal. Although there are many different kinds of cell surface receptors, which vary in both structure and function, most can be divided into three common types: ligand-gated ion channels, G protein-coupled receptors, and receptor tyrosine kinases.

(i) Ligand-gated ion channels: Pore-forming receptors that gate the flow of ions across the cell membrane (and the membrane of various intracellular organelles) [373, 431], see Fig. 3.2. In order to create a pore, the cell surface receptor has a membrane-spanning region with a hydrophilic channel through the center of it. The channel allows specific ions to cross the membrane by excluding the hydrophobic core of the phospholipid bilayer. When a ligand binds to the extracellular region of the receptor, the protein undergoes a conformational change that opens the channel. Ligand-gated ion channels play a crucial role in cell-to-cell communication in the nervous system, where neurotransmitters released from the axon terminal of one neuron bind to receptors in the dendritic membrane of another neuron, thus mediating the flow of ions across the synapse. Note that the opening and closing of ion channels can also be mediated by other types of signal. For example, the opening and closing of voltage-gated ion channels depends on the voltage gradient across the plasma membrane (Sect. 3.4), while mechanically gated ion channels allow sound, pressure, or movement to cause a change in the excitability of specialized sensory cells and sensory neurons. The stimulation of a mechanoreceptor causes mechanically sensitive ion channels to open and produce a transduction current that changes the membrane potential of the cell—a process known as mechano-transduction. An important example of mechano-transduction will be considered in Chap. 10, where we describe models of active process in hair cells of the inner ear. Another will be analyzed in Chap. 14 within the context of focal adhesions and cell motility.

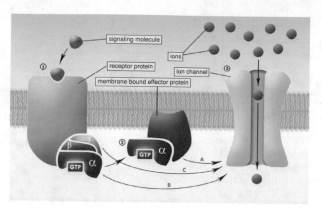

Fig. 3.3: Schematic diagram of a G protein-gated ion channel. (1) Binding of a ligand to the G protein-coupled receptor activates the associated G protein through binding to GTP. (2) The G protein dissociates from the receptor, and separates into an α subunit and a $\beta\gamma$-complex. (3) This can lead to the opening of an ion channel via several possible pathways: (A) interaction of the GTP-bound α-subunit with the ion channel; (B) interaction of the $\beta\gamma$-complex with the ion channel; (C) A signaling cascade triggered by the release of the G protein. [Public domain figure downloaded from Wikipedia Commons.]

(ii) G protein-coupled receptors (GPCRs): A large family of cell surface receptors that share a common structure and method of signaling. When a G protein-coupled receptor is activated by binding of a ligand, it undergoes a conformational change, causing activation of an associated G protein. G proteins bind the nucleotide guanosine triphosphate (GTP), which they can break down by hydrolysis to form guanosine diphosphate (GDP). A G protein attached to GTP is active, while a G protein that is bound to GDP is inactive. The G proteins associated with GPCRs consist of three subunits. Activation of the GPCR by ligand binding switches the G protein to its active state, and the latter separates into two pieces, one consisting of the α subunit, and the other consisting of the β and γ subunits. These subunits dissociate from the GPCR and then interact with other proteins, triggering a signaling cascade. In certain cases, this can lead to the opening of an ion channel that is not directly linked to the GPCR, see Fig. 3.3. (Receptors that directly form a channel pore are called ionotropic, whereas GPCRs are called metabotropic, since they indirectly gate ion channels.) Eventually, the α subunit hydrolyzes GTP back to GDP, at which point the G protein becomes inactive, and reassembles as a three-piece unit associated with a GPCR. Cell signaling using G protein-coupled receptors is a cycle, that can repeat multiple times in response to ligand binding. One particularly interesting class of GPCRs is the odorant (scent) receptors associated with the sense of smell.

(iii) Receptor tyrosine kinases (RTKs): A class of enzyme-linked receptors that bind various types of growth factor and hormones in humans and other animals. Not only are they key regulators of normal cellular processes but also play a critical role in the development and progression of many types of cancer. Typically, when signaling molecules bind to the extracellular domains of two neighboring RTKs,

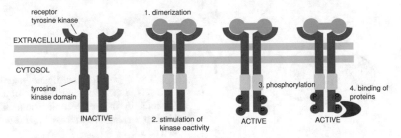

Fig. 3.4: Enzyme-coupled receptor. A pair of neighboring receptors are inactive in the absence of ligand. When ligands bind to the pair they tend to dimerize. This stimulates kinase activity, resulting in phosphorylation of tyrosines and the subsequent binding of intracellular proteins.

they dimerize. Each receptor then attaches phosphates to amino acids known as tyrosines, which are located in the intracellular domain of the other receptor (trans-phosphorylation). The phosphorylated tyrosine then triggers downstream signaling cascades, see Fig. 3.4.

3.2 Cooperative binding

Consider a set of ligand-gated receptors R in the plasma membrane of a cell that can bind to ligand L in the extracellular environment, see Fig. 3.5. For the moment we assume that each receptor has a single binding site, and that the ion channel is only open when the binding site is occupied by a ligand. The corresponding reaction scheme is taken to be

$$R + L \underset{k_-}{\overset{k_+}{\rightleftharpoons}} RL,$$

where k_+ and k_- are the binding and unbinding rates, respectively. If the number of ligands in the environment is large relative to the number of receptors, then $c = [L]$ is approximatively constant and this reaction scheme is well approximated by

$$R \underset{k_-}{\overset{ck_+}{\rightleftharpoons}} RL.$$

If the number of receptors is sufficiently large and fluctuations in the ligand concentration are negligible, then one can use equilibrium thermodynamics and the law of mass action to determine the fraction of open ion channels (see Sect. 1.3). In the case of the simple reaction scheme $R + L \rightleftharpoons LR$, the law of mass action gives

$$\frac{[LR]}{[R][L]} = \frac{1}{K_d},$$

where $K_d = k_-/k_+$ is the dissociation constant. Assuming that the total number of receptors is fixed, $[R] + [LR] = [R_{tot}]$, we have

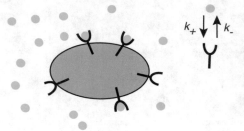

Fig. 3.5: Population of ligand-gated ion channels in the plasma membrane of a cell. Ligands in the extracellular domain can bind to a receptor at a rate k_+ and unbind at a rate k_-.

$$\frac{[LR]}{(R_{\text{tot}}] - [LR])[L]} = \frac{1}{K_d},$$

which on rearranging gives

$$\frac{[LR]}{[R_{\text{tot}}]} = \frac{[L]}{[L] + K_d}. \tag{3.2.1}$$

The fraction of bound receptors increases linearly with $[L]$ at low ligand concentrations but saturates at high concentrations for which $[L] \gg K_d$.

A sharper dependence on $[L]$ can be obtained if there is some form of cooperative binding [670]. The latter refers to situations in which a receptor has multiple binding sites, which can influence each other. An extreme example is when a receptor has n binding sites such that mutual interactions force all of the binding sides to be either simultaneously occupied or simultaneously empty. This can be represented by the reaction scheme

$$R_0 + nL \rightleftharpoons R_n,$$

where R_0 denotes a receptor with empty binding sites and R_n denotes a receptor with all sites filled. The law of mass action shows that at equilibrium

$$\frac{[R_n]}{[L]^n[R_0]} = \frac{1}{K_n},$$

Fig. 3.6: Cooperative binding model. Plot of fraction of open receptors as a function of ligand concentration for various n, where n is the number of binding sites.

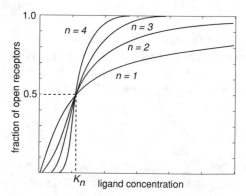

where K_n is an effective dissociation rate. Note that since the forward reaction involves n ligands, one has to include the factor $[L]^n$. Again setting $[R_n] + [R_0] = [R_{tot}]$ and rearranging gives

$$\frac{[R_n]}{[R_{tot}]} = \frac{[L]^n}{[L]^n + K_n}. \tag{3.2.2}$$

The dependence of the fraction of open ion channels as a function of $[L]$ and n is illustrated in Fig. 3.6.

3.2.1 Monod-Wyman-Changeux (MWC) model

The above model of receptor-ligand binding is unrealistic in at least two aspects. First, the binding to multiple sites is not all-or-none, that is, a fraction of sites can be occupied at any one time. Second, it is possible for the ion channel to be either open or closed in each binding state—changes in binding state shift the balance between open and closed. A more realistic model of a ligand-gated ion channel with cooperative binding has been developed for the nicotinic acetylcholine receptor, which is found at the neuromuscular junction, and also for cGMP-gated ion channels that enable photoreceptors to amplify their response to light [560] (see also Chap. 10). It is analogous to the classical Monod-Wyman-Changeux (MWC) model of dimoglobin [589]. The nicotinic receptor has two binding sites for acetylcholine and the equilibrium between the open and closed state of the channel is shifted to the open state by the binding of acetylcholine. A schematic illustration of the different receptor states together with a reaction diagram are shown in Fig. 3.7(a,b). In the diagram, T_j denotes a closed receptor with j occupied sites and R_j denotes a receptor in the corresponding open state. Also shown is the equilibrium constant (inverse of the dissociation constant) for each of the reversible reactions. In particular, K_T and K_R are the equilibrium constants for binding of an acetylcholine molecule to an individual site of a closed and an open receptor, respectively. The additional factor of 2 takes into account the fact that there are two unoccupied sites in the forward reaction $T_0 \rightarrow T_1$, whereas the additional factor of 1/2 takes into account the fact that there are two occupied sites in the backward reaction $T_2 \rightarrow T_1$ (and similarly for R_j). Finally, Y_j is the equilibrium constant associated with the opening and closing of a receptor with j occupied sites.

Applying the law of mass action to each of the reversible reactions leads to the following set of equations for the concentrations:

$$\frac{[R_i]}{[T_i]} = Y_i, \quad \frac{[T_1]}{[L][T_0]} = 2K_T, \quad \frac{[R_1]}{[L][R_0]} = 2K_R, \tag{3.2.3a}$$

$$\frac{[T_2]}{[L][T_1]} = K_T/2, \quad \frac{[R_2]}{[L][R_1]} = K_R/2. \tag{3.2.3b}$$

We are interested in the fraction of receptors that are in the open state, which is

$$p_{\text{open}} = \frac{[R_0] + [R_1] + [R_2]}{[R_0] + [R_1] + [R_2] + [T_0] + [T_1] + [T_2]}.$$

Equations (3.2.3a, 3.2.3b) can be used to express $[T_j]$ and $[R_j]$ in terms of $[T_0]$ and $[R_0]$:

$$[T_1] = 2K_T[L][T_0], \quad [T_2] = (K_T[L])^2[T_0], \quad [R_1] = 2K_R[L][R_0], \quad [R_2] = (K_R[L])^2[R_0].$$

Substituting these results into the formula for p_{open} and using (3.2.3a) gives

$$p_{\text{open}} = \frac{Y_0(1 + K_R[L])^2}{Y_0(1 + K_R[L])^2 + (1 + K_T[L])^2}. \tag{3.2.4}$$

We now observe that when $[L] = 0$,

$$p_{\text{open}}(0) = \frac{1}{1 + 1/Y_0},$$

whereas when $[L]$ is large

$$p_{\text{open}}([L]) \approx \frac{1}{1 + (K_T/K_R)(1/Y_0)}.$$

It follows that if the open receptor has a higher affinity for binding acetylcholine than the closed receptor ($K_R > K_T$) then $p_{\text{open}}([L]) > p_{\text{open}}(0)$. An interesting feature of MWC type models is that activation of the receptor, as specified by p_{open}, is a sigmoidal function of ligand concentration $[L]$. This is illustrated in Fig. 3.7(c) for a curve fitted to data from cGMP ion channels. Thus binding is effectively cooperative even though there are no direct interactions between binding sites.

Finally, note that it is straightforward to generalize the MWC model to the case of n binding sites, see Ex. 3.1. Defining the fraction of open receptors according to

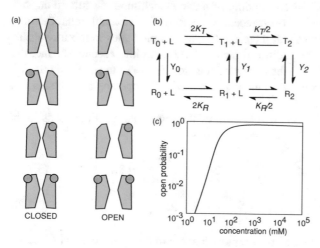

Fig. 3.7: The MWC receptor model with two binding sites. (a) Schematic illustration of different conformational states distinguished by the number of occupied binding sites and whether the ion channel is open or closed. (b) Reaction diagram. (c) Fraction of open receptors as a function of ligand concentration for the MWC model with two binding sites. Sketch of a typical curve for a cGMP ion channel.

$$p_{\text{open}} = \frac{\sum_{j=0}^{n}[R_j]}{\sum_{j=0}^{n}[R_j] + \sum_{j=0}^{n}[T_j]},$$

the law of mass action gives

$$p_{\text{open}} = \frac{Y_0(1 + K_R[L])^n}{Y_0(1 + K_R[L])^n + (1 + K_T[L])^n}. \tag{3.2.5}$$

The MWC model has emerged as a general mechanism for receptor-ligand interactions within a diverse range of applications, including ion channel gating, chemotaxis, and gene regulation (see the review by Marzen et al. [560]). We will consider the particular example of bacterial chemosensing in Chap. 10.

3.2.2 Ising model of cooperative binding

Consider a 1D array of N receptors such that neighboring receptor-ligand complexes can interact, see Fig. 3.8. If a receptor is bound by a ligand but the adjacent receptors are unoccupied, then there is a reduction in free energy given by $-a - \mu$, where $-a$ is the binding energy and μ is the chemical potential of the ligand. However, due to ligand interactions, when two adjacent receptors are simultaneously occupied by ligands, there is an additional reduction in free energy given by $-J$. We can express this mathematically by introducing the occupation states σ_n, $n = 1, \ldots, N$, with $\sigma_n = 1$ if the nth receptor is occupied and $\sigma_n = 0$ if it is unoccupied. The total free energy for a given configuration $\sigma = (\sigma_1, \ldots, \sigma_N)$ is then given by

$$E[\sigma] = -(a + \mu) \sum_{n=1}^{N} \sigma_n - J \sum_{n=1}^{N-1} \sigma_n \sigma_{n+1}.$$

The corresponding Boltzmann–Gibbs distribution is

$$p(\sigma) = Z^{-1} e^{-E[\sigma]/k_B T}, \tag{3.2.6}$$

and the partition function is

$$Z = \sum_{\sigma_1 = 0,1} \cdots \sum_{\sigma_N = 0,1} e^{\alpha \sum_{j=1}^{N} \sigma_j + \gamma \sum_{j=1}^{N-1} \sigma_j \sigma_{j+1}}, \tag{3.2.7}$$

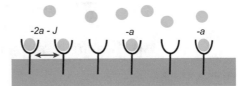

Fig. 3.8: 1D array of membrane receptors with nearest neighbor interactions. An isolated bound receptor has lower free energy $-a$, whereas a pair of adjacent bound receptors each have a lower free energy $-2a - J$ with J representing the strength of cooperative interactions.

with $\alpha = (a+\mu)/k_BT$ and $\gamma = J/k_BT$. The partition function can be treated as a generating function for the mean occupancy, that is,

$$\langle\sigma\rangle := \sum_\sigma p(\sigma)\left(\frac{1}{N}\sum_{n=1}^N \sigma_n\right) = \frac{1}{N}\frac{d}{d\alpha}\ln Z[\alpha]. \tag{3.2.8}$$

The above model is equivalent to the classical Ising model of a 1D lattice of magnetic spins $s_n = \sigma_n - 1/2 = \pm 1/2$ (see Ex. 1.2). The latter can also be interpreted as a model of a 1D polymer (Chap. 12). A well-known result from statistical mechanics is that the 1D Ising model can be solved exactly. In particular, one can derive an exact expression for Z using transfer matrices [163]. First, it is convenient to take the sites to be arranged on a ring by making the identification $\sigma_{N+1} = \sigma_1$. (This yields a good approximation for large N, since boundary effects at the ends of a 1D array are negligible.) We then rewrite Z in the more suggestive form

$$Z = \sum_{\sigma_1}\cdots\sum_{\sigma_N}\left[e^{\alpha(\sigma_1+\sigma_2)/2+\gamma\sigma_1\sigma_2}\right]\left[e^{\alpha(\sigma_2+\sigma_3)/2+\gamma\sigma_2\sigma_3}\right]\cdots\left[e^{\alpha(\sigma_N+\sigma_1)/2+\gamma\sigma_N\sigma_1}\right].$$

We can view each term on the right-hand side as the element of a matrix $\mathbf{T}$ with matrix elements labeled by σ_1,σ_2 etc., that is, $T_{\sigma_1\sigma_2} = e^{\alpha(\sigma_1+\sigma_2)/2+\gamma\sigma_1\sigma_2}$. Hence.

$$\mathbf{T} = \begin{pmatrix} T_{11} & T_{10} \\ T_{01} & T_{00} \end{pmatrix} = \begin{pmatrix} e^{\alpha+\gamma} & e^{\alpha/2} \\ e^{\alpha/2} & 1 \end{pmatrix}.$$

In terms of the transfer matrix $\mathbf{T}$

$$Z = \sum_{\sigma_1}\cdots\sum_{\sigma_N} T_{\sigma_1\sigma_2}T_{\sigma_2\sigma_3}\cdots T_{\sigma_N\sigma_1} = \text{Tr}[\mathbf{T}^N],$$

where we have used the standard rules of matrix multiplication. The eigenvalues of $\mathbf{T}$ satisfy the characteristic equations

$$(e^{\alpha+\gamma} - \lambda)(1-\lambda) - e^\alpha = 0,$$

which gives the pair of eigenvalues $\lambda = \lambda_\pm$ with

$$\lambda_\pm = \frac{1}{2}\left[(e^{\alpha+\gamma}+1) \pm \sqrt{(e^{\alpha+\gamma}+1)^2 + 4e^\alpha}\right].$$

Since $\mathbf{T}$ is a symmetric matrix, there exists a unitary matrix $\mathbf{U}$ such that

$$\mathbf{U}\mathbf{T}\mathbf{U}^{-1} = \mathbf{T}_d \equiv \text{diag}(\lambda_+,\lambda_-).$$

Moreover,

$$\text{Tr}[\mathbf{T}^N] = \text{Tr}[[\mathbf{U}^{-1}\mathbf{T}_d\mathbf{U}]^N] = \text{Tr}[\mathbf{U}^{-1}\mathbf{T}_d^N\mathbf{U}] = \text{Tr}[\mathbf{T}_d^N].$$

Finally, we note that

$$Z = \mathrm{Tr}[\mathbf{T}_d^N] = \lambda_+^N + \lambda_-^N \approx \lambda_+^N$$

for large N. Hence,

$$\langle \sigma \rangle = \frac{1}{N} \frac{d}{d\alpha} \ln \lambda_+^N = \frac{d}{d\alpha} \ln \lambda_+ = \frac{1}{\lambda_+} \frac{d\lambda_+}{d\alpha}.$$

A plot of $\langle \sigma \rangle$ against $k_B T \alpha = a + \mu$ yields a sigmoid function, indicative of cooperativity—recall from Sect. 1.3 that $e^{\mu/k_B T}$ is proportional to the ligand concentration. An analogous model of cooperativity arises in gene regulatory networks involving transcription factor proteins binding to multiple sites of a promoter (Chap. 5).

3.3 Markov chain models of ion channel kinetics

3.3.1 Single ion channel models

Recordings from single ion channels show that their dynamics is highly stochastic, see Fig. 1.2. Typically, the opening of an ion channel is modeled as a continuous-time Markov chain, the general theory of which is presented in Sect. 3.6. Here we consider some simple low-dimensional models of stochastic ion channels.

Two-state model. Consider a simple two-state model of an ion channel that can exist either in a closed state (C) or an open state (O). Transitions between the two states are governed by a continuous-time Markov process or Markov chain

$$C(\text{closed}) \underset{\beta}{\overset{\alpha}{\rightleftharpoons}} O(\text{open}), \tag{3.3.1}$$

with transition rates α, β. (In the case of voltage-gated ion channels, α, β will depend on the local membrane potential, see Sect. 3.4, whereas in the case of ligand-gated ion channels, they will depend on the local concentration of the ligand.) In order to understand what such a process means, let $N(t)$ be a discrete random variable with $N(t) = 0$ if the channel is closed and $N(t) = 1$ if it is open. Introduce the probability distribution $P_n(t) = \mathbb{P}[N(t) = n]$. From conservation of probability, $P_0(t) + P_1(t) = 1$. The transition rates then determine the probability of jumping from one state to the other in a small interval Δt:

$$\alpha \Delta t = \mathbb{P}[N(t + \Delta t) = 1 | N(t) = 0], \quad \beta \Delta t = \mathbb{P}[N(t + \Delta t) = 0 | N(t) = 1].$$

It follows that there are two possible ways for the ion channel to be in the closed state at time $t + \Delta t$, depending on whether it is closed or open at time t:

$$P_0(t + \Delta t) = \mathbb{P}[C \to C] P_0(t) + \mathbb{P}[O \to C] P_1(t)$$
$$= [1 - \alpha \Delta t] P_0(t) + \beta \Delta t P_1(t).$$

Writing down a similar equation for the open state, dividing by Δt, and taking the limit $\Delta t \to 0$ leads to the pair of equations

$$\frac{dP_0}{dt} = -\alpha P_0 + \beta P_1 \quad \frac{dP_1}{dt} = \alpha P_0 - \beta P_1, \tag{3.3.2}$$

which are equivalent, since $P_0(t) + P_1(t) = 1$.

Equation (3.3.2) has the unique stable steady state

$$P_0^* = \frac{\beta}{\alpha + \beta}, \quad P_1^* = \frac{\alpha}{\alpha + \beta}. \tag{3.3.3}$$

Such a steady state has to be consistent with equilibrium statistical mechanics. For a single ion channel maintained at a fixed temperature T, the open and closed probabilities are determined by the Boltzmann–Gibbs distribution, see Sect. 1.3. In particular, we find that $P_1^* = P_0^* e^{-\Delta E / k_B T}$, where $\Delta E = E_1 - E_0$ is the difference in free energy between the open and closed states. It follows that

$$\frac{\alpha}{\beta} = e^{-\Delta E / k_B T}. \tag{3.3.4}$$

(In the case of voltage-gated ion channels, see Sect. 3.4, ΔE is a function of membrane voltage v. Typically, $\Delta E(v) = qv$, where the constant q is determined by the displacement of charge when the ion channel changes its conformational state.)

Now consider the time-dependent solution of equation (3.3.2) under the initial condition $P_n(0) = \delta_{n,n_0}$. Using the fact that $P_0(t) + P_1(t) = 1$ we can obtain the solution

$$P_{0n_0}(t) = \delta_{0,n_0} e^{-t/\tau_c} + \beta \tau_c (1 - e^{-t/\tau_c}),$$
$$P_{1n_0}(t) = \delta_{1,n_0} e^{-t/\tau_c} + \alpha \tau_c (1 - e^{-t/\tau_c}),$$

with $\tau_c = 1/(\alpha + \beta)$ and $P_{nn_0}(t) = \mathbb{P}[N(t) = n | N(0) = n_0]$. A number of results follow from this. First τ_c is the relaxation time of the Markov chain, with $P_m(t) \to P_m^*$ in the limit $t \to \infty$. Second, in the large time limit, $\langle N(t) \rangle = P_1^*$ and

$$\langle N(t + \tau)N(t) \rangle = \mathbb{P}[N(t + \tau) = 1 | N(t) = 1]\mathbb{P}[N(t) = 1]$$
$$= P_{11}(\tau)P_1^* = (P_1^*)^2 + P_0^* P_1^* e^{-\tau/\tau_c}. \tag{3.3.5}$$

Note that the Markov process is stationary at large times, since the statistics only depends on time differences.

Multi-state models. The above two-state model is a simplification of more detailed Markov models, in which there can exist inactivated states and multiple subunits [767]. For example, the Na^+ channel inactivates as well as activates, see Fig. 3.9. Moreover, both K^+ and Na^+ channels consist of multiple subunits, each of which can be in an open state, and the channel only conducts when all subunits are open. For example, suppose that a channel consists of two identical, independent subunits, each of which can be open or closed, and that an ionic current can only flow through the channel if both the subunits are open. Let S_n denote the state in which n subunits are open. The transitions between the different states of the ion channel are governed by the reaction scheme

$$S_0 \underset{\beta}{\overset{2\alpha}{\rightleftharpoons}} S_1 \underset{2\beta}{\overset{\alpha}{\rightleftharpoons}} S_2,$$

where α, β are the rates of opening and closing of a single subunit. The factors of two take into account the fact that the state S_0 (S_2) has two closed (open) states either of which can open (close). Introduce the discrete random variable $N(t)$ with $N(t) = n$ if the current state of the ion channel is n. The corresponding master equation for $P_n(t) = \mathbb{P}[N(t) = n]$ takes the form

$$\frac{dP_0}{dt} = \beta P_1 - 2\alpha P_0, \quad \frac{dP_2}{dt} = \alpha P_1 - 2\beta P_2, \qquad (3.3.6)$$

with $P_0(t) + P_1(t) + P_3(t) = 1$. The steady-state solution satisfies $P_0^* = \beta P_1^*/(2\alpha)$ and $P_2^* = \alpha P_1^*/(2\beta)$, which implies

$$P_1^* \left(1 + \frac{\beta}{2\alpha} + \frac{\alpha}{2\beta} \right) = 1.$$

Hence,

$$P_0^* = \frac{\beta^2}{(\alpha+\beta)^2}, \quad P_1^* = \frac{2\alpha\beta}{(\alpha+\beta)^2}, \quad P_2^* = \frac{\alpha^2}{(\alpha+\beta)^2}.$$

It is straightforward to show that the steady-state solution is stable by linearizing the kinetic equations. That is, setting $p_n(t) = P_n(t) - P_n^*$ and using $\sum_n p_n(t) = 0$, we have

$$\frac{dp_0}{dt} = -\beta(p_0 + p_2) - 2\alpha p_0, \quad \frac{dp_2}{dt} = -\alpha(p_0 + p_2) - 2\beta p_2.$$

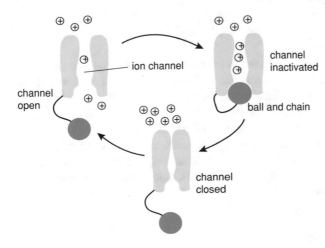

channel open

ion channel

channel inactivated

ball and chain

channel closed

Fig. 3.9: Schematic diagram of the opening/-closing of a single sodium ion channel described by a 3-state model. The reduced 2-state model ignores the inactivated state.

Introducing the vector $\mathbf{p} = (p_0, p_2)^\top$, this pair of equations can be rewritten in the matrix form

$$\frac{d\mathbf{p}}{dt} = \mathbf{Mp}, \quad \mathbf{M} = \begin{pmatrix} -\beta - 2\alpha & -\beta \\ -\alpha & -\alpha - 2\beta \end{pmatrix}.$$

The linear system has solutions of the form $\mathbf{p} = \mathbf{v}e^{\lambda t}$ with $(\lambda, \mathbf{v})$ satisfying the eigenvalue equations $\mathbf{Mv} = \lambda \mathbf{v}$. This only has nontrivial solutions if $\mathbf{M} - \lambda \mathbf{I}$ is not invertible, where $\mathbf{I}$ is the unit matrix. We thus obtain the characteristic equation

$$0 = \det(\mathbf{M} - \lambda \mathbf{I}) \equiv (\lambda + \beta + 2\alpha)(\lambda + \alpha + 2\beta) - \alpha\beta.$$

Rearranging, we have

$$(\lambda + \alpha + \beta)(\lambda + 2[\alpha + \beta]) = 0,$$

and so $\lambda = \lambda_{1,2}$ with $\lambda_1 = -(\alpha + \beta)$ and $\lambda_2 = -2(\alpha + \beta)$. Since $\lambda_{1,2} < 0$, it follows that $\mathbf{p}(t) \to 0$ as $t \to \infty$ and the steady state is stable.

Model reduction. One of the interesting features of ion channel models with two or more subunits is that the kinetic equations can sometimes be reduced to a lower dimensional set of equations due to the existence of a stable invariant manifold— solutions that start in the manifold cannot leave it, and other solutions exponentially converge to the invariant manifold. In the case of the two-subunit model, this can be shown by direct substitution. That is, setting

$$P_0 = (1 - p)^2, \quad P_1 = 2p(1 - p), \quad P_2 = p^2, \tag{3.3.7}$$

and substituting into the kinetic equation (3.3.6) gives

$$-2(1 - p)\frac{dp}{dt} = 2\beta p(1 - p) - 2\alpha(1 - p)^2, \quad 2p\frac{dp}{dt} = 2\alpha p(1 - p) - 2\beta p^2,$$

which are both satisfied provided that

$$\frac{dp}{dt} = \alpha(1 - p) - \beta p. \tag{3.3.8}$$

Thus, if the initial state can be expressed in terms of the single variable p according to equation (3.3.7), then the solution remains in this one-dimensional space with the dynamics described by the single kinetic equation (3.3.8). Moreover, the stability of the unique steady state implies that the invariant manifold is stable. Since the conducting state of each ion channel corresponds to S_2 (both subunits in the open state), it follows that the fraction of conducting ion channels at time t is $p(t)^2$. Thus the expected conductance of ion channels is proportional to p^2. Such a result generalizes to more complex ion channel models such as Hodgkin–Huxley. (The extension to multiple subunits is considered in Ex. 3.2. Further examples of multi-state ion channel models are explored in Ex. 3.3 and 3.4.)

3.3.2 Birth–death master equation for an ensemble of ion channels

Recall from Chap. 2 that the diffusion equation either determines the probability density $p(x,t)$ for the location of a single Brownian particle, or the concentration $u(x,t)$ of a large ensemble of independent particles. In the first case there exists a corresponding continuous stochastic variable $X(t)$, which is the position of the particle for a single realization of a trajectory. A similar distinction arises in the case of ion channels. Equations (3.3.2) and (3.3.6), for example, are master equations describing the probability distribution of a single ion channel, associated with a discrete random variable $N(t)$. In the case of a large number of independent and identical ion channels, we can reinterpret $P_n(t)$ as the fraction of ion channels in state n at time time t. For example, on setting $P_1 = x$ and $P_0 = 1 - x$ in equation (3.3.2), with $x(t)$ the fraction of open ion channels at time t, we obtain the kinetic equation

$$\frac{dx}{dt} = -\beta x + \alpha(1 - x).$$
(3.3.9)

We want to investigate fluctuations that arise when there is a finite number of ion channels.

Suppose that there are N_0 identical, independent two-state ion channels evolving according to the simple Markov process (3.3.2). In order to take into account fluctuations in the case of finite N_0, it is necessary to keep track of the probability $P_n(t)$ that there are n open channels at time t, $0 \leq n \leq N_0$. More precisely,

$$P_n(t) = \mathbb{P}[N(t) = n | N(0) = n_0],$$

where $N(t)$, $0 \leq N(t) \leq N_0$, is a discrete random variable specifying the number of open channels at time t (rather than the conformational state of a single ion channel). For ease of notation, we drop the explicit dependence on the initial condition $N(0) = n_0$. Consider a time interval $[t, t + \Delta t]$ with Δt sufficiently small so that only one channel has a significant probability of making a $C \to O$ or $O \to C$ transition. There are four possible events that can influence $P_n(t)$ during this interval, two of which involve transitions into the state of n open ion channels, and two of which involve transitions out of the state. Collecting these terms and taking the limit $\Delta t \to 0$ leads to the master equation

$$\frac{d}{dt}P_n(t) = \alpha(N_0 - n + 1)P_{n-1}(t) + \beta(n + 1)P_{n+1}(t) - [\alpha(N_0 - n) + \beta n]P_n(t)$$
(3.3.10a)

for $0 < n < N_0$. The first term on the right-hand side represents the probability flux that one of $N_0 - (n - 1)$ closed channels undergoes the transition $C \to O$, whereas the second term represents the probability flux that one of $n + 1$ open channels undergoes the transition $O \to C$. The last two terms represent transitions $n \to n \pm 1$. At the boundaries $n = 0, N_0$, we have the modified equations

$$\frac{d}{dt}P_0(t) = \beta P_1(t) - \alpha N_0 P_1(t),$$ (3.3.10b)

$$\frac{d}{dt}P_{N_0}(t) = \alpha P_{N-1}(t) - \beta N_0 P_{N_0}(t).$$ (3.3.10c)

(If there are $N(t)$ open channels then it immediately follows that there are $N_0 - N(t)$ closed channels, so we don't need to keep track of the latter as well). Define the mean number of open channels at time t by

$$\bar{n}(t) = \sum_{n=0}^{N_0} n P_n(t).$$

By differentiating both sides of these equations with respect to t and using the master equation (3.3.10), we recover the kinetic equation (3.3.9) with $x = \bar{n}/N_0$.

The above population ion channel model is an example of a finite birth–death process described by a master equation of the general form

$$\frac{d}{dt}P_n(t) = \omega_+(n-1)P_{n-1}(t) + \omega_-(n+1)P_{n+1}(t) - [\omega_+(n) + \omega_-(n)]P_n(t)$$

(3.3.11)

for $0 \leq n \leq N_0$ and reflecting boundary conditions at $n = 0, N_0$, that is, $\omega_-(0) = 0, \omega_+(N_0) = 0$. Note that $P_{-1} = P_{N_0+1} \equiv 0$ in equation (3.3.11). In the case of the simple ion channel model with constant transition rates α, β, we have

$$\omega_+(n) = (N_0 - n)\alpha, \quad \omega_-(n) = n\beta.$$ (3.3.12)

However, as we shall see later, more general ion channel models can have transition rates $\omega_\pm(n)$ that are nonlinear functions of n. In the latter case, multiplying both sides of the more general master equation (3.3.11) by n/N_0 and summing over n gives

$$\frac{d\langle n/N_0 \rangle}{dt} = \langle \Omega_+(n/N_0) \rangle - \langle \Omega_-(n/N_0) \rangle,$$ (3.3.13)

where $\omega_\pm(n) = N_0 \Omega_\pm(n/N_0)$, and the brackets $\langle \ldots \rangle$ denote a time-dependent ensemble averaging over realizations of the stochastic dynamics, that is,

$$\langle A(n/N_0) \rangle = \sum_{n=0}^{N_0} P_n(t) A(n/N_0)$$

for any function of state $A(n/N_0)$. If the transition rates in (3.3.11) are nonlinear functions of n, then there is coupling between different order moments resulting in a moment closure problem. That is, $\langle \Omega_\pm(n/N_0) \rangle \neq \Omega_\pm(\langle n \rangle/N_0)$ for finite N_0. However, in the thermodynamic limit $N_0 \to \infty$, statistical correlations can be ignored so that one can take the mean-field limit

$$\langle \Omega_\pm(n/N_0) \rangle \to \Omega_\pm(\langle n/N_0 \rangle).$$

This then yields a deterministic equation for the fraction x of open ion channels:

$$\frac{dx}{dt} = \Omega_+(x) - \Omega_-(x). \tag{3.3.14}$$

A steady-state solution P_n^* of the master equation (3.3.11) satisfies $J(n) = J(n+1)$ with

$$J(n) = \omega_-(n)P_n^* - \omega_+(n-1)P_{n-1}^*.$$

Using the fact that n is a nonnegative integer, that is, $P_n^* = 0$ for $n < 0$, it follows that $J(n) = 0$ for all n. Hence, by iteration,

$$P_n^* = P_0^* \prod_{m=1}^{n} \frac{\omega_+(m-1)}{\omega_-(m)}, \tag{3.3.15}$$

with

$$P_0^* = \left(1 + \sum_{n=1}^{N_0} \prod_{m=1}^{n} \frac{\omega_+(m-1)}{\omega_-(m)}\right)^{-1}.$$

For finite N_0, such a solution exists provided that all birth/death rates are positive definite. It can then be shown that (see Sect. 3.6) $\lim_{t\to\infty} P_n(t) = P_n^*$.

In the particular case of the transition rates (3.3.12), we have

$$P_n^* = P_0^* \left[\frac{\alpha}{\beta}\right]^n \frac{N_0!}{n!(N_0-n)!}. \tag{3.3.16}$$

After calculating P_0^*, we obtain the binomial distribution

$$P_n^* = p_0^n (1-p_0)^{N_0-n} \frac{N_0!}{n!(N_0-n)!}, \tag{3.3.17}$$

where $p_0 = \alpha/(\alpha+\beta)$. The mean and variance of the binomial distribution can be obtained using generating functions. That is,

$$\Gamma(z) \equiv \sum_{n=0}^{N_0} z^n P_n^* = \sum_{n=0}^{N_0} \frac{N_0!}{n!(N_0-n)!} (zp_0)^n (1-p_0)^{N_0-n}$$
$$= (zp_0 + 1 - p_0)^{N_0}.$$

It follows that

$$\langle n \rangle = \Gamma'(1) = N_0 p_0 (zp_0 + 1 - p_0)^{N_0-1}\big|_{z=1} = N_0 p_0,$$

and

$$\langle n(n-1) \rangle = \Gamma''(0) = N_0(N_0-1)p_0^2(zp_0 + 1 - p_0)^{N_0-2}\big|_{z=1} = N_0(N_0-1)p_0^2.$$

Hence, the mean and variance of the Binomial distribution are

$$\langle n \rangle = N_0 p_0, \quad \text{Var}[n] = N_0 p_0 (1 - p_0). \tag{3.3.18}$$

This model is explored further in Ex. 3.5 and 3.13. Other examples of birth–death processes are considered in Ex. 3.6–3.8.

System-size expansion of the birth–death master equation. A useful diffusion approximation of the birth–death master equation (3.3.11) for large but finite N_0 can be obtained by carrying out a Kramers–Moyal or system-size expansion to second order in N_0^{-1} [300, 822], which was originally applied to ion channel models by Fox and Lu [285]. This yields a Fokker–Planck (FP) equation describing the evolution of the probability density of a corresponding continuous stochastic process that is the solution to a stochastic differential equation (SDE). A rigorous analysis of the diffusion approximation has been carried out by Kurtz [475]. First, introduce the rescaled variable $x = n/N_0$ and transition rates $N_0 \Omega_\pm(x) = \omega_\pm(N_0 x)$. Equation (3.3.11) can then be rewritten in the form

$$\frac{dp(x,t)}{dt} = N_0 [\Omega_+(x - 1/N_0) p(x - 1/N_0, t) + \Omega_-(x + 1/N_0) p(x + 1/N_0, t)$$
$$- (\Omega_+(x) + \Omega_-(x)) p(x,t)],$$

where $p(n/N_0, t) = P_n(t)$. Treat x as a continuous variable and Taylor expand terms on the right-hand side to second order in N_0^{-1}. In particular, setting $f_\pm(x,t) = \widehat{\omega}_\pm(x) p(x,t)$, we have

$$f_+(x - 1/N_0) = f_+(x,t) - \frac{1}{N_0} \frac{\partial f_+(x,t)}{\partial x} + \frac{1}{2N_0^2} \frac{\partial^2 f_+(x,t)}{\partial x^2} + O(1/N_0^3),$$

and

$$f_-(x + 1/N_0) = f_-(x,t) + \frac{1}{N_0} \frac{\partial f_-(x,t)}{\partial x} + \frac{1}{2N_0^2} \frac{\partial^2 f_-(x,t)}{\partial x^2} + O(1/N_0^3).$$

This leads to the FP equation

$$\frac{\partial p(x,t)}{\partial t} = -\frac{\partial}{\partial x} [A(x) p(x,t)] + \frac{1}{2N_0} \frac{\partial^2}{\partial x^2} [D(x) p(x,t)], \tag{3.3.19}$$

with

$$A(x) = \Omega_+(x) - \Omega_-(x), \quad D(x) = \Omega_+(x) + \Omega_-(x). \tag{3.3.20}$$

The FP equation is supplemented by the no-flux or reflecting boundary conditions at the ends $x = 0, 1$ and a normalization condition:

$$J(0,t) = J(1,t) = 0, \quad \int_0^1 p(x,t) dx = 1. \tag{3.3.21}$$

Here $J(x,t)$ is the probability flux,

$$J(x,t) = -\frac{1}{2N_0}\frac{\partial}{\partial x}[D(x)p(x,t)] + A(x)p(x,t). \tag{3.3.22}$$

Linear noise approximation. Recall from Sect. 2.3 that the solution to the FP equation (3.3.19) determines the probability density function for a corresponding stochastic process $X(t)$, which evolves according to the SDE [300]

$$dX = A(X)dt + \frac{1}{\sqrt{N_0}}B(X)dW(t), \tag{3.3.23}$$

with $B(x)^2 = D(x)$. Here $W(t)$ denotes a Wiener process with $dW(t)$ distributed according to a Gaussian process with mean and covariance

$$\langle dW(t)\rangle = 0, \quad \langle dW(t)dW(s)\rangle = \delta(t-s)dtds. \tag{3.3.24}$$

Note that the noise term in (3.3.23) is multiplicative, since it depends on the current state $X(t)$. It is well known that there is an ambiguity in how one integrates multiplicative noise terms, which relates to the issue of Ito versus Stratonovich versions of stochastic calculus [300], see Sect. 2.2. However, for this particular example, based on the reduction of a master equation, the explicit form of the corresponding FP equation (3.3.19) ensures that the noise should be interpreted in the sense of Ito. In the limit $N_0 \to \infty$, we recover the deterministic equation (3.3.14).

In the particular case of the two-state ion channel model with transition rates (3.3.12), we have

$$A(x) = \alpha - (\alpha + \beta)x, \quad D(x) = \alpha + (\beta - \alpha)x.$$

Moreover, the solution $x(t)$ of the deterministic rate equation converges to the unique stable fixed point $x^* = \alpha/(\alpha + \beta)$. One can thus view the corresponding SDE (3.3.23) as describing a stochastic path in phase space that involves Gaussian-like fluctuations of order $1/\sqrt{N_0}$ about the deterministic trajectory. Substituting $X - x^* = Y/\sqrt{N_0}$ into the SDE (3.3.23) and formally Taylor expanding to lowest order in $1/\sqrt{N_0}$ yields the so-called linear noise approximation

$$dY = -kYdt + B(x^*)dW(t), \tag{3.3.25}$$

with

$$k \equiv -A'(x^*) = \alpha + \beta, \quad B(x^*) = \sqrt{D(x^*)} = \sqrt{\frac{2\alpha\beta}{\alpha+\beta}}.$$

This takes the form of an Ornstein–Uhlenbeck equation [300]; see equation (2.2.32). Hence, in the stationary limit $t \to \infty$,

$$\langle Y(t)\rangle \to 0, \quad \langle Y(t)^2\rangle \to \frac{B(x^*)^2}{2k}\left[1 - e^{-2kt}\right].$$

Since $Y(t) = \sqrt{N_0}(X(t) - x^*)$, $X(t) = n(t)/N_0$ and $x^* = p_0$, we recover the results of (3.3.18).

Steady-state solution and the quasi-potential. The FP equation (3.3.19) has a unique steady-state solution obtained by setting $J = J(x) = 0$ for all $0 \le x \le 1$. The resulting first-order ODE can be solved to give a steady-state probability density of the form

$$P_{\text{FP}}(x) = \mathcal{N}\frac{e^{-N_0\Psi(x)}}{D(x)}, \qquad (3.3.26)$$

with the quasi-potential (see Sect. 2.4.2)

$$\Psi(x) \equiv -2\int^x \frac{A(x')}{D(x')}dx' = -2\int^x \frac{\Omega_+(x') - \Omega_-(x')}{\Omega_+(x') + \Omega_-(x')}dx'. \qquad (3.3.27)$$

Here $\mathcal{N}$ is a normalization factor. How does the resulting steady-state density given by (3.3.26) with quasi-potential (3.3.27) compare to the steady-state solution of the corresponding master equation (3.3.11) in the large N_0 limit? In order to answer this question, let us consider the particular transition rates (3.3.12). Taking logarithms of both sides of equation (3.3.16) and using Stirling's formula $\ln(n!) \approx n\ln n - n$ yields a steady-state density similar in form to (3.3.26) but with a different quasi-potential: $P(x) \sim e^{-N_0\Phi(x)}$, with

$$\Phi(x) = -x\ln(\alpha/\beta) + x\ln(x) + (1-x)\ln(1-x) = \int^x \ln\frac{\Omega_-(x')}{\Omega_+(x')}dx'. \qquad (3.3.28)$$

Since $\Phi(x) \ne \Psi(x)$, we see that the steady-state probability density under the diffusion approximation can deviate significantly from the effective potential obtained directly from the master equation. However, this discrepancy is not much of an issue for the simple two-state system, since the underlying kinetic equation has a unique fixed point. Indeed, both potentials have the same global minimum at $x = x^*$, $\Phi'(x^*) = \Psi'(x^*) = 0$. Moreover, we find that $\Psi''(x^*) = \Phi''(x^*)$. Since N_0 is large, we can make the Gaussian approximation

$$P(x) \approx p(x^*)\exp\left[-N_0\Phi(x^*) - N_0\Phi''(x^*)(x - x^*)^2/2\right].$$

and similarly for $P_{\text{FP}}(x)$. Under this approximation, the mean and variance of the fraction of open channels are given by

$$\frac{\bar{n}}{N_0} = x^* = \frac{\alpha}{\alpha + \beta}, \qquad \frac{\langle(n - \bar{n})^2\rangle}{N_0^2} = \frac{1}{N_0\Phi''(x^*)} = \frac{x^*(1 - x^*)}{N_0}, \qquad (3.3.29)$$

and we obtain the same results using the Gaussian approximation of $P_{\text{FP}}(x)$. Thus the diffusion approximation accounts well for the Gaussian-like fluctuations around a globally stable fixed point.

Remark 3.1. The diffusion approximation can lead to exponentially large errors when there are multiple stable fixed points. Two known examples of bistability in a population of ion channels are (i) stochastic calcium release in ooycytes and cardiac myocytes [377, 378], and (ii) membrane voltage fluctuations underlying the initiation of spontaneous action potentials. In the first case, there is bistability in the fraction x of open ion channels arising from the fact that the transition rates in the birth-death process are nonlinear functions of x. This is due to a feedback mechanism involving calcium-induced-calcium release. On the other hand, bistability in the membrane voltage of a neuron occurs under the assumption that the kinetics of calcium or sodium ion channels is relatively fast and potassium kinetics are frozen.

3.3.3 First passage times for a birth-death process

Suppose that we wanted to determine the first time that all of the ion channels are closed (or open) given that $N(0) = m$. This is a first passage time problem that can be analyzed by imposing an absorbing boundary condition at $n = 0$ (or $n = N_0$). This means imposing the condition $\omega_+(0) = 0$ (or $\omega_-(N_0) = 0$). For the sake of illustration, we will consider an absorbing boundary at $n = a - 1$ and a reflecting boundary at $n = N_0$. Setting $\omega_+(n) = \lambda_n$ and $\omega_-(n) = \mu_n$ in equation (3.3.11), the corresponding backward master equation for $Q_m(t) = \mathbb{P}[N(t) = n | N(0) = m]$ and fixed n takes the form (see Box 5G)

$$\frac{d}{dt}Q_m(t) = \lambda_m[Q_{m+1}(t) - Q_m(t)] + \mu_m[Q_{m-1}(t) - Q_m(t)]$$

for $a \leq m \leq N_0$ with boundary conditions $Q_{a-1}(t) = 0$ and $Q_{N_0+1}(t) = Q_{N_0}(t)$. The latter reflecting boundary condition is equivalent to $\lambda_{N_0} = 0$. Following along analogous lines to continuous Markov processes (see Sect. 2.4.2), introduce the survival probability

$$G_m(t) = \sum_{n=a}^{N} \text{Prob}[N(t) = n | N(0) = m],$$

and let T denote the first time that the system reaches the state $n = a - 1$. Then

$$\mathbb{P}[T \geq t | N(0) = m] = G_m(t),$$

and the MFPT given the initial state $N(0) = m$ is

$$\tau_m = -\int_0^\infty t \frac{dG_m(t)}{dt} dt = \int_0^\infty G_m(t) dt. \tag{3.3.30}$$

Since each term in the sum defining $G_m(t)$ satisfies the backward master equation, it follows that

$$\frac{d}{dt}G_m(t) = \lambda_m[G_{m+1}(t) - G_m(t)] + \mu_m[G_{m-1}(t) - G_m(t)], \tag{3.3.31}$$

with boundary conditions $G_{a-1}(t) = 0$ and $G_{N_0+1}(t) = G_{N_0}(t)$.

Integrating the backwards equation for G_m with respect to t yields an iterative equation for the MFPT:

$$\lambda_m[\tau_{m+1} - \tau_m] + \mu_m[\tau_{m-1} - \tau_m] = -1, \quad a \leq m \leq N \tag{3.3.32}$$

with $\tau_{a-1} = 0$, $\tau_{N_0+1} = \tau_{N_0}$. Set $U_m = \tau_{m+1} - \tau_m$, $U_{N_0} = 0$, so that the iterative equation (3.3.32) becomes

$$\lambda_m U_m - \mu_m U_{m-1} = -1.$$

Define ϕ_m according to the recursion formula

$$\lambda_m \phi_m = \mu_m \phi_{m-1}, \quad a \leq m \leq N_0,$$

with $\phi_{a-1} = 1$, which implies that

$$\phi_m = \prod_{j=a}^{m} \frac{\mu_j}{\lambda_j}. \tag{3.3.33}$$

Setting $U_m = \phi_m S_m$, we then have

$$\mu_m \phi_{m-1}[S_m - S_{m-1}] = -1, \quad S_{N_0} = 0.$$

This has the solution, for $a \leq m \leq N - 1$,

$$S_m = \sum_{j=m}^{N_0-1} \frac{1}{\mu_{j+1} \phi_j}.$$

Hence, using the definitions of U_m amd S_m,

$$\tau_{m+1} - \tau_m = \phi_m \sum_{j=m}^{N_0-1} \frac{1}{\mu_{j+1} \phi_j}.$$

Iterating this equation by starting at $m = a - 1$ with $\tau_{a-1} = 0$, we obtain the final result

$$\tau_m = \sum_{l=a-1}^{m-1} \phi_l \sum_{j=l}^{N_0-1} \frac{1}{\mu_{j+1} \phi_j}, \quad a \leq m \leq N_0. \tag{3.3.34}$$

Similarly, if there is a reflecting boundary condition at $n = 0$ and an absorbing boundary condition at $n = b + 1$, we have (see Ex. 3.6)

$$\tau_m = \sum_{l=m}^{b} \phi_l \sum_{j=0}^{l} \frac{1}{\lambda_j \phi_{j+1}}, \quad \phi_m = \prod_{j=m}^{b} \frac{\lambda_j}{\mu_j}, \quad \phi_{b+1} = 1. \tag{3.3.35}$$

3.4 Voltage-gated ion channels

Conductance-based models of the Hodgkin–Huxley type have been used to describe many important features of the electrophysiology of neurons and other secretory cells [431]. It is typically assumed that the number of voltage-gated ion channels is sufficiently large so that one can determine the average transmembrane ionic currents based on the opening probabilities of individual channels, which is an application of the law of large numbers. However, the resulting deterministic equations cannot account for spontaneous events driven by ion channel fluctuations. For example, ion channel noise can induce spontaneous action potentials (SAPs), which can have a large effect on a neuron's function [856]. If SAPs are too frequent, a neuron cannot reliably perform its computational role. Hence, ion channel noise imposes a fundamental limit on the density of neural tissue. Smaller neurons must function with fewer ion channels, making ion channel fluctuations more significant and more likely to cause an SAP. In this section, we describe how to couple the voltage dynamics to the stochastic opening and closing of a finite number of ion channels, and use this to analyze the effects of ion channel fluctuations on neural excitability.

3.4.1 Conductance-based model of a neuron

A neuron typically consists of a cell body (or soma) where the nucleus is located, a branching output structure known as the axon and a branching input structure known as the dendritic tree, see Fig. 3.10. Neurons mainly communicate with each other by sending electrical impulses or spikes (action potentials) along their axons. (Some neurons are also coupled diffusively via gap junctions, see Sect. 6.7.) These axons make contacts on the dendrites of other neurons via microscopic junctions known as synapses. The basic components of synaptic processing induced by the arrival of an action potential are shown in the inset of Fig. 3.10. Depolarization of the presynaptic axon terminal causes voltage-gated Ca^{2+} channels within an active zone to open. The influx of Ca^{2+} produces a high concentration of Ca^{2+} near the active zone, which in turn causes vesicles containing neurotransmitter to fuse with the presynaptic cell membrane and release their contents into the synaptic cleft (exocytosis). The released neurotransmitter molecules then diffuse across the synaptic cleft and bind to specific receptors on the postsynaptic membrane. These receptors cause ion channels to open, thereby changing the membrane conductance and membrane potential of the postsynaptic cell.

    The opening of synaptic ion channels results in the flow of electrical current along the dendritic tree of the stimulated neuron. If the total synaptic current from all of the activated synapses forces the electrical potential within the cell body to cross some threshold, then the neuron fires a spike. The standard biophysical model for describing the dynamics of a single neuron with somatic membrane potential V is based upon conservation of electric charge:

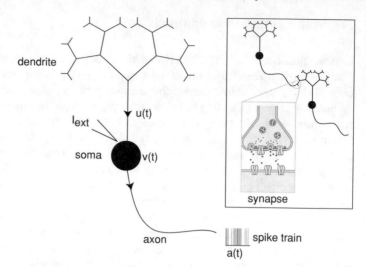

Fig. 3.10: Basic structure of a neuron. [Inset shows a synaptic connection from an upstream or presynaptic neuron and a downstream or postsynaptic neuron]. See textfor details.

$$C\frac{dV}{dt} = -I_m - I_{syn} + I_{ext}, \qquad (3.4.1)$$

where C is the cell capacitance, I_m is the membrane current, I_{syn} denotes the sum of synaptic currents entering the cell body and I_{ext} describes any externally injected currents. Ions can diffuse in and out of the cell through ion-specific channels embedded in the cell membrane. Ion pumps within the cell membrane maintain concentration gradients, such that there is a higher concentration of Na^+ and Ca^{2+} outside the cell and a higher concentration of K^+ inside the cell. The membrane current through a specific channel varies approximately linearly with changes in the potential V relative to some equilibrium or reversal potential, which is the potential at which there is a balance between the opposing effects of diffusion and electrical forces. Summing over all channel types, the total membrane current (flow of positive ions) leaving the cell through the cell membrane is

$$I_m = \sum_s g_s(V - V_s), \qquad (3.4.2)$$

where g_s is the conductance due to channels of type s and V_s is the corresponding reversal potential. In the case of a channel selective to a single ion, V_s satisfies the Nernst equation (see Ex. 3.9)

$$V_s = \frac{k_B T}{q} \ln\left(\frac{[outside]_s}{[inside]_s}\right), \qquad (3.4.3)$$

where q is the charge of the ion, k_B is the Boltzmann constant, T is temperature (in degrees Kelvin), and $[outside]_s, [inside]_s$ denote the extracellular and intracellular concentrations of the given ion. Typical values for the common ion species are

$V_K \approx -75\text{mV}$, $V_{Na} \approx 50\text{mV}$, $V_{Ca} \approx 150\text{mV}$, and $V_{Cl} \approx -60\text{mV}$ (which is close to the resting potential of the cell).

The generation and propagation of an action potential arises from nonlinearities associated with active membrane conductances. Recordings of the current flowing through single channels indicate that channels fluctuate rapidly between open and closed states in a stochastic fashion. Nevertheless, most models of a neuron use deterministic descriptions of conductance changes, under the assumption that there are a large number of approximately independent channels of each type. It then follows from the law of large numbers that the fraction of channels open at any given time is approximately equal to the probability that any one channel is in an open state, which is voltage dependent. The conductance g_s for ion channels of type s is thus taken to be the product $g_s = \bar{g}_s P_s(V)$ where $\bar{g}_s$ is equal to the density of channels in the membrane multiplied by the conductance of a single channel and P_s is the fraction of open channels.

3.4.2 Stochastic Morris–Lecar model

Let $V(t)$ denote the membrane voltage of some subcellular compartment, which could represent the site of action potential initiation in a neuron [431]. Suppose that there are two types of voltage-gated ion channel, which for concreteness we identify with calcium (Ca^{2+}) and potassium (K^+), respectively. Take the total number of Ca^{2+} and K^+ ion channels to be N_0 and M_0, respectively. For simplicity, we model the opening and closing of each ion channel as a two-state Markov process

$$C \underset{\beta_i(V)}{\overset{\alpha_i(V)}{\rightleftharpoons}} O, \quad i = \text{Ca, K,} \tag{3.4.4}$$

with the transition rates taken to be voltage dependent. When an ion channel of type i is open it allows the flow of an ion-current I_i that is based on Ohm's law (3.4.2) with $s = $ Ca, K. The steady state or resting voltage V^* is assumed to satisfy $V_{Ca} < V^* < V_K$ so that Ca^{2+} ions tend to flow into the compartment (depolarization), whereas K ions tend to flow out of the compartment (hyperpolarization).

Let $N(t)$ and $M(t)$ denote the number of open Ca^{2+} and K^+ ion channels at time t. An application of Kirchoff's law implies that if $M(t) = m, N(t) = n$, then

$$C\frac{dV}{dt} = F(V, m, n) \equiv \frac{n}{N_0} f_{Ca}(V) + \frac{m}{M_0} f_K(V) + f_L(V) + I_{app}, \tag{3.4.5}$$

where C is a membrane capacitance, I_{app} represents any externally applied current,

$$f_{Ca}(V) = N_0 g_{Ca}(V_{Ca} - V), \quad f_K(V) = M_0 g_K(V_K - V), \quad f_L(V) = \bar{g}_L(V_L - V),$$

and $f_L(V)$ represents a leak current. The discrete variables $N(t)$ and $M(t)$ each evolve according to a birth–death process with associated master equation of the form (3.3.10), except that now the birth and death rates are voltage dependent. For example, $P_n(t) = \mathbb{P}[N(t) = n]$ evolves according

$$\frac{d}{dt}P_n(t) = \alpha_{Ca}(V)(N_0 - n + 1)P_{n-1}(t) + \beta_{Ca}(V)(n + 1)P_{n+1}(t)$$
$$- [\alpha_{Ca}(V)(N_0 - n) + \beta_{Ca}(V)n]P_n(t), \tag{3.4.6}$$

and similarly for $P_m(t) = \mathbb{P}[M(t) = m]$. It is important to note that equation (3.4.5) only holds between jumps in the discrete variables.

Remark 3.2. The stochastic process defined by (3.4.4) and (3.4.5) is an example of a stochastic hybrid system, since it involves the coupling of a piecewise deterministic continuous dynamical system with a discrete Markov process. There has been a lot of recent interest in such systems, particularly within the context of conductance-based models [116, 148, 433, 614, 627, 767, 836]. They also arise in models of polymerization (Chap. 4), gene networks (Chap. 5), diffusion processes in stochastically gated environments (Chap. 6), active motor transport (Chap. 7), bacterial chemotaxis (Chap. 10), and cell adhesion (Chap. 14). Some of the basic properties of stochastic hybrid systems will be described in Box 5D.

Morris–Lecar model and phase-plane analysis. Suppose that we take the thermodynamic limits $N_0, M_0 \to \infty$ with $\bar{g}_{Ca} := N_0 g_{Ca}$ and $\bar{g}_K := M_0 g_K$ fixed. Setting $x = N/N_0$ and $w = M/M_0$, we then obtain the deterministic kinetic equations

$$\frac{dx}{dt} = \alpha_{Ca}(V)(1 - x) - \beta_{Ca}(V)x,$$
$$\frac{dw}{dt} = \alpha_K(V)(1 - w) - \beta_K(V)w.$$

Exploiting the fact that the dynamics of the Ca^{2+} channels is much faster than the voltage and K^+ dynamics, we assume that the fraction of open Ca^{2+} ion channels is in quasi-equilibrium ($dx/dt \approx 0$):

$$x = a_{Ca}(V) = \frac{\alpha_{Ca}(V)}{\alpha_{Ca}(V) + \beta_{Ca}(V)}.$$

We then obtain the Morris–Lecar (ML) conductance-based model of a neuron [592]:

$$C\frac{dV}{dt} = a_\infty(V)f_{Ca}(V) + wf_K(V) + f_L(V) + I_{app},$$
$$\frac{dw}{dt} = (1 - w)\alpha_K(v) - w\beta_K(V) = \frac{\phi}{\tau(V)}[w_\infty(V) - w]. \tag{3.4.7}$$

Standard formulations of the ML model take

$$\alpha_{Ca}(V) = \beta_{Ca}\exp\left(\frac{2[V - v_{Ca,1}]}{v_{Ca,2}}\right), \tag{3.4.8}$$

with $\beta_{Ca}, v_{Ca,1}, v_{Ca,2}$ constant. Moreover, the transition rates $\alpha_K(v)$ and $\beta_K(v)$ are chosen such that

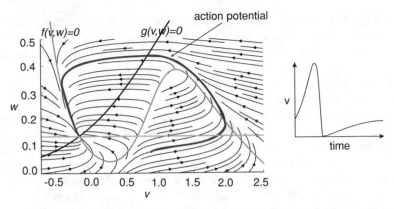

Fig. 3.11: Deterministic phase plane dynamics of the ML model. Nullclines: $\dot{v} = 0$ (gray) and $\dot{w} = 0$ (black). Point of intersection of two nullclines is the resting state (v^*, w^*). Black streamlines represent deterministic trajectories. Red curve represents an action potential. Inset shows the action potential as a function of time. (Intersection of the horizontal line with the v-nullcline indicates fixed points of the reduced system obtained by fixing $w = w^*$). Parameter values are $C = 20$ mF, $V_{Ca} = 120$ mV, $V_K = -84$ mV, $V_L = -60$ mV, $g_{Ca} = 4.4$ mS, $g_K = 8$ mS, $g_L = 2.2$ mS, $\beta_{Ca} = 0.8$ s^{-1}, $v_{Ca,1} = -1.2$ mV, $v_{Ca,2} = 18$ mV, $v_{K,1} = 2$ mV, $v_{K,2} = 30$ mV, and $\phi = 0.04$ ms^{-1}.

$$w_\infty(v) = \frac{1}{2}\left(1 + \tanh\left[\frac{v - v_{K,1}}{v_{K,2}}\right]\right), \quad \tau(v) = \cosh\left[\frac{v - v_{K,1}}{2v_{K,2}}\right]. \qquad (3.4.9)$$

The deterministic ML model can be analyzed using phase plane analysis. The voltage V has a cubic-like nullcline (along which $\dot{V} = 0$) and the recovery variable has a monotonically increasing nullcline (along which $\dot{w} = 0$), see Fig. 3.11. First, suppose that the applied current I_{app} is sufficiently small so that the nullclines have a single intersection point at (V^*, w^*) on the left branch of the cubic nullcline. This

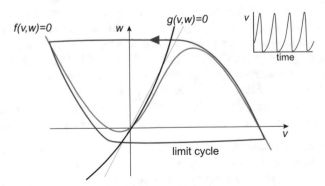

Fig. 3.12: Schematic diagram illustrating the trajectory of a globally stable periodic orbit in the phase-plane for the FitzHugh–Nagumo equations. The unique rest point is unstable. Inset shows the voltage as a function of time.

corresponds to a fixed point of the system, which we identify with the resting state. The neuron is said to be excitable, since small perturbations of the resting state decay to zero, whereas sufficiently large perturbations result in a time-dependent trajectory taking a prolonged excursion through state space before returning to the resting state. In the latter case, the time-dependent plot of the variable V can be interpreted as an action potential. There is effectively a threshold phenomenon in which sub-threshold perturbations result in a simple return to the resting state, whereas super-threshold perturbations generate an action potential. It is also possible to convert the ML model from an excitable to an oscillatory regime by modifying the external current I_{app}. For an intermediate range of values of I_{app} one finds that the fixed point shifts to the middle branch of the cubic nullcline where it is unstable. The fixed point now coexists with a stable limit cycle, along which the trajectory alternates periodically between the left and right branches, resulting in a periodic train of action potentials, see Fig. 3.12.

Potassium channel fluctuations. Now suppose that the fraction of open Ca^{2+} ion channels is in quasi-equilibrium, whereas we keep track of fluctuations in the number $M(t)$ of open K^+ channels. Equation (3.4.5) then reduces to the simpler stochastic hybrid system

$$C\frac{dV}{dt} = F(V(t), m/M) \equiv a_{Ca}(v)f_{Ca}(V) + \frac{m}{M_0}f_K(V) + f_L(V) + I_{app}. \quad (3.4.10)$$

for $M(t) = m$, and $M(t)$ evolving according to a birth–death process. Fig. 3.13 shows Morris–Lecar simulations that include stochastic voltage-dependent gating of 100 K^+ ion channels. Spontaneous action potentials are induced by this simulated model when the system is in an excitable regime. The discreteness and stochasticity of the gating variable $w = m/M_0$ allows trajectories to fluctuate around the fixed point of the deterministic model. Occasionally, K^+ ion channels spontaneously inactivate (w fluctuates toward 0) and a regenerative Ca^{2+} current leads to an action potential. This type of spontaneous activity has been observed in stochastic versions of the Hodgkin–Huxley equations, and is thought to influence subthreshold membrane potential oscillations and excitability of stellate neurons of the medial entorhinal cortex of the hippocampal region [855]. Also illustrated in Fig. 3.13 are the effects of noise when the underlying deterministic system is in an oscillatory or bistable regime. It can be seen that in the former case, noise results in irregular oscillations, whereas in the latter case stochastic bistability is observed. That is, channel noise allows the alternate sampling of two stable fixed points in the (V, w) phase plane.

Calcium channel fluctuations. In the excitable regime, one can exploit the fact that the initiation of an action potential typically occurs without w deviating much from the resting value w^*. Therefore suppose, that we fix $w = w^*$ and consider the stochastic opening and closing of the Ca^{2+} channels. The stochastic membrane voltage then evolves according to the stochastic hybrid system

$$C\frac{dV(t)}{dt} = G(V(t), n/N_0) \equiv \frac{n(t)}{N_0}f(V(t)) - g(V(t)) \quad (3.4.11)$$

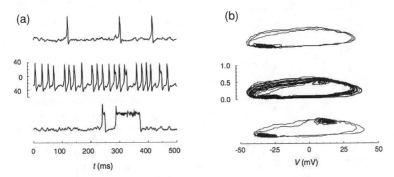

Fig. 3.13: Simulations of the stochastic ML model with the calcium channels in quasi-equilibrium and 100 randomly opening and closing potassium channels. (a) Output voltage trace. (b) Trajectories in the $V - w$ plane, where $w = m/M$. There are three different regimes shown, which can be obtained by varying I_{app}: excitable (upper plots); oscillatory (middle plots); bistable (lower plots). [Adapted from Smith (2002).]

for $N(t) = n$, where $f(V) = f_{Ca}(V)$ and $g(V) = -[w^* f_K(V) + f_L(V) + I_{app}]$ represents the sum of effective leakage currents and external inputs. Since the right-hand side of equation (3.4.11) is negative (positive) for large (small) V, it follows that there exists an invariant interval for the voltage dynamics. In particular, let v_0 denote the voltage for which $\dot{V} = 0$ when $n = 0$ and let v_{N_0} be the corresponding voltage when $n = N_0$. That is, $g(v_0) = 0$ and $f_{Ca}(v_N) - g(v_N) = 0$. Then $V(t) \in [v_0, v_{N_0}]$ if $V(0) \in [v_0, v_{N_0}]$. Equation (3.4.11) only holds between jumps in the number of open ion channels, with the latter described by the master equation (3.4.6). In the following, we drop the Ca subscript on the transition rates α, β appearing in (3.4.6).

In the absence of noise ($N_0 \to \infty$), the system (3.4.11) evolves according to the deterministic equation

$$\frac{dv}{dt} = A(v) = \frac{\alpha(v)}{\alpha(v) + \beta(v)} f(v) - g(v) \equiv -\frac{dU(v)}{dv}, \qquad (3.4.12)$$

where $U(v)$ is a deterministic potential. In Fig. 3.14, we plot $U(v)$ as a function of v for various values of the external input current and the particular transition rates

$$\alpha(v) = \beta \exp\left(\frac{2(v - v_1)}{v_2}\right), \qquad \beta = \text{constant}.$$

The minima and maxima of the potential correspond to stable and unstable fixed points of the deterministic dynamics, respectively. It can be seen that below some threshold applied current $I_{app} < I^*$, there exist two stable fixed points $v_{\pm}$ (minima) separated by an unstable fixed point at v_0 (maximum), that is, the system exhibits bistability. The left-hand fixed point represents the resting state, whereas the right-hand fixed point corresponds to an excited state. Thus, in the bistable regime the

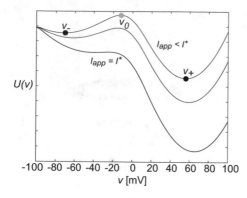

Fig. 3.14: Plot of deterministic potential $U(v)$ as a function of voltage v for different values of the external stimulus current I_{app}. Parameter values are $N = 10$, $v_{Na} = 120$ mV, $v_{eff} = -62.3$ mV, $g_{Na} = 4.4$ mS/cm^2, $g_{eff} = 2.2$ mS/cm^2, $\beta = 0.8$ s^{-1}, and $\alpha(v) = \beta \exp[(v + 1.2)/9]$.

deterministic system requires an external perturbation in order to generate an action potential starting from the resting state. On the other hand, for the stochastic system it is possible that fluctuations in the opening and closing of Ca^{2+} ion channels induce a transition from the resting state to the excited state by crossing over the potential hill at v_0 (as determined by the corresponding quasi-potential, see Sect. 2.4.2). Of course, once such an event occurs, one has to take into account the K$^+$ dynamics in order to incorporate the effects of repolarization that return the system to the resting state. If one includes the slow opening and closing of these channels, then the underlying deterministic system becomes excitable rather than bistable. For simplicity, we will assume that this does not significantly affect the noise-induced initiation of an action potential. It turns out that such an assumption breaks down if fluctuations in the opening and closing of K$^+$ channels become significant [614].

In order to investigate action potential initiation, we will make the following approximations:

(i) For sufficiently large N_0, we approximate the jump Markov process for the ion channels by a continuous Markov process using a linear noise diffusion approximation as outlined in Sect. 3.3.2;

(ii) The transitions between different discrete states are much faster than the voltage dynamics so that, for fixed V, the fraction of open ion channels is close to the quasi-equilibrium $x^*(V) = a(V) \equiv \alpha(V)/(\alpha(V) + \beta(V))$. This limiting case was originally considered by Chow and White [181].

Under these approximations, we set

$$\frac{N(t)}{N_0} = X(t) = x^*(V) + \frac{Y(t)}{\sqrt{N_0}},$$

with $Y(t)$ evolving according to the OU process (3.3.25). Combining with the voltage dynamics, we obtain the pair of SDEs

$$dV(t) = \left[f(V(t))(a(V(t)) + Y(t)/\sqrt{N_0}) - g(V(t)) \right] dt, \tag{3.4.13a}$$

$$dY(t) = -k(V(t))Y(t)dt + b(V(t))dW(t), \tag{3.4.13b}$$

with

$$k(V) = \alpha(V) + \beta(V), \quad b(V) = \sqrt{\frac{2\alpha(V)\beta(V)}{\alpha(V) + \beta(V)}}.$$

Thus the stochastic voltage is coupled to a fast Ornstein–Uhlenbeck process $Y(t)$.

In the case of fast Ca^{2+} channel kinetics (approximation (ii)), we can take $Y(t)$ to be in quasi-equilibrium for a given V, that is, $Y(t)dt \approx k(V)^{-1}b(V)dW(t)$. This then yields a scalar SDE for the voltage:

$$dV = [f(V)a(V) - g(V)]dt + \frac{1}{\sqrt{N_0}}\sigma(V)f(V)dW(t), \tag{3.4.14}$$

where

$$\sigma(V) = \frac{b(V)}{k(V)} = \frac{1}{\alpha(V) + \beta(V)}\sqrt{\frac{2\alpha(V)\beta(V)}{\alpha(V) + \beta(V)}}. \tag{3.4.15}$$

In deriving (3.4.14), we have effectively taken a zero correlation limit of an Ornstein–Uhlenbeck process. It can be shown that the multiplicative noise term should be interpreted in the sense of Stratonovich [433], see also Sect. 2.3.5. However, for large N_0 this yields an $O(1/N_0)$ correction to the drift term in the FP equation, which can be dropped. We thus obtain the FP equation

$$\frac{\partial p(v,t)}{\partial t} = -\frac{\partial}{\partial v}[A(v)p(v,t)] + \frac{1}{2N_0}\frac{\partial^2}{\partial v^2}[D(v)p(v,t)], \tag{3.4.16}$$

with

$$A(v) = f(v)a(v) - g(v), \quad D(v) = [\sigma(v)f(v)]^2. \tag{3.4.17}$$

The FP equation is supplemented by reflecting boundary conditions at $v = V_1, V_2$:

$$J(V_1,t) = J(V_2,t) = 0, \tag{3.4.18}$$

with

$$J(v,t) = A(v)p(v,t) - \frac{1}{2N_0}\frac{\partial}{\partial v}D(v)p(v,t). \tag{3.4.19}$$

Note that equation (3.4.16) is identical in form to (3.3.19) except we now have a stochastic process with respect to membrane voltage rather than fraction of open ion channels; the latter is slaved to the voltage.

The calculation of the rate of escape from the resting state based on the diffusion approximation proceeds along identical lines to Sect. 2.4.2 for a Brownian particle subject to multiplicative noise, see Fig. 2.8 with $x \to v$. That is,

$$\lambda_- = \frac{D(v_-)}{4\pi}\sqrt{|\Phi''(v_0)||\Phi''(x_-)|}e^{-N_0[\Phi(v_0) - \Phi(v_-)]}, \tag{3.4.20}$$

where

$$\Phi(v) = -2 \int^v \frac{A(y)}{B(y)} dy = -2 \int^v \frac{f(y)a(y) - g(y)}{\sigma(y)f(y)} dy \qquad (3.4.21)$$

is the quasi-potential. Comparison with equations (2.4.26) and (2.4.27a) shows that the quasi-potential Φ is scaled by a factor of 2, and the exponential term has a factor of N_0, which acts as an "inverse temperature." Since the diffusion approximation of the underlying master equation requires N_0 to be large, it follows that the system operates in a weak noise (low temperature) regime. Moreover, the rate of escape is exponentially small for large N_0 as $\Phi(v_0) > \Phi(v_-)$. Keener and Newby [433] explicitly calculated the MFPT $\tau_- = 1/\lambda_-$ and compared it with Monte Carlo simulations of the full stochastic model (3.4.11). They showed that the analytical result deviates significantly from the full model, where it overestimates the mean time to spike. This is related to the fact that the quasi-potential of the steady-state density under the diffusion approximation generates exponentially large errors in the MFPT.

> *Remark 3.3. Breakdown of the diffusion approximation.* The diffusion approximation still accounts for the effects of fluctuations well within the basin of attraction of each metastable fixed point, but there is now a small probability that there is a noise–induced transition to the basin of attraction of the other fixed point. Since the probability of such a transition is usually of order $e^{-\tau N}$ with $\tau = O(1)$, except close to the boundary of the basin of attraction, such a contribution cannot be analyzed accurately using standard FP methods [822]. That is, in the case of weak noise, the transitions between different metastable states typically involve rare transitions (large fluctuations) that lie in the tails of the associated probability distribution, where the Gaussian approximation no longer necessarily holds. These exponentially small transitions play a crucial role in allowing the system to approach the unique stationary state (if it exists) in the asymptotic limit $t \to \infty$. In other words, for bistable (or multistable) systems, the limits $t \to \infty$ and $N_0 \to \infty$ do not commute [47, 356, 826]. In order to ensure accurate estimates of transition rates, one has to use alternative approximation schemes based on large deviation theory and perturbation methods, which are applied directly to the underlying master equation (Chap. 8). The same issue will arise in the analysis of genetic switches (Chap. 5).

3.5 Stochastic models of Ca^{2+} release

Ca^{2+} is one of the most important and well-studied cellular signaling molecules. From a modeling perspective, it attracts a great deal of interest due to the fact that calcium signaling often involves complex spatiotemporal dynamics, including oscillations and waves. For reviews on the modeling of calcium dynamics within cells, see Chap. 7 of Keener and Sneyd [431], Falcke [267] and Chap. 4 of Bressloff [115]. In vertebrates, most of the Ca^{2+} is stored in bones, from where it can be released by hormonal stimulation to maintain a high extracellular Ca^{2+} concentration (around 1mM). On the other hand, active ion pumps and exchangers maintain the cytoplasmic Ca^{2+} concentration at relatively low levels (around 10–100 nM). The resulting steep concentration gradient across the plasma membrane means that cells are able to increase their cytoplasmic Ca^{2+} concentration rapidly by opening either voltage-gated or ligand-gated Ca^{2+} ion channels. Here we will consider another major mechanism for controlling intracellular Ca^{2+} based on the action of protein receptors embedded in the surface membrane of intracellular stores. Some of the main features of such receptors are as follows:

1. Cells can regulate their cytoplasmic Ca^{2+} concentration via the intracellular supply of Ca^{2+} from internal stores such as the endoplasmic reticulum (ER) and mitochondria. Inositol (1,4,5)-trisphosphate (IP$_3$) receptors (IP$_3$Rs) and ryanodine receptors (RyRs) distributed throughout the ER, for example, mediate the release of Ca^{2+} into the cytoplasm, whereas Ca^{2+} ion pumps maintain the relatively high Ca^{2+} concentration within the ER. The RyR plays a critical role in excitation-contraction coupling in skeletal and cardiac muscle cells, but is also found in non-muscle cells such as neurons. One important feature of RyRs is that they can undergo Ca^{2+}-induced Ca^{2+} release (CICR), in which elevated cytoplasmic Ca^{2+} activates RyRs that release further Ca^{2+}, which then activates other Ry receptors, resulting in a nonlinear regenerative feedback mechanism. The IP$_3$R is similar in structure to the RyR, but is found predominantly in non-muscle cells and is sensitive to the second messenger IP$_3$. The binding of an extracellular ligand such as a hormone or a neurotransmitter to a metabotropic receptor results in the activation of a G protein and the subsequent activation of phospholipase C (PLC). This then cleaves phosphotidylinositol biphosphate (PIP$_2$) into diacylglycerol (DAG) and IP$_3$. The water soluble IP$_3$ is free to diffuse throughout the cell cytoplasm and bind to IP$_3$Rs located on the ER membrane, which then open and release Ca^{2+} from the ER. The opening and closing of an IP$_3$R is also modulated by the concentration of cytoplasmic Ca^{2+}, so it too can undergo CICR.

2. Another mechanism for controlling cytoplasmic Ca^{2+} is through buffering (binding) to large proteins. It is estimated that at least 99% of the total cytoplasmic Ca^{2+} is bound to buffers. A summary of the basic extracellular and intracellular mechanisms for controlling cytoplasmic Ca^{2+} is shown in Fig. 3.15.

3. One of the most dramatic consequences of CICR is the propagation of intracellular Ca^{2+} waves mediated primarily by the opening of IP$_3$Rs. These waves were first observed in nonneuronal cells such as *Xenopus laevis* oocytes [501, 637], where the resulting changes in Ca^{2+} concentration across the whole cell provided a developmental signal.

4. Many cell types exhibit spontaneous localized Ca^{2+} release events known as sparks or puffs [165, 166]. The fluorescent imaging of Ca^{2+} puffs and sparks has established that Ca^{2+} release is a stochastic process that occurs at spatially discrete sites consisting of clusters of IP$_3$Rs and RyRs, respectively. Ca^{2+} puffs are found in *Xenopus laevis* oocytes, and have an amplitude ranging from around 50–600nM, a spatial spread of approximately 6μm and a typical duration of 1s [637, 641, 874]. For sufficiently high levels of IP$_3$ concentration, the amplification of Ca^{2+} puffs by CICR can lead to the formation of Ca^{2+} waves [637, 641, 874]. Calcium sparks, which are thought to be the building blocks of

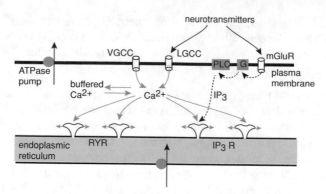

Fig. 3.15: Calcium signaling. The entry of Ca^{2+} from outside the cell is mediated by voltage-gated (VGCC) and ligand-gated (LGCC) calcium channels. Stimulation of metabotropic glutamate receptors (mGluRs) produces IP_3 second messengers that bind to IP_3 receptors (IP_3Rs), which subsequently release Ca^{2+} from the endoplasmic reticulum. Both IP_3Rs and Ryanodine receptors (RyRs) are sensitive to Ca^{2+}, resulting in calcium-induced calcium release (CICR). The latter can sometimes result in the propagation of a Ca^{2+} wave along the dendrites.

the large regenerative Ca^{2+} signal that controls contraction in cardiac and skeletal muscle cells, arise from the opening of clusters of RyRs by local CICR. The frequency of calcium spark events is sensitive to changes in membrane potential, although they rarely induce calcium waves due to shorter duration and less spatial spread.

3.5.1 Stochastic model of Ca^{2+} puffs in a cluster of IP_3Rs

Stochastic models of Ca^{2+} puffs typically treat a cluster of IP_3Rs as a set of N_0 channels that open and close independently, but are indirectly coupled by the common cytoplasmic Ca^{2+} concentration [265, 266, 754, 790]. Models differ in the level of detail regarding individual receptors. The first deterministic kinetic model of Ca^{2+}-gated IP_3Rs was proposed by De Young and Keizer [878], in their study of agonist-induced Ca^{2+} oscillations. This model assumes that the IP_3R consists of three equivalent receptor subunits, all of which have to be in a conducting state in order to generate a Ca^{2+} flux. Each subunit is taken to have an IP_3 binding site, an activating Ca^{2+} binding site, and an inactivating Ca^{2+} binding site; the conducting state corresponds to the state in which all subunits have the first two binding sites occupied but the third unoccupied, see Fig. 3.16(a). Although the De Young–Keizer model is simple to describe, it involves a relatively large number of variables that have to be coupled to the Ca^{2+} and IP_3 concentrations. A simplified version of the model was subsequently developed by Li and Rinzel [517]. They exploited the fact that the binding of IP_3 and activating Ca^{2+} are fast relative to inactivating Ca^{2+},

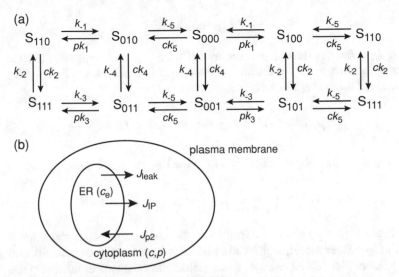

Fig. 3.16: IP$_3$R model. (a) Reaction diagram of the De Young–Keizer model [878] of an IP$_3$R subunit. The state of the subunit is denoted by S_{ijk}, where $i = 0, 1$ denotes whether the IP$_3$ binding site is unoccupied ($i = 0$) or occupied ($i = 1$), and j, k denote the corresponding status of the activating and inactivating Ca²⁺ binding sites, respectively. Although there are 24 separate single-step reactions, the model only has 10 independent rate constants. This is a consequence of equilibrium thermodynamics and two additional constraints: (i) the rate constants are taken to be independent of whether or not the Ca²⁺ activating binding site is occupied, and (ii) the kinetics of Ca²⁺ activation are assumed to be independent of IP$_3$ binding and Ca²⁺ inactivation. (b) Schematic diagram of fluxes in the Li-Rinzel model [517]. Here c and p denote the concentration of Ca²⁺ and IP$_3$ in the cytoplasm, and c_e is the concentration in the endoplasmic reticulum (ER). Both p and c_e are held fixed.

and used a quasi-steady-state argument to reduce the eight-state subunit model to a model that simply keeps track of whether or not the inactivating Ca²⁺ binding site of a subunit is occupied. There is then a single gating variable

$$h = x_{000} + x_{010} + x_{100} + x_{110},$$

where x_{ijk} denotes the fraction of subunits in state S_{ijk}, see Fig. 3.16(a). Thus, h^3 is the fraction of receptors in a cluster not inactivated by Ca²⁺. One finds that h evolves according to an equation of the form (Ex. 3.10)

$$\frac{dh}{dt} = \alpha_h(p)(1 - h) - \beta_h(p)ch, \tag{3.5.1a}$$

where c is the cytoplasmic Ca²⁺ concentration and p is the IP$_3$ concentration, which is taken to be fixed. It is assumed that there are three fluxes contributing to the change in Ca²⁺ concentration:

$$\frac{dc}{dt} = J_{IP} + J_{leak} - J_{p2}, \tag{3.5.1b}$$

where J_{IP} is the flux through the cluster of IP$_3$Rs, J_{leak} is a leakage flux from the ER to the cytoplasm, and J_{p2} is the flux pumped back into the ER, see Fig. 3.16(b). The expressions for the various fluxes are

$$J_{IP} = f(c,p)^3 h^3 [c_e - c], \quad J_{leak} = v_0 [c_e - c], \quad J_{p2} = \frac{v_1 c^2}{k_3^2 + c^2}, \tag{3.5.2}$$

where c_e is the fixed Ca^{2+} concentration in the ER, v_0, v_1 are constants, and

$$f(c,p) = \left(\frac{p}{p + K_1}\right) \cdot \left(\frac{c}{c + K_5}\right), \quad K_j = k_{-j}/k_j. \tag{3.5.3}$$

The function $f(c,p)$ can be derived from the quasi-steady-state reduction of the De Young–Keizer model, which shows that the fraction of open subunits is $x_{110} = f(c,p)h$, see Ex. 3.10. The cubic terms reflect the existence of three subunits. Parameter values of the model can be found in [517]. Note that the simplified model resembles a conductance-based model of a neuron, see Sect. 3.4, after replacing Ca^{2+} concentration c by membrane voltage v and c_e by a reversal potential.

We now describe a stochastic version of the Li-Rinzel model for a cluster of IP$_3$Rs due to Shuai and Jung [754]. For stochastic versions of the full De Young–Keizer model see, for example, [265, 266, 344, 790]. The deterministic equations (3.5.1) describe the mean behavior of a large cluster of Ca^{2+} channels, just as the Hodgkin–Huxley equations for membrane voltage apply to a large number of voltage-gated ion channels. If the number of channels is relatively small, then it is necessary to take into account thermally driven fluctuations in the opening and closing of individual channels. In the case of the Li-Rinzel model, one only needs to consider the state of the Ca^{2+}-inactivating binding site of each subunit. The latter is modeled as the two-state Markov process

$$A^* \underset{\beta_h c}{\overset{\alpha_h}{\rightleftharpoons}} A, \tag{3.5.4}$$

where A (A^*) denotes the unbound (bound) state. Suppose that there are N_0 independent IP$_3$Rs, each with three independent subunits labeled $i = 1, 2, 3$ that are described by the above two-state Markov process. Let $N_i(t)$ ($i = 1, 2, 3$) denote the number of receptors at time t that have the ith subunit in state A. Under the adiabatic assumption that the Ca^{2+} concentration c evolves much more slowly than the state transitions of the channels, we can write down a master equation for the probability $P(n_i, t) = \text{Prob}[N_i(t) = n_i | N_i(0) = \bar{n}_i]$ according to

$$\frac{dP(n_i, t)}{dt} = (N_0 - n_i + 1)\alpha_h P(n_i - 1, t) + (n_i + 1)c\beta_h P(n_i + 1, t) \tag{3.5.5}$$

$$- (n_i c\beta_h + (N_0 - n_i)\alpha_h)P(n_i, t), \quad i = 1, 2, 3.$$

As with voltage-gated ion channels (see Sect. 3.4), we have a stochastic hybrid system, since the $A \rightarrow A^*$ transition rate depends on the Ca^{2+} concentration $c(t)$, which evolves according to a piecewise deterministic equation of the form (3.5.1a). The latter, in turn, couples to the discrete stochastic variables $N_i(t)$ through the flux

$$J_{IP} = f(c(t), p)[c_e - c(t)] \prod_{i=1}^{3} \frac{N_i(t)}{N_0}. \tag{3.5.6}$$

Finally, for large N_0, one can obtain a further simplification by carrying out a system-size expansion of the master equation (3.5.5) along identical lines to Sect. 3.3.2. This yields the following SDE for $H_i(t) = N_i(t)/N_0$ with H_i treated as a continuous stochastic variable:

$$dH_i = \alpha_h(1 - H_i) - c\beta_h H_i + \frac{1}{\sqrt{N_0}} b(H_i) dW_i, \tag{3.5.7}$$

where

$$b(H_i) = \sqrt{\alpha_h(1 - H_i) + c\beta_h H_i},$$

and $W_i(t)$ is an independent Wiener process with

$$\langle dW_i(t) \rangle = 0, \quad \langle dW_i(t) dW_j(t') \rangle = \delta(t - t') dt \, dt' \delta_{i,j}.$$

Shuai and Jung [754] simulated the stochastic Li-Rinzel model in order to investigate the effects of noise on Ca^{2+} oscillations in a space-clamped model. They assumed that the deterministic system (3.5.1) was monostable at low and high IP_3 concentrations and exhibited limit cycle oscillations (occurring via a Hopf bifurcation, see Box 5A) at intermediate concentrations. They showed that noise can enlarge the range of IP_3 concentrations over which oscillations occur—an effect known as coherence resonance. They also found a broad distribution of puff amplitudes, lifetimes, and inter-puff intervals. In particular, at low IP_3 concentrations, the amplitude distribution is a monotonically decaying function, whereas at higher concentrations it is unimodal. This suggests that Ca^{2+} puffs become more significant as IP_3 concentration is increased, and hence could impact the spontaneous generation of Ca^{2+} waves. This issue was investigated numerically by Falcke [266] using a stochastic version of the De Young–Keizer model that was incorporated into a reaction-diffusion model of spatially distributed channel clusters. He showed that there is indeed a transition from Ca^{2+} puffs to waves as the IP_3 concentration is increased. At low concentrations, only puffs occur, since there is not enough Ca^{2+} released to stimulate neighboring clusters, which means that the response is purely local. However, as IP_3 concentration increases, global Ca^{2+} waves can emerge from local nucleation sites of high Ca^{2+} concentration. At intermediate levels of IP_3, global events are rare and waves only progress a short distance before dying out. On the other hand, for higher IP_3 concentrations, global waves occur regularly with a well-defined period. Again this oscillatory-like behavior can occur in parameter regimes for which the deterministic model is non-oscillatory.

3.5.2 *Stochastic model of* Ca^{2+} *sparks in cardiac myocytes*

We now turn to a stochastic model of Ca^{2+} sparks in cardiac myocytes [377], which includes details of the geometry of Ca^{2+} release units, in particular, the narrow junctional gap known as the diadic space that separates the sarcoplasmic reticulum (SR) from the plasma membrane, see Fig. 3.17. (In smooth muscle cells the smooth ER is referred to as the sarcoplasmic reticulum.) In a typical myocyte, there could be up to 10,000 Ca^{2+} release units, each one containing a cluster of around 50 RyRs on the surface of the SR. The cluster of RyRs is apposed to L-type Ca^{2+} channels located on so-called t-tubules, which are invaginations of the plasma membrane into the myocyte. (For simplicity, we assume that the L-type Ca^{2+} channels are not involved in the spontaneous generation of Ca^{2+} sparks so are ignored in the model.) The diadic space separating the SR from the t-tubules is a region of the mytoplasm (intracellular fluid of myocytes), which is approximately cylindrical in shape with height 10 nm and radius 100 nm. Since the diadic space is a small enclosed volume, it supports an elevation in Ca^{2+} concentration relative to the bulk mytoplasm following the release of Ca^{2+} from an RyR. Such a local elevation plays a crucial role in the calcium-induced-calcium release (CICR) that results in a Ca^{2+} spark. The SR in a neighborhood of the RyRs is known as the junctional SR (JSR), which may have a different Ca^{2+} concentration from the bulk or network SR (NSR).

We present the model in non-dimensional form; details of model approximations and estimates of experimentally based model parameters can be found in [377]. First, the diadic space is modeled as a single compartment with Ca^{2+} concentration c satisfying the current conservation equation

$$\tau_D \frac{dc}{dt} = J_{RyR} - J_D. \tag{3.5.8}$$

Here τ_D is a time constant, J_{RyR} is the total Ca^{2+} current through the RyRs, and J_D is the diffusive current from the diadic space to the bulk mytoplasm. The latter

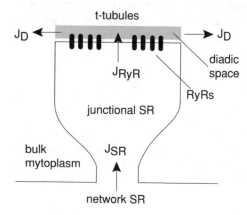

Fig. 3.17: Schematic diagram of a Ca^{2+} release unit in a cardiac myocyte. It is divided into four compartments: the network SR, the junctional SR, the diadic space, and the bulk mytoplasm. See text for details.

is modeled as the Fickian flux $J_D = c - c_m$, where c_m is the bulk mytoplasm Ca^{2+} concentration. The total flux through the RyRs is taken to be proportional to the number n of open RyRs times the Ca^{2+} concentration c_{sr} in the JSR:

$$J_{RyR} = c_{sr}x, \quad x = \frac{n}{N_0}, \tag{3.5.9}$$

with N_0 the total number of RyRs in the cluster. Each RyR has Ca^{2+} binding sites, which can be activating or deactivating. When an RyR is in an activated state it is promoted to a mode where it continuously open and closes according to a Markov process, with a mean open time of 1 ms [885]. The opening of an RyR channel results in an extra Ca^{2+} current flowing into the diadic space, which increases the rate at which Ca^{2+} binds to the other RyRs via CICR, thus creating a positive feedback loop. This feedback loop provides a mechanism for bistability. Note that the RyRs also contain inactivating Ca^{2+} binding sites, but these do not play a role in initiating a Ca^{2+} spark so are not included in the model. For simplicity, the RyRs are modeled using a two-state Markov process involving a single closed state and a single open state (see also [435]):

$$C(\text{closed}) \underset{k_-}{\overset{k_+(c)}{\rightleftharpoons}} O(\text{open}), \tag{3.5.10}$$

with transition rates

$$k_+(c) = \frac{1}{k\tau_o} \frac{c^\alpha}{c^\alpha + 1}, \quad k_- = \frac{1}{\tau_o}. \tag{3.5.11}$$

Here τ_o is the mean open time of a RyR, α is the number of Ca^{2+} ions that are needed to open a RyR, and k determines the proportion of time the RyRs are open. It is assumed that the RyRs are gated independently of each other. They are, however, indirectly coupled via the Ca^{2+} concentration in the diadic space. The time constant τ_D of diffusive flux from the diadic space is several orders of magnitude smaller than the mean open time τ_o of a RyR, that is, $\tau_D \sim 3\mu s$, whereas $\tau_o \sim 1$ ms. Therefore, the Ca^{2+} concentration in the diadic space can be taken to be in quasi-equilibrium, $\tau_D \rightarrow 0$, so that $c = c_m + c_{sr}x$. It follows that the transition rate can be re-expressed as a function of the fraction of open channels and the Ca^{2+} concentration in the SR, $k_+ = k_+(c_m + c_{sr}x)$.

Now consider N_0 independent RyRs within a Ca^{2+} release unit, each described by the above two-state Markov process. Let $N(t)$ be the number of open channels at time t and set $P(n,t) = \text{Prob}[N(t) = n | N(0) = n_0]$. The distribution $P(n,t)$ evolves according to the birth–death master equation (3.3.11) with transitions rates $\omega_\pm(n/N_0) : n \rightarrow n \pm 1$:

$$\omega_+(x) = N_0(1-x)\frac{(c_m + c_{sr}x)^\alpha}{k((c_m + c_{sr}x)^\alpha + 1)}, \quad \omega_-(x) = N_0 x. \tag{3.5.12}$$

For the moment, it is assumed that c_m and c_{sr} are fixed so that ω_+ can be treated as a function of x alone. (Later the dynamics of c_{sr} following initiation of a Ca^{2+} spark will also be taken into account.) The units of time are fixed by setting $\tau_o = 1$. In the deterministic limit $N_0 \to \infty$, we obtain the kinetic equation (3.3.14), which takes the explicit form

$$\frac{dx}{dt} = \Omega_+(x) - \Omega_-(x) = (1-x)\frac{(c_m + c_{sr}x)^\alpha}{k((c_m + c_{sr}x)^\alpha + 1)} - x. \tag{3.5.13}$$

It can be shown that, for physiologically reasonable parameter values, this equation exhibits bistablity [377], that is, there exists a pair of stable fixed points $x_\pm$ separated by an unstable fixed point x_0. The fixed point $x_- \approx 0$ represents a quiescent state, whereas the other fixed point x_+ represents a Ca^{2+} spark in which a significant fraction of RyRs are in the active mode and can be interpreted as a burst phase.

Noise-induced transitions from x_- to x_+ determine the distribution of inter-spark intervals, just as noise-induced transitions from x_+ to x_- determine the distribution of spark lifetimes. Hence, estimating the mean time for the occurrence of a spark event reduces to the problem of calculating the MFPT to reach x_+, starting from a neighborhood of x_-, by crossing x_0. This calculation can be carried out by performing a system-size expansion of the BD master equation given by equations (3.3.11) and (3.5.12), which yields an FP equation with multiplicative noise; see equation (3.3.19). The inverse of Kramer's escape rate formulae (2.4.27a) can then be used to determine the mean time τ_i to initiate a Ca^{2+} spark starting from the quiescent state x_- after substituting for the transition rates using equation (3.5.12). Similarly, the mean duration τ_f of a spark, which corresponds to the mean time to transition back from x_+ to x_-, is given by the inverse of equation (2.4.27b). It turns out that in the case of Ca^{2+} release, the quiescent state x_- is in an $O(1/N_0)$ neighborhood of the boundary $x = 0$, so that the prefactor of the MFPT has to be modified accordingly, see [377, 378] for details. Nevertheless, the leading order exponential is unchanged. Hinch compared the theoretical prediction with Monte Carlo simulations of the full system for various c_{st}, and found that the spark time increases with the number of receptors in the cluster and the mean spark length increases rapidly with c_{sr}.

Irrespective of the particular method used to solve the FPT problem, it was assumed above that the concentration c_{sr} in the junctional SR (JSR) is held fixed. This is a reasonable approximation when considering the initiation of a Ca^{2+} spark. However, following Ca^{2+} release from the RyRs, the Ca^{2+} concentration c_{sr} slowly changes according to

$$\tau_{sr}\frac{dc_{sr}}{dt} = -c_{sr}x + k_{sr}[c_0 - c_{sr}], \tag{3.5.14}$$

where $\tau_{sr} \gg \tau_o \gg \tau_D$. The first term on the right-hand side is the loss of Ca^{2+} through the RyRs, whereas the second terms is the influx J_{SR} of Ca^{2+} from the network SR with fixed Ca^{2+} concentration c_0, see Fig. 3.17. The variation of c_{sr} means that one has to modify the analysis of the time to terminate the Ca^{2+} spark. Following Hinch [377], this can be achieved by combining the theory of stochastic transitions with the classical phase-plane analysis of slow-fast excitable systems, see [431] and Sect. 3.4. That is, (3.5.13) and (3.5.14) form an excitable system with the fraction x of open RyRs acting as the fast variable and c_{sr} acting as the

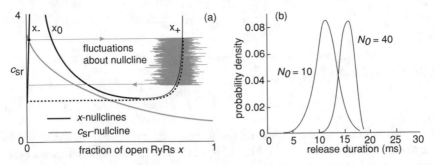

Fig. 3.18: (a) Sketch of nullclines in the deterministic planar Ca^{2+} spark model with x denoting the fraction of open RyRs and c_{SR} the Ca^{2+} concentration in the junctional SR. The c_{sr} nullcline is a monotonically decreasing function $x(c_{sr})$, whereas the x nullcline is cubic-like with three branches $x_{\pm}(c_{sr})$ and $x_0(c_{sr})$. (Note that the branch $x_-(c_{sr}) \approx 0$; we have moved it away from the vertical axis for the sake of illustration.) In the given diagram there is a single, stable fixed point on the left-hand branch. In the stochastic version of the model a Ca^{2+} spark initiates a jump to the right-hand branch $x_+(c_{sr})$. This is followed by a stochastic trajectory in which the slow variable $c_{sr}(t)$ moves down the nullcline until it undergoes a noise-induced transition back to the left-hand branch before the knee at $x = x_c$. In the deterministic case, the return transition occurs at the knee (dashed curve). (b) Distribution of spark durations for clusters containing 10 RyRs and 40 RyRs. Model parameters are $\alpha = 4$, $c_m = 0$, $k = 0.2$, $c_0 = 3.5$, and $k_{sr} = 0.3$. [Adapted from Hinch [377].]

slow variable. In Fig. 3.18(a) we sketch the nullclines of the deterministic system in a parameter regime where there is a single, stable fixed point (x^*, c_{sr}^*). In the full stochastic model, the initiation of a Ca^{2+} spark induces a transition to the right-hand x-nullcline according to $x_-(c_{sr}^*) \to x_+(c_{sr}^*)$. The slow variable then moves down the right-hand nullcline $x_+(c_{sr})$ according to the equation

$$\tau_{sr} \frac{dc_{sr}}{dt} = -c_{sr}x_+(c_{sr}) + k_{sr}[c_0 - c_{sr}]. \tag{3.5.15}$$

That is, although x is a stochastic variable, it fluctuates much faster than the dynamics of c_{sr} so one can substitute a time-averaged value of x in (3.5.14).

Suppose that $c_{sr}(t)$ is the solution of (3.5.15) with $c_{sr}(0) = c_{sr}^*$, that is, the Ca^{2+} spark occurs at $t = 0$. In principle, the spark can terminate at any time $t > 0$ due to fluctuations in the number of open RyRs. Using a separation of time-scales, we can estimate the rate of transition λ back to the left-hand branch at time t by solving the FPT problem using a diffusion approximation with $c_{sr}(t)$ fixed. Since λ depends on c_{sr}, we have a time-dependent transition rate $\lambda(t) = \lambda(c_{sr}(t))$. One can now calculate the distribution of spark durations T. Let $P(\tau) = \text{Prob}(T > \tau)$ and introduce the spark duration probability density $p(\tau) = -dP/d\tau$. The probability that a spark terminates in an infinitesimal time interval $\delta\tau$ is $\lambda(\tau)\delta\tau$, so that

$$P(\tau + \delta\tau) = P(\tau)(1 - \lambda(\tau)\delta\tau).$$

Dividing both sides by $\delta\tau$ and taking the limit $\delta\tau \to 0$ gives

$$\frac{dP}{d\tau} = -\lambda(\tau)P(\tau),$$

which can be integrated to yield $P(\tau) = \exp\left(-\int_0^\tau \lambda(t)dt\right)$. Note that by definition $P(0) = 1$. It follows that

$$p(\tau) = \lambda(\tau)\exp\left(-\int_0^\tau \lambda(t)dt\right). \tag{3.5.16}$$

An illustration of the distribution of spark durations obtained from the stochastic phase-plane analysis is shown in Fig. 3.18(b); the results are consistent with experimental data of Wang et al. [843].

Finally, note that one of the major simplifications of the Hinch model [377] is that the background Ca^{2+} concentrations in the mytoplasm (c_m) and the network SR (c_{nsr}) are held fixed. It thus fails to capture the collective behavior of a large population of Ca^{2+} release units (CaRUs), which are coupled via global changes in these background concentrations (assuming diffusion can be neglected on the relevant time-scales). This has motivated the development of a whole-cell model of calcium-induced calcium release in cardiac myocytes, based on a system of $\mathcal{N}$ globally coupled CaRUs [860, 861].

3.6 Markov chains

The various stochastic ion channel models presented in previous sections, including birth–death processes, are examples of continuous-time Markov chains. The latter play a major role in a wide range of discrete stochastic processes in biology. For example, genetic and biochemical signaling networks are typically modeled in terms of a class of Markov chains known as chemical reaction networks (Chap. 5). Another important class of Markov chain is the Poisson process, which is often used to model the sequence of discrete events characterizing the signal received by a biochemical sensor. In this section, we introduce some of the basic theory of both discrete-time (Box 3A) and continuous-time Markov chains, and then focus on the particular example of Poisson processes. Chemical reaction networks will be discussed at length in Chap. 5. For a more extensive treatment of discrete Markov processes see [343].

3.6.1 Master equation of a continuous-time Markov chain

Let $X = \{X(t); t \geq 0\}$ be a family of random variables taking values in some finite subset Γ of the integers. (In the case of an ion channel, $X(t)$ would represent the conformational state of the ion channel, e.g., whether it is open or closed.) X is called a continuous-time Markov chain if it satisfies the Markov property

$$\mathbb{P}(X(t_l) = n | X(t_1), \dots, X(t_{l-1})) = \mathbb{P}(X(t_l) = n | X(t_{l-1}))$$

for any $n \in \Gamma$ and any sequence $t_1 < t_2 < \dots < t_l$ of times. Introduce the transition rates

$$\mathbb{P}(X(t + \Delta t) = n | X(t) = m) = W_{nm}\Delta t + O(\Delta t^2) \text{ for } n \neq m,$$
$$\mathbb{P}(X(t + \Delta t) = n | X(t) = n) = 1 - \sum_{k \neq n} W_{kn}\Delta t + O(\Delta t^2),$$

with $W_{nm} \geq 0$ for $n \neq m$. Suppose that the chain is in state $X(t) = m$ at time t. One of two distinct events can then occur in the small time interval $(t, t + \Delta t)$:

1. The chain is in the same state at time $t + \Delta t$ with probability $1 - \sum_{k \neq n} W_{kn}\Delta t + O(\Delta t^2)$.
2. The chain jumps to a new state $n \neq m$ with probability $W_{nm}\Delta t + O(\Delta t^2)$.

The probability of two or more transitions in the small time interval is negligible. It follows that the probability distribution $P_n(t) = \mathbb{P}(X(t) = n)$ satisfies

$$P_n(t + \Delta t) = \sum_{m \neq n} W_{nm}\Delta t P_m(t) + \left(1 - \sum_{k \neq n} W_{kn}\Delta t \right) P_n(t).$$

Rearranging this equation, dividing through by Δt and taking the limit $\Delta t \to 0$ yields the master equation

$$\frac{dP_n(t)}{dt} = \sum_{m \neq n} W_{nm}P_m(t) - \left(\sum_{k \neq n} W_{kn} \right) P_n(t), \tag{3.6.1}$$

where $\mathbf{W}$ is known as the transition matrix. This version of the master equation can be interpreted as follows: the first sum on the right-hand side involves all transitions $m \to n$, whereas the second sum involves all transitions $n \to k$ for $k \neq n$. The master equation of a continuous-time Markov chain is the analog of the FP equation for SDEs, and plays a major role in the analysis of discrete processes in cell biology. The master equation can also be rewritten in the more compact form

$$\frac{dP_n(t)}{dt} = \sum_{m \in \Gamma} A_{nm}P_m(t), \tag{3.6.2}$$

where the matrix $\mathbf{A}$ is known as the generator of the Markov chain with

$$A_{nm} = W_{nm} - \delta_{n,m} \sum_k W_{kn}. \tag{3.6.3}$$

In the case of the two-state ion channel model, the generator is

$$\mathbf{A} = \begin{pmatrix} -\alpha & \beta \\ \alpha & -\beta \end{pmatrix}. \tag{3.6.4}$$

Consider a set C of states. The set C is said to be closed if $A_{nm} = 0$ for all $m \in C$ and $n \notin C$. A closed set containing exactly one state is said to be absorbing, that is, if the Markov chain reaches the absorbing state then it stays there. The set C is said to be irreducible if any two states $m, n \in C$ can be connected together via a set of jumps. If the whole state space Γ is irreducible, then the Markov chain is said to be irreducible. An important issue is whether or not a Markov chain converges to a unique stationary distribution, assuming the latter exists. A stationary distribution of a Markov chain is a vector $\mathbf{P}^*$ whose components satisfy

(i) $P_n^* \geq 0$ for all $n \in \Gamma$, $\quad \sum_{n \in \Gamma} P_n^* = 1$,

(ii) $\sum_{m \in \Gamma} A_{nm} P_m^* = 0$ for all $n \in \Gamma$.

The following can be shown for an irreducible continuous-time Markov chain:

1. If there exists a stationary distribution $\mathbf{P}^*$ then it is unique and

$$P_n(t) \rightarrow P_n^* \text{ as } t \rightarrow \infty.$$

2. If there is no stationary distribution then $P_n(t) \rightarrow 0$ as $t \rightarrow \infty$ for all n. (This can only occur if the number of discrete states is infinite.)

One can establish the above results using the theory of discrete-time Markov chains (see Box 3A). For simplicity, assume that the Markov chain is finite. The basic idea is as follows. Formally, we can integrate the master equation (3.6.2) to obtain the solution

$$P_n(t) = \sum_{m \in \Gamma} \left[e^{\mathbf{A}t} \right]_{nm} P_m(0). \tag{3.6.5}$$

Now suppose that we sample the distribution at discrete times $t = lh, l = 0, 1, \ldots$ for some positive constant h. We then have a discrete-time Markov chain

$$\Pi_n(l) = \sum_{m \in \Gamma} [\mathbf{K}^l]_{nm} \Pi_m(0),$$

where $\Pi_n(l) = P_n(lh)$ and $\mathbf{K} = e^{\mathbf{A}h}$. If the continuous-time Markov chain is irreducible then the corresponding discrete-time Markov chain is irreducible and aperiodic (see Box 3A). Hence,

$$\lim_{l \rightarrow \infty} \Pi_n(l) = \Pi_n^{(h)},$$

where $\Pi_n^{(h)}$ is unique (given h). If this argument is repeated for any two rational numbers h_1 and h_2, then $P_n(h_1 l)$ and $P_n(h_2 l)$ will have an infinite number of times in common, which implies $\Pi^{(h_1)} = \Pi^{(h_2)}$. Continuity of $P_n(t)$ as a function of time thus establishes the existence of a unique stationary distribution $\mathbf{P}^*$.

Another way to relate continuous-time and discrete-time Markov chains is to decompose the transition rate matrix $\mathbf{W}$ as $W_{nm} = K_{nm} \lambda_m$, with $\sum_{n \neq m} K_{nm} = 1$. Here λ_m determines the jump times from the state m based on an exponential distribution, whereas K_{nm} is the probability that when it jumps, the new state is n for $n \neq m$. The evolution of the continuous-time Markov chain for $N(t)$ can then be described as follows. Suppose the system starts at time zero in the state $N(0) = n_0$. Let t_1 be the random variable (stopping time) such that

$$\mathbb{P}(t_1 < t) = 1 - \exp\left(-t\lambda_{n_0}\right).$$

Draw a value of t_1 from $\mathbb{P}(t_1 < t)$, choose an internal state $n_1 \in \Gamma$ with probability $K_{n_1 n_0}$, and set $N(t_1^+) = n_1$. Iterating this procedure, one can construct a sequence of increasing jumping times $(t_k)_{k \geq 0}$ (setting $t_0 = 0$) and a corresponding sequence of discrete states $(n_k)_{k \geq 0}$.

Box 3A. Discrete-time Markov chains and the Perron–Frobenius theorem.

Markov chains. Let $\{X_0, X_1, \ldots\}$ be a sequence of discrete random variables that take one of N_0 values in some countable set Γ with $N_0 = |\Gamma|$. (Note that it is possible for $N_0 = \infty$.) Without loss of generality, we take $\Gamma = \{0, 1, \ldots, N_0 - 1\}$. The stochastic process X is said to be a Markov chain if it satisfies the Markov property

$$\mathbb{P}(X_l = n | X_0, \ldots, X_{l-1}) = \mathbb{P}(X_l = n | X_{l-1})$$

for all discrete times $l \geq 1$ and $n \in \Gamma$. The evolution of the chain is described by its transition probabilities

$$K_{nm} = \mathbb{P}(X_{l+1} = n | X_l = m)$$

for all n, m, l. (Probabilists usually write the transition matrix in the adjoint form, that is, $K_{mn} = \mathbb{P}(X_{l+1} = n | X_l = m)$.) For simplicity, we assume that the Markov chain X is homogeneous, that is, K_{nm} is independent of the discrete time l. $\mathbf{K}$ is a stochastic matrix, since it satisfies the following two properties:

(i) $\mathbf{K}$ has nonnegative entries, $K_{nm} \geq 0$

(ii) $\sum_{n \in \Gamma} K_{nm} = 1$.

Introduce the l-step transition matrix $\mathbf{K}(l)$ with components

$$K_{nm}(l) = \mathbb{P}(X_{l_0+l} = n | X_{l_0} = m).$$

Clearly $\mathbf{K}(1) = \mathbf{K}$. We can now derive a discrete version of the Chapman–Kolomogorov equation:

$$
\begin{aligned}
K_{nm}(l + l') = \mathbb{P}(X_{l+l'} = n | X_0 = m) &= \sum_{k \in \Gamma} \mathbb{P}(X_{l+l'} = n, X_l = k | X_0 = m) \\
&= \sum_{k \in \Gamma} \mathbb{P}(X_{l+l'} = n | X_l = k, X_0 = m) \mathbb{P}(X_l = k | X_0 = m) \\
&= \sum_{k \in \Gamma} \mathbb{P}(X_{l+l'} = n | X_l = k) \mathbb{P}(X_l = k | X_0 = m),
\end{aligned}
$$

from the Markov property, and hence

$$K_{nm}(l + l') = \sum_{k \in \Gamma} K_{nk}(l') K_{km}(l). \tag{3.6.6}$$

In matrix form, the Chapman–Kolmogorov equation (3.6.6) implies that $\mathbf{K}(l + l') = \mathbf{K}(l)\mathbf{K}(l')$ and hence $\mathbf{K}(l) = \mathbf{K}^l$. Finally, introducing the l-dependent probability $P_n(l) = \mathbb{P}(X_l = n)$, we have

$$P_n(l) = \mathbb{P}(X_l = n) = \sum_{m \in \Gamma} \mathbb{P}(X_l = n | X_0 = m) P_m(0)$$

$$= \sum_{m \in \Gamma} K_{nm}(l) P_m(0) = \sum_{m \in \Gamma} [\mathbf{K}^l]_{nm} P_m(0).$$

This establishes that the probability vector $\mathbf{P}(l) = (P_0(l), \ldots P_{N_0-1}(l))$ can be determined from the initial probability vector $\mathbf{P}(0)$ using $\mathbf{P}(l) = \mathbf{K}^l \mathbf{P}(0)$.

Recurrent and transient states. A state n of a Markov chain is called recurrent or persistent if

$$\mathbb{P}(X_l = n \text{ for some } l \geq 1 | X_0 = n) = 1.$$

That is, the probability of eventual return to n, having started from n is unity. If the probability of return is strictly less than one, then n is called transient. Let

$$f_{nm}(l) = \mathbb{P}(X_1 \neq n, X_2 \neq n, \ldots, X_{l-1} \neq n, X_l = n | X_0 = m)$$

be the probability that the first visit to state n, starting from m, occurs at the lth step. This is analogous to the first passage time distribution. It follows that

$$f_{nm} = \sum_{l=1}^{\infty} f_{nm}(l)$$

is the probability that the chain ever visits n starting from m. We see that the state n is persistent if and only if $f_{nn} = 1$. In order to derive a criterion for persistence, it is useful to introduce the generating functions

$$G_{nm}(z) = \sum_{l=0}^{\infty} z^l K_{nm}(l), \quad F_{nm}(z) = \sum_{l=1}^{\infty} z^l f_{nm}(l).$$

Here $G_{nm}(0) = \delta_{nm}$ and $F_{nm}(0) = 0$ for all n, m. Clearly $f_{nm} = F_{nm}(1)$. We now derive an important relationship between the two generating functions. Let A be the event that $X_l = n$ given $X_0 = m$, and let $B_{nm}(s)$ be the event that the first arrival at n occurs at the sth step. Then

$$\mathbb{P}(A) = \sum_{s=1}^{l} \mathbb{P}(A | B_{nm}(s)) \mathbb{P}(B_{nm}(s)).$$

Since $\mathbb{P}(B_{nm}(s)) = f_{nm}(s)$ and $\mathbb{P}(A | B_{nm}(s)) = K_{nn}(l-s)$, we have

$$K_{nm}(l) = \sum_s f_{nm}(s) K_{nn}(l-s).$$

Multiplying both sides by z^l and summing over l shows that

$$\sum_{l=1}^{\infty} z^l K_{nm}(l) = \sum_{l=1}^{\infty} \sum_{s=1}^{l} (z^s f_{nm}(s))(z^{l-s} K_{nn}(l-s))$$

$$= \sum_{s=1}^{\infty} \sum_{l=s}^{\infty} (z^s f_{nm}(s))(z^{l-s} K_{nn}(l-s)),$$

that is,

$$G_{nm}(z) = \delta_{nm} + F_{nm}(z) G_{nn}(z). \tag{3.6.7}$$

Abel's theorem. Equation (3.6.7) allows us to derive a condition for persistence of a state n. If $|z| < 1$, then $G_{nn}(z) = (1 - F_{nn}(z))^{-1}$. Hence, if $f_{nn} = F_{nn}(1) = 1$ (persistence), then as $z \to 1^-$, we have $G_{nn}(z) \to \infty$. We now use a well-known theorem from power series known as Abel's Theorem.

Theorem 3.1. *Let* $\{a_k\}$ *be a sequence of real or complex numbers and let*

$$G(z) = \sum_{k=0}^{\infty} a_k z^k,$$

with z real (or in a restricted region of the complex plane). If the series $\sum_k a_k$ is convergent, then

$$\lim_{z \to 1^-} G(z) = \sum_{k=0}^{\infty} a_k.$$

An application of Abels' Theorem establishes that n is persistent if $\sum_l K_{nn}(l) = \infty$. It then follows from equation (3.6.7) for $m \neq n$ that $\sum_l K_{nm}(l) = \infty$ for all m such that $f_{nm} > 0$. An immediate consequence of this result is that n is transient if $\sum_l K_{nn}(l) < \infty$ and if this holds then $\sum_l K_{nm}(l) < \infty$ for all m. Moreover, if n is transient then $K_{nm}(l) \to 0$ as $l \to \infty$.

Mean recurrence time. Given that $X_0 = m$, let

$$T_{nm} = \min\{l \geq 1 : X_l = n\}$$

be the time of the first visit to n, with the convention that $T_{nm} = \infty$ if the visit never occurs. Define the mean recurrence time μ_n of the state n as

$$\mu_n = \mathbb{E}[T_{nn}] = \begin{cases} \sum_l l f_{nn}(l) & \text{if } n \text{ is persistent} \\ \infty & \text{if } n \text{ is transient} \end{cases}.$$

Note that μ_n may be infinite even when n is persistent, so a persistent state n is called null if $\mu_n = \infty$ and non-null or positive if $\mu_n < \infty$. It can be shown that a persistent state is null if and only if $K_{nn}(l) \to 0$ as $l \to \infty$ [343].

Irreducibility. Consider a set C of states. The set C is said to be closed if $K_{nm} = 0$ for all $m \in C$ and $n \notin C$. A closed set containing exactly one state is said to be absorbing, that is, if the Markov chain reaches the absorbing state then it stays there. The set C is said to be irreducible if any two states $m, n \in C$ can be connected together in a finite number of steps. That is, there exist $l, l \geq 0$ for which $K_{nm}(l)K_{mn}(l) > 0$. If the whole state space Γ is irreducible, then the Markov chain is said to be irreducible. Any state space Γ can be partitioned uniquely as $\Gamma = T \cup C_1 \cup C_2 \cup \ldots$, where T is a set of transient states and the C_r are irreducible closed sets of persistent states. In the case of a finite state space, not all states can be transient. For if they were, then $\sum_{n \in \Gamma} K_{nm}(l) = 1$ but $K_{nm}(l) \to 0$ as $l \to \infty$ leading to the contradiction

$$1 = \lim_{l \to \infty} \sum_{n \in \Gamma} K_{nm}(l) = 0.$$

One final definition is useful: a state n is said to be periodic if there exists an integer $d(n) > 1$ such that $K_{nn}(l) = 0$ unless l is a multiple of $d(n)$. If $d(n) = 1$ then the state n is said to be aperiodic.

Stationary distributions and the limit theorem. An important issue is whether or not a Markov chain converges to a unique stationary distribution, assuming the latter exists. A stationary distribution of a Markov chain is a vector $\mathbf{p}^*$ whose components satisfy

(i) $p_n^* \geq 0$ for all n, $\quad \sum_{n \in \Gamma} p_n^* = 1$
(ii) $p_n^* = \sum_{m \in \Gamma} K_{nm} p_m^*$ for all n.

It is stationary since $\mathbf{p}^* = \mathbf{K}^n \mathbf{p}^*$. Note that in the case of an irreducible chain, $p_n^* > 0$ for all n. We then have the following theorem [343]:

Theorem 3.2. *An irreducible, aperiodic Markov chain has a stationary distribution $\mathbf{p}^*$ if and only if all the states are non-null persistent: $\mathbf{p}^*$ is the unique stationary distribution and $p_n^* = \mu_n^{-1}$ for each $n \in \Gamma$, where μ_n is the mean recurrence time of n.*

In the case of a finite state space irreducibility is sufficient to establish existence of a unique stationary state. The next theorem establishes the link between the existence of a stationary distribution and the limiting behavior of $K_{nm}(l)$.

Theorem 3.3. *For an irreducible, aperiodic Markov chain (all states n have period $d(n) = 1$), we have that*

$$K_{nm}(l) \to \frac{1}{\mu_n} \text{ as } l \to \infty \text{ for all } m, n.$$

In the case of a transient or null persistent chain, the theorem is satisfied, since $K_{nm}(l) \to 0$ for all n, m and $\mu_n = \infty$. In the case of a non-null persistent Markov chain, we have

$$\lim_{l \to \infty} K_{nm}(l) \to p_n^* = \frac{1}{\mu_n},$$

where $\mathbf{p}^*$ is the unique stationary distribution. It immediately follows that the limiting transition probability does not depend on its starting point $X_0 = m$ and

$$P_n(l) = \mathbb{P}(X_l = n) = \sum_{m \in \Gamma} K_{nm}(l) P_m(0) \to \frac{1}{\mu_n} \text{ as } l \to \infty.$$

Theorem 3.3 is an example of an ergodic theorem for Markov chains.

Perron–Frobenius theorem. The theory of Markov chains is considerably simplified if Γ is finite. For if Γ is irreducible then it is necessarily non-null persistent, and the probability distribution converges to a unique stationary distribution when it is aperiodic. This can also be established using the well-known Perron–Frobenius theorem for finite square matrices.

Theorem 3.4. *If* **K** *is the transition matrix of a finite, irreducible chain with period d then*

1. *$\lambda_1 = 1$ is an eigenvalue of* **K**
2. *the d complex roots of unity, $\lambda_k = e^{2\pi i k/d}$ for $k = 0, 1, \ldots, d-1$, are eigenvalues of* **K**
3. *the remaining eigenvalues $\lambda_{d+1}, \ldots, \lambda_N$ satisfy $|\lambda_j| < 1$.*

The left eigenvector corresponding to λ_1 is $\boldsymbol{\psi} = (1, 1, \ldots, 1)$, whereas the right-eigenvector is a stationary state $\mathbf{p}^*$. In the aperiodic case, there exists a simple real eigenvalue $\lambda_1 = 1$ and all other eigenvalues satisfy $|\lambda_j| < 1$. One can thus establish that the probability distribution converges to $\mathbf{p}^*$ in the large-n limit.

3.6.2 Poisson processes

There are a number of systems where the signal received by a biochemical sensor involves a sequence of discrete events that can be modeled as a Poisson process. Examples include the arrival of photons at photoreceptors of the retina (Chap. 10), and the arrival of action potentials (spikes) at the synapse of a neuron. Another type of event is the random arrival of customers at some service station, resulting in the formation of a queue. In Sect. 5.4, queuing theory will be used to study transcriptional bursting in gene networks. However, these processes are typically non-Markovian.

A Poisson process is an example of a counting process that is also a Markov renewal process (see Box 3B). Let $\lambda > 0$ be fixed. The counting process $\{N(t), t \in [0, \infty)\}$ is called a Poisson process with rate λ if the following conditions hold:

1. $N(0) = 0$;
2. $N(t)$ has independent, stationary increments;
3. The number n of events in any interval of length τ has the Poisson distribution

$$P_n(\tau) = \frac{(\lambda \tau)^n}{n!} e^{-\lambda \tau}. \tag{3.6.8}$$

Another classical way to obtain a Poisson process is to consider the occurrence of events in infinitesimal time intervals (a renewal process). That is, suppose we divide a given time interval τ into M bins of size $\Delta \tau = \tau/M$ and assume that $\Delta \tau$ is small enough so that the probability of finding two events within any one bin can be neglected. Given a rate λ, take the probability of finding one event in a given bin to be $\lambda \Delta \tau$. The probability $P_n(\tau)$ of finding $N(\tau) = n$ independent events in the interval τ is then given by the binomial distribution

$$P_n(\tau) = \lim_{\Delta \tau \to 0} \frac{M!}{(M-n)!n!} (\lambda \Delta \tau)^n (1 - \lambda \Delta \tau)^{M-n}.$$

This consists of the probability $(\lambda \Delta \tau)^n$ of finding n events in n specific bins multiplied by the probability $(1 - \lambda \Delta \tau)^{M-n}$ of not finding events in the remaining bins. The binomial factor is the number of ways of choosing n out of M bins that contain an event. Using the approximation $M - n \approx M = \tau/\Delta \tau$ for large M, and defining $\varepsilon = -\lambda \Delta \tau$, we have that

$$\lim_{\Delta \tau \to 0} (1 - \lambda \Delta \tau)^{M-n} = \lim_{\varepsilon \to 0} \left((1+\varepsilon)^{1/\varepsilon} \right)^{-\lambda \tau} = e^{-\lambda \tau}.$$

For large M, $M!/(M-n)! \approx M^n = (\tau/\Delta \tau)^n$, so that we recover the Poisson distribution (3.6.8).

A simple method for calculating the moments of the Poisson distribution is to introduce the moment generating function

$$G(s; \tau) = \sum_{n=0}^{\infty} P_n(\tau) e^{sn}. \tag{3.6.9}$$

Differentiating with respect to s shows that

$$\frac{d^k G(s; \tau)}{ds^k} \bigg|_{s=0} = \sum_{n=0}^{\infty} P_n(\tau) n^k \equiv \langle n^k(\tau) \rangle. \tag{3.6.10}$$

The generating function for the Poisson process can be evaluated explicitly as

$$G(s; \tau) = \exp(-\lambda \tau) \exp(\lambda \tau e^s), \tag{3.6.11}$$

from which we deduce that the mean and variance are

$$\langle N(\tau) \rangle = \lambda \tau, \quad \sigma_n^2(\tau) = \lambda \tau. \tag{3.6.12}$$

The ratio of the standard deviation to the mean is called the coefficient of variation $C_V = \sigma_\tau / \langle N(\tau) \rangle$, whereas the Fano factor F is the ratio of the variance to the mean. It follows that for a homogeneous Poisson process $C_V = 1/\sqrt{\lambda \tau}$ and $F = 1$.

Another useful quantity of a Poisson process is the distribution of inter-arrival times $\tau_n = T_n - T_{n-1}$, where T_n is the nth arrival time, which can be defined iteratively according to

$$T_n = \inf\{t \geq 0 | N(t + T_{n-1}) = n\}, \quad T_0 = 0.$$

Suppose that an event last occurred at time T_n. The probability that the next event occurs in the interval $T_n + \tau \leq T_{n+1} \leq T_n + \tau + \Delta\tau$ is equal to the probability that no event occurs for a time τ, which is $p = e^{-\lambda\tau}$ multiplied by the probability $\lambda\Delta\tau$ of an event within the following interval $\Delta\tau$:

$$\mathbb{P}(\tau \leq T_{n+1} - T_n \leq \tau + \Delta\tau) = \lambda\Delta\tau e^{-\lambda\tau}.$$

The inter-arrival time probability density is thus an exponential, $\rho(\tau) = \lambda e^{-\lambda\tau}$. It follows that the mean is

$$\langle\tau\rangle = \int_0^\infty \lambda e^{-\lambda\tau}\tau d\tau = \frac{1}{\lambda},$$

and the variance is

$$\sigma_\tau^2 = \int_0^\infty \lambda e^{-\lambda\tau}\tau^2 d\tau - \langle\tau\rangle^2 = \frac{1}{\lambda^2}.$$

Note that the fastest way to generate a Poisson sequence of events for constant λ is to iterate the arrival times according to $T_{n+1} = T_n - \ln(x_{rand})/\lambda$ with x_{rand} uniformly distributed over $[0, 1]$.

Inhomogeneous Poisson process. It is possible to generalize the above Poisson model to the case of a time-dependent rate $\lambda(t)$. The simplest way to analyze this inhomogeneous Poisson process is to consider the joint probability density of n arrival times, $\rho(T_1, \ldots T_n)$. In particular, $\rho(T^1, \ldots T^n)(\Delta t)^n$ is given by the product of the probabilities $\lambda(T_j)\Delta t$ that the jth event occurs in the time interval $T_j \leq t \leq T_j + \Delta t$ and the probabilities of not firing during the inter-arrival intervals. The latter is given by

$$\mathbb{P}(\text{no events in}(T_j, T_{j+1})]) = \prod_{m=1}^M (1 - \lambda(T_j + m\Delta t)\Delta t),$$

where we have partitioned the interval (T_j, T_{j+1}) into M bins of size Δt. Taking the logarithm,

$$\ln\mathbb{P}(\text{no events in}(T_j, T_{j+1})]) = \sum_{m=1}^M \ln(1 - \lambda(T_j + m\Delta t)\Delta t) \approx -\sum_{m=1}^M \lambda(T_j + m\Delta t)\Delta t.$$

Taking the limit $\Delta t \to 0$ and exponentiating again shows that

$$\mathbb{P}(\text{no events in}(T_j, T_{j+1})]) = \exp\left(-\int_{T_j}^{T_{j+1}} \lambda(t)dt\right).$$

Hence, the probability density that during the time interval $[0, T]$ there are exactly n events at times $T_1, \ldots, T_n$ is

$$\rho(T_1, \ldots T_n; T) = \left(\prod_{i=1}^{n} \lambda(T_i) \right) \exp \left(- \int_0^T \lambda(t) dt \right). \tag{3.6.13}$$

We now note that the probability of n events occurring in $[0, T]$ in particular time bins can also be written as

$$\rho(T_1, \ldots T_n; T)(\Delta t)^n = n! \frac{\prod_{i=1}^{n} \lambda(T_i) \Delta t}{\int_0^T \lambda(t) dt} P_n[T],$$

which establishes that the distribution for an inhomogeneous Poisson process is given by

$$P_n(T) = \frac{\Lambda(T)^n}{n!} e^{-\Lambda(T)}, \quad \Lambda(T) = \int_0^T \lambda(t) dt. \tag{3.6.14}$$

Some further examples of more complicated Poisson processes are considered in Ex. 3.11 and 3.12.

Box 3B. Counting and renewal processes

A random process $\{N(t), t \in [0, \infty)\}$ is said to be a counting process if $N(t)$ is the number of events that have occurred in the interval $[0, t]$ such that

1. $N(0) = 0$;
2. $N(t) \in \{0, 1, 2, \ldots\}$ for all $t \geq 0$;
3. for $0 \leq s < t$, $N(t) - N(s)$ is the number of events in the interval $(s, t]$.

Let T_I be the occurrence or arrival times of the ith event. This leads to two further definitions useful in characterizing counting processes. First, let $\{X(t), t \in [0, \infty)\}$ be a continuous-time stochastic process. We say that $X(t)$ has independent increments if, for all $0 \leq t_1 < t_2 \cdots < t_n$, the random variables $X(t_j) - X(t_{j-1})$, $j = 2, \ldots, n$, are independent. In the case of a counting process this means that the numbers of arrivals in non-overlapping time intervals are independent. Following on from this, $X(t)$ is said to have stationary increments if, for all $t_2 > t_1 \geq 0$ and all $r > 0$, the random variables $X(t_2) - X(t_1)$ and $X(t_2 + r) - X(t_1 + r)$ have the same distributions. In particular, a counting process has stationary increments if, for all $t_2 > t_1 \geq 0$, $N(t_2) - N(t_1)$ has the same distribution as $N(t_2 - t_1)$. In other words, the distribution of the number of events in an interval depends only on the length of the interval. In general, counting processes are non-Markovian, with the notable exception of a Poisson process.

An equivalent way to define a Poisson process is as a counting process whose inter-arrival times τ_n are independent identically distributed random variables whose probability density is an exponential. This naturally leads to the following generalization of a Poisson process known as a renewal process. A renewal process $N(t)$ is a counting process for which

$$N(t) = \max\{n : T_n \leq t\}, T_0 = 0, \quad T_n = \tau_1 + \ldots + \tau_n$$

for $n \geq 1$, where $\{\tau_j\}$ is a sequence of independent identically distributed nonnegative random variables. It immediately follows that

$$N(t) \geq n \text{ if and only if } T_n \leq t.$$

It turns out that the Poisson process is the only renewal process that is also a Markov chain. Finally, note that if τ_j is strictly positive, then $\mathbb{P}[N(t) < \infty] = 1$ for all t.

3.7 Exercises

Problem 3.1. Monod-Wyman-Changeux (MWC) model. Generalize the MWC model from two binding sites to n binding sites.

(a) Let R_j, T_j, $j = 0, 1, \ldots, n$, denote the global states (open or closed) with j ligands bound to the allosteric site. Also define K_T and K_R to be the equilibrium constants for binding of an acetylcholine molecule to an individual site of an closed and open receptor, respectively. Finally, take Y_j to be the equilibrium constant associated with the opening and closing of a receptor with j occupied sites. Defining the fraction of open receptors according to

$$p_{\text{open}} = \frac{\sum_{j=0}^{n}[R_j]}{\sum_{j=0}^{n}[R_j] + \sum_{j=0}^{n}[T_j]},$$

use the law of mass action to derive the sigmoidal function

$$p_{\text{open}} = \frac{Y_0(1 + K_R[L])^n}{Y_0(1 + K_R[L])^n + (1 + K_T[L])^n}.$$

Hint: care needs to be taken in working out the combinatorial factors multiplying K_T and K_R in the reaction diagram.

(b) Rederive the result for the MWC model in part (a) using the Boltzmann–Gibbs distribution (Sect. 1.3). Use the following observations. A microstate is specified by the number of occupied binding sites m, $0 \leq m \leq n$, and whether the channel is open or closed. The free energy of a given microstate is

$$E_m^r = m\left[\varepsilon_b^r - \mu_0 - k_B T \ln(c/c_0)\right],$$

where ε_b^r is the binding energy when the channel is open $(r = R)$ or closed $(r = T)$, and the chemical potential $\mu = \mu_0 - k_B T \ln(c/c_0)$ takes into account that ligands are being taken out of solution (see Sect. 1.3). Evaluate the partition function

$$Z = \sum_{r=R,T} \frac{n!}{m!(n-m)!} e^{-E_m^r/k_B T},$$

and hence determine p_{open}. Explain the presence of the combinatorial factor in the definition of Z.

Problem 3.2. Ion channel with multiple subunits. Consider an ion channel with $k, k > 2$, identical subunits, each of which can be open or closed, and a current only passes if all k subunits are open. Let S_j denote the state in which j subunits are open and let α, β denote the rates of opening and closing of a single subunit. Write down the corresponding reaction scheme
(a) Derive the kinetic equations for x_j, which is the fraction of channels in state j such that $\sum_{j=0}^{k} x_j = 1$.

(b) By direct substitution, show that

$$x_j = \frac{k!}{(k-j)!j!} n^j (1-n)^{k-j}$$

is an invariant manifold of the dynamics, provided that

$$\frac{dn}{dt} = \alpha(1-n) - \beta n.$$

Problem 3.3. Chain of ion channel states. The time course of the opening and closing of some ion channels seems to follow a power law rather than an exponential law at large times. One way to understand such power law behavior is to consider an ion channel with N closed states such that the transition to an open state can only take place from state 1 at one end of a chain

$$0 \overset{\alpha}{\leftarrow} 1 \underset{\beta_1}{\overset{\gamma_1}{\rightleftharpoons}} 2 \ldots \underset{\beta_{N-1}}{\overset{\gamma_{N-1}}{\rightleftharpoons}} N.$$

(a) Write down the corresponding set of kinetic equations. Hence, show that when $\gamma_n = \beta_n = 1$ for all n and $\alpha = 1$, we obtain the discrete diffusion equation along a chain with a reflecting boundary at $n = N$ and an absorbing boundary at $n = 0$:

$$\frac{dp_1}{dt} = p_2 - 2p_1,$$

$$\frac{dp_n}{dt} = p_{n-1} + p_{n+1} - 2p_n, \quad 1 < n < N,$$

$$\frac{dp_N}{dt} = p_{N-1} - p_N,$$

where $p_n(t)$ is the probability that the channel is in state n at time t.

(b) Given the initial condition $p_n(0) = \delta_{n,1}$, show that in the large N limit, the exact solution is

$$p_n(t) = e^{-2t}[I_{n-1}(2t) - I_{n+1}(2t)],$$

where $I_n(z)$ is the modified Bessel function of integer order:

$$I_n(z) = \int_{-\pi}^{\pi} e^{ink} e^{z\cos(k)} \frac{dk}{2\pi}.$$

Hint: Use discrete Fourier transforms to solve the discrete diffusion equation on the infinite lattice (Box 2A), and then use the method of images to write down the solution for a semi-infinite lattice.

(c) When $2t \gg n$, the modified Bessel function has the asymptotic expansion

$$I_n(2t) = \frac{e^{2t}}{\sqrt{4\pi t}} \left[1 - \frac{4n^2 - 1}{16t} + \cdots\right].$$

Use this to show that, for large t,

$$p_n(t) \approx \frac{n}{2\pi^{1/2} t^{3/2}}.$$

(d) Define $F(t)$ to be the total probability of finding the system in a closed state:

$$F(t) = \sum_{n=1}^{\infty} p_n(t).$$

Show that $dF/dt = -p_1$ and, hence, $F(t) \approx (\pi t)^{-1/2}$ for large N, t.

Problem 3.4. Ligand-gated ion channel. Consider the following second-order kinetic scheme for a ligand-gated ion channel:

$$C \underset{r_2}{\overset{r_1(T)}{\rightleftharpoons}} C_1$$

$$\nwarrow_{r_4} \qquad \swarrow_{r_3}$$

$$O$$

Here C and C_1 are the closed forms of the receptor, O is the open (conducting) form, and the r_i are voltage independent transition rates. The transition rate r_1 for $C \rightarrow C_1$

depends on the concentration of ligand $\mathcal{T}$. Suppose that we make the following approximations: (i) The transmitter concentration $\mathcal{T}$ occurs as a pulse $\delta(t - t_0)$ for a release event occurring at time $t = t_0$, that is, $r_1(\mathcal{T}) = r_1\delta(t - t_0)$; (ii) The fraction of channels in state C is taken to be fixed at unity—this is reasonable if the number of channels in state C is much larger than the number of channels in states C_1 or O.

(a) Write down the kinetic equations for the fraction of receptors in the states C_1 and O, which are denoted by z and s respectively.

(b) Solve the resulting pair of inhomogeneous linear equations assuming that $z(0) = s(0) = 0$. In particular show that the fraction of open channels is given by

$$s(t) = r_1 r_3 \left(\frac{1}{\tau_2} - \frac{1}{\tau_1} \right)^{-1} (e^{-(t-t_0)/\tau_1} - e^{-(t-t_0)/\tau_2}), \quad t > t_0,$$

with $\tau_1 = 1/(r_2 + r_3)$, $\tau_2 = 1/r_4$.

(c) Show that in the limit $\tau_2 \to \tau_1 = \tau_s$ this reduces to the so-called alpha function

$$s(t) = r_1 r_3 (t - t_0)e^{-(t-t_0)/\tau_s}, \quad t > t_0.$$

Such a response function is often used to model the response of synaptic receptors following the release of neurotransmitter.

Problem 3.5. Master equation for an ensemble of ion channels. Consider the master equation for the two-state ion channel model:

$$\frac{d}{dt}P_n(t) = \alpha(N_0 - n + 1)P_{n-1}(t) + \beta(n+1)P_{n+1}(t)$$
$$- [\alpha(N_0 - n) + \beta n]P_n(t).$$

(a) By multiplying both sides by n and summing over n, derive the following kinetic equation for the mean $\bar{n} = \sum_{n=0}^{N_0} nP(n,t)$:

$$\frac{d\bar{n}}{dt} = \alpha(N_0 - \bar{n}) - \beta\bar{n}.$$

(b) Derive a corresponding equation for the variance $\sigma^2 = \langle n^2 \rangle - \langle n \rangle^2$. That is, multiply both sides of the master equation by n^2 and sum over n to determine an equation for the second moment, and then use part (a). Show that the variance decays exponentially at a rate $2(\alpha + \beta)$ to the steady-state value

$$\sigma^2 = \frac{\alpha\beta}{(\alpha + \beta)^2},$$

and hence deduce that fluctuations become negligible in the large N_0 limit.

(c) Compare the results obtained from the master equation with the analysis based on the linear noise approximation.

(d) Construct the master equation for an ensemble of N_0 identical, independent channels each of which has two subunits. That is, determine an equation for the evolution of the probability distribution $P_{n_0,n_2}(t)$ that there are n_j ion channels with j open subunits such that $N_0 = n_0 + n_1 + n_2$.

Problem 3.6. Birth–death process with an absorbing state.

(a) Consider the birth–death process (3.3.11) with $\omega_+(n) = \lambda_n$ and $\omega_-(n) = \mu_n$. Suppose that there is an absorbing state at $n = b+1 \leq N_0$ and a reflecting boundary at $n = 0$. Following the analysis of Sect. 3.3.2) shown that the MFPT starting from the state $m \in [0,b]$ is

$$\tau_m = \sum_{l=m}^{b} \phi_l \sum_{j=0}^{l} \frac{1}{\lambda_j \phi_{j+1}}, \quad \phi_m = \prod_{j=m}^{b} \frac{\lambda_j}{\mu_j}, \quad \phi_{b+1} = 1.$$

(b) Now suppose there is an absorbing state at $n = 0$ and take $N_0 \to \infty$. It is not *a priori* obvious that absorption is now certain, since the system could either drift to infinity or wander forever between the nonzero states. Let $\pi_m, m > 0$, be the probability of absorption at zero given the initial state m. Since the transitions $m \to m+1$ and $m \to m-1$ occur with probability $\lambda_m/(\lambda_m + \mu_m)$ and $\mu_m/(\lambda_m + \mu_m)$, respectively, we have the iterative equation

$$\pi_m = \frac{\lambda_m}{\lambda_m + \mu_m} \pi_{m+1} + \frac{\mu_m}{\lambda_m + \mu_m} \pi_{m-1},$$

with $\pi_0 = 1$ and $0 \leq \pi_m \leq 1$ for $m \geq 1$. Show that this iterative equation has a solution of the form

$$\pi_{m+1} - \pi_1 = (\pi_1 - 1) \sum_{n=1}^{m} \left(\prod_{k=1}^{n} \frac{\lambda_k}{\mu_k} \right).$$

Hence, show that absorption is certain starting from any state if

$$\sum_{n=1}^{\infty} \left(\prod_{k=1}^{n} \frac{\lambda_k}{\mu_k} \right) = \infty.$$

Using the MFPT formula (3.3.34) with $m = 1$ and $N_0 \to \infty$, derive a condition for the MFPT $\tau_1 < \infty$.

(c) Assume that absorption is not certain,

$$\sum_{n=1}^{\infty} \left(\prod_{k=1}^{n} \frac{\lambda_k}{\mu_k} \right) < \infty.$$

Using the fact that $\pi_m \to 0$ as $m \to \infty$ obtain the result

$$\pi_m = \frac{\sum_{n=m}^{\infty}\left(\prod_{k=1}^{n}\frac{\lambda_k}{\mu_k}\right)}{1+\sum_{n=1}^{\infty}\left(\prod_{k=1}^{n}\frac{\lambda_k}{\mu_k}\right)}.$$

Problem 3.7. Non-homogeneous birth–death process. Consider the master equation for a non-homogeneous birth–death process of the form

$$\frac{d}{dt}P_n(t) = \omega_+(n-1,t)P_{n-1}(t) + \omega_-(n+1,t)P_{n+1}(t)$$
$$- [\omega_+(n,t) + \omega_-(n,t)]P_n(t),$$

with transition rates $\omega_+(n,t) = \lambda(t)n$, $\omega_-(t) = \mu(t)n$. Let $P_n(0) = \delta_{n,N}$.

(a) Show that the generating function $\Gamma(z,t) = \sum_{m=0}^{\infty} z^m P_m(t)$ satisfies the PDE

$$\frac{\partial\Gamma}{\partial t} = [\lambda(t)(z^2-z) + \mu(t)(1-z)]\frac{\partial\Gamma}{\partial z}, \quad \Gamma(z,0) = z^N.$$

(b) The characteristic curves parameterized by s,τ are given by

$$\frac{dt}{d\tau} = 1, \quad \frac{dz}{d\tau} = (1-z)[\lambda(t)z - \mu(t)], \quad \frac{d\Gamma}{d\tau} = 0,$$

with initial conditions

$$t(s,0) = 0, \quad z(s,0) = s, \quad \Gamma(s,0) = s^N.$$

Solve the equation for z by setting $t = \tau$ and making the change of variables $1 - z = 1/y$. Hence show that

$$\Gamma(z,t) = \left(1 + \left[\frac{e^{\rho(t)}}{z-1} - \int_0^t \lambda(t')e^{\rho(t')}dt'\right]^{-1}\right)^N,$$

where

$$\rho(t) = \int_0^t [\mu(\tau) - \lambda(\tau)]d\tau.$$

(c) Setting $z = 0$ and using the expression for $d\rho(t)/dt$ in part (b), show that

$$P_0(t) = \left[\frac{\int_0^t \mu(t')e^{\rho(t')}dt'}{1 + \int_0^t \mu(t')e^{\rho(t')}dt'}\right]^N.$$

Hence derive a condition for extinction to occur, that is,

$$\lim_{t\to\infty} P(0,t) = 1.$$

Compare this with the condition for extinction in the deterministic kinetic model

$$\frac{dn}{dt} = [\lambda(t) - \mu(t)]n, \quad n(0) > 0.$$

Problem 3.8. Linear growth model. Consider a birth–death process with

$$\omega_+(n) = \lambda n + a, \quad \omega_-(n) = \mu n,$$

and $\lambda, \mu, a > 0$. The term λn represent the natural rates of growth within a population, whereas a could denote a rate of immigration.

(a) Write down the explicit birth–death master equation for $P_0(t)$ and $P_n(t), n > 0$, using the given transition rates.

(b) Multiplying the nth equation by n and summing over n, derive an ODE for the expectation $M(t) = \sum_{n=1}^{\infty} nP_n(t)$. Hence, show that

$$M(t) = at + n_0, \quad \lambda = \mu,$$

and

$$M(t) = \frac{a}{\lambda - \mu}\left[e^{(\lambda-\mu)t} - 1\right] + n_0 e^{(\lambda-\mu)t}, \quad \lambda \neq \mu,$$

where $P_n(0) = \delta_{n,n_0}$. What happens in the limit $t \to \infty$?

(c) Now suppose $a = 0$ so that $n = 0$ is an absorbing state. Let π_m denote the probability of absorption at zero given that the system is initially at state m. Using Ex. 3.6(b), show that $\pi_m = \mu^m/\lambda^m$ when $\mu < \lambda$ and $\pi_m = 1$ when $\mu \geq \lambda$ for all $m \geq 1$.

Problem 3.9. Electrodiffusion. The flow of ions through channels in the cell membrane is driven by the combination of concentration gradients and electric fields. If interactions between the ions are ignored then each ion can be treated as an independent Brownian particle moving under the influence of the electric force $-q\nabla\phi$, where ϕ is the electrical potential and q is the charge on the ion. Multiplying the corresponding FP equation by the number N of ions and using an Einstein relation, we obtain the Nernst–Planck equation

$$\frac{\partial c(\mathbf{x},t)}{\partial t} = -\nabla \cdot \mathbf{J}, \quad \mathbf{J}(\mathbf{x},t) = -D\left(\nabla c + \frac{qc}{k_B T}\nabla\phi\right),$$

where c denotes ion concentration. Treating an ion channel as a quasi-one-dimensional domain, this reduces to the 1D equation

$$\frac{\partial c(x,t)}{\partial t} = -\frac{\partial J}{\partial x}, \quad J(x,t) = -D\left(\frac{\partial c}{\partial x} + \frac{qc}{k_B T}\frac{\partial\phi}{\partial x}\right).$$

(a) Suppose that the cell membrane extends from $x = 0$ (inside) to $x = L$ (outside) and denote the extracellular and intracellular ion concentrations by c_e and c_i, respectively. Solve the 1D steady-state Nernst–Planck equation to show that there is zero flux through the membrane if the potential difference $V = \phi_i - \phi_e$ across the membrane is given by the Nernst potential

$$V_R = \frac{k_B T}{q} \ln\left(\frac{c_e}{c_i}\right).$$

(b) Now suppose that there is a constant nonzero flux J of ions through the channel, and assume for simplicity that the electric field is uniform, that is, $\partial\phi/\partial x = -V/L$. Solving the steady-state Nernst–Planck equation with boundary conditions $c(0) = c_i, c(L) = c_e$, derive the Goldman–Hodgkin–Katz equation for the current density:

$$I \equiv qJ = \frac{D}{L}\frac{q^2 V}{k_B T}\frac{c_i - c_e \exp(-qV/k_B T)}{1 - \exp(-qV/k_B T)}.$$

Check that the Nernst potential is recovered when $J = 0$.

(c) Consider two ion species with opposite charges $q_1 = -q_2 = q$. Applying part (b) to the current for each ion species, derive an expression for the membrane voltage V at which the total ionic current is zero.

Problem 3.10. De Young Keizer model. Carry out the reduction of the De Young–Keizer model introduced in Sect. 3.5

(a) First write down the kinetic equations for the four states without calcium bound to the inactivating site, $(S_{000}, S_{010}, S_{100}, S_{110})$, using the reaction diagram Fig. 3.16a.

(b) Perform the quasi-steady-state approximation by setting all time derivatives to zero and dropping all slow transitions involving binding/unbinding of the calcium inactivating binding site, that is, set $k_{\pm 2} = k_{\pm 4} = 0$. Show that

$$x_{100} = \frac{K_5 h}{c + K_5} - x_{000}, \quad x_{010} = \frac{K_1 h}{p + K_1} - x_{000},$$

where $K_i = k_{-i}/k_i$ and

$$x_{000} + x_{010} + x_{100} + x_{110} = h.$$

Hence, determine x_{000} and x_{110}.

(c) Show that the corresponding quasi-steady-state solutions for the states with calcium bound to the inactivating site, $(S_{001}, S_{011}, S_{101}, S_{111})$, are obtained from (b) by taking $K_1 \to K_3$ and $h \to 1 - h$. Note that

$$x_{001} + x_{011} + x_{101} + x_{111} = 1 - h.$$

(d) Finally, add together the four kinetic equations of part (a) and substitute for the x_{ijk} using their quasi-steady-state solutions. Hence derive the equation

$$\frac{dh}{dt} = -\frac{(k_{-4} K_2 K_1 + k_{-2} p K_4)c}{K_4 K_2 (p + K_1)} h + \frac{k_{-2} p + k_{-4} K_3}{p + K_3}(1 - h).$$

Problem 3.11. Mixed Poisson process. Consider a Poisson process whose time-independent rate λ is itself a random variable with probability density $p(\lambda)$. For a given choice of λ, the generating function of the Poisson process is

$$G_\lambda(s,t) \equiv \sum_{n=0}^{\infty} P_n(t)e^{ns} = e^{\lambda t(e^s-1)}.$$

Let

$$G(s,t) = \mathbb{E}[G_\lambda(s,t)] = \int_0^{\infty} p(\lambda)e^{\lambda t(e^s-1)}d\lambda.$$

By performing a Taylor expansion in s, show that the mean and variance of $N(t)$ after taking expectation with respect to the Poisson process and λ are

$$\mathbb{E}[N(t)] = \langle\lambda\rangle t, \quad \text{Var}[N(t)] = \sigma_\lambda^2 t^2 + \mathbb{E}[N(t)].$$

Here $\langle\lambda\rangle = \int_0^\infty p(\lambda)d\lambda$ etc. Hence show that the variance of a mixed Poisson process is greater than a standard Poisson process (assuming $\sigma_\lambda^2 > 0$), which is an example of over-dispersion.

Problem 3.12. Compound Poisson process. Let $Y_1, Y_2, \ldots$, be independent, identically distributed random variables with probability generating function $g(z) = \mathbb{E}[e^{sY}]$. Assume that they are independent of some Poisson process $N(t)$ with constant rate λ. Then

$$\widehat{N}(t) = \sum_{i=1}^{N(t)} Y_i$$

defines the counting function of a compound Poisson process. Show the following:

(a) $\mathbb{E}[e^{s\widehat{N}(t)}] = e^{-\lambda t(1-g(s))}.$

(b) $\mathbb{E}[\widehat{N}(t)] = \lambda\langle Y\rangle t.$

(c) $\text{Var}[\widehat{N}(t)] = \text{Var}[N(t)]\langle Y\rangle^2 + \mathbb{E}[N(t)]\sigma_Y^2.$

Problem 3.13. Computer simulations: Two-state ion channels. In this problem, we investigate the diffusion approximation of the master equation (3.3.10) for an ensemble of two-state ion channels:

$$\frac{d}{dt}P(n,t) = \alpha(N-n+1)P(n-1,t) + \beta(n+1)P(n+1,t)$$
$$- [\alpha(N-n) + \beta n]P(n,t).$$

Take $\alpha = 1, \beta = 2$ and $N = 100$.

(a) Numerically solve the master equation using Euler's direct method for $t \in [0,1]$ and $\Delta t = 0.01$. Plot the histogram of $P_n(T)$ for $T = 1$ and compare with the steady-state distribution (3.3.17).

(b) Use the Gillespie algorithm (Chap. 5) to generate sample paths for the number $n(t)$ of open ion channels for $t \in [0,10]$. The two reactions are $n \to n+1$ at a rate $\alpha(N-n)$ and $n \to n-1$ at a rate βn. By averaging over sample paths, compare the histogram of $n(T)$ with the distribution $P_n(T)$ for $T = 1$.

(c) Use Euler's direct method (see Sect. 2.6) to simulate the Langevin equation

$$dX(t) = [\alpha(1-X) - \beta X]dt + \frac{1}{\sqrt{N}}\sqrt{\alpha(1-X) + \beta X}dW(t),$$

obtained by carrying out a Kramers–Moyal expansion of the master equation. Here $X(t)$ is the fraction of open ion channels at time t. Construct a histogram of $X(T)$ for $T = 1$ and compare with the results of part (b). Repeat for $N = 10$ and $N = 1000$ and comment on the differences.

Chapter 4
Molecular motors

The cytoskeleton within the cytoplasm plays important roles in maintaining the structural integrity of a cell, intracellular transport, cell motility, and cell division. In eukaryotes, the cytoskeleton consists of three types of protein filaments—microtubules, intermediate filaments, and actin filaments, see Fig. 4.1. Actin filaments are the thinnest structures (around 6 nm) whose basic building block is the globular protein G-actin. These can assemble into a long filamentous chain known as F-actin, which has the superficial appearance of two interlocked strands. Actin filaments are relatively flexible and strong. Actin dynamics plays a major role in cell motility, where one end (the + or barbed end) elongates due to polymerization while the other end (the - or pointed end) contracts due to a combination of depolymerization and myosin motors. F-actin also serves as a tensile platform for myosin motors involved in the pulling action of muscle contraction. Actin filaments are themselves assembled into two general types of structures: bundles called filopodia, which consist of parallel arrays of filaments, and cross-linked networks called lamellipodia. Microtubules are hollow cylinders around 23 nm in diameter, which typically consist of 13 protofilaments each of which is a polymer made up of heterodimers of alpha and beta tubulin. Microtubules project radially from organizing centers known as centrosomes, and play a key role in cell division via the mitotic spindle. Finally, intermediate filaments average 10 nm in diameter are more strongly bound than F-actin, and act to maintain the structural integrity of a cell. Actin and tubulin filaments are assembled via the polymerization of subunits, which change their chemical state when incorporated into a filament. For example, actin monomers contain an adenosine triphosphate (ATP) molecule that rapidly hydrolyzes to adenosine diphosphate (ADP) following polymerization. Similarly, the β unit of the tubulin heterodimer contains a guanosine triphosphate (GTP) molecule that hydrolyzes to guanosine diphosphate (GDP) after polymerization. These chemical transformations can lead to more complex phenomena than observed in simple polymers, such as treadmilling and dynamical instabilities.

The degree of flexibility of a polymer can be characterized in terms of the so-called persistence length ξ_p, which specifies the length scale over which correla-

© Springer Nature Switzerland AG 2021

P. C. Bressloff, *Stochastic Processes in Cell Biology*, Interdisciplinary Applied Mathematics 41, https://doi.org/10.1007/978-3-030-72515-0_4

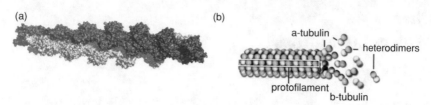

Fig. 4.1: Cell cytoskeletal filaments. (a) Computer reconstruction of the atomic structure of an actin filament with 13 subunits by Thomas Splettstoesser using open-source software PyMOL. (b) Schematic illustration of helical structure of a microtubule [Public domain figures downloaded from Wikipedia Commons.]

tions in the orientation of the polymer persist. (Persistence length is defined more precisely in Sect. 12.1.) if $\xi_p \gg L$, where L is the total length of the polymer, then the polymer is said to be rigid, whereas if $\xi_p \ll L$ then it acts like a random coil. F-actin and microtubules are rigid or semirigid polymer filaments. On the other hand, biopolymers such as DNA and proteins are much more flexible relative to their lengths, which means one has to take into account the wide range of different configurations that can occur through bending and folding of the polymers, as well as the associated energetics.

Another major function of actin and microtubular polymer filaments is that they act as effective 1D tracks for the active movement of molecular motor proteins. Diffusion inside the cytosol or within the plasma membrane of a cell is a means by which dissolved macromolecules can be passively transported without any input of energy. However, there are two main limitations of passive diffusion as a mechanism for intracellular transport:

(i) It can take far too long to travel the long distances necessary to reach targets within a cell, which is particularly acute in the case of the axons and dendrites of neurons.

(ii) Diffusive transport tends to be unbiased, making it difficult to target resources to specific areas within a cell.

Active intracellular transport can overcome these difficulties so that movement is both faster and direction-specific, but does so at a price. Active transport cannot occur under thermodynamic equilibrium, which means that energy must be consumed by this process, typically via the hydrolysis of ATP. The main types of active intracellular transport involve the molecular motors kinesin and dynein carrying resources along microtubular filament tracks, and myosin V motors transporting cargo along actin filaments. As we have already highlighted, microtubules and actin filaments are polarized polymers with biophysically distinct $(+)$ and $(-)$ ends. It turns out that this polarity determines the preferred direction in which an individual molecular motor moves. For example, kinesin moves toward the $(+)$ end whereas dynein moves toward the $(-)$ end of a microtubule. Each motor protein undergoes a sequence of conformational changes after reacting with one or more ATP molecules, causing it to step forward along a filament in its preferred direction, see Fig. 4.2.

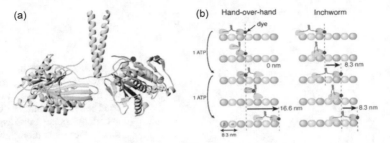

Fig. 4.2: (A) The kinesin molecule consists of two motor units (analogous to feet) that are linked together via a common stalk that attaches to cargo to be transported. (B) Previous studies had shown that the kinesin molecule moves along the microtubule in steps of 8 nm. Kinetic analysis of the dwell time between steps shows that there is an alternation of displacement from one step to the other, supporting a hand-over-hand model rather than an inchworm model. In the latter case, both feet would move only in 8nm steps as the kinesin molecule's center of mass moves. However, if the kinesin molecule moves in a hand-over-hand motion, then the "rear" foot should take a 16nm step forward during one cycle, and then 0 nm during the next cycle [Adapted from A. Yildiz *et al.* [877].]

Thus, ATP provides the energy necessary for the molecular motor to do work in the form of pulling its cargo along a filament in a biased direction. When modeling active transport, one usually neglects the dynamics of microtubules and actin filaments and simply treats them as static 1D tracks with periodic structure. On the other hand, the regulation of polymerization and depolymerization by molecular motors plays an important role in the formation and maintenance of certain cytoskeletal structures such as the mitotic spindle (see Chap. 14).

The movement of molecular motors such as kinesin occurs over several length- and time scales [421, 436, 461, 525]. In the case of a single motor, there are at least three regimes:

(a) The mechanico-chemical energy transduction process that generates a single step of the motor. In the case of dimeric or double-headed kinesin, a single step is of length 8nm and the total conformational cycle takes around 10 ms.

(b) The effective biased random walk along a filament during a single run, in which the motor takes multiple steps before dissociating from the filament. For example, kinesin takes around 100 steps in a single run, covering a distance of around 1μm. Walking distances can be substantially increased if several molecular motors pull the cargo.

(c) The alternating periods of directed motion along the filament and diffusive or stationary motion when the motor is unbound from the filament. In the unbound state, a motor diffuses in the surrounding aqueous solution with a diffusion coefficient of the order 1 μm^2/s. However, molecular crowding tends to confine the motor so that it stays close to its detachment point. In the case of multiple molecular motors transporting cargo, the resulting complex can exhibit bidirectional motion [345, 473, 741, 854]; see Sect. 4.5.2.

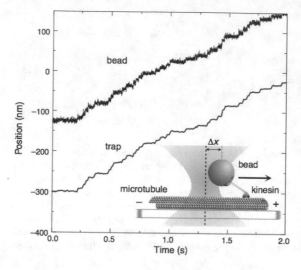

Fig. 4.3: Operation of an optical tweezer or trap. Experimental traces show kinesin-driven bead movement and the corresponding optical trap displacement; the separation between bead and trap was maintained at around 175 nm. The movement can be characterized in terms of a stochastic sequence of steps of average length 8 nm. Inset: schematic representation of the experimental setup [Adapted from M. Schnitzer *et al.* [736].]

Remark 4.1. Optical tweezers or traps. In recent years there have been spectacular advances in single-molecule techniques for measuring the force-dependent dynamics of molecular motors, DNA and other macromolecules vital for cell function. In particular, the use of an optical tweezer (or trap) allows piconewton forces to be applied to molecules over nanometer length scales. The basic idea of an optical tweezer is to use radiation pressure from individual photons emitted by a laser to generate forces on a micron-sized glass bead. This then imparts a force on the macromolecule of interest via a velcro-like link between the molecule and the glass bead. By applying known forces to the bead, it is possible to study the mechano-chemistry of the attached macromolecule as a function of the applied force. An example of such an experiment is illustrated in Fig. 4.3. Here a kinesin molecular motor attached to a silica bead moves along a clamped microtubule filament [736]. The bead is placed inside an optical trap such that the force on the bead is proportional to its displacement Δx from the center of the trap. The applied force is maintained at a constant level by using feedback to move the trap in sync with the bead so that Δx is kept constant. Such an experiment can be used to determine the probability distribution of motor step lengths, for example. Another common application of the optical tweezer involves attaching the glass bead to one end of a DNA strand. This can then be used to determine the force-extension curve of a DNA polymer, measure the force produced by RNA polymerase during the transcription of DNA, or measure the force necessary to pack viral DNA into the capsid (protein shell) of a bacteriophage (virus that infects and replicates within bacteria).

Advances in experimental techniques have generated considerable information about the structural properties of molecular motors and their dynamics. For example, optical traps have been used to measure how changes in ATP concentration affect the force–displacement properties of both kinesin [736, 829] and dynein [297, 446]. A sketch of typical results obtained for kinesin is shown in Fig. 4.4. Such data can be incorporated into models at levels (b) and (c). On the other hand, information about the energetics of the various conformational states and the rates of transitions between them are not yet sufficient to develop detailed biophysical models of motors. Hence, it is not possible to generate realistic velocity–force curves, for example, without considerable data fitting. Thus much of the work on molecular motors at the smallest scale (a) is of a more qualitative nature, in which one tries to understand the basic principles that allow nanoscale machines to do useful work in the presence of thermal noise—so-called Brownian ratchet models.

In addition to intracellular transport, molecular motors perform many other functions within a cell:

(i) Muscle contraction and cell locomotion due to the collective action of multiple myosin II motor heads (cross bridges) interacting with actin filaments (see Chap. 15 of Keener and Sneyd [431]).

(ii) The reversible action of rotary motor ATP synthase, which either produces ATP using ion gradients or acts as an ion pump fueled by ATP hydrolysis [248].

(iii) The swimming and tumbling of bacteria such as *E. coli* driven by flagella rotary motors (Chap. 10).

(iv) Transcription of RNA from DNA via RNA polymerase [841] (Sect. 5.5).

(v) The action of viral DNA packaging motors that inject viral genomic DNA into the protein shell (capsid) of a bacteriophage (a virus that infects and replicates within bacteria) as part of its replication cycle [879].

In this chapter, we introduce some basic stochastic dynamical models of molecular motors, including both polymerizing filaments and processive motors. The role of processive motors in intracellular transport will be considered in Chap. 7, whereas the mechanical and dynamical properties of cytoskeletal elements will be explored in Chap. 14. (For a comprehensive introduction to the mechanics of motor proteins and the cytoskeleton, see the book by Howard [397].) In Sect. 4.1, we consider a simple 1D stochastic model of actin polymerization and depolymerization, which neglects molecular details such as the structure of heterodimers and helical protofilaments. The model takes the form of a birth–death process that keeps track of the addition or removal of monomers from one or both ends of the polymer. We then consider the well-known Dogterom–Leibler model of microtubule (MT) catastrophe [223]; see Sect. 4.2. This takes the form of a two-state velocity-jump process

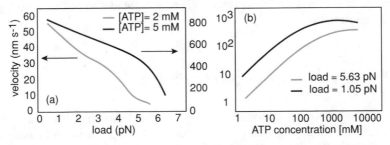

Fig. 4.4: Sketch of results from optical trap measurements of kinesin. (a) Variation of velocity with load for high and low ATP concentrations. (b) Variation of velocity with ATP concentration for low and high loads [Redrawn from Visscher et al. [829].]

for the length of the MT. We also include the effects of nucleation and constrained growth, analyze first passage time problems, and consider a more detailed model of the effects of ATP hydrolysis and cap formation. In Sect. 4.3, we introduce the theory of Brownian ratchets by considering the Fokker–Planck equation for a Brownian particle moving in a periodic ratchet (asymmetric) potential. We show that the mean velocity of the Brownian particle is zero, which implies that such a potential cannot provide a mechanism for a molecular motor to do useful work against an applied load. One mechanism for breaking the periodicity is to rectify the motion, as exemplified by polymerization ratchets and translocation ratchets; energy is provided by the binding of a molecule to the polymer or protein. Polymerization ratchets play a major role in cell motility [208, 392, 587], where the force exerted by an actin filament extrudes the cell membrane in a particular direction (Chap. 14). On the other hand, the translocation ratchet is used to model the transport of a polymer through a membrane pore. On one side of the membrane, proteins known as chaperones, which are too large to pass through the pore, bind the polymer and thus rectify its motion through the pore (see also Sect. 12.1).

In Sect. 4.4, we describe a qualitative model of processive molecular motors such as kinesin and dynein that is based on a two-state Brownian ratchet (flashing ratchet). The basic idea is that the motor has to negotiate a periodic potential energy landscape based on its interactions with the MT filaments, and the form of the landscape depends on the conformational state of the motor. (The idea of representing a molecular motor in terms of several conformational states that depend on interactions with a filament was first introduced by Huxley in his theoretical study of muscles [405].) We show that useful work can be generated provided that the transition rates between the different conformational states do not satisfy detailed balance, which is achieved via the hydrolysis of ATP. We end by briefly describing an alternative, kinetic approach to modeling the stepping of molecular motors, based on a discrete Markov process. The state transition diagram includes both jumps between conformational states and jumps between neighboring sites on the filament. We further develop the theory of molecular motors in Sect. 4.5, where we consider two examples of the collective motion of an ensemble of molecular motors: (i) the tug-of-war model of bidirectional vesicular transport by opposing groups of processive motors; (ii) a model of interacting motors attached to a rigid cytoskeletal backbone. The second model supports collective oscillations consistent with those seen experimentally.

4.1 Simple model of actin polymerization

Consider, for simplicity, monomers binding or unbinding at the + end of a single-stranded filament, see Fig. 4.5. (For an extension to multi-stranded filaments, see [786].) Suppose that the minimum length of the polymer is either a single monomer or a critical nucleus of M monomers, which for the moment is considered stable. Let $N(t)$, $N(t) \geq 0$, denote the number of monomers added to this critical nucleus, and

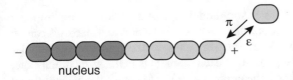

Fig. 4.5: Simple model of F-actin undergoing polymerization at one end.

nucleus

take the rate of monomer binding and unbinding to be π and ε, respectively. The probability $P_n(t)$ that the filament contains $N(t) = n$ additional monomers at time t satisfies the master equation

$$\frac{dP_n}{dt} = \varepsilon P_{n+1}(t) + \pi P_{n-1}(t) - [\varepsilon + \pi]P_n(t) \qquad (4.1.1)$$

for $n > 0$, supplemented by the reflecting boundary condition

$$\frac{dP_0}{dt} = \varepsilon P_1(t) - \pi P_0(t), \qquad (4.1.2)$$

and the normalization condition $\sum_{n=0}^{\infty} P_n(t) = 1$. We are assuming that there is an unlimited supply of monomers. First, note that if we multiply both sides of (4.1.1) by n and sum over n then we obtain a mean-field equation for the mean change in length $\langle N(t) \rangle = \sum_{n=0}^{\infty} n P_n(t)$, namely

$$\frac{d\langle N(t) \rangle}{dt} = \pi - \varepsilon + \varepsilon P_0. \qquad (4.1.3)$$

If the master equation has a stationary solution then

$$\varepsilon P_{n+1} + \pi P_{n-1} - [\varepsilon + \pi]P_n = 0, \quad n > 0, \quad \varepsilon P_1 = \pi P_0.$$

Since the binding/unbinding rates are n-independent, the solution is of the form $P_n = C\lambda^n$. Substituting this into the stationary equation for $n > 0$ gives

$$\varepsilon \lambda^{n+1} + \pi \lambda^{n-1} - [\varepsilon + \pi]\lambda^n = 0, \quad n > 0,$$

which reduces to the quadratic equation

$$\lambda^2 - (1+r)\lambda + r = 0, \quad r = \frac{\pi}{\varepsilon}.$$

This has the solutions $\lambda = 1$ or $\lambda = r$. The normalization condition $C\sum_{n=0}^{\infty} \lambda^n = 1$ requires that $\lambda < 1$ so that there exists a unique stationary solution provided that $\pi/\varepsilon < 1$. Solving for C using the normalization condition then gives

$$P_n = \left(1 - \frac{\pi}{\varepsilon}\right)\left(\frac{\pi}{\varepsilon}\right)^n, \quad n \geq 0. \qquad (4.1.4)$$

It also follows that $d\langle n \rangle/dt = 0$.

So far we have considered a stochastic model of a single filament. Each model simulation generates a sample path of the stochastic behavior, and statistics can be extracted by running many trials. An alternative picture is to consider a large population of N_0 identical filaments. Suppose that each filament is in the same initial state. For sufficiently large N_0, we expect the number $X_n(t)$ of filaments having additional length n to be $X_n(t) = N_0 P_n(t)$. Since the transition rates are n-independent, we can simply multiply the master equation by N_0 to obtain corresponding kinetic equations for the X_n with $\sum_{n=0}^{\infty} X_n(t) = N_0$. As we have discussed in the context of ion channel population models, Sect. 3.3, the kinetic equations for a population of filaments are deterministic. If one wanted to take into account fluctuations due to intrinsic noise, then one would have to consider the master equation for the probability distribution $\mathbb{P}(M_0, M_1, \ldots, M_{N_0}, t)$ where M_n is the number of filaments in state n.

The deterministic population model has been extended to take into account the disappearance and production of critical nuclei of size M [237, 400]. Taking X_n to denote the fraction of filaments of length n, the kinetic equations are

$$\frac{dX_n}{dt} = \varepsilon X_{n+1}(t) + \pi X_{n-1}(t) - [\varepsilon + \pi]X_n(t), \quad n > M, \tag{4.1.5}$$

and

$$\frac{dX_M}{dt} = \varepsilon X_{M+1}(t) - (\pi + \varepsilon)X_M(t) + \sigma, \tag{4.1.6}$$

where nuclei can disappear (convert back to M monomers) at a rate ε and are produced at a rate σ. Assuming a fixed background monomer concentration a, the binding and production rates are taken to be

$$\pi = \pi_0 a, \quad \sigma = \sigma_0 a^M, \tag{4.1.7}$$

with σ_0, π_0 independent of a. One no longer has a conservation condition for the total number of filaments. However, if $\pi/\varepsilon < 1$ then one can still construct a steady-state solution of the form $X_n = C(\pi/\varepsilon)^n$. Substituting into the steady-state solution for $n = M$, we have

$$C\varepsilon\left(\frac{\pi}{\varepsilon}\right)^{M+1} - C(\varepsilon + \pi)\left(\frac{\pi}{\varepsilon}\right)^M + \sigma = 0,$$

which implies that $C = (\sigma/\varepsilon)(\varepsilon/\pi)^M$ and

$$X_n = \frac{\sigma}{\varepsilon}\left(\frac{\pi}{\varepsilon}\right)^{n-M} = \frac{\sigma_0}{\varepsilon}\left(\frac{\pi_0}{\varepsilon}\right)^{n-M} a^n. \tag{4.1.8}$$

It immediately follows that the mean filament length L in the population is

$$L = M + \frac{\displaystyle\sum_{n=M}^{\infty}(n-M)X_n}{\displaystyle\sum_{n=M}^{\infty} X_n} = M + \frac{\displaystyle\sum_{n=0}^{\infty} n(\pi/\varepsilon)^n}{\displaystyle\sum_{n=0}^{\infty}(\pi/\varepsilon)^n} = M + \frac{\pi/\varepsilon}{1 - \pi/\varepsilon}.$$

Hence, the mean length diverges as $\pi \to \varepsilon$ from below.

It is also possible to analyze the stability of the steady state in the case that the polymers have a maximum size $n = J$. After dividing the kinetic equation by ε and rescaling time, we obtain the matrix equation

$$\frac{d\mathbf{X}}{dt} = \mathbf{MX} + \mathbf{s},$$

where $\mathbf{X} = (X_M, X_{M+1}, \ldots, X_J)^T$, $\mathbf{s} = (\sigma/\varepsilon, 0, \ldots, 0)^T$ and $\mathbf{M}$ is the tridiagonal matrix

$$\mathbf{M} = \begin{pmatrix} -(1+r) & 1 & 0 & 0 \cdots 0 \\ r & -(1+r) & 1 & 0 \cdots 0 \\ 0 & r & -(1+r) & 1 \cdots 0 \\ \vdots & \vdots & \vdots & \vdots \vdots \vdots \\ 0 & \cdots & \cdots & 0 \ r \ 1 \end{pmatrix}.$$

This linear system has a general solution of the form

$$\mathbf{X}(t) = \mathbf{X}_0 + \sum_{j=1}^{J-M+1} \mathbf{v}_j e^{\lambda_j t},$$

where $\mathbf{X}_0$ is the steady-state solution, and $0, \lambda_j, j = 1, J+1-M$ are the eigenvalues of $\mathbf{M}$. We can now use same basic results from linear algebra. First, since $\mathbf{M}$ is tridiagonal with $M_{m,n+1}M_{n+1,n} = r > 0$, it follows that the eigenvalues are real and simple. In particular, none of the eigenvalues λ_j vanish. The Gershgorin disc theorem can then be used to establish that none of the eigenvalues λ_j are positive definite, and are thus negative definite. The theorem states that the eigenvalues of the tridiagonal matrix $\mathbf{M}$ are contained in the union of discs D_n in the complex λ-plane with

$$D_n = \{|\lambda - M_{nn}| \le \sum_{k \ne n} |M_{nk}|\}.$$

The first disc D_1 has center at $\lambda = -(1+r)$ and radius 1, whereas all the discs $D_2, \ldots, D_{J-1}$ have centers at $\lambda = -(1+r)$ and radii $1+r$. All of these lie in the left-half complex plane. Finally, D_J has a center at $\lambda = -1$ and radius r, which also lies the left-half complex plane provided that $r \le 1$.

Now suppose that actin monomers can bind or unbind at both ends with rates $k_{on}^{\pm}$ and $k_{off}^{\pm}$, as shown in Fig. 4.6. The binding rate is multiplied by a fixed background monomer concentration a. (The spatial effects of a nonuniform monomer concentration are considered by Edelstein-Keshet and Ermentrout [238]; see also Ex. 4.1.)

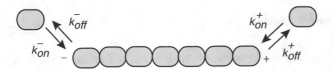

Fig. 4.6: Model of F-actin undergoing polymerization at both ends

The difference between the two ends is due to the fact the ATP-actin quickly hydrolyzes to ADP-actin so that the tip consists of ATP-actin and the tail consists of ADP-actin. Rather than writing down the master equation for the system, let us consider the equations for the mean number of monomers $n_\pm$ added at each end. Assuming that the filament is sufficiently long, we have

$$\frac{dn_+}{dt} = k_{on}^+ a - k_{off}^+, \quad \frac{dn_-}{dt} = k_{on}^- a - k_{off}^-. \tag{4.1.9}$$

It is clear that the $\pm$ end grows provided that $a > a_c^\pm$, where $a_c^\pm = k_{off}^\pm / k_{on}^\pm$. If $a_c^+ \approx a_c^-$, then both ends shrink or grow simultaneously. On the other hand, if $a_c^+ < a < a_c^-$ then the plus end grows at the same time the minus end shrinks. Finally, adding the pair of equations in (4.1.9) shows that

$$\frac{dn}{dt} = k_{on} a - k_{off},$$

with $n = n_+ + n_-$, $k_{off} = k_{off}^+ + k_{off}^-$, and $k_{on} = k_{on}^+ + k_{on}^-$. Hence, if the monomer concentration $a = a_0$, where

$$a_0 = \frac{k_{off}^+ + k_{off}^-}{k_{on}^+ + k_{on}^-} \equiv \frac{k_{off}}{k_{on}},$$

then the total filament length remains constant even though monomers are constantly moving along its length—treadmilling.

4.2 Microtubule catastrophes

An interesting aspect of MTs is that they undergo periods of persistent MT growth interrupted by occasional switching to rapid shrinkage know as "microtubule catastrophe" [582]. MTs grow by the attachment of guanosine triphosphate (GTP)-tubulin complexes at one end. In order to maintain growth, the end of the MT must consist of a "cap" of consecutive GTP-tubulin monomers. However, each polymerized complex can hydrolyze into guanosine diphosphate (GDP)-tubulin such that if all the monomers in the cap convert to GDP, then the MT is destabilized, and there is rapid shrinkage due to detachment of the GDP-tubulin monomers. The competition between the attachment of GTP-tubulin and hydrolysis from GTP to GTD is thought to be the basic mechanism for alternating periods of growth and shrinkage.

4.2.1 Dogterom–Leibler model

One approach to modeling catastrophe is to assume that the MT exists either in a growing phase ($n = +$) or a shrinking phase ($n = -$), and to treat stochastic transitions between the two states as a continuous-time Markov process [96, 223] (see Sect. 3.6). The microscopic details of cap formation and hydrolysis are not modeled explicitly. It is assumed that one end of an MT is fixed and the position of the other end is taken to be a stochastic variable $X(t)$, which can also be identified as the variable length of the MT. Let $p_n(x,t)$ be the probability density that at time t the end of the MT is at $X(t) = x$ and in the discrete state $N(t) = n \in \{+, -\}$. That is, for a given set of initial conditions, $p_n(x,t) = p(x,n,t|y,m,0)$, where

$$p(x,n,t|y,m,0)dx = \mathbb{P}[x \leq X(t) \leq x + dx, N(t) = n | X(0) = y, N(0) = m],$$

and $p_n(x,0) = \delta(x - y)\delta_{n,m}$. The Dogterom–Leibler model of MT catastrophe takes the form [223]

$$\frac{\partial p_+}{\partial t} = -v_+ \frac{\partial p_+}{\partial x} - k_c p_+ + k_r p_-, \tag{4.2.1a}$$

$$\frac{\partial p_-}{\partial t} = v_- \frac{\partial p_-}{\partial x} - k_r p_- + k_c p_+ \tag{4.2.1b}$$

for $x \in (0, \infty)$. Here v_+ and v_- are the average speeds of growth and shrinking, k_c is the catastrophe rate, and k_r is the rescue rate. Equation (4.2.1) is supplemented by a reflecting boundary condition at $x = 0$:

$$v_+ p_+(0,t) = v_- p_-(0,t). \tag{4.2.2}$$

That is, if the MT hits the boundary at $x = 0$, then it immediately undergoes a forced rescue. (One can modify the model to incorporate finite nucleation times; see Sect. 4.2.2.) Note that equation (4.2.1) is equivalent to the differential Chapman–Kolmogorov (CK) equation for a simple velocity-jump process. This equation and various extensions are also used to model promoter noise in gene networks (Sect. 5.3), bidirectional motor transport (Sect. 7.2), and bacterial chemotaxis (Chap. 10).

We can determine a condition for the existence of a steady-state solution by adding equations (4.2.1a) and (4.2.1b) and setting $\partial_t p_\pm = 0$. This shows that

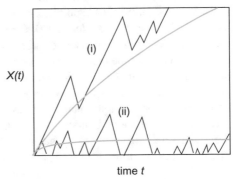

Fig. 4.7: Schematic diagram illustrating two phases of MT growth depending on the sign of the mean growth rate V. (i) For $V > 0$ the MT undergoes unbounded growth. (ii) For $V < 0$ the frequency of catastrophes increases so that there is bounded growth. The mean length as a function of time is shown by gray curves.

$$v_+ \frac{dp_+}{dx} - v_- \frac{\partial p_-}{\partial x} = 0,$$

and thus $v_+ p_+ - v_- p_- = \text{constant}$. Assuming a semi-infinite domain for x, normalizability of $p_\pm$ implies that the constant must be zero. It follows that

$$\frac{dp_+(x)}{dx} = \left[\frac{k_r}{v_-} - \frac{k_c}{v_+}\right] p_+(x) := \frac{V}{D} p_+(x), \qquad (4.2.3)$$

where

$$V = \frac{k_r v_+ - k_c v_-}{k_c + k_r}, \qquad D = \frac{v_+ v_-}{k_c + k_r}. \qquad (4.2.4)$$

Here V is the mean speed of microtubular growth (based on the steady-state solution of the two-state Markov process describing the switching between growth and shrinking phases), and D is an effective diffusivity. It immediately follows that there exists a steady-state solution, $p_+(x) = p_+(0)e^{Vx/D}$, $0 < x < \infty$, if and only if $V < 0$. In the regime $V > 0$, catastrophe events are relatively rare and the MT continuously grows with mean speed V, whereas, for $V < 0$ the catastrophe events occur much more frequently so that there is a balance between growth and shrinkage that results in a steady-state distribution of MT lengths, see Fig. 4.7. In the latter case, the total probability density is

$$p(x) := p_+(x) + p_-(x) = \mathcal{N}\left(1 + \frac{v_-}{v_+}\right)e^{xV/D}, \qquad (4.2.5)$$

with

$$\mathcal{N}^{-1} = \left(1 + \frac{v_-}{v_+}\right)\int_0^\infty e^{xV/D}dx = \frac{D}{|V|}\left(1 + \frac{v_-}{v_+}\right). \qquad (4.2.6)$$

This follows from conservation of probability, $\int_0^\infty p(x)dx = 1$. Finally, note that typical parameter values for MT growth and shrinkage are as follows [888]: $k_c = 3 \text{ s}^{-1}$, $k_r = 0.03 - 0.2 \text{ s}^{-1}$, $v_+ = 0.05 \,\mu\text{m/s}$, and $v_- = 0.5 \,\mu\text{m/s}$.

Constrained microtubular growth and sticky boundaries. A number of studies have extended the original Dogterom–Leibler model to the case of bounded domains [330, 594, 887, 888]. For example, consider a single MT in a 1D domain of fixed length L. Assume that for lengths $0 < x < L$, it undergoes transitions between a growing and a shrinking phase according to equation (4.2.1). A variety of boundary conditions can then be imposed whenever the MT hits the wall at $x = L$ or shrinks back to zero length at $x = 0$. For example, suppose that whenever the MT hits the boundary at $x = L$, its growth velocity v_+ drops to zero and it sticks to the wall until it transitions to a shrinkage state at a rate r_b. Mathematically speaking, one can incorporate a sticky boundary at $x = L$ by introducing a boundary state B, and introducing the probability $P_b(t)$ that the MT is in this state at time t, see Fig. 4.8(a). The boundary condition for (4.2.1) at $x = L$ takes the form

$$v_- p_-(L,t) = r_b P_b(t), \qquad (4.2.7)$$

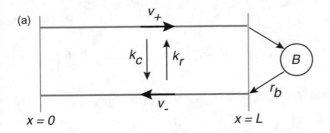

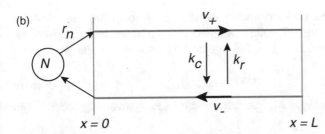

Fig. 4.8: Schematic representation of catastrophe and rescue of a confined MT. (a) A growing MT can stick to the wall at $x = L$ and subsequently switch to the shrinkage state at a rate r_b. (b) One can incorporate the effects of nucleation by introducing a sticky boundary at $x = 0$, with r_n the nucleation rate for switching to a growth state.

with $P_b(t)$ evolving according to the equation

$$\frac{dP_b}{dt} = v_+ p_+(L,t) - r_b P_b(t).$$ (4.2.8)

The steady-state analysis for $p(x)$ proceeds along identical lines to the standard Dogterom–Leibler model, yielding the solution (4.2.5). However, the normalization factor $\mathcal{N}$ is now determined from the modified conservation equation

$$\int_0^L p(x)dx + P_b = 1,$$ (4.2.9)

with

$$P_b = \frac{v_+}{r_b} p_+(L) = \mathcal{N}\frac{v_+}{r_b} e^{LV/D}.$$

We thus find that

$$\mathcal{N}^{-1} = \frac{D}{V}\left(1 + \frac{v_+}{v_-}\right)\left(e^{LV/D} - 1\right) + \frac{v_+}{r_b} e^{LV/D}.$$ (4.2.10)

For the sake of illustration, suppose $V > 0$ so that $p(x)$ is an exponentially increasing function of x. In the steady state, the average length of an MT under confined growth is [888]

$$\langle X \rangle = \int_0^L x p(x)\,dx + P_b L \qquad (4.2.11)$$

$$= \mathcal{N}\left\{ \left(1 + \frac{v_+}{v_-}\right) \left(\frac{D}{V}\right)^2 \left[1 + e^{LV/D}(LD/V - 1)\right] + \frac{Lv_+}{r_b} e^{LV/D} \right\}.$$

In the limit $r_b \to \infty$ (instant wall-induced catastrophes), we see that

$$\frac{\langle X \rangle}{L} \approx \frac{1}{1 - e^{-VL/D}} - \frac{D}{VL}.$$

It is also possible to use a sticky boundary to take into account the finite time for nucleation of a new growing MT, following the collapse of a previous MT. Now the reflecting boundary condition at $x = 0$ is replaced by a sticky boundary condition [330], see Fig. 4.8(b),

$$v_+ p_+(0,t) = r_n P_n(t), \qquad (4.2.12)$$

with the probability of being in the nucleating state N, $P_n(t)$, evolving according to the equation

$$\frac{dP_n}{dt} = v_- p_-(0,t) - r_n P_n(t). \qquad (4.2.13)$$

(Note that sticky boundary conditions will also arise in the analysis of bacterial chemotaxis [22] (Chap. 10), cell mitosis (Chap. 14), and cytoneme-based morphogenesis (Chap. 14).)

Another type of nontrivial confinement arises when the fixed right-hand boundary is replaced by an elastically coupled barrier [888], see Fig. 4.9. If the barrier is displaced from its equilibrium position L by the growing MT with length $x > L$, then it generates an elastic force $F(x) = \kappa(x - L)$ that resists additional growth. It is assumed that the force is zero when $x < L$. The spring constant is taken to lie in the range $10^{-7} - 10^{-5}$ N/m and $L = 10\,\mu$m. Under the applied force F, the growth rate

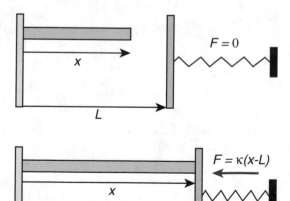

Fig. 4.9: Schematic representation of a single MT growing against an elastic wall. If $x < L$ then the MT tip experiences no force, whereas if $x > L$ and in the growing phase it experiences a force $F = \kappa(x - L)$.

decreases due to the fact an additional free energy Fd is needed in order to incorporate a single tubulin dimer of size $d = 0.6$ nm against the force F. If the on and off rates for tubulin binding in the absence of a force are k_{on} and k_{off}, respectively, then the force-dependent growth velocity takes the form

$$v_+(F) = \left(k_{on} e^{-Fd/k_B T} - k_{off} \right) d, \tag{4.2.14}$$

with $k_{on} \sim 50$ s^{-1} and $k_{off} \sim 6$ s^{-1}. It turns out that the catastrophe rate is itself a decreasing function of v_+ [888], which implies that $k_c = k_c(F)$ with k_c an increasing function of F. If F were a constantly applied force at the tip of the MT and the right-hand boundary were removed, then the steady-state density would be of the form

$$p(x) := \mathcal{N} \left(1 + \frac{v_-}{v_+(F)} \right) e^{x/\lambda(F)}, \quad \lambda(F) = \frac{v_+(F)v_-}{v_+(F)k_r - v_- k_c(F)}. \tag{4.2.15}$$

In this case, the transition between bounded and unbounded growth would be determined by the force-dependent parameter $\lambda^{-1}(F)$, which is a monotonically decreasing function of F. Suppose that $\lambda^{-1}(F)$ switches sign from positive to negative values as F increases beyond a critical force F_c for which $\lambda^{-1}(F_c) = 0$. Then F_c is the critical constant force for the transition from unbounded to bounded growth. On the other hand, when there is an elastic force, one typically finds that $p(x)$ is a unimodal function of x with a maximum at some x_m [888].

4.2.2 First passage times

A natural issue for constrained MT growth is how to determine the mean first passage time (MFPT) to hit the boundary at $x = 0$ or $x = L$ given an initial length of the MT. This requires extending the theory of first passage time (FPT) problems for stochastic differential equations (SDEs), see Sect. 2.4, to velocity-jump process such as the Dogterom–Leibler model. One of the additional challenges of constrained MT growth is how to handle sticky boundaries. Suppose that we wish to determine the MFPT to hit the wall at $x = L$, say, given a sticky (nucleating) boundary at $x = 0$. The presence of the latter means that it is now necessary to keep track of each time the MT hits $x = 0$ before eventually exiting at $x = L$, since the MT spends an exponentially distributed time $\tilde{\tau}$ in the nucleated state before re-entering the growth phase. One approach is to analyze the forward CK equation using Laplace transforms and summing over all possible paths that eventually escape at $x = L$, after indexing them according to the number of times they visit the sticky boundary [330]. An alternative method is to introduce an appropriate set of splitting probabilities and conditional MFPTs and to calculate them explicitly by solving a corresponding set of integral equations [594]. Although the latter direct method neatly avoids the need to sum over paths in the case of sticky boundaries, the analysis is still quite involved, particularly when extended to more complicated FPT problems. Here we consider a more efficient method for solving FPT problems in the presence of sticky boundaries, which uses some classical concepts from probability theory, namely, conditional expectations and the strong Markov property [132]. In order to proceed, however, it is first necessary to solve the FPT problem without a sticky boundary (Figure 4.10).

MFPT with an absorbing boundary at $x = 0, L$. In order to determine the splitting probability and conditional MFPT to hit the wall at $x = L$ before ever reaching $x = 0$, and the corresponding quantities to reach $x = 0$ first, we impose absorbing boundaries at $x = 0, L$,

$$p_+(0,t) = p_-(L,t) = 0.$$

The basic setup is similar to the FPT problem considered in Sect. 2.4 and Fig. 2.6 with an absorbing boundary at both ends. In particular, we have to determine the splitting probabilities and conditional MFPTs. Define the conditional FPT to reach $x = L$ by

$$\mathcal{T}_m(y) = \inf\{t \geq 0; X(t) = L | X(0) = y, N(0) = m\}, \qquad (4.2.16)$$

with $\mathcal{T}_m(y) = \infty$ if the particle (MT tip) hits $x = 0$ first. The probability flux through the end $x = L$ is

$$J_m(y,t) = v_+ p_+(L,t|y,m,0).$$

It follows that for $0 < y < L$

$$\Pi_m(y,t) := \mathbb{P}[\mathcal{T}_m(y) > t] = \int_t^\infty J_m(y,t')dt', \qquad (4.2.17)$$

where $\Pi_m(y,t)$ denotes the probability that the particle exits at $x = L$ after time t, having started in state (y,m). In particular, $\pi_m(y) = \Pi_m(y,0)$ is the total splitting probability for such an event. In determining the corresponding MFPT, it is necessary that the expectation is only taken with respect to trajectories that never reach $x = 0$. Denote the set of events in which the MT first reaches $x = L$ without ever returning to the origin by Ω_0. Then the conditional MFPT is

absorbing BC   absorbing BC

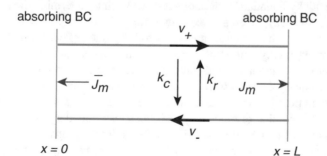

Fig. 4.10: Simplified FPT problem with absorbing boundaries at $x = 0, L$. The splitting probabilities and conditional MFPTs are obtained from the respective probability fluxes J_m and $\bar{J}_m$.

$$T_m(y) = \mathbb{E}[\mathcal{T}_m(y) \,|\, 1_{\Omega_0}] = \frac{\mathbb{E}[\mathcal{T}_m(y) 1_{\Omega_0}]}{\mathbb{P}[1_{\Omega_0}]}$$

$$= -\frac{1}{\pi_m(y)} \int_0^\infty t \frac{\partial \mathbb{P}[\mathcal{T}_m(y) > t]}{\partial t} dt = \pi_m^{-1}(y) \int_0^\infty \Pi_m(y,t) dt, \qquad (4.2.18)$$

where 1_{Ω_0} is an indicator function that ensures that expectation is only taken with respect to events that lie in Ω_0. Similarly, let

$$\overline{\mathcal{T}}_m(y) = \inf\{t \geq 0; X(t) = 0 \,|\, X(0) = y, N(0) = m\} \qquad (4.2.19)$$

be the FPT to exit the interval at $x = 0$, and set

$$\overline{\Pi}_m(y,t) := \mathbb{P}[\overline{\mathcal{T}}_m(y) > t] = \int_t^\infty \overline{J}_m(y,t') dt', \qquad (4.2.20)$$

where $\overline{J}_m(y,t)$ is the probability flux through the end $x = 0$,

$$\overline{J}_m(y,t) = v_- p_-(0,t|y,m,0).$$

The corresponding splitting probability and conditional MFPT are

$$\overline{\pi}_m(y) = \overline{\Pi}_m(y,0) \quad \overline{\pi}_m(y)\overline{T}_m(y) = \int_0^\infty \overline{\Pi}_m(y,t) dt, \qquad (4.2.21)$$

One way to determine the splitting probabilities and conditional MFPTs is to use the backward CK equation. Setting $q_m(y,t) = p_n(x,t|y,m,0)$ for fixed x,n, the backward CK equation takes the form

$$\frac{\partial q_+}{\partial t} = v_+ \frac{\partial q_+}{\partial y} - k_c[q_+ - q_-], \qquad (4.2.22\text{a})$$

$$\frac{\partial q_-}{\partial t} = -v_- \frac{\partial q_-}{\partial y} + k_r[q_+ - q_-], \qquad (4.2.22\text{b})$$

with the boundary conditions

$$q_-(0,t) = 0, \quad q_+(L,t) = 0. \qquad (4.2.23)$$

Set $x = L$, $n = +$, and integrate both sides with respect to t. This yields

$$\frac{\partial \Pi_+}{\partial t} = v_+ \frac{\partial \Pi_+}{\partial y} - k_c[\Pi_+ - \Pi_-], \qquad (4.2.24\text{a})$$

$$\frac{\partial \Pi_-}{\partial t} = -v_- \frac{\partial \Pi_-}{\partial y} + k_r[\Pi_+ - \Pi_-]. \qquad (4.2.24\text{b})$$

Taking the limit $t \to 0$ with $\partial \Pi_m(y,t)/\partial t \to 0$ for $0 < y < L$ gives the steady-state equations

$$0 = v_+ \frac{\partial \pi_+}{\partial y} - k_c[\pi_+ - \pi_-], \qquad (4.2.25a)$$

$$0 = -v_- \frac{\partial \pi_-}{\partial y} + k_r[\pi_+ - \pi_-], \qquad (4.2.25b)$$

which are the analog of equation (2.4.18) for diffusion. In order to determine the boundary conditions at $x = 0, L$ note that if the MT starts out in the shrinking phase at $x = 0$, it never reaches $x = L$, whereas if it starts out at $x = L$ in the growing phase it is immediately absorbed. Hence, $\pi_-(0) = 0$ and $\pi_+(L) = 1$. Integrating equations (4.2.24) with respect to t then gives

$$-\pi_+ = v_+ \frac{\partial \pi_+ T_+}{\partial y} - k_c[\pi_+ T_+ - \pi_- T_-], \qquad (4.2.26a)$$

$$-\pi_- = -v_- \frac{\partial \pi_- T_-}{\partial y} + k_r[\pi_+ T_+ - \pi_- T_-], \qquad (4.2.26b)$$

with boundary conditions $\pi_-(0)T_-(0) = \pi_+(L)T_+(L) = 0$. These are the analogs of equation (2.4.20). Identical equations hold for exit at $x = 0$, with the modified boundary conditions $\overline{\pi}_-(0) = 1$ and $\overline{\pi}_+(L) = 0$. Explicit expressions for the solutions can be found in [594]. A related FPT problem is considered in Ex. 4.2.

MFPT with a nucleating boundary at $x = 0$. We now describe a probabilistic method for incorporating the effects of a sticky boundary at $x = 0$, which is based on taking conditional expectations and applying the strong Markov property [132]; see Box 4A for a definition of the latter. The basic method will appear in various subsequent chapters, including the analysis of stochastically gated diffusion (Sect. 6.7), search processes with stochastic resetting (Sect. 7.5), and cytoneme-based morphogen transport (Chap. 14). We begin by defining the following set of FPTs, see Fig. 4.11:

$$\mathcal{T} = \inf\{t \geq 0; X(t) = L\},$$
$$\mathcal{S} = \inf\{t \geq 0; X(t) = 0\},$$
$$\mathcal{R} = \inf\{t \geq 0; X(\mathcal{S} + \mathcal{N} + t) = L\},$$

where we have suppressed the dependence on the initial condition (y, m) and $\mathcal{N}$ is the first nucleation time. Introducing the set

$$\Omega = \{\mathcal{S} < \mathcal{T}\},$$

we can decompose the MFPT to escape at $x = L$ according to

$$\tau := \mathbb{E}[\mathcal{T}] = \mathbb{E}[\mathcal{T}1_{\Omega^c}] + \mathbb{E}[\mathcal{T}1_\Omega]$$
$$= \mathbb{E}[\mathcal{T}1_{\Omega^c}] + \mathbb{E}[(\mathcal{S} + \mathcal{N} + \mathcal{R})1_\Omega]. \qquad (4.2.27)$$

Here Ω^c is the complementary set of Ω and 1_Ω again denotes the indicator function that ensures expectation is only taken with respect to events that lie in Ω. Note that

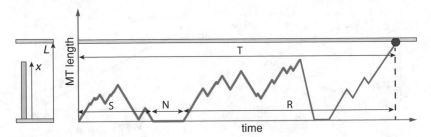

Fig. 4.11: Decomposition $\mathcal{T} = \mathcal{S} + \mathcal{N} + \mathcal{R}$ of the FPT to hit the wall at $x = L$ starting at $x = 0$ in the growing phase. Here $\mathcal{S}$ is the FPT to return to zero without reaching $x = L$, $\mathcal{N}$ is the first nucleation time, and $\mathcal{R}$ is the FPT to hit the wall after the first nucleation event.

$\mathbb{E}[\mathcal{T}1_{\Omega^c}] = \pi T$, where T is the conditional FPT that the MT hits $x = L$ before ever hitting $x = 0$, and $\mathbb{E}[\mathcal{S}1_{\Omega}] = \overline{\pi}\overline{T}$, where $\overline{T}$ is the conditional FPT that the MT hits $x = 0$ before ever hitting $x = L$. Moreover,

$$\mathbb{E}[\mathcal{N}1_{\Omega}] = \mathbb{E}[\mathcal{N}|\mathcal{S} < \mathcal{T}]\mathbb{P}[\Omega] = r_n^{-1}\mathbb{P}[\Omega],$$

where r_n is the nucleation rate and $\mathbb{P}[\Omega]$ is the splitting probability $\overline{\pi}$ to hit $x = 0$ before $x = L$. Incorporating the explicit dependence on the initial conditions, we thus find

$$\tau_m(y) = \pi_m(y)T_m(y) + \overline{\pi}_m(y)\left[\overline{T}_m(y) + \frac{1}{r_n}\right] + \mathbb{E}[\mathcal{R}1_{\Omega}].$$

We now exploit an important property of the velocity-jump process, namely, it satisfies the strong Markov property; see Box 4A and Chap. 9. In terms of our current example, the strong Markov property implies that even though the stopping times are random, the stochastic process $\widehat{X}(t') = X(t - \mathcal{S} - \mathcal{N})$ with times $t' \geq 0$ is identical to the original stochastic process $X(t)$ with the initial condition $\widehat{X}(0) = X(\mathcal{S} + \mathcal{N})$. In particular, the MFPT for $\widehat{X}$ to reach the boundary at $x = L$ is simply $\tau_+(0)$, so that

$$\mathbb{E}[\mathcal{R}1_{\Omega}] = \overline{\pi}_m(y)\tau_+(0).$$

Hence,

$$\tau_m(y) = \pi_m(y)T_m(y) + \overline{\pi}_m(y)\left[\overline{T}_m(y) + \frac{1}{r_n} + \tau_+(0)\right]. \tag{4.2.28}$$

Equation (4.2.28) is identical to the result derived in Ref. [594] using integral equations. Finally, the unknown constant $\tau_+(0)$ can be determined self-consistently by setting $y = 0$ and $m = +$:

$$\tau_+(0) = \pi_+(0)T_+(0) + \overline{\pi}_+(0)\left[\overline{T}_+(0) + \frac{1}{r_n} + \tau_+(0)\right].$$

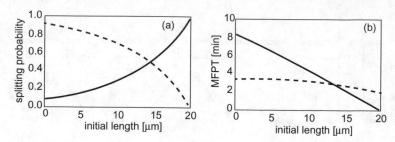

Fig. 4.12: (a) Splitting probability for hitting the wall at $x = L$ (solid curve) and shrinking to zero (dashed curve) as a function of the initial length. (b) Conditional MFPT for hitting $x = L$ before shrinking to zero as a function of the initial length. MT is assumed to start in the growth phase. Parameter values are $L = 20\,\mu$m, $v_+ = 2.4\,\mu$m/min, $v_- = 9.6\,\mu$m/min, $k_c = 0.3$/min, $k_r = 0.42$/min, and $r_n = 0.15$/min.

Rearranging the equation and using $\pi_+(0) + \overline{\pi}_+(0) = 1$ yields

$$\tau_+(0) = T_+(0) + \frac{\overline{\pi}_+(0)}{\pi_+(0)}\left[\overline{T}_+(0) + \frac{1}{r_n}\right].$$

Example theoretical plots are shown in Fig. 4.12 using parameter values for yeast cells taken from [594].

Box 4A. Stopping times and the strong Markov property.

Here we give heuristic definitions of stopping times and the strong Markov property. First, it is useful to introduce the notion of a natural filtration (see Chap. 9 for more details). Given a stochastic process $\{X(t) : t \in [0,T]\}$, the natural filtration $\mathcal{F}_t$ is taken to be the set of sample paths generated by the stochastic process, that is,

$$\mathcal{F}_t = \{X(s), s \leq t\}.$$

(More precisely, $\mathcal{F}_t$ belongs to a σ-algebra; see Sect. 9.1.) In other words, $\mathcal{F}_t$ contains the history of the stochastic process up to time t. A random variable is said to be measurable with respect to the natural filtration $\mathcal{F}_t$ if its probability density can be determined from the distribution of sample paths up to time t.

Stopping times. Let T denote some subinterval of $\mathbf{Z}^+$ (discrete time) or $\mathbb{R}^+$ (continuous time), and consider a stochastic process $\{X(t)\}_{t \in T}$ in $\mathbb{R}^d$. A random variable $\tau \in \mathbb{R}^+$, which is defined on the same probability space as $\mathbf{X}$, is a stopping time if for every $t \in T$ we can completely determine whether or not τ has occurred before time t using the information contained in all the events that have occurred up to time t, that is, the filtration $\mathcal{F}_t$. A classical example of a stopping time is the FPT τ_A for a stochastic process $\{X(t)\}_{t \in T}$ to reach the set $A \subset \mathbb{R}^d$,

$$\tau_A = \inf\{t \geq 0 : X(t) \in A\}.$$

Strong Markov property. A stochastic process $\{X(t)\}_{t \in T}$ is said to have the Markov property if the conditional probability distribution of future states of the process (conditional on both past and present states) depends only upon the present state, not on the sequence of events that preceded it. That is, for all $t' > t$ we have

$$\mathbb{P}[X_{t'} \leq x | X_s, s \leq t] = \mathbb{P}[X_{t'} \leq x | X_t].$$

The strong Markov property is similar to the Markov property, except that the "present" is defined in terms of a stopping time. That is, given any finite-valued stopping time τ with respect to the natural filtration of X, if the stochastic process $Y(t) = X(t + \tau) - X(\tau)$ is independent of $\{X(s), s < \tau\}$ and has the same distribution as $\widehat{Y}(t) = X(t) - X(0)$ then X is said to satisfy the strong Markov property.

4.2.3 Model of hydrolysis and caps

Another approach to modeling MT dynamics is to include a simplified description of hydrolysis and cap formation that involves only a few model parameters [25, 277, 278]. Here we will describe in some detail the model of Antal *et. al* [25]. The MT is taken to consist of a mixture of GTP-tubulin complexes (GTP-T) and GTD-tubulin complexes (GTD-T). A given configuration is represented by a string of $+$ and $-$ symbols corresponding to GTP-T and GTD-T, respectively. Three basic processes are considered:

1. Attachment: Growth of an MT occurs via the attachment of a GTP-T monomer at one end, with the attachment rate depending on the identity of the current monomer at the tip. That is

$$|\ldots+\rangle \implies |\ldots++\rangle \text{ at rate } \lambda, \quad |\ldots-\rangle \implies |\ldots-+\rangle \text{ at rate } p\lambda,$$

with $p \leq 1$.

2. Conversion: Once incorporated into the MT, each GTP-T can independently convert by hydrolysis to GTD-T:

$$|\ldots+\ldots\rangle \implies |\ldots-\ldots\rangle \text{ at rate } \gamma.$$

3. Detachment: Shrinkage of an MT occurs via the detachment of a GTD-T monomer from the end of the MT

$$|\ldots -\rangle \implies |\ldots\rangle \text{ at rate } \mu.$$

In general, one finds that there are two phases in the parameter space (λ, μ, p), one corresponding to a growing phase with average growth rate $V(\lambda, \mu, p)$, and the other to a bounded phase. The two phases are separated by a phase boundary $\mu = \mu_*(\lambda, p)$ along which $V = 0$. Following Antal *et. al* [25], we will develop the stochastic analysis of the model by considering various limiting cases.

Unconstrained growth. First, suppose that there is unrestricted growth ($\mu = 0$) and the attachment rate is independent of the end monomer ($p = 1$). The speed of growth is then simply $V = \lambda$. The kinetic equation for the mean number x of GTP-T monomers in the chain is

$$\frac{dx}{dt} = \lambda - \gamma x.$$

The probability $P_m(t)$ that there are m GTP-T monomers at time t evolves according to a birth–death master equation of the form

$$\frac{dP_m}{dt} = \gamma(m+1)p_{m+1}(t) + \lambda P_{m-1}(t) - (\lambda + \gamma m)P_m(t). \tag{4.2.29}$$

Equation (4.2.29) is a rare example of a master equation that can be solved exactly, and one finds that $P_n(t)$ is given by a Poisson distribution. (A master equation identical in form to (4.2.29) will arise in a model of unregulated gene expression; see Sect. 5.2.) The simplest way to see this is to introduce the generating function

$$G(z,t) = \sum_{n \geq 0} z^n P_n(t),$$

and use equation (4.2.29) to show that

$$\frac{\partial G}{\partial t} + \gamma(z-1)\frac{\partial G}{\partial z} = \lambda(z-1)G.$$

This is a linear first-order PDE with nonconstant coefficients, which can be solved using the method of characteristics (see Box 2C). Recall that the basic idea is to construct characteristic curves $z = z(t)$ along which $G(t) \equiv G(z(t), t)$ satisfies

$$\frac{dG}{dt} = \frac{\partial G}{\partial t} + \frac{dz}{dt}\frac{\partial G}{\partial z},$$

such that the evolution of G is consistent with the original PDE. This then yields the characteristic equations

$$\frac{dz}{dt} = \gamma(z-1), \quad \frac{dG}{dt} = \lambda(z-1)G.$$

Solving for $z(t)$, we have $z(t) = 1 + se^{\gamma t}$, where s parameterizes the initial data. Then

$$\frac{dG}{dt} = \lambda s e^{\gamma t} G, \quad G(t) = F(s) \exp\left(\lambda s e^{\gamma t} / \gamma\right)$$

for some function F determined by the initial data. In order to obtain the solution $G(z,t)$ we eliminate s in terms of z, which gives

$$G(z,t) = F([z-1]e^{-\gamma t}) \exp\left(\lambda(z-1)/\gamma\right). \tag{4.2.30}$$

Since $G(1,t) = 1$, we require $F(0) = 1$. Moreover, given the initial condition $p_n(0) = \delta_{n,0}$, we have $G(z,0) = 1$ and $F(z) = e^{-\lambda z/\gamma}$. It follows that

$$G(z,t) = e^{\lambda\left(1-e^{-\gamma t}\right)(z-1)/\gamma}. \tag{4.2.31}$$

If we now Taylor expand $G(z,t)$ in powers of z, we find that

$$P_n(t) = e^{-\lambda(t)} \frac{\lambda(t)^n}{n!}, \quad \lambda(t) = \frac{\lambda}{\gamma}(1 - e^{-\gamma t}), \tag{4.2.32}$$

which is a time-dependent Poisson distribution of rate $\lambda(t)$. It immediately follows that

$$\langle N(t) \rangle = \lambda(t), \quad \mathrm{var}[N(t)] = \lambda(t).$$

In the more general case $n_0 \neq 0$, the mean and variance can be calculated from the formulae

$$\langle N(t) \rangle = \left.\frac{\partial G(z,t)}{\partial z}\right|_{z=1}, \quad \langle (N^2(t) - N(t)) \rangle = \left.\frac{\partial^2 G(z,t)}{\partial z^2}\right|_{z=1}.$$

Calculating these derivatives yields

$$\langle N(t) \rangle = (n_0 - \lambda/\gamma)e^{-\gamma t} + \lambda/\gamma, \; \mathrm{var}[N(t)] = \langle n(t) \rangle - n_0 e^{-2\gamma t}.$$

Cap length. The conversion of GTP-T to GTD-T means that more recently attached monomers around the tip region are more likely to be GTP-T, whereas monomers in the tail are predominantly GTD-T. The cap is defined to be the region from the end of the microtubule to the first GTD-T monomer, see Fig. 4.13. Let π_k be the probability that the cap is of length k and consider the associated master equation

GTP-T region

GTD-T region

k

λ

cap

Fig. 4.13: Schematic diagram illustrating a cap of GTP-T monomers and additional islands of GTP-T.

$$\frac{d\pi_k}{dt} = \lambda(\pi_{k-1} - \pi_k) - k\pi_k + \sum_{s \geq k+1} \pi_s \qquad (4.2.33)$$

for $k \geq 0$ and $\pi_{-1} \equiv 0$. We are assuming each GTP-T monomer in the cap is equally likely to hydrolyze and we have set $\gamma = 1$. The last term on the right-hand side represents the probability that a cap of length greater then k hydrolyzes at the $(k+1)$th site. Adding the first $k-1$ equations gives (in steady state)

$$\pi_1 + 2\pi_2 + \ldots + (k-1)\pi_{k-1} + \lambda\pi_{k-1} = N_1 + N_2 + \ldots + N_k,$$

where $N_k = \sum_{s \geq k} \pi_s$. Using the fact that $N_j = N_k + \pi_{k-1} + \ldots + \pi_j$ for $j < k$, we see that

$$\pi_{k-1} = \frac{k}{\lambda} N_k.$$

From the identity $N_{k-1} - N_k = \pi_{k-1}$, it follows that

$$N_k = \frac{\lambda}{k+\lambda} N_{k-1}.$$

Iterating this equation and using $N_0 \equiv 1$ gives

$$N_k = \frac{\lambda^k \Gamma(1+\lambda)}{\Gamma(k+1+\lambda)},$$

where $\Gamma(z)$ is the gamma function

$$\Gamma(z) = \int_0^\infty t^{z-1} e^{-t} dt, \qquad (4.2.34)$$

with

$$\frac{\Gamma(1+\lambda)}{\Gamma(k+1+\lambda)} = \frac{1}{(k+\lambda)(k-1+\lambda)\ldots(1+\lambda)}.$$

Thus, the stationary cap length distribution is

$$\pi_k = \frac{(k+1)\lambda^k \Gamma(1+\lambda)}{\Gamma(k+2+\lambda)}. \qquad (4.2.35)$$

We can now calculate the mean cap length using

$$\langle k \rangle = \sum_{k \geq 1} k\pi_k = \sum_{k \geq 1} k[N_k - N_{k+1}] = \sum_{k \geq 1} N_k = -1 + \sum_{k \geq 0} N_k.$$

Given the solution for N_k and the properties of confluent hypergeometric functions $F(a,c;x)$, see Box 4B, we have

$$\langle k \rangle = -1 + F(1,\lambda+1;\lambda) = -1 + \lambda e^\lambda \lambda^{-\lambda} \gamma(\lambda,\lambda),$$

where $\gamma(a,x)$ is the incomplete gamma function. Finally, using the asymptotic result

$$\gamma(\lambda,\lambda) \rightarrow \sqrt{\frac{\pi}{2\lambda}}\lambda^\lambda e^{-\lambda}, \quad \lambda \gg 1,$$

we see that

$$\langle k \rangle \rightarrow \sqrt{\pi\lambda/2}. \tag{4.2.36}$$

Thus a growing MT with λ GTP-T monomers has a cap size that scales as $\sqrt{\lambda}$.

Box 4B. Hypergeometric series.

The confluent hypergeometric function is defined according to the infinite series

$$F(a,c;x) = \sum_{k=0}^{\infty} \frac{(a)_n x^n}{(c)_n n!},$$

where we have used the Pochhammer symbol

$$(a)_n = a(a+1)(a+2)\ldots(a+n-1), \quad (a)_0 = 1.$$

The congruent hypergeometric function has the integral representation

$$F(a,c;x) = \frac{\Gamma(c)}{\Gamma(a)\Gamma(c-a)} \int_0^1 e^{xt}t^{a-1}(1-t)^{c-a-1}dt$$

for $\mathrm{Re}[c] > \mathrm{Re}[a] > 0$. An important special case is

$$F(1,1+\lambda;x) = \int_0^1 e^{xt}(1-t)^{\lambda-1}dt = e^x \int_0^1 e^{-ux}u^{\lambda-1}du$$

$$= e^x x^\lambda \int_0^x e^{-u}u^{\lambda-1}du$$

$$= \lambda e^x x^{-\lambda}\gamma(\lambda,x),$$

where $\gamma(\lambda,x)$ is the incomplete gamma function.

Constrained growth. There are two mechanisms for slowing the growth rate: conversion from GTP-T to GTD-T at the tip resulting in a reduced rate of attachment (for $p < 1$) and detachment of GTD-T at the tip ($\mu > 0$). First, consider the effect of having $p < 1$ but no detachment. In determining the rate of growth, it is now necessary to keep track of the hydrolysis state of the end monomer. Thus, the kinetic equation for the number of GTP-T monomers in the chain becomes (for $\gamma = 1$)

$$\frac{dx}{dt} = -x + p\lambda\pi_0 + \lambda(1 - \pi_0),$$

where π_0 is the probability that there is no cap. Extending equation (4.2.33) to the case $p < 1$ gives

$$\frac{d\pi_0}{dt} = -p\lambda\pi_0 + (1-\pi_0).$$

This pair of equations yields the steady-state solution

$$n_0 = p\lambda\frac{1+\lambda}{1+p\lambda}.$$

The steady-state speed of growth of the MT is

$$V(p,\lambda) = p\lambda n_0 + \lambda(1-n_0) = p\lambda\frac{1+\lambda}{1+p\lambda}.$$

Let us now calculate the probability distribution of MT lengths $P(L,t)$. Let $X(L,t)$ and $Y(L,t)$ denote the conditional probabilities that the length equals L and the end monomer is GTP-T and GTD-T, respectively. These probabilities evolve according to

$$\frac{dX(L,t)}{dt} = \lambda X(L-1,t) + p\lambda Y(L-1,t) - (1+\lambda)X(L),$$
$$\frac{dY(L,t)}{dt} = X(L) - p\lambda Y(L).$$

Adding this pair of equations and using $P = X+Y$,

$$\frac{dP(L,t)}{dt} = \lambda X(L-1,t) + p\lambda Y(L-1,t) - \lambda X(L) - p\lambda Y(L).$$

For sufficiently long filaments, the state of the last monomer does not depend on polymer length so that

$$X(L) = (1-\pi_0)P(L), \quad Y(L) = \pi_0 P(L).$$

Substituting into the equation for dP/dt gives

$$\frac{dP(L,t)}{dt} = V(p,\lambda)[P(L-1,t) - P(L,t)]. \tag{4.2.37}$$

This represents a Poisson process with rate V so that

$$P(L,t) = \frac{(Vt)^L}{L!}e^{-Vt}, \tag{4.2.38}$$

and

$$\langle L\rangle = Vt, \quad \mathrm{var}[L] = Vt.$$

Finally, consider the general case $\mu > 0$ and $p < 1$. The probability that there is no cap (the end monomer is GTD-T) evolves as

$$\frac{d\pi_0}{dt} = -p\lambda\pi_0 + (1-\pi_0) - \mu\mathcal{N}_0,$$

where $\mathcal{N}_0 = \Pr\{+-\rangle\}$, and the corresponding speed of growth is

$$V(\lambda, \mu, p) = p\lambda\pi_0 + \lambda(1 - \pi_0) - \mu\pi_0.$$

The difficulty in analyzing the general case is due to the fact that one has to solve an infinite hierarchy of higher-order correlations in order to determine $\mathcal{N}_0$. However, progress can be made in certain limiting cases [25]. For example, suppose that $\lambda, \mu \ll 1$ so that rate of conversion GTP-T $\to$ GTD-T is much faster than the other processes. Consequently, hydrolysis occurs as soon as a monomer attaches, that is, $\pi_0 \approx 1$. Hence, the growth phase occurs when $p\lambda > \mu$ and $V = p\lambda - \mu$. On the other hand, when $\lambda \gg 1$, there is a high probability that the end monomer is GTP-T so that $\pi_0 \approx \mathcal{N}_0$. Consequently, the steady-state distribution is

$$\pi_0 = \frac{1}{1 + p\lambda + \mu}, \quad V = \lambda - \frac{(1-p)\lambda + \mu}{1 + p\lambda + \mu}.$$

4.3 Brownian motion in a periodic potential

One qualitative method for modeling the stepping of molecular motors is based on the theory of Brownian ratchets; see the extensive review by Reimann [683]. Here, we develop the theory by considering the classical problem of how to solve the Fokker–Planck equation for a Brownian particle in a periodic potential. Therefore, consider the 1D FP equation

$$\frac{\partial p}{\partial t} = D_0 \left[\frac{1}{k_B T} \frac{\partial [V'(x) - F_0]p}{\partial x} + \frac{\partial^2 p}{\partial x^2} \right], \tag{4.3.1}$$

where $V(x)$ is an L-periodic potential, $V(x+L) = V(x)$ for all x, and F_0 is a constant external force, see Fig. 4.14. We begin by describing the standard Stratonovich-based calculation of the mean velocity [357, 683, 782, 890], and show that it is zero when $F_0 = 0$, i.e., the motor cannot do any useful work. We then consider one mechanism for breaking periodicity that is based on rectification. An alternative mechanism, involving ATP hydrolysis and the breaking of detailed balance, will be the subject of Sect. 4.4. The first step is to introduce the effective potential or free energy $\mathcal{V}(x) = V(x) - F_0 x$ and to note that $\mathcal{V}'(x)$ is periodic even though $\mathcal{V}$ is not. Next we define the reduced probability density and currents

$$\widehat{p}(x,t) = \sum_{n=-\infty}^{\infty} p(x+nL,t), \quad \widehat{J}(x,t) = \sum_{n=-\infty}^{\infty} J(x+nL,t), \tag{4.3.2}$$

with

$$J(x,t) = -D_0 \left[\frac{1}{k_B T} \mathcal{V}'(x)p + \frac{\partial p}{\partial x} \right].$$

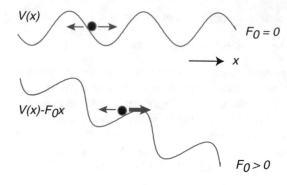

Fig. 4.14: Brownian particle moving in a periodic potential $V(x)$. In the absence of tilt ($F_0 = 0$), the mean velocity in the long time limit is zero. On the other hand, in the presence of a tilt ($F_0 \neq 0$), the net motion of the particle is in the direction of the force.

It immediately follows that

$$\widehat{p}(x+L,t) = \widehat{p}(x,t), \quad \int_0^L \widehat{p}(x,t)dx = 1. \tag{4.3.3}$$

The periodicity of $\mathcal{V}'(x)$ implies that if $p(x,t)$ is a solution of the FP equation, then so is $p(x+nL,t)$. (Note that $p(x,t)$ itself is not periodic, otherwise it would not be possible to satisfy the normalization condition $\int_{\infty}^{\infty} p(x,t)dx = 1$.) The principle of superposition for a linear PDE then shows that $\widehat{p}$ satisfies the FP equation

$$\frac{\partial \widehat{p}(x,t)}{\partial t} + \frac{\partial \widehat{J}(x,t)}{\partial x} = 0, \tag{4.3.4}$$

with

$$\widehat{J}(x,t) = -D_0 \left[\frac{1}{k_B T} \mathcal{V}'(x)\widehat{p} + \frac{\partial \widehat{p}}{\partial x} \right], \tag{4.3.5}$$

and periodic boundary conditions at $x = 0, L$. There exists a stationary solution $\widehat{p}_0$ of the reduced FP equation with constant flux $\widehat{J}_0$ such that

$$\frac{d}{dx}\left(e^{\mathcal{V}(x)/k_B T} \widehat{p}_0(x) \right) = -\frac{\widehat{J}_0}{D_0} e^{\mathcal{V}(x)/k_B T}. \tag{4.3.6}$$

(The full FP equation does not have a nontrivial steady state, since $p(x,t) \to 0$ pointwise as $t \to \infty$.) Integrating this equation from x to $x+L$ and using periodicity yields the stationary solution

$$\widehat{p}_0(x) = \frac{\widehat{J}_0 \mathcal{N}(x)}{\left[1 - e^{-F_0 L/k_B T} \right]}, \tag{4.3.7}$$

where

$$\mathcal{N}(x) = \frac{1}{D_0} e^{-\mathcal{V}(x)/k_B T} \int_x^{x+L} e^{\mathcal{V}(y)/k_B T} dy. \tag{4.3.8}$$

Finally, $\widehat{J}_0$ is determined by imposing the normalization condition on $\widehat{p}_0$.

A quantity of particular interest is the ensemble averaged velocity v with $v dt = \langle dX(t) \rangle$. It turns out that this is equal to the rate of change of the ensemble averaged position [683]. Recall that the solution of the FP equation (4.3.1), $p(x,t)$, is the probability density on the sample space Ω of solutions to the Langevin equation

$$dX(t) = \frac{F_0 - V'(x)}{\gamma} + \sqrt{2D_0} dW(t),$$

where γ is the drag coefficient with $D_0 \gamma = k_B T$. (We assume some fixed initial condition $X(0) = x_0$.) The connection between the two paradigms can be expressed as $p(x,t) = \langle \delta(x - X(t)) \rangle$, where for fixed (x,t), $\langle \ldots \rangle$ denotes averaging with respect to realizations of the Wiener process. (This should be contrasted with the definition $\langle x(t) \rangle = \int x p(x,t) dx$.) Taking differentials of both sides with respect to time gives

$$\partial_t p(x,t) dt = -\langle \delta'(x - X(t)) dX(t) \rangle,$$

and, since $\partial_t p = -\partial_x J(x,t)$ implies that

$$J(x,t) dt = \langle \delta(x - X(t)) dX(t) \rangle.$$

Integrating both sides with respect to x yields the result

$$\langle dX(t) \rangle = \left[\int_{-\infty}^{\infty} J(x,t) dx \right] dt. \tag{4.3.9}$$

The right-hand side of (4.3.9) can be rewritten as

$$-\left[\int_{-\infty}^{\infty} x \partial_x J(x,t) dx \right] dt = \left[\int_{-\infty}^{\infty} x \partial_t p(x,t) dx \right] dt = \frac{d\langle x(t) \rangle}{dt} dt.$$

We deduce the important result that [683]

$$\langle dX(t) \rangle = \frac{d\langle x(t) \rangle}{dt} dt.$$

Equation (4.3.9) thus implies that

$$v = \int_{-\infty}^{\infty} J(x,t) dx = \int_0^L \widehat{J}(x,t) dx. \tag{4.3.10}$$

Since $v = L\widehat{J}_0$ for constant current, it follows that

$$v = L \frac{1 - e^{-F_0 L / k_B T}}{\int_0^L \mathcal{N}(x) dx}. \tag{4.3.11}$$

It can be seen that there is no net motion in a purely periodic potential, since the numerator vanishes when $F_0 = 0$. Moreover the net direction of motion for $F_0 \neq 0$ is in the direction of the applied force. Note that in the case of a space-dependent

Fig. 4.15: Brownian particle moving in a periodic
ratchet potential $V(x)$.

diffusion coefficient $D(x)$, the above analysis is easily extended with $\mathcal{N}(x)$ now given by [150]

$$\mathcal{N}(x) = e^{-V(x)/k_BT} \int_x^{x+L} \frac{1}{D(y)} e^{V(y)/k_BT} dy.$$

The result that there is no net motion in a periodic potential ($F = 0$) can be counter-intuitive when considering ratchet potentials as shown in Fig. 4.15, since one might think that it is more difficult to move backward and cross the steep slope.

4.3.1 Polymerization ratchet

One interesting application of ratchet potentials is to the so-called polymerization ratchet [656], which is a simplified model of how actin polymerization changes the shape of a cell's membrane during cell motility [584, 587]. Suppose that a section of cell membrane wall is undergoing Brownian motion in the presence of a resistive force F due to stretching, see Fig. 4.16(a). This motion is rectified by the addition of actin monomers to the end of an actin polymer filament, whenever the gap x between membrane wall and filament is sufficiently large. Assume that in the absence of a load force, actin monomers are added at a rate k_+m and lost at a rate k_-, where m is the background concentration of monomers. First, consider the limiting case in which the mean time between attachments is sufficiently large so that the Brownian particle reaches thermal equilibrium. This means that the probability density for a gap of size x is given by the Boltzmann–Gibbs distribution (Sect. 1.3)

$$p(x) = \frac{F}{k_BT} e^{-Fx/k_BT}.$$

An estimate of the mean polymerization velocity is then $v = a\left[k_+mP(x > a) - k_-\right]$, where a is the size of a monomer and

$$P(x > a) = \int_a^\infty p(x)dx = e^{-Fa/k_BT}.$$

Finally, using detailed balance,

$$\frac{k_+m}{k_-} = e^{\Delta G/k_BT},$$

where ΔG is the binding energy, we have

$$v = ak_-\left[e^{[\Delta G - Fa]/k_BT} - 1\right], \tag{4.3.12}$$

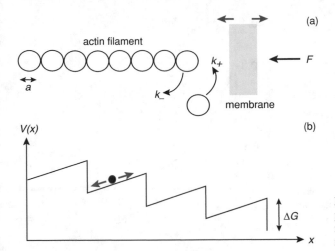

actin filament

(a)

k_+

F

k_-

membrane

$V(x)$

(b)

ΔG

x

Fig. 4.16: (a) Cartoon of polymerization ratchet model. (b) Simplified ratchet model.

which suggests that growth stops when the resistive force F becomes sufficiently large such that $F \geq F_S$, where the stall force $F_S = \Delta G/a$. A sketch of the velocity–load curve for typical values of k_+, m, k_- and a is shown in Fig. 4.17.

Let us now turn to the diffusion-limited case, which has been analyzed by Peskin *et al* [656] using a Fokker–Planck description of the process shown in Fig. 4.16(a). Here we will consider a reduced model, consisting of a Brownian particle moving in a ratchet potential, see Fig. 4.16(b). This is obtained by ignoring spontaneous unbinding of monomers ($k_- = 0$) and assuming that as soon as the distance between the polymer and the wall is equal to a, a new monomer is immediately inserted, resulting in a sudden drop in energy by an amount ΔG. However, it is possible to reverse the direction by jumping over a free energy barrier of height ΔG—this represents the dislodging of a monomer due to wall motion. The analysis of the reduced model proceeds along similar lines to the general motion of a Brownian particle in a tilted potential with

$$\mathcal{V}(x) = Fx - n\Delta G, \quad na < x < (n+1)a.$$

However, one now needs to take into account the discontinuities in $V(x)$ at the points $x = na$, integer n. Thus, equation (4.3.6) still holds, but care must be taken when integrating this equation with respect to $x \in (0, a]$. That is, it is necessary to introduce the matching condition

$$\lim_{x \to a^+} \widehat{p}_0(x) e^{\mathcal{V}(x)} = \lim_{x \to a^-} \widehat{p}_0(x) e^{\mathcal{V}(x)}.$$

One finds that (see Ex. 4.3)

$$v = \frac{2D_0}{a} \frac{\omega^2/2}{\mathcal{A}(1 - e^{-\omega}) - \omega}, \quad \omega = \frac{Fa}{k_B T}, \tag{4.3.13}$$

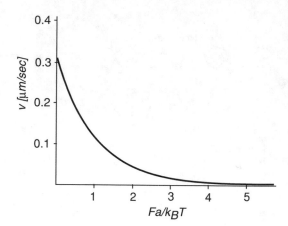

Fig. 4.17: Sketch of typical velocity–force curve based on equation (4.3.9).

with

$$\mathcal{A} = \frac{e^{\Delta G/k_B T} - 1}{e^{(\Delta G - Fa)/k_B T} - 1}. \tag{4.3.14}$$

Note that $v \to 0$ as $Fa \to \Delta G$, since $\mathcal{A} \to \infty$. On the other hand, in the regime $\Delta G \gg Fa$ and $k_B T \gg Fa$, we have $v \approx 2D_0/a$. This latter result can be understood as follows: in the absence of a force F, the mean time for a diffusive displacement of size a is $T = a^2/2D_0$ so that the mean speed is $v = a/T$. In Ex. 4.4, we consider an alternative formulation of the polymerization ratchet based on a population model; see also Ref. [656].

In Chap. 14, we will consider a microscopic model of cell protrusion based on an extension of the polymerization ratchet model known as the tethered ratchet model. This was developed by Mogilner and Oster within the context of the simpler problem of *Listeria* propulsion [584, 586], and has subsequently been incorporated into more complex models of cell crawling [392]. Such a model takes into account thermal bending fluctuations of a semi-stiff actin filament, which generates the gap necessary for insertion of an additional monomer rather than membrane diffusion (the elastic ratchet model). More specifically, the tethered ratchet model proposes that there are two classes of filament: some are attached, under tension and nucleating rather than growing, while others are unattached and pushing via an elastic ratchet mechanism.

4.3.2 Translocation ratchet

Following gene expression, many proteins have to translocate into or across a cellular membrane. Examples include translocation through nuclear pores and through pores in the endoplasmic reticulum. It has been suggested that translocation may be driven by a Brownian ratchet [304, 656, 760]. The basic mechanism is illustrated in Fig. 4.18. Once the protein chain enters a pore, thermal fluctuations cause it to diffuse back and forth through the pore without any net displacement. However, suppose that the protein has ratchet sites that are equally spaced along the chain with nearest neighbor separation δ. In the case of a perfect ratchet, it is assumed that once a ratchet site has passed through the pore it cannot re-enter the pore, that is, it is reflected. On the other hand, for an imperfect ratchet there is a certain probability π of reflection. The latter could be due to the binding of a macromolecule (chaperonin) to the ratchet site on the distal side of the pore.

Consider a translocation ratchet and let $p(x,t)$ be the probability density that $X(t) = x$, where $X(t)$, $0 < X(t) < \delta$, is the position of the first ratchet site to the right of the pore exit. Let F be the net force resisting translocation of the protein. The FP equation takes the form

$$\frac{\partial p}{\partial t} + \frac{\partial J}{\partial x} = 0, \quad J = -\frac{DF}{k_B T}p - D\frac{\partial p}{\partial x}. \tag{4.3.15}$$

The corresponding boundary conditions for a perfect ratchet are

$$J(0,t) = J(\delta,t), \quad p(\delta,t) = 0. \tag{4.3.16}$$

The periodic flux condition expresses the fact that as soon as one ratchet site crosses $x = \delta$, another site appears at $x = 0$, with $x = \delta$ treated as an absorbing boundary. The steady-state solution satisfies the constant flux condition

$$-\frac{DF}{k_B T}p - D\frac{\partial p}{\partial x} = J_0.$$

Multiplying both sides by $D^{-1}e^{Fx/k_B T}$, integrating from x to δ, and using the absorbing boundary condition yield

$$p(x) = \frac{k_B T J_0}{DF}\left[e^{F(\delta-x)/k_B T} - 1\right].$$

Imposing the normalization condition $\int_0^1 p(x)dx = 1$ then determines J_0 according to

$$1 = \frac{J_0 \delta^2}{D}\frac{1}{\omega}[e^\omega - 1 - \omega], \quad \omega = \frac{F\delta}{k_B T}.$$

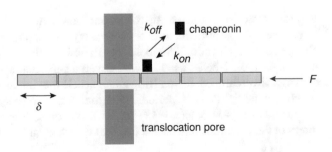

Fig. 4.18: Cartoon of a translocation ratchet.

It follows that the average speed of the perfect translocation ratchet is [656]

$$v = \delta J_0 = \frac{2D}{\delta} \frac{\omega^2}{e^\omega - 1 - \omega}. \tag{4.3.17}$$

Now suppose that each ratchet site can exist in two states that are in equilibrium

$$S_0 \underset{k_{\text{off}}}{\overset{k_{\text{on}}}{\rightleftharpoons}} S_1,$$

with only S_1 ratcheted. Hence S_0 passes freely through the pore in both directions, whereas S_1 is reflected. The probability of being in the ratcheted state is then

$$\pi = \frac{k_{\text{on}}}{k_{\text{on}} + k_{\text{off}}}.$$

The only modification of the perfect ratchet equations is that the absorbing boundary condition is replaced by [656]

$$p(\delta) = (1 - \pi)p(0). \tag{4.3.18}$$

Repeating the above calculation yields the modified velocity; see Ex. 4.5

$$v = \delta J_0 = \frac{2D}{\delta} \left[\frac{\omega^2/2}{\frac{e^\omega - 1}{1 - K(e^\omega - 1)} - \omega} \right]. \tag{4.3.19}$$

A simplified, discrete translocation model is considered in Ex. 4.6.

Note that one major simplification of the above model is that it treats the translocating polymer as rigid. However, a polymer such as a protein or DNA tends to be highly coiled (small persistence length) so that one has to take into account an effective entropic force, reflecting the fact that a free polymer has many more configurational states than one that is threaded through a pore [184, 599, 635, 789]. The statistical mechanics of polymers is considered in Chap. 12.

4.4 Brownian ratchet model of a processive molecular motor

In performing a single step along a filament track, a molecular motor cycles through a sequence of conformational states before returning to its initial state (modulo the change in spatial location). Suppose that there is a total of M conformational states in a single cycle labeled $i = 1, \ldots, M$. Given a particular state i, the motor is modeled as an overdamped, driven Brownian particle moving in an asymmetric periodic (ratchet) potential $V_i(x)$. A periodic potential is said to be symmetric if there exists Δx such that $V_i(-x) = V_i(x + \Delta x)$ for all x, otherwise it is asymmetric. The asymmetry of the potentials reflects the fact that cytoskeletal filaments are polarized. The

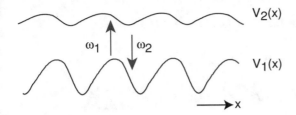

Fig. 4.19: Brownian ratchet model of a molecular motor that can exist in two internal states with associated l-periodic ratchet potentials $V_1(x)$ and $V_2(x)$. State transition rates are denoted by ω_1 and ω_2.

Langevin equation for the location of the particle $X(t)$, assuming that it remains in a given conformational state, is

$$dX = -\frac{V_i'(X)}{\gamma}dt + dW_i(t), \qquad (4.4.1)$$

with $\langle dW_i(t)\rangle = 0$ and $\langle dW_i(t)dW_j(t')\rangle = 2D\delta_{i,j}\delta(t-t')dt\,dt'$. The corresponding FP equation is

$$\frac{\partial p_i(x,t)}{\partial t} = -\frac{\partial J_i(x,t)}{\partial x}, \qquad (4.4.2)$$

where $p_i(x,t)$ is the probability density that the motor particle is in internal state i and at location x at time t, and $J_i(x,t)$ is the probability flux

$$J_i(x,t) = \frac{1}{\gamma}\left[-V_i'(x) - k_B T\frac{\partial}{\partial x}\right]p_i(x,t), \qquad (4.4.3)$$

where $D\gamma = k_B T$. If the state transitions between the conformational states are now introduced according to a discrete Markov process, then it is necessary to add source terms to the FP equation:

$$\frac{\partial p_i(x,t)}{\partial t} = -\frac{\partial J_i(x,t)}{\partial x} + \sum_{j\neq i}[\omega_{ij}(x)p_j(x,t) - \omega_{ji}(x)p_i(x,t)],$$

where $\omega_{ij}(x)$ is the rate at which the motor switches from state j to state i.

In order to develop the basic theory, consider the simple case of two internal states $N = 2$ following along the lines of [421, 638, 657, 667]. Then

$$\frac{\partial p_1(x,t)}{\partial t} + \frac{\partial J_1(x,t)}{\partial x} = -\omega_1(x)p_1(x,t) + \omega_2(x)p_2(x,t), \qquad (4.4.4a)$$

$$\frac{\partial p_2(x,t)}{\partial t} + \frac{\partial J_2(x,t)}{\partial x} = \omega_1(x)p_1(x,t) - \omega_2(x)p_2(x,t). \qquad (4.4.4b)$$

Note that adding the pair of equations together and setting $p = p_1 + p_2$, $J = J_1 + J_2$ leads to the conservation equations $\partial_t p + \partial_x J = 0$. An example of l-periodic ratchet potentials $V_1(x), V_2(x)$ is shown in Fig. 4.19, with l the basic step length of a cycle along the filament track. The analysis of the two-state model proceeds along similar lines to the one-state model considered in Sect. 4.3. That is, set

$$\widehat{p}_j(x,t) = \sum_{n=-\infty}^{\infty} p_j(x+nl,t), \quad \widehat{J}_j(x,t) = \sum_{n=-\infty}^{\infty} J_j(x+nl,t). \tag{4.4.5}$$

The total probability flux can then be written as

$$\widehat{J}(x,t) = -\frac{1}{\gamma}\left[V_1'(x)\widehat{p}_1(x,t) + V_2'(x)\widehat{p}_2(x,t) + k_BT\frac{\partial \widehat{p}(x,t)}{\partial x}\right].$$

Consider the steady-state solution for which there is a constant total flux $\widehat{J}_0$ so that

$$V_1'(x)\widehat{p}_1(x) + V_2'(x)\widehat{p}_2(x) + k_BT\frac{\partial \widehat{p}(x)}{\partial x} = -\widehat{J}_0\gamma.$$

Defining $\lambda(x) = \widehat{p}_1(x)/\widehat{p}(x)$, this equation can be rewritten as

$$V_{\mathrm{eff}}'(x)\widehat{p}(x) + k_BT\frac{\partial \widehat{p}(x)}{\partial x} = -\widehat{J}_0\gamma, \tag{4.4.6}$$

where

$$V_{\mathrm{eff}}(x) = \int_0^x \left[\lambda(y)V_1'(y) + (1-\lambda(y))V_2'(y)\right]dy. \tag{4.4.7}$$

Suppose that the system is in thermodynamic equilibrium. The state transition rates and steady-state probabilities then satisfy the detailed balance condition (see Sect. 1.3)

$$\frac{\omega_1(x)}{\omega_2(x)} = e^{[V_1(x)-V_2(x)]/k_BT} = \frac{\widehat{p}_2(x)}{\widehat{p}_1(x)}. \tag{4.4.8}$$

Therefore,

$$\lambda(x) = \frac{1}{1+e^{-[V_1(x)-V_2(x)]/k_BT}}, \tag{4.4.9}$$

and, in particular, $\lambda(x)$ reduces to an l-periodic function. It follows that $V_{\mathrm{eff}}(x)$ in equation (4.4.6) is also an l-periodic potential and hence there is no net motion in a particular direction (in the absence of an external force or tilt); see Sect. 4.3. In conclusion, in order for a molecular motor to sustain directed motion that can pull against an applied load, we require a net positive supply of chemical energy that maintains the state transition rates away from detailed balance—this is the role played by ATP.

Therefore, consider the situation in which transitions between the two states occur as a result of chemical reactions involving ATP hydrolysis. Denoting the two conformational states of the motor by M_1, M_2, the scheme is taken to be [638]

$$\mathrm{ATP} + M_1 \underset{\alpha_2}{\overset{\alpha_1}{\rightleftharpoons}} M_2 + \mathrm{ADP} + P$$

$$\mathrm{ADP} + P + M_1 \underset{\gamma_2}{\overset{\gamma_1}{\rightleftharpoons}} M_2 + \mathrm{ATP}$$

$$M_1 \underset{\beta_2}{\overset{\beta_1}{\rightleftharpoons}} M_2,$$

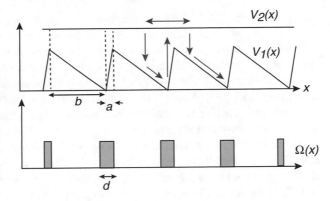

Fig. 4.20: Simplified Brownian ratchet model in which $V_1(x)$ is a periodic ratchet potential and $V_2(x)$ is constant. Also shown is an example of a localized function $\Omega(x)$ that signals regions where there is a breakdown of detailed balance.

with $\alpha_j, \gamma_j, \beta_j$ x-dependent. The first reaction pathway involves ATP hydrolysis with chemical free energy gain $\Delta\mu$ and a corresponding transition from state 1 to state 2, the second involves hydrolysis in the opposite direction, while the third involves thermal state transitions without any change in chemical free energy. Basic chemical kinetics implies that

$$\frac{\alpha_1}{\alpha_2} = e^{(V_1-V_2+\Delta\mu)/k_BT}, \quad \frac{\gamma_1}{\gamma_2} = e^{(V_1-V_2-\Delta\mu)/k_BT},$$

$$\frac{\beta_1}{\beta_2} = e^{(V_1-V_2)/k_BT}. \tag{4.4.10}$$

It follows that the net transition rates between the two conformational states are

$$\omega_1 = \alpha_2 e^{(V_1-V_2+\Delta\mu)/k_BT} + \gamma_2 e^{(V_1-V_2-\Delta\mu)/k_BT} + \beta_2 e^{(V_1-V_2)/k_BT}, \tag{4.4.11}$$

$$\omega_2 = \alpha_2 + \gamma_2 + \beta_2. \tag{4.4.12}$$

Clearly detailed balance no longer holds. In general, it is now necessary to determine the steady-state solution of the pair of equations in (4.4.4) numerically. Given such a solution, the efficiency of the motor doing work against a load F may be determined as follows. First the flux (4.4.3) has an additional term of the form $Fp_i(x,t)/\gamma$. The mechanical work done per unit time against the external force is then $\dot{W} = Fv$ where $v = l\widehat{J_0}$ is the velocity of the motor. On the other hand, the chemical energy consumed per unit time is $\dot{Q} = r\Delta\mu$, where r is the steady-state rate of ATP consumption:

$$r = \int_0^l [(\alpha_1(x) - \gamma_1(x))\widehat{p}_1(x) - (\alpha_2(x) - \gamma_2(x))\widehat{p}_2(x)]dx.$$

The efficiency of the motor is then defined to be [421] $\eta = Fv/r\Delta\mu$.

A major mathematical challenge is to determine how the effective speed of the molecular motor depends on the asymmetries of the potentials and/or the transition rates. (If these functions are symmetric, then there is no net polarization and the speed is zero.) In the weak diffusion limit, analytical tools from PDE theory and transport processes have been used to prove the existence of a steady-state solution of equation (4.4.4) in a bounded domain with no-flux boundary conditions. Moreover, for certain classes of potential function and transition rates, the steady-state density localizes to one end or other of the domain—the so-called motor effect [171, 655, 833]. A simple heuristic argument for directed motion can be given [421, 667] by considering the switch between an asymmetric ratchet potential $V_1(x)$ and a uniform potential $V_2(x) = \text{const}$ for which pure diffusion occurs, see Fig. 4.20. Suppose that the motor starts at a minimum of the potential $V_1(x)$ and is excited to state 2. In this state it undergoes diffusion, which generates a Gaussian probability density with a width $\sqrt{2Dt}$ at time t. The motor should spend sufficient time τ_2 in state 2 so that it has a reasonable chance to jump down to the well of the next minimum on the right, and yet not enough time to jump too far to the left. This suggests that $\tau_2 \sim a^2/D$ where a is the width of the steep part of the potential. The motor also needs just enough time in state 1 in order to move down the shallow part of the potential to the next minimum. If the width of the shallow part is b and the maximum potential is V_1, then the net drift is $V_1/\gamma b$, Assuming that the drift induced by the force dominates diffusion in state 1, we have $\tau_1 \sim b^2\gamma/V_1 \ll \tau_2$. Such a condition violates detailed balance. One way to measure the deviation from detailed balance is to introduce the quantity

$$\Omega(x) = \omega_1(x) - \omega_2(x)\mathrm{e}^{[V_1(x)-V_2(x)]/k_BT}. \qquad (4.4.13)$$

One finds that the mean velocity v depends on the amplitude $\Omega(x)$ and whether it is homogeneous (x-independent), or localized as shown in Fig. 4.20. In the homogeneous case, the motor speed is a unimodal function of the amplitude $\Omega_0 = \max_x |\Omega(x)|$, with the maximum at $\Omega_0 \sim 1/\tau_1$ and $v \to 0$ as $\Omega_0 \to \infty$. On the other hand, the speed is monotonically increasing for a localized perturbation, see Fig. 4.21.

The Brownian ratchet is one important example of a stochastic system with a nonequilibrium steady state (NESS). An NESS has a number of characteristics that distinguish it from an equilibrium state: irreversibility, breakdown of detailed balance, and free energy dissipation. In particular, it is a steady state in which there are constant nonzero fluxes or currents. (Another type of stochastic process that exhibits an NESS is diffusion with stochastic resetting, see Sect. 7.5.) For a recent

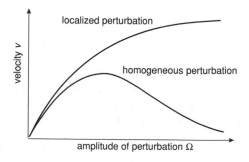

Fig. 4.21: Schematic diagram of the mean velocity v (for zero external force) of a two-state Brownian ratchet as a function of Ω, which measures the departure from equilibrium and is related to the fuel concentration [Redrawn from [421].]

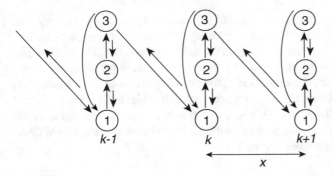

Fig. 4.22: State transition diagram for a discrete Brownian ratchet that cycles through $M = 3$ internal states and makes a single step of length Δx.

review of the theory and applications of nonequilibrium steady states, see Zhang et al. [365, 890]. Although Brownian ratchet models provide important insights into the mechanisms underlying molecular motor dynamics, they have certain limitations. First, as we have already highlighted, it is difficult to obtain analytical solutions of the full equations in order to construct velocity–force curves, for example. Second, there is currently not enough experimental data regarding the potentials $V_j(x)$ and transition rates $\omega_{ij}(x)$ to sufficiently constrain models. Moreover, a large number of model parameters are needed to specify these functions, making data fitting problematic.

The above motivates an alternative approach to modeling molecular motors, based on a discrete Markov process [249, 461, 520, 525]. The basic idea is to take the transition rate functions to be localized at a discrete set of spatial positions $x = x_k$, $k = 1, \ldots, K$, and to replace the continuum diffusion and drift terms by hopping rates between nearest lattice sites. The resulting discrete Brownian ratchet model can be mapped on to a stochastic network of KM states as shown in Fig. 4.22. The stochastic dynamics is now described by a master equation, an example of which is

$$\frac{dP_{km}(t)}{dt} = \sum_{n \neq m} [P_{kn}(t)W_{km;kn} - P_{km}(t)W_{kn;km}] + P_{k+1,1}(t)W_{kM;k+1,1} \qquad (4.4.14)$$

$$+ P_{k-1,M}(t)W_{k1;k-1,M} - P_{k,1}(t)W_{k-1,M;k,1} - P_{k,M}(t)W_{k+1,1;k,M},$$

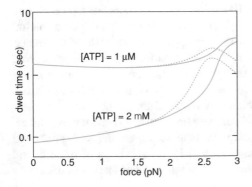

Fig. 4.23: Mean dwell times of myosin V as a function of external load at different ATP concentrations. The solid lines correspond to theoretical predictions from a discrete stochastic model analyzed by Kolomeisky and Fisher [459], whereas the dashed lines correspond to experimental data of Mehta et al. [571], whereas t [Redrawn from Kolomeisky and Fisher [459].]

where $P_{km}(t) = p_m(x_k, t)$ and for "vertical" transitions $W_{km;kn} = \omega_{mn}(x_k)$. In this example steps along the filament (power strokes) only occur between states $m = 1$ and $m = M$. One can then use methods developed by Derrida [215] to calculate the effective diffusion and velocity of a particle whose state probability evolves according to such a master equation. One of the advantages of the discrete models is that the relatively small number of parameters makes it easier to fit to experimental data as shown in Fig. 4.4 [461]. This is illustrated in Fig. 4.23, which shows the results of fitting a discrete model to data on the ATP dependence of myosin V dwell times, that is, the times between successive steps.

4.5 Collective effects of multiple molecular motors

In many cases, molecular motors work in groups rather than in isolation [346]. For example, the number of myosin motors involved in muscle contraction can be around 10^{19}, the number of dynein motors responsible for the beating of cilia or flagella is roughly 10^4, and up to 10 coordinated motors can be involved in the intracellular transport of vesicles. A useful characterization of the collective behavior in motor assemblies distinguishes between rowers and porters [503]. Rowers spend most of their time unbound from their cytoskeletal filaments so they need to operate as part of a large assembly in order to produce sufficiently high velocities. This is exemplified by various classes of myosin motors. On the other hand, porters such as kinesin need to be more carefully coordinated, since the presence of other motors can impede the motion of any individual motor within the assembly. A variety of theoretical models have shown that there are interesting collective effects in motor ensembles, including bidirectional motion, spontaneous oscillations, hysteresis, and the formation of self-organizing structures.

4.5.1 Cooperative cargo transport by multiple motors

Often intracellular cargo such as a vesicle is transported by multiple motors forming a motor/cargo complex. Here we will consider a simple model of a motor/cargo complex due to Klumpp et al. [452], in which there is only a single type of motor, kinesin say, responsible for transport. We will keep track of the state of the motor, but not its position along an MT; the coupling between the two will be considered in Chap. 7, when we consider active transport models. Suppose that there are N identical motors irreversibly attached to the cargo particle, but which can bind to and unbind from the filament along which they move. Thus, the number n of motor molecules that are bound to the filament can vary between $n = 0$ and $n = N$. Hence, there are $N + 1$ different states of the cargo particle corresponding to the unbound state with $n = 0$ and to N bound states with $n = 1, 2, \ldots, N$. Each of these bound states contains $N!/(N-n)!n!$ substates corresponding to the different combinations of connecting n motor molecules to the filament. The dynamics of the motor complex depends on the properties of individual motors combined with the observation that an applied force is shared equally between the motors.

(a) When bound to an MT, the velocity of a single molecular motor decreases approximately linearly with the force applied against the movement of the motor [829]. Thus, each motor is assumed to satisfy the linear force–velocity relation

$$v(F) = \begin{cases} v_f(1 - F/F_s) & \text{for } F \leq F_s \\ v_b(1 - F/F_s) & \text{for } F \geq F_s, \end{cases} \tag{4.5.1}$$

where F is the applied force, F_s is the stall force satisfying $v(F_s) = 0$, v_f is the forward motor velocity in the absence of an applied force in the preferred direction of the particular motor, and v_b is the backward motor velocity when the applied force exceeds the stall force.

(b) The binding rate is independent of the applied force, whereas the unbinding rate is taken to be an exponential function of the applied force:

$$\pi(F) = \bar{\pi}, \quad \gamma(F) = \bar{\gamma} e^{\frac{F}{F_d}}, \tag{4.5.2}$$

where F_d is the experimentally measured force scale on which unbinding occurs. The force dependence of the unbinding rate is based on measurements of the walking distance of a single motor as a function of load [736], in agreement with Kramer's rate theory [357] (see also Sect. 2.4.2).

(c) Now suppose that the externally applied load or force F acts on a motor/cargo complex with n independent molecular motors. If the motors are not directly coupled to each other then they act independently and share the load. It follows that a single motor feels the force F/n. Hence, the velocity of the cargo when there are n bound motors is

$$v_n = v(F/n).$$

Moreover, (4.5.2) implies that the population binding and unbinding rates take the form

$$\gamma_n = n\gamma(F/n), \quad \pi_n = (N-n)\bar{\pi}. \tag{4.5.3}$$

Birth–death master equation. One can model transitions between the different internal states of the cargo complex using a birth–death master equation, see Fig. 4.24. Let $P_n(t)$ be the probability that there are n bound motors at time t. Then

$$\frac{dP_n}{dt} = \gamma_{n+1} P_{n+1} + \pi_{n-1} P_{n-1} - (\gamma_n + \pi_n) P_n, \tag{4.5.4}$$

with reflecting boundaries at $n = 0, N$. There are a number of quantities that characterize the properties of the transport process. We begin by calculating the steady-state distribution of bound motors, which satisfies the equation $J_n = J_{n+1}$ with

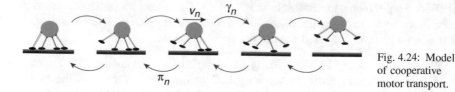

Fig. 4.24: Model of cooperative motor transport.

$$J_n = \gamma_n P_n - \pi_{n-1} P_{n-1}.$$

Since n is nonnegative, we have $P_n = 0$ for all $n < 0$, which means that $J_n = 0$ for all $n \geq 0$. Hence,

$$\gamma_{n+1} P_{n+1} = \pi_n P_n,$$

so that by iteration,

$$P_n = P_0 \prod_{i=0}^{n-1} \frac{\pi_i}{\gamma_{i+1}}, \qquad (4.5.5)$$

with P_0 determined from the normalization $\sum_{n=0}^{N} P_n = 1$. The correctly normalized probability distribution of the bound states is then

$$\widetilde{P}_n = \frac{P_n}{1 - P_0}.$$

It follows that the mean number of bound motors in steady state is

$$N_b = \sum_{n=1}^{N} \frac{n P_n}{1 - P_0}, \qquad (4.5.6)$$

and the mean cargo velocity is

$$v_{\text{eff}} = \sum_{n=1}^{N} v_n \frac{P_n}{1 - P_0}. \qquad (4.5.7)$$

The steady-state distribution of the number of bound motors also yields an explicit expression for the effective cargo detachment rate. In steady state, the effective cargo attachment and detachment rates satisfy

$$\gamma_{\text{eff}}(1 - P_0) = \pi_{\text{eff}} P_0,$$

where $1 - P_0$ is the probability that the cargo is bound to the filament via at least one motor. The effective binding rate is $\pi_{\text{eff}} = \bar{\pi}$, since attachment is established as soon as one motor binds to the filament. Thus

$$\gamma_{\text{eff}} = \frac{\bar{\pi} P_0}{1 - P_0} = \frac{\gamma_1 P_1}{1 - P_0}.$$

Using

$$P_{n+1} = P_1 \prod_{i=1}^{n} \frac{\pi_i}{\gamma_{i+1}}, \quad n \geq 1,$$

and $1 - P_0 = P_1 + \sum_{n=1}^{N-1} P_{n+1}$, we have

$$\gamma_{\text{eff}} = \gamma_1 \left[1 + \sum_{n=1}^{N-1} \prod_{i=1}^{n} \frac{\pi_i}{\gamma_{i+1}} \right]^{-1}. \tag{4.5.8}$$

Explicit formulae can be obtained in the absence of a load force, for which $\gamma_n = n\bar{\gamma}$ and $\pi_n = (N-n)\bar{\pi}$ (see Ex. 4.7). For example, the steady-state distribution is

$$P_n = P_0 \frac{N!}{(N-n)!n!} \left(\frac{\bar{\pi}}{\bar{\gamma}} \right)^n, \quad P_0 = \left(1 + \frac{\bar{\pi}}{\bar{\gamma}} \right)^{-N}, \tag{4.5.9}$$

and for large N,

$$N_b \approx N \frac{\bar{\pi}/\bar{\gamma}}{1 + \bar{\pi}/\bar{\gamma}}. \tag{4.5.10}$$

First passage time density. The unbinding rate (4.5.8) can be interpreted as the inverse of a MFPT for a birth–death process. Following Sect. 3.3.3, consider an absorbing boundary at $n = 0$ and a reflecting boundary at $n = N$. Using equation (3.3.34) with $a = 1$ shows that the MFPT for all motors to be unbound, given that there were n bound motors at $t = 0$, is

$$\tau_n = \sum_{l=0}^{n-1} \phi_l \sum_{m=l}^{N-1} \frac{1}{\gamma_{m+1}\phi_m}, \quad \phi_m = \prod_{j=1}^{m} \frac{\gamma_j}{\pi_j}, \quad \phi_0 = 1. \tag{4.5.11}$$

In particular,

$$\tau_1 = \sum_{m=1}^{N-1} \frac{1}{\gamma_{m+1}} \left(\prod_{j=1}^{m} \frac{\gamma_j}{\pi_j} \right)^{-1} + \frac{1}{\gamma_1} = \gamma_{\text{eff}}^{-1}.$$

For the given birth–death process, it is possible to determine the full FPT density rather than just the mean [452]. (Similar methods arise in the analysis of continuous-time random walks; see Sect. 6.1.2). First, let $\psi_m(t)$ denote the probability density for the FPT to go from m bound motors to the unbound state at time t. The FPT density satisfies a recursion formula of the form

$$\psi_m(t) = \int_0^t \rho_m(\tau) \left[\frac{\pi_m}{\gamma_m + \pi_m} \psi_{m+1}(t - \tau) + \frac{\gamma_m}{\gamma_m + \pi_m} \psi_{m-1}(t - \tau) \right] d\tau \tag{4.5.12}$$

for $m \neq 0, N$, where $\rho_m(\tau)$ is the waiting time density to make a jump from state m to $m \pm 1$. In order to understand the origin of this formula, let $m \xrightarrow{t} m'$ denote the possibly multistep change of state $m \to m'$ occurring over the time interval t. It follows that $m \xrightarrow{t} 0$ can be decomposed as $m \xrightarrow{\tau} m \pm 1 \xrightarrow{t-\tau} 0$ with $\pi_m/(\gamma_m + \pi_m)$ the probability that the first transition is to $m+1$ and $\gamma_m/(\gamma_m + \pi_m)$ the probability that the first transition is to $m-1$. Integrating over all intermediate times $\tau \in [0,t]$ yields equation (4.5.12). In the case of a homogeneous continuous-time Markov process, the waiting time density is an exponential (see Sect. 5.2.3),

$$\rho_m(\tau) = \alpha_m e^{-\alpha_m \tau}, \quad \alpha_m = \pi_m + \gamma_m. \tag{4.5.13}$$

Hence, equation (4.5.12) becomes

$$\psi_m(t) = \int_0^t e^{-(\pi_m + \gamma_m)\tau} \left[\pi_m \psi_{m+1}(t-\tau) + \gamma_m \psi_{m-1}(t-\tau) \right] d\tau. \tag{4.5.14}$$

Since $\pi_N = 0$, we have

$$\psi_N(t) = \int_0^t e^{-\gamma_N \tau} \gamma_N \psi_{N-1}(t-\tau) d\tau.$$

Moreover, $\psi_0(y) = \delta(t)$ as the unbinding time is zero. In order to solve the recursion equation (4.5.14), we use Laplace transforms:

$$\widetilde{\psi}_m(s) = \frac{1}{\pi_m + \gamma_m + s} \left[\pi_m \widetilde{\psi}_{m+1}(s) + \gamma_m \widetilde{\psi}_{m-1}(s) \right], \quad m = 0, N, \tag{4.5.15a}$$

and

$$\widetilde{\psi}_0(s) = 1, \quad \widetilde{\psi}_N(s) = \frac{\gamma_N}{\gamma_N + s} \widetilde{\psi}_{N-1}(s). \tag{4.5.15b}$$

Setting $m = N - 1$ in equation (4.5.15a) yields

$$\widetilde{\psi}_{N-1}(s) = \frac{1}{\pi_{N-1} + \gamma_{N-1} + s} \left[\pi_{N-1} \widetilde{\psi}_N(s) + \gamma_{N-1} \widetilde{\psi}_{N-2}(s) \right],$$

which on substituting for $\widetilde{\psi}_N(s)$ and rearranging shows that

$$\widetilde{\psi}_{N-1}(s) = \frac{\gamma_{N-1}}{\gamma_{N-1} + s + \pi_{N-1} \left(1 - \frac{\gamma_N}{\gamma_N + s} \right)} \widetilde{\psi}_{N-2}(s).$$

Similarly, setting $m = N - 2$ in equation (4.5.15a) gives

$$\widetilde{\psi}_{N-2}(s) = \frac{\gamma_{N-2}}{\gamma_{N-2} + s + \pi_{N-2} \left(1 - \frac{\gamma_{N-1}}{\gamma_{N-1} + s + \pi_{N-1} \left(1 - \frac{\gamma_N}{\gamma_N + s} \right)} \right)} \widetilde{\psi}_{N-3}(s).$$

This procedure can be iterated until $m = 1$ and we impose $\widetilde{\psi}_0(s) = 1$. One thus obtains finite continued fractions that can be rewritten as polynomial quotients in s and thus inverted to obtain $\psi_m(t)$ as a sum of decaying exponentials in time t [452].

Distribution of walking distances. A similar analysis can be carried out for the probability density $\varphi_m(x)$ that the complex travels a distance x before unbinding, starting out in the state m. By the same reasoning as for FPTs,

$$\varphi_m(x) = \int_0^x \widehat{\rho}_m(y) \left[\frac{\pi_m}{\gamma_m + \pi_m} \varphi_{m+1}(x-y) + \frac{\gamma_m}{\gamma_m + \pi_m} \varphi_{m-1}(x-y) \right] dy, \quad (4.5.16)$$

where $\widehat{\rho}_m(y)$ is the distribution of walking distances y in state m before the jump to $m \pm 1$. It follows that

$$\widehat{\rho}_m(y)dy = \rho_m(\tau)d\tau, \quad \tau = y/v_m,$$

where v_m is the velocity in state m. Therefore,

$$\widehat{\rho}_m(y) = \frac{\gamma_m + \pi_m}{v_m} e^{-(\gamma_m + \pi_m)y/v_m}, \quad (4.5.17)$$

and

$$\varphi_m(x) = \int_0^x e^{-(\pi_m + \gamma_m)y/v_m} \left[\frac{\pi_m}{v_m} \varphi_{m+1}(x-y) + \frac{\gamma_m}{v_m} \varphi_{m-1}(x-y) \right] dy. \quad (4.5.18)$$

Hence, the moments of the walking distance are given by the same expressions as the moments of the FPT with $\gamma_m, \pi_m \to \gamma_m/v_m, \pi_m/v_m$. In particular, the mean walking distance is

$$\langle x \rangle = \int_0^\infty x\varphi_1(x)dx = \frac{v_1}{\gamma_1} \left[1 + \sum_{n=1}^{N-1} \prod_{i=1}^n \frac{v_{i+1}\pi_i}{v_i\gamma_{i+1}} \right]. \quad (4.5.19)$$

The mean distance for the specific choices $of \gamma_m = m\varepsilon$, $\pi_m = (N-m)\pi$, and $v_m = v$ is calculated in Ex. 4.7.

4.5.2 Tug-of-war model

Using single-particle tracking experiments, trajectories of individual motor–cargo complexes can be recorded, and shown to exhibit many random turning events [345, 473, 741, 854]. This immediately raises the issue of how bidirectional transport is achieved, given that motors such as kinesin and dynein are unidirectional. Recall that MTs are polarized filaments with biophysically distinct + and − ends, and the polarity determines the preferred direction of motion of individual motors: kinesin (dynein) moves toward the + (−) end. There is considerable debate in the literature regarding the most likely mechanism for bidirectional transport. Several different scenarios have been proposed including those shown in Fig. 4.25: (a) An asymmetric tug-of-war model involving the joint action of multiple kinesin and dynein motors pulling in opposite directions; (b) A symmetric tug-of-war model where all the motors are of the same type, but they are distributed on MTs of opposite polarity; (c) A hopping model, in which the whole motor–cargo complex hops between MTs of opposite polarity. Yet another suggested mechanism (not shown) is some form of coordination complex that controls the switching between different motor species. It might be possible to apply statistical methods to single-particle tracking data which, when combined with knowledge of the individual motor dynamics, could identify the underlying mechanism(s) for bidirectional transport [32]. However, as far as we are aware, the debate continues! For the sake of illustration, we will focus on the first scenario here.

Suppose that a certain vesicular cargo is transported along a one-dimensional track via N_+ right-moving (anterograde) motors and N_- left-moving (retrograde motors), see Fig. 4.25(a). At a given time t, the internal state of the cargo–motor complex is fully characterized by the numbers n_+ and n_- of anterograde and retrograde motors that are bound to an MT and thus actively pulling on the cargo. Assume that over the time scales of interest, all motors are permanently bound to the cargo so that $0 \leq n_\pm \leq N_\pm$. The tug-of-war (ToW) model of Muller *et. al.* [596, 597] assumes that the motors act independently other than exerting a load on motors with the opposite directional preference. (However, some experimental work suggests that this is an oversimplification, that is, there is some direct coupling between motors [228].) Thus the properties of the motor complex can be determined from the corresponding properties of the individual motors together with a specification of the effective load on each motor.

The ToW model can be constructed as a generalization of the cooperative motor model. At the single motor level, equations (4.5.1) and (4.5.2) still hold, except that the value of parameters such as the stall force will differ for kinesin and dynein. However, even in the absence of an externally applied load, there is now an effective

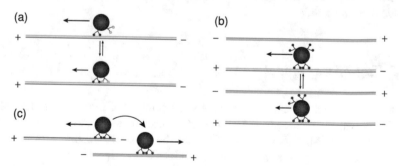

Fig. 4.25: Molecular motor-assisted models of bidirectional transport. (a) Asymmetric tug-of-war model: Opposing groups of motors (e.g., dynein and kinesin) compete to transport along a single polarized MT track. (b) Symmetric tug-of-war model: Groups of motors of the same directional preference are distributed among two parallel MTs of opposite polarity. (c) Hopping model in which a motor–cargo complex jumps between MTs of opposite polarity.

force on one class of motor due to the opposing action of the other class. Let F_c denote the net load on the set of anterograde motors, which is taken to be positive when pointing in the retrograde direction. If the molecular motors are not directly coupled to each other, then a single anterograde motor feels the force F_c/n_-, whereas a single retrograde motor feels the opposing force $-F_c/n_+$. At the population level, the binding and unbinding rates for the two types of motor are

$$\gamma_j(n_j) = n_j \bar{\gamma}_j e^{F_c/F_{D,j} n_j}, \quad \pi_j(n_j) = (N_j - n_j) \bar{\pi}_j \quad j = \pm. \tag{4.5.20}$$

The cargo force F_c is determined self-consistently by the condition that all the motors move with the same cargo velocity v_c. Suppose that $N_+ \geq N_-$ so that the net motion is in the anterograde direction, which is taken to be positive. In this case, the forward motors are stronger than the backward motors so that $n_+ F_{s+} > n_- F_{s-}$. Equation (4.5.1) implies that

$$v_c = v_{f+}(1 - F_c/(n_+ F_{s+})) = -v_{b-}(1 - F_c/(n_- F_{s-})). \tag{4.5.21}$$

This generates a unique solution for the load F_c and cargo velocity v_c:

$$F_c(n_+, n_-) = (\mathcal{F} n_+ F_{s+} + (1 - \mathcal{F}) n_- F_{s-}), \tag{4.5.22}$$

where

$$\mathcal{F} = \frac{n_- F_{s-} v_{f+}}{n_- F_{s-} v_{f+} + n_+ F_{s+} v_{b-}}, \tag{4.5.23}$$

and

$$v_c(n_+, n_-) = \frac{n_+ F_{s+} - n_- F_{s-}}{n_+ F_{s+}/v_{f+} + n_- F_{s-}/v_{b-}}. \tag{4.5.24}$$

The corresponding expressions when the backward motors are stronger, $n_+ F_{s+} < n_- F_{s-}$, are found by interchanging v_f and v_b.

The original study of [596, 597] considered the stochastic dynamics associated with transitions between different internal states (n_+, n_-) of the motor complex, without specifying the spatial position of the complex along a 1D track. This defines a Markov process with a corresponding master equation for the time evolution of the probability distribution $P(n_+, n_-, t)$. They determined the steady-state probability distribution of internal states and found that the motor complex exhibited at least three different modes of behavior, see Fig. 4.26, which were consistent with experimental studies of motor transport using single-particle tracking; the transitions between these modes of behavior depend on motor strength, which primarily depends upon the stall force.

(i) The motor complex spends most of its time in states with approximately zero velocity.

(ii) The motor complex alternates between fast backward and forward movements, so that there is a bimodal velocity distribution with peaks close to the single motor velocities of $1\,\mu$ m/s.

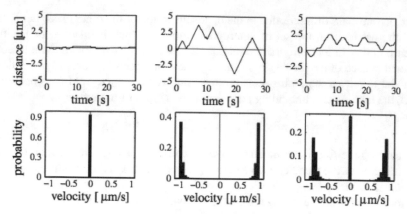

Fig. 4.26: Motility states for the symmetric tug-of-war model consisting of $N = 4$ plus and $N = 4$ minus motors. The three columns correspond to the three listed motility states (i), (ii), and (iii), respectively. The top row shows a typical trajectory for given motility state, whereas the bottom row shows a histogram of the distribution of velocities. The different motility behavior is obtained by taking different stall forces F_S and unbinding rates $\bar{\varepsilon}$: (i) $F_S = 2$ pN, $\bar{\varepsilon} = 0.4$ s^{-1}, (ii) $F_S = 6$ pN, $\varepsilon_0 = 1$ s^{-1}, (iii) $F_S = 4.75$ pN, $\varepsilon_0 = 0.4$ s^{-1}. These and other single motor parameter values are based on experimental data from kinesin 1: detachment force $F_d = 3$ pN, binding rate $\bar{\pi} = 5$ s^{-1}, forward velocity $v_f = 1$ μm/s, and backward velocity $v_b = 6$ nm/s [Adapted from Muller et al. [596].]

(iii) The motor complex exhibits fast backward and forward movement interrupted by stationary pauses, which is consistent with experimental studies of bidirectional transport. The velocity distribution now has three peaks.

One of the useful features of the tug-of-war model is that it allows various biophysical signaling mechanisms to be incorporated into the model [609, 610, 664]. This will be exploited in Sect. 7.3.4, when we use the tug-of-war model to study the effects of local chemical signaling on intracellular cargo transport.

ATP signaling. Experimentally, it is found that [ATP] primarily affects the stall force, forward motor velocity, and unbinding rate; see for example Fig. 4.4. There are a number of models of the [ATP] and force-dependent motor parameters that closely match experiments for both kinesin [276, 585, 736, 829] and dynein [297, 446]. We give some examples of [ATP]-dependent parameters. First, the forward velocity can be modeled using Michaelis–Menten kinetics (see Sect. 1.3)

$$v_f([ATP]) = \frac{v_f^{max}[ATP]}{[ATP] + K_v}, \qquad (4.5.25)$$

where $v_f^{max} = 1$ μm/s, $K_v = 79.23$ μM for kinesin, and $v_f^{max} = 0.7$ μm/s, $K_v = 38$ μM for dynein. (The backward velocity of both kinesin and dynein is small, $v_b \approx \pm 0.006$ μm/s, so that the [ATP] dependence can be neglected.) The binding rate is determined by the time necessary for an unbound motor to diffuse within range of the MT and bind to it, which is assumed to be independent of [ATP]. The unbinding rate of a single motor under zero load can be determined using the [ATP] dependent average run length $L_k([ATP]) = L_k^{max}/([ATP] + K_u)$. The mean time to detach from the MT is $v_f([ATP])/L_k([ATP])$ so that

$$\bar{\gamma}([ATP]) = \frac{v_f^{max}([ATP] + K_u)}{L_k^{max}([ATP] + K_v)}, \qquad (4.5.26)$$

where $L_k^{max} = 0.86\,\mu\text{m}$, $K_u = 3.13\,\mu\text{M}$ for kinesin and $L_k^{max} = 1.5\,\mu\text{m}$, $K_u = 1.5\,\mu\text{M}$ for dynein. Finally, a model for the [ATP]-dependent stall force of kinesin is

$$F_s([ATP]) = F_s^0 + \frac{(F_s^{max} - F_s^0)[ATP]}{K_s + [ATP]}, \qquad (4.5.27)$$

where $F_s^0 = 5.5$ pN, $F_s^{max} = 8$ pN, $K_s = 100\,\mu\text{M}$ for kinesin and $F_s^0 = 0.22$ pN, $F_s^{max} = 1.24$ pN, $K_s = 480\,\mu\text{M}$ for dynein.

Tau signaling. The second signaling mechanism involves MT associated proteins (MAPs). These molecules bind to MTs and effectively modify the free energy landscape of motor–microtubule interactions [800]. For example, tau is a MAP found in the axon of neurons and is known to be a key player in Alzheimer's disease [463]. Experiments have shown that the presence of tau on the MT can significantly alter the dynamics of kinesin; specifically, by reducing the rate at which kinesin binds to the MT [828]. It has also been shown that, at the low tau concentrations affecting kinesin, dynein is relatively unaffected by tau. Thus tau signaling can be incorporated into the tug-of-war model by considering a tau concentration-dependent kinesin binding rate of the form [611]

$$\bar{\pi}(\tau) = \frac{\bar{\pi}^{max}}{1 + e^{-\gamma(\tau_0 - \tau)}}, \qquad (4.5.28)$$

where τ is the dimensionless ratio of tau per MT dimer and $\bar{\pi}^{max} = 5$ s^{-1}. The remaining parameters are found by fitting the above function to experimental data [828], so that $\tau_0 = 0.19$ and $\gamma = 100$.

4.5.3 Rigidly linked molecular motors

As our second example of a motor assembly, consider the case where N motors are rigidly coupled to a common backbone, which is connected to a fixed cytoskeletal structure via a spring K [419–421], see Fig. 4.27. Each motor is treated as a two-state Brownian ratchet, such that the motor sees an asymmetric, l-periodic potential $V_1(x)$ when bound to a cytoskeletal filament and a flat potential V_2 when unbound. The motors switch between the two states with position-dependent transition rates $\omega_{1,2}(x)$, which do not satisfy detailed balance due to ATP hydrolysis. Cooperativity

arises due to the global motion of the backbone relative to the filament, which simultaneously modifies the positions of all the motors. Such a configuration mimics the
experimental protocol known as a motility assay as well as the coupling of myosin
motor cross bridges in muscles.

Suppose that the displacement of the backbone at time t is $Y(t)$, and the center-
of-masses of the motors are separated by a uniform spacing q on the backbone. It
follows that the position of the nth motor at time t is $x_n(t) = Y(t) + nq$. Each motor
is either in the bound state ($\sigma = 1$) or the unbound state ($\sigma = 2$) and the energy of
the nth motor in state σ is $V_\sigma(x_n(t))$. However, since $V_\sigma(x+l) = V_\sigma(x)$ for all x, we
need only specify the position of the motor using the cyclic coordinate $\xi_n(t) = x_n(t)$
mod l with $0 < \xi_n < l$. Let $P_\sigma(\xi,t)$ be the probability density that there exists a
motor in state σ at cyclic position ξ at time t. It follows that

$$P(\xi,t) \equiv P_1(\xi,t) + P_2(\xi,t) = \frac{1}{N} \sum_{n=1}^{N} \delta(\xi - \xi_n(t)).$$

A major simplification occurs if the periodic spacing q of the motors is incommen-
surate with the period l of the filament track, that is, l/q is an irrational number.
In the limit $N \to \infty$, the motor positions ξ_n form a dense subset on $[0,l]$ such that
$P(x,t) \to 1/l$. The equations of motion for P_σ are [419]

$$\frac{\partial P_1}{\partial t} + v \frac{\partial P_1}{\partial \xi} = -\omega_1 P_1 + \omega_2 P_2, \tag{4.5.29a}$$

$$\frac{\partial P_2}{\partial t} + v \frac{\partial P_2}{\partial \xi} = \omega_1 P_1 - \omega_2 P_2, \tag{4.5.29b}$$

where $v = dY/dt$ is the velocity of the backbone and the transition rates satisfy
equation (4.4.13). It is convenient to set $\Omega(\xi) = \Omega\theta(\xi)$ and $\int_0^l \theta(\xi)d\xi = 1$. For
concreteness, ω_2 is a constant and $\omega_1(\xi)$ is specified by taking $\theta(\xi)$ to be a periodic sequence of square pulses of width d, see Fig. 4.27. Finally, the velocity v is

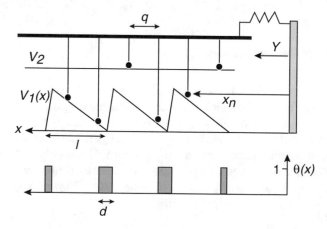

Fig. 4.27: Ensemble of
molecular motors cou-
pled rigidly to a moving
backbone. See text for
details.

determined by the force balance equation (per motor)

$$F_{\text{ext}} = \mu v(t) + KY(t) + \int_0^l (P_1 \partial_\xi V_1 + P_2 \partial_\xi V_2) d\xi, \qquad (4.5.30)$$

where F_{ext} is an externally applied force. The first term on the right-hand side is a frictional force with damping coefficient μ, the second term is the elastic force from the spring with elastic modulus NK, and the third term is the force due to the potentials.

For an incommensurate system with $P_2 = l^{-1} - P_1$, and a soft spring ($K = 0$), there exists a steady state that satisfies the pair of equations

$$v \frac{dP_1}{d\xi} = -(\omega_1 + \omega_2)P_1 + \frac{\omega_2}{l}, \qquad (4.5.31)$$

and (since V_2 is constant)

$$F_{\text{ext}} = \mu v + \int_0^l P_1 \partial_\xi V_1 d\xi. \qquad (4.5.32)$$

Let us non-dimensionalize the system by fixing the length and time units such that $l, q, \omega_2 = O(1)$ and suppose that $v \ll 1$. The solution for P_1 may then be expanded as a Taylor series in v:

$$P_1(\xi) = \sum_{n=0}^{\infty} v^n P_1^{(n)}(\xi).$$

Substituting into equation (4.5.31) gives

$$P_1^{(0)}(\xi) = \frac{1}{l} \frac{\omega_2(\xi)}{\omega_1(\xi) + \omega_2(\xi)}, \quad P_1^{(n)}(\xi) = -\frac{1}{\omega_1(\xi) + \omega_2(\xi)} \partial_\xi P_1^{(n-1)}(\xi). \quad (4.5.33)$$

Substituting the Taylor expansion of P_1 into (4.5.32) then yields

$$F_{\text{ext}} = F^{(0)} + (\mu + F^{(1)})v + \sum_{n=2}^{\infty} v^n F^{(n)} \qquad (4.5.34)$$

for

$$F^{(n)} = \int_0^l P_1^{(n)} \partial_\xi V_1 d\xi.$$

It follows from equation (4.4.13) that

$$P_1^{(0)}(\xi) = l^{-1} \frac{1}{e^{(V_1(\xi) - V_2)/k_B T} + 1 + \Omega \theta(\xi)/\omega_2(\xi)}.$$

Hence, if detailed balance holds ($\Omega = 0$) then from the periodicity of the potentials,

$$F^{(0)} = -\frac{k_B T}{l} \int_0^l \partial_\xi \ln\left[1 + e^{-(V_1(\xi) - V_2)/K_B T}\right] d\xi = 0.$$

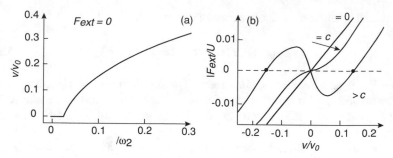

Fig. 4.28: Ensemble of molecular motors coupled rigidly to a moving backbone. (a) Sketch of non-dimensionalized velocity v/v_0 against Ω/ω_2, with $v_0 = \omega_2 l$. There exists a critical amplitude $\Omega = \Omega_c$, where Ω measures the size of the deviation from detailed balance, beyond which the motor assembly has a nonzero velocity in the absence of an external force. (b) Sketch of force–velocity curves for different amplitudes Ω and a symmetric potential $V_1(x)$ of height U and periodicity l, illustrating spontaneous symmetry breaking. For the dimensionless parameter values $d/l = 0.1$ and $\mu v_0 l/U = 0.1$, one finds that $\Omega_c/\omega_2 \approx 0.026$ [Redrawn from Julicher and Prost [419].]

One also finds that $F^{(1)} > 0$, that is, there is an effective increase in friction so $v = 0$ in the absence of an external force. On the other hand, when detailed balance is broken, two important features arise [419]:

(i) If $\Omega \neq 0$ and the potential $V_1(\xi)$ is asymmetric, then $F^{(0)} \neq 0$ and the system can do work against an external load as previously found for a single motor. One subtle point is that we still have $F^{(0)} = 0$ if the transition rates are homogeneous, which differs from the single motor case.

(ii) If $\Omega \neq 0$ then $F^{(1)}$ may become negative as an emergent feature of cooperativity. This can overcome the effects of external damping, resulting in a nonzero velocity in the absence of an external force. Note that in contrast to a single motor, it is no longer necessary for asymmetric potentials or transition rates.

In order to explore feature (ii) in more detail, suppose that both the potential $V_1(\xi)$ and $\theta(\xi)$ are symmetric functions, with $\theta(\xi)$ determining the deviation from detailed balance. The perturbation analysis implies that $P^{(n)}(\xi)$ is symmetric for even n and antisymmetric for odd n; see (4.5.33). Since taking a derivative converts a symmetric function to an antisymmetric one and vice versa, we see from (4.5.34) that $F^{(n)} = 0$ for all even n. Therefore, we have

$$F_{\text{ext}} = (\mu + F^{(1)})v + v^3 F^{(3)} + O(v^5). \qquad (4.5.35)$$

This represents the normal form of an imperfect pitchfork bifurcation. First consider the case $F_{\text{ext}} = 0$. Suppose that $F^{(1)}$ is a monotonically decreasing function of Ω such that $\mu + F^{(1)} = 0$ at a critical value $\Omega = \Omega_c$. Then for $\Omega < \Omega_c$, the only solution is $v = 0$ and the system does not move. However at the critical point $\Omega = \Omega_c$, the zero velocity state loses stability and two new stable solutions $v_\pm$ emerge via spontaneous symmetry breaking, with $v_\pm \sim \sqrt{\Omega - \Omega_c}$ close to the bifurcation point, see

Fig. 4.28(a). One can also determine the relationship between an external force and velocity for different values of Ω as illustrated in Fig. 4.28(b).

Finally, suppose that the spring constant K of the spring connecting the backbone to the cytoskeleton is nonzero. It is now possible for the system to exhibit spontaneous oscillations [420]. In order to show this, consider the linear stability of the zero velocity steady-state solution of equations (4.5.29) and (4.5.30) of zero external force. The steady-state solution is given by

$$P_1(\xi) = R(\xi) \equiv \frac{1}{l} \frac{\omega_2(\xi)}{\omega_1(\xi) + \omega_2(\xi)}, \quad Y = Y_0 \equiv -\frac{1}{K} \int_0^l R(\xi) \partial_\xi V_1(\xi) d\xi.$$

Linearize about the steady-state solution by setting

$$P_1(\xi,t) = R(\xi) + p(\xi)e^{\lambda t}, \quad Y(t) = Y_0 + ye^{\lambda t},$$

and $v(t) = \partial_t Y(t) = \lambda y e^{\lambda t}$, we find

$$\lambda p(\xi) + \lambda y R'(\xi) = -(\omega_1(\xi) + \omega_2(\xi))p(\xi)$$

$$0 = \mu \lambda y + Ky + \int_0^l p(\xi) V_1'(\xi) d\xi.$$

Expressing $p(\xi)$ in terms of y and eliminating y yields the eigenvalue equation

$$\mu + \frac{K}{\lambda} = \int_0^l \frac{R'(\xi) V_1'(\xi)}{\lambda + \omega_1(\xi) + \omega_2(\xi)} d\xi. \tag{4.5.36}$$

The zero velocity state is stable if $\mathrm{Re}\,\lambda < 0$. An instability will occur if there exists a critical value $\Omega = \Omega_c(K)$ for which $\lambda = i\omega$. Note that if $K = 0$ then the left-hand side is real, which implies that λ is real and destabilization of the zero velocity state cannot occur via a Hopf bifurcation. (See Box 5A for a definition of a Hopf bifurcation.) On the other hand, if $K > 0$ then a Hopf bifurcation can occur resulting in a limit cycle oscillation with frequency $\omega(K)$.

4.6 Exercises

Problem 4.1. Spatial polymerization of a filament. Suppose that a polymer filament is placed in a cylinder with uniform cross section A. Also assume that the monomers within the tube can undergo diffusion along the axis of the tube, which is taken to be the x-axis. Let $x_\pm(t)$ denote the positions of the $\pm$ ends of the filament within the tube. The apparent velocities of these ends due to polymerization/depolymerization are

$$\frac{dx_+}{dt} = v_+ = l[k_{on}^+ a(x_+,t) - k_{off}^+],$$

$$\frac{dx_-}{dt} = v_- = -l[k_{on}^- a(x_-,t) - k_{off}^-],$$

where l is the size of a monomer. The ends of the filament act as sources or sinks for monomer, so that the monomer concentration $a(x,t)$ along the axis satisfies the inhomogeneous diffusion equation

$$\frac{\partial a}{\partial t} = D\frac{\partial a^2}{\partial x^2} - \gamma[\delta(x-x_+)v_+ - \delta(x-x_-)v_-], \quad \gamma = \frac{1}{Al}.$$

(a) Derive the diffusion equation by considering conservation of monomer passing through an infinitesimal volume $A\Delta x$ centered about either end of the filament. Explain the minus sign in the definition of v_-.

(b) Suppose that the tube is infinitely long and

$$a(x,t) \to \alpha, \quad x \to \pm\infty.$$

Look for a traveling wave solution in which the filament maintains a fixed length L and $v_\pm = v$, where v is the speed of the wave. That is, set $x_+ = vt, v_- = vt - L$ and go to a moving frame $z = x - vt$ with $a(x,t) = \mathcal{A}(z)$ such that

$$-v\frac{d\mathcal{A}}{dz} = D\frac{d^2\mathcal{A}}{dz^2} + v\gamma[\delta(z+L) - \delta(z)].$$

Explicitly solve this equation by matching the solution at the points $z = -L, 0$. In particular, show that

$$\mathcal{A}(-L) = \alpha, \quad \mathcal{A}(0) = \alpha - 1 + e^{-\gamma v L/D}.$$

(c) Substituting for $\mathcal{A}$ in the expressions for $v_\pm$ and setting $v_+ = v_- = v$, determine v and L. Show that a physical solution only exists if

$$\alpha > \frac{k_{off}^+ + k_{off}^-}{k_{on}^+ + k_{on}^-}.$$

Problem 4.2. FPT problem for confined microtubular growth. Consider the Dogterom–Leibler model for an MT confined to a domain of size L. Suppose that there is a reflecting boundary at $x = 0$ and an absorbing boundary at $x = L$, that is,

$$v_+p_+(0,t) = v_-p_-(0,t), \quad p_-(L,t) = 0.$$

The FPT for the MT to reach the boundary at $x = L$, having started in state (y,m) at time $t = 0$ with $m \in \{+,-\}$, is defined according to

$$T_m(y) = \inf\{t \geq 0; X(t) = L | X(0) = y, N(0) = m\}.$$

Let $S_m(y,t)$ be the survival probability

$$S_m(y,t) = \int_0^L p(x,+,t|y,m,0)dx.$$

(a) Using the same arguments as the analysis of MFPT problems for SDEs, see Sect. 2.4, show that

$$\tau_m(y) := \mathbb{E}[T_m(y)] = \int_0^\infty S_m(y,t)dt.$$

Starting from the backward CK equation, derive the following pair of equations for the MFPTs $\tau_m = \mathbb{E}[T_m(y)]$:

$$-1 = v_+ \frac{d\tau_+}{dy} - k_c(\tau_+ - \tau_-), \quad 0 = -v_- \frac{d\tau_-}{dy} + k_r[\tau_+ - \tau_-],$$

together with the boundary conditions

$$\tau_+(0) = \tau_-(0), \quad \tau_+(L) = 0.$$

(b) Defining

$$\widehat{\tau}(y) = \frac{v_+}{k_c}\tau_+(y) - \frac{v_-}{k_r}\tau_-(y),$$

use the pair of equations in part (a) to obtain the solution $\widehat{\tau}(y) = -y/k_c + A$, where A is a constant. Use the boundary condition at $y = 0$ to show that

$$A = \frac{v_- v_+}{k_r k_c} \frac{V}{D}\tau_-(0),$$

with the mean velocity V and diffusivity D defined by

$$V = \frac{k_r v_+ - k_c v_-}{k_c + k_r}, \quad D = \frac{v_+ v_-}{k_c + k_r}.$$

(c) Use the result of part (b) and the boundary condition $\tau_+(L) = 0$ to express $\tau_-(L)$ in terms of $\tau_-(0)$. Eliminate $\tau_+(y)$ in the equation for $d\tau_-/dy$ by a second application of part (b) and solve the resulting boundary value problem. Hence, obtain the solution

$$\tau_-(y) = \frac{k_c}{v_+}[I(L)/v_+ + L/k_c] - \frac{k_r I(y)}{v_+ v_-},$$

where

$$I(y) := \frac{Dy}{V} - \left(\frac{D}{V}\right)^2 \left(1 - e^{-Vy/D}\right).$$

Problem 4.3. Polymerization ratchet I. Consider a Brownian particle moving in the ratchet potential

$$V(x) = Fx - n\Delta G, \quad na < x < (n+1)a.$$

Following the analysis of Sect. 4.3, we obtain the equation

$$\frac{d}{dx}\left(e^{\mathcal{V}(x)/k_BT}\widehat{p}_0(x)\right) = -\frac{\widehat{J}_0}{D_0}e^{\mathcal{V}(x)/k_BT}.$$

for the stationary distribution $\widehat{p}_0(x) = \sum_{n=-\infty}^{\infty} p_0(x+na)$.

(a) Integrate the above equation from 0^+ to x, $0 < x < a$, and impose the matching condition

$$\lim_{x\to a^+}\widehat{p}_0(x)e^{\mathcal{V}(x)} = \lim_{x\to a^-}\widehat{p}_0(x)e^{\mathcal{V}(x)},$$

together with periodicity $\widehat{p}_0(a^+) = \widehat{p}_0(0^+)$. Hence show that

$$\widehat{p}_0(x) = \frac{\widehat{J}_0 k_B T}{F D_0}\left[\mathcal{A}e^{-Fx/k_BT} - 1\right],$$

with

$$\mathcal{A} = \frac{e^{\Delta G/k_BT} - 1}{e^{(\Delta G - Fa)/k_BT} - 1}.$$

(b) Explain the matching condition used in part (a).

(c) Determine the constant flux $\widehat{J}_0$ using the normalization condition $1 = \int_0^a \widehat{p}_0(x)dx$. Hence show that the speed of growth $v = \widehat{J}_0 a$ is given by

$$v = D_0\frac{F^2 a}{(k_BT)^2}\left[\mathcal{A}\left(1 - e^{-Fa/k_BT}\right) - \frac{Fa}{k_BT}\right]^{-1}.$$

(d) Show that in the regime $\Delta G \gg Fa$ and $k_BT \gg Fa$,

$$v \approx 2D_0/a.$$

Problem 4.4. Polymerization ratchet II. Let $c(x,t)$ denote the density of a population of polymerizing filaments whose tips are at a distance x from some membrane

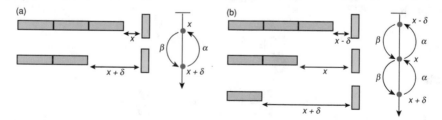

Fig. 4.29: Possible jumps in the distance x of a polymer tip from a membrane barrier due to the addition (subtraction) of a monomer of size δ at a rate α (β). (a) Case $x < \delta$ for which $x \rightleftharpoons x+\delta$. (b) Case $x > \delta$ for which $x - \delta \rightleftharpoons x \rightleftharpoons x+\delta$.

barrier, and whose other ends are fixed. The relative position of each tip undergoes Brownian motion (due to the fluctuating barrier) together with jumps due to
the addition or removal of a monomer of size δ, as illustrated in Fig. 4.29. Write
$c(x,t) = c_-(x,t)$ for $x < \delta$ and $c(x,t) = c_+(x,t)$ for $x > \delta$. In the case of zero depolymerization ($\beta = 0$),

$$\frac{\partial c_-}{\partial t} = D\frac{\partial^2 c_-}{\partial x^2} + \varepsilon D\frac{\partial c_-}{\partial x} + \alpha c_+(x+\delta,t), \quad x < \delta,$$

$$\frac{\partial c_+}{\partial t} = D\frac{\partial^2 c_+}{\partial x^2} + \varepsilon D\frac{\partial c_+}{\partial x} + \alpha[c_+(x+\delta,t) - c_+(x,t)], \quad x > \delta.$$

Here $\varepsilon = f/k_B T$ with f the applied force. The corresponding boundary conditions
are

$$J_-(0,t) = 0, \quad J_-(\delta,t) = J_+(\delta,t), \quad c_-(\delta,t) = c_+(\delta,t),$$

with

$$J(x,t) = -D\left[\frac{\partial c}{\partial x} + \varepsilon c(x,t)\right].$$

Calculate the velocity v of the ratchet in the steady state using the following steps:

(a) Assume a solution of the form $c_+(x) = Ae^{-\mu x}$ with μ to be determined later by
the boundary conditions. The constant A is determined by the total number of filaments. Substitute into the steady-state equation for c_- and solve using the boundary
condition $J_-(0,t) = 0$. In particular, show that

$$c_-(x) = [c_-(0) + \mu\Gamma]e^{-\varepsilon x} - \Gamma\left[\varepsilon e^{-\mu x} - \varepsilon + \mu\right],$$

where

$$\Gamma = \frac{\alpha A}{D\varepsilon\mu(\mu - \varepsilon)}e^{-\mu\delta}.$$

(b) Using continuity of the current J at $x = \delta$, show that μ satisfies the transcendental
equation

$$(\mu - \varepsilon)\mu = \frac{\alpha}{D}\left[1 - e^{-\mu\delta}\right].$$

(c) Using continuity of c at $x = \delta$, show that

$$c_-(0) + \mu\Gamma = \mu\Gamma e^{\varepsilon\delta}.$$

(d) Calculate the velocity v according to the formula

$$v = \left(\frac{N_+}{N_+ + N_-}\right)\alpha\delta,$$

where N_+ is the number of filaments with gap $x > \delta$ and N_- is the number of
filaments with gap $x < \delta$,

$$N_- = \int_0^\delta c_-(x)dx, \quad N_+ = \int_\delta^\infty c_+(x)dx.$$

That is, derive the result

$$v = \frac{D\delta\varepsilon^2(\mu - \varepsilon)}{\delta\varepsilon^2 + (e^{\varepsilon\delta} - \delta\varepsilon - 1)\mu},$$

Finally, fixing spatial units by setting $\delta = 1$ and taking $\alpha/D \ll 1$ in the result of part (b), obtain an approximation for μ in terms of ε, and thus show that

$$v \approx \delta\alpha e^{-f\delta/k_B T}.$$

This recovers the form of the growth rate in equation (4.3.12).

Problem 4.5. Translocation ratchet. The FP equation for the translocation ratchet takes the form

$$\frac{\partial p}{\partial t} + \frac{\partial J}{\partial x} = 0, \quad J = -\frac{DF}{k_B T}p - D\frac{\partial p}{\partial x}.$$

Suppose that each ratchet site can exist in two states that are in equilibrium

$$S_0 \underset{k_{\mathrm{off}}}{\overset{k_{\mathrm{on}}}{\rightleftharpoons}} S_1,$$

with only S_1 ratcheted. The FP equation is then supplemented by the boundary conditions

$$J(0,t) = J(\delta,t), \quad p(\delta) = (1 - \pi)p(0), \quad \pi = \frac{k_{\mathrm{on}}}{k_{\mathrm{on}} + k_{\mathrm{off}}}.$$

Show that the velocity of translocation is

$$v = \delta J_0 = \frac{2D}{\delta} \left[\frac{\omega^2/2}{\frac{e^\omega - 1}{1 - K(e^\omega - 1)} - \omega} \right].$$

Problem 4.6. Discrete model of polymer chaperone-assisted translocation. Following Krapivsky and Mallick [467], consider a (rigid) polymer chain that passes through a membrane nanopore as shown in Fig. 4.30. Take the pore to be located at $x = 0$ and focus on the polymer segment to the right of the pore. At any given time t, the segment consists of L monomer units, each of size a. In the absence of any chaperones, the polymer executes an unbiased random walk, hopping by one monomer unit to the right or left at equal rates α. In the following we set $\alpha = a = 1$. Now suppose that the region on the right side of the membrane has a fixed density of chaperones that absorb irreversibly at a rate λ onto unoccupied monomeric sites of the polymer. A chaperone is assumed to be larger than the pore, so that it rectifies the polymer diffusion when bound to the site immediately to the right of the pore (see also Sect. 4.3), resulting in a nonzero speed V.

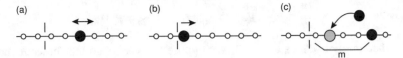

Fig. 4.30: Schematic of chaperone-assisted translocation model of [467]. (a) If the site adjacent to pore is unoccupied then the polymer can hop in either direction. (b) The polymer can hop only to the right because an adsorbed chaperone is next to the pore and is too large to enter. (c) Adsorption of a new chaperone (shaded) at a site on the leftmost chaperone-free segment of length m.

(a) Let E_m denote the probability that an interval of length m to the right of the pore is chaperone-free. It follows that the mean speed of translocation is $V = 1 - E_1$, where $1 - E_1$ is the probability that there is a chaperone immediately to the right of the pore. Define the segment probability $S_m = E_m - E_{m+1}$, which is the probability that the leftmost chaperone is at a distance $m + 1$ from the pore. The probabilities E_m then evolve according to

$$\frac{dE_m}{dt} = S_{m-1} - S_m - \lambda m E_m, \quad m \geq 1.$$

Using the definitions of E_m and S_m and the scenarios shown in Fig. 4.30, explain the meaning of each term on the right-hand side of this equation.

(b) Using the definition of S_m and the normalization condition $\sum_{m \geq 0} S_m = 1$ show that

$$\frac{dE_m}{dt} = E_{m-1} + E_{m+1} - 2E_m - \lambda m E_m, \quad m \geq 1,$$

with boundary condition $E_0 = 1$.

(c) Using the well-known identity of Bessel functions,

$$J_{v-1}(x) + J_{v+1}(x) - \frac{2v}{x} J_v(x) = 0,$$

and the boundary condition, show that the steady-state probabilities are given by

$$E_m = \frac{J_{m+2/\lambda}(2/\lambda)}{J_{2/\lambda}(2/\lambda)},$$

and hence that the expected speed is

$$V = 1 - \frac{J_{1+2/\lambda}(2/\lambda)}{J_{2/\lambda}(2/\lambda)}.$$

Plot V as a function of the absorption rate λ.

Problem 4.7. Cooperative model of motor transport. Consider the cooperative model of motor transport in the absence of a load force, whose birth–death master equation is given by equation (4.5.4):

$$\frac{dP_n}{dt} = \gamma_{n+1}P_{n+1} + \pi_{n-1}P_{n-1} - (\gamma_n + \pi_n)P_n,$$

with reflecting boundaries at $n = 0, N$. Suppose that the cargo attachment/detach-ment rates and velocity are given by

$$\gamma_n = n\gamma, \quad \pi_n = (N - n)\pi, \quad v_n = v.$$

(a) Show that the steady-state distribution is

$$P_n = P_0 \frac{N!}{(N-n)!n!} \left(\frac{\pi}{\gamma}\right)^n, \quad P_0 = \left(1 + \frac{\pi}{\gamma}\right)^{-N}.$$

(b) By constructing the generating function $G(s) = \sum_{n=1}^{N} e^{sn}P_n/(1 - P_0)$, derive an expression for the mean number of bound motors using $N_b = G'(0)$ and show that for large N,

$$N_b \approx N \frac{\pi/\gamma}{1 + \pi/\gamma}.$$

(c) Show that the mean walking distance is

$$\langle x \rangle = \frac{v}{N\pi}\left[\left(1 + \frac{\pi}{\gamma}\right)^N - 1\right].$$

Problem 4.8. Computer simulations: polymerization.
(a) Use Euler's direct method to solve the ODE corresponding to the master equation (4.1.1) for polymerization at one end:

$$\frac{dP_n}{dt} = \varepsilon P_{n+1}(t) + \pi P_{n-1}(t) - [\varepsilon + \pi]P_n(t), \quad n > 0,$$

with $P_n(0) = \delta_{n,10}$, $\varepsilon = 0.7$, and $\pi = 0.4$. Take $t = [0, 2]$.

(b) Use the Gillespie algorithm (Sect. 5.2) to generate sample paths for the length $N(t)$ of the polymer. The two reactions are $n \to n + 1$ at rate π and $n \to n - 1$ at rate ε. By averaging over sample paths, compare the histogram of $N(T)$ with the distribution $P_n(T)$ for $T = 2$.

(c) Does the histogram of $N(T)$ appear to converge to a stationary distribution for large T and $\varepsilon = 0.7$, $\pi = 0.4$? What about the case $\varepsilon = 0.4$, $\pi = 0.7$?

Chapter 5
Stochastic gene expression and regulatory networks

Genetically identical cells exposed to the same environmental conditions can show significant variation in molecular content and marked differences in phenotypic characteristics. This intrinsic variability is linked to the fact that many cellular events at the genetic level involve small numbers of molecules (low copy numbers). Although stochastic gene expression was originally viewed as having detrimental effects on cellular function, with various implications for disease, it is now seen as being potentially advantageous. For example, intrinsic noise can provide the flexibility needed by cells to adapt to fluctuating environments or respond to sudden stresses, and can also support a mechanism by which population heterogeneity is established during cell differentiation and development. Since establishing a functional role for stochastic gene expression in λ-phage [26], there has been an explosion of studies focused on investigating the origins and consequences of noise in gene expression, see the reviews [131, 425, 550, 645, 671, 720, 816]. This typically involves identifying the molecular mechanisms of noise generation at the single gene level, and then building on this knowledge to test and predict its effects on larger regulatory networks. Gene regulation refers to the cellular processes that control the expression of proteins, dictating under what conditions specific proteins should be produced from their parent DNA. This is particularly crucial for multicellular organisms, where all cells share the same genomic DNA, yet do not express the same proteins. That is, selective gene expression allows the cells to specialize into different phenotypes (cell differentiation), resulting in the development of different tissues and organs with distinct functional roles.

In this chapter, we explore the effects of noise on gene expression at the single-cell level. We begin in Sect. 5.1 by describing various deterministic rate models of gene regulatory networks, including autoregulatory networks, the toggle switch, the *lac* operon, the repressilator, NK-βB oscillations, and the circadian clock. We also briefly review linear stability analysis, Hopf bifurcation theory, and the occurrence of oscillations in delay differential equations. We then turn to the analysis of molecular noise associated with low copy numbers, based on the chemical master equation (Sect. 5.2). Since chemical master equations are difficult to analyze

© Springer Nature Switzerland AG 2021
P. C. Bressloff, *Stochastic Processes in Cell Biology*, Interdisciplinary
Applied Mathematics 41, https://doi.org/10.1007/978-3-030-72515-0_5

directly, we show how they can be approximated by a corresponding Fokker-Planck (FP) equation using a system-size expansion analogous to the birth–death processes considered in Chap. 3. We also describe an alternative, numerical approach based on Gillespie's stochastic simulation algorithm for generating exact sample paths of a continuous-time Markov process. We then explore various affects of molecular noise on gene expression, including translational bursting, noise-induced switching, and noise-induced oscillations.

One of the assumptions of many stochastic models of gene networks is that the binding/unbinding of transcription factors at promoter sites is faster than the rates of synthesis and degradation. If this assumption is relaxed, then there exists another source of intrinsic noise known as promoter noise. The latter is modeled in terms of a stochastic hybrid system (see Box 5D). This involves the coupling between a continuous-time Markov chain and a continuous process that may be deterministic or stochastic. The evolution of the system is now described by a differential Chapman-Kolmogorov (CK) equation, which is a mixture of a master equation and an FP equation. In the limit of fast switching, a quasi-steady-state (QSS) approximation is used to reduce the CK equation to an effective FP equation. This is analogous to the system-size expansion of chemical master equations. Various examples of networks with promoter noise are presented in Sect. 5.3, including a two-state regulatory network with or without feedback, which switches between an active and inactive state, and a stochastic version of the toggle switch. One of the effects of promoter noise is transcriptional bursting. In Sect. 5.4, we show how the latter can be formulated in terms of queuing theory. In Sect. 5.5, we look at some time-limiting steps in gene regulation, including kinetic proofreading based on enzymatic reactions, and DNA transcription times. In Sect. 5.6, we provide a brief introduction to epigenetics. The latter refers to phenotypic states that are not encoded as genes, but as inherited patterns of gene expression originating from environmental factors, and maintained over multiple cell generations when the original environmental stimuli have been removed. A number of epigenetic mechanisms are discussed, including the infection of *E. coli* by the λ phage DNA virus, and local mechanisms such as DNA methylation and gene silencing by nucleosome modifications. Finally, in Sect. 5.7, we consider the role of gene expression in the downstream interpretation of spatially distributed morphogen gradients during early development.

5.1 Gene regulatory networks

In Fig. 5.1, we show the two main stages in the expression of a single gene according to the central dogma.

1. **Transcription (DNA $\rightarrow$ RNA).** The first major stage of gene expression is the synthesis of a messenger RNA molecule (mRNA) with a nucleotide sequence complementary to the DNA strand from which it is copied - this serves as the template for protein synthesis. Transcription is mediated by a molecular machine known as RNA polymerase. The key steps in transcription are binding of RNA polymerase (P) to the relevant promoter region of DNA (D) to form a closed

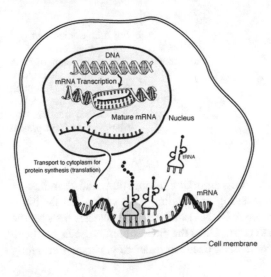

Fig. 5.1: Schematic diagram of a eukaryotic cell showing the main steps of protein synthesis according to the central dogma. See text for details. [Public domain figure downloaded from Wikipedia Commons.]

complex (PD_c), the unzipping of the two strands of DNA to form an open complex (PD_o), and finally promoter escape, when RNA polymerase reads one of the exposed strands

$$P + D \underset{k_-}{\overset{k_+}{\rightleftharpoons}} PD_c \overset{k_{open}}{\longrightarrow} PD_0 \overset{k_{escape}}{\longrightarrow} \text{transcription}.$$

Once the RNAP is reading the strand, the promoter is unoccupied and ready to accept a new polymerase. The binding/unbinding of polymerase is very fast, $k_\pm \gg k_{open}$ so that the first step happens many times before formation of an open complex. Hence, one can treat the RNA polymerase as in quasi-equilibrium with the promoter characterized by an equilibrium constant $K_P = k_+/k_-$. The rate of transcription will thus be proportional to the fraction of bound RNA polymerase, $k_+/(k_+ + k_-)$.

2. **Translation (RNA →protein).** The second major stage is synthesis of a protein from mRNA. In the case of eukaryotes, transcription takes place in the cell nucleus, whereas subsequent protein synthesis takes place in the cytoplasm, which means that the mRNA has to be exported from the nucleus as an intermediate step. Translation is mediated by a macromolecule known as a ribosome, which produces a string of amino acids (polypeptide chains), each specified by a codon (represented by three letters) on the mRNA molecule. Since there are four nucleotides (A, U, C, G), there are 64 distinct codons, e.g., AUG, CGG, most of which code for a single amino acid. The process of translation consists of ribosomes moving along the mRNA without backtracking (from one end to the other, technically known as the 5' end to the 3' end) and is conceptually divided into

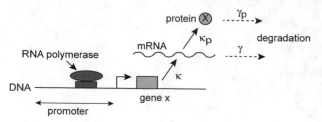

Fig. 5.2: Unregulated transcription of a gene x following binding of RNA polymerase to the promoter region. The resulting mRNA exits the nucleus and is then translated by ribosomes to form protein X.

three major stages (as is transcription): initiation, elongation, and termination. Each elongation step involves translating or "reading" of a codon and the binding of a freely diffusing transfer RNA (tRNA) molecule that carries the specific amino acid corresponding to that codon. Once the chain of amino acids has been generated, a number of further processes occur in order to generate a correctly folded protein.

A very simple model of (unregulated) gene expression is shown in Fig. 5.2. Let a and x denote the concentrations of mRNA and protein, respectively, due to the activity of a single gene. The kinetic equations take the simple form

$$\frac{da}{dt} = \kappa - \gamma a, \quad \frac{dx}{dt} = \kappa_p a - \gamma_p x, \tag{5.1.1}$$

where κ is the transcription rate of mRNA by the gene, κ_p is the translation rate of protein by each mRNA molecule, and γ, γ_p are degradation rates. At steady state, the protein concentration is

$$x^* = \frac{\kappa_p}{\gamma_p} a^* = \frac{\kappa_p \kappa}{\gamma_p \gamma}.$$

Remark 5.1. Low copy numbers. The production of mRNA from a typical gene in *E. coli* occurs at a rate of approximately 10nM per minute, while the average lifetime of mRNA due to degradation is around a minute. This implies that on average $a^* = 1 - 10$ nM. The translation rate is a few proteins per minute and the protein lifetime tends to be at least several hours. Hence $x^* \approx 100 - 1000$ nM. Given that the cell volume of *E. coli* is around 10^{-15} L, it follows that 1 nM corresponds to one molecule per cell. This implies that the steady-state levels of mRNA and protein are typically $\ll 10^4$ molecules per cell! Deterministic chemical kinetic equations based on the law of mass action assume that the concentrations of the various reactants are continuously differentiable. For molar concentrations with molecule numbers of order 10^{23} this is a reasonable approximation. On the other hand, the number of mRNA and protein molecules within a cell are much smaller, and one has to take into account the discrete nature of chemical reactions (molecular noise). Since molecular noise is inherent to the given system, rather than arising from external factors, it is an example of intrinsic rather than extrinsic noise.

The above simplified picture ignores a major feature of cellular processing, namely, gene regulation. Individual cells frequently have to make "decisions," that is, to express different genes at different spatial locations and times, and at different activity levels. One of the most important mechanisms of genetic control is tran-

scriptional regulation, that is, determining whether or not an mRNA molecule is
made. The control of transcription (switching a gene on or off) is mediated by proteins known as transcription factors, see Fig. 5.3. Negative control (or repression) is
mediated by repressors that bind to a promotor region along the DNA where RNA
polymerase has to bind in order to initiate transcription—it thus inhibits transcription. On the other hand, positive control (activation) is mediated by activators that
increase the probability of RNA polymerase binding to the promoter. The presence
of transcription factors means that cellular processes can be controlled by extremely
complex gene networks, in which the expression of one gene produces a repressor
or activator, which then regulates the expression of the same gene or another gene.
This can result in many negative and positive feedback loops, the understanding of
which lies at the heart of systems biology [11]. In addition to transcriptional regulation, there are a variety of other mechanisms that can control gene expression,
including mRNA and protein degradation, and translational regulation.

Once feedback and nonlinearities are included in gene networks, a rich reper-
toire of dynamics can occur, including bistability (genetic switches) and oscillations. One important tool for understanding the regulatory mechanisms of gene networks, both quantitatively and qualitatively, is the construction of synthetic gene
networks [158, 246, 360, 623, 669, 785]. The latter typically consist of a small
number of interacting genes that can be inserted into organisms with minimal in-
terference from the hosts' own regulatory processes. The simplicity of synthetic

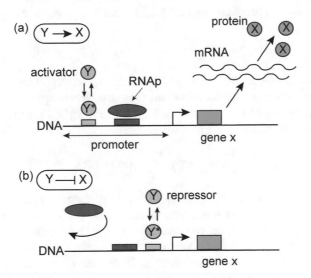

Fig. 5.3: Transcriptional regulation due to the binding of a repressor or activator protein to a promoter region along the DNA. (a) Increased transcription due to the binding of an activator protein
Y to the promoter. An activator typically transitions between inactive and active forms; the active
form Y* has a high affinity to the promoter binding site. An external chemical signal can regulate
transitions between the active and inactive states. (c) Transcription can be stopped (or reduced) by
a repressor protein Y binding to the promoter and blocking the binding of RNA polymerase.

networks compared to naturally occurring gene networks provides a more practical framework for exploring regulatory mechanisms via mathematical models, as well as a potential source of new engineered technologies. For example, it is possible to construct coupled systems of genetic oscillators using a synthetic version of a natural form of bacterial communication known as quorum sensing (QS), with the goal of enhancing the oscillatory response via synchronization [207, 299, 567, 895]. In the following, we describe various example of synthetic and naturally occurring gene networks with nonlinear feedback. Coupled genetic oscillators and QS will be explored in Chap. 15.

5.1.1 Autoregulatory gene network

One of the simplest examples of a gene network with regulatory feedback occurs when a gene is directly regulated by its own gene product (autoregulation), see Fig. 5.4(a). A simple kinetic model of autoregulatory feedback is

$$\frac{da}{dt} = \kappa g(x) - \gamma a, \quad \frac{dx}{dt} = \kappa_p a - \gamma_p x. \tag{5.1.2}$$

The function $g(x)$ represents the nonlinear feedback effect of the protein on the transcription of mRNA. A typical choice for g in the case of an activator (+) or repressor (-) is a Hill function, see Fig. 5.4(b). That is, $g = g_-$ or $g = g_+$ with

$$g_+(x) = 1 + \sigma \frac{x^2}{K^2 + x^2}, \quad g_-(x) = \frac{K^2}{K^2 + x^2} + \sigma. \tag{5.1.3}$$

Here σ determines the asymptotic rate of production at large protein concentrations, and K^2 is an effective dissociation constant. We can further simplify the dynamics by assuming that the mRNA is in quasi-equilibrium, $m \simeq \kappa g(x)/\gamma$, so that the protein concentration evolves according to

$$\frac{dx}{dt} = -\gamma_p x + \frac{\kappa_p \kappa}{\gamma} g(x) \equiv f(x). \tag{5.1.4}$$

A fixed point x^* of the kinetic equation (5.1.4) is given by a solution of the algebraic equation $f(x^*) = 0$. One can thus determine the number of fixed points and their stability using a graphical construction, see Fig. 5.4(c). In the case of negative feedback, there is a single stable fixed point, whereas with positive feedback, the network can be monostable or bistable. In the latter case, the autoregulatory network can act as a genetic switch.

Promoter binding/unbinding. The form of the nonlinear function in equation (5.1.2) is determined by the binding/unbinding kinetics of proteins interacting with operator sites on the promoter, under the adiabatic assumption that binding/unbinding rates are much faster than other processes such as protein production and degradation. Suppose that the promoter of the autoregulatory repressor network has a

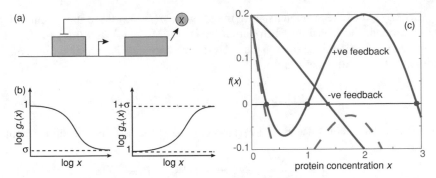

Fig. 5.4: (a) An autoregulatory network. A gene x is repressed (or activated) by its own protein product. (b) Promoter activity functions $g_-(x)$ and $g_+(x)$ for repression and activation, respectively, as functions of protein concentration x. (c) Fixed points of a deterministic autoregulatory network. The network is monostable in the case of negative feedback (red curve) and weak positive feedback (dashed blue curve), but can exhibit bistability in the case of strong positive feedback (solid blue curve).

single operator site OS_1 for binding protein X, see Fig. 5.5(a). The gene is assumed to be OFF when X is bound to the promoter and ON otherwise. If O_0 and O_1 denote the unbound and bound promoter states, then the corresponding reaction scheme is

$$O_0 + X \underset{k_-}{\overset{k_+}{\rightleftharpoons}} O_1, \tag{5.1.5}$$

where $k_\pm$ are the binding/unbinding rates. If the number of X proteins is sufficiently large, then we have the reduced reaction scheme

$$O_0 \underset{k_-}{\overset{k_+[X]}{\rightleftharpoons}} O_1,$$

where $[X]$ is the concentration of X.

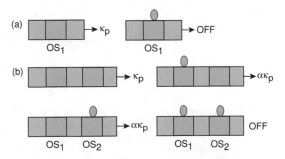

Fig. 5.5: Autoregulatory feedback circuit. (a) Two-state promoter binding/unbinding model. (b) Multiple operator sites. The synthesis rate κ_p is reduced by a factor α, $0 < \alpha < 1$, if one of the sites is occupied, and ceases completely if both sites are occupied.

Let $M(t)$ label the promoter state at time t, with $M(t) = m$ if the state is currently O_m. Introduce the probability distribution $P_m(t) = \mathbb{P}[M(t){=}m]$ with $P_0(t) + P_1(t){=}1$. Following along identical lines to the analysis of a two-state ion channel (Sect. 3.3), we have the pair of equations

$$\frac{dP_0}{dt} = -k_+[X]P_0 + k_-P_1 \qquad \frac{dP_1}{dt} = k_+[X]P_0 - k_-P_1, \qquad (5.1.6)$$

which are equivalent, since $P_0(t) + P_1(t) = 1$. Equation (5.1.6) has the unique stable steady state

$$P_0^* = \frac{K}{[X]+K}, \qquad P_1^* = \frac{[X]}{[X]+K}, \qquad K = \frac{k_-}{k_+}. \qquad (5.1.7)$$

Here K is a dissociation constant (inverse of the equilibrium constant). Assuming that the gene only produces X in the unbound state, the kinetic equation for $x = [X]$, under the adiabatic assumption, is

$$\frac{dx}{dt} = \kappa_p P_0^* - \gamma_p x = \frac{\kappa_p K}{x+K} - \kappa_p x. \qquad (5.1.8)$$

Similarly, if X acts as its own activator, then the gene only produces protein in the bound state, and

$$\frac{dx}{dt} = \kappa_p P_1^* - \gamma_p x = \frac{\kappa_p x}{x+K} - \kappa_p x. \qquad (5.1.9)$$

Dimerization. A simple generalization of the previous example is to assume that the operator site only binds dimers (cooperative binding, see Sect. 3.2). The corresponding reaction schemes are now

$$X + X \underset{k_{-1}}{\overset{k_1}{\rightleftharpoons}} 2X, \quad OS_1 + 2X \underset{k_-}{\overset{k_+}{\rightleftharpoons}} OS_1^*. \qquad (5.1.10)$$

We assume that dimerization is the fastest reaction, so that the concentration of dimers is in quasi-equilibrium, that is, from the equilibrium law of mass action,

$$[X^2] = \frac{k_1}{k_{-1}}[X]^2 = x^2/K_1,$$

where K_1 is the dimerization constant. The reduced reaction scheme for the two promoter states is

$$O_0 \underset{k_-}{\overset{k_+[X]^2/K_1}{\rightleftharpoons}} O_1,$$

and the corresponding master equation for the probabilities $P_m(t)$ is

$$\frac{dP_0}{dt} = -\frac{k_+}{K_1}x^2 P_0 + k_-P_1, \qquad \frac{dP_1}{dt} = \frac{k_+}{K_1}x^2 P_0 - k_-P_1.$$

It follows that

$$P_0^* = \frac{K^2}{x^2 + K^2}, \quad P_1^* = \frac{x^2}{x^2 + K^2}, \quad K^2 = \frac{k_-}{k_+}K_1. \tag{5.1.11}$$

Multiple operator sites. Now consider a promoter with two nonoverlapping operator sites OS_1 and OS_2, each of which can independently bind a protein X, see Fig. 5.5(b). The two independent binding reactions reactions are

$$OS_1 + X \underset{k_{-1}}{\overset{k_1}{\rightleftharpoons}} OS_1^*, \quad OS_2 + X \underset{k_{-2}}{\overset{k_2}{\rightleftharpoons}} OS_2^*. \tag{5.1.12}$$

We have four promoter states $O_m, m = 0, 1, 2, 3$ as follows:

$$O_0 : OS_1 \text{ and } OS_2 \text{ unoccupied,}$$
$$O_1 : OS_1 \text{ occupied by } X, OS_2 \text{ unoccupied,}$$
$$O_2 : OS_1 \text{ unoccupied }, OS_2 \text{ occupied by } X,$$
$$O_3 : OS_1 \text{ and } OS_2 \text{ each occupied by } X.$$

The reaction scheme for the promoter states is

$$O_0 \underset{k_{-1}}{\overset{k_1[X]}{\rightleftharpoons}} O_1, \quad O_0 \underset{k_{-2}}{\overset{k_2[X]}{\rightleftharpoons}} O_2, \quad O_1 \underset{k_{-2}}{\overset{k_2[X]}{\rightleftharpoons}} O_3, \quad O_2 \underset{k_{-1}}{\overset{k_1[X]}{\rightleftharpoons}} O_3. \tag{5.1.13}$$

The corresponding steady-state probability distributions for the different promoter states are then

$$P_0^* = \frac{1}{1 + [X]/K_1 + [X]/K_2 + [X]^2/K_1 K_2},$$

$$P_1^* = \frac{[X]/K_1}{1 + [X]/K_1 + [X]/K_2 + [X]^2/K_1 K_2},$$

$$P_2^* = \frac{[X]/K_2}{1 + [X]/K_1 + [X]/K_2 + [X]^2/K_1 K_2},$$

$$P_3^* = \frac{[X]^2/K_1 K_2}{1 + [X]/K_1 + [X]/K_2 + [X]^2/K_1 K_2},$$

with dissociation constants

$$K_1 = \frac{k_{-1}}{k_1}, \quad K_2 = \frac{k_{-2}}{k_2}.$$

We assume that promoter state O_0 produces protein X at a rate κ_p, states O_1, O_2 produce X at the reduced rate $\alpha \kappa_p$, $0 < \alpha < 1$, and state O_3 is off. It follows that the protein concentration evolves according to the kinetic equation

$$\frac{dx}{dt} = \kappa_p P_0^* + \alpha \kappa_p (P_1^* + P_2^*) - \gamma_p x. \tag{5.1.14}$$

Remark 5.2. Promoter noise. In order to derive an explicit expression for nonlinear feed-back terms in gene regulatory networks, one assumes that transcription factor/DNA binding is faster than other processes, such as the rate of synthesis and degradation (adiabatic assumption). One can then use the steady-state probabilities for the different promoter states in the kinetic equations for protein production and degradation. If the adiabatic assumption breaks down, then one has another source of intrinsic noise due to the stochastic nature of transcription factor binding (promoter noise), see Sect. 5.3.

5.1.2 Mutual repressor model (toggle switch)

Considerable insight into genetic switches has been obtained by constructing a synthetic version of a switch in *E. coli*, in which the gene product of the switch is a fluorescent reporter protein [301]. This allows the flipping of the switch to be observed by measuring the fluorescent level of the cells. The underlying gene circuit is based on a mutual repressor model, see Fig. 5.6. It consists of two repressor proteins X and Y whose transcription is mutually regulated. That is, the protein product of one gene binds to the promoter of the other gene and represses its output.

For simplicity, the explicit dynamics of transcription and translation are ignored so that we only model the mutual effects of the proteins on protein production. Denoting the concentrations of the proteins by $x(t), y(t)$, and assuming the two proteins act symmetrically, the resulting kinetic equations are

$$\frac{dx}{dt} = -\gamma_p x + \frac{\kappa_p K^2}{K^2 + y^2}, \quad \frac{dy}{dt} = -\gamma_p y + \frac{\kappa_p K^2}{K^2 + x^2}. \tag{5.1.15}$$

Here γ_p is the rate of protein degradation, κ_p is the rate of protein production in the absence of repression, and K is a binding constant for the repressors. (We have absorbed a factor of κ/γ into κ_p, see equation (5.1.4).)

It is convenient to rewrite the equations in non-dimensional form by measuring x and y in units of K^{-1} and time in units of γ_p^{-1}:

$$\frac{du}{dt} = -u + \frac{\alpha}{1 + v^2}, \quad \frac{dv}{dt} = -v + \frac{\alpha}{1 + u^2}, \tag{5.1.16}$$

with $\alpha = \kappa_p / K\gamma_p$. Analysis of the fixed point solutions of this pair of equations establishes that the mutual repressor model acts as a bistable switch. The fixed point equation for u is

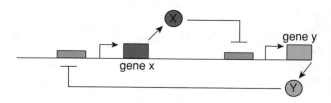

Fig. 5.6: Mutual repressor model of a genetic switch. A gene x expresses a protein X that represses the transcription of gene y, and the protein Y represses the transcription of gene x.

$$u = \alpha \left[1 + \left(\frac{\alpha}{1+u^2} \right)^2 \right]^{-1},$$

which can be rearranged to yield a product of two polynomials:

$$(u^2 - \alpha u + 1)(u^3 + u - \alpha) = 0.$$

The cubic is a monotonically increasing function of u and thus has a single root given implicitly by

$$u = \frac{\alpha}{1+u^2} = v.$$

This solution is guaranteed by the exchange symmetry of the underlying equations. The roots of the quadratic are given by

$$u = U_{\pm} \equiv \frac{1}{2} \left[\alpha \pm \sqrt{\alpha^2 - 4} \right],$$

with $v = U_{\mp}$. It immediately follows that there is a single fixed point when $\alpha < 2$ and three fixed points when $\alpha > 2$. Moreover, linear stability analysis establishes that the symmetric solution is stable when $\alpha < 2$, and undergoes a pitchfork bifurcation at the critical value $\alpha_c = 2$ where it becomes unstable and a pair of stable fixed points emerge.

Modified mutual repressor model. Consider a simplified mutual repressor model consisting of a single promoter with two operator sites OS_1 and OS_2 that bind to dimers of protein Y and protein X, respectively (see Fig. 5.7). If the dimer of one protein is bound to its site, then this represses the expression of the other protein. However, both sites cannot be occupied at the same time. Hence, the promoter can be in three states O_m, $m = 0, 1, 2$: no dimer is bound to the promoter (O_0); a dimer of protein Y is bound to the promoter (O_1); a dimer of protein X is bound to the promoter (O_2). Assuming that the number of proteins is sufficiently large, we have the transition scheme

$$O_1 \underset{\beta y^2}{\overset{\beta K}{\rightleftharpoons}} O_0 \underset{\beta K}{\overset{\beta x^2}{\rightleftharpoons}} O_2,$$

Fig. 5.7: (a) Simplified mutual repressor with a single promoter having two operator sites OR_1 and OR_2. (b) A dimer of protein X can bind to OR_2 and a dimer of protein Y can bind to OR_1, but they cannot both be bound at the same time.

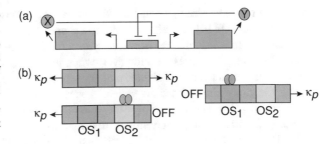

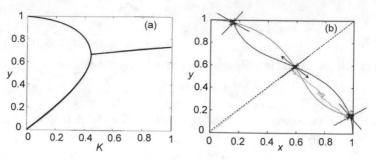

Fig. 5.8: Modified mutual repressor model (a) Bifurcation diagram. (b) Phase-plane dynamics. The black curve shows the x_2-nullcline and the gray curve shows the x_1-nullcline. The open circles show the stable fixed points, the filled circle shows the unstable saddle. The irregular curve shows a stochastic trajectory leaving the lower basin of attraction to reach the curve (separatrix) that separates the basins of attraction of the two fixed points.

where β is a transition rate and K is a non-dimensional dissociation constant. Protein X (Y) is produced at a rate κ_p when the promoter is in the states $O_{0,2}$ $(O_{0,1})$, and both proteins are degraded at a rate γ_p in all three states. The kinetic equations become, see Ex. 5.1,

$$\frac{dx}{dt} = -\gamma_p x + \kappa_p \frac{K+x^2}{x^2+y^2+K}, \quad \frac{dy}{dt} = -\gamma_p y + \kappa_p \frac{K+y^2}{x^2+y^2+K}. \qquad (5.1.17)$$

For sufficiently small K one finds that the system exhibits bistability. This is illustrated in Fig. 5.8.

> Remark 5.3. Noise-induced switching. In the absence of molecular or promoter noise, the bistable system will settle into one of the stable fixed points, depending on the initial condition. All initial conditions that approach a given fixed point form the basin of attraction. However, when intrinsic noise is included, it is possible for a stochastic trajectory to leave a fixed point and cross the boundary (separatrix) between the basins of attraction, resulting in a switch to the other fixed point. At the population level, stochastic genetic switching can produce phenotypic heterogeneity. Phenotypic heterogeneity is expected to be particularly beneficial to microbial cells that need to adapt efficiently to sudden changes in environmental conditions. Fluctuations in gene expression provide a mechanism for sampling distinct physiological states, and could, therefore, increase the probability of survival during times of stress, without the need for genetic mutation (Chap. 15).

The *lac* operon. The idea of a genetic switch was first proposed over 40 years ago by Jacob and Monod [410], in their study of the *lac* operon. When there is an abundance of glucose, *E. coli* uses glucose exclusively as a food source irrespective of whether or not other sugars are present. However, when glucose is unavailable, *E. coli* can feed on other sugars such as lactose, and this occurs via the *lac* operon switch that induces the expression of various genes. A variety of mathematical models of the *lac* operon have been developed over the years [341, 342, 722, 865, 875, 876]. Here we briefly describe a simplified model presented in Chap. 10 of Keener and Sneyd [431]. The basic feedback control mechanism is illustrated in Fig. 5.9. There are

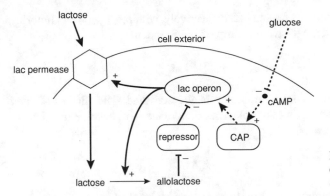

Fig. 5.9: Feedback control circuit of the *lac* operon. See text for details.

two control sites on the *lac* operon, see Fig. 5.10, a repressor site that blocks RNA polymerase from binding to the promoter site, and a preceding control site to which a dimeric CAP (catabolic activator protein) molecule can bind provided it forms a complex with cyclic AMP (cAMP). Bound CAP promotes the binding of RNA polymerase to the promoter region. When there is sufficient glucose in the cell exterior, the action of cAMP is inhibited so that CAP cannot bind and the *lac* operon is repressed. On the other hand, when glucose is removed, the CAP-cAMP complex can bind to the activator site and activate the *lac* operon. The latter consists of several genes that code for the proteins responsible for lactose metabolism. One of these proteins is *lac* permease, which allows the entry of lactose into the cell that is enhanced by a positive feedback loop. The feedback mechanism involves another protein, β-galactosidase, which converts lactose into allolactose. Allolactose can bind to the repressor protein and prevent it binding to the repressor binding site. This further activates the *lac* operon, resulting in the further production of allolactose and increased entry of lactose via the *lac* permease.

Suppose that the CAP dynamics is ignored, so that we can focus on the positive feedback loop indicated in Fig. 5.9 by solid arrows. Let A denote the concentration of allolactose and similarly for lactose (L), the permease (P), the protein product β-galactosidase (B), mRNA (M), and the repressor (R). Let p_{on} and p_{off} denote the

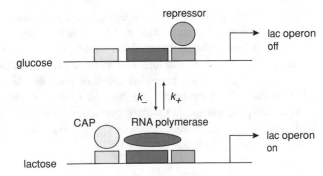

Fig. 5.10: Repressor and CAP sites for the *lac* operon.

probabilities that the operon is on and off, respectively, with $p_{on} + p_{off} = 1$. Ignoring the effect of the CAP site, we have the simple kinetic scheme

$$\frac{dp_{on}}{dt} = k_{-r}(1 - p_{on}) - k_r R^* p_{on},$$

where R^* is the concentration of repressor in the activated state. Each activated repressor protein interacts with two molecules of allolactose to become inactivated, so from mass-action kinetics,

$$\frac{dR^*}{dt} = k_{-a}R - k_a A^2 R^*,$$

where the binding/unbinding of a single repressor molecule to the operon has a negligible effect on the total concentration $R_T = R + R^*$. The next simplification is to take these reactions to be much faster than those associated with gene expression so that p_{on} and R^* take the steady-state values

$$R^* = \frac{R_T}{1 + K_a A^2}, \quad p_{on} = \frac{1}{1 + K_r R^*},$$

with $K_a = k_a/k_{-a}$ and $K_r = k_r/k_{-r}$. Combining these two results gives the steady-state probability

$$p_{on} = \frac{1 + K_a A^2}{1 + K_r R_T + K_a A^2} \equiv \Gamma(A).$$

It follows that the concentration of mRNA is determined by the equation

$$\frac{dM}{dt} = \alpha_M \Gamma(A) - \gamma_M M, \qquad (5.1.18a)$$

where α_M and γ_M are the rates of mRNA production and degradation. This is the first of the model equations. The next two equations represent the dynamics of the enzymes directly produced by the on state of the operon, namely, permease and β-galactosidase

$$\frac{dP}{dt} = \alpha_P M - \gamma_P P, \quad \frac{dB}{dt} = \alpha_B M - \gamma_B B. \qquad (5.1.18b)$$

Note that although both enzymes are produced by different parts of the same mRNA, the effective production rates differ due to different times of production (permease is produced after β-galactosidase), and the time delay associated with permease migrating to the cell membrane. The final two equations specify the dynamics of lactose and allolactose based on Michaelis-Menten kinetics. Let L_e be a fixed concentration of lactose exterior to the cell. Lactose enters the cell at a Michaelis-Menten rate proportional to the permease concentration P, where it is converted to allolactose via the enzymatic action of β-galactosidase; the latter also breaks down allolactose into glucose and galactose. Thus

$$\frac{dL}{dt} = \alpha_L P \frac{L_e}{K_{Le} + L_e} - \alpha_A B \frac{L}{K_L + L} - \gamma_L L, \qquad (5.1.18c)$$

$$\frac{dA}{dt} = \alpha_A B \frac{L}{K_L + L} - \beta_A B \frac{A}{K_A + A} - \gamma_A A. \qquad (5.1.18d)$$

Keener and Sneyd [431] show that for physiologically based parameter values, equation (5.1.18) exhibits bistability in the interior lactose concentration as a function of the exterior lactose concentration L_e.

5.1.3 Repressilator and degrade-and-fire models of a genetic oscillator

Let us return to the autoregulatory network of equation (5.1.2), which for negative feedback takes the form

$$\frac{dx_1}{dt} = \frac{\kappa}{1 + (x_2/K)^n} - \gamma x_1, \qquad \frac{dx_2}{dt} = \kappa_p x_1 - \gamma_p x_2, \qquad (5.1.19)$$

where x_1 and x_2 are the mRNA and protein concentrations, respectively. As we illustrated in Fig. 5.4(c), the network has a single fixed point. A closer look at the trajectories of the system reveal that this fixed point is a stable focus, which means that it supports damped oscillations as it approaches the fixed point. That is, linearization about the fixed point yields a Jacobian with a complex pair of imaginary eigenvalues, see Box 5A. However, it is straightforward to convert the network to an oscillator by adding an additional, intermediate chemical species Z. This is exemplified by the model of a clock gene (see below), where the intermediate species is the nuclear protein. There are a number of other possible choices for Z.

One interpretation of the intermediate species Z is that, it is another transcription factor. This can be generalized to the notion of multi-gene negative feedback circuits. The most studied of these is the repressilator, which was the first reported experimental realization of a synthetic gene oscillator [246]. The basic circuit consists of three genes encoding three transcriptional repressors: the lactose repressor LacI, the tetracycline repressor TetR, and the cI repressor from bacteriophage λ. The synthetic circuit consists of a ring, in which the protein of a given gene acts as a repressor of the next gene in the ring, see Fig. 5.11. Representing the action of each repressor as a Hill function, we obtain the following system of equations for the concentrations x_1, x_2, x_3 of LacI, TetR and cI, respectively

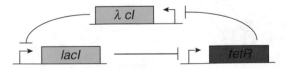

Fig. 5.11: Schematic diagram of repressilator circuit.

$$\frac{dx_1}{dt} = \kappa_1 + \frac{\sigma_1}{1+(x_3/K_3)^{n_3}} - \gamma_1 x_1, \tag{5.1.20a}$$

$$\frac{dx_2}{dt} = \kappa_2 + \frac{\sigma_2}{1+(x_1/K_1)^{n_1}} - \gamma_2 x_2, \tag{5.1.20b}$$

$$\frac{dx_3}{dt} = \kappa_3 + \frac{\sigma_3}{1+(x_2/K_2)^{n_2}} - \gamma_3 x_3, \tag{5.1.20c}$$

where $\kappa_j, j = 1,2,3$ are background production rates. If one also includes the mRNA dynamics and sets $\kappa_j = \kappa$, $\sigma_j = \sigma$, $K_j = K$, $n_j = n$, and $\gamma_i = \gamma$, then one can write down a dimensionless version of the resulting system of equations in the compact form

$$\frac{dm_i}{dt} = -m_i + \frac{\sigma}{1+p_j^n} + \kappa, \tag{5.1.21a}$$

$$\frac{dp_i}{dt} = -\beta(p_i - m_i), \tag{5.1.21b}$$

where $i \in \{1,2,3\}$ and $j \in \{3,1,2\}$. The parameter β denotes the ratio of the protein and mRNA degradation rates, time is rescaled in units of the mRNA lifetime, protein concentrations are in units of K, and mRNA concentrations are rescaled by their translation efficiency (the average number of proteins produced per mRNA molecule). One finds that the system (5.1.21) can exhibit oscillations as β is increased from zero. (Note that for nonlinear systems is is not usually possible to construct an explicit oscillatory solutions. Instead, one determines conditions under which a fixed point becomes unstable due to a pair of complex conjugate eigenvalues of the associated Jacobian crossing the imaginary axis as a parameter is varied, and then uses the Hopf bifurcation theorem, see Box 5A.) There have been several extensions of the repressilator model, both deterministic [155, 595, 784] and stochastic [442, 530, 578, 665].

Another interpretation of the intermediate species Z is that it could be some intermediate, non-active (unfolded) version of the repressor X_2. This then leads to the concept of a chain of intermediate steps associated with the sequential assembly of the mRNA and mature protein [785]. Suppose that there are $N-1$ intermediate

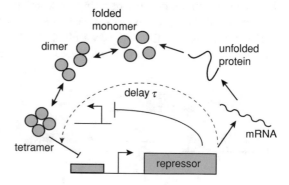

Fig. 5.12: Degrade-and-fire model of an autorepressor genetic oscillator. Various stages of transcription and translation are shown, which contribute an effective delay τ to the negative feedback.

protein states denoted by Z_i, $i \in \{1,\dots,N-1\}$, evolving according to the system of equations

$$\frac{dx_1}{dt} = \frac{\kappa}{1+(x_2/K)^n} - \gamma x_1,$$

$$\frac{dz_i}{dt} = P_i(z_{i-1}) - P_{i+1}(z_i) - \gamma_z z_i, \quad i = 1,\dots,N-1,$$

$$\frac{dx_2}{dt} = \kappa_p z - \frac{V x_2}{K'+x_2} - \gamma_p x_2, \tag{5.1.22}$$

where $z_0 \equiv x_1$. The production of Z_i is taken to be irreversible and given by some production function $P_i(z_{i-1})$. The mature protein X_2 is taken to undergo enzymatic degradation. From a stochastic perspective, one can treat the time from initiation of translation to the production of the fully formed protein as a sequence of exponentially distributed waiting times. If each reaction is irreversible and has rate λ, then the mean delay time is $\tau = N/\lambda$. Moreover, if the number of reactions N increases for fixed τ, then the fluctuations in the delay time decrease, and one can treat τ as a deterministic delay. One can thus reduce the system (5.1.22) to a scalar delay equation for the mature protein concentration $x = x_2$, see Fig. 5.12 [561, 562]:

$$\frac{dx}{dt} = \frac{\sigma_{\text{eff}}}{1+(x_\tau/K)^n} - \frac{Vx}{K'+x} - \gamma_p x, \tag{5.1.23}$$

where $x_\tau = x(t-\tau)$. It is well known that a one-dimensional ODE with a discrete delay can support oscillations, see [769] and Box 5B. One finds that if $\sigma_{\text{eff}} > V \gg K/\tau > K'/\tau$, then the system exhibits strongly nonlinear degrade-and-fire oscillations, in which the number of proteins rises rapidly from zero, after which it slowly decays back to zero (approximate sawtooth function). This can be idealized using the following piecewise dynamical equation:

$$\frac{dx}{dt} = -\gamma_p, \quad T_j < t < T_{j+1}, \quad x(T_j) = 0, \quad \lim_{\varepsilon \to 0^+} x(T_j+\varepsilon) = b_j. \tag{5.1.24}$$

Here T_j is the time of the j-th protein burst of size b_j, which is represented as an instantaneous jump in the concentration whenever x reaches zero. It follows immediately that $T_{j+1} - T_j = b_j/\gamma_p$.

The circadian clock gene. The circadian rhythm plays a key physiological role in the adaptation of living organisms to the alternation of night and day [306, 634]. Experimental studies of a wide range of plants and animals has established that in almost all cases, autoregulatory feedback on gene expression plays a central role in the molecular mechanisms underlying circadian rhythms [446, 643]. Based on experimental data, a variety of models of increasing complexity have been developed, which show how regulatory feedback loops in circadian gene networks generate sustained oscillations under conditions of continuous darkness [281, 317, 505, 506, 619, 770, 818]. The resulting circadian oscillator has a natural period of approximately 24 hours, which can be entrained to the external light-dark cycle. Here we consider a minimal model of the circadian clock in the fungus *Neurospora* [505]. A related model for circadian rhythms in *Drosophila* was previously introduced by Golbeter [317]. A schematic diagram of the basic model is shown in Fig. 5.13. A clock gene X (*frq* in *Neurospora*, *per* in Drosophila) is transcribed to form mRNA (M), which exits the nucleus and is subsequently translated into cytoplasmic clock protein (X_C). The resulting protein either degrades or enters the nucleus (X_N) where it inhibits its own gene expression. The governing equations for the concentrations m, x_C, x_N of mRNA, cytosolic protein and nuclear protein, respectively, are

$$\frac{dm}{dt} = \kappa \frac{K_m^n}{K_m^n + x_N^n} - \gamma \frac{m}{K_m' + m}, \tag{5.1.25a}$$

$$\frac{dx_C}{dt} = \kappa_p m - \gamma_p \frac{x_C}{K_p + x_C} - k_1 x_C + k_2 x_N, \tag{5.1.25b}$$

$$\frac{dx_N}{dt} = k_1 x_C - k_2 x_N. \tag{5.1.25c}$$

Here, κ is the unregulated rate of transcription, κ_p is the rate of translation, and γ, γ_p are the rates of mRNA and protein degradation; degradation is assumed to obey Michaelis-Menten kinetics (see Sect. 1.3.5). The negative regulation of transcription is taken to be cooperative with a Hill coefficient of n. Finally, the rate constants k_1, k_2 characterize the transport of protein into and out of the nucleus. It can be shown that the above model exhibits limit cycle oscillations in physiologically reasonable parameter regimes, and thus provides a molecular basis for the sustained oscillations of the circadian clock under constant darkness [505].

Note that in more complex organisms such as mammals, the circadian system exhibits a hierarchical structure that spans the whole organism down to the cellular level [217, 889]. The complex network of rhythmic elements is coordinated centrally by the suprachiasmatic nucleus (SCN), which is a small region in the hypothalamus above the optic chiasm. SCN neurons, which show robust circadian activity on the cellular and molecular levels, receive light-dependent input from the retina via the optic chiasm, as well as other sources of environmental input carrying informa-

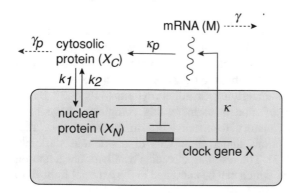

Fig. 5.13: Minimal model for a negative auto regulation network underlying circadian rhythms. Transcription of a clock gene (X) produces mRNA (M), which is transported outside the nucleus and then translated into cytosolic clock protein (X_C). The protein is either degraded or transported into the nucleus (X_N) where is exerts negative feedback on the gene expression.

tion about temperature, food, and noise levels, for example. The SCN functions as a central pacemaker and synchronizer that acts on the circadian rhythms in the other parts of the body, where cell-autonomous, peripheral circadian clocks can be found. The core regulatory network underlying the mammalian circadian clock involves at least 14 genes that are interconnected via transcriptional/translational feedback loops.

Oscillations in NF-κB signaling. There are a number of other cellular oscillators that are based on gene regulatory circuits, most of which have periods shorter than the circadian 24-hour cycle (ultradian oscillators). One well-known example is found in the nuclear localization of the NF-κB transcription factor, a key player in the response of mammalian cells to pathogens and stress [27, 382, 404, 588, 714]. In the absence of stimulation, NF-κB resides in the cytosol, since nuclear localization of NF-κB is inhibited by binding to IκB. On the other hand, following stimulation by cytokines, which are substances secreted by certain cells of the immune system, IκB kinase activity induces NF-κB to unbind, thus allowing it to enter the nucleus and initiate the expression of target genes, including its own inhibitor IκB. This generates a negative feedback loop that has a time delay between IκB transcription and translation. When this is combined with rapid IκB degradation, oscillations of NF-κB nuclear localization arise with a period in the range of 2–4 hours, depending on the cell type. The NF-κB pathway exhibits a range of behaviors upon stimulation, including both damped and persistent oscillations, which can be understood in terms of the action of three distinct forms of the inhibitor (IκBα, IκBβ and IκBε), see Fig. 5.14 and Ex. 5.2. When all three of these forms are present in the cell, the pathway exhibits damped oscillations in response to stimulation. However, when cells are modified so that certain IκB proteins are absent, the response changes. When IκBα is absent, cells show pathologically high activity, whereas when both IκBβ and IκBε are absent, cells respond to stimuli with sustained oscillations in NF-κB activity. This difference in behavior is a consequence of the fact that, of the three IκBα forms, only IκBα receives positive feedback from NF-κB.

Finally, note that yet another well-studied ultradian oscillator involves the tumor suppressor protein p53, which is phosphorylated in response to γ-irradiation-induced DNA damage [834]. Phosphorylated p53 induces the expression of several

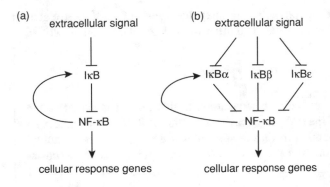

Fig. 5.14: (a) The main components of the NF-κB signaling pathway involving the action of the inhibitor IκB. (b) More detailed digram that includes three different types of IκB inhibitor. See text for details.

genes, including genes of its own regulators. A negative feedback loop results in oscillations of p53 activity with a period similar to NF-κB localization.

Box 5A. Linear stability analysis and Hopf bifurcations.

Stability of equilibria. Consider the d-dimensional differential equation

$$\frac{dx_i}{dt} = F_i(\mathbf{x}), \quad \mathbf{x} \in \mathbb{R}^d, \tag{5.1.26}$$

and suppose there is a fixed point $\mathbf{x}^*$ for which $F_i(\mathbf{x}^*) = 0$ for all $i = 1, \ldots, d$. Imagine that one is in a local neighborhood of the fixed point by writing $\mathbf{x}(t) = \mathbf{x}^* + \mathbf{y}(t)$. Substituting into the differential equation and Taylor expanding to first order in $\mathbf{y}$, we have

$$\frac{dy_j}{dt} = \frac{dx_j}{dt} = F_j(\mathbf{x}^* + \mathbf{y}) = F_j(\mathbf{x}^*) + \sum_{k=1}^d \frac{\partial F_j}{\partial x_k}(\mathbf{x}^*) y_k + \text{h. o. t.}.$$

Imposing the fixed point conditions and dropping the higher-order terms then yields the linear equation

$$\frac{d\mathbf{y}}{dt} = \mathbf{M}\mathbf{y}, \quad M_{jk} = \frac{\partial F_j}{\partial x_k}(\mathbf{x}^*),$$

where $\mathbf{M}$ is the Jacobian matrix. Trying a solution of the form $\mathbf{y} = \mathbf{v}e^{\lambda t}$ generates the eigenvalue equation $\mathbf{M}\mathbf{v} = \lambda \mathbf{v}$. This will have a nontrivial solution provided that $\mathbf{M} - \lambda \mathbf{I}$ is non-invertible, where $\mathbf{I}$ is the $d \times d$ unit matrix. This yields the characteristic equation

$$\text{Det}[\mathbf{M} - \lambda \mathbf{I}] = 0,$$

which reduces to a d-th order polynomial for λ. The latter can be factorized as

$$\prod_{j=1}^d (\lambda - \lambda_j) = 0,$$

where the roots λ_j are the (possibly complex) eigenvalues of $\mathbf{M}$. If these eigenvalues are distinct, then the general solution to the linear ODE can be written as

$$\mathbf{y}(t) = \sum_{j=1}^d c_j \mathbf{v}_j e^{\lambda_j t},$$

where $\mathbf{v}_j$ are the corresponding eigenvectors. In this case, it is clear that the fixed point is stable if the real part of each eigenvalue of the Jacobian matrix $\mathbf{M}$ is negative. It turns out that this result still holds if the eigenvalues are not all distinct. In the one-dimensional case $\dot{x} = F(x)$, a fixed point x^* is stable (unstable) if $F'(x^*) < 0$ ($F'(x^*) > 0$). Introducing the "potential" $U(x)$ according to $F(x) = -U'(x)$, it immediately follows that fixed points of $F(x)$ are stationary points of $U(x)$, and stable (unstable) fixed points correspond to minima (maxima) of $U(x)$. In higher dimensions, one cannot generally express the functions $F_j(\mathbf{x})$ in terms of a scalar potential, $\mathbf{F} = -\nabla U(\mathbf{x})$, unless the system corresponds to a gradient dynamical system. (In classical mechanics, $\mathbf{F}$ would then represent a conservative force.)

Hopf bifurcation. Any differential equation describing the dynamics of some biological system will depend on one or more parameters μ. This can be made explicit by writing

$$\frac{d\mathbf{x}}{dt} = \mathbf{F}(\mathbf{x}(t);\mu), \quad \mathbf{x} \in \mathbb{R}^d, \mu \in \mathbb{R}^m.$$

For simplicity, suppose that only one parameter is varied and set $m = 1$. The dynamical system is said to have a bifurcation at the critical value $\mu = \mu_c$ if there is a change in the (topological) structure of trajectories as μ crosses μ_c. (For a detailed introduction to bifurcation theory see the book *Elements of applied bifurcation theory* by Kuznetsov [483].) In the case of local bifurcations, this means that there is a change in the number and/or stability of equilibria or fixed points. A Hopf bifurcation occurs when a fixed point changes stability as μ crosses μ_c resulting in the emergence of a small amplitude limit cycle. If the limit cycle is stable and surrounds an unstable fixed point, then the Hopf bifurcation is said to be supercritical, whereas, if the limit cycle is unstable and surrounds a stable fixed point then it is said to be subcritical. Before stating the general conditions for the occurrence of a Hopf bifurcation in a planar system ($d = 2$), it is useful to consider an explicit example:

$$\frac{dx}{dt} = \mu x - \omega y + \sigma x(x^2 + y^2), \quad \frac{dy}{dt} = \omega x + \mu y + \sigma y(x^2 + y^2), \quad (5.1.27)$$

with $\sigma = \pm 1$.

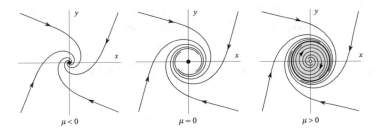

Fig. 5.15: Supercritical Hopf bifurcation. Phase portraits for system $\dot{x} = \mu x + y - x(x^2 + y^2)$ and $\dot{y} = -x + \mu y - y(x^2 + y^2)$ for the three cases $\mu < 0$, $\mu = 0$ and $\mu > 0$.

This pair of equations has a single fixed point at the origin. Transforming to polar coordinates by setting $x = r\cos\theta, y = r\sin\theta$, we have

$$\frac{dr}{dt} = r(\mu + \sigma r^2), \quad \frac{d\theta}{dt} = -\omega, \tag{5.1.28}$$

with $r \geq 0$. It immediately follows that all nonzero solutions rotate clockwise at the same angular frequency ω. Suppose that $\sigma = -1$. If $\mu \leq 0$ then $\dot{r} < 0$ for all $r > 0$ and trajectories converge to the fixed point at the origin, which is stable. On the other hand, if $\mu > 0$ then $\dot{r} < 0$ for $r \in (\sqrt{\mu}, \infty)$ and $\dot{r} > 0$ for $r \in (0, \sqrt{\mu})$. Hence, the origin is now an unstable fixed point, whereas there is a stable periodic orbit at $r = \sqrt{\mu}$, see Fig. 5.15. In other words, the system undergoes a supercritical Hopf bifurcation at the critical value $\mu = \mu_c = 0$. Similarly, if $\sigma = +1$ then the fixed point undergoes a subcritical Hopf bifurcation with an unstable limit cycle existing for $\mu < 0$.

Hopf bifurcation theorem (2D). Consider the planar dynamical system

$$\frac{dx}{dt} = f(x, y; \mu), \quad \frac{dy}{dt} = g(x, y; \mu)$$

for the single parameter μ. Suppose that there exists a fixed point at $(x, y) = (0, 0)$, say. Linearizing about the origin gives the linear system

$$\frac{d}{dt}\begin{pmatrix} x \\ y \end{pmatrix} = \mathbf{M}(\mu)\begin{pmatrix} x \\ y \end{pmatrix},$$

where $\mathbf{M}(\mu)$ is the Jacobian

$$\mathbf{M}(\mu) = \begin{pmatrix} \partial_x f(0, 0; \mu) & \partial_y f(0, 0; \mu) \\ \partial_x g(0, 0; \mu) & \partial_y g(0, 0; \mu) \end{pmatrix}.$$

Let the eigenvalues of the Jacobian be given by the complex conjugate pair $\lambda(\mu), \overline{\lambda}(\mu) = \alpha(\mu) \pm i\beta(\mu)$. Furthermore, suppose that at a certain value $\mu = \mu_c$ (with $\mu_c = 0$, say) the following conditions hold:

1. $\alpha(0) = 0$, $\beta(0) = \omega \neq 0$ with $\text{sign}(\omega) = \text{sign}[\partial_\mu g(0, 0; 0)]$.

2. $\left.\dfrac{d\alpha(\mu)}{d\mu}\right|_{\mu=0} \neq 0$ (transversality condition).

3. The so-called first Liapunov coefficient l_1 doesn't vanish, $l_1 \neq 0$ (genericity condition), where

$$l_1 = \frac{1}{16}(f_{xxx} + f_{xyy} + g_{xxy} + g_{yyy})$$
$$+ \frac{1}{16\omega}(f_{xy}[f_{xx} + f_{yy}] - g_{xy}[g_{xx} + g_{yy}] - f_{xx}g_{xx} + f_{yy}g_{yy}),$$

with $f_{xy} = \partial_x \partial_y f(0,0;0)$ etc.

Then a unique curve of periodic solutions bifurcates from the origin as μ crosses zero. The amplitude of the limit cycle grows like $\sqrt{|\mu|}$ and the period tends to $2\pi/\omega$ as $|\mu| \to 0$. Suppose for the sake of illustration that $\alpha'(\mu) > 0$, and set $\sigma = \text{sign}(l_1)$.

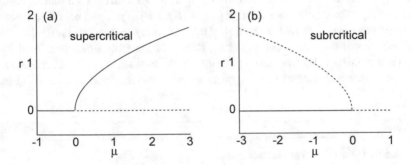

Fig. 5.16: Bifurcation diagrams for (a) supercritical ($\sigma = -1$) and (b) subcritical ($\sigma = +1$) Hopf bifurcations in the same dynamical system as Fig. 5.15.

If $\sigma = -1$ then the fixed point is asymptotically stable for $\mu \leq 0$ and unstable for $\mu > 0$. Moreover, there is a unique and stable periodic orbit that exists for $\mu > 0$, corresponding to the case of a supercritical Hopf bifurcation, see Fig. 5.16(a).

If $\sigma = +1$ then the fixed point is asymptotically stable for $\mu < 0$ and unstable for $\mu \geq 0$. Moreover, and unstable limit cycle exists for $\mu < 0$, and we have a subcritical Hopf bifurcation, see Fig. 5.16(b).

Finally, sufficiently close to the bifurcation point, the dynamical system is locally topologically equivalent to the normal form given by equation (5.1.27).

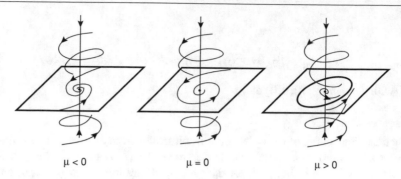

$$\mu < 0 \qquad\qquad \mu = 0 \qquad\qquad \mu > 0$$

Fig. 5.17: Hopf bifurcation in 3D.

There is also an n-dimensional version of the Hopf bifurcation theorem for $n \geq 3$. In particular, suppose that the $n \times n$ Jacobian matrix has a pair of complex conjugate eigenvalues $\lambda(\mu), \overline{\lambda}(\mu) = \alpha(\mu) \pm i\beta(\mu)$, and all other eigenvalues have negative real parts, $\text{Re}[\lambda_j(\mu)] < 0$ for $j = 3,\ldots,n$. It can then be shown that the system converges to a family of smooth two-dimensional invariant manifolds W_c (trajectories starting in W_c stay within W_c), and the above Hopf bifurcation theorem then applies to the effective dynamics on W_c (Fig. 5.17).

Box 5B. Delay differential equations.

We illustrate how oscillations can arise in simple delay differential equations (DDEs). (For a more detailed introduction, with applications to the life sciences, see the book by Smith [769].)

Simple linear DDE. Consider the DDE

$$\frac{dx}{dt} = -\alpha x(t - \tau) \tag{5.1.29}$$

for real α and $\tau > 0$. Clearly, when $\tau = 0$ and $\alpha > 0$, the fixed point $u = 0$ is asymptotically stable, that is, $x(t) = x(0)e^{-\alpha t}$. When $\tau > 0$, we need to consider initial data on the interval $-\tau \leq t \leq 0$. For example, suppose that $x(t) = 1$ for $t \in [-\tau, 0]$. We can then generate the solution by iteration. First, for $t \in [0, \tau]$ we have $\dot{x} = -x(t - \tau) = -1$ so that

$$x(t) = x(0) + \int_0^t (-1)ds = 1 - t, \quad t \in [0, \tau].$$

Similarly, on $t \in [\tau, 2\tau]$ we have $\dot{x} = -[1 - (t - \tau)]$ and thus

$$x(t) = x(\tau) - \int_\tau^t [1 - (s - \tau)]ds = 1 - t + \frac{1}{2}(t - \tau)^2, \quad t \in [\tau, 2\tau].$$

Repeating this analysis, we find that $x(t)$ is a polynomial of degree n on each subinterval $t \in [(n-1)\tau, n\tau]$, and is a smooth function except at each $n\tau$, $n \geq 0$. Simulating the DDE for $\alpha = 1$, shows that the steady-state solution $x = 0$ is stable for $\tau < \pi/2$ and unstable when $\tau > \pi/4$; in the latter case the system supports undamped oscillations.

In order to understand the occurrence of oscillations, we look at the spectrum of the linear operator

$$\mathbb{L}(X) = \frac{dX}{d\tau} + \alpha X(t - \tau).$$

In particular

$$\mathbb{L}(e^{\lambda t}) = e^{\lambda t}(\lambda + \alpha e^{-\lambda \tau}). \tag{5.1.30}$$

It follows that $e^{\lambda t}$ is a solution of equation (5.1.29) if and only if $\lambda \in \mathbb{C}$ is a root of the characteristic equation

$$h(\lambda) := \lambda + \alpha e^{-\lambda \tau} = 0. \tag{5.1.31}$$

Note that λ is said to be root of order l, $l \geq 1$, if

$$h(\lambda) = h'(\lambda) = h''(\lambda) = \cdots h^{(l-1)}(\lambda) = 0, \quad h^{(l)}(\lambda) \neq 0.$$

Differentiating both sides of equation (5.1.30) with respect to λ establishes that $t^j e^{\lambda t}$, $j = 0, 1, \ldots, k$, are solutions of (5.1.29) if and only if λ is a root of order at least $k + 1$. Analogous to standard ODEs, it can be shown that the solution $x = 0$ is asymptotically stable if $\text{Re}[\lambda] < 0$ for all roots of the characteristic equation, and is unstable if there exists a root with positive root.

We begin by listing some general properties of the roots, which follow from the fact that $h(\lambda)$ is an analytic function:

1. If λ is a root, then it has finite order.

2. If λ is a root, then so is its complex conjugate $\overline{\lambda}$.

3. The set of roots can have no accumulation points in $\mathbb{C}$, which implies that, for each $R > 0$, the set of roots satisfying $|\lambda| < R$ is finite. Hence, the set of roots is a countable (possibly finite) set.

4. If the set of roots is infinite, which we denote by $\{\lambda_n\}_{n=1}^\infty$, then $|\lambda_n| \to \infty$ as $n \to \infty$. Since $|\alpha| e^{-\tau \text{Re}[\lambda_n]} = |\lambda_n|$, we see that $\text{Re}[\lambda_n] \to -\infty$. Hence, for each $a \in \mathbb{R}$, the number of roots satisfying $\text{Re}[\lambda] \geq a$ is finite.

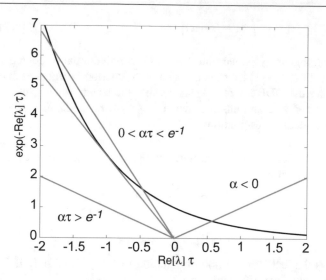

Fig. 5.18: Graphical construction of real roots of characteristic function $h(\lambda)$, which are given by intersections of curve e^{-x} with the straight lines $-x/\alpha\tau$.

Further analysis can be performed by decomposing each root into real and imaginary parts, $\lambda = u + iv$. This yields the pair of equations

$$u = -\alpha e^{-\tau u}\cos(\tau v), \quad v = \alpha e^{-\tau u}\sin(\tau v). \qquad (5.1.32)$$

The existence of real roots, satisfying $-x/(\alpha\tau) = e^{-x}$ with $x = \tau u$, may be explored graphically, see Fig. 5.18. One finds that there is one positive real root for $\alpha < 0$ (origin is unstable for positive feedback), a pair of negative real roots when $0 < \alpha < \tau e^{-1}$ and no real roots when $\alpha > \tau e^{-1}$. Therefore, in the case of negative feedback ($\alpha > 0$), the origin can only become unstable due to at least one pair of complex conjugate eigenvalues crossing the imaginary axis. One obtains the following results:

1. If $0 < \alpha\tau < \pi/2$, then there exists $\delta > 0$ such that $\text{Re}(\lambda) \leq -\delta$ for all roots (origin is stable).

2. If $\alpha\tau = \pi/2$ then $\lambda = \pm i\pi\tau/2$ are roots of order one.

3. If $\alpha\tau > \pi/2$, then there are roots $\lambda = x \pm iy$ with $u > 0$, $v \in (\pi\tau/2, \pi\tau)$ (origin is unstable).

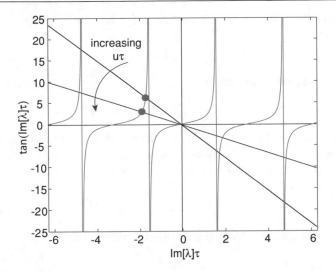

Fig. 5.19: Graphical construction of complex root with positive real part.

We now prove these results. Since $\alpha > 0$, it follows that if $u + iv$ is a root with $u \geq 0$ and $v > 0$, then $\cos(v\tau) \leq 0 < \sin(v\tau)$ and thus

$$v\tau \in S := \cup_{n=0}^{\infty}\{[\pi/2, \pi) + 2n\pi\}.$$

As $\sin(v\tau)/v\tau$ is strictly monotonically decreasing within each disconnected domain of $v\tau \in S$, and $\sin(v\tau)/v\tau = 2/\pi$ when $v\tau = \pi/2$, we conclude that $\sin(v\tau)/(v\tau) \leq 2/\pi$ for all $v\tau \in S$. Hence, if $u > 0$ and $v \geq 0$ then the following inequalities hold

$$\frac{1}{\alpha\tau} \leq \frac{e^{\tau u}}{\alpha\tau} = \frac{\sin(v\tau)}{v\tau} \leq \frac{2}{\pi}.$$

In other words, if $\alpha\tau < \pi/2$ then $\mathrm{Re}(\lambda) < 0$ for every root λ. Finally, we can establish the existence of roots with positive real part when $\alpha\tau > \pi/2$ using another graphical construction, see Fig. 5.19. Equation (5.1.32) implies that

$$\tan(v\tau) = -\frac{v\tau}{u\tau}.$$

For each $u > 0$ there is a solution for v in the domain $v\tau \in (\pi/2, \pi)$, which is given by the intercept of the straight line having negative slope with the function $\tan(v\tau)$. This generates a one parameter family of solutions $v = v(u)$ with u then determined from the equation

$$u = -\alpha e^{-\tau u} \cos(\tau v(u)).$$

Nonlinear DDE with negative feedback. Let us now consider a nonlinear DDE of the form

$$\frac{dx}{dt} = -f(x(t-\tau)),\tag{5.1.33}$$

where f is a smooth function satisfying

$$f(0) = 0, \quad f'(0) = 1 \quad f''(0) = A, \quad f'''(0) = B.$$

It follows that $x = 0$ is a fixed point and the negative sign means that the nonlinear function represents delayed negative feedback. It is convenient to move the delay parameter outside of the nonlinear function by rescaling time. This gives

$$\frac{dx}{dt} = -\tau f(x(t-1)).\tag{5.1.34}$$

Linearizing about $x = 0$ leads to the characteristic equation

$$\lambda + \tau e^{-\lambda} = 0.$$

We thus obtain equation (5.1.31) under the mapping $\alpha \to \tau$ and $\tau \to 1$. It follows from the previous analysis that the origin is stable for $0 \le \tau < \pi/2$ and $\lambda = \pm i\pi/2$ are roots at $\tau = \pi/2$. Writing $\tau = \pi/2 + \mu$, the characteristic function can be written as

$$F(\lambda;\mu) := \lambda + (\pi/2 + \mu)e^{-\lambda},$$

with μ treated as a bifurcation parameter. Note that

$$F(i\pi/2,0) = 0, \quad \partial_\lambda F(i\pi/2,0) = 1 + i\pi/2, \quad \partial_\mu F(i\pi/2.0) = -i.$$

The implicit function theorem ensures that in a neighborhood of $\mu = 0$, we can solve $F = 0$ for $\lambda = \lambda(\mu) = \alpha(\mu) + i\omega(\mu)$ with $\lambda(0) = i\pi/2$ and

$$\frac{d\lambda}{d\mu}(0) = -\partial_\mu F/\partial_\lambda F = \frac{\pi/2 + i}{1 + (\pi/2)^2}.$$

In particular, $\alpha'(0) > 0$ (transversality condition). Continuity of the roots as a function of μ implies that there exists $\mu_0 > 0$ such that for $|\mu| < \mu_0$, the roots consist of the conjugate pair $\alpha(\mu) \pm i\omega(\mu)$ and all other roots satisfy $\text{Re}[\lambda], -\delta/2$ for some $\delta > 0$. (This is a consequence of the fact that all roots have negative real part bounded away from zero when $\mu < 0$.) The existence of a periodic solution now follows from the Hopf bifurcation theorem for nonlinear DDEs [769]. It can also be shown that the Hopf bifurcation is supercritical if

$$A^2\left(\frac{11\pi - 4}{5\pi}\right) > B,$$

and is subcritical if the reverse inequality holds.

5.1.4 Activator-repressor relaxation oscillator

Another important class of synthetic gene oscillator is based on the combination of both negative and positive feedback loops to create a relaxation oscillator [35, 361, 362, 785]. One of the potential advantages of such oscillators is that they appear more robust. For the sake of illustration, we will consider the particular synthetic network developed by Hasty et al. [362], see Fig. 5.20. It consists of two plasmids, both of which contain the same promoter denoted by P_{RM}^*. (A plasmid is a genetic structure in a cell that can replicate independently of the chromosomes, typically a small circular DNA strand in the cytoplasm of a bacterium or protozoan. Plasmids are often used for the experimental manipulation of genes.) On the first plasmid, the promoter controls the λ phage cI gene, whereas on the second plasmid, the promoter regulates the lac gene. Each copy of the promoter P_{RM}^* consists of three operator sites $OS1, OS2$ and $OS3^*$. CI dimers can independently bind to $OS1$ and $OS2$, whereas a Lac tetramer can bind to $OS3^*$. (The P_{RM}^* promoter is a mutant of the wild type promoter P_{RM}, consisting of sites $OS1, OS2, OS3$, all of which bind CI dimers.) The gene is OFF if $OS3^*$ is occupied, irrespective of the states of the other two sites, and is ON otherwise. The production rate of protein in the on state is enhanced by a factor α if both sites $OS1$ and $OS2$ are occupied.

Let X and Y denote CI and Lac proteins, respectively, and let D_i denote the promoter region of plasmid i. The set of binding reactions are as follows:

$$2X \underset{k_{-1}}{\overset{k_1}{\rightleftharpoons}} X_2, \quad 4Y \underset{k_{-2}}{\overset{k_2}{\rightleftharpoons}} 2Y_2 \underset{k_{-3}}{\overset{k_3}{\rightleftharpoons}} Y_4,$$

$$D_i + X_2 \underset{k_{-4}}{\overset{k_4}{\rightleftharpoons}} D_iX_2, \quad D_iX_2 + X_2 \underset{k_5}{\overset{k_{-5}}{\rightleftharpoons}} D_iX_2X_2,$$

$$D_i + Y_4 \underset{k_{-6}}{\overset{k_6}{\rightleftharpoons}} D_iY_4, \quad D_iX_2 + Y_4 \underset{k_{-7}}{\overset{k_7}{\rightleftharpoons}} D_iX_2Y_4, \quad D_iX_2X_2 + Y_4 \underset{k_{-8}}{\overset{k_8}{\rightleftharpoons}} D_iX_2X_2Y_4.$$

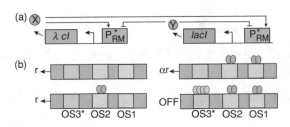

Fig. 5.20: Synthetic relaxation oscillator. (a) Circuit diagram consisting of an activator gene cI and a repressor gene $lacI$. (b) Different states of the mutant promoter P_{RM}^* indicating the production rate in each state. The gene is OFF if a Lac tetramer is bound to site $OS3^*$ irrespective of the occupancy of sites $OS1$ and $OS2$.

For each reversible reaction a, $a = 1, \ldots, 8$, we have an equilibrium constant $K_a = k_a/k_{-a}$ (in units of inverse Molars). These are further constrained by taking $K_5 = \sigma K_4$ and $K_6 = K_7 = K_8$. Similarly, the production of the proteins are given by the irreversible reactions

$$D_1 \underset{r}{\rightarrow} D_1 + X, \quad D_1 X_2 \underset{r}{\rightarrow} D_1 X_2 + X, \quad D_1 X_2 X_2 \underset{\alpha r}{\rightarrow} D_1 X_2 X_2 + X,$$

$$D_2 \underset{r}{\rightarrow} D_2 + Y, \quad D_2 X_2 \underset{r}{\rightarrow} D_2 X_2 + Y, \quad D_2 X_2 X_2 \underset{\alpha r}{\rightarrow} D_2 X_2 X_2 + Y.$$

Given these reactions, the kinetic equations for the monomer CI concentration x and monomer Lac concentration y take the form

$$\frac{dx}{dt} = -2k_1 x^2 + 2k_{-1} x_2 + r([D_1] + [D_1 X_2] + \alpha[D_1 X_2 X_2]) - \gamma_x x, \qquad (5.1.35a)$$

$$\frac{dy}{dt} = -2k_2 y^2 + 2k_{-2} y_2 + r([D_2] + [D_2 X_2] + \alpha[D_2 X_2 X_2]) - \gamma_y y. \qquad (5.1.35b)$$

Here $x_2 = [X_2]$ and $y_2 = [Y_2]$ denote the concentrations of CI and Lac dimers, respectively, and $[D_1]$ denotes the concentration of D_1 etc..

As it stands, the pair of equations does not form a closed system due to coupling with the concentrations x_2 and y_2. These in turn couple to the concentrations of other protein multimers and DNA-protein complexes. However, it is known that the multimerization and DNA binding processes are much faster than the rates of transcription and degradation so that one can carry out an adiabatic approximation. That is, x_2, y_2, etc., are taken to be in quasi-equilibrium. Introduce the dimensionless variables

$$\tilde{x} = \sqrt{K_1 K_4} x, \quad \tilde{y} = (K_2^2 K_3 K_6)^{1/4} y, \quad \tilde{t} = \sqrt{K_1 K_4} r \mathcal{M}_1 t, \qquad (5.1.36)$$

where $\mathcal{M}_i$ is the copy number concentration of plasmid i. Eliminating the slow variables and dropping the tildes, we obtain the equations

$$\frac{dx}{dt} = \frac{1 + x^2 + \alpha \sigma x^4}{(1 + x^2 + \sigma x^4)(1 + y^4)} - \gamma_x x, \qquad (5.1.37a)$$

$$\tau_y \frac{dy}{dt} = \frac{1 + x^2 + \alpha \sigma x^4}{(1 + x^2 + \sigma x^4)(1 + y^4)} - \gamma_y x, \qquad (5.1.37b)$$

where the degradation rates have also been non-dimensionalized, and

$$\tau_y = \left(\frac{K_1^2 K_4^2}{K_2^2 K_3 K_6} \right)^{1/4} \frac{\mathcal{M}_1}{\mathcal{M}_2}. \qquad (5.1.38)$$

We sketch the derivation of equation (5.1.37). Writing down the kinetic equations for the concentrations

$$[X_2], [Y_2], [Y_4], [D_i], [D_i X_2], [D_i X_2 X_2], [D_i Y_4], [D_i X_2 Y_4], [D_i X_2 X_2 Y_4],$$

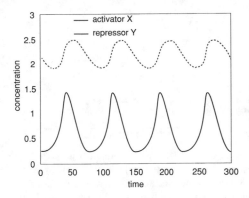

Fig. 5.21: Illustration of periodic time variation of activator and repressor concentrations in relaxation oscillator of Hasty et al. [362], with the same parameter values as used in Ref. [406]: $\alpha = 11, \sigma = 2, \gamma_x = 0.2, \gamma_y = 0.3$ and $\tau_y = 5$. (Units are arbitrary.)

and setting time derivatives to zero for these fast variables shows that

$$[X_2] = K_1 x^2, \quad [Y_4] = K_3 K_2^2 y^4,$$
$$[D_i X_2] = K_4 [D_i][X_2] = K_1 K_4 x^2 [D_i],$$
$$[D_i X_2 X_2] = K_5 [D_i X_2][X_2] = \sigma(K_1 K_4)^2 x^4 [D_i],$$
$$[D_i Y_4] = K_6 [D_i][Y_4] = K_2^2 K_3 K_6 y^4 [D_i],$$
$$[D_i X_2 Y_4] = K_7 [D_i X_2][Y_4] = K_6 K_4 [D_i][X_2][Y_4] = (K_1 K_4 x^2)(K_2^2 K_3 K_6 y^4)[D_i],$$
$$[D_i X_2 X_2 Y_4] = K_8 [D_i X_2 X_2][Y_4] = \sigma(K_1 K_4 x^2)^2 (K_2^2 K_3 K_6 y^4)[D_i].$$

Substituting these quasi-equilibrium expressions into equation (5.1.35) yields

$$\frac{dx}{dt} = r[D_1]\left(1 + K_1 K_4 x^2 + \alpha\sigma(K_1 K_4 x^2)^2\right) - \gamma_x x, \tag{5.1.39a}$$

$$\frac{dy}{dt} = r[D_2]\left(1 + K_1 K_4 x^2 + \alpha\sigma(K_1 K_4 x^2)^2\right) - \gamma_y y. \tag{5.1.39b}$$

Conservation of the total number $\mathcal{M}_i$ of i plasmids implies that

$$[D_i] + [D_i X_2] + [D_i X_2 X_2] + [D_i Y_4] + [D_i X_2 Y_4] + [D_i X_2 X_2 Y_4] = \mathcal{M}_i,$$

which implies that

$$\left(1 + K_1 K_4 x^2 + \sigma(K_1 K_4 x^2)^2\right)\left(1 + K_6 K_3 K_2^2 y^4\right)[D_i] = \mathcal{M}_i.$$

Substituting this equation into (5.1.39) and performing the rescalings (5.1.36) yields equation (5.1.37).

Since the plasmid copy numbers can be chosen by the designer of the construct, it follows that τ_y is a design parameter. Note that Hasty et al. [362] took a high copy plasmid for the CI gene ($\mathcal{M}_1 \approx 50$) and a low copy number for the lac gene ($\mathcal{M}_2 \approx 1$), which yielded $\tau_y \approx 5$. The degradation rates were taken as experimental control parameters. An illustration of the model's oscillatory behavior is shown in

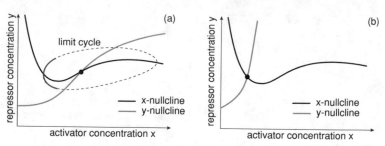

Fig. 5.22: Phase-plane analysis of activator-repressor model. The nullcline of the activator is cubic-like, whereas the nullcline of the repressor is monotonic. Point(s) of intersection of the two nullclines are fixed points. (a) If the degradation rate of the activator is around 5 times faster then the repressor, then the repressor nullcline intersects the middle branch of the activator nullcline. The fixed point is unstable and the system exhibits a limit cycle, with most of the time spent in the repressed regime (indicated by red curve). (b) If the degradation rates are comparable, then the repressor nullcline intersects the first branch of the activator nullcline. The fixed point is now stable and the system is excitable. This means that for small perturbations of the fixed point, trajectories make small excursions before relaxing back to the fixed point. On the other hand, large perturbations result in large excursions in the phase plane before returning to a neighborhood of the fixed point, resulting in a single burst of gene expression. This is analogous to the firing of a neuronal action potential.

Fig. 5.21. Both the activator X and the repressor Y are expressed periodically. The concentrations grow until there is a sufficient amount of repressor to cut off expression; the concentrations then decrease until repression is released and the next burst of expression begins. The sharp peaks in the concentration profiles, particularly for X, is indicative of a relaxation oscillator. A characteristic feature of a relaxation oscillator is that there is a separation of time scales in which a single cycle can be partitioned into regions of rapid changes in the activator concentration separated by one or more regions of slow variation. (The repressilator is a more regular oscillator with sinusoidal like oscillations and no separation of time scales.) In the activator-repressor model, the slow response occurs in the repressed regime (low x). This can be further understood using phase-plane analysis, as detailed in the caption of Fig. 5.22.

Remark 5.4. *Effects of noise on genetic oscillators.* There are at least two different issues that arise when noise is included. First, to what extent is a genetic oscillator robust to noise arising from low copy numbers, for example? This has been explored in some detail within the context of circadian rhythms [48, 282, 283, 324, 325, 505]. In the absence of noise, one can represent the dynamics in terms of motion around a limit cycle. Each point on the limit cycle can be assigned a phase $\theta \in [0, 2\pi]$ such that the dynamics is given by

$$\frac{d\theta}{dt} = \omega, \tag{5.1.40}$$

where $\omega = 2\pi/\Delta$ and Δ is the period of the oscillation. When noise is included, one tends to observe an irregular trajectory around the limit cycle. If the trajectory wanders too far from the limit cycle then it may escape completely, and the notion of phase breaks down. However, even if the trajectory remains close to the limit cycle, the resulting stochastic phase of the oscillator undergoes diffusion around the limit cycle, so all phase information is eventually lost. The effective diffusion coefficient of this process is one measure of robustness [324, 325].

A second effect of noise is that it can extend the parameter regime over which oscillations can occur. That is, even though the deterministic system converges to a fixed point, the system exhibits oscillatory behavior when noise is included. This can be established by looking at the power spectrum of the protein concentration, see Sect. 5.2.6.

5.2 Molecular noise and the chemical master equation

In the previous section, we identified at least two possible sources of noise: molecular noise arising from the low copy number of mRNA and proteins, and promoter noise associated with the binding/unbinding of transcription factors. In both cases, it is necessary to replace the deterministic kinetic equations based on the law of mass action by stochastic models. In this section, we focus on chemical master equations that take into account fluctuations in the number of molecules of each species, whereas in Sect. 5.3, we use stochastic hybrid systems to model promoter noise. The latter is neglected here by assuming that the binding/unbinding of transcription factors is sufficiently fast.

Intrinsic versus extrinsic noise. Following Swain et al. [247], it is useful to distinguish between contributions arising from fluctuations that are inherent to a given system of interest (intrinsic noise) from those arising from external factors (extrinsic noise). As we have already indicated, intrinsic noise is due to fluctuations generated by the binding/unbinding of a repressor or activator, and mRNA and protein production and decay—these can be significant due to the small number of molecules involved. Extrinsic-noise sources are defined as fluctuations and population variability in the rate constants associated with these events. The classification of a noise source as intrinsic rather than extrinsic is context-dependent, so that intrinsic noise at one level can act as extrinsic noise at another level. Gene-intrinsic noise refers to the variability generated by molecular-level noise in the reaction steps that are intrinsic to the process of gene expression. Network-intrinsic noise is generated by fluctuations and variability in signal transduction and includes gene-intrinsic noise in the expression of regulatory genes. Cell-intrinsic noise arises from gene-intrinsic noise and network-intrinsic noise, as well as fluctuations and variability in cell-specific factors, such as the activity of ribosomes and polymerases, metabolite concentrations, cell size, cell age, and stage of the cell cycle.

    An operational definition of gene-intrinsic noise is the difference in the expression of two almost identical genes from identical promoters in single cells averaged over a large cell population. This definition is based on the assumptions that the two genes are affected identically by fluctuations in cell-specific factors and that their expression is perfectly correlated if these fluctuations are the only source of population heterogeneity. The contribution of gene-intrinsic noise can then be investigated experimentally using two-reporter assays [247]. (A biochemical assay is an experimental procedure for quantitatively measuring the presence or amount of one or more target molecular constituents.) These assays evaluate, in single cells, the difference in the abundances of two equivalent reporters, such as red and green fluorescent protein, expressed from identical promotors, located at equivalent chromosomal positions. Cells with the same amount of each protein appear yellow, whereas cells expressing more of one fluorescent protein rather than the other appear green or red, see Fig. 5.23a,b. In the absence of intrinsic noise, the expression of the two reporter proteins should be strongly correlated. On the other hand, since the expression of the two reporters are independent, any intrinsic stochasticity in gene expression will be manifested as differences in expression levels within the same cell. By considering the spread of the expression levels across a population of cells, it is possible to separate out the noise contribution generated by the biochemical reaction steps that are intrinsic to the process of gene expression from extrinsic environmental noise. There are, however, some potential limitations of this approach. For example, contributions from extrinsic factors, such as imperfect timing in replication and intracellular heterogeneity might be measured as gene-intrinsic noise. Moreover, because increased variability in regulatory signals might cause cells to adapt distinct expression states, the measured population-average gene-intrinsic noise and the extrinsic regulatory noise might not always be independent.

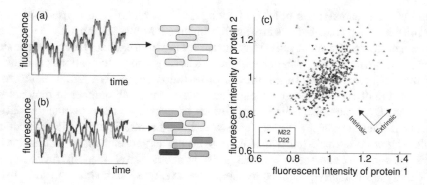

Fig. 5.23: Measuring intrinsic and extrinsic noise in gene expression. Two almost identical genes, which encode red and green fluorescent proteins, are expressed from identical promotors, and are influenced identically by cell-specific factors, such as gene regulatory signals. (a) Cells with equal amounts of the two proteins appear yellow, indicating that the level of intrinsic noise is low. Noise fluctuations of the two proteins in the same cell appear correlated over time. (b) If intrinsic noise is significant then the expression of the two genes becomes uncorrelated in individual cells, giving rise to a cell population in which some cells express more of one fluorescent protein than the other. (c) Plot of fluorescence in two strains (M22 and D22) of the bacterium *Escherichia coli*. Each point represents the mean fluorescence intensities from one cell. Spread of points perpendicular to the diagonal line on which the two fluorescent intensities are equal corresponds to intrinsic noise, whereas the spread parallel to this line corresponds to extrinsic noise. [Adapted from Elowitz *et al.* [247].]

At the population level, gene-intrinsic noise leads to intercellular variability in gene expression that is uncorrelated between cells. However, cells are also subjected to time-varying inputs of a stochastic (extrinsic noise) or deterministic nature, either from the extracellular environment or from upstream processes within a cell that may be entrained to the cell cycle or circadian rhythms. One of the important consequences of time-varying inputs is that they can induce varying degrees of synchrony or correlations across a population of cells [210]. Population gene expression is explored further in Chap. 15.

5.2.1 Master equation for chemical reaction networks

In the case of low copy numbers, one has to keep track of fluctuations in the number of each chemical species. As a simple example consider, unregulated transcription of mRNA by a single gene, see Fig. 5.2. Suppose that at time t, the number of mRNA molecules is $N(t)$. In an infinitesimal time interval Δt, there are three possible events:

1. $N(t + \Delta t) = N(t) + 1$ due to the gene transcription of one new mRNA molecule, which occurs with probability $\Omega \kappa \Delta t$.
2. $N(t + \Delta t) = N(t) - 1$ due to the degradation of one existing mRNA molecule, which occurs with probability $\gamma N(t) \Delta t$.
3. $N(t + \Delta t) = N(t)$ with probability $1 - \Omega \kappa \Delta t - \gamma N(t) \Delta t$.

The probability of two or more transitions in the small time interval is negligible. Let $P_n(t) = \mathbb{P}[N(t) = n]$. It follows that

$$P_n(t + \Delta t) = [\Omega \kappa P_{n-1}(t) + \gamma(n+1)P_{n+1}(t)] \Delta t + [1 - (\Omega \kappa + \gamma n)\Delta t]P_n(t).$$

Rearranging this equation, dividing through by Δt and taking the limit $\Delta t \to 0$ yields the birth–death master equation

$$\frac{dP_n}{dt} = \Omega \kappa P_{n-1}(t) + \gamma(n+1)P_{n+1}(t) - (\Omega \kappa + \gamma n)P_n(t) \tag{5.2.1}$$

for $n \geq 0$ and $P_{-1}(t) \equiv 0$.

Equation (5.2.1) is identical in form to the master equation (4.2.29) for the growth of a microtubule under GTP hydrolysis. It immediately follows that $P_n(t)$ is given by

$$P_n(t) = e^{-\lambda(t)} \frac{\lambda(t)^n}{n!}, \quad \lambda(t) = \frac{\kappa \Omega}{\gamma}(1 - e^{-\gamma t}), \tag{5.2.2}$$

which is a time-dependent Poisson distribution of rate $\lambda(t)$. Hence, in the limit $t \to \infty$ we obtain the time-independent Poisson distribution

$$P_n^* = e^{-\lambda_0} \frac{\lambda_0^n}{n!}, \quad \lambda_0 = \kappa \Omega / \gamma. \tag{5.2.3}$$

It follows that $\langle N \rangle = \lambda_0$ and $\mathrm{Var}[N] = \lambda_0$. This is an important result because both the mean and variance in the number of protein molecules can be measured experimentally. One commonly used measure of the level of noise in a regulatory networks is the so-called Fano factor

$$\text{Fano factor} = \frac{\langle N^2 \rangle - \langle N \rangle^2}{\langle N \rangle}. \tag{5.2.4}$$

For the unregulated process, the Fano factor is one. This is a baseline value for quantifying the effects of gene regulation on the level of noise.

In order to extend the above analysis to more complex gene regulatory networks, we need to have a systematic method for constructing the corresponding master equation. This can be achieved by first formulating the deterministic mass-action kinetics in terms of stoichiometric coefficients and propensities. Consider a set of chemical species $i = 1, \ldots, K$ with concentrations x_i undergoing R single-step reactions labeled $a = 1, \ldots, R$. Recall from Sect. 1.3 that the kinetic law of mass action for the concentrations x_i takes the general form

$$\frac{dx_i}{dt} = \sum_{a=1}^{R} S_{ia} f_a(\mathbf{x}), \quad f_a(\mathbf{x}) = \kappa_a \prod_{j=1}^{K} x_j^{s_j} \tag{5.2.5}$$

for $i = 1, \ldots, K$,, where $\mathbf{S}$ is the $K \times R$ stoichiometric matrix and the functions f_a are the propensities. In the case of the autoregulatory network (5.1.2), we have $K = 2$ with x_1 and x_2 the concentrations of mRNA and protein, respectively. There are $R = 4$ reactions. For $a = 1, 2$ (mRNA production and degradation), we have

$$S_{i,1} = \delta_{i,1}, \quad S_{i,2} = -\delta_{i,1}, \quad f_1(\mathbf{x}) = \kappa g(x_1), \quad f_2(\mathbf{x}) = \gamma x_1.$$

Similarly, for $a = 3, 4$ (protein production and degradation), we have

$$S_{i,3} = \delta_{i,2}, \quad S_{i,4} = -\delta_{i,2}, \quad f_3(\mathbf{x}) = \kappa_p x_1, \quad f_4(\mathbf{x}) = \gamma_p x_2.$$

Here $\mathbf{x} = (x_1, x_2)^\top$.

Given the kinetic equation (5.2.5), the corresponding chemical master equation takes the form

$$\frac{dP(\mathbf{n}, t)}{dt} = \Omega \sum_{a=1}^{R} \left(\prod_{i=1}^{K} \mathbb{E}^{-S_{ia}} - 1 \right) f_a(\mathbf{n}/\Omega) P(\mathbf{n}, t). \tag{5.2.6}$$

Here $\mathbb{E}^{-S_{ia}}$ is a step or ladder operator such that for any function $g(\mathbf{n})$,

$$\mathbb{E}^{-S_{ia}} g(n_1, \ldots, n_i, \ldots, n_N) = g(n_1, \ldots, n_i - S_{ia}, \ldots, n_N). \tag{5.2.7}$$

One point to note is that when the number of molecules is sufficiently small, the characteristic form of a propensity function $f(\mathbf{x})$ in equation (5.2.6) has to be modified:

$$\left(\frac{n_j}{\Omega}\right)^{s_j} \rightarrow \frac{1}{\Omega^{s_j}} \frac{n_j!}{(n_j - s_j)!}.$$

Although the chemical master equation is linear in $P(\mathbf{n}, t)$, the transition rates can be nonlinear functions of the copy numbers n_i; the master equation is then said to be nonlinear. In general, it is not possible to obtain exact solutions of the master equation (5.2.6) even in the case of a stationary solution. (Note, however, that recent progress has been made in generalizing the theory of deterministic chemical reaction networks to stochastic models, using notions of weak reversibility and deficiency zero, see Box 5C and [15].) Therefore, one often resorts to some form of approximation scheme.

Box 5C. Weak reversibility and deficiency zero.

Analyzing a system of nonlinear ODEs is generally a daunting task, even with regards finding a general set of conditions for the existence and stability of fixed points. However, in the case of the mass-action kinetic equation (5.2.5), there is an underlying network structure that makes the task much easier. This network structure is associated with a directed graph whose nodes are complexes (sets of chemical species that occur on the left-hand (reactants) or right-hand (products) side of single-step reactions), and whose links represent these reactions [271, 395, 396]. One of the most important theorems regarding deterministic mass-action kinetics is the deficiency zero theorem of Feinberg [272, 273]. Roughly speaking, this theorem states that if certain easily verifiable properties of the network hold, then within each invariant manifold of the dynamics (compatibility class) there is exactly one fixed point with strictly positive components, and the fixed point is locally asymptotically stable. For a review of deterministic chemical reaction network theory, see [351]. Here we briefly summarize some of the main ideas needed to state the zero deficiency theorem.

Weak reversibility. Consider a well-mixed system with K chemical species, $\mathcal{X} = \{X_1, \ldots, X_K\}$, undergoing R chemical reactions labeled $a = 1, \ldots, R$. Let $\mathbf{s}_a$ be the vector specifying the number of molecules of each species forming the a-th reactant complex, and let $\mathbf{r}_a$ be the corresponding vector of the a-th product complex. Let $\mathcal{C} = \{\mathbf{r}_a, \mathbf{s}_a; a = 1, \ldots R\}$ and $\mathcal{R} = \{\mathbf{s}_a \rightarrow \mathbf{r}_a; a = 1, \ldots R\}$ denote the sets of complexes and reactions, respectively. (Note that the same complex can act as a product or reactant in more than one reaction.) The triple $\{\mathcal{X}, \mathcal{C}, \mathcal{R}\}$ is called a chemical reaction network. Corresponding to each reaction network is a unique, directed graph constructed as follows. The nodes of the graph are given by the set of distinct complexes $\mathbf{z} \in \mathcal{C}$. A directed edge is then placed from a complex $\mathbf{z}$ to a complex $\mathbf{z}'$ if and only if $\mathbf{z} \rightarrow \mathbf{z}' \in \mathcal{R}$. Each connected component of the graph is called a linkage class of the graph, with the number of linkage classes denoted by l. A network is said to be reversible if for every forward reaction $\mathbf{z} \rightarrow \mathbf{z}' \in \mathcal{R}$ there is a corresponding backward reaction $\mathbf{z}' \to \mathbf{z} \in \mathcal{R}$. A network is said to be weakly reversible if for any reaction $\mathbf{z} \to \mathbf{z}' \in \mathcal{R}$, there is a sequence of directed reactions starting with $\mathbf{z}'$ as a reactant complex and ending with $\mathbf{z}$ as a product complex. As an illustration, consider the following set of chemical reactions:

There are $K = 4$ chemical species (A, B, C, D), $m = 5$ complexes $(A, B, C + D, 2A, 2C)$, and $R = 5$ reactions. There are two disconnected graphs so the number of linkages $l = 2$. The top graph is reversible, whereas the second is weakly reversible.

Deficiency. Another important notion is the span of the stoichiometric vectors $\mathbf{v}_a = \mathbf{r}_a - \mathbf{s}_a$, that is,

$$S = \text{span}_{a=1,\dots,R}\{\mathbf{v}_a\} \subset \mathbb{R}^K.$$

In general, S will be a proper subset of $\mathbb{R}^K$ so that $s \equiv \dim[S] < K$. For the above example

$$S = \text{span}\left\{ \begin{bmatrix} -1 \\ 1 \\ 0 \\ 0 \end{bmatrix}, \begin{bmatrix} 0 \\ -1 \\ 1 \\ 1 \end{bmatrix}, \begin{bmatrix} 1 \\ 0 \\ -1 \\ -1 \end{bmatrix}, \begin{bmatrix} -2 \\ 0 \\ 2 \\ 0 \end{bmatrix}, \begin{bmatrix} 2 \\ 0 \\ -2 \\ 0 \end{bmatrix} \right\}$$

$$= \text{span}\left\{ \begin{bmatrix} -1 \\ 1 \\ 0 \\ 0 \end{bmatrix}, \begin{bmatrix} 0 \\ -1 \\ 1 \\ 1 \end{bmatrix}, \begin{bmatrix} -2 \\ 0 \\ 2 \\ 0 \end{bmatrix} \right\},$$

so that $s = 3$. Since the vector space S specifies the possible combinations of changes in the number of molecules of each species over the set of reactions, it follows that if the initial concentrations are given by the vector $\mathbf{x}(0) = x_0$, then the solution $\mathbf{x}(t)$ of the mass-action kinetic equation (5.2.5) will lie in the space $x_0 + S$. This invariant manifold of the dynamics is called a stoichiometric compatability class, and since the concentration (number) of each chemical species is positive, any solution lies in the positive stoichiometric compatability class. We need one final definition to state the zero deficiency theorem, namely, the deficiency of a chemical reaction network is

$$\delta = m - l - s,$$

where m is the number of complexes, l is the number of linkage classes, and s is the dimension of the stoichiometric subspace of the network. For example, the above reaction network has $\delta = 0$, since $m = 5, l = 2, s = 3$.

Theorem 5.1. (The deficiency zero theorem [273]). *Consider a weakly reversible, deficiency zero chemical reaction network* $(\mathcal{X}, \mathcal{C}, \mathcal{R})$ *with dynamics given by the mass-action kinetics (5.2.5). Then for any choice of rate constants, within each positive stoichiometric compatability class there is precisely one fixed point, which is locally asymptotically stable with respect to that compatability class.*

A crucial step in the proof of the deficiency zero theorem and its extensions is to reformulate the mass-action kinetics in terms of the space of complexes $\mathbb{R}^m$ rather than the space of chemical species $\mathbb{R}^n$. As a first step, we rewrite equation (5.2.5) as

$$\frac{dx_i}{dt} = \sum_{\mathbf{z} \to \mathbf{z}'} \kappa_{\mathbf{z} \to \mathbf{z}'} \left(\prod_{j=1}^K x_j^{s_j} \right) (z_i' - z_i).$$

That is, we specify the reactions in terms of the reactant and product complexes rather than the reaction label a. We construct a new vector space $\mathbb{R}^m$ of complexes by introducing the set of basis vectors $\{\omega_{\mathbf{z}}, \mathbf{z} \in \mathcal{C}\}$ and decomposing any $v \in \mathbb{R}^m$ as $v = \sum_{\mathbf{z} \in \mathcal{C}} v_{\mathbf{z}} \omega_{\mathbf{z}}$. Let $Z : \mathbb{R}^m \to \mathbb{R}^K$ be the linear map defined by $Z(\omega_{\mathbf{z}}) = \mathbf{z}$. Let $\Psi : \mathbb{R}^K \to \mathbb{R}^m$ be the nonlinear map given by

$$\Psi(\mathbf{x}) = \sum_{\mathbf{z} \in \mathcal{C}} \left(\prod_{j=1}^K x_j^{z_j} \right) \omega_{\mathbf{z}}. \tag{5.2.8}$$

Finally, introduce the map $A_\kappa : \mathbb{R}^m \to \mathbb{R}^m$,

$$A_\kappa(v) = \sum_{\mathbf{z} \to \mathbf{z}'} \kappa_{\mathbf{z} \to \mathbf{z}'} v_{\mathbf{z}} (\omega_{\mathbf{z}'} - \omega_{\mathbf{z}}). \tag{5.2.9}$$

The mass-kinetics can now be written in the form

$$\frac{d\mathbf{x}}{dt} = Z(A_\kappa(\Psi(\mathbf{x}))). \tag{5.2.10}$$

From equation (5.2.5), a fixed point $\mathbf{x}_*$ exists if and only if $\sum_a S_{ia} f_a(\mathbf{x}_*) = 0$ for all $i = 1, \dots, K$. In terms of the decomposition (5.2.10), it is sufficient to show that $A_\kappa(\Psi(\mathbf{x}_*)) = 0$. This is equivalent to the so-called complex balance condition: for each $\mathbf{z} \in \mathcal{C}$

$$\sum_{a : \mathbf{r}_a = \mathbf{z}} \kappa_a \left(\prod_{j=1}^K x_j^{r_{ja}} \right) = \sum_{b : \mathbf{s}_b = \mathbf{z}} \kappa_b \left(\prod_{j=1}^K x_j^{s_{jb}} \right). \tag{5.2.11}$$

Complex balance corresponds to the condition that for each $\mathbf{z} \in \mathcal{C}$, the rates of all reactions into the complex $\mathbf{z}$ are balanced by the rates of all reactions out of $\mathbf{z}$. This is distinct from the standard condition of detailed balance in reversible networks (see Sect. 1.3). The requirement of zero deficiency and weak reversibility establishes that there exists at least one fixed point for which $A_\kappa(\Psi(\mathbf{x}^*)) = 0$. One can establish that this fixed point is unique and asymptotically stable.

Extensions. There have been a number of more recent results in chemical reaction network theory. For example, in the case of deterministic models, mathematical analysis has been used to investigate conditions for the existence of multiple equilibria [202, 203], as well as dynamical properties such as persistence and global stability [16, 23]; persistence is the condition that if all chemical species are initially present then none of them completely disappear during the time-evolution of the network. Another important development has been to show that a product-form stationary solution of a chemical master equation exists for each closed irreducible subset of the state-space, provided that corresponding deterministic mass-action kinetics admits a complex balanced equilibrium [15]. More explicitly, there exists a stationary solution $P(\mathbf{n})$ of equation (5.2.6) consisting of a product of Poisson distributions

$$P(\mathbf{n}) = \prod_{i=1}^{K} \frac{c_i^{n_i}}{n_i!} e^{-c_i}, \quad A_\kappa(\Psi(\mathbf{c})) = 0. \tag{5.2.12}$$

5.2.2 System-size expansion of chemical master equation

The most common approximation scheme for chemical master equations is obtained by carrying out a Taylor expansion in the inverse of the system size Ω [242, 300, 822], along analogous lines to the system-size expansion of birth–death processes, see Sect. 3.3.2. The basic idea is to set $f_a(\mathbf{n}/\Omega)P(\mathbf{n},t) \rightarrow f_a(\mathbf{x})p(\mathbf{x},t)$ with $\mathbf{x} = \mathbf{n}/\Omega$ treated as a continuous vector so that

$$\prod_{i=1}^{K} \mathbb{E}^{-S_{ia}} h(\mathbf{x}) = h(\mathbf{x} - \mathbf{S}_a/\Omega)$$

$$= h(\mathbf{x}) - \Omega^{-1} \sum_{i=1}^{K} S_{ia} \frac{\partial h}{\partial x_i} + \frac{1}{2\Omega^2} \sum_{i,j=1}^{K} S_{ia} S_{ja} \frac{\partial^2 h(\mathbf{x})}{\partial x_i \partial x_j} + O(\Omega^{-3}).$$

Carrying out a Taylor expansion of the master equation to second order thus yields the multivariate FP equation of the Ito form

$$\frac{\partial p}{\partial t} = -\sum_{i=1}^{K} \frac{\partial A_i(\mathbf{x}) p(\mathbf{x},t)}{\partial x_i} + \frac{1}{2\Omega} \sum_{i,j=1}^{K} \frac{\partial^2 D_{ij}(\mathbf{x}) p(\mathbf{x},t)}{\partial x_i \partial x_j}, \tag{5.2.13}$$

where

$$A_i(\mathbf{x}) = \sum_{a=1}^{R} S_{ia} f_a(\mathbf{x}), \quad D_{ij}(\mathbf{x}) = \sum_{a=1}^{R} S_{ia} S_{ja} f_a(\mathbf{x}). \tag{5.2.14}$$

The FP equation (5.2.13) corresponds to the Ito SDE

$$dX_i = A_i(\mathbf{X}) dt + \frac{1}{\sqrt{\Omega}} \sum_{a=1}^{R} B_{ia}(\mathbf{X}) dW_a(t), \tag{5.2.15}$$

where $W_a(t)$ are independent Wiener processes [300],

$$\langle dW_a(t) \rangle = 0, \quad \langle dW_a(t) dW_b(t') \rangle = \delta_{a,b} \delta(t-t') dt\, dt', \tag{5.2.16}$$

and $\mathbf{D} = \mathbf{B}\mathbf{B}^T$, that is

$$B_{ia} = S_{ia} \sqrt{f_a(\mathbf{x})}. \tag{5.2.17}$$

Note that equation (5.2.15) is often referred to as the chemical Langevin equation.

As we highlighted in Sect. 3.3.2, the system-size expansion works particularly well when the underlying deterministic system (5.2.5) has a unique stable fixed point since, after a transient phase, the dynamics consists of Gaussian-like fluctuations about the fixed point. In particular, suppose that the deterministic system (5.2.5), written as

$$\frac{dx_i}{dt} = A_i(\mathbf{x}),$$

has a unique stable fixed point $\mathbf{x}_*$ for which $A_i(\mathbf{x}_*) = 0$, and introduce the Jacobian matrix $\mathbf{M}$ with

$$M_{ij} = \left. \frac{\partial A_i}{\partial x_j} \right|_{\mathbf{x}=\mathbf{x}_*}. \tag{5.2.18}$$

The chemical Langevin equation suggests that, after a transient phase, the stochastic dynamics is characterized by Gaussian fluctuations about the fixed point, see Fig. 5.24. Substituting $X_i(t) = x_i^* + Y_i(t)/\sqrt{\Omega}$ into the Langevin equation (5.2.15) and keeping only lowest order terms in $\Omega^{-1/2}$ yields the Ornstein-Uhlenbeck (OU) process [300]

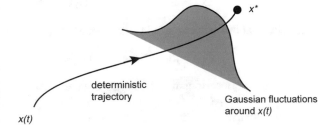

Fig. 5.24: Linear noise approximation, in which fluctuations in a neighborhood of a fixed point x^* are treated as Gaussian.

deterministic trajectory

x^*

Gaussian fluctuations around $x(t)$

$x(t)$

$$dY_i = \sum_{j=1}^{K} M_{ij}Y_j dt + \sum_{a=1}^{R} B_{ia}(\mathbf{x}_*)dW_a(t). \tag{5.2.19}$$

Introducing the stationary covariance matrix

$$\Sigma_{ij} = \langle [Y_i(t) - \langle Y_i(t) \rangle][Y_j(t) - \langle Y_j(t) \rangle] \rangle,$$

it immediately follows from the analysis of the multivariate OU process, see Sect. 2.3, that

$$\mathbf{M}\Sigma + \Sigma\mathbf{M}^T = -\mathbf{B}\mathbf{B}^T. \tag{5.2.20}$$

Equation (5.2.20), which is a form of fluctuation-dissipation theorem, is often used to estimate the size of protein fluctuations in gene networks due to intrinsic noise.

Stochastic autoregulatory network. As an illustration of the system-size expansion, we consider a stochastic version of the autoregulatory network shown in Fig. 5.4. The corresponding kinetic equations are

$$\frac{dx_1}{dt} = -\gamma x_1 + \kappa g(x_2), \quad \frac{dx_2}{dt} = \kappa_p x_1 - \gamma_p x_2, \tag{5.2.21}$$

where $x_1(t)$ and $x_2(t)$ denote the concentrations of mRNA and protein molecules at time t, respectively. The parameters γ, γ_p represent the degradation rates, κ_p represents the translation rate of proteins, and $\kappa g(x)$ represents the nonlinear feedback effect of the protein on the transcription of mRNA. Suppose that there are $N_1(t) = n_1$ mRNA and $N_2(t) = n_2$ proteins at time t. The corresponding master equation for the probability distribution $P = P(n_1, n_2, t)$ is

$$\frac{dP(n_1, n_2, t)}{dt} = \kappa\Omega g(n_2/\Omega)P(n_1 - 1, n_2, t) + \gamma(n_1 + 1)P(n_1 + 1, n_2, t)$$

$$+ \kappa_p n_1 P(n_1, n_2 - 1, t) + \gamma_p(n_2 + 1)P(n_1, n_2 + 1, t)$$

$$- [\Omega\kappa g(n_2/\Omega) + \gamma n_1 + \kappa_p n_1 + \gamma_p n_2] P(n_1, n_2, t) \tag{5.2.22}$$

for $n_1, n_2 \geq 0$. In order to avoid writing down the equations for dP/dt when $n_1 = 0$ and/or $n_2 = 0$, we impose the conditions $P(n_1, -1, t) = P(-1, n_2, t) \equiv 0$.

Expressing the above chemical master equation in the canonical form (5.2.6), we can use a system-size expansion to derive an FP equation of the form (5.2.13) with drift terms

$$A_1(\mathbf{x}) = \kappa g(x_2) - \gamma x_1, \quad A_2(\mathbf{x}) = \kappa_p x_1 - \gamma_p x_2,$$

and a diagonal diffusion matrix $\mathbf{D}$ with nonzero components

$$D_{11} = \kappa g(x_2) + \gamma x_1, \quad D_{22} = \kappa_p x_1 + \gamma_p x_2.$$

Linearizing the corresponding Langevin equation about the unique fixed point by setting $X_i(t) = x_i^* + \Omega^{-1/2}Y_i(t)$, shows that $Y_i(t)$ is an OU process, whose stationary covariance matrix Σ satisfies the matrix equation

$$\mathbf{M}\Sigma + \Sigma\mathbf{M}^T = -\mathbf{B}\mathbf{B}^T \equiv -\mathbf{D},$$

with

$$\mathbf{M} = \begin{pmatrix} -\gamma & \mu \\ \kappa_p & -\gamma_p \end{pmatrix}, \quad \mathbf{D} = \begin{pmatrix} \kappa g(x_2^*) + \gamma x_1^* & 0 \\ 0 & \kappa_p x_1^* + \gamma_p x_2^* \end{pmatrix}, \quad \mu = \kappa_p g'(x_2^*).$$

Note that $\mu > 0$ for an activator and $\mu < 0$ for a repressor.

The solution of the matrix equation yields, see Ex. 5.3

$$\Sigma_{22} = \left[1 + \frac{b}{1+\eta} \frac{1-\phi}{1+b\phi} \right] x_2^*,$$

where

$$b = \frac{\kappa_p}{\gamma}, \quad \phi = -\frac{\mu}{\gamma_p}, \quad \eta = \frac{\gamma_p}{\gamma}.$$

Here, η is the ratio of degradation rates, and ϕ describes the strength and sign of the feedback. The parameter b can be interpreted as a mean burst size, and is motivated by the observation that a single mRNA generates a burst of protein production before it decays, which is known as translational bursting, see Sect. 5.2.4. Finally, making the identifications $\text{var}[N_2] = \Omega\Sigma_{22}$ and $\langle N_2 \rangle = \Omega x_2^*$, the Fano factor for proteins is

$$\frac{\text{var}[N_2]}{\langle N_2 \rangle} = 1 + \frac{b}{1+\eta} \frac{1-\phi}{1+b\phi}.$$

There are two distinct contributions to the deviation from Poisson-like statistics. First, the factor $b/(1+\eta)$, which implies that that protein bursting ($b > 1$) tends to increase fluctuations in the absence of feedback ($\phi = 0$). Second, the factor

$$\frac{1-\phi}{1+b\phi} = \frac{\gamma_p + g'(x_2^*)}{\gamma_p - bg'(x_2^*)},$$

which shows that negative feedback ($g'(x^*) < 0$) suppresses fluctuations.

A simplified version of a stochastic autoregulatory network is considered in Ex. 5.4, where the number of mRNA molecules is taken to be in quasi-equilibrium so that the only random variable is the number of protein molecules n. Another application of the linear noise approximation is explored in Ex. 5.5, and a spectral analysis of protein bursting in the absence of autoregulatory feedback is considered in Ex. 5.6.

5.2.3 The stochastic simulation algorithm (SSA)

The SSA, which was originally developed by Gillespie [310–312], is an efficient numerical scheme for generating exact sample paths of a continuous-time Markov process whose probability distribution evolves according to the chemical master equation (5.2.6). In the following, we assume units have been chosen so that the system size $\Omega = 1$. We can thus identify the components of $\mathbf{x}$ as the molecular counts for each chemical species.

The starting point for constructing the SSA is to define a new probability function $p(\tau, a | \mathbf{x}, t)$, which is the probability, given $\mathbf{X}(t) = \mathbf{x}$, that the next reaction in the system will occur in the time interval $[t + \tau, t + \tau + \Delta\tau)$ and will be the reaction a. From this perspective, both τ and a are random variables conditioned on $\mathbf{X}(t) = \mathbf{x}$. An analytical expression for $p(\tau, a | \mathbf{x}, t)$ can be obtained by introducing another probability function $P_0(\tau | \mathbf{x}, t)$, which is the probability, given $\mathbf{X}(t) = \mathbf{x}$, that no reaction of any kind occurs in the time interval $[t, t + \tau)$. It follows from the definitions of P_0 and the propensities f_a that P_0 satisfies the equation

$$P_0(\tau + d\tau | \mathbf{x}, t) = P_0(\tau | \mathbf{x}, t)\left[1 - \sum_{a=1}^{R} f_a(\mathbf{x}) d\tau\right],$$

which is the product of the probability that no reaction occurs in $[t, \tau)$ and the probability that there are no transitions in the infinitesimal interval $[t + \tau, t + \tau + d\tau)$. Rearranging and taking the limit $d\tau \to 0$ yields

$$\frac{dP_0(\tau | \mathbf{x}, t)}{d\tau} = -F(\mathbf{x}) P_0(\tau | \mathbf{x}, t), \quad F(\mathbf{x}) = \sum_{a=1}^{R} f_a(\mathbf{x}).$$

Under the initial condition $P(0 | x, t) = 1$, we have the solution

$$P_0(\tau | \mathbf{x}, t) = \exp(-F(\mathbf{x})\tau).$$

We now note

$$p(\tau, a | \mathbf{x}, t) d\tau = P_0(\tau | \mathbf{x}, t) f_a(\mathbf{x}) d\tau,$$

which implies that p can be written in the form

$$p(\tau, a | \mathbf{x}, t) = F(\mathbf{x}) \exp(-F(\mathbf{x})\tau) \frac{f_a(\mathbf{x})}{F(\mathbf{x})}. \tag{5.2.23}$$

Hence, τ is an exponential random variable with mean and standard deviation $1/F(\mathbf{x})$, while a is a statistically independent integer random variable with $\mathbf{x}$-dependent probability $f_a(\mathbf{x})/F(\mathbf{x})$.

One exact Monte Carlo method for generating samples of the random variables τ, a is to draw two random numbers r_1, r_2 from the uniform distribution on $[0, 1]$ with

$$\tau = -\frac{1}{F(\mathbf{x})} \ln r_1, \tag{5.2.24a}$$

$$a = \text{ the smallest integer for which } \sum_{s=1}^{a} f_a(\mathbf{x}) > r_2 F(\mathbf{x}). \tag{5.2.24b}$$

The direct method of implementing the SSA is as follows:

1. Initialize the time $t = t_0$ and the chemical state $\mathbf{x} = \mathbf{x}_0$

2. Given the state $\mathbf{x}$ at time t, determine the $f_a(\mathbf{x})$ for $a = 1, \ldots, R$ and their sums $F(\mathbf{x})$

3. Generate values for τ and a using equations (5.2.24a) and (5.2.24b)

3. Implement the next reaction by setting $t \rightarrow t' = t + \tau$ and $x_j \rightarrow x'_j = x_j + S_{ja}$.

4. Return to step 2 with $(\mathbf{x}, t)$ replaced by $(\mathbf{x}', t')$, or else stop.

There have been a variety of subsequent algorithms that differ in the implementation of step 2, including the next reaction method [309] and the modified next reaction method [14]. The latter is based on the random time change representation of Kurtz [17]

In many applications the mean time between reactions, $1/F(\mathbf{x})$, is very small so that simulating every reaction becomes computationally infeasible, irrespective of the version of the SSA is chosen. Gillespie [311] introduced tau-leaping in order to address this problem by sacrificing some degree of exactness of the SSA in return for a gain in computational efficiency. The basic idea is to "leap" the system forward by a pre-selected time τ (distinct from the τ of the SSA), which may include several reaction events. Given $\mathbf{X}(t) = \mathbf{x}$, τ is chosen to be large enough for efficient computation but small enough so that

$$f_a(\mathbf{x}) \approx \text{constant in } [t, t + \tau) \text{ for all } a.$$

Let $\mathcal{N}(\lambda)$ denote a Poison counting process with mean λ. During the interval $[t, t + \tau)$ there will be approximately $\mathcal{N}(\lambda_a)$ reactions of type a with $\lambda_a = f_a(\mathbf{x})\tau$. Since each of these reactions increases increases x_j by S_{ja}, the state at time $t + \tau$ will be

$$X_j(t + \tau) = \mathbf{x} + \sum_{a=1}^{R} \mathcal{N}_a(f_a(\mathbf{x})\tau)S_{ja}, \qquad (5.2.25)$$

where the $\mathcal{N}_a$ are independent Poisson processes. This equation is known as the tau-leaping formula. However, there are two fundamental problems with the original formulation of tau-leaping. First, it is difficult to choose the appropriate value of τ at each iteration of the algorithm - occasionally large changes in propensities occur that cause one or more components x_j to become negative. Second, although tau-leaping becomes exact in the limit $\tau \rightarrow 0$, the inefficiency becomes prohibitive since the R generated Poisson random numbers will be zero most of the time resulting in no change of state. These two issues have been addressed in various modifications in the tau-leaping procedure, see for example Cao et al. [161]. Some computer problems using the SSA are explored in Ex. 5.7.

5.2.4 Translational protein bursting

Advances in experimental techniques have allowed the observation of real-time fluctuations in gene expression within individual cells. For example, by trapping individual *E. coli* cells in a microfluidic chamber, it is possible to observe the step-wise increase of β-gal of the *lacZ* gene due to translational bursting [291]. That is, each mRNA molecule is translated into a few protein molecules before it degrades, resulting in a sequence of protein bursts. Assuming that the lifetime of mRNA is short compared to the lifetime of the protein and there is no regulatory feedback, fluctuations in the level of mRNA can be integrated out, and the production of proteins can be treated as random uncorrelated events, in which the number of proteins is exponentially distributed. Here we describe some stochastic models of translational bursting.

Stochastic model of translational bursting with multiple mRNAs. First, consider a single mRNA molecule with a degradation rate γ, which starts synthesizing a protein at time $t = 0$. Let $P_0(n,t)$ ($P_c(n,t)$) denote the probability that there are n proteins at time t and the mRNA has not (has) decayed. Neglecting protein degradation, we have the master equation [86, 156, 564]

$$\frac{dP_0(n,t)}{dt} = -\gamma P_0(n,t) + \kappa_p[P_0(n-1,t) - P_0(n,t)], \qquad (5.2.26a)$$

$$\frac{dP_c(n,t)}{dt} = \gamma P_0(n,t), \qquad (5.2.26b)$$

where κ_p is the rate of protein production and we have set $P_0(-1,t) \equiv 0$. Let $P(n) = \lim_{t\to\infty} P_c(n,t)$. Note that $\lim_{t\to\infty} P_0(n,t) = 0$ due to the decay of mRNA. Integrating (5.2.26b) with respect to time gives

$$P(n) = \gamma \int_0^\infty P_0(n,t)dt,$$

since $P_c(n,0) = 0$. In order to compute $P_0(n,t)$, integrate equation (5.2.26a) with respect to time using $P_0(n,0) = \delta_{n,0}$:

$$-\delta_{n,0} = -P(n) + \frac{\kappa_p}{\gamma}[P(n-1) - P(n)].$$

Setting $n = 0$ gives $P(0) = \gamma/(\kappa_p + \gamma)$, and for $n \geq 1$, we have the recurrence relation

$$P(n) = \frac{\kappa_p}{\kappa_p + \gamma}P(n-1) \to P(n) = \left(\frac{\kappa_p}{\kappa_p + \gamma}\right)^n \frac{\gamma}{\kappa_p + \gamma}.$$

Introducing the generating function

$$G(z) \equiv \sum_{n\geq 0} z^n P(n) = \frac{\gamma}{\kappa_p + \gamma}\frac{1}{1 - z\kappa_p/(\kappa_p + \gamma)} = \frac{\gamma}{\kappa_p + \gamma - z\kappa_p},$$

the burst size is given by

$$b = \sum_{n \geq 0} nP(n) = G'(1) = \left. \frac{\kappa_p \gamma}{[\kappa_p + \gamma - z\kappa_p]^2} \right|_{z=1} = \frac{\kappa_p}{\gamma}.$$

Now suppose that there are m mRNA molecules, and that translation of each mRNA proceeds independently. The probability of producing N proteins due to bursts from each mRNA molecule can be expressed as a multiple convolution [658]. For example, if $m = 2, 3$ then

$$P_2(N) = \sum_{n=0}^{N} P(n)P(N-n), \quad P_3(N) = \sum_{n=0}^{N} P(n) \sum_{n'=0}^{N-n} P(n')P(N-n-n').$$

Assume that the number of proteins is sufficiently large so that we can approximate the sums by integrals, for example,

$$P_2(N) = \int_0^N P(n)P(N-n)dn.$$

The advantage of the integral formulation is that one can use Laplace transforms and the convolution theorem. In particular,

$$\widetilde{P}_m(s) \equiv \int_0^\infty P_m(n)e^{-sn}dn = \left[\widetilde{P}(s)\right]^m,$$

where

$$\widetilde{P}(s) = \int_0^\infty \frac{b^n}{(1+b)^{n+1}} e^{-sn}dn = \frac{1}{1+b} \int_0^\infty \left(\frac{b}{1+b}e^{-s}\right)^n dn$$

$$= \frac{1}{1+b} \int_0^\infty \exp\left(n\ln\left(\frac{b}{1+b}e^{-s}\right)\right) dn = -\frac{1}{1+b} \left[\ln\left(\frac{b}{1+b}\right) - s\right]^{-1}.$$

It follows that

$$\widetilde{P}_m(s) = \left(\frac{-1}{1+b}\right)^m \left[\ln\left(\frac{b}{1+b}\right) - s\right]^{-m} = \left(\frac{1}{1+b}\right)^m [s - \beta]^{-m},$$

where $\beta = \ln(b/(1+b))$. Now use the following Laplace identities:

$$\mathscr{L}(e^{\beta n}) = \frac{1}{s-\beta}, \quad \left(\frac{d}{d\beta}\right)^k \mathscr{L}(e^{\beta n}) = \mathscr{L}(n^k e^{\beta n}) = \frac{k!}{[s-\beta]^{k+1}}.$$

These establish that

$$\widetilde{P}_m(s) = \left(\frac{1}{1+b}\right)^m \frac{1}{(m-1)!} \mathscr{L}(n^{m-1}e^{\beta n}).$$

Hence,

$$P_m(n) = \left(\frac{1}{1+b}\right)^m \frac{1}{(m-1)!} n^{m-1} e^{\beta n} = \left(\frac{b}{1+b}\right)^n \left(\frac{1}{1+b}\right)^m \frac{n^{m-1}}{\Gamma(m)}.$$

For $n, b \gg 1$, we can make the approximation

$$\left(\frac{b}{1+b}\right)^n = e^{-n\ln(1+b^{-1})} \approx e^{-n/b},$$

which leads to the gamma distribution for n with m fixed

$$P_m(n) \equiv F(n; m, b^{-1}) = \frac{n^{m-1} e^{-n/b}}{b^m \Gamma(m)}. \tag{5.2.27}$$

From properties of the gamma distribution, we immediately note that for a given number of mRNA molecules $\langle n \rangle = mb$ and $\text{var}(n) = mb^2$. Hence, under the various approximations the Fano factor is of the order of the burst size b. Finally, an estimate for m is $m \approx k/\gamma_0$ where k is the rate of production of mRNAs and γ_0 is the frequency of the cell cycle (assuming that it is higher than the rate of protein degradation).

Chapman-Kolmogorov equation for population bursting. An alternative approach to analyzing protein bursting is to start from the Chapman-Kolmogorov (CK) equation [291]

$$\frac{\partial p(x,t)}{\partial t} = \frac{\partial}{\partial x}[\gamma_p x p(x)] + k \int_0^x w(x - x') p(x', t) dx', \tag{5.2.28}$$

where $p(x,t)$ is the probability density for x protein molecules (treating x as a continuous variable) at time t, and

$$w(x) = \frac{1}{b} e^{-x/b} - \delta(x). \tag{5.2.29}$$

The first term on the right-hand side of the CK equation represents protein degradation, whereas the second term represents the production of proteins from exponentially distributed bursts. The gamma distribution (5.2.27) with $n \to x$ is obtained as the stationary solution of the CK equation, which can be established using Laplace transforms (see below). It is also possible to incorporate autoregulatory feedback into the CK equation by allowing the burst rate to depend on the current level of protein x, which acts as its own transcription factor [291, 544, 545, 884]:

$$\frac{\partial p(x,t)}{\partial t} = \frac{\partial}{\partial x}[\gamma_p x p(x,t)] + k \int_0^x w(x - x') c(x') p(x', t) dx'. \tag{5.2.30}$$

One possible form of the response function $c(x)$ is a Hill function $c(x) = k^s/(k^s + x^s)$, with $s > 0$ ($s < 0$) corresponding to negative (positive) feedback. In this case, the stationary density takes the form

$$p(x) = A x^{m(1+\varepsilon)-1} e^{-x/b} [1 + (x/k)^s]^{-m/s}.$$

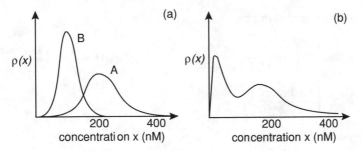

Fig. 5.25: Sketch of typical steady-state probability densities obtained in a model of protein autoregulation [291]. Parameter values are $m = 10$, $b = 20$ and $k = 70$ nM. (a) Negative feedback. Curves A and B correspond to the two cases of no regulation ($c \equiv 1$) and regulation $c = k^s/(k^s + x^s) + \varepsilon$ with $s = 1$ and $\varepsilon = 0.05$. (b) Positive feedback with $\varepsilon = 0.2$ and $s = -4$.

Plots of the stationary density show that negative autoregulation sharpens the density (noise reduction), whereas positive feedback broadens the density and can lead to bistability, see Fig. 5.25.

Suppose that $c(x) = 1$ (no autoregulatory feedback). Laplace transforming with respect to the protein number x, $\mathscr{L}(p) = \int_0^\infty e^{-zx} p(x,t)dx = \widetilde{p}(z,t)$, and using the convolution theorem gives

$$\frac{\partial \widetilde{p}(z,t)}{\partial t} = \gamma_p z \mathscr{L}[xp](z,t) + k\widetilde{w}(z)\widetilde{p}(z,t).$$

Now

$$\int_0^\infty e^{-zx} xp(x,t)dx = -\frac{\partial}{\partial z}\int_0^\infty e^{-zx} p(x,t)dx = -\frac{\partial}{\partial z}\widetilde{p}(z,t),$$

and

$$\widetilde{w}(z) = \int_0^\infty e^{-zx}\left[\frac{1}{b}e^{-x/b} - \delta(x)\right]dx = \frac{1}{1+bz} - 1 = -\frac{bz}{1+bz}.$$

Therefore,

$$\frac{\partial \widetilde{p}(z,t)}{\partial t} + \gamma_p z\frac{\partial}{\partial z}\widetilde{p}(z,t) = k\widetilde{w}(z)\widetilde{p}(z,t).$$

This is a quasilinear equation, which can be solved using the method of characteristics [717]. That is, the corresponding characteristic equations are

$$\frac{dz}{dt} = \gamma_p z, \quad \frac{d\widetilde{p}}{dt} = k\widetilde{w}(z)\widetilde{p}.$$

Solving for z, we have $z(t) = z_0 e^{\gamma_p t}$ where z_0 parameterizes the initial data. Hence

$$\frac{d\widetilde{p}}{dt} = -k\frac{bz_0 e^{\gamma_p t}}{1+bz_0 e^{\gamma_p t}}\widetilde{p},$$

which can be integrated to give

$$\widetilde{p}(t) = F(z_0)\exp\left(-k\int_0^t \frac{bz_0 e^{\gamma_p t'}}{1+bz_0 e^{\gamma_p t'}}dt'\right) = F(z_0)\exp\left(-\frac{k}{\gamma_p}\ln\left[\frac{1+bz_0 e^{\gamma_p t}}{1+bz_0}\right]\right)$$

$$= F(z_0)\left[\frac{1+bz_0 e^{\gamma_p t}}{1+bz_0}\right]^{-k/\gamma_p}.$$

Setting $z_0 = ze^{-\gamma_p t}$ then yields

$$\widetilde{p}(z,t) = F(ze^{-\gamma_p t})\left[\frac{1+bz}{1+bze^{-\gamma_p t}}\right]^{-k/\gamma_p}.$$

Since $\int_0^\infty p(x,t)dx = 1$ for all $t > 0$, we see that $\widetilde{p}(0,t) = 1$ for all $t > 0$. In the limit $t \to \infty$,

$$\widetilde{p}(z,t) \to \widetilde{p}(z) = F(0)[1+bz]^{-k/\gamma_p}.$$

Since $\int_0^\infty p(x)dx = 1$, we see that $\widetilde{p}(0) = 1$, which implies that $F(0) = 1$. Using the inverse Laplace transform

$$\mathcal{L}^{-1}(b^{-1}+z)^{-m} = e^{-x/b}\frac{x^{m-1}}{\Gamma(m)},$$

we obtain the result that the stationary density is

$$p(x) = \frac{1}{b^m \Gamma(m)}x^{m-1}e^{-x/b}, \quad m = \frac{k}{\gamma_p}.$$

Now suppose that $c(x)$ is given by the Hill function

$$c(x) = \frac{k^s}{k^s + x^s} + \varepsilon.$$

Laplace transforming the steady-state equation gives

$$\frac{\partial\widetilde{p}}{\partial z} = -\frac{k}{\gamma_p}\frac{1}{b^{-1}+z}\int_0^s \widetilde{c}(z-s)\widetilde{p}(s)ds,$$

where we have applied the convolution theorem. Multiplying both sides by $z+b^{-1}$ and taking the inverse Laplace transform gives

$$\frac{\partial xp}{\partial x} + \frac{xp}{b} = \frac{k}{\gamma_p}c(x)p(x).$$

This can be solved for $p(x)$:

$$p(x) = Ax^{-1}e^{-x/b}\exp\left(\frac{k}{\gamma_p}\int\frac{c(y)}{y}dy\right),$$

where A is a normalization factor. Performing the integral with respect to y for the given form of $c(y)$ finally yields the result

$$p(x) = Ax^{m(1+\varepsilon)-1}e^{-x/b}[1+(x/k)^s]^{-m/s}.$$

Finally, note that a rigorous analysis of the existence of a stationary density and convergence to the stationary density has been carried out for a more general class of integral operator equations than (5.2.30) [544, 545, 884].

5.2.5 Noise-induced switching

Another consequence of molecular noise is noise-induced switching. Consider a bistable autoregulatory network with positive nonlinear feedback. For simplicity, we assume that the concentration of mRMA is in quasi-equilibrium so that the deterministic rate equation is given by

$$\frac{dx}{dt} := A(x) = -\gamma_p x + \kappa_p g(x), \quad g(x) = 1 + \sigma\frac{x^2}{K^2+x^2}. \tag{5.2.31}$$

(We have absorbed a factor κ/γ into κ_p.) Let us introduce the deterministic potential $U(x)$ by setting $A(x) = -dU/dx$. In the absence of molecular noise, the dynamics can then be written as

$$\frac{dx}{dt} = -\frac{dU}{dx}. \tag{5.2.32}$$

The minima and maxima of the potential $U(x)$ correspond to stable and unstable fixed points of the deterministic dynamics, respectively. Hence, we can represent the deterministic dynamics in terms of motion down a double well potential with minima $x_\pm$ separated by a maximum x_0, along analogous lines to Fig. 2.7. Now, however, the variable x represents protein concentration rather than position of a Brownian particle.

In the presence of molecular noise, the number of proteins $N(t)$ evolves according to a birth–death process with the following master equation for $P_n(t) = \mathbb{P}[N(t) = n]$:

$$\frac{dP_n}{dt} = \Omega\kappa_p g(n/\Omega)P_{n-1}(t) + \gamma_p(n+1)P_{n+1}(t) - (\Omega\kappa_p g(n/\Omega) + \gamma_p n)P_n(t). \tag{5.2.33}$$

The presence of fluctuations in the number of proteins means that there can be noise-induced transitions between the two metastable states $x_\pm$. These can be studied by considering the diffusion approximation of the birth–death master equation (5.2.33), which yields the FP equation (see Sect. 3.3.2)

$$\frac{\partial p}{\partial t} = -\frac{\partial A(x)p(x,t)}{\partial x} + \frac{1}{2\Omega}\frac{\partial^2 D(x)p(x,t)}{\partial x^2}, \tag{5.2.34}$$

where

$$A(x) = \kappa_p g(x) - \gamma_p x, \quad D(x) = \kappa_p^2 g(x) + \gamma_p x.$$

Applying the theory of noise-induced escape for Brownian particles moving in a double well potential and subject to multiplicative noise, see Sect. 2.4.2 and Fig. 2.8, then yields the following formula for the rate of escape from the state x_-:

$$\lambda_- = \frac{D(x_-)}{2\pi} \sqrt{|\Phi''(x_0)|\Phi''(x_-)} e^{-\Omega[\Phi(x_0)-\Phi(x_-)]}, \tag{5.2.35}$$

with

$$\Phi(x) = -2\int^x \frac{A(y)}{D(y)} dy. \tag{5.2.36}$$

Comparison with equations (2.4.26) and (2.4.27a) shows that the quasipotential Φ is scaled by a factor of 2, and the exponential term has a factor of Ω, which acts as an "inverse temperature." Since the diffusion approximation of the underlying master equation requires Ω to be large, it follows that the system operates in a weak noise (low temperature) regime. Moreover, the rate of escape is exponentially small for large Ω as $\Phi(x_0) > \Phi(x_-)$. (An example of bistability in a chemical reaction network that can be analyzed exactly is considered in Ex. 5.8.) Finally, note the same issue regarding possible exponential errors in the escape rate under the diffusion approximation hold for gene networks as they do for ion channel models (Chap. 3). More accurate approximation schemes will be considered in Chap. 8.

5.2.6 Noise-induced oscillations

Stochastic biochemical and gene networks can exhibit noise-induced oscillations (quasi-cycles) in parameter regimes for which the underlying deterministic kinetic equations have only fixed point solutions. These quasi-cycles are characterized by a peak in the power spectrum obtained using a linear noise approximation of the chemical master equation. As an illustrative example, consider the Brusselator [104, 105]. Here we sketch the analysis, with further details left to Ex. 5.9. The Brusselator is an idealized model of an autocatalytic reaction, in which at least one of the reactants is also a product of the reaction [313]. The model consists of two chemical species X and Y interacting through the following reaction scheme:

$$\emptyset \xrightarrow{a} X, \quad X \xrightarrow{b} Y, \quad 2X + Y \xrightarrow{c} 3X, \quad X \xrightarrow{d} \emptyset.$$

These reactions describe the production and degradation of an X molecule, an X molecule spontaneously transforming into a Y molecule, and two molecules of X reacting with a single molecule of Y to produce three molecules of X. The corresponding mass-action kinetic equations for $u_1 = [X], u_2 = [Y]$ are (after rescaling so that $c = d = 1$)

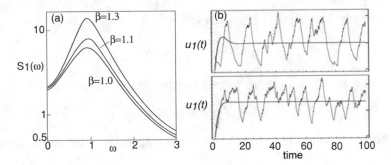

Fig. 5.26: (a) Power spectrum $S_1(\omega)$ of fluctuations in the concentration of X_1 molecules in Brusselator system for parameter values in the fixed point regime of the kinetic equation (5.2.37): $a = 1$ and $b = 1.0, 1.1, 1.3$. (b) Relaxation to a stable fixed point (red curve) in the deterministic system is replaced by a noisy oscillation in the stochastic case (black curves). Two trials are shown.

$$\frac{du_1}{dt} = a - (b+1)u_1 + u_1^2 u_2, \tag{5.2.37a}$$

$$\frac{du_2}{dt} = bu_1 - u_1^2 u_2. \tag{5.2.37b}$$

The system has a fixed point at $u_1^* = a, u_2^* = b/a$, which is stable when $b < a^2 + 1$ and unstable when $b > a^2 + 1$ (see below). Moreover, the fixed point undergoes a Hopf bifurcation at the critical value $b = a^2 + 1$ for fixed a, leading to the formation of a stable limit cycle.

It is straightforward to write down the stoichiometric coefficients and propensities, and thus construct a stochastic version of the Brusselator that keeps track of the number of molecules $n_1(t)$ and $n_2(t)$. Suppose that the system size Ω is sufficiently large so that we can set

$$\frac{n_j}{\Omega} = u_j^* + \frac{1}{\sqrt{\Omega}} v_j,$$

and carry out a linear noise approximation of the associated master equation. This yields a multivariate OU process whose chemical Langevin equation is of the form

$$\frac{dv_j(t)}{dt} = \sum_{j'} M_{jj'} v_{j'}(t) + \sum_{j'} D_{jj'} \eta_{j'}(t), \tag{5.2.38}$$

with white noise terms satisfying

$$\langle \eta_j(t) \rangle = 0, \quad \langle \eta_j(t)\eta_{j'}(t') \rangle = \delta_{j,j'}\delta(t - t').$$

One can show that, see Ex. 5.9

$$\mathbf{M} = \begin{pmatrix} b-1 & a^2 \\ -b & -a^2 \end{pmatrix}, \quad \mathbf{D} = \begin{pmatrix} 2(b+1)a & -2ba \\ -2ba & 2ba \end{pmatrix}. \tag{5.2.39}$$

Fourier transforming the Langevin equation with respect to time,

$$V_i(\omega) = \int_{-\infty}^{\infty} e^{i\omega t} v_i(t) dt,$$

and defining the power spectrum of the i-th chemical species by

$$\langle V_i(\omega) V_i(\omega') \rangle = S_i(\omega) \delta(\omega + \omega'),$$

one obtains the following expressions for the components of the power spectrum [104]

$$S_1(\omega) = 2a((1+b)\omega^2 + a^4)\Gamma(\omega)^{-1}, \quad S_2(\omega) = 2ab(\omega^2 + 1 + b)\Gamma(\omega)^{-1},$$
$$(5.2.40)$$

where $\Gamma(\omega) = (a^2 - \omega^2)^2 + (1 + a^2 - b)^2 \omega^2$. In Fig. 5.26 we plot the power spectrum $S_1(\omega)$ in a parameter regime where the deterministic kinetic equation (5.2.37) supports a stable fixed point. It can be seen that there is a peak in the power spectrum at $\omega = \omega_c \neq 0$, indicating the presence of stochastic oscillations (quasi-cycles) even though the deterministic system operates below the Hopf bifurcation point. Moreover the frequency ω_c is approximately equal to the Hopf frequency of limit cycle oscillations beyond the bifurcation point. Thus, intrinsic noise can extend the parameter regime over which a biochemical system can exhibit oscillatory behavior. Another example of noise-induced oscillations is considered in Ex. 5.10.

5.3 Promoter noise and stochastic hybrid systems

So far we have assumed that the promoter state of a gene is in quasi-equilibrium so that the effects of promoter noise can be ignored. We now relax this assumption; the resulting network is then modeled as a stochastic hybrid system.

5.3.1 Two-state gene network with promoter and protein noise

Consider the simple two-state gene network shown in Fig. 5.27, which consists of a gene that can be in one of two states, active or inactive, depending on whether or not a protein Y is bound to a promoter site. (For the moment there is no feedback regulation.) In the active state the gene produces protein X at a rate κ_p, which subsequently degrades at a rate γ_p, whereas no protein is produced in the inactive state. For simplicity, the stages of transcription and translation are lumped together so we do not keep track of the amount of mRNA. The reaction scheme of the regulatory network is

$$G_{\text{off}} + Y \underset{k_-}{\overset{k_+}{\rightleftharpoons}} G_{\text{on}}, \quad G_{\text{on}} \overset{\kappa_p}{\longrightarrow} G_{\text{on}} + X, \quad X \overset{\gamma}{\longrightarrow} \emptyset,$$

where G_{on} and G_{off} denote the active and inactive states of the gene. If the number of Y molecules is large, then the first reaction can be rewritten as

$$G_{\text{off}} \underset{k_-}{\overset{ck_+}{\rightleftharpoons}} G_{\text{on}},$$

where c is the concentration of transcription factor.

Let $M(t)$ denote the current state of the gene with $M(t) = 1$ (active) or $M(t) = 0$ (inactive), and let $N(t)$ be the number of protein molecules. If we include both promoter and protein fluctuations, then we have to deal with the master equation for the joint probability distribution $P(m,n,t) = \mathbb{P}[M(t) = m, N(t) = n]$, where $m \in \{0,1\}$ and $n \geq 0$, which takes the form

$$\frac{dP(m,n,t)}{dt} = [ck_+ m + k_-(1-m)]P(1-m,n,t) + \kappa_p m P(m,n-1,t) \tag{5.3.1}$$
$$+ \gamma_p(n+1)P(m,n+1,t) - [ck_+(1-m) + k_- m + \kappa_p m + \gamma_p n]P(m,n,t),$$

with $P(m,-1,t) = 0$. Since the transition rates are linear in n and m, one could determine the means and variances by taking moments of equation (5.3.1). Here, we consider an alternative method, in which we treat the activation state $M(t)$ of the gene as an upstream drive of gene expression (extrinsic noise), and use the theory of conditional expectations, whereby one first averages with respect to the intrinsic noise and then with respect to the extrinsic noise. The advantage of this method is that it can be applied to more general forms of time-dependent inputs [210]. Furthermore, one is usually interested in the statistics of $N(t)$ rather than $M(t)$.

Suppose we condition on a particular realization of the stochastic process $M(t)$, $\sigma_t = \{M(\tau), 0 \leq \tau \leq t\}$. (Mathematically speaking, the information generated by a given realization forms a σ-algebra, see Chap. 9.) First, define the conditional probability distribution $\widehat{P}_n(t) := \mathbb{P}[N(t) = n \cup N(0) = 0|\sigma_t]$, which satisfies the master equation

$$\frac{d}{dt}\widehat{P}_n(t) = M(t)\kappa_p \widehat{P}_{n-1}(t) + \gamma_p(n+1)\widehat{P}_{n+1}(t) - [M(t)\kappa_p + \gamma_p n]\widehat{P}_n(t) \tag{5.3.2}$$

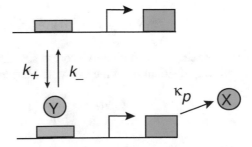

Fig. 5.27: Simple example of a two-state gene network. The promoter transitions between an active state (bound by a transcription factor protein Y) and an inactive state with rates $k_{\pm}$. The active state produces protein X at a rate κ_p and protein X degrades at a rate γ_p.

for $n \geq 0$ and $\widehat{P}_{-1}(t) = 0$. Since $M(t)$ is a discrete stochastic variable, it follows that equation (5.3.2) is defined between jumps in $M(t)$. More specifically, if $P_{mm_0}(t) = \mathbb{P}[M(t) = m|M(0) = n_0]$ then the master equation for $M(t)$ takes the form

$$\frac{dP_{mm_0}}{dt} = \sum_{n=0,1} A_{mn} P_{nm_0}, \quad \mathbf{A} = \begin{pmatrix} -ck_+ & k_- \\ ck_+ & -k_- \end{pmatrix}.$$

Using the fact that $P_{0m_0}(t) + P_{1m_0}(t) = 1$ we can solve this pair of equations to give

$$P_{1m_0}(t) = \delta_{1,m_0} e^{-t/\tau_c} + \frac{ck_+}{\tau_c}(1 - e^{-t/\tau_c}), \quad \tau_c = \frac{1}{k_- + ck_+}.$$

A number of results follow from this. First τ is the relaxation time of the two-state Markov chain with $P_{mm_0}(t) \to \rho_m$ in the limit $t \to \infty$ and

$$\rho_0 = \frac{k_-}{k_- + ck_+}, \quad \rho_1 = \frac{ck_+}{k_- + ck_+}. \tag{5.3.3}$$

In the stationary state we have $\langle M(t) \rangle = \rho_1$, and the stationary autocorrelation function is given by

$$\langle (M(t) - \rho_1)(M(t') - \rho_1) \rangle = \rho_0 \rho_1 e^{-|t-t'|/\tau_c}. \tag{5.3.4}$$

Hence, the two-state Markov process provides an alternative form of colored noise to an Ornstein-Uhlenbeck process [300]. In the physics literature it is often referred to as dichotomous noise [63].

Comparing equation (5.3.2) with (5.2.1), we immediately see that $\widehat{P}_n(t)$ is given by the inhomogeneous Poisson distribution

$$\widehat{P}_n(t) = e^{-\chi(t)} \frac{\chi(t)^n}{n!}, \quad \chi(t) = \kappa_p \int_0^t M(\tau) e^{-\gamma_p(t-\tau)} d\tau. \tag{5.3.5}$$

It follows that the mean and variance of $N(t)$ condition on a particular realization σ_t are

$$\mathbb{E}[N(t)|\sigma_t] = \chi(t), \quad \text{Var}[N(t)|\sigma_t] = \chi(t).$$

Given that $\chi(t)$ is itself stochastic, the full system is an example of a doubly stochastic Poisson process [199, 338], see Sect. 13.1. (There is also a binomial component if $N(0) > 0$.) From the tower property of expectation (see Sect. 1.2), we see that

$$\mathbb{E}[N(t)] = \mathbb{E}[\mathbb{E}[N(t)|\sigma_t]] = \kappa_p \int_0^t \mathbb{E}[M(\tau)] e^{-\gamma_p(t-\tau)}. \tag{5.3.6}$$

Suppose $M(t)$ is a stationary process so that $\mathbb{P}[M(t) = m] = \rho_m$ where ρ_m is the stationary distribution of the two-state Markov chain, see equation (5.3.14). Then

$$\mathbb{E}[N(t)] = \frac{\kappa_p \rho_1}{\gamma_p}(1 - e^{-\gamma_p t}) \xrightarrow[t \to \infty]{} \frac{\kappa_p \rho_1}{\gamma_p}. \tag{5.3.7}$$

Similarly, from the law of total variance,

$$\text{Var}[N(t)] = \mathbb{E}[\text{Var}[N(t)|\sigma_t]] + \text{Var}[\mathbb{E}[N(t)|\sigma_t]] \tag{5.3.8}$$

$$= \mathbb{E}[N(t)] + \kappa_p^2 \int_0^t \int_0^t \mathbb{E}[(M(\tau) - \rho_1)(M(\tau') - \rho_1)]e^{-\gamma_p(2t-\tau-\tau')}d\tau'd\tau.$$

From equation (5.3.4), we see that for $\gamma_p \neq 1/\tau_c$

$$\int_0^t \int_0^t \mathbb{E}[(M(\tau) - \rho_1)\mathbb{E}[M(\tau') - \rho_1)]e^{-\gamma_p(2t-\tau-\tau')}d\tau'd\tau$$

$$= \rho_0\rho_1 \int_0^t \int_0^t e^{-\gamma_p(2t-\tau-\tau')}e^{-|\tau-\tau'|/\tau_c}d\tau'd\tau$$

$$= 2\rho_0\rho_1 \int_0^t \int_0^\tau e^{-\gamma_p(2t-\tau-\tau')}e^{-(\tau-\tau')/\tau_c}d\tau'd\tau$$

$$= 2\rho_0\rho_1 \frac{e^{-2\gamma_p t}}{\gamma_p + \tau_c^{-1}} \left[\frac{e^{2\gamma_p t} - 1}{2\gamma_p} + \frac{1 - e^{(\gamma_p - \tau_c^{-1})t}}{(\gamma_p - \tau_c^{-1})} \right] \underset{t\to\infty}{\longrightarrow} \frac{\rho_0\rho_1}{\gamma_p(\gamma_p + \tau_c^{-1})}.$$

We thus find that

$$\lim_{t\to\infty} \text{Var}[N(t)] = \frac{\kappa_p\rho_1}{\gamma_p} + \frac{\rho_0\rho_1}{\gamma_p(\gamma_p + ck_+ + k_-)}. \tag{5.3.9}$$

The first term on the right-hand side is the expected contribution from the Poisson process associated with protein fluctuations, whereas the second term is the contribution from promoter noise.

5.3.2 Two-state gene network as a stochastic hybrid system

Now suppose that the switching rates $k_\pm$ are finite but the expected number of proteins is sufficiently large (thermodynamic limit) so that we can represent the dynamics in terms of a continuous-valued protein concentration x [430, 886]. Again let $M(t)$ denote the current state of the gene with $M(t) = 1$ (active) or $M(t) = 0$ (inactive). The concentration evolves according to the (piecewise) deterministic equation

$$\frac{dx}{dt} = \kappa_p m - \gamma_p x, \tag{5.3.10}$$

for $M(t) = m$. One could also have a nonzero protein production rate in both states [766]. Note that equation (5.3.10) is defined between jumps in $M(t)$, that is, it is piecewise deterministic. Incorporating promoter noise into models of gene regulation naturally leads to the coupling between a piecewise continuous differential equation and a continuous-time Markov chain, also known as a piecewise deterministic Markov process (PDMP) [131, 211, 274, 440]. The continuous variables are the concentrations of various proteins and the discrete variable represents the activation state of the gene [430, 438, 613, 617, 766]. We have already encountered several examples of PDMPs, including conductance-based models of voltage fluctuations in neurons, see Sect. 3.4, and the Dogterom-Leibler model of microtubule catastrophes, see Sect. 4.2. The latter is also an example of a velocity jump process, see Sect. 7.2. Note that a PDMP is a special case of a more general class of processes involving a coupling between discrete and continuous stochastic variables known as stochastic hybrid systems. For example, between jumps in the discrete variables, the continuous variables could evolve according to a stochastic differential equation rather than deterministically, or reset their values at each jump. We will analyze the effects of promoter noise in the two-state gene regulatory network by using some basic features of stochastic hybrid systems, which are summarized in Box 5D.

Given the initial conditions $X(0) = x_0, M(0) = m_0$, we introduce the probability density $p_m(x,t|x_0,m_0,0)$ according to

$$p_m(x,t|x_0,m_0,0)dx = \mathbb{P}\{X(t) \in (x,x+dx), M(t) = m|x_0,m_0),$$

which evolves according to the differential Chapman-Kolmogorov (CK) equation

$$\frac{\partial p_0}{\partial t} = -\frac{\partial}{\partial x}(-\gamma_p x p_0(x,t)) + k_- p_1(x,t) - k_+ p_0(x,t), \tag{5.3.11a}$$

$$\frac{\partial p_1}{\partial t} = -\frac{\partial}{\partial x}([\kappa_p - \gamma_p x]p_1(x,t)) + k_+ p_0(x,t) - k_- p_1(x,t). \tag{5.3.11b}$$

(For simplicity, we have absorbed the concentration c of transcription factor into k_+.) The CK equation is supplemented by the no-flux boundary conditions $J(x,t) = 0$ at $x = 0, \kappa_p/\gamma_p$, where $J(x,t) = F_0(x)p_0(x,t) + F_1(x)p_1(x,t)$. That is, $p_1(0,t) = 0$ and $p_0(\kappa_p/\gamma_p,t) = 0$. In the limit that the switching between active and inactive states is much faster than the protein dynamics, the probability that the gene is active rapidly converges to the steady state $k_+/(k_+ + k_-)$, and we obtain the deterministic equation

$$\frac{dx}{dt} = \kappa_p\langle M(t)\rangle - \gamma_p x = \frac{\kappa_p k_+}{k_+ + k_-} - \gamma_p x. \tag{5.3.12}$$

Previously, we have absorbed the factor $k_+/(k_- + k_+)$ into the protein production rate κ_p.

Following Ref. [430], we characterize the long-time behavior of the system in terms of the steady-state solution, which satisfies

$$\frac{d}{dx}(-\gamma_p x p_0(x)) = k_- p_1(x) - k_+ p_0(x), \tag{5.3.13a}$$

$$\frac{d}{dx}([\kappa_p - \gamma_p x]p_1(x)) = k_+ p_0(x) - k_- p_1(x). \tag{5.3.13b}$$

The no-flux boundary conditions imply that $p_0(\kappa_p/\gamma_p) = 0$ and $p_1(0) = 0$. First, note that we can take $x \in [0, r/\gamma]$ and impose the normalization condition

$$\int_0^{\kappa_p/\gamma_p} [p_0(x) + p_1(x)]dx = 1.$$

Integrating equations (5.3.13a) and (5.3.13b) with respect to x then leads to the constraints

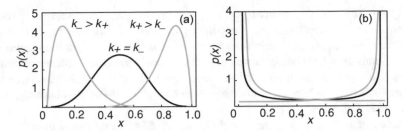

Fig. 5.28: Sketch of steady-state protein density $p(x)$ for a simple regulated network in which the promoter transitions between an active and inactive state at rates $k_\pm$. (a) Case $k_\pm/\gamma_p > 1$: there is a graded density that is biased towards $x = 0,1$ depending on the ratio k_+/k_-. (b) Case $k_\pm/\gamma_p < 1$: there is a binary density that is concentrated around $x = 0,1$ depending on the ratio k_+/k_-.

$$\int_0^{\kappa_p/\gamma_p} p_0(x)dx = \frac{k_-}{k_- + k_+} = \rho_0, \quad \int_0^{\kappa_p/\gamma_p} p_1(x)dx = \frac{k_+}{k_- + k_+} = \rho_1. \quad (5.3.14)$$

Here ρ_n is the stationary distribution of the two-state Markov chain. Adding equations (5.3.13a) and (5.3.13b), we can solve $p_0(x)$ in terms of $p_1(x)$ and then generate a closed differential equation for $p_1(x)$, see the derivation of equation (5.3.32). We thus obtain a solution of the form (see Ex. 5.11)

$$p_0(x){=}C(\gamma_p x)^{-1+k_+/\gamma_p}(\kappa_p - \gamma_p x)^{k_-/\gamma_p}, \quad p_1(x){=}C(\gamma_p x)^{k_+/\gamma_p}(\kappa_p - \gamma_p x)^{-1+k_-/\gamma_p}$$

$$(5.3.15)$$

for some constant C. Imposing the normalization conditions, then determines C as

$$C = \gamma\left[\kappa_p^{(k_++k_-)/\gamma_p} B(k_+/\gamma_p, k_-/\gamma_p)\right]^{-1},$$

where $B(\alpha, \beta)$ is the Beta function:

$$B(\alpha, \beta) = \int_0^1 t^{\alpha-1}(1-t)^{\beta-1}dt.$$

Finally, setting $\kappa_p/\gamma_p = 1$, the total probability density $p(x) = p_0(x) + p_1(x)$ is given by [430]

$$p(x) = \frac{x^{k_+/\gamma_p-1}(1-x)^{k_-/\gamma_p-1}}{B(k_+/\gamma_p, k_-/\gamma_p)}. \quad (5.3.16)$$

In Fig. 5.28, we sketch $p(x)$, $0 < x < \kappa_p/\gamma_p$ for various values of $K_\pm = k_\pm/\gamma_p$. It can be seen that when the rates $k_\pm$ of switching between the active and inactive gene states are faster than the rate of protein degradation γ_p then the steady-state density is unimodal (graded), whereas if the rate of degradation is faster then the density tends to be concentrated around $x = 0$ or $x = 1$, consistent with a binary process. In other words, if switching between promoter states is much slower than other processes then one can have a transcriptional contribution to protein burst-

ing [430]. This scenario tends to occur in eukaryotic gene expression, for which
the presence of nucleosomes and the packing of DNA-nucleosome complexes into
chromatin generally make promotors inaccessible to the transcriptional machinery.
Hence, transitions between open and closed chromatin structures, corresponding to
active and repressed promotor states, can be quite slow. The bimodal distribution
corresponds to time-dependent fluctuations that exhibit translational bursting, as illustrated in Fig. 5.29.

Hufton et al. [402] have generalized the two-state switching model to include the
effects of molecular noise and nonlinearities. These can be included by carrying out
a system-size expansion of the master equation (5.3.1) after setting $x(t) = n(t)/\Omega$,
where $n(t)$ is the number of proteins and Ω is the system size. This leads to a
CK equation that describes the time evolution of the probability densities for the
piecewise stochastic differential equation

$$dX(t) = F_m(X)dt + \sqrt{\frac{D_m(X)}{\Omega}}dW(t) \qquad (5.3.17)$$

for $M(t) = m \in \{0,1\}$ and

$$F_m(x) = \kappa_p m - \gamma_p x, \quad D_m(x) = \kappa_p m + \gamma_p x. \qquad (5.3.18)$$

We now have a combination of discrete promoter noise and continuous intrinsic
molecular noise. Hufton *et al.* show how to approximate the steady-state densities
by carrying out a linear noise approximation, which can be applied even when F_m,
σ_m and $k_{\pm}$ are nonlinear functions of x (autoregulatory feedback), see also [805].
A number of other papers have also analyzed the joint effects of promoter noise
and protein molecular noise in gene networks. For example, the doubly stochastic
Poisson process for time-dependent production and degradation of mRNA is the
basis of a study concerned with the effects of upstream drives on gene transcription
[210]. These authors focus on how a distribution of production and degradation rates

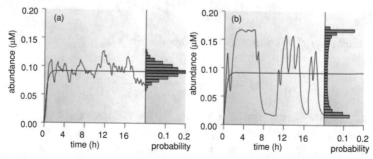

Fig. 5.29: (a) Unimodal distribution for relatively fast switching (1 minute). Bimodal distribution
and translational bursting for slow switching (1 hour). [Adapted from Kaern et al. (2005).]

affects population heterogeneity. Yet other work has focused on understanding the role of fluctuations in enzyme degradation and gene regulation [191, 353]. Finally, Sanchez et al. [719], consider the effects of promoter architecture on the cell-to-cell variability in gene expression.

Box 5D. Stochastic hybrid systems.

Here we give a general definition of a stochastic hybrid system and summarize some of its basic properties.

Definition of a stochastic hybrid system. Consider a system whose states are described by a pair of stochastic variables $(X(t), M(t)) \in \Sigma \times \{0, \cdots, M_0\}$, where $X(t)$ is a continuous variable in a connected, bounded domain $\Sigma \subset \mathbb{R}^d$ with a regular boundary $\partial \Sigma$, and $M(t)$ a discrete stochastic variable taking values in the finite set $\Gamma \equiv \{0, \cdots, M_0\}$. When the internal state is $M(t) = m$, the system evolves according to the ordinary differential equation (ODE)

$$\frac{dX}{dt} = F_m(X), \qquad (5.3.19)$$

where the vector field $F_m : \mathbb{R}^r \to \mathbb{R}^d$ is a sufficiently smooth function. Assume that the dynamics of x is confined to the domain Σ so that existence and uniqueness of a trajectory holds for each m. (It is possible to have a countable set of discrete variables, although one can always relabel the internal states so that they are effectively indexed by a single integer. One could also consider generalizations of the continuous process, in which the ODE (5.3.19) is replaced by a stochastic differential equation (SDE) or even a partial differential equation (PDE). In order to allow for such possibilities, we will refer to all of these processes as examples of a stochastic hybrid system.) For fixed x, the discrete stochastic variable evolves according to a homogeneous, continuous-time Markov chain (Sect. 3.6) with transition matrix $\mathbf{W}(x)$ and corresponding generator $\mathbf{A}(x)$, which are related according to

$$A_{mm'}(x) = W_{mm'}(x) - \delta_{m',m} \sum_{k \in \Gamma} W_{km}(x). \qquad (5.3.20)$$

In the case of the two-state gene regulatory network, we have $x \in \mathbb{R}, m \in \{0, 1\}$,

$$F_m(x) = \kappa_p m - \gamma_p x, \quad \mathbf{A}(x) = \mathbf{A} = \begin{pmatrix} -k_+ & k_- \\ k_+ & -k_- \end{pmatrix}. \qquad (5.3.21)$$

In the various applications considered in this book, the corresponding Markov chain is irreducible for all $x \in \Sigma$, that is, for fixed x there is a nonzero probability of transitioning, possibly in more than one step, from any state to any other state of the Markov chain (see Sect. 3.3). This implies the existence of a unique invariant probability distribution on Γ for fixed $x \in \Sigma$, denoted by $\rho(x)$, such that

$$\sum_{m' \in \Gamma} A_{mm'}(x)\rho_{m'}(x) = 0, \quad \forall m \in \Gamma. \tag{5.3.22}$$

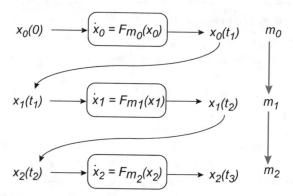

Fig. 5.30: Schematic illustration of a piecewise deterministic Markov process.

Iterative formulation. Suppose that we decompose the transition matrix of the Markov chain as

$$W_{m'm}(x) = P_{m'm}(x)\lambda_m(x),$$

with $\sum_{m' \neq m} P_{m'm}(x) = 1$ for all x. Hence $\lambda_m(x)$ determines the jump times from the state m whereas $P_{m'm}(x)$ determines the probability distribution that when it jumps the new state is m' for $m' \neq m$. The hybrid evolution of the system with respect to $X(t)$ and $M(t)$ can then be described as follows, see Fig. 5.30. Suppose the system starts at time zero in the state (x_0, m_0). Call $x_0(t)$ the solution of (5.3.19) with $m = m_0$ such that $x_0(0) = x_0$. Let t_1 be the random variable (stopping time) such that

$$\mathbb{P}(t_1 < t) = 1 - \exp\left(-\int_0^t \lambda_{m_0}(x_0(t'))dt'\right).$$

Then in the random time interval $s \in [0, t_1)$ the state of the system is $(x_0(s), m_0)$. Now draw a value of t_1 from $\mathbb{P}(t_1 < t)$, choose an internal state $m_1 \in \Gamma$ with probability $P_{m_1 m_0}(x_0(t_1))$, and call $x_1(t)$ the solution of the following Cauchy problem on $[t_1, \infty)$:

$$\begin{cases} \dot{x}_1(t) = F_{m_1}(x_1(t)), & t \geq t_1 \\ x_1(t_1) = x_0(t_1). \end{cases}$$

Iterating this procedure, one can construct a sequence of increasing jumping times $(t_k)_{k \geq 0}$ (setting $t_0 = 0$) and a corresponding sequence of internal states $(m_k)_{k \geq 0}$. The evolution $(X(t), M(t))$ is then defined as

$$(X(t), M(t)) = (x_k(t), m_k) \quad \text{if } t_k \leq t < t_{k+1}. \tag{5.3.23}$$

Note that the path $X(t)$ is continuous and piecewise C^1.

The above formulation is the basis of a simulation algorithm for PDMPs [886]. If $\lambda_m(x) = \kappa_m$ is independent of x for all m, then the stochastic jump time from state m is given by

$$\tau = \frac{1}{\kappa_m} \ln(1/u),$$

where $u \in [0,1]$ is a realization of a uniform random variable. The simulation of the PDMP is then very similar to the SSA for a chemical master equation. On the other hand, in the x-dependent case one has to numerically solve the integral equation

$$\int_0^\tau \lambda_{m_k}(x_k(t))dt = \ln(1/u) \tag{5.3.24}$$

at the k-th iteration. This can be achieved by solving the ODE system

$$\frac{dx_k}{dt} = F_{m_k}(x_k(t)),$$
$$\frac{dR_k}{dt} = \lambda_{m_k}(x_k(t))[1 - R_k(t)],$$

and setting $R_k(t) = u$. In contrast to discrete Markov processes, the time spent in a discrete state (the sojourn time τ) of a PDMP may become infinite. That is,

$$\mathbb{P}(\tau = \infty) = c > 0, \quad \lim_{t \to \infty} \mathbb{P}(\tau < t) \to 1 - c.$$

Chapman-Kolmogorov equation. Given the initial conditions $X(0) = x_0, N(0) = n_0$, we introduce the probability density $p_n(x,t|x_0,n_0,0)$ with

$$\mathbb{P}[X(t) \in (x, x+dx), M(t) = m|x_0, m_0] = p_m(x,t|x_0,m_0,0)dx.$$

It can be shown that p evolves according to the forward differential Chapman-Kolmogorov (CK) equation [116, 300]

$$\frac{\partial p_m}{\partial t} = -\nabla \cdot [F_m(x)p_m(x,t)] + \sum_{m' \in \Gamma} A_{mm'} p_{m'}(x,t). \tag{5.3.25}$$

For notational convenience, we have dropped the explicit dependence on initial conditions. The first term on the right-hand side represents the probability flow associated with the piecewise deterministic dynamics for a given n, whereas the second term represents jumps in the discrete state n. It remains to specify boundary conditions for the CK equation. For the sake of illustration, suppose that $d = 1$ (one-dimensional continuous dynamics) with $\Sigma = [0, L]$. No-flux boundary conditions at the ends $x = 0, L$ take the form $J(0,t) = J(L,t) = 0$ with

$$J(x,t) = \sum_{m=0}^{M_0} F_m(x)p_m(x,t).$$ (5.3.26)

On the other hand, an absorbing boundary condition at $x = L$, say, is

$$p_m(L,t) = 0, \quad \forall n \text{ such that } F_m(L) < 0.$$

In general, it is difficult to obtain an analytical steady-state solution of (5.3.25), assuming it exists, unless $d = 1$ and $M_0 = 1$, which holds for the two-state gene regulatory network. The CK equation then reduces to

$$\frac{\partial p_m}{\partial t} = -\frac{\partial}{\partial x}[F_m(x)p_m(x,t)] + \sum_{m' \in \Gamma} A_{mm'} p_{m'}(x,t),$$ (5.3.27)

with

$$\mathbf{A}(x) = \begin{pmatrix} -\alpha(x) & \beta(x) \\ \alpha(x) & -\beta(x) \end{pmatrix}$$

for a pair of transition rates $\alpha(x), \beta(x)$. The steady-state version of (5.3.27) reduces to the pair of equations

$$0 = -\frac{d}{dx}(F_0(x)p_0(x)) - \alpha(x)p_0(x) + \beta(x)p_1(x),$$ (5.3.28a)

$$0 = -\frac{d}{dx}(F_1(x)p_1(x)) + \alpha(x)p_0(x) - \beta(x)p_1(x).$$ (5.3.28b)

Adding the pair of equations yields

$$\frac{d}{dx}(F_0(x)p_0(x)) + \frac{d}{dx}(F_1(x)p_1(x)) = 0,$$ (5.3.29)

that is,

$$F_0(x)p_0(x) + F_1(x)p_1(x) = c$$

for some constant c. The reflecting boundary conditions imply that $c = 0$. Since $F_n(x)$ is nonzero for all $x \in \Sigma$, we can express $p_1(x)$ in terms of $p_0(x)$

$$p_1(x) = -\frac{F_0(x)p_0(x)}{F_1(x)}.$$ (5.3.30)

Substituting into equation (5.3.28a) gives

$$0 = \frac{d}{dx}(F_0(x)p_0(x)) + \left(\frac{\beta(x)}{F_1(x)} + \frac{\alpha(x)}{F_0(x)}\right)F_0(x)p_0(x),$$ (5.3.31)

which yields the solutions

$$p_m(x) = \frac{1}{Z|F_m(x)|} \exp\left(-\int_{x_*}^{x} \left(\frac{\beta(y)}{F_1(y)} + \frac{\alpha(y)}{F_0(y)}\right) dy\right). \qquad (5.3.32)$$

where $x_* \in \Sigma$ is arbitrary and assuming that the normalization factor Z exists.

Fast switching limit. In many of the applications in cell biology, one finds that the transition rates between the discrete states $m \in \Gamma$ are much faster than the relaxation rates of the piecewise deterministic dynamics for $x \in \mathbb{R}^d$. Thus, there is a separation of time scales between the discrete and continuous processes, so that if t is the characteristic time scale of the relaxation dynamics then εt is the characteristic time scale of the Markov chain for some small positive parameter ε. Assuming that the Markov chain is ergodic, in the limit $\varepsilon \to 0$ one obtains a deterministic dynamical system in which one averages the piecewise dynamics with respect to the corresponding unique stationary measure. This then raises the important problem of characterizing how the law of the underlying stochastic process approaches this deterministic limit in the case of weak noise, $0 < \varepsilon \ll 1$. Fast switching can be incorporated into the model by introducing a small positive parameter ε and rescaling the transition matrix so that equation (5.3.25) becomes

$$\frac{\partial p_m}{\partial t} = -\nabla \cdot [F_m(x)p_m(x,t)] + \frac{1}{\varepsilon} \sum_{m' \in \Gamma} A_{mm'} p_{m'}(x,t), \qquad (5.3.33)$$

with F_m and $\mathbf{A}$ independent of ε (at least to lowest order). The fast switching limit then corresponds to the case $\varepsilon \to 0$. Let us now define the averaged vector field $\overline{F} : \mathbb{R}^d \to \mathbb{R}^d$ by

$$\overline{F}(x) = \sum_{m \in \Gamma} \rho_m(x)F_m(x). \qquad (5.3.34)$$

Intuitively speaking, one would expect the stochastic hybrid system (5.3.19) to reduce to the deterministic dynamical system

$$\begin{cases} \dot{x}(t) = \overline{F}(x(t)) \\ x(0) = x_0 \end{cases} \qquad (5.3.35)$$

in the fast switching limit $\varepsilon \to 0$. That is, for sufficiently small ε, the Markov chain undergoes many jumps over a small time interval Δt during which $\Delta x \approx 0$, and thus the relative frequency of each discrete state m is approximately $p_m^*(x)$. This can be made precise in terms of a law of large numbers for stochastic hybrid systems [264, 440].

Quasi-steady-state diffusion approximation. For small but nonzero ε, one can use perturbation theory to derive lowest order corrections to the deterministic mean-field equation, which leads to a Langevin equation with noise amplitude $O(\sqrt{\varepsilon})$. More specifically, perturbations of the mean-field equation (5.3.35) can be analyzed using a quasi-steady-state (QSS) diffusion or adiabatic approximation, in which the CK equation (5.3.33) is approximated by a Fokker-Planck (FP) equation for the total density $C(x,t) = \sum_m p_m(x,t)$. The QSS approximation was first developed from a probabilistic perspective by Papanicolaou [636]. It has subsequently been applied to a wide range of problems in biology, including models of intracellular transport in axons [289, 675] and dendrites [608–610] and bacterial chemotaxis [374, 375, 626]. There have also been more recent probabilistic treatments of the adiabatic limit, which have been applied to various stochastic neuron models [628]. Finally, note that it is also possible to obtain a diffusion limit by taking the number of discrete states $M_0 + 1$ to be large [148, 627]. The basic steps of the QSS reduction are as follows:

a) Decompose the probability density as

$$p_m(x,t) = C(x,t)\rho_m(x) + \varepsilon w_m(x,t),$$

where $\sum_m p_m(x,t) = C(x,t)$ and $\sum_m w_m(x,t) = 0$. Substituting into equation (5.3.33) yields

$$\rho_m(x)\frac{\partial C}{\partial t} + \varepsilon \frac{\partial w_m}{\partial t} = -\nabla \cdot (F_m(x)[\rho_m(x)C + \varepsilon w_m])$$

$$+ \frac{1}{\varepsilon}\sum_{m' \in \Gamma} A_{mm'}(x)[\rho_{m'}(x)C + \varepsilon w_{m'}].$$

Summing both sides with respect to m then gives

$$\frac{\partial C}{\partial t} = -\nabla \cdot [\overline{F}(x)C] - \varepsilon \sum_{m \in \Gamma} \nabla \cdot [F_m(x)w_m], \qquad (5.3.36)$$

where $\overline{F}(x)$ is the mean vector field of equation (5.3.34).

b) Using the equation for C and the fact that $\mathbf{A}(x)\rho(x) = 0$, we have

$$\varepsilon \frac{\partial w_m}{\partial t} = \sum_{m' \in \Gamma} A_{mm'}(x)w_{m'} - \nabla \cdot [F_m(x)\rho_m(x)C] + \rho_m(x)\nabla \cdot [\overline{F}(x)C]$$

$$- \varepsilon \left[\nabla \cdot (F_m(x)\omega_n) - \rho_m(x) \sum_{m' \in \Gamma} \nabla \cdot [F_{m'}(x)w_{m'}] \right].$$

c) Introduce the asymptotic expansion

$$w_m \sim w_m^{(0)} + \varepsilon w_m^{(1)} + \varepsilon^2 w_m^{(2)} + \dots$$

and collect $O(1)$ terms:

$$\sum_{m'\in\Gamma} A_{mm'}(x)w_{m'}^{(0)} = \nabla \cdot [\rho_m(x)F_m(x)C(x,t)] - \rho_m(x)\nabla \cdot [\overline{F}(x)C].$$

(5.3.37)

The Fredholm alternative theorem (see Box 2E) ensures that this has a solution for fixed x,t and $A_{nm}^\dagger = A_{mn}$. More specifically, $\mathbf{A}$ has a one-dimensional null-space spanned by the vector with components $u_m = 1$, since $\sum_m u_m A_{mn} = \sum_m A_{nm}^\dagger u_m = 0$. Hence, equation (5.3.37) has a solution provided that

$$0 = \sum_m \left(\nabla \cdot [\rho_m(x)F_m(x)C(x,t)] - \rho_m(x)\nabla \cdot [\overline{F}(x)C]\right).$$

This equation holds since $\sum_m \rho_m(x) = 1$ and $\sum_m \rho_m(x)F_m(x) = \overline{F}(x)$ for all x. The solution is unique on imposing the condition $\sum_m w_m^{(0)}(x,t) = 0$, so that we can set

$$w_m^{(0)}(x) = \sum_{m'\in\Gamma} A_{mm'}^\ddagger(x)\left(\nabla \cdot [\rho_{m'}(x)F_{m'}(x)C(x,t)] - \rho_{m'}(x)\nabla \cdot [\overline{F}(x)C]\right),$$

(5.3.38)

where $\mathbf{A}^\ddagger$ is the pseudo-inverse of the generator $\mathbf{A}$. One typically has to determine the pseudo-inverse of $\mathbf{A}$ numerically.

d) Combining equations (5.3.37) and (5.3.36) shows that C evolves according to the Ito Fokker-Planck (FP) equation

$$\frac{\partial C}{\partial t} = -\nabla \cdot [\overline{F}(x)C] - \varepsilon\nabla \cdot [\mathcal{V}(x)C] + \varepsilon\sum_{i,j-1}^d \frac{\partial^2 D_{ij}(x)C}{\partial x_i\partial x_j},$$

(5.3.39)

where the $O(\varepsilon)$ correction to the drift, $\mathcal{V}(x)$, and the diffusion matrix $D(x)$ are given by

$$\mathcal{V} = \sum_{m,m'}\left\{(\rho_m F_m)\nabla \cdot (F_{m'}A_{m'm}^\ddagger) - \overline{F}\nabla \cdot (F_{m'}A_{m'm}^\ddagger\rho_m)\right\},$$

(5.3.40a)

and

$$D_{ij}(x) = \sum_{m,m'\in\Gamma} F_{m,i}(x)A_{mm'}^\ddagger(x)\rho_{m'}(x)[\overline{F}_j(x) - F_{m',j}(x)].$$

Using the fact that $\sum_{m'} A_{mm'}^\ddagger = 0$ we, can rewrite the diffusion matrix as

$$D_{ij}(x) = \sum_{m,m'\in\Gamma} [F_{m,i}(x) - \overline{F}_i(x)]A_{mm'}^\ddagger(x)\rho_{m'}(x)[\overline{F}_j(x) - F_{m',j}(x)].$$

(5.3.40b)

One-dimensional case. In the one-dimensional case, the CK equation (5.3.33) reduces to the one-dimensional Ito FP equation

$$\frac{\partial C}{\partial t} = -\frac{\partial}{\partial x}([\overline{F}(x) + \varepsilon \mathcal{V}(x)]C) + \varepsilon \frac{\partial^2}{\partial x^2}(D(x)C),\tag{5.3.41}$$

with the diffusion coefficient $D(x)$ given by

$$D(x) = \sum_{m \in \Gamma} Z_m(x)F_m(x),\tag{5.3.42}$$

where $Z_m(x)$ is the unique solution to

$$\sum_{m' \in \Gamma} A_{mm'}(x)Z_{m'}(x) = [\overline{F}(x) - F_m(x)]\rho_m(x), \quad \sum_m Z_m(x) = 0.\tag{5.3.43}$$

For $M_0 > 1$ one typically has to solve equation (5.3.43) numerically in order to find the pseudo-inverse of $\mathbf{A}$. However, in the special case of a two-state discrete process ($m = 0, 1$), one has the explicit solution

$$D(x) = \frac{\beta(x)[F_0(x) - \overline{F}(x)]F_0(x) + \alpha(x)[F_1(x) - \overline{F}(x)]F_1(x)}{[\alpha(x) + \beta(x)]^2}.\tag{5.3.44}$$

At a fixed point x_* of the deterministic equation $\dot{x} = \overline{F}(x)$, we have $\overline{F}(x_*) = 0$ and $\beta(x_*)F_0(x_*) = -\alpha(x_*)F_1(x_*)$. This gives the reduced expression

$$D(x_*) = \frac{|F_0(x_*)F_1(x_*)|}{\alpha(x_*) + \beta(x_*)}.\tag{5.3.45}$$

Remark 5.5. One subtle point is the nature of boundary conditions under the QSS reduction, since the FP equation is a second-order parabolic PDE, whereas the original CK equation is an $(M_0 + 1)^{\text{th}}$-order hyperbolic PDE. It follows that for $M_0 > 1$, there is a mismatch in the number of boundary conditions between the CK equation and the FP equation. This implies that the QSS reduction may break down in a small neighborhood of the boundary, as reflected by the existence of boundary layers [900]. One way to eliminate the existence of boundary layers is to ensure that the boundary conditions of the CK equation are compatible with the QSS reduction. For example, at $x = L$,

$$p_m(L, t) = p_m^*(L)C(L, t) + O(\varepsilon).$$

This holds for the various problems considered in this book.

5.3.3 Noisy mutual repressor model

The modeling of promoter noise in a two-state gene regulatory network can be extended to more complicated gene regulatory networks [438]. For example, consider the mutual repressor model of Fig. 5.6. Suppose that the number of proteins X and Y are n_1 and n_2, respectively, and assume that they are finite. The state transition diagram for the three promoter states is now

$$
O_1 \underset{n_1(n_1-1)\beta}{\overset{\beta K'}{\rightleftharpoons}} O_0 \underset{\beta K'}{\overset{n_2(n_2-1)\beta}{\rightleftharpoons}} O_2,
$$

where β is a transition rate and K' is a non-dimensional dissociation constant. Let $p_m(n_1,n_2,t)$, $m = 0,1,2$, be the probability that there are $N_1(t) = n_1$ proteins Y, $N_2(t) = n_2$ proteins X, and the promoter is in state $M(t) = m$ at time t. The master equation for the full system takes the form

$$
\frac{d}{dt}p_m(n_1,n_2,t) = \sum_{l=0,1,2} A_{ml}\,p_l(n_1,n_2,t) + \sum_{n_1',n_2'} W^m_{n_1 n_2, n_1' n_2'} p_m(n_1',n_2',t), \quad (5.3.46)
$$

where

$$
\mathbf{A} = \beta \begin{pmatrix} -n_1(n_1-1)-n_2(n_2-1) & K' & K' \\ n_1(n_1-1) & -K' & 0 \\ n_2(n_2-1) & 0 & -K' \end{pmatrix}, \quad (5.3.47)
$$

and

$$
\sum_{n_1',n_2'} W^0_{n_1 n_2, n_1' n_2'} p_0(n_1',n_2',t)
$$
$$
= \gamma_p[(n_1+1)p_0(n_1+1,n_2,t)+(n_2+1)p_0(n_1,n_2+1,t)-(n_1+n_2)p_0(n_1,n_2,t)]
$$
$$
+ \kappa_p(p_0(n_1-1,n_2,t)+p_0(n_1,n_2-1,t)-2p_0(n_1,n_2,t)), \quad (5.3.48a)
$$

$$
\sum_{n_1',n_2'} W^1_{n_1 n_2, n_1' n_2'} p_1(n_1',n_2',t)
$$
$$
= \gamma_p[(n_1+1)p_1(n_1+1,n_2,t)+(n_2+1)p_1(n_1,n_2+1,t)-(n_1+n_2)p_1(n_1,n_2,t)]
$$
$$
+ \kappa_p(p_1(n_1-1,n_2,t)-p_1(n_1,n_2,t)), \quad (5.3.48b)
$$

$$
\sum_{n',m'} W^2_{n_1 n_2, n_1' n_2'} p_1(n_1',n_2',t)
$$
$$
= \gamma_p[(n_1+1)p_2(n_1+1,n_2,t)+(n_2+1)p_2(n_1,n_2+1,t)-(n_1+n_2)p_2(n_1,n_2,t)]
$$
$$
+ \kappa_p(p_2(n_1,n_2-1,t)-p_2(n_1,n_2,t)). \quad (5.3.48c)
$$

The sums $\sum_{n'_1, n'_2} W^m_{n_1 n_2, n'_1 n'_2} p_m(n'_1, n'_2, t)$ for $m = 0, 1, 2$, are determined by noting that protein X is produced at a rate κ_p when $m = 0, 2$ and protein Y is produced at a rate κ_p when $m = 0, 1$. Both proteins degrade at a rate γ_p irrespective of the state m.

As in the case of the two-state gene network, we can consider various approximations of the full master equation (5.3.46) [438, 613]. For example, suppose that we carry out a system-size expansion of the sums

$$\sum_{n'_1, n'_2} W^m_{n_1 n_2, n'_1 n'_2} p_m(n'_1, n'_2, t)$$

with respect to the mean number $N = \kappa_p / \gamma_p$ of proteins. That is, introducing the rescaling $t \to t\gamma_p$, setting $y = n_1 / N$, $x = n_2 / N$, and keeping only the first terms in the Taylor expansions, we obtain the CK equation (see Ex. 5.12)

$$\frac{\partial p_m}{\partial t} = -\frac{\partial F_m(x) p_m}{\partial x} - \frac{\partial G_m(y) p_m}{\partial y} + \frac{1}{\varepsilon} \sum_{l=0,1,2} A_{ml}(x,y) p_l \qquad (5.3.49)$$

for $p_m(x, y, t)$, with

$$F_0(x) = 1 - x, \quad F_1(x) = -x, \quad F_2(x) = 1 - x,$$
$$G_0(y) = 1 - y, \quad G_1(y) = 1 - y, \quad G_2(y) = -y,$$

$$\mathbf{A} = \begin{pmatrix} -x^2 - y^2 & K & K \\ y^2 & -K & 0 \\ x^2 & 0 & -K \end{pmatrix}, \qquad (5.3.50)$$

and $\varepsilon = \gamma_p^3 / \beta \kappa_p^2$ and $K = K' \gamma_p^2 / \kappa_p^2$ are dimensionless parameters. Sample paths of the resulting stochastic system evolve according to a 3-state stochastic hybrid system in $\mathbb{R}^2$

$$\frac{dx}{dt} = F_m(x), \quad \frac{dy}{dt} = G_m(y), \quad \text{for } M(t) = m, \qquad (5.3.51)$$

with jumps in the state of the operator generated by the matrix $\mathbf{A}(x, y)$.

In the deterministic limit $N \to \infty$ and $\varepsilon \to 0$, we obtain the deterministic equations

$$\frac{dx}{dt} = \overline{F}(x, y), \quad \frac{dy}{dt} = \overline{G}(x, y), \qquad (5.3.52)$$

where

$$\overline{F}(x, y) = \sum_{m=0,1,2} F_m(x) \rho_m(x, y), \quad \overline{G}(x, y) = \sum_{m=0,1,2} G_m(x) \rho_m(x, y),$$

and $\rho_m(x, y)$ is the stationary distribution of the matrix $\mathbf{A}(x, y)$ for fixed x, y. It can be shown that these equations are identical to (5.1.17). Recall that the deterministic mutual repressor model can exhibit bistability. Numerical simulations of the full discrete Markov process for $N_1(t)$ and $N_2(t)$ with master equation (5.3.46) show that for finite switching rates, promoter noise can induce transitions between the two metastable states. Noise-induced transitions can also be observed in the stochastic hybrid system (5.3.51). In order to derive a Kramer's escape rate formula for these transitions requires developing methods for analyzing metastability for continuous-time Markov chains and stochastic hybrid systems (Chap. 8).

5.4 Transcriptional bursting and queuing theory

In Sect. 5.3.1, we considered a simple two-state gene network, where the gene switches between an on and off state and mRNA is only produced during the on-phase. We showed that promoter noise could generate a form of transcriptional bursting. However, in eukaryotic cells, transcription appears to follow an ordered, multistep, and cyclic process, which involves transitions between distinct chromatin states [90, 847]. Schwabe et al. [742] refer to this complex transcription mechanism as a molecular ratchet model, in which the time a single gene spends in the off- and on-phases, and the time between consecutive transcription initiation events are random variables with corresponding waiting time densities $f(t), g(t)$, and $h(t)$, see Fig. 5.31(a). The complexity of the underlying molecular mechanisms means that these waiting time densities are not necessarily exponential, as assumed in the previous model.

Suppose that an on state persists for some random time $T_{on} = t$ and let $B(t)$ be the number of transcription initiation events (number of mRNA) that occur over this time interval. (For simplicity, we assume that each mRNA generates the same number of proteins so that burst size is measured in terms of the number of mRNA produced during an on-phase.) Let $F(b,t) = \mathbb{P}[B(t) \geq b | T_{on} = t]$. If the waiting time

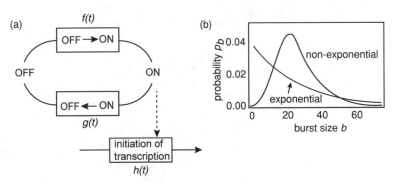

Fig. 5.31: (a) Schematic illustration of molecular ratchet model of transcription at the single gene level, where the waiting time distribution of the off state, on state, and transcription initiation are denoted by $f(t), g(t), and h(t)$. (b) Schematic illustration of a typical burst-size distribution for exponential and non-exponential on state time distributions.

density between events is $h(\tau)$ and τ_j, $j = 1,\ldots,b$ denotes the time of the j-th event, then

$$F(b,t) = \int_0^t h(t - \tau_b) \int_0^{\tau_b} h(\tau_b - \tau_{b-1}) \cdots \int_0^{\tau_2} h(\tau_2 - \tau_1) d\tau_1 \cdots d\tau_b$$
$$= \int_0^t F(b-1, \tau_b) h(t - \tau_b) d\tau_b. \tag{5.4.1}$$

The probability that exactly b initiation events occur in the time interval t is

$$P(b,t) = F(b,t) - F(b+1,t),$$

and the mean burst size given t is

$$\langle b(t) \rangle = \sum_{b=0}^{\infty} bP(b,t).$$

Laplace transforming equation (5.4.1) and using the convolution theorem shows that

$$\widetilde{F}(b,s) = \widetilde{h}(s)\widetilde{F}(b-1,s) = [\widetilde{h}(s)]^b \widetilde{F}(0,s).$$

Since $F(0,t) = 1$, it follows that $\widetilde{F}(0,s) = s^{-1}$. Hence,

$$\widetilde{P}(b,s) = \widetilde{F}(b,s) - \widetilde{F}(b+1,s) = s^{-1}(1 - \widetilde{h}(s))\widetilde{h}(s)^b,$$

which means that

$$\langle \widetilde{b}(s) \rangle = s^{-1}(1 - \widetilde{h}(s)) \sum_{b=0}^{\infty} b\widetilde{h}(s)^b = s^{-1}\widetilde{h}(s)(1 - \widetilde{h}(s)) \frac{d}{d\widetilde{h}(s)} \sum_{b=0}^{\infty} \widetilde{h}(s)^b$$
$$= s^{-1}\widetilde{h}(s)(1 - \widetilde{h}(s)) \frac{d}{d\widetilde{h}(s)} \frac{1}{1 - \widetilde{h}(s)} = s^{-1} \frac{\widetilde{h}(s)}{1 - \widetilde{h}(s)}.$$

Hence

$$\langle b(t) \rangle = \mathscr{L}^{-1}\left[\frac{1}{s} \frac{\widetilde{h}(s)}{1 - \widetilde{h}(s)} \right](t). \tag{5.4.2}$$

A similar analysis establishes that

$$\langle b(t)^2 \rangle = \mathscr{L}^{-1}\left[\frac{1}{s} \frac{\widetilde{h}(s)(1 + \widetilde{h}(s))}{(1 - \widetilde{h}(s))^2} \right](t). \tag{5.4.3}$$

One now obtains the moments of the burst-size distribution by integrating with respect to all times $T_{\text{on}} = t$

$$\langle b \rangle = \int_0^{\infty} \langle b(t) \rangle g(t) dt, \quad \langle b^2 \rangle = \int_0^{\infty} b(t)^2 g(t) dt. \tag{5.4.4}$$

In the case of exponential waiting times $g(t) = k_g e^{-k_g t}$ and $h(t) = k_h e^{-k_h t}$, we have $\widetilde{h}(s) = k_h/(k_h + s)$, so that

$$\langle b(t) \rangle = \mathscr{L}^{-1} \left[\frac{k_h}{s^2} \right](t) = k_h t, \quad \langle b(t)^2 \rangle = \mathscr{L}^{-1} \left[\frac{k_h}{s^2} \frac{2k_h + s}{s} \right](t) = k_h^2 t^2 + k_h t.$$

It follows that

$$\langle b \rangle = \frac{k_h}{k_g}, \quad \langle b^2 \rangle = \langle b \rangle + 2\langle b \rangle^2, \tag{5.4.5}$$

and thus the Fano factor is

$$Q_b \equiv \frac{\text{Var}[b]}{\langle b \rangle} = 1 + \langle b \rangle. \tag{5.4.6}$$

The corresponding distribution of burst sizes is a geometric distribution,

$$p_b \equiv \int_0^\infty P(b,t) g(t) dt = p(1-p)^b, \quad p = \frac{k_g}{h_h + k_g}.$$

Schwabe et al. [742] explore the effects of non-exponential on state time distributions, and show that they can lead to a burst-size distribution that is peaked (rather than exponentially decreasing) and this reduces transcriptional noise, see Fig. 5.31(b).

Queuing model of transcriptional bursting. Now suppose that one combines the stochastic process that generates a distribution of burst sizes with the waiting time density $f(\tau)$ of being in the off-phase. The resulting system can be mapped into one of the classical models of queuing theory [472], see Fig. 5.32. Queuing theory concerns the mathematical analysis of waiting lines formed by customers randomly arriving at some service station, and staying in the system until they receive service from a group of servers. Different types of queuing process are defined in terms of (i) the stochastic process underlying the arrival of customers, (ii) the distribution of the number of customers (batches) in each arrival, (iii) the stochastic process underlying the departure of customers (service-time distribution), and (iv) the number of servers. The above model of transcriptional bursting can be mapped onto a queuing process as follows: individual mRNAs are analogous to customers, transcriptional bursts correspond to customers arriving in batches, and the degradation of mRNAs is the analog of customers exiting the system after being serviced. Thus, the waiting time density for mRNA degradation is the analog of the service-time distribution. Finally, since the mRNAs are degraded independently of each other, the effective number of servers in the corresponding queuing model is infinite, that is, the presence of other customers does not affect the service time of an individual customer.

The particular queuing model that maps onto the model of transcriptional bursting is the $GI^X/M/\infty$ system, Here the symbol G denotes a general waiting time density for the arrival process, that is, the waiting time density $f(\tau)$ for the gene to switch from the off-phase to the on-phase, and I^X refers to the fact that customers

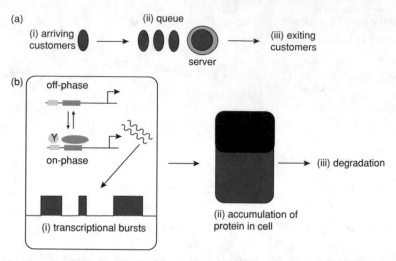

Fig. 5.32: Diagram illustrating the mapping between queuing theory and transcriptional bursting. (a) Example of a single-server queue. (b) Transcriptional bursting. The stochastic switching of a gene between an on- phase and an off-phase generates a sequence of transcriptional bursts that is analogous to the arrival of customers in the queuing model. This results in the accumulation of proteins within the cell, which is the analog of a queue. Degradation corresponds to exiting of customers after being serviced by an infinite number of servers.

(mRNAs) arrive in batches of independently distributed random sizes X (mRNA burst sizes). The symbol M stands for a Markovian or exponential service-time density (mRNA degradation-time density), and "∞" denotes infinite servers. Exploiting this mapping, exact results for the moments of the mRNA steady-state distribution can be obtained from the corresponding known expressions for the moments of the steady-state distribution of the number of current customers in the $GI^X/M/\infty$ model [472]. The latter were originally derived by Liu et al. [528], and we will follow their analysis in Box 5E. Here we simply quote the results for the mean and variance of the steady-state mRNA copy number N

$$\langle N \rangle = \frac{\lambda}{\mu}\langle b\rangle, \quad \mathrm{Var}[N] = \langle N \rangle - \langle N \rangle^2 + \frac{\mu}{\lambda}\langle N \rangle^2 \left(\frac{\widetilde{f}(\mu)}{1-\widetilde{f}(\mu)} + \frac{\langle b^2\rangle - \langle b\rangle}{2\langle b\rangle^2} \right). \quad (5.4.7)$$

Here $\widetilde{f}(s)$ is the Laplace transform of the waiting time density $f(\tau)$, μ is the degradation rate, λ is the mean rate at which transcription bursts occur with mean burst size $\langle b\rangle$. In the special case that the arrival times are also exponentially distributed, $\widetilde{f}(s) = \lambda/(s+\lambda)$, and the variance simplifies to

$$\mathrm{Var}[N] = \frac{\lambda}{\mu}\langle b^2\rangle. \quad (5.4.8)$$

Box 5E. Queuing theory

Single-server queues. A single-server queuing system is characterized by two sequences $\{T_n, n \geq 1\}$ and $\{S_n, n \geq 1\}$ of independent positive random variables. The first is the inter-arrival times of customers with common distribution function F and the second is the service times with common distribution function H. Each arriving customer joins the line of customers who are waiting to receive attention from the single server. When the n-th customer reaches the head of the line, she is served for a period S_n and then immediately leaves the system. Let $Q(t)$ denote the number of waiting customers at time t, including any customer currently receiving service. Then $\{Q(t) : t \geq 0\}$ is itself a stochastic process whose statistics is determined by F and H. Some of the general questions regarding queuing theory include the following: (i) Under what conditions is $Q(t)$ a Markov process or $Q(t)$ at least contains an embedded Markov chain? (ii) When is $Q(t)$ asymptotically stationary? (iii) When does the queue length grow beyond all bounds, in the sense that the single server cannot cope with the traffic density? Queues are typically labeled by the triplet $A/B/s$, where A describes F, B describes H, and s is the number of servers. Common choices for A and B are either $M(\lambda)$ or G, where $M(\lambda)$ is an exponential distribution (Markovian) and G is a general non-Markovian distribution. Two examples of single-server queues are $M(\lambda)/M(\mu)/1$, which is the only system whose queue lengths are homogeneous Markov chains, and $M(\lambda)/G/1$. Although Q is not a Markov chain in the latter case, there exists an imbedded discrete-time Markov chain, as explored in Ex. 5.13. In contrast to single-server queues, the model of transcriptional bursting involves an infinite number of servers and customers arrive in randomly varying batches rather than as individuals.

Moments of $GI^X/M/\infty$ queuing model. Let $F(t)$ and $H(t)$ be the probability distributions for the inter-arrival times and the service times, respectively. It is assumed that the mean arrival rate λ and the mean service rate μ are finite. Since the distribution of service times is taken to be exponential, we have $H(t) = 1 - e^{-\mu t}$. The arrivals occur in batches of variable size X, where $\mathbb{P}[X = b] = p_b$ for $b \geq 1$, with finite mean and variance, and generating function

$$A(z) = \sum_{r=1}^{\infty} p_b z^b.$$

It follows that for any positive integer k, the k-th factorial moment of X is

$$A_k = \left.\frac{d^k A(z)}{dz^k}\right|_{z=1} = \sum_{r=k}^{\infty} b(b-1)\cdots(b-k+1)p_b. \qquad (5.4.9)$$

Let T_n be the time of the n-th group arrival, with X_n denoting the size of the batch and S_{ni} the service time of the i-th member of the batch. The number of busy servers at time is then

$$N(t) = \sum_{0 \leq T_n \leq t} \chi(t - T_n, X_n), \qquad (5.4.10)$$

$$\chi(t - T_n, X_n) = \sum_{i=1}^{X_n} I(t - T_n, S_{ni}), \quad I(t - T_n, S_{ni}) = \begin{cases} 1 \text{ if } t - T_n \leq S_{ni} \\ 0 \text{ if } t - T_n > S_{ni} \end{cases}.$$

In other words, we are keeping track of all batches that arrived prior to time t and the number of customers in each batch that haven't yet exited the system. Introduce the generating function

$$G(z, t) = \sum_{k=0}^{\infty} z^k \mathbb{P}[N(t) = k], \qquad (5.4.11)$$

and the binomial moments

$$B_r(t) = \sum_{k=r}^{\infty} \frac{k!}{(k-r)! r!} \mathbb{P}[N(t) = k], \quad r = 1, 2, \ldots. \qquad (5.4.12)$$

Integral equation. Suppose that the system is empty at time $t = 0$. We will derive an integral equation for the generating function $G(z, t)$. Conditioning on the first arrival time by setting $T_1 = y$, it follows that $N(t) = \chi(t - y, X_1) + N^*(t - y)$ if $y \leq t$ and zero otherwise, where $\chi(t - y, X_1)$ and $N^*(t - y)$ are independent of each other, and $N^*(t)$ has the same distribution as $N(t)$. Moreover

$$\mathbb{P}[I(t - y, S_{1i}) = k] = \mathbb{P}[t - y \leq S_{1i}] \delta_{k,1} + \mathbb{P}[t - y > S_{1i}] \delta_{k,0}$$
$$= [1 - H(t - y)] \delta_{k,1} + H(t - y) \delta_{k,0},$$

so that

$$\sum_{k=0}^{\infty} z^k P[I(t - y, S_{1i}) = k] = z + (1 - z) H(t - y).$$

Since $I(t - y, S_{1i})$ for $i = 1, 2, \ldots$ are independent and identically distributed, the law of total expectation, see equation (1.2.6), implies that

$$\mathbb{E}[z^{\chi(t-y,X_1)}] = \mathbb{E}[\mathbb{E}[z^{\chi(t-y,X_1)}|X_1]] = \sum_{b=1}^{\infty} p_b \mathbb{E}[z^{\sum_{i=1}^{b} I(t-y,S_{1i})}]$$

$$= \sum_{b=1}^{\infty} p_b \prod_{i=1}^{b} \mathbb{E}[z^{I(t-y,S_{1i})}] = \sum_{b=1}^{\infty} p_b \prod_{i=1}^{b} \left(\sum_{k=0}^{\infty} z^k P[I(t-y,S_{1i}) = k] \right)$$

$$= \sum_{b=1}^{\infty} p_b [z + (1 - z) H(t - y)]^b = A[z + (1 - z) H(t - y)].$$

Another application of the law of total expectation gives

$$
\begin{aligned}
G(z,t) = \mathbb{E}[z^{N(t)}] &= \mathbb{E}[\mathbb{E}[z^{N(t)}|y]] \\
&= \int_t^\infty \mathbb{E}[z^0]dF(y) + \int_0^t \mathbb{E}[z^{\chi(t-y,X_1)}]\mathbb{E}[z^{N^*(t-y)}]dF(y) \\
&= 1 - F(t) + \int_0^t G(z,t-y)A[z+(1-z)H(t-y)]dF(y). \quad (5.4.13)
\end{aligned}
$$

One can now obtain an iterative equation for the binomial moments by differentiating equation (5.4.13) multiple times with respect to z and using

$$
B_r(t) = \frac{1}{r!}\frac{d^r G(z,t)}{dz^r}\bigg|_{z=1}.
$$

Noting that

$$
\frac{d^r}{dz^r}A[z+(1-z)H(t-y)]\bigg|_{z=1} = [1-H(t-y)]^r A_r, \quad r=1,2,\dots
$$

we obtain the integral equation

$$
B_r(t) = \int_0^t B_r(t-y)dF(y) + \sum_{k=1}^r \frac{A_k}{k!}\int_0^t B_{r-k}(t-y)[1-H(t-y)]^k dF(y).
$$

This can be written in the more compact form

$$
B_r(t) = \int_0^t B_r(t-y)dF(y) + \int_0^t \mathcal{H}_r(t-y)dF(y), \quad (5.4.14)
$$

where

$$
\mathcal{H}_r(t) = \sum_{k=1}^r \frac{A_k}{k!}B_{r-k}(t)[1-H(t)]^k. \quad (5.4.15)
$$

Equation (5.4.14) is an example of a renewal-type equation for $C(t) = B_r(t) + \mathcal{H}_r(t)$, see equation (5.4.25) of Box 5F, which has the unique solution

$$
B_r(t) = \int_0^t \mathcal{H}_r(t-y)dm(y), \quad (5.4.16)
$$

where

$$
m(y) = \sum_{n=1}^\infty F_n(y), \quad F_n(y) = \mathbb{P}[T_n \le y] \quad (5.4.17)
$$

is the so-called renewal function.

Steady-state binomial moments. The steady-state binomial moments can now be obtained by taking the limit $t \to \infty$ in equation (5.4.16) and using the key renewal theorem (Box 5F)

$$B_r^* \equiv \lim_{t\to\infty} B_r(t) = \lambda \sum_{k=1}^{r} \frac{A_k}{k!} \int_0^\infty B_{r-k}(t)[1-H(t)]^k dt, \qquad (5.4.18)$$

where $\lambda^{-1} = \mathbb{E}[T_1]$. In the case of an exponential service-time distribution, $1 - H(t) = e^{-\mu t}$ so that

$$B_r^* = \lambda \sum_{k=1}^{r} \frac{A_k}{k!} \widetilde{B}_{r-k}(k\mu), \qquad (5.4.19)$$

where $\widetilde{B}_r(s)$ is the Laplace transform of $B_r(t)$. An iterative equation for the Laplace transforms can be obtained by Laplace transforming equations (5.4.14) and (5.4.15), and using the convolution theorem

$$\widetilde{B}_r(s) = \widetilde{B}_r(s)\widetilde{f}(s) + \widetilde{\mathcal{H}}_r(s)\widetilde{f}(s),$$

which, on rearranging, yields

$$\widetilde{B}_r(s) = \frac{\widetilde{f}(s)}{1-\widetilde{f}(s)} \widetilde{\mathcal{H}}_r(s), \quad \widetilde{\mathcal{H}}_r(s) = \sum_{k=1}^{r} \frac{A_k}{k!} \widetilde{B}_{r-k}(s+k\mu). \qquad (5.4.20)$$

Note that $B_0(t) = 1$ so $\widetilde{B}_0(s) = 1/s$ and, hence,

$$\widetilde{B}_1(s) = \frac{\widetilde{f}(s)}{1-\widetilde{f}(s)} \frac{A_1}{s+\mu}.$$

Equations (5.4.19) and (5.4.20) determine completely the steady-state binomial moments. In particular,

$$B_1^* \equiv \langle N \rangle = \frac{\lambda}{\mu} A_1, \qquad (5.4.21a)$$

$$B_2^* \equiv \frac{1}{2}\left(\langle N^2 \rangle - \langle N \rangle\right) = \lambda \left(A_1 \widetilde{B}_1(\mu) + \frac{A_2}{4\mu}\right)$$

$$= \frac{\lambda A_1}{2\mu}\left(\frac{\widetilde{f}(\mu)}{1-\widetilde{f}(\mu)} A_1 + \frac{A_2}{2A_1}\right). \qquad (5.4.21b)$$

Box 5F. Renewal theory

The renewal equation. Consider a general renewal process (see Box 3B). Introduce the inter-arrival distribution $F(t) = \mathbb{P}[\tau_1 \leq t]$ and let $F_k(t) = \mathbb{P}[T_k \leq t]$

be the distribution function of the k-th arrival time T_k. Clearly $F_1 = F$. From the identity $T_{k+1} = T_k + \tau_{k+1}$, one has the iterative equation

$$F_{k+1}(t) = \int_0^t F_k(t-y)dF(y), \quad k \geq 1. \tag{5.4.22}$$

Moreover

$$\mathbb{P}[N(t) = k] = \mathbb{P}[N(t) \geq k] - \mathbb{P}[N(t) \geq k+1] = \mathbb{P}[T_k \leq t] - \mathbb{P}[T_{k+1} \leq t]$$
$$= F_k(t) - F_{k+1}(t).$$

An important object in renewal theory is the renewal function m

$$m(t) \equiv \mathbb{E}[N(t)] = \sum_{k=1}^{\infty} F_k(t). \tag{5.4.23}$$

The last expression results by expressing $N(t)$ as a sum of Heaviside functions

$$N(t) = \sum_{k=1}^{\infty} H(t - T_k),$$

so that

$$m(t) = \mathbb{E}\left[\sum_{k=1}^{\infty} H(t-T_k)\right] = \sum_{k=1}^{\infty} \mathbb{E}[H(t-T_k)] = \sum_{k=1}^{\infty} F_k(t).$$

A fundamental result of renewal theory is that m satisfies the renewal equation

$$m(t) = F(t) + \int_0^t m(t-x)dF(x). \tag{5.4.24}$$

This can be established using the theorem of total expectation:

$$|m(t) = \mathbb{E}[N(t)] = \mathbb{E}[\mathbb{E}[N(t)|\tau_1]] = \int_0^{\infty} \mathbb{E}[N(t)|\tau_1 = x]dF(x)$$
$$= \int_0^t \mathbb{E}[N(t)|\tau_1 = x]dF(x) + \int_t^{\infty} \mathbb{E}[N(t)|\tau_1 = x]dF(x)$$
$$= \int_0^t (1 + \mathbb{E}[N(t-x)])dF(x) = \int_0^t (1 + m(t-x))dF(x).$$

We have used the fact that $\mathbb{E}[N(t)|\tau_1 = x] = 0$ if $t < x$. It can also be shown that $m(t) = \sum_{k=1}^{\infty} F_k(t)$ is the unique solution to the renewal equation (5.4.24) that is bounded on finite intervals. An important generalization of (5.4.24) is the renewal-type equation

$$C(t) = \mathcal{H}(t) + \int_0^t C(t-x)dF(x), \tag{5.4.25}$$

where $\mathcal{H}$ is a uniformly bounded function. The solution of the latter equation is specified by the following theorem (see [343]):

Theorem 5.1. *The function*

$$C(t) = \mathcal{H}(t) + \int_0^t \mathcal{H}(t-y)dm(y) \tag{5.4.26}$$

is a solution of the renewal-type equation (5.4.25). Moreover, if $\mathcal{H}$ is bounded on finite intervals then so is C, and the solution is unique.

Proof. If $h : [0,\infty) \to \mathbb{R}$, then define the functions $h * m$ and $h * F$ by

$$(h*m)(t) = \int_0^t h(t-x)dm(x), \quad (h*F)(t) = \int_0^t h(t-x)dF(x),$$

assuming the integrals exist. From the definitions, we have

$$m*F = F*m, \quad (h*m)*F = h*(m*F).$$

Moreover, the iterative equation (5.4.22), the renewal equation (5.4.24) and the solution (5.4.26) can be written in the compact forms

$$F_{k+1} = F_k * F = F * F_k, \quad m = F + F*m, \quad C = \mathcal{H} + \mathcal{H}*m.$$

Convolving C with respect to F gives

$$C*F = \mathcal{H}*F + \mathcal{H}*m*F = \mathcal{H}*F + \mathcal{H}*(m-F) = \mathcal{H}*m = C - \mathcal{H},$$

which establishes that the function C is a solution of the renewal-type equation (5.4.25). It remains to prove that if $\mathcal{H}$ is bounded on finite intervals then the solution is unique.

If $\mathcal{H}$ is bounded on finite intervals $[0,T]$, then from equation (5.4.26)

$$\sup_{0 \le t \le T} |C(t)| \le \sup_{0 \le t \le T} |H(t)| + \sup_{0 \le t \le T} \left| \int_0^t \mathcal{H}(t-y)dm(y) \right|$$
$$\le (1 + m(T)) \sup_{0 \le t \le T} |H(t)| < \infty.$$

Finiteness of the renewal function means that the solution C is also bounded on finite intervals. Now suppose that $\widehat{C}$ is another bounded solution of (5.4.25) and set $\Delta(t) = C(t) - \widehat{C}(t)$. Equations (5.4.22) and (5.4.25) imply that

$$\Delta = \Delta * F = \Delta * F * F = \Delta * F_2 = \Delta * F * F_2 = \Delta * F_3 \cdots,$$

that is, $\Delta = \Delta * F_k$, $k \geq 1$. Therefore,

$$|\Delta(t)| \leq F_k(t) \sup_{0 \leq u \leq t} |\Delta(u)|, \quad k \geq 0.$$

Finally, taking the limit $k \to \infty$ and noting that

$$F_k(t) = \mathbb{P}[N(t) \geq k] \to 0 \text{ as } k \to \infty,$$

we conclude that $|\Delta(t)| = 0$ for all t.

We end our brief discussion of renewal theory by stating several limit theorems. Set $\mu = \mathbb{E}[\tau_1] < \infty$.

Theorem 5.2. (*Elementary Renewal Theorem*).

$$\frac{N(t)}{t} \overset{\text{a.s.}}{\Rightarrow} \frac{1}{\mu} \text{ as } t \to \infty.$$

Theorem 5.3. (*Key Renewal Theorem*). *Define a random variable Y and its distribution F_Y to be arithmetic with span $\lambda, \lambda > 0$, if Y takes values in the set $\{m\lambda, m \in \mathbf{Z}\}$ with probability one, and λ is the maximal value with this property. Suppose that the first inter-arrival time τ_1 is not arithmetic. If $g : [0, \infty) \to [0, \infty)$ is such that g is monotone decreasing and $\int_0^\infty g(t)dt < \infty$, then*

$$\int_0^t g(t - x)dm(x) \to \frac{1}{\mu} \int_0^\infty g(x)dx \text{ as } t \to \infty.$$

5.5 Time-limiting steps in gene regulation

5.5.1 Kinetic proofreading in protein synthesis

A major requirement for proper cell function is that the genetic code is "read" with few mistakes during protein synthesis or DNA replication. For example, both transcription and translation involve the incorporation of specific molecular substrates at particular times, namely, a specific mRNA nucleotide during the production of mRNA or a specific amino acid during production of a protein. The incorporation of each substrate involves some recognition site within an RNA polymerase or a ribosome, respectively, that is more energetically disposed to bind the correct substrate C, say, rather than an incorrect substrate D. In a simple reaction scheme, the frequency of errors is of the order $e^{-\Delta G_{CD}/k_B T}$, where ΔG_{CD} is the smallest difference in binding energies between the correct substrate and an incorrect substrate. The basic problem is that typical values of ΔG_{CD} cannot account for the small error rates observed in protein synthesis. For example, the maximum frequency at which a wrong but similar amino acid is inserted during protein translation is 10^{-4}, which means that even smaller error rates must occur in each recognition step. The error rates are smaller still in the case of DNA transcription, taking values around 10^{-9}. *Kinetic proofreading* is a mechanism for error correction in biochemical processes, which was first introduced by Hopfield [394] and independently by Ninio [618]. The proofreading mechanism increases specificity of biochemical interactions by including a number of intermediate steps that can undo errors at the cost of increased reaction time and free energy expenditure.

Consider the binding interaction between a codon E of mRNA and the anti-codon of a tRNA during protein synthesis, see Fig. 5.33. Let C denote the correct tRNA and D an incorrect tRNA. Here E may be viewed as an enzyme acting on a substrate C or D according to a classical Michaelis-Menten scheme (see Sect. 1.3). That is,

$$E \underset{k_{\text{off}}}{\overset{k_{\text{on}}[C]}{\rightleftharpoons}} EC \overset{W}{\rightarrow} E + \text{correct amino acid incorporated} \tag{5.5.1a}$$

$$E \underset{k'_{\text{off}}}{\overset{k'_{\text{on}}[D]}{\rightleftharpoons}} ED \overset{W}{\rightarrow} E + \text{incorrect amino acid incorporated.} \tag{5.5.1b}$$

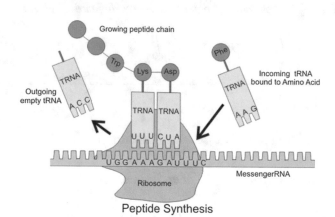

Peptide Synthesis

Fig. 5.33: Ribosomes can bind to an mRNA chain and use it as a template for determining the correct sequence of amino acids in a particular protein. Amino acids are selected, collected and carried to the ribosome by transfer RNA (tRNA molecules), which enter one part of the ribosome and bind to the messenger RNA chain. The attached amino acids are then linked together by another part of the ribosome. Once the protein is produced, it can then "fold" to produce a specific functional three-dimensional structure. Specificity is achieved through the interaction between the codon (triplet of nucleotides) in mRNA and the anti-codon in the tRNA. [Public domain figure downloaded·from Wikipedia Commons.]

(It is assumed that the catalytic step has no selectivity, that is, the rate of catalysis W is the same for both substrates.) The corresponding kinetic equations are

$$\frac{d[EC]}{dt} = k_{on}[E][C] - (k_{off} + W)[EC], \tag{5.5.2a}$$

$$\frac{d[ED]}{dt} = k'_{on}[E][D] - (k'_{off} + W)[ED], \tag{5.5.2b}$$

$$[E]_{Total} = [EC] + [ED] + [E]. \tag{5.5.2c}$$

The last equation ensures that the total concentration of ribosomes or enzymes is fixed. At steady state, we have

$$[EC] = [E]\frac{k_{on}[C]}{k_{off} + W}, \quad [ED] = [E]\frac{k'_{on}[D]}{k'_{off} + W}.$$

It follows that the rates of correct and incorrect translation are

$$R_{correct} = W[EC], \quad R_{incorrect} = W[ED],$$

and the error rate is

$$F_0 = \frac{R_{incorrect}}{R_{correct}} = \left[\frac{k'_{on}[D]}{k'_{off} + W}\right]\left[\frac{k_{on}[C]}{k_{off} + W}\right]^{-1}. \tag{5.5.3}$$

Typically, one finds that the on rates are approximately the same for all tRNAs (being diffusion limited) and the tRNAs have similar concentrations, that is, $k_{on} \approx k'_{on}$ and $[D] \approx [C]$. Hence, the error rate F_0 is minimized by taking the catalytic rate W to be much smaller than the off rates. Introducing the dissociation constants $K_C = k_{off}/k_{on}, K_D = k'_{off}/k'_{on}$, we have

$$F_0 \approx \frac{K_C}{K_D} = e^{-\Delta G_{CD}/k_B T}. \tag{5.5.4}$$

The above simple binding model neglects the fact that when tRNA binds to a codon, it is chemically altered via the hydrolysis of GTP (guanosine triphosphate); an analogous process occurs during the polymerization of microtubules (see Sect. 4.1). The transition to the new state is irreversible and in this state, the tRNA can also dissociate from the mRNA. This leads to the new reaction scheme

$$E \underset{k_{off}}{\overset{k_{on}[C]}{\rightleftharpoons}} EC \overset{r}{\rightarrow} EC^* \overset{W}{\rightarrow} E + \text{correct amino acid incorporated} \tag{5.5.5a}$$
$$\downarrow$$
$$E$$

$$E \underset{k'_{off}}{\overset{k'_{on}[D]}{\rightleftharpoons}} ED \overset{r}{\rightarrow} ED^* \overset{W}{\rightarrow} E + \text{incorrect amino acid incorporated}. \tag{5.5.5b}$$
$$\downarrow$$
$$E$$

Let the off rates of the modified substrates C^* and D^* be q_{off} and q'_{off}, respectively, and assume that $[C^*], [D^*] \approx 0$ so that direct formation of a complex EC^* or ED^* can be neglected. The steady-state concentrations of the modified substrate then satisfy, see Ex. 5.14

$$[EC^*] = \frac{1}{q_{\text{off}} + W} \frac{rk_{\text{on}}}{k_{\text{off}} + r} [E][C].$$

Again assuming that the rates of catalysis W, r are much smaller than the on and off rates, and taking the concentrations of all tRNAs to be the same, we obtain the new error rate

$$F \approx \frac{q_{\text{off}}}{q'_{\text{off}}} \frac{k'_{\text{on}}}{k'_{\text{off}}} \frac{k_{\text{off}}}{k_{\text{on}}}.$$

Finally, taking the on rates to be tRNA nonspecific

$$F = \frac{Q_C}{Q_D} \frac{K_C}{K_D} = e^{-\Delta G_{CD}/k_B T} e^{-\Delta G_{C^* D^*}/k_B T} < F_0. \tag{5.5.6}$$

In particular, if the difference in biding energies of the two substrates are the same for the native and modified states, then

$$F = \left[e^{-\Delta G_{CD}/k_B T} \right]^2. \tag{5.5.7}$$

In summary, the inclusion of an irreversible step into the kinetic scheme, $EA \rightarrow EA^*$, which necessitates the expenditure of energy, provides an additional opportunity for the incorrect substrate to dissociate, and leads to a reduction in the error rate. An even higher level of accuracy can be achieved by having a sequence of n irreversible proofreading stages

$$E \underset{k_{\text{off}}}{\overset{k_{\text{on}}[C]}{\rightleftharpoons}} EC \overset{r_1}{\rightarrow} EC^* \overset{r_2}{\rightarrow} EC^{**} \ldots \overset{r_m}{\rightarrow} EC^{**\ldots *} \overset{W}{\rightarrow} E + \text{product}.$$
$$\hspace{3.5cm} \downarrow \hspace{1.3cm} \downarrow \hspace{2.2cm} \downarrow$$
$$\hspace{3.5cm} E \hspace{1.3cm} E \hspace{2.2cm} E$$

(See Ex. 5.15 for another system that appears to exploit kinetic proofreading, namely T cells of the adaptive immune system.)

5.5.2 Fluctuations in DNA elongation times

One of the simplifications in the models of gene expression discussed so far is that the multistage processes underlying transcription and translation have been reduced to single-step processes with exponential waiting times (Poisson approximation). However, transcription (and also translation) can be broken up into at least three stages called initiation, elongation, and termination [197, 339]. During the initiation stage, RNA polymerase (RNAP) binds to a promoter site on the DNA and unzips

Fig. 5.34: Schematic illustration of the TEC. In the absence of fluctuations, the active transcription site in the pretranslocated state ($m = 0$) takes one step beyond the nascent mRNA to enter the posttranslocated state ($m = 1$). Translocation is a reversible reaction with transition rates a and b. The site can then add one nucleotide to the mRNA via polymerization so that $n \to n + 1$ and $m = 1 \to m = 0$. This step is also reversible with forward and backward transition rates k_f, k_b, respectively. [Redrawn from [832].]

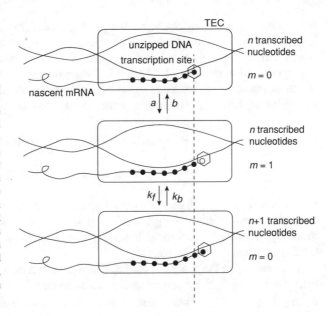

the double helix so that the strand of DNA to be transcribed is made accessible. Following the transcription of the first few nucleotides, the so-called transcription elongation complex (TEC) is formed, which consists of the RNAP, the DNA, and the emerging mRNA. This signals the beginning of the elongation phase where the TEC slides along the DNA, extending the transcript one nucleotide at a time. The process is terminated when a specific site is reached, and the nascent mRNA is released. An implicit assumption of the single-step Poisson approximation of gene transcription is that the rate-limiting step is initiation. However, there is growing evidence from single-molecule experiments that initiation can be much faster than elongation [41]. Moreover, *in vitro* studies of *E. coli* RNAP have established that processive mRNA synthesis is often disrupted by transcriptional pauses that can last anything from 1s to more than 20s [369, 607]. In some cases, the pauses are linked with the reverse translocation of the RNAP along the DNA, a process known as backtracking [622]. These observations suggest that the distribution of transcription times might be non-exponential with heavy-tails arising from the long transcriptional pauses.

Recently, there have been a number of stochastic models of the elongation stage and backtracking [556, 695, 715, 832, 870]. For the sake of illustration, we will focus on the model of Voliotis et al. [832], which is based on a master equation description of the dual processes of the TEC translocating along the DNA and the extension of the nascent mRNA via polymerization. A schematic diagram of the basic kinetics is shown in Fig. 5.34 in the simpler case that backtracking cannot occur. Suppose that n nucleotides have been transcribed and the active site of RNA polymerase is at the end of the precursor mRNA chain. Denote this so-called pre-translocated state of the active site by $m = 0$. The active site can then shift one step beyond the precursor mRNA to form a posttranslocated state denoted by $m = 1$. It is now in a position to add the next nucleotide to the precursor mRNA by polymerization so that $n \rightarrow n + 1$ and m resets to 0. The rates of polymerization and depolymerization are given by k_f and k_b, while the rates of forward and backward translocation are given by a and b. Let $P_{n,m}(t)$ be the probability of finding the TEC in state (n, m) at time t. The corresponding master equation is given by [832]

$$\frac{dP_{n,0}}{dt} = k_f P_{n-1,1} + b P_{n,1} - (k_b + a) P_{n,0}, \tag{5.5.8a}$$

$$\frac{dP_{n,1}}{dt} = k_b P_{n+1,0} + a P_{n,0} - (k_f + b) P_{n,1}, \tag{5.5.8b}$$

with $n = 0, 1, \ldots N_0 - 1$. There is a reflecting boundary condition at $n = 0$, which can be implemented by introducing a fictitious state $n = -1$ and setting $k_b P_{0,0} = k_f P_{-1,1}$. Similarly, there is an absorbing boundary condition $P_{N_0,0} = 0$, since the process terminates when $n = N_0$ is reached.

The first step in the analysis is to introduce the mean occupancy for each translocation step $(m = 0, 1)$ by summing over all nucleotide positions $n = 0, \ldots, N_0 - 1$. Setting $\Pi_m(t) = \sum_{n=0}^{N_0-1} P_{n,m}(t)$ and using the boundary conditions, we have

$$\frac{d\Pi_0}{dt} = (k_f + b)\Pi_1 - (k_b + a)\Pi_0, \quad \Pi_1 = 1 - \Pi_0,$$

with the initial condition $\Pi_0(0) = 1$, There is convergence to the steady-state solution

$$\Pi_0^* = (k_f + b)\tau, \quad \Pi_1^* = (k_b + a)\tau, \quad \tau = \frac{1}{k_f + k_b + a + b},$$

with τ the relaxation time. Assuming polymerization/depolymerization is much slower than translocation ($k_f, k_b \ll a, b$), we can make the quasi-steady-state approximation $P_{m,n}(t) = \Pi_m^* P_n(t)$ (see also Sect. 5.3.2), with $P_n(t)$ satisfying the birth–death master equation

$$\frac{dP_n}{dt} = \omega_- P_{n+1} + \omega_+ P_{n-1} - (\omega_+ + \omega_-) P_n, \tag{5.5.9}$$

and the effective polymerization/depolymerization rates are

$$\omega_+ = k_f(k_b + a)\tau \approx \frac{k_f a}{a+b}, \quad \omega_- = k_b(k_f + b)\tau \approx \frac{k_b b}{a+b}.$$

The boundary conditions become $\omega_- P_0 = \omega_+ P_{-1}$ (reflecting) and $P_{N_0} = 0$ (absorbing). The elongation time is defined as the time for the TEC to reach position $n = N_0$ starting from $n = 0$. In terms of the mean-field model given by the birth–death process, calculating the mean and variance of the elongation time requires solving a first passage time for the discrete Markov process. This can be achieved by following analogous steps to the analysis of continuous processes in Sect. 2.4.2. The analysis for a general master equation is presented in Box 5G.

Suppose that the TEC starts at position $n(0) = n_0$. Define the survival probability that the TEC has not yet reached the absorbing boundary at $n = N_0$ by

$$S(n_0, t) = \sum_{n=0}^{N_0-1} P(n, t | n_0, 0), \qquad (5.5.10)$$

where we have made the initial condition explicit by setting $P_n(t) \to P(n, t | n_0, 0)$. If T is the (stochastic) elongation time, then $S(n_0, t)$ is the probability that $T \geq t$. This implies that the cumulative distribution function of the elongation time is $1 - S(n_0, t)$. Hence, the first and second moments of the elongation time are

$$T(n_0) = \langle T \rangle = \int_0^\infty t \frac{\partial S(n_0, t)}{\partial t} dt = \int_0^\infty S(n_0, t) dt, \qquad (5.5.11)$$

and

$$T_2(n_0) = \langle T^2 \rangle = \int_0^\infty t^2 \frac{\partial S(n_0, t)}{\partial t} dt = 2 \int_0^\infty t S(n_0, t) dt. \qquad (5.5.12)$$

Equations for S and the moments of T can be obtained by considering the backward master equation

$$\frac{dP(n, t | n_0, 0)}{dt} = \omega_+ \left[P(n, t | n_0 + 1, 0) - P(n, t | n_0, 0) \right]$$
$$+ \omega_- \left[P(n, t | n_0 - 1, 0) - P(n, t | n_0, 0) \right]. \qquad (5.5.13)$$

The backward equation follows from differentiating with respect to t both sides of the Chapman-Kolmogorov equation

$$P(n_1, s | n_0, 0) = \sum_n P(n_1, s | n, t) P(n, t | n_0, 0).$$

Summing equation (5.5.13) from $n = 0$ to $n = N_0 - 1$ shows that

$$\frac{dS(n_0, t)}{dt} = \omega_+ \left[S(n_0 + 1, t) - S(n_0, t) \right] + \omega_- \left[S(n_0 - 1, t) - S(n_0, t) \right], \qquad (5.5.14)$$

supplemented by the boundary conditions $S(N_0, t) = 0$ and $S(0, t) = S(-1, t)$, and the initial condition $S(n_0, 0) = 1$.

Let us now calculate the mean elongation time. Integrating equation (5.5.14) with respect to t gives

$$-1 = \omega_+ \left[T(n_0 + 1) - T(n_0) \right] + \omega_- \left[T(n_0 - 1) - T(n_0) \right], \qquad (5.5.15)$$

with $T(N_0) = 0$ and $T(0) = T(-1)$. Setting $U(n_0) = T(n_0) - T(n_0 - 1)$, we obtain the first-order difference equation

$$\omega_+ U(n_0 + 1) - \omega_- U(n_0) = -1,$$

which has the solution $U(0) = 0$, $U(1) = -1/\omega_+$, $U(2) = -1/\omega_+ - \omega_-/\omega_+^2$, etc., that is

$$U(n) = -\frac{1}{\omega_+}\left[1 + \frac{\omega_-}{\omega^+} + \left(\frac{\omega_-}{\omega^+}\right)^2 + \ldots + \left(\frac{\omega_-}{\omega^+}\right)^{n-1}\right].$$

Since $-T(n) = U(N_0) + U(N_0 - 1) + \ldots + U(n+1)$, we deduce that

$$T(n_0) = \sum_{n_0+1}^{N_0} \frac{1}{\omega_+} \sum_{m=0}^{n-1} \left(\frac{\omega_-}{\omega^+}\right)^m. \tag{5.5.16}$$

Introducing $K = \omega_-/\omega_+$ and noting that $0 \leq K < 1$, we can sum the geometric series to give [300, 832]

$$T(n_0) = \frac{1}{\omega_+}\sum_{n=n_0+1}^{N_0} \frac{1 - K^n}{1 - K} = \frac{1}{\omega_+(1-K)}\left[N_0 - n_0 - \frac{K^{n_0+1} - K^{N_0+1}}{1 - K}\right].$$

Finally, setting $n_0 = 0$ we obtain the mean elongation time $\mu = T(0)$ with

$$\mu = \frac{1}{\omega_+(1-K)}\left[N_0 - \frac{K(1 - K^{N_0})}{1 - K}\right]. \tag{5.5.17}$$

The variance can be calculated in a similar fashion (see Ex. 5.16). Here we simply note that when chain lengthening is dominant, $K \ll 1$, both the mean and variance are linear functions of the chain length N_0

$$\mu = \frac{N_0}{\omega_+} + K\frac{N_0 - 1}{\omega_+} + O(K^2), \tag{5.5.18}$$

and

$$\sigma^2 = \frac{N_0}{\omega_+^2} + 4K\frac{N_0 - 1}{\omega_+^2} + O(K^2). \tag{5.5.19}$$

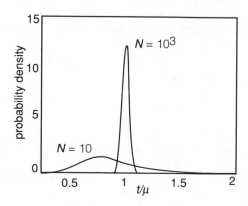

Fig. 5.35: Sketch of the density of elongation times (in units of the mean elongation time) for the TEC mean-field model without backtracking. Examples of small and large template sizes are shown. Also, $K = 0.01$ and $\omega_+ = 20$ s^{-1}. [Based on results from Voliotis et al. [832].]

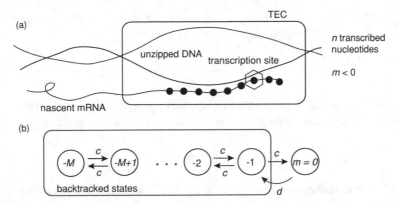

Fig. 5.36: Schematic illustration of TEC backtracking. (a) Example of a backtracked state of the transcription elongation complex (TEC) with $m = -2$. (b) Unbiased random walk model of backtracking. Transitions between backtracked states occur at a rate c. The TEC enters the backtracking regime from the state $m = 0$ at a rate d and exits the backtracking regime at the rate c. [Redrawn from [832].]

For sufficiently long sequences $N_0 \gg 1$, one finds that the distribution of elongation times is given by a narrow Gaussian with fluctuations scaling as $1/\sqrt{N_0}$. This adds a characteristic delay to the Poisson-like distribution of initiation times, see Fig. 5.35. Moreover, if initiation is much faster than elongation, then the transcription time is much more regular than if initiation dominates.

Voliotis et al. [832] show that the above picture persists when backtracking pauses are included, provided that they are sufficiently rare. However, the distribution of elongation times is drastically altered when backtracking become significant. The simplest way to incorporate backtracking into the model is to treat it as a separate process. That is, one can introduce additional translocation states of the TEC given by $m = -1, \ldots, -M$, which represent backtracked states shifted by $|m|$ steps from state $m = 0$. The duration of backtracking pauses can also be analyzed in terms of a first passage time problem, in this case, a random walk on a finite lattice with a reflecting boundary at $m = -M$ and an absorbing boundary at $m = 0$, see Fig. 5.36. One finds that there is a broad distribution of pause durations that exhibits power law behavior at intermediate duration times. Hence, the distribution of elongation times is significantly altered. Numerical simulations of the full model also suggest that the distribution of elongation times with long pauses naturally exhibits switching between high and low mRNA product rates, resulting in transcriptional bursting.

Box 5G. First passage times for master equations.

Backward master equation. Consider a discrete Markov process $N(t) \in \Gamma$ whose probability distribution $P(n, t | n_0, t_0) = \mathbb{P}[N(t) = n | N(t_0) = n_0]$ evolves according to the forward master equation (Chap. 3)

$$\frac{dP(n,t|n_0,t_0)}{dt} = \sum_m W_{nm} P(m,t|n_0,t_0) - P(n,t|n_0,t_0) \sum_m W_{mn}, \quad t > t_0,$$

where $\mathbf{W}$ is the transition matrix. Setting $P_n(t) = P(n,t|n_0,t_0)$ and $\mathbf{P} = (P_n)_{n\in\Gamma}$, the master equation can be written compactly in terms of the matrix equation

$$\frac{d\mathbf{P}(t)}{dt} = \mathbf{A}\mathbf{P}(t),$$

where $\mathbf{A}$ is the generator with $A_{nm} = W_{nm} - \delta_{n,m} \sum_k W_{km}$. In order to analyze first passage time problems, it is convenient to consider the corresponding backward master equation. One way to derive the latter is to consider the first step out of the initial state n_0 at time t_0 rather than the last step of the trajectory that enters the final state n at time t. Over an infinitesimal interval dt, we have two mutually exclusive events such that

$$P(n,t|n_0,t_0) = \left(1 - \sum_m W_{mn_0} dt\right) P(n,t|n_0,t_0+dt)$$
$$+ \sum_m [W_{mn_0} dt] P(n,t|m,t_0+dt).$$

The first term on the right-hand side represents the case that the system is still in the same state n_0 at time $t_0 + dt_0$ and subsequently evolves to the state n at time t, whereas the second term includes all transitions $n_0 \to m$ during the time interval dt followed by evolution to state n at time t. Assuming that the Markov process is stationary, we can can set $P(n,t|S_0,t_0+dt) = P(n,t-dt|n_0,t_0)$. Now dividing through by dt and taking the limit $dt \to 0$ yields the backward master equation

$$\frac{dP(n,t|n_0,t_0)}{dt} = \sum_m W_{mn_0} P(n,t|m,t_0) - P(n,t|n_0,t_0) \sum_m W_{mn_0}. \quad (5.5.20)$$

Setting $Q_m(t) = P(n,t|m,t_0)$ for given n, we can write the backward master equation in the matrix form

$$\frac{d\mathbf{Q}}{dt} = \mathbf{A}^\top \mathbf{Q}. \quad (5.5.21)$$

Survival probability and first passage times. Analogous to continuous Markov processes, the first passage time to reach a particular target state n_f, say, can be determined by imposing an absorbing boundary at n_f. This is achieved by setting the transition rates out of n_f to zero, that is, $W_{mn_f} = 0$ for all m. The next step is to introduce the survival probability that the system has not yet reached n_f at time t, having started at n_0 at time t_0:

$$S(n_0,t) = \sum_{n \neq n_f} P(n,t|n_0,0),$$

where we have set $t_0 = 0$ for convenience. The survival probability is related to the first passage time distribution $F(n_0, \tau)$ according to

$$S(n_0,t) = 1 - \int_0^t F(n_0, \tau) d\tau,$$

which means that S is the cummulative distribution for F and

$$F(n_0, \tau) = - \left. \frac{dS(n_0,t)}{dt} \right|_{t=\tau}.$$

From the forward master equation with $A_{nn_f} = 0$, we have

$$\frac{dS(n_0,t)}{dt} = \sum_{n \neq n_f} \frac{dP(n,t|n_0,0)}{dt} = \sum_{n \neq n_T} \sum_{m \neq n_f} A_{nm} P(m,t|n_0,0)$$

$$= - \sum_{m \neq n_f} W_{n_f m} P(m,t|n_0,0) = -J(n_f,t|n_0,0),$$

where $J(n_T,t|n_0,0)$ is the probability current into the absorbing state. We have used the fact that $\sum_n A_{nm} = 0$ so that $\sum_{n \neq n_f} A_{nm} = -A_{n_f m} = -W_{n_f m}$ for $m \neq n_f$. Alternatively, we can obtain an equation for the first passage time density using the backward master equation

$$\frac{dS(n_0,t)}{dt} = \sum_{n \neq n_f} \frac{dP(n,t|n_0,0)}{dt} = \sum_{n \neq n_f} \sum_m A_{mn_0} P(n,t|m,0)$$

$$= \sum_m A_{mn_0} \left(\sum_{n \neq n_f} P(n,t|m,0) \right) = \sum_m A_{mn_0} S(m,t).$$

That is, the survival probability satisfies the backward master equation. Now differentiating both side with respect to t establishes that the first passage time density also satisfies the backward master equation

$$\frac{dF(n_0,t)}{dt} = \sum_m A_{mn_0} F(m,t). \qquad (5.5.22)$$

Although the backward equation isn't necessarily simpler to solve than the forward equation, the former is a particularly convenient starting point for calculating moments of F. For example, the MFPT, which is defined as

$$T(n_0) = \int_0^\infty F(n_0, \tau) \tau d\tau, \qquad (5.5.23)$$

can be determined by applying the backward master operator to both sides

$$\sum_m A_{mn_0} T(m) = \int_0^\infty \sum_m A_{mn_0} F(m,\tau)\tau d\tau = \int_0^\infty \partial_\tau F(n_0,\tau)\tau d\tau$$

$$= \int_0^\infty \partial_\tau S(n_0,\tau)d\tau = S(n_0,\infty) - S(n_0,0) = -1.$$

In matrix form, with $\mathbf{T} = (T(m))_{m\in\Gamma}$, we have $\mathbf{A}^\top \mathbf{T} = -1$.

5.6 Epigenetics

Epigenetics concerns phenotypic states that are not encoded as genes, but as inherited patterns of gene expression originating from environmental factors [196, 773], and maintained over multiple cell generations when the original environmental stimuli have been removed. Examples of environmental influences range from changes in the supply of nutrients in bacteria to stress in humans. The simplest and best studied example of inherited gene expression involves the infection of *E. coli* by the λ phage DNA virus [668]. Following infection, the λ phage either multiplies and ultimately kills the host (lysis), or it integrates its DNA into that of the host (lysogeny) and is passively replicated over many generations, see Fig. 5.37. These two distinct

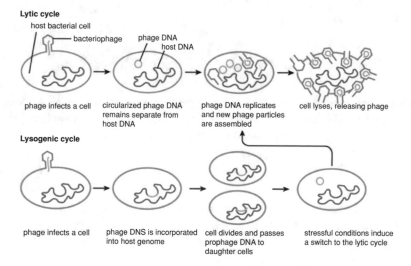

Fig. 5.37: Major transitions during lambda infection of a bacterial cell showing both lysogenic and lytic cycles. [Public domain figure downloaded from Wikipedia Commons.]

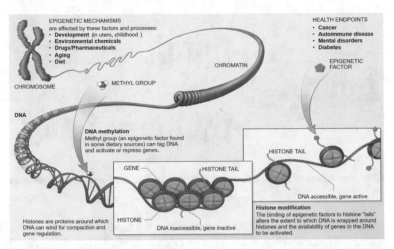

Fig. 5.38: Illustration of two major epigenetic mechanisms for *cis*-acting gene regulation. (i) DNA methylation occurs when methyl groups tag DNA and repress or activate genes. (ii) Nucleosome or histone modification occurs when the binding of epigenetic factors to histone tails alters the extent to which DNA is wrapped around histones and the availability of genes in the DNA to be activated. (A nucleosome is a basic unit of DNA packaging in eukaryotes, consisting of a segment of DNA wound in sequence around eight histone protein cores. This structure is often compared to thread wrapped around a spool.) [Public domain figure from the National Institutes of Health (http://commonfund.nih.gov/epigenomics/figure.aspx).]

scenarios involve the expression of different genes. In wild-type, one finds that the lyosgenic state is extremely stable with spontaneous loss occurring around 10^{-5} per cell and generation, which corresponds to a lifetime of five years. Hence, the spontaneous escape from the metastable lysogenic state is an excellent example of a rare event (Chap. 8), which allows the environmentally induced phenotype to persist over multiple generations. A model of lysogeny in λ-phage will be considered in Sect. 5.6.1.

The epigenetic mechanism in λ phage involves proteins produced by one gene diffusing across the cell and binding to operator sites of another gene. This has been the main form of gene regulation considered in this chapter, and is known as *trans*-acting gene regulation. However, there are also more local mechanisms that maintain epigenetic states locally on the genome, which belong to the class of so-called *cis*-acting gene regulation. Examples in eukaryotes include DNA methylation [98, 542] and gene silencing by nucleosome modifications [340, 806, 868], see Fig. 5.38. A model of the latter will be considered in Sect. 5.6.2 and applied to a problem in olfactory sensing.

DNA methylation [9]. DNA methylation (addition of the methyl group CH_3) in vertebrate DNA occurs on cytosine (C) nucleotides predominantly in the sequence CG (G for guanine). This is base paired to exactly the same sequence (in opposite orientation) on the other strand of the DNA helix. A simple mechanism then allows an existing pattern of DNA methylation to be inherited directly by a daughter cell, as illustrated in Fig. 5.39. An enzyme known as maintenance methyl transferase acts

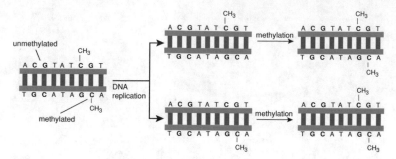

Fig. 5.39: Replication of DNA methylation patterns. See text for details. [Redrawn from [9].]

preferentially on unmethylated CG sequences that are base paired with a methylated CG sequence. In other words, the pattern of DNA methylation on the parental DNA strand provides a template for the methylation of the daughter DNA strands. One major role of DNA methylation, in conjunction with other regulatory mechanisms, is to provide a particularly efficient form of gene repression. This is achieved via a number of mechanisms. First, the methyl groups associated with cytosines are located in the major groove of DNA, which allows them to interfere with the binding of transcription factors. Second, there are a variety of proteins within a vertebrate cell that bind specifically to methylated DNA. Probably the best characterized are those associated with histone-modifying enzymes, which generate a repressive chromatin state that prevents regions of DNA to be accessed for transcription. Finally, note that DNA methylation patterns are dynamic during mammalian development, and incorrect patterns have a widespread involvement in cancer progression.

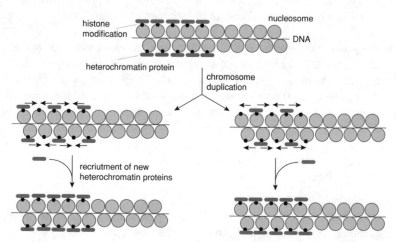

Fig. 5.40: Replication of nucleosome modification patterns. See text for details. [Redrawn from [9].]

Histone modification [9]. The packaging of DNA in chromatin is heterogeneous due to the fact that histone modification can lead to regions of more tightly wound DNA that is inaccessible for transcription. The regions of denser packaging are maintained by so-called heterochromatin proteins. During chromosome replication, the modified chromatin components are shared between the sister chromosomes. Following DNA replication, the inherited modified histones and heterochromatin proteins act in concert to replace the missing components, resulting in reproduction of the original pattern of nucleosome modification, see Fig. 5.40.

5.6.1 Regulatory network of λ phage lysogeny

Lysogeny in λ phage is maintained by a regulatory network involving λ phage DNA, a pair of regulatory proteins CI and Cro, and an operator complex OR consisting of three binding sites OR_j, $j = 1, 2, 3$, overlapping with two promoter sites P_{RM} and P_R, see Fig. 5.41. Either CI or Cro can bind to the operators OR_j. The protein CI has the highest affinity for OR_1, and when it is bound it blocks RNA polymerase from binding to the promoter P_R and initiating transcription of the gene cro. It also binds cooperatively to OR_2, resulting in autoactivation of its own production by a factor of 10. Finally, if CI is bound to all three sites, then it represses its own production. On the other hand, the protein Cro mainly binds to OR_3, consequently blocking the promoter P_{RM} and the synthesis of CI. If it also binds to OR_1 or OR_2 then it represses its own production. Note that the native state of both proteins is a dimer, so we take $x = [CI]$ and $y = [Cro]$ to be the concentrations of dimers. (The regulatory network is a more complicated version of the mutual repressor model of a bistable switch, see Sect. 5.1.)

Suppose that the following simplifications are imposed: (i) Cro and CI never bind the operator complex simultaneously; (ii) strong cooperativity causes the binding of CI to OR_1 and OR_2 to be concurrent; (iii) the states in which Cro is bound to OR_1 or OR_2 can be lumped together. There are then five promoter states O_j, $j = 0, 1, 2, 3, 4$:

O_0 : all operator sites are unoccupied

O_1 : OR_1 and OR_2 both occupied by a CI dimer

O_2 : all operator sites are occupied by a CI dimer

O_3 : OR_3 occupied by a Cro dimer

O_4 : OR_3 and at least one of the other operator sites is occupied by a Cro dimer

The reaction scheme for the promoter states is

$$O_0 \underset{k_{-1}}{\overset{k_1 x^2}{\rightleftharpoons}} O_1, \quad O_1 \underset{k_{-2}}{\overset{k_2 x}{\rightleftharpoons}} O_2, \quad O_0 \underset{k_{-3}}{\overset{k_3 y}{\rightleftharpoons}} O_3, \quad O_3 \underset{k_{-4}}{\overset{k_4 y}{\rightleftharpoons}} O_4. \tag{5.6.1}$$

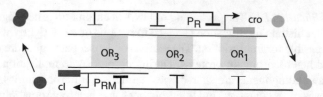

Fig. 5.41: Schematic diagram of operator complex OR with three binding sites OR_j, $j = 1, 2, 3$. The gene cI is transcribed when OR_3 is free and OR_2 is occupied by the protein CI, whereas the gene cro is transcribed when both OR_2 and OR_1 are free. Dimers of CI bind cooperatively to OR_1 and OR_2.

The law of mass action can be used to determine the steady-state probabilities P_j^* for promoter state O_j with $\sum_{j=0}^{4} P_j^* = 1$.

Finally, we assume that the state O_0 produces CI at a rate R_{RM} and the state O_1 produces CI at a rate $\widehat{R}_{RM}$, with $\widehat{R}_{RM} \gg R_{RM}$; no other state produces CI. Similarly, we assume that states O_0 and O_3 produce Cro at a rate R_R, and Cro production is off in the other states. The kinetic equations for the concentrations of CI and Cro then take the form, see Ex. 5.17

$$\frac{dx}{dt} = \phi_{CI}(x, y) - \gamma_{CI} x, \quad \frac{dy}{dt} = \phi_{Cro}(x, y) - \gamma_{Cro} y, \tag{5.6.2}$$

where the net production rates are

$$\phi_{Cro}(x, y) = \frac{R_{RM} + \widehat{R}_{RM} x^2 / K_1}{1 + x^2/K_1 + x^2/K_1 K_2 + y/K_3 + y^2/K_3 K_4}, \tag{5.6.3}$$

$$\phi_{CI}(x, y) = \frac{R_M(1 + y/K_3)}{1 + x^2/K_1 + x^2/K_1 K_2 + y/K_3 + y^2/K_3 K_4}. \tag{5.6.4}$$

Here $K_j = k_{-j}/k_j$.

Lysogeny is maintained during bacterial growth provided that the number of CI molecules per bacterial cell is sufficiently high (200-350 per cell). However, if the CI concentration becomes too low, then increased activation of cro increases the concentration of Cro protein and decreases cI activation. Thus, lysogeny ends and lysis begins. This can be modeled in terms of a bistable switch, in which one of the metastable fixed points is the lysogenic state, and termination of lysogenesis occurs when there is a noise-induced path that exits the basin of attraction of this state; this tends to occur when the bacteria are under environmental stress. The analysis of such noise-induced transitions has been developed in terms of a first passage time problem, after carrying out a system-size expansion of the associated chemical master equation [37, 38].

5.6.2 Two-state epigenetic model of nucleosome modification

We now consider a simplified model of nucleosome-based epigenetics in fission yeast [220, 577]. Suppose that each nucleosome exists in only two mutually exclusive chemical states termed modified (M) and anti-modified (A), see Fig. 5.42. Each nucleosome type recruits a modifying enzyme that converts the other type to its own type. In order for robust epigenetic memory, it is necessary to have some form of cooperativity in the recruitment process. This is implemented by requiring that two M (A) nucleosomes are needed to induce a transition $A \to M$ ($M \to A$). It is also possible for nucleosomes to spontaneously switch to the other state. At each update, a nucleosome site n is chosen at random. With probability $1 - \alpha$ nucleosome n randomly set to an A or M state, or with probability α two other nucleosomes are chosen and if these are in the same state, then the nucleosome n is set to this state.

Ignoring the effects of molecular noise (large number of sites N), the kinetic equation for the fraction x of M-state sites is

$$\frac{dx}{dt} = R_+(x) - R_-(x), \tag{5.6.5}$$

where

$$R_+(x) = \alpha(1-x)x^2 + (1-\alpha)(1-x), \quad R_-(x) = \alpha x(1-x)^2 + (1-\alpha)x. \tag{5.6.6}$$

The first terms in the definitions of $R_\pm$ represent the rate of recruitment, while the second terms represent spontaneous switching. Recruitment of any site occurs with probability α/N and spontaneous switching with probability $(1-\alpha)/N$. The factors of x and $(1-x)$ are based on the law of mass action, and the quadratic terms reflect cooperativity. The kinetic equation can be simplified as

$$\frac{dx}{dt} = A(x) := \alpha(2x-1)[x(1-x) - 1/F], \quad F = \frac{\alpha}{1-\alpha}. \tag{5.6.7}$$

Here F is the ratio of recruitment to spontaneous switching. It can be shown that the system is bistable if $F > 4$.

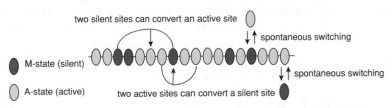

Fig. 5.42: Schematic illustration of two-state model of nucleosome modification with N sites introduced in Ref. [577]. Each site represents a nucleosome that can be active/anti-modified (A) or silent/modified (M). Transitions between these two states are in part random, and in part autoregulated by recruitment of histone-modifying enzymes by local nucleosomes.

In the case of low copy numbers, we have to replace the deterministic kinetic equation by a birth–death master equation for the probability $P_m(t)$ that there are m sites of type N

$$\frac{dP_m}{dt} = N[R_-([m+1]/N)P_{m+1} + R_+([m-1]/N)P_{m-1}(t)$$
$$- (R_+(m/N) + R_-(m/N))P_m(t)]. \tag{5.6.8}$$

Carrying out a system-size expansion along the lines of Sect. 3.3.2, we obtain the following FP equation for $P(x,t)$

$$\frac{\partial P}{\partial t} = -\frac{\partial A(x)P(x,t)}{\partial x} + \frac{1}{2N}\frac{\partial^2}{\partial x^2}D(x)P(x,t), \tag{5.6.9}$$

with

$$A(x) = R_+(x) - R_-(x), \quad D(x) = R_+(x) + R_-(x) = \alpha[x(1-x)+1/F].$$

We can characterize the stochastic dynamics in terms of trajectories along a one-dimensional landscape determined by the quasipotential, see Sect. 5.2.5

$$\Phi(x) = -2\int_x \frac{A(y)}{D(y)}dy. \tag{5.6.10}$$

Bistability of the deterministic system means that the quasipotential will be a double well potential, and the epigenetic state of the nucleosomes will be maintained via multiple generations if the potential barrier between the two metastable states is sufficiently high.

Differentiation of olfactory neurons. The olfactory system integrates signals from receptors expressed in olfactory sensory neurons. Each sensory neuron expresses only one of many similar olfactory receptors (ORs). The choice of receptor is made stochastically early in the differentiation process and is maintained throughout the life of the neuron. With thousands of possible OR genes to select, the system requires a local mechanism that couples expression of each gene to itself while still leaving similar genes non-activated. In addition, activation of a gene needs to project a form of negative feedback that keeps other OR genes from being activated. In other words, the system involves a combination of local positive feedback and global negative feedback [275, 538]. The basic model is shown in Fig. 5.43 [12].

Let x_j denote the fraction of active nucleosome sites of the j-th gene. As a generalization of the two-state model, we take

$$\frac{dx_j}{dt} = \Gamma(\mathbf{x})(1-x_j)x_j^2 - \alpha x(1-x)^2 + \beta\Gamma(\mathbf{x})(1-x_j) - \beta x_j. \tag{5.6.11}$$

Here $\alpha < 1$ represents an inherent bias resulting in a weaker recruitment from the silenced state, and β represents spontaneous conversions. The major modification is that the fraction of silenced sites available for activation of spontaneous switching is reduced by the factor $\Gamma(\mathbf{x})$, due to binding of a represser protein P to some sub-population of silenced sites. The repressor is assumed to be produced by all active genes, such that the repressor concentration $[P]$ is given by

$$[P] = \kappa_p \sum_j x_j^h,$$

for some hill coefficient h. Let $s_j = 1 - x_j$ denote the fraction of silenced nucleosomes of the j-th gene, and introduce the decomposition $s_j = s_j^u + s_j^b$, where s_j^u is the fraction of nucleosomes that are silenced and not bound by P, whereas s_j^b is the fraction of nucleosomes that are silenced and bound by P. Only the former are available for activation. The kinetic equation for conversion is of the form

$$\frac{ds_j^b}{dt} = k_+[P]s_j^u - k_- s_j^b.$$

Assuming fast rates of conversion between P bound and P unbound silenced nucleosomes, it follows that

$$s_j^b = \frac{k_+[P]}{k_-} s_j^u = r \sum_j x_j^h, \qquad r = \frac{k_+ \kappa_p}{k_-}.$$

Hence

$$s_j^u = \frac{1}{1 + r \sum_j x_j^h} s_j, \qquad (5.6.12)$$

which implies that $\Gamma(\mathbf{x}) = (1 + r \sum_j x_j^h)^{-1}$.

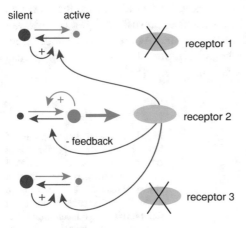

Fig. 5.43: Multistability in an olfactoric neuronal cell that randomly selects and expresses one of around 1000 olfactoric receptors genes [80]. The two-state epigenetic model of nucleosome modification (via positive feedback) in one gene is supplemented by a global negative feedback associated to the activity of all olfactoric genes. The combination of local activation and global repression leads to the expression of receptor 2 which, in turn, favors the silenced nucleosome state for all other olfactoric genes.

5.7 Morphogen gradients and gene expression

It has been known for some time that morphogen gradients play a crucial role in the spatial regulation of patterning during development [487, 755, 799, 846, 863, 864]. That is, a spatially varying concentration of a morphogen protein drives a corresponding spatial variation in gene expression through some form of concentration thresholding mechanism. For example, in regions where the morphogen concentration exceeds a particular threshold, a specific gene is activated. Hence, a continuously varying morphogen concentration can be converted into a discrete spatial pattern of differentiated gene expression across a cell population, see Fig. 5.44.

Within developmental cell biology, the first morphogens were identified in experimental studies of *Drosophila* embryos. *Drosophila* is particularly suitable for investigating early development due to the relative simplicity of its anatomy and the applicability of powerful genetic methods. One finds that the patterning of a spatially uniform arrangement of identical cells is driven by graded distributions of various transcription factors, see Fig. 5.45. Examples of transcription factors that play a key role in establishing the adult body plan are as follows: (i) The activator Bicoid (Bcd), whose anterior-posterior concentration gradient specifies anterior segments of the body [663]; (ii) The activator/repressor Dorsal (Dl), whose ventral-to-dorsal nuclear localization gradient specifies the arrangement of the muscle, nerve and skin tissues [393]; (iii) The repressor Capicua (Cic), whose concentration profile has minima at both the anterior and posterior ends, plays an important role in determining the terminal structures of the embryo [294]. Each of these gradients arises from asymmetries in the unfertilized egg. For example, the Bcd protein concentration gradient arises from the localized spatial pattern of bcd mRNA, which is established during egg formation (oogenesis). The egg chamber of drosophila develops through 14 stages to form the mature egg. During oogenesis, the so-called nurse cells produce bcd mRNA, and others such as oskar (osk) and gurken (gcd) are delivered to the developing oocyte. The different mRNAs are localized at different parts of oocyte boundary depending on the polarization of microtubules, properties of the molecular motors, and mRNA anchoring proteins [58].

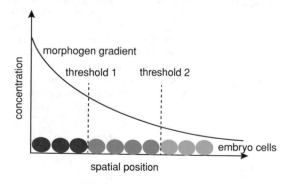

Fig. 5.44: Schematic diagram of how a morphogen gradient acts on embryo cells. Cells exposed to the morphogen concentration above threshold 1 will activate a "red" gene; cells exposed to the morphogen concentration below threshold 1 but above threshold 2 will activate a "green" gene; and cells exposed to the morphogen concentration below threshold 2 will activate a "blue" gene.

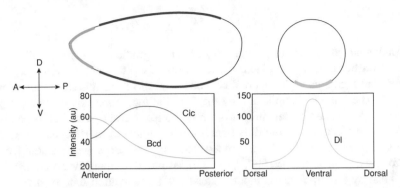

Fig. 5.45: Graded distributions of transcription factors in the early Drosophila embryo. The high concentration regions of Bicoid (Bcd) and Capicua (Cic) are shown on a sideview of the embryo, whereas the high concentration region of Dorsal (Dl) is shown on a cross-sectional view. Also shown are sketches of typical fluorescence intensity profiles of the signals from an embryo stained with the appropriate antibody.

Diffusion-based morphogen gradients. The most common mechanism for morphogen gradient formation, as exemplified by the Bcd gradient of *Drosophila*, is thought to involve a localized source of protein production within the embryo, combined with diffusion away from the source and subsequent degradation [486, 487, 755, 799, 846]. The latter can arise either from degradation within the extracellular domain or by binding to membrane bound receptors and subsequent removal from the diffusing pool by endocytosis. The rates of binding and internalization thus control the effective degradation rate. (An alternative mechanism of morphogenesis that involves direct contacts between cells mediated by thin actin-rich filaments or cytonemes will be considered in Chap. 14.) The basic reaction-diffusion equation for diffusion-based morphogenesis takes the form

$$\frac{\partial C(x,t)}{\partial t} = D\frac{\partial^2 C(x,t)}{\partial x^2} - kC(x,t), \quad D\frac{\partial C(0,t)}{\partial x} = 0, \quad -D\frac{\partial C(0,t)}{\partial x} = Q_0, \quad (5.7.1)$$

where $C(x,t)$ denotes the concentration of protein, t is the time from the onset of morphogen synthesis, and x, $0 < x < L$, is the distance from the anterior pole of the embryo whose size is L. The total rate of degradation is given by k, and Q_0 represents the rate of protein synthesis at $x = 0$. Experimentally, it is observed that the protein concentration decays to zero before reaching the posterior end, so the solution is approximately independent of L. The steady-state solution takes the form

$$C^*(x) = C^*(0)\left(\frac{e^{x/\lambda} + e^{2L/\lambda}e^{-x/\lambda}}{1 + e^{2L/\lambda}}\right), \quad \lambda = \sqrt{\frac{D}{k}}. \quad (5.7.2)$$

The constant $C^*(0)$ can be determined from the boundary condition at $x = 0$. In particular, when $L \gg \lambda$, the concentration decays exponentially with length constant λ

$$C^*(x) = \frac{Q_0}{D\lambda}e^{-x/\lambda}, \quad \lambda = \sqrt{D/k}. \tag{5.7.3}$$

Mathematical aspects of protein concentration gradient formation will be considered further in Chap. 11. Here we focus on how different genes are expressed at different locations within an embryo, based on different concentrations of morphogen. The issue is complicated by the fact that the expression of a single gene is typically regulated by networks involving multiple transcription factors and feedback. These can involve both direct and indirect targets of morphogens, as well as auxiliary signaling factors that are independent of the morphogen gradient [28]. In the following illustrative examples, we will assume that the activation/inactivation of a gene by a morphogen is much faster than the relaxation time of the concentration gradient and the synthesis of target proteins. Hence, the transcription rate of mRNA at location x at time t is an algebraic function of the local concentration $C(x,t)$.

Monotonic patterning. As a simple example, suppose that a morphogen A activates a particular gene by binding to a single promoter site of the gene. The rate of transcription will then depend on $\mathbb{P}[A] = \mathbb{P}[A\,\text{bound}] = C_A/(K_A + C_A)$, where $C_A = C_A(x,t)$ is the local morphogen concentration at time t, and K_A is the corresponding dissociation constant, see Chap. 3. Let c_m and c_p denote the concentrations of mRNA and protein of a gene controlled by the morphogen gradient at (x,t). The law of mass action then yields the kinetic equations

$$\frac{dc_m}{dt} = \kappa_1 \frac{C_A}{K_A + C_A} - \gamma_m c_m, \tag{5.7.4a}$$

$$\frac{dc_p}{dt} = \kappa_2 c_m - \gamma_p c_p. \tag{5.7.4b}$$

Here κ_1, κ_2 are transcription and translation rates, and γ_m, γ_p are degradation rates. If $\gamma_m \gg \gamma_p$ then we obtain a reduced equation for c_p given by

$$\frac{dc_p}{dt} = \frac{\kappa_1 \kappa_2}{\gamma_m} \frac{C_A}{K_A + C_A} - \gamma_p c_p. \tag{5.7.5}$$

Under the further assumption that protein degradation is faster than the time scale of morphogen gradient formation, we can take c_p to be in the quasi-steady state

$$c_p(x,t) \approx c_p^{max} \frac{C_A(x,t)}{K_A + C_A(x,t)}, \quad c_p^{max} = \frac{\kappa_1 \kappa_2}{\gamma_m \gamma_p}. \tag{5.7.6}$$

Thus, the protein concentration at location x at time t is proportional to the probability that the gene is activated by the local morphogen concentration. This result extends to more complicated cases, where the activation of the gene depends on additional promoter binding sites and/or other transcription factors. One simple generalization is to include cooperative binding of the morphogen A to n binding sites (see Chap. 3), in which case the probability of gene activation is given by the Hill function $C_A^n/(K_A^n + C_A^n)$.

Non-monotonic patterning. An interesting example of a more complex gene network is shown in Fig. 5.46. It consists of a morphogen A that activates a gene B and a gene C, with the proteins produced by gene B acting as repressors of gene C. This network can convert a monotonic morphogen gradient into a unimodal pattern [755]. It is assumed that gene C has two independent promoter binding sites, one with an affinity for morphogen A and the other with an affinity for protein B. Gene C is only active if the promoter is bound by an activator and not bound by a repressor. Under these assumptions the probability that gene C is active is given by

$$\mathbb{P}[A\,\text{bound}, B\,\text{not bound}] = \mathbb{P}[A](1 - \mathbb{P}[B]) = \frac{C_A}{K_A^C + C_A}\frac{K_B^C}{K_B^C + C_B}.$$

Here K_A^C and K_B^C are the dissociation constants for the repressor and activator within the promoter region of gene C. Given that gene B is also activated by the morphogen A, it follows that

$$C_B = C_B^{\max}\frac{C_A}{K_A^B + C_A}.$$

Substituting this expression into the equation for $\mathbb{P}[A\,\text{bound}, B\,\text{not bound}]$, and assuming $K_A^B \gg C_B^{\max}$, we have

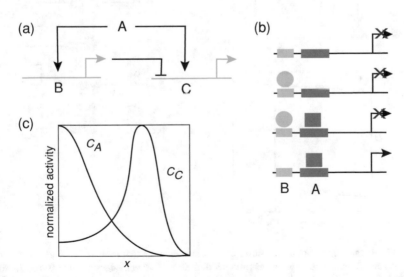

Fig. 5.46: Gene circuit for non-monotonic patterning. (a) Basic gene circuit showing that morphogen A acts as an activator of genes B and C, and gene B produces proteins that suppress gene C. (b) Possible promoter states of gene C. The latter is only active if site A is bound. and site B is vacant. (c) The activation of gene C is a non-monotonic function of the morphogen concentration C_A, which leads to a non-monotonic spatial distribution C_C of protein C.

$$\mathbb{P}[A \text{ bound}, B \text{ not bound}] = \frac{C_A}{K_A^C + C_A} \frac{K_B^C}{K_B^C + C_B^{\max} C_A / K_A^B}.$$

The activation function on the right-hand side is a unimodal function of the morphogen concentration C_A with a maximum at

$$C_A^* = \sqrt{K_A^C K_B^C K_A^B / C_B^{\max}}.$$

Hence, if this value lies within the range of the morphogen concentration gradient, then the activity of gene C will have a maximum at the location x^* for which $C_A(x^*) = C_A^*$. It is thought that a mechanism analogous to this circuit plays a role in the dorsal-ventral patterning of the *Drosophila* embryo.

Positive feedback and bistability. Recall from Sect. 5.1 that a gene regulatory network with positive feedback can exhibit bistability. Here we consider this type of circuit within the context of morphogen gradients. For simplicity, consider a gene B that is activated both by the morphogen A and its own protein product B (positive feedback). We assume that there are two cooperative promoter sites for B, which are independent of the binding site for A, see Fig. 5.47. The probability that gene B is

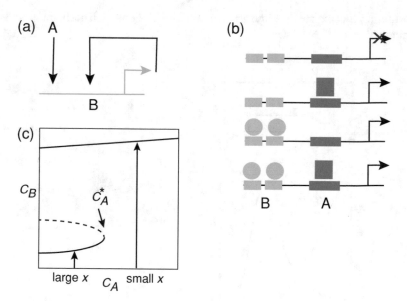

Fig. 5.47: Bistable gene circuit. (a) Basic gene circuit showing that morphogen A acts as an activator of gene B, and gene B produces proteins that also act as activators of gene B. (b) Possible promoter states of gene B, in which activator B acts cooperatively. (c) Bifurcation diagram illustrating bistability of the concentration C_B with the morphogen concentration acting as a bifurcation parameter. Stable and unstable steady states are indicated by solid and dashed lines, respectively. As C_A crosses a critical value or threshold, the concentration of protein B switches from a small to a high value. This results in a sharp transition in the spatial variation in C_B.

activated is given by

$$\mathbb{P}[A \text{ or } B \text{ bound}] = 1 - (1 - \mathbb{P}[A])(1 - \mathbb{P}[B]) = \mathbb{P}[B] + \mathbb{P}[A](1 - \mathbb{P}[B]).$$

By analogy with equation (5.7.5), the concentration C_B of protein B evolves according to the kinetic equation

$$\frac{dC_B}{dt} = \gamma_B \left[C_B^{\max} \left(\frac{C_B^2}{K_B^2 + C_B^2} + \frac{C_A}{K_A + C_A} \left(1 - \frac{C_B^2}{K_B^2 + C_B^2} \right) \right) - C_B \right], \qquad (5.7.7)$$

where γ_B is the degradation rate of protein B. Solutions to this equation can exhibit bistability, as illustrated in Fig. 5.47(c). First, suppose $C_A = 0$. It can be shown graphically that when $C_B^{\max} > 2K_B$ there exist two stable fixed points, corresponding to low and high concentration levels, separated by an unstable fixed point. As C_A increases, the unstable fixed point and the low concentration or "off" state annihilate in a saddle-node bifurcation at a critical value C_A^*, leaving only the high concentration or "on" state.

The above steady-state behavior can now be used to determine the spatial-dependence of C_B in response to a steady-state morphogen gradient, represented by a monotonically decreasing function $C_{A,s}(x)$. Assuming that the system starts in a state for which $C_B = 0$ for all x, it follows that there exists a point x^* with $C_A(x^*) = C_A^*$ such that C_B is in the on state for $x < x^*$ and in the off state when $x > x^*$, with a sharp jump in $C_B(x)$ at $x = x^*$. Finally, note that if the morphogen signal is subsequently switched off ($C_A = 0$), the basic patterning will persist due to the fact that the on branch still exists at $C_A = 0$. A similar analysis can be carried out in mutual repressor networks exhibiting bistability.

Effects of intrinsic noise on gene expression boundaries. The above example shows how a regulatory gene network with positive feedback, possibly via mutual inhibition, provides a mechanism for forming sharp boundaries in the absence of noise. We would then expect intrinsic noise fluctuations to smooth these boundaries leading to less precision. However, it has been shown, counterintuitively, that noise-induced switching in such networks can actually sharpen tissue boundaries [653, 732, 891]. Here we explore this issue along the lines of Ref. [653].

For the sake of illustration, consider a genetic switch consisting of two genes A and B that repress each other, together with a morphogen M that activates gene A, see Fig. 5.48. We will assume that each of the repressors has two independent binding sites on the promoter of the opposite gene, and when these are both occupied they prevent the binding of RNA polymerase. The morphogen acts independently of the repressor B and is also assumed to have two independent binding sites. The kinetic equations for the concentration C_A and C_B of the repressors, given a concentration C_M for the activator are taken to be [653]

$$\frac{dC_A}{dt} = \gamma_A p_A(C_B, C_M) - k_A x_A, \qquad \frac{dC_B}{dt} = \gamma_B p_B(C_A) - k_B x_B, \qquad (5.7.8)$$

with

$$p_A = \left(1 + \rho_A \left(\frac{1 + C_M/K_M}{1 + \kappa C_M/K_M} \right)^2 \left(1 + \frac{C_B}{K_B} \right)^2 \right)^{-1},$$

$$p_B = \left(1 + \rho_B \left(1 + \frac{C_A}{K_A} \right)^2 \right)^{-1}.$$

Here, ρ_A, ρ_B determine the basal level of gene activation, κ controls the strength of the activator signal with $\kappa > 1$, and K_A, K_B, K_M are dissociation constants. It can be shown that there exist parameter regimes within which the system exhibits bistability as the morphogen concentration is varied [653], see Fig. 5.48(b). In the bistable regime, one stable fixed point has $C_A \gg C_B$ and the other has $C_A \ll C_B$. Thus, we will refer to the two fixed points as the A state and the B state, respectively.

As in our discussion of Fig. 5.47, bistability implies that the final steady state at location x will depend on both the steady-state morphogen concentration $C_M(x)$ and the initial condition of the system. Suppose that the dynamical system is bistable in the range $C_* < C_M < C^*$, see Fig. 5.48(b). Suppose that, before the application of the morphogen signal, the "prepattern" gene B is active uniformly across the domain. It follows that in the presence of the morphogen gradient, the system will switch to "target" gene A at positions x where $C_M(x) > C^*$. Let $T(x)$ denote the time for a cell to change its expression state from B to A at x. Using the well known phenomena of critical slowing down close to a saddle-node bifurcation, we note that $T(x)$ becomes very large as $x \to x^*$ from above, where $C(x^*) = C^*$.

In order to explore the effects of noise, Perez-Carrasco et al. [653] assume that the stochastic version of the model can be treated as a birth–death process for the production and degradation of the proteins, and then carry out a system-size expan-

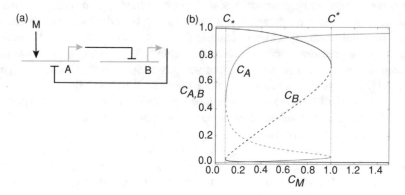

Fig. 5.48: Bistability in a mutual repressor gene network. (a) Gene circuit consisting of a pair of mutually repressing genes A and B, and a a morphogen M that activates gene A. (b) Bifurcation diagram of deterministic kinetic equations. Parameter values are $\gamma_A = \gamma_B = k_A = k_B = 1$, $\rho_A = 1$, $\rho_B = 1.75 \times 10^{-4}$, $K_A = 10^{-3}$, $K_B = 3 \times 10^{-2}$, $K_M = 1$, $\kappa = 10$. [Adapted from Perez-Carrasco et al. [653].]

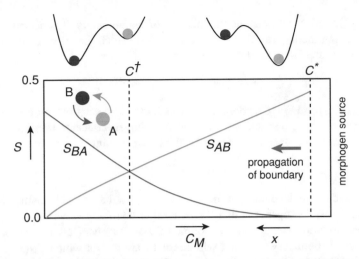

Fig. 5.49: Shift in patterning boundary due to stochastic switching in the bistable domain. Schematic illustration of the variation in the switching terms $S = \Omega^{-1} \ln \tau$ for the transitions $A \to B$ and $B \to A$ as a function of the morphogen concentration C_M. One finds that S_{BA} decreases with C_M, whereas S_{BA} increases with C_M. The stochastic boundary occurs at the concentration $C^\dagger$ at which the transition rates balance. In terms of energy landscapes, the dominant state is the one with a deeper energy potential.

sion to obtain a chemical Langevin approximation. This takes the form

$$\frac{dC_A}{dt} = \gamma_A p_A - k_A C_A + \sqrt{v_A \gamma_A p_A + k_A x_A} \xi_A(t), \qquad (5.7.9a)$$

$$\frac{dC_B}{dt} = \gamma_B p_B - k_B C_B + \sqrt{v_B \gamma_B p_B + k_B x_B} \xi_B(t), \qquad (5.7.9b)$$

where $\xi_j(t)$, $j = A, B$, are Gaussian white noise terms with

$$\langle \xi_j(t) \rangle = 0, \quad \langle \xi_i(t) \xi_j(t') \rangle = \Omega^{-1} \delta(t - t') \delta_{i,j}$$

and Ω denotes the system size. Note the introduction of the additional parameters v_A, v_B, which are used to approximate the effects of expression bursts without explicitly modeling the mRNA dynamics. One can view Ω, v_A, v_B as noise control parameters that do not affect the deterministic dynamics.

There are two major effects of intrinsic noise on the formation of the morphogen gradient [653]. First, noise can speed up the transition to the target gene in the monostable domain, thus counteracting the critical slowing down around x^*. Second, noise-induced switching between the metastable fixed points of the bistable domain moves the deterministic boundary at x^* into the interior of the domain, with the new boundary location $x^\dagger$ determined by the morphogen concentration $C^\dagger$ at which the switching rates $\tau_{A \to B}$ and $\tau_{B \to A}$ between the two metastable states are equal. This point is independent of Ω, v_A and v_B and results in a robust threshold boundary. In order to determine the location of the boundary, it is necessary to use large deviation theory (see Chap. 8). Here we just sketch the basic idea. To leading order in the system size Ω, the transition rates can be written as

$$\tau_{A\to B} \sim e^{\Omega S_{AB}(C_M)}, \quad \tau_{B\to A} \sim e^{\Omega S_{BA}(C_M)}, \tag{5.7.10}$$

where $S_{AB}(C_M)$ and $S_{BA}(C_M)$ denote the so-called minimal actions for the transitions $A \to B$ and $B \to A$, respectively, and the morphogen concentration C_M acts as a parameter. Hence, to leading order in Ω, the balance point is given by

$$S_{AB}(C^\dagger) = S_{BA}(C^\dagger), \tag{5.7.11}$$

with S_{AB} and S_{BA} independent of the system size. The basic mechanism is illustrated in Fig. 5.49. The upshot is that the patterning boundary slowly moves from C^* to $C^\dagger$ in a direction away from the source of morphogens. Roughly speaking, the position of the boundary ay time t will be the location x for which $C_M(x,t) = C(t)$, where $S_{BA}(C(t)) \approx \Omega^{-1}\ln(t/\Gamma)$ and Γ is a prefactor. It follows that the speed of propagation of the barrier in concentration space is

$$v(t) := \frac{dC}{dt} = \left(\frac{dt}{dC}\right)^{-1} \approx \frac{1}{\Omega t}\left(\frac{dS_{BA}}{dC}\right)^{-1} \tag{5.7.12}$$

for large Ω. It can be seen from Fig. 5.49 that dS_{BA}/dC increases as C decreases, so that $v(t)$ decreases monotonically with time t. One word of caution is that since the transitions times are exponentially large, the system may not reach its stochastic equilibrium position over time scales relevant to developmental processes. Finally, note that another possible mechanism for sharpening the boundary in the presence of noise is cell sorting, based on mechanical cell-cell interactions involving adhesion and repulsion [844].

5.8 Exercises

Problem 5.1. Modified mutual repressor model. Consider a simplified mutual repressor model consisting of a single promoter with two operator sites OS_1 and OS_2 that bind to dimers of protein Y and protein X, respectively (see Fig. 5.6). If the dimer of one protein is bound to its site, then this represses the expression of the other protein. However, both sites cannot be occupied at the same time. Hence, the promoter can be in three states O_m, $m = 0, 1, 2$: no dimer is bound to the promoter (O_0); a dimer of protein X is bound to the promoter (O_1); a dimer of protein Y is bound to the promoter (O_2). Assuming that the number of proteins is sufficiently large, we have the transition scheme

$$O_1 \underset{\beta x^2}{\overset{\beta k}{\rightleftharpoons}} O_0 \underset{\beta k}{\overset{\beta y^2}{\rightleftharpoons}} O_2,$$

where β is a transition rate and k is a non-dimensional dissociation constant. Protein X (Y) is produced at a rate κ_p when the promoter is in the states $O_{0,1}$ ($O_{0,2}$), and both proteins are degraded at a rate γ_p in all three states.

(a) Write down the kinetic equations for the probability $P_j(t)$ that the promoter is in state O_j, $j = 0, 1, 2$ at time t.

(b) Assuming that the binding reactions are much faster than the rates of protein production and degradation, write down the quasi- steady-state solutions for P_j^*.

(c) Under the assumption of part (b), show that the kinetic equations for the protein concentrations x and y are given by

$$\frac{dx}{dt} = -\gamma_p x + \kappa_p \frac{k + x^2}{x^2 + y^2 + k},$$

$$\frac{dy}{dt} = -\gamma_p y + \kappa_p \frac{k + y^2}{x^2 + y^2 + k}.$$

(d) Set $\gamma_p = \kappa_p = 1$. Show numerically that the deterministic system is bistable for $0 < k < k_c$ and bistable for $k > k_c$ for $k_c = 4/9$, by plotting the nullclines $\dot{x} = 0$ and $\dot{y} = 0$ in the $x - y$ plane.

Problem 5.2. Oscillations in NF-κB signaling. Consider a model involving a single form of IκB. There are three state variables: x (concentration of nuclear NF-κB), y (concentration of cytosolic IκB), and y_m (concentration of IκB mRNA). Conservation of NF-κB means that the cytosolic concentration of NF-κB is $1 - x_n$. The dimensionless kinetic equations are

$$\frac{dx}{dt} = A \frac{1 - x(t)}{\varepsilon + y(t)} - \frac{Bx(t)y(t)}{\delta + x(t)},$$

$$\frac{dy}{dt} = y_m(t) - I(t) \frac{(1 - x(t))y(t)}{\varepsilon + y(t)},$$

$$\frac{dy_m}{dt} = x(t)^2 - y_m(t),$$

where $I(t)$ represents an extracellular signal.

(a) Numerically simulate the above system of ODES using the following parameter values: $A = 0.007, B = 954.5, \delta = 0.029, \varepsilon = 5 \times 10^{-5}$ and constant input $I = 0.035$. Verify that $x(t)$ exhibits spike-like oscillations. Given that the model is expressed in time units 0.017^{-1} minutes (determined by degradation rate of NF-κB), what is the period of oscillations?

(b) Show that the oscillations become smoother when I is tripled.

(c) The pair of isoforms IκBβ and IκBε that are not dependent on NF-κB for synthesis, can be incorporated into the model by including a constant background rate c_0 of IκB production, that is,

$$\frac{dy_m}{dt} = c_0 + x(t)^2 - y_m(t).$$

Verify that when $c_0 = 0.005$, the system still exhibits sustained oscillations but with diminished amplitude. What happens when c_0 is increased to 0.02?

Problem 5.3. Stochastic model of autoregulation I. Carry out the steps in the analysis of protein bursting in an autoregulatory repressor network. The matrix equation for the covariance matrix is

$$\mathbf{M}\Sigma + \Sigma\mathbf{M}^T = -\mathbf{D},$$

with

$$\mathbf{M} = \begin{pmatrix} -\gamma & \mu \\ \kappa_p & -\gamma_p \end{pmatrix}, \quad \mathbf{D} = \begin{pmatrix} \kappa g(x_2^*) + \gamma x_1^* & 0 \\ 0 & \kappa_p x_1^* + \gamma_p x_2^* \end{pmatrix}, \quad \mu = \kappa_p g'(x_2^*).$$

Note that $\mu > 0$ for an activator and $\mu < 0$ for a repressor.

1. Write down the equations in component form. Since $\Sigma_{12} = \Sigma_{21}$, there are three independent equations.

2. Solve for Σ_{11} and Σ_{22} in terms of x_2^* and Σ_{12} and substitute into the remaining equation. (Use the fixed point equations $\kappa g(x_2^*) = \gamma x_1^*$ and $\kappa_p x_1^* = \gamma_p x_2^*$.)

3. After some algebra, express Σ_{12} in terms of x_2^*, and hence show that

$$\Sigma_{22} = \left[1 + \frac{b}{1+\eta} \frac{1-\phi}{1+b\phi} \right] x_2^*,$$

where

$$b = \frac{\kappa_p}{\gamma}, \quad \phi = -\frac{\mu}{\gamma_p}, \quad \eta = \frac{\gamma_p}{\gamma}.$$

Problem 5.4. Stochastic model of autoregulation II. Consider a stochastic version of the autoregulatory network shown in Fig. 5.4, in which the number of mRNA molecules is in quasi-equilibrium so that the only dynamical variable is the number of protein molecules n. There are then two basic reactions (production and degradation of protein), which we model as

$$n \xrightarrow{f_1} n+b, \quad f_1 = g(n/\Omega), \quad S_1 = b,$$
$$n \xrightarrow{f_2} n-1, \quad f_2 = \gamma_p n/\Gamma, \quad S_2 = -1.$$

Here γ_p is the protein degradation rate, the function $g(n/\Omega)$ represents the effects of feedback, and it is assumed that each synthesis event produces b proteins. The integer b can be interpreted as a mean burst size.

(a) Write down the master equation that takes into account the effects of molecular noise. Then perform a system-size expansion to obtain the following FP equation for $x = n/\Omega$:

$$\frac{\partial p}{\partial t} = -\frac{\partial A(x)p(x,t)}{\partial x} + \frac{1}{2\Omega}\frac{\partial^2 D(x)p(x,t)}{\partial x^2},$$

where

$$A(x) = bg(x) - \gamma_p x, \quad D(x) = b^2 g(x) + \gamma_p x.$$

(b) Let x^* denote the single stable fixed point x^* for which $bg(x^*) = \gamma_p x^*$. Linearizing about the fixed point by setting $x(t) = x^* + y(t)/\sqrt{\Omega}$, derive differential equations for the mean $\langle y \rangle$ and variance σ_y^2

$$\frac{d\langle y \rangle}{dt} = (bg'(x^*) - \gamma_p)\langle y \rangle, \quad \frac{d\sigma_y^2}{dt} = 2(bg'(x^*) - \gamma_p)\sigma_y^2 + (b^2 g(x^*) + \gamma_p x^*).$$

(c) Hence show that the stationary Fano factor of the autoregulatory feedback model is

$$\frac{\sigma_x^2}{x^*} = \frac{b+1}{2}\frac{\gamma_p}{|bg'(x^*) - \gamma_p|}.$$

Interpret the result.

Problem 5.5. Linear noise approximation for a population of cells. Consider a population of n_{max} identical cells each with a single copy of the same gene. Let $n_1(t)$ denote the number of active genes and $n_2(t)$ the number of proteins. Setting $x_j = \langle n_j \rangle / \Omega$, where Ω is the system size, the various reactions and the corresponding rate equations based on mass action (valid in the limit $\Omega \to \infty$) are as follows:

1. *Gene activation and inactivation*

$$n_1 \xrightarrow{k_+(n_{max}-n_1)} n_1 + 1, \quad n_1 \xrightarrow{k_- n_1} n_1 - 1,$$

with

$$\frac{dx_1}{dt} = k_+(x_{max} - x_1) - k_- x_1.$$

2. *Protein production and degradation*

$$n_2 \xrightarrow{\kappa_p n_1} n_2 + 1, \quad n_2 \xrightarrow{\gamma_p n_2} n_2 - 1,$$

with

$$\frac{dx_2}{dt} = \kappa_p x_1 - \gamma_p x_2.$$

(a) Write down the master equation for the joint probability distribution $P(n_1, n_2, t)$.

(b) Identify the stoichiometric coefficients and the propensities for the four single-step reactions, and hence write down the 2D Fokker-Planck equation for $p(x_1, x_2, t)$ in the large Ω regime.

(c) Write down the corresponding chemical Langevin equation. Verify that the deterministic rate equations are obtained in the thermodynamic limit $\Omega \to \infty$. Show that there is a unique stable fixed point (x_1^*, x_2^*).

(d) Linearize the Langevin equation about the fixed point by setting $X_i(t) = x_i^* + \Omega^{-1/2}Y_i(t)$, and obtain a corresponding OU process for $Y_i(t)$. Hence, establish that the stationary covariance matrix Σ satisfies the equation

$$M\Sigma + \Sigma M^T = -D,$$

with

$$M = \begin{pmatrix} -(k_+ + k_-) & 0 \\ \kappa_p & -\gamma_p \end{pmatrix}, \qquad D = \begin{pmatrix} 2k_- x_1^* & 0 \\ 0 & 2\kappa_p x_1^* \end{pmatrix}.$$

(e) By solving the matrix equation for the covariance, obtain the following Fano factors:

$$\frac{\text{var}[n_1]}{\langle n_1 \rangle} = \frac{k_-}{k_+ + k_-} = 1 - \langle n_1 \rangle / n_{\max},$$

$$\frac{\text{var}[n_2]}{\langle n_2 \rangle} = 1 + \frac{\kappa_p}{k_+ + k_- + \gamma_p} \frac{\text{var}[n_1]}{\langle n_1 \rangle}.$$

Note that $\langle n_j \rangle = \Omega x_j^*$ and $\text{var}[n_j] = \Omega \Sigma_{jj}$. We immediately see that the presence of a transcription factor increases the Fano factor of the protein above one.

Problem 5.6. Frequency domain analysis of a simple gene network. Consider a simple model of protein translation given by the stochastic kinetic equations

$$\frac{dX}{dt} = \kappa - \gamma X + \eta(t), \qquad \frac{dY}{dt} = \kappa_p x - \gamma_p Y + \eta_p(t),$$

where $X(t)$ and $Y(t)$ are the concentrations of mRNA and protein, γ, γ_p are degradation rates, κ is the rate of mRNA production, and κ_p is the rate of protein production. Moreover, $\eta(t)$ and $\eta_p(t)$ are independent white noise terms with $\langle \eta \rangle = \langle \eta_p \rangle = 0$, and

$$\langle \eta(t)\eta(t') \rangle = q\delta(t - t'), \quad \langle \eta_p(t)\eta_p(t') \rangle = q_p\delta(t - t'), \quad \langle \eta(t)\eta_p(t') \rangle = 0.$$

(a) Introducing the Fourier transforms

$$\widehat{\eta}(\omega) = \int_{-\infty}^{\infty} e^{i\omega t} \eta(t)dt, \quad , \quad \eta(t) = \int_{-\infty}^{\infty} e^{-i\omega t} \widehat{\eta}(\omega) \frac{d\omega}{2\pi},$$

show that

$$\langle \eta(\omega)\eta(\omega') \rangle = 2\pi q\delta(\omega + \omega').$$

(b) By linearizing about the steady state $x^* = \kappa/\gamma, y^* = \kappa\kappa_p/(\gamma\gamma_p)$ and using Fourier transforms show that the power spectra of the fluctuations $\widehat{X}(t) = X(t) - x^*$ and $\widehat{Y}(t) = Y(t) - y^*$ are given by (after dropping the tildes)

$$S_{XX}(\omega) = \frac{q}{\omega^2 + \gamma^2}, \quad S_{YY}(\omega) = \frac{q_p}{\omega^2 + \gamma_p^2} + \frac{\kappa_p^2 q}{(\omega^2 + \gamma^2)(\omega^2 + \gamma_p^2)}.$$

(c) Using the definition of the power spectrum, written in the form

$$\langle X(t)^2 \rangle = \int_{-\infty}^{\infty} S_{XX}(\omega) \frac{d\omega}{2\pi},$$

show that

$$\langle X(t)^2 \rangle = \frac{q}{2\gamma}.$$

Similarly, show that

$$\langle Y(t)^2 \rangle = \frac{q_p}{2\gamma} + \frac{\kappa_p^2 q}{2\gamma^2 \gamma_p} + O(\gamma^{-3}).$$

Hint: You should assume that $\gamma \gg \gamma_p$ and use the result

$$\int_{-\infty}^{\infty} \frac{1}{\omega^2 + a^2} \frac{d\omega}{2\pi} = \frac{1}{2a}.$$

(d) From the linear noise approximation, one obtains the following Fano factors for the number $M(t)$ of mRNA and number $N(t)$ of proteins:

$$\frac{\mathrm{var}[M]}{\langle M \rangle} = 1, \quad \frac{\mathrm{var}[N]}{\langle N \rangle} = 1 + b,$$

where $b = \kappa_p/\gamma$. Use this to determine q and q_p.

Problem 5.7. Computer simulations. Use the SSA to numerically investigate the following systems:

(a) The model of regulated transcription considered in problem [5.7]. There are two discrete variables (number of active genes n_1, and number of mRNA molecules n_2) and four reactions (gene activation and deactivation, mRNA production and degradation). Take the parameter values $k_+ = 0.03\,\mathrm{min}^{-1}, k_- = 0.2\,\mathrm{min}^{-1}, \kappa = 10\,\mathrm{min}^{-1}, \gamma = 0.2\,\mathrm{min}^{-1}$ and consider the two cases $n_{\max} = 10$, $n_{\max} = 100$. Run the simulations for sufficient time to reach steady state. Plot a histogram of $n_1(T)$ and $n_2(T)$ based on 100 simulations, say, where T is the final time. Determine the mean and variance, and compare the numerical Fano factor with the theoretical expressions based on the diffusion approximation. [In this problem all factors of the system size Ω cancel, that is, $\Omega f_a(\mathbf{n}/\Omega) = f_a(\mathbf{n})$.]

(b) Consider a toggle switch with promoters states in quasi-equilibrium, whose deterministic kinetic equations are given by

$$\frac{dx_1}{dt} = -x_1 + \frac{\kappa_p}{1+x_2^4}, \quad \frac{dx_2}{dt} = -x_2 + \frac{\kappa_p}{1+x_1^4}. \tag{5.8.13}$$

The latter is bistable for any $\kappa_p > 1$. What is the origin of the quartic powers? Determine the stoichiometric coefficients and the propensities. Run simulations of a stochastic version of the model, in which fluctuations in the number of protein molecules are taken into account. (Set the system size $\Omega = 1$.) Consider the cases $\kappa_p = 5, 50, 500, 5000$, (and at least 10,000 reaction steps for each). Verify that the system exhibits bistability for $\kappa_p = 5000$, whereas noise dominates at κ_p so there is no bistability. What happens for $\kappa_p = 50, 500$?

(c) The circadian clock model with stoichiometry and propensities are listed in Table 1. Use the following parameter values: $\kappa = 0.5\,\mathrm{nM}\,\mathrm{h}^{-1}, \gamma = 0.3\,\mathrm{nM}\,\mathrm{h}^{-1}, K_m = 2.0\,\mathrm{nM}, K_m' = 0.2\,\mathrm{nM}, \kappa_p = 2.0\,\mathrm{h}^{-1}, \gamma_p = 1.5\,\mathrm{nM}\,\mathrm{h}^{-1}, K_p = 0.1\,\mathrm{nM}, k_1 = k_2 = 0.2\,\mathrm{h}^{-1}$. Take the system size $\Omega = 100$. Plot a sample trajectory of the number of mRNA M and the number of cytosolic clock proteins X_C as a function of time, and check that the oscillation period is around 22 hours. Compare with solutions of the deterministic kinetic rate equations. Also plot several sample trajectories in the (M, X_C)-phase-plane superimposed on the deterministic limit-cycle.

Problem 5.8. Bistability in an autocatalytic reaction. Consider the following nonlinear, autocatalytic reaction scheme for a protein that can exists in two states X and Y:

$$X \underset{k_2}{\overset{k_1}{\rightleftharpoons}} Y, \quad X + 2Y \overset{k_3}{\rightarrow} 3Y.$$

Let $[X]$ and $[Y]$ denotes the concentrations of the molecule in each of the two states such that $[X] + [Y] = Y_{\mathrm{tot}}$ fixed. The kinetic equation for $[Y]$ is

$$\frac{d[Y]}{dt} = -k_2[Y] + k_1[X] + k_3 V^2 [Y]^2 [X],$$

where V is cell volume.

(a) Let $y = [Y]/Y_{\mathrm{tot}}$. Show that after an appropriate rescaling of time, the corresponding kinetic equation for y is

$$\frac{dy}{dt} = y(\mu(1-y)y - 1) + \lambda(1-y),$$

where $\mu = k_3 Y_{\mathrm{tot}}^2/k_2, \lambda = k_1/k_2$. Determine the existence and stability of the fixed points for y. Plot the bifurcation diagram with μ treated as a bifurcation parameter and $\lambda = 0.03$. Hence, show that the system is bistable over a range of values of μ.

(b) Suppose that there are N molecules, that is, $N = V Y_{\mathrm{tot}}$, where V is cell volume. Construct the master equation for the probability $P(n,t)$ that there are $n(t) = n$ molecules in state Y at time t.

(c) Using equation (3.3.15), show that the steady-state distribution is

$$P_s(n) = \frac{C_N N!}{n!(N-n)!} \prod_{m=0}^{n-1} \left[\lambda + \frac{\mu}{N^2} m(m-1)\right].$$

Plot $P_s(n)$ as a function of n (treated as a continuous variable over the range $[0,400]$) for $N = 400$, $\mu = 4.5$ and $\mu = 6$ with $\lambda = 0.03$. Comment on the location of the peaks in terms of fixed points of the deterministic system.

(d) Derive the corresponding Fokker-Planck equation using a Kramers-Moyal expansion, and determine the steady-state solution. Calculate the steady-state solution and compare with the exact solution of part (c) for $N = 40$ and $N = 400$.

Problem 5.9. The Brusselator. The Brusselator is an idealized model of an autocatalytic reaction, in which at least one of the reactants is also a product of the reaction. The model consists of two chemical species X and Y interacting through the following reaction scheme:

$$\emptyset \xrightarrow{a} X, \quad X \xrightarrow{b} Y, \quad 2X + Y \xrightarrow{c} 3X, \quad X \xrightarrow{d} \emptyset.$$

These reactions describe the production and degradation of an X molecule, an X molecule spontaneously transforming into a Y molecule, and two molecules of X reacting with a single molecule of Y to produce three molecules of X. The corresponding mass-action kinetic equations for $u_1 = [X], u_2 = [Y]$ are (after rescaling so that $c = d = 1$)

$$\frac{du_1}{dt} = a - (b+1)u_1 + u_1^2 u_2,$$

$$\frac{du_2}{dt} = bu_1 - u_1^2 u_2.$$

(a) Determine the stability of the fixed point at $u_1^* = a, u_2^* = b/a$. Hence establish that the fixed point undergoes a Hopf bifurcation at the critical value $b = a^2 + 1$.

(b) Let $n_1(t)$ and $n_2(t)$ denote the number of X and Y molecules at time t, respectively, and take Ω to be cell volume. Determine the stoichiometric coefficients and propensities of the four single-step reactions, and use this to construct the Brusselator master equation.

(c) Use Gillespie's SSA to simulate the stochastic Brusselator. Take parameter values (in time^{-1}) $a = 50$, $b = 50$, $c = 1$ and $d = 5$. Set the system size $\Omega = 100$. Take the initial conditions $n_1 = 1000, n_2 = 2000$. Plot both the solutions $n_1(t), n_2(t)$ in the time domain, and in the phase-plane.

(d) Now suppose that Ω is sufficiently large so that we can carry out a system-size expansion of the master equation. Set $c = d = 1$. Linearize the resulting FP equation about the fixed point (u_1^*, u_2^*), with $u_1^* = a, u_2^* = b/a$, to obtain the multivariate FP equation for an OU process with

$$\mathbf{M} = \begin{pmatrix} b-1 & a^2 \\ -b & -a^2 \end{pmatrix}, \quad \mathbf{D} = \begin{pmatrix} 2(b+1)a & -2ba \\ -2ba & 2ba \end{pmatrix}.$$

(e) Using the results of part (d), write down the Langevin equation for the stochastic Brusselator under the linear noise approximation. Show that the components of the spectrum are given by

$$S_1(\omega) = 2a((1+b)\omega^2 + a^4)\Gamma(\omega)^{-1}, \quad S_2(\omega) = 2ab(\omega^2 + 1 + b)\Gamma(\omega)^{-1},$$

where

$$\Gamma(\omega) = (a^2 - \omega^2)^2 + (1 + a^2 - b)^2\omega^2.$$

Problem 5.10. Noise-induced oscillations in a relaxation oscillator. Consider a stochastic relaxation oscillator involving an activator and repressor. The activator enhances expression of both proteins , whereas the repressor acts by binding to the activator to form an inert complex. Let N_A, N_R, and N_C denote the number of activator A, repressor R, and complex C molecules, respectively. Denote the corresponding concentrations by $x_A = N_A/\Omega$, etc. There are four reactions involving production and degradation of proteins with burst sizes b_A and b_R:

$$N_A \xrightarrow{f_A} N_A + b_A, \quad N_R \xrightarrow{f_R} N_R + b_R$$
$$N_A \xrightarrow{g_A} N_A - 1, \quad N_A \xrightarrow{g_A} N_A - 1,$$

with propensities

$$f_A = \kappa_A \frac{\alpha_0 + x_A/K_A}{1 + x_A/K_A}, \quad f_R = \kappa_R \frac{x_R/K_R}{1 + x_R/K_R}, \quad g_A = \gamma_A x_A, \quad g_R = \gamma_R x_R.$$

There are two more reactions involving the formation of the complex, and degradation of the complex to R due to degradation of A within the complex:

$$A + R \rightarrow C, \quad C \rightarrow R,$$

with propensities $k_C x_A x_R$ and $\gamma_A x_C$, respectively. It follows that the kinetic equations are:

$$\frac{dx_A}{dt} = b_A f_A(x_A) - k_C x_A x_R - k_A x_A,$$
$$\frac{dx_R}{dt} = b_R f_R(x_R) - k_C x_A x_R + k_A x_C - \gamma_R x_R,$$
$$\frac{dx_A}{dt} = k_C x_A x_R - k_A x_C.$$

(a) Numerically simulate the deterministic system for the following parameter values: $\kappa_A = 50, b_A = 5, \gamma_R = 5, b_R = 10, K_A = 0.5, K_R = 1, \gamma_A = 1, \gamma_R = 0.1, \alpha_0 = 0.1, k_C = 200$. Show that the system is excitable by running two simulations, one from the initial conditions $(x_A, x_R, x_C) = (0, 10, 35)$, and another from initial conditions $(x_A, x_R, x_C) = (5, 10, 35)$. What changes about the behavior if $\gamma_R = 0.2$?

(b) Set the system size $\Omega = 1$. For the same parameter values as a) with $\gamma_R = 0.1$ use Gillespie's SSA to numerically run simulations of the stochastic version of the model based on the chemical master equation. Explain the observed differences in behavior between the deterministic and stochastic versions of the model.

Problem 5.11. Binary response in stochastic gene expression. Consider the stochastic model of gene expression in which the gene randomly switches between an active and inactive state. The steady-state probability densities $p_{0,1}(x)$ for protein concentration x when the gene is in an active ($j = 1$) or inactive ($j = 0$) state satisfy the pair of equations

$$\frac{d}{dx}(-\gamma_p x p_0(x)) = k_- p_1(x) - k_+ p_0(x),$$

$$\frac{d}{dx}([\kappa_p - \gamma_p x] p_1(x)) = k_+ p_0(x) - k_- p_1(x),$$

with boundary conditions $p_0(\kappa_p/\gamma_p) = 0$ and $p_1(0) = 0$.

(a) Derive the normalization conditions

$$\int_0^{\kappa_p/\gamma_p} p_0(x) dx = \frac{k_-}{k_- + k_+}, \quad \int_0^{\kappa_p/\gamma_p} p_1(x) dx = \frac{k_+}{k_- + k_+}.$$

(b) By adding the pair of steady-state equations show that one solution is

$$p_0(x) = \frac{\kappa_p - \gamma_p x}{\gamma_p x} p_1(x).$$

(c) Substituting for $p_0(x)$, solve the resulting differential equation for $P(x) = (\kappa_p - \gamma_p x) p_1(x)$, and thus obtain the solution

$$p_0(x) = C(\gamma_p x)^{-1+k_+/\gamma_p}(\kappa_p - \gamma_p x)^{k_-/\gamma_p}, \quad p_1(x) = C(\gamma_p x)^{k_+/\gamma_p}(\kappa_p - \gamma_p x)^{-1+k_-/\gamma_p}.$$

(d) Using part (c), show that

$$\int_0^{\kappa_p/\gamma_p} p_0(x) dx = \frac{C}{\gamma_p} \kappa_p^{(k_+ + k_-)/\gamma_p} B(k_+/\gamma_p, 1 + k_-/\gamma_p),$$

$$\int_0^{\kappa_p/\gamma_p} p_1(x) dx = \frac{C}{\gamma_p} \kappa_p^{(k_+ + k_-)/\gamma_p} B(1 + k_+/\gamma_p, k_-/\gamma_p),$$

where $B(\alpha, \beta)$ is the Beta function:

$$B(\alpha, \beta) = \int_0^1 t^{\alpha-1}(1-t)^{\beta-1} dt.$$

(e) Using the standard property

$$B(\alpha, \beta) = \frac{\Gamma(\alpha)\Gamma(\beta)}{\Gamma(\alpha+\beta)},$$

show that the solution in part (c) satisfies the normalization conditions provided that

$$C = \gamma \left[\kappa_p^{(k_+ + k_-)/\gamma_p} B(k_+/\gamma_p, k_-/\gamma_p) \right]^{-1}.$$

Problem 5.12. Promoter noise in a mutual repressor network. Consider the mutual repressor network of Fig. 5.7. Suppose that the number of proteins X and Y are n_1 and n_2, respectively, and assume that they are finite. The state transition diagram for the three promoter states is now

$$O_1 \underset{n_1(n_1-1)\beta}{\overset{\beta K'}{\rightleftharpoons}} O_0 \underset{\beta K'}{\overset{n_2(n_2-1)\beta}{\rightleftharpoons}} O_2,$$

where β is a transition rate and K' is a non-dimensional dissociation constant. Let $p_m(n_1, n_2, t)$, $m = 0, 1, 2$, be the probability that there are $N_1(t) = n_1$ proteins Y, $N_2(t) = n_2$ proteins X, and the promoter is in state $M(t) = m$ at time t. The master equation for the full system takes the form

$$\frac{d}{dt} p_m(n_1, n_2, t) = \sum_{l=0,1,2} A_{ml} p_l(n_1, n_2, t) + \sum_{n_1', n_2'} W_{n_1 n_2, n_1' n_2'}^m p_m(n_1', n_2', t).$$

(a) Carry out a system-size expansion of the sums $\sum_{n_1', n_2'} W_{n_1 n_2, n_1' n_2'}^m p_m(n_1', n_2', t)$, $m = 0, 1, 2$, with respect to the mean number $N = \kappa_p/\gamma_p$ of proteins. First, introduce the rescaling $t \to t\gamma_p$ and set $y = n_1/N, x = n_2/N$. Keeping only the first terms in the Taylor expansions, derive the CK equation

$$\frac{\partial p_m}{\partial t} = -\frac{\partial F_m(x) p_m}{\partial x} - \frac{\partial G_m(y) p_m}{\partial y} + \frac{1}{\varepsilon} \sum_{l=0,1,2} A_{ml}(x, y) p_l$$

for $p_m(x, y, t)$ with

$$F_0(x) = 1 - x, \quad F_1(x) = -x, \quad F_2(x) = 1 - x,$$
$$G_0(y) = 1 - y, \quad G_1(y) = 1 - y, \quad G_2(y) = -y,$$

$$A = \begin{pmatrix} -x^2 - y^2 & K & K \\ y^2 & -K & 0 \\ x^2 & 0 & -K \end{pmatrix},$$

and $\varepsilon = \gamma_p^3/\beta \kappa_p^2$ and $K = K' \gamma_p^2/\kappa_p^2$ are dimensionless parameters.

(b) Using part (a), show that we recover the deterministic rate equation (5.1.17) in the fast switching limit.

Problem 5.13. Single-server queue $M/G/1$. Consider the queue $M(\lambda)/G/1$ consisting of a single server, in which individual customers arrive according to a Poisson process with intensity λ and the waiting time to service a customer is more general than exponentially distributed (non-Markovian). Let τ_n be the time of departure of the n-th customer and let $Q(\tau_n)$ be the number of customers waiting in line on the n-th customer's departure. If $Q(\tau_n) > 0$ then the $(n+1)$-th customers begins to be serviced immediately at time τ_n. During the corresponding random service time S_{n+1} another R customers arrive and join the waiting line. Hence,

$$Q(\tau_{n+1}) = R + Q(\tau_n) - 1, \quad \text{if } Q(\tau_n) > 0.$$

On the other hand, if $Q(\tau_n) = 0$, then the server has to wait for the $(n+1)$-th customer to arrive before starting work. Again the service takes a random time S_{n+1}, during which another R customers arrive, so that

$$Q(\tau_{n+1}) = R \quad \text{if } Q(\tau_n) = 0.$$

Combining these two results, we have the iterative equation

$$Q(\tau_{n+1}) = R + Q(\tau_n) - H(Q(\tau_n)),$$

where $H(q)$ is a Heaviside function and the number of arrivals R is generated according to the given Poisson process. Since R only depends on the service time S_{n+1} and is independent of $Q(\tau_n)$, it follows that $Q(\tau_n)$ evolves according to a Markov chain (see Box 3A). The Markov chain is irreducible and aperiodic. Recall that a Markov chain X is completely determined by its transition probabilities

$$K_{ji} = \mathbb{P}(X_{n+1} = j | X_l = i).$$

Given a service time $S_{n+1} = S$, we have

$$K_{j0} = \mathbb{P}[Q(\tau_{n+1}) = j | Q(\tau_n) = 0] = \mathbb{E}[\mathbb{P}[R = j | S]] = \mathbb{E}\left[\frac{(\lambda S)^j}{j!} e^{-\lambda S}\right] := \omega_j.$$

Similarly, if $i \geq 1$ then

$$K_{ji} = \mathbb{P}[Q(\tau_{n+1}) = j | Q(\tau_n) = i] = \mathbb{E}[\mathbb{P}[R = j - i + 1 | S]] = \mathbb{E}\left[\frac{(\lambda S)^{j-i+1}}{(j-i+1)!} e^{-\lambda S}\right]$$

$$= \omega_{j-i+1}$$

for $j - i + 1 \geq 0$ and zero if $j - i + 1 < 0$. That is $\omega_{-k} = 0$ for $k > 0$.

(a) A stationary distribution p_j^* exists if it satisfies

$$p_j^* = \sum_{i \geq 0} K_{ji} p_i^*,$$

with $p_j^* \geq 0$ for all j and $\sum_j p_j^* = 1$. Substituting for K_{ji} show by iteration that if p_0^* is known, then there exists a unique solution for $p_j^*, j > 0$.

(b) Summing the stationary equation for $j = 0, \ldots, n$ and rearranging, show that

$$p_{n+1}^* \omega_0 = \varepsilon_n p_0^* + \sum_{i=1}^{n} \varepsilon_{n+i-1} p_i^*, \quad n \geq 0,$$

where $\varepsilon_n = 1 - \omega_0 - \omega_2 - \ldots - \omega_n$. Also show that $\sum_i \omega_i = 1$ and hence $\varepsilon_n > 0$. Finally, use proof by induction to establish that if $p_0^* \geq 0$ then $p_j^* \geq 0$ for all $j \geq 0$.

(c) Introducing the generating functions

$$G(s) = \sum_j s^j p_j^*, \quad \Delta(s) = \sum_j s^j \omega_j,$$

use equation (5.4.13) to show that

$$G(s) = \frac{p_0^*(s-1)\Delta(s)}{s - \Delta(s)}.$$

The existence of a stationary solution then reduces to the conditions $p_0^* > 0$ and $G(s) \to 1$ as $s \to 1^-$. Using L'Hopital's rule, show that both conditions are satisfied if $p_0^* = 1 - \Delta'(1) > 0$, that is, $\Delta'(1) < 1$.

(d) Show that

$$\Delta(s) = \mathbb{E}[\exp(\lambda S(s-1))] = \Gamma(\lambda(s-1)),$$

where Γ is the moment generating function for the process S. Hence, deduce that a stationary density exists if and only if $\lambda \mathbb{E}[S] := \rho < 1$.

Problem 5.14. Kinetic proofreading in protein synthesis. Consider the kinetic proofreading model given by the modified Michaelis-Menten reaction kinetics of equation (5.5.5).

(a) Write down the kinetic equations for the evolution of the concentrations $[EC]$ and $[EC^*]$.

(b) Show that the steady-state concentration of the modified enzyme-substrate complex $[EC^*]$ is

$$[EC^*] = \frac{1}{q_{\text{off}} + W} \frac{r k_{\text{on}}}{k_{\text{off}} + r} [E][C].$$

(c) Repeating the analysis for the incorrect substrate D, derive the following approximation for the error rate:

$$F = \frac{Q_C}{Q_D} \frac{K_C}{K_D} = e^{-\Delta G_{CD}/k_B T} e^{-\Delta G_{C^* D^*}/k_B T}.$$

Problem 5.15. Kinetic proofreading in a T-cell receptor. T cells, which mature in the thymus, are one of two key cell types of the adaptive immune system, and whose basic function is the detection and destruction of intracellular pathogens such as certain bacteria and all viruses [192, 322, 508]. (The other cell type consists of B cells, which mature in the bone marrow, and are mainly concerned with the detection and destruction of extracellular pathogens.) In order to execute their function, T cells

(a)

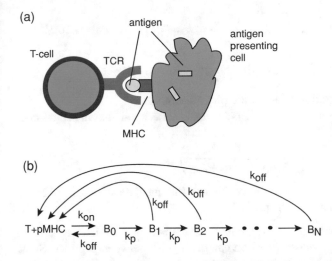

Fig. 5.50: Kinetic proof-reading model of T-cell activation. (a) Schematic diagram of a T-cell receptor (TCR) binding to an antigen that is attached to a major histocompatibility complex molecule (MHC) in the surface of an antigen presenting cell. (b) Reaction diagram, see text for details.

scan the surfaces of cells for molecular markers of infection. Detection of the appropriate marker activates the T cell which then responds to the pathogen, either by killing the infected cell (effector T cells) or by signaling other parts of the immune system such as B cells (helper T cells). Since T cells only scan the surface of other cells, it is necessary that some cells are able to present information regarding their internal contents to the surface. This is achieved by cutting intracellular proteins into peptide fragments, and transporting these fragments to the surface for surveillance by T cells. If a pathogen is present within the cell, then signature peptide groups known as antigens will be made accessible. A major challenge for the pathogen recognition machinery is that the vast majority of peptides on a given antigen presenting cell do not signify the presence of a pathogen. Thus, a T cell has to recognize an antigen against a noisy background of these so-called self- peptides, just as a ribosome has to recognize the correct tRNA during each stage of protein synthesis. Both processes implement a kinetic proofreading mechanism for error correction. Here we consider a kinetic proofreading model for T-cell activation introduced by McKeithan [566].

The model of McKeithan considers the interaction of a T-cell receptor (TCR) with a ligand consisting of a peptide fragment that is bound to a specialized molecule in the surface of an antigen presenting cell, known as a major histocompatibility complex molecule (MHC), see Fig. 5.50(a). The peptide-MHC complex that forms the ligand is denoted by pMHC. There are two basic assumptions of the model: (i) In order to respond to an antigen, a TCR in an inactive state T has to undergo a sequence of N modifications to form the activated state B_N. (ii) Dissociation of pMHC from the TCR can occur at any stage, after which the receptor quickly returns to its inactive state, see Fig. 5.50(b). Suppose that the off rate back to the inactive state T is the same for all intermediate states. We then have the following hierarchy of kinetic equations for the concentrations $[T], [B_j], j = 0, \ldots, N$:

$$\frac{d[T]}{dt} = -k_{on}[T][P] + k_{off}\sum_{i=0}^{N}[B_i], \quad \frac{d[B_0]}{dt} = k_{on}[T][P] - k_{off}[B_0] - k_p[B_0],$$

$$\frac{d[B_i]}{dt} = k_p([B_{i-1}] - [B_i]) - k_{off}[B_i], \quad \frac{d[B_N]}{dt} = k_p[B_{N-1}] - k_{off}[B_N],$$

where $[P]$ is the concentration of a specific pMHC complex.

(a) Solving the above system of equations in steady state, show that the fraction of activated complexes is

$$\frac{[B_N]}{\sum_{i=0}^{N}[B_i]} = \left(\frac{k_p}{k_p + k_{off}}\right)^N.$$

Note that $k_p/(k_p + k_{off})$ is the probability that any intermediate step i, the T cell is modified before dissociation of the pMHC.

(b) Suppose that k_p is independent of the particular substrate, such that the off rate k_{off} is the only parameter whose variation can distinguish between peptides. Show that as the number of steps N increases, the fraction of activated T cells becomes more sensitive to small changes in k_{off} (increased selectivity), reflecting the objective of the kinetic proofreading mechanism. Does an increase in selectivity come at a cost?

Problem 5.16. Model of transcriptional elongation. Consider the birth–death master equation for the elongation phase of transcription in the absence of backtracking.

(a) Starting from the backward master equation (5.5.14) derive a difference equation for the second moment $T_2(n_0)$ of the elongation time, where n_0 is the starting position along the chain, analogous to the difference equation (5.5.15) for the first moment.

(b) Solve the difference equation in part (a) recursively by introducing the variable $U_2(n_0) = T_2(n_0) - T_2(n_0 - 1)$.

(c) Using the result from part (b) and the formula (5.5.17) for the mean elongation time, determine the variance σ^2 of the elongation time in terms of $K = \omega_-/\omega_+$ and ω_+, where $\omega_\pm$ are the effective polymerization/depolymerization rates, and show that when $K \ll 1$,

$$\sigma^2 = \frac{N}{\omega_+^2} + 4K\frac{N-1}{\omega_+^2} + O(K^2).$$

Problem 5.17. Lambda phage. Recall from Sect. 5.6 that lysogeny is maintained by a regulatory network involving λ phage DNA, a pair of regulatory proteins CI and Cro, and an operator complex OR consisting of three binding sites OR_j, $j = 1, 2, 3$, overlapping with two promoter sites P_{RM} and P_R. Suppose that there are five promoter states O_j, $j = 0, 1, 2, 3, 4$:

O_0 : all operator sites are unoccupied

O_1 : OR_1 and OR_2 both occupied by a CI dimer

O_2 : all operator sites are occupied by a CI dimer

O_3 : OR_3 occupied by a Cro dimer

O_4 : OR_3 and at least one of the other operator sites is occupied by a Cro dimer

The reaction scheme for the promoter states is

$$O_0 \underset{k_{-1}}{\overset{k_1 x^2}{\rightleftharpoons}} O_1, \quad O_1 \underset{k_{-2}}{\overset{k_2 x}{\rightleftharpoons}} O_2, \quad O_0 \underset{k_{-3}}{\overset{k_3 y}{\rightleftharpoons}} O_3, \quad O_3 \underset{k_{-4}}{\overset{k_4 y}{\rightleftharpoons}} O_4.$$

(a) Use mass-action kinetics to derive the master equation for the probabilities $P_m(t)$, where $P_m(t) = \mathbb{P}[\text{promoter in state } O_m \text{ at time } t]$. Hence, determine the steady-state distributions P_m^* with $\sum_{j=0}^4 P_j^* = 1$.

(b) Suppose that the state O_0 produces CI at a rate R_{RM} and the state O_1 produces CI at a rate $\widehat{R}_{RM}$, with $\widehat{R}_{RM} \gg R_{RM}$; no other state produces CI. Similarly, suppose that states O_0 and O_3 produce Cro at a rate R_R, and Cro production is off in the other states. Derive the kinetic equations for the concentrations of CI and cro

$$\frac{dx}{dt} = \phi_{CI}(x, y) - \gamma_{CI} x, \quad \frac{dy}{dt} = \phi_{Cro}(x, y) - \gamma_{Cro} y,$$

where the net production rates are

$$\phi_{Cro}(x, y) = \frac{R_{RM} + \widehat{R}_{RM} x^2 / K_1}{1 + x^2/K_1 + x^2/K_1 K_2 + y/K_3 + y^2/K_3 K_4},$$

$$\phi_{CI}(x, y) = \frac{R_M (1 + y/K_3)}{1 + x^2/K_1 + x^2/K_1 K_2 + y/K_3 + y^2/K_3 K_4}.$$

Here $K_j = k_{-j}/k_j$.

Problem 5.18. Differentiation of olfactory neurons. Consider the model given by equation (5.6.11). Suppose that there are 10 genes, each with $N = 50$ nucleosomes. Use the Gillespie SSA to investigate a stochastic version of the model with system size $N = 50$ and parameter values $\alpha = 0.3$, $r = 1$, $h = 4$, and $\beta = 0.03$.

Chapter 6
Diffusive transport

The efficient delivery of proteins and other molecular products to their correct location within a cell (intracellular transport) is of fundamental importance to normal cellular function and development [9]. Moreover, the breakdown of intracellular transport is a major contributing factor to many degenerative diseases. Broadly speaking, there are four basic mechanisms for intracellular transport [114], see Fig. 6.1:

(i) *Passive diffusion* within the cytosol or the surrounding plasma membrane of the cell. Since the aqueous environment of a cell is highly viscous at the length and velocity scales of macromolecules (low Reynolds number), a diffusing particle can be treated as an overdamped Brownian particle where inertial effects are ignored (Chap. 2).

(ii) *Facilitated diffusion through membrane pores and channels.* Facilitated diffusion involves the passive transport of small solute molecules and ions across a biological membrane via specific transmembrane integral proteins such as a channel or pore. (A pore refers to the case of a thin membrane, whereas a channel refers to the case of a thick membrane, i.e., a long pore.) It is passive in the sense that facilitated transport does not involve the use of chemical energy; rather, molecules and ions move down their concentration gradient. However, it is distinct from diffusion, since it would be difficult for molecules or ions to cross the membrane without assistance from the transmembrane proteins. (There is also an active form of membrane transport involving ion pumps, which are rotary motors that move ions across a membrane against their concentration gradient.)

(iii) *Active motor-driven transport* along polymerized filaments such as microtubules and F-actin that comprise the cytoskeleton [397] (Chap. 4). Newly synthesized products from the nucleus are mainly transported to other intracellular compartments or the cell membrane via a microtubular network that projects radially from organizing centers (centrosomes). The same network is used to

© Springer Nature Switzerland AG 2021
P. C. Bressloff, *Stochastic Processes in Cell Biology*, Interdisciplinary
Applied Mathematics 41, https://doi.org/10.1007/978-3-030-72515-0_6

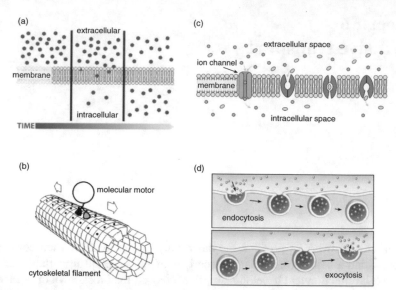

Fig. 6.1: Different mechanisms of intracellular transport: (a) Passive diffusion through a cell membrane. (b) Active motor-driven transport. (c) Facilitated diffusion. (d) Exocytosis and endocytosis. [Public domain figures downloaded from Wikipedia Commons.]

transport degraded cell products back to the nucleus and is also exploited by various animal viruses including HIV in order to reach the nucleus from the cell surface and release their genome through nuclear pores. Active transport is faster and more easily regulated than passive diffusion, but requires a constant supply of energy to do useful work.

(iv) *Exo/endocytosis and secretory trafficking.* An important mechanism for regulating the distribution of proteins and lipids in the plasma membrane and intracellular compartments is through endocytosis and exocytosis [352]. Endocytosis is the physical process whereby a vesicle forms within a membrane (budding) and is then released into the cytoplasm, whereas exocytosis is the complementary process in which an intracellular vesicle fuses with the membrane and releases its contents into a particular compartment or secretes its contents externally (as in the release of neurotransmitters at a synapse). One important role of exo/endocytosis is to regulate protein receptors in the plasma membrane [239, 492], see Fig. 6.2. That is, molecular motors transport internalized receptors from the plasma membrane to intracellular compartments that either recycle the receptors to the cell surface (early endosomes and recycling endosomes) or sort them for degradation (late endosomes and lysosomes). When this is coupled with membrane diffusion, it is possible to regulate the number of receptors within the membrane, which plays an important role in synaptic plasticity. Another important role of exo/endocytosis occurs in the early secretory pathway of eukaryotic cells, which is a critical system for the maturation and transportation of newly synthesized lipids and proteins to specific target sites within the cell membrane [526].

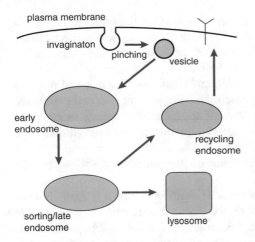

Fig. 6.2: Cartoon sketch of the endocytic pathway. A small region of the plasma membrane is invaginated and pinched off into a vesicle. The internalized vesicle fuses with larger, early endosomes and is then trafficked to sorting endosomes. From there, internalized material either is sent to recycling endosomes followed by reinsertion into the cell surface or is targeted for lysosomal degradation.

In this chapter we focus on passive and facilitated diffusive transport in cells, mechanisms (i) and (ii), whereas active motor transport, mechanism (iii), is covered in Chap. 7. An extensive treatment of secretory receptor trafficking, mechanism (iv), can be found in the book by Lauffenburger and Linderman [492]. Diffusive transport within living cells has a number of characteristic features that reflects the complex nature of the cellular environment:

1. The intracellular environment is extremely crowded with macromolecules, subcellular compartments, and confinement domains, suggesting that anomalous subdiffusion is likely to occur [219]. The plasma membrane is also a complex heterogeneous environment [480, 827]. Many papers model diffusion in such environments in terms of continuous-time random walks and fractional Brownian motion. However, it is still unclear to what extent intracellular diffusion is anomalous in the long-time limit rather than just at intermediate times. This motivates studying diffusion in the presence of obstacles and transient traps whereby normal diffusion is recovered asymptotically [620, 725, 727, 729].

2. Molecules inside the cell are often confined to a domain with small exits on the boundary of the domain. Examples include an ion searching for an open ion channel within the cell membrane, the transport of newly transcribed mRNA from the nucleus to the cytoplasm via nuclear pores, the confinement of neurotransmitter receptors within a synapse of a neuron, and the confinement of calcium and other signaling molecules within subcellular compartments such as dendritic spines of neurons. This has led to recent interest in using Green's function and asymptotic methods to solve the so-called narrow escape problem [68, 385, 389, 390, 660, 738]. A related class of problems involves the search for a small target within the interior of a cellular domain. This requires finding corrections to the classical Smoluchowski theory of diffusion-limited (or diffusion-controlled) chemical reactions [188, 434, 674, 696, 771], which are reactions in which the rate-limiting step is determined by the diffusion time to bring reactants

within sufficient proximity for a reaction to occur. One important example is a promotor protein searching for a specific binding site on DNA, which is facilitated by an intermittent search process in which the particle switches between 3D and 1D diffusion [87, 88, 194, 354, 462, 580, 749].

3. Another issue is how the random opening and closing of channels and pores in the plasma membrane of a cell or subcellular compartment affects the diffusive exchange of molecules between the interior and exterior of the cell or compartment. A related problem concerns the intercellular transport of molecules through cells coupled by stochastically gated gap junctions. Many of these examples can be formulated in terms of diffusion in a domain with randomly switching boundaries or partial boundaries.

4. One of the characteristic features of channel diffusion is that it is spatially confined, which leads to strong entropic effects due to the reduction in the available degrees of freedom [151] (see Sect. 1.3 for a definition of entropy). Moreover, various mechanisms of facilitated diffusion can occur through interactions between a diffusing particle and proteins within the channel, as exemplified by nuclear pore complexes, which are the sole mediators of exchange between the nucleus and cytoplasm [59]. When a channel becomes sufficiently narrow, particles are no longer able to pass each other (single-file diffusion)—one then finds that a tagged particle exhibits anomalous subdiffusion on long time scales [49]. Entropic effects also play a major role in the translocation of biological polymers such as DNA through nanopores.

We begin by considering the anomalous effects of molecular crowding and trapping, where the differences in diffusive behavior at multiple time scales are highlighted (Sect. 6.1). In particular, we describe continuous-time random walk models of diffusion–trapping, molecular crowding and homogenization theory, percolation theory (Box 6B), and diffusion in the plasma membrane. In Sect. 6.2, we consider one important application of diffusion–trapping models, namely to the diffusive transport of protein receptors in dendritic membranes, which plays a major role in determining the strength of a neuron's synapses. We also introduce the notion of an accumulation time and show how to construct 1D Green's function (Box 6C). In Sect. 6.3 we consider the classical Smoluchowski theory of diffusion-limited reactions. We briefly describe one application of the theory to chemoreception, which will be developed more fully in Chap. 10. We then consider various extensions of Smoluchowski reaction rate theory, including mechanisms underlying protein search for specific targets on DNA, stochastically gated diffusion-limited reactions and the problem of multi-particle correlations, and the role of ligand rebinding in enzymatic reactions with multiple binding sites. In Sect. 6.4 we describe how 2D and 3D Green's functions (Box 6D) and singular perturbation theory (matched asymptotics) can be used to analyze narrow escape and narrow capture problems. The former concerns the escape of a particle from a bounded domain through small openings in the boundary of the domain, whereas the latter refers to a diffusion–trapping problem in which the interior traps are much smaller than the size of the domain. Asymptotic methods can also be used to extend the Smoluchowski theory of diffusion-limited reactions to bounded domains. In Sect. 6.5 we consider an alternative measure of the time scale for diffusive search processes, based on the FPT of the fastest particle to find a target among a large population of independent Brownian particles, which is an example of an extreme statistic. This leads to the so-called "redundancy principle," which provides a possible explanation for the apparent redundancy in the number of molecules involved in various cellular processes, namely, that it accelerates search processes [740]. In many cases the boundary regions through which diffusing particles can escape randomly switch between open and closed states, which requires the analysis of PDEs in randomly switching environments. The latter topic is introduced in Sect. 6.6, where we highlight the connection to stochastic hybrid systems (Box 6E) and show how a switching environment induces statistical correlations, analogous to stochastically gated diffusion-limited reactions. We also consider the problem of proteins within the cell randomly switching between conformational states with different diffusivities. In Sect. 6.7 the analysis of randomly switching environments is extended to the case of molecular diffusion between cells, which are coupled by stochastically gated gap junctions. In particular, we calculate the effective permeability of the gap junctions. The diffusive transport through narrow membrane pores and channels is explored in Sect. 6.8, including the Fick–Jacobs equation for confined diffusion, single-file diffusion, and the translocation of a flexible polymer. Entropic effects play a major role in cases of restricted diffusive transport. This aspect is considered further in Sect. 6.8.3, where we describe various models for the facilitated diffusion of macromolecules through nuclear pores. Finally, in Sect. 6.9, we consider an example of diffusive transport on a tree-like structure. In particular, we calculate the probability that a particle is absorbed at the terminal node of a semi-infinite Cayley tree. This depends on the Laplace transform of the flux through the terminal node, which can be calculated using Green's functions and a recursion method.

6.1 Anomalous diffusion

Anomalous diffusion is a vast subject area that has many applications in the physical and biological sciences. We will only be able to touch on various aspects of the theory here. One major topic that we do not address (other than a brief discussion of fractional Brownian motion) is fractional diffusion equations, since this is covered extensively elsewhere [252, 572, 573]. Other subjects that are not covered include diffusion in disordered media, fractals, and percolation theory, see for example [106, 363]. Some recent surveys of anomalous diffusion with an emphasis on biological applications include [383, 466, 574].

6.1.1 Molecular crowding, diffusion–trapping, and long-time correlations

In normal (unobstructed) diffusion in d dimensions, the mean-square displacement (MSD) of a Brownian particle is proportional to time, $\langle R^2 \rangle = 2dDt$, which is a consequence of the central limit theorem. A general signature of anomalous diffusion is the power law behavior [106, 383, 572], see Fig. 6.3,

$$\langle R^2 \rangle = 2dDt^\alpha, \tag{6.1.1}$$

corresponding to either subdiffusion ($\alpha < 1$) or superdiffusion ($\alpha > 1$). Due to recent advances in single-particle tracking methods (see below), subdiffusive behavior has been observed for a variety of biomolecules and tracers within living cells. Examples include messenger RNA molecules [319] and chromosomal loci [848] moving within the cytoplasm of bacteria, lipid granule motion in yeast cells [417], viruses [745], telomeres in cell nuclei [142], and protein channels moving within the plasma membrane [850]. There are a number of subcellular mechanisms thought to generate subdiffusive motion of particles in cells, each with its own distinct type of physical model:

(i) *Molecular crowding.* One of the characteristic features of the interior aqueous environment of cells (cytoplasm) and intracellular compartments such as the endoplasmic reticulum and mitochondria is that they are crowded with small solutes, macromolecules, and skeletal proteins, which occupy 10% to 50% of the volume [219]. Cell membranes are also crowded environments containing lipids (molecules consisting of nonpolar, hydrophobic hydrocarbon chains that end in a polar hydrophilic head), which are often organized into raft structures, and various mobile and immobile proteins [480]. If the concentration of obstacles is sufficiently high, then subdiffusive behavior occurs [853], in which the domain of free diffusion may develop a fractal-like structure [725]. Diffusion on a fractal is a stationary process and is thus ergodic. There is an ongoing debate whether molecular crowding results in anomalous diffusion or leads to a simple reduction in the normal diffusion coefficient on long time scales. One way to determine the effective diffusion coefficient in the latter case is to use homogenization theory, see Sect. 6.1.3.

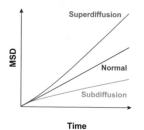
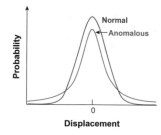

Fig. 6.3: Anomalous diffusion. Two characteristic features of anomalous diffusion are shown: superlinear or sublinear variation of the mean-square displacement (MSD) with time; large tails in the probability density.

(ii) *Diffusion–trapping.* There are many examples of intracellular transport where there is transient trapping of diffusing particles, resulting in anomalous diffusion on intermediate time scales and normal diffusion on long time scales. One important example is the surface diffusion and trapping of neurotransmitter receptors in the dendrites of a neuron, see Sect. 6.2. The switch between anomalous and normal diffusion has been explored in some detail by Saxton [727, 729], who carried out Monte Carlo simulations of random walks on a 2D lattice with a finite hierarchy of binding sites, that is, binding sites with a finite set of energy levels. This means that there are no traps that have an infinite escape time so that diffusing particles ultimately equilibrate with the traps and diffusion becomes normal. On the other hand, in the case of infinite hierarchies, arbitrarily deep traps exist but are very rare, resulting in a nonequilibrium system in which anomalous subdiffusion occurs at all times [106, 363]. The latter process can be modeled in terms of a so-called continuous-time random walk (CTRW), see Sect. 6.1.2. In addition to having a heavy-tailed waiting time distribution, the CTRW is weakly nonergodic; the temporal average of a long particle trajectory differs from the ensemble average over many diffusing particles [364, 416, 417, 574, 850].

(iii) *Long-time correlations.* This mechanism involves the viscoelastic properties of the cytoplasm due to the combined effects of macromolecular crowding and the presence of elastic elements such as nucleic acids and cytoskeletal filaments. As a particle moves through the cytoplasm, the latter "pushes back", thus generating long-time correlations in the particle's trajectory. This memory effect can lead to subdiffusive behavior that can be modeled in terms of fractional Brownian motion (FBM) or the fractional Langevin equation (FLE) [153, 555, 839]. In contrast to CTRWs, the probability density for unconfined subdiffusion in FBM/FLE is a Gaussian (with a time-dependent diffusivity). Moreover, FBM/FLE are ergodic systems although, under confinement, time-averaged quantities behave differently from their ensemble-averaged counterparts [416].

Determining which type of stochastic model best fits experimental data is a nontrivial task, particularly since CTRW, diffusion on fractals, and FBM/FLE generate similar scaling laws for ensemble-averaged behavior in the long-time limit. Thus other measures such as ergodicity (equivalence of time averages and ensemble averages) are being used to help identify which model provides the best characterization for anomalous diffusion in living cells [793, 848, 850]. A more fundamental difficulty in experimentally establishing the existence of anomalous diffusion is that the behavior of $\langle R^2 \rangle$ can depend on the spatial or temporal scale over which observations are made. Yet another difficulty in interpreting experimental data is that there are certain practical limitations of current methods [219]. The most effective method for describing membrane diffusion is single-particle tracking (SPT), in which one images the trajectory of a marker attached to a diffusing molecule [479, 728, 734]. This involves the selective labeling of proteins or lipids with fluorophores such as quantum dots, green fluorescent protein (GFP), or organic dyes so that continuous high-resolution tracking of individual molecules can be carried out. SPT can yield nanometer spatial resolution and submillisecond temporal res-

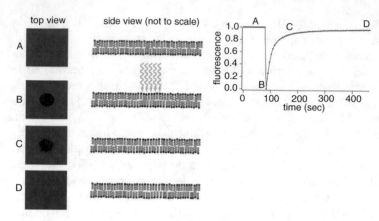

Fig. 6.4: Schematic illustration of fluorescence recovery after photobleaching (FRAP). (A) A membrane bilayer is uniformly labeled with a fluorescent tag. (B) This label is selectively photobleached by a small (30 μm) fast light pulse. (C) The intensity of the fluorescent signal within the bleached area is monitored as the bleached dye diffuses out and new dye diffuses in (D). Eventually uniform intensity is restored. [Public domain figure downloaded from Wikipedia Commons.]

olution of individual trajectories. Various transport properties of the particle are then derived through a statistical analysis of the trajectory, including a measurement of the mean-square displacement. Visualization of the diffusive behavior of single-membrane proteins in living cells has revealed that these molecules undergo a variety of stochastic behaviors including normal and anomalous diffusion and confinement within subcellular compartments. SPT also provides information on the structure of the surrounding membrane and the molecular interactions. The rapid increase in the range of applications of STP to cell biology has been driven by major improvements in the visualization of trajectories combined with new strategies for labeling proteins with nanoprobes. However, single-molecule approaches still have their own limitations, such as the shortness of observation times and the possibility that identified molecules are not representative of the population, which can lead to sampling errors. Moreover, it is not currently suitable for measuring diffusion in three dimensions due to the relatively rapid speed of 3D diffusion and the problems of imaging in depth. Hence, in the case of diffusion within the cytosol, it is necessary to use a method such as fluorescence recovery after photobleaching (FRAP) [39, 687], see Fig. 6.4. Here fluorescently labeled molecules are introduced into the cell and those in some specified volume are bleached by a brief intense laser pulse. The diffusion of unbleached molecules into the bleached volume is then measured. FRAP is limited because it only provides ensemble-averaged information of many fluorescent particles, and it also has a restricted measurement time, making it difficult to capture long-tail phenomena expected in anomalous subdiffusion.

6.1.2 Continuous-time random walks

An important generalization of the standard random walk, known as the continuous-time random walk (CTRW), is often used to model anomalous diffusion arising from trapping processes. The basic idea is that trapping can increase the time between jumps (waiting times) of a random walk, so that jumps no longer occur at fixed discrete-time steps. A CTRW is typically written in the form [403]

$$R_n(\ell,t) = \sum_{\ell' \in \Gamma} p(\ell|\ell') \int_0^t \psi(t-t') R_{n-1}(\ell',t') dt', \qquad (6.1.2)$$

where $R_n(\ell,t)$ is the probability for a walker to just arrive at site ℓ at time t in n steps, $p(\ell|\ell')$ is the probability of the jump $\ell \to \ell'$, and $\psi(t)$ is the waiting time density for a single step over a time interval of length t. Thus steps can now take place at different times. In general, a CTRW has a memory of previous time steps so is non-Markovian. (In fact it can often be reformulated in terms of a continuous-time master equation with memory, see Ex. 6.1.) How does one recover the standard Markovian random walk? First, we consider the exponential waiting time density $\psi(t)$ given by $\psi(t) = \Lambda e^{-\Lambda t}$. Substituting into the CTRW and differentiating both sides with respect to t show that

$$\frac{1}{\Lambda} \frac{dR_n}{dt} + R_n(\ell,t) \approx R_n(\ell,t+\varepsilon) = \sum_{\ell'} p(\ell|\ell') R_{n-1}(\ell',t)$$

for $\varepsilon = \Lambda^{-1} \ll 1$. This is the standard recursive equation after setting $t = n\varepsilon$.

In order to analyze equation (6.1.2) we will make extensive use of transform methods. It is useful to keep track of the different transform pairs, which are as follows:

$n \leftrightarrow z\,(z-\text{transform})$, $\ell \leftrightarrow \mathbf{k}\,(\text{discrete Fourier transform})$, $t \leftrightarrow s\,(\text{Laplace transform})$.

First, taking the Laplace transform of (6.1.2) and using the convolution theorem give

$$\widetilde{R}_n(\ell,s) = \widetilde{\psi}(s) \sum_{\ell' \in \Gamma} p(\ell|\ell') \widetilde{R}_{n-1}(\ell',s). \qquad (6.1.3)$$

This has the solution (for $n \geq 1$)

$$\widetilde{R}_n(\ell,s) = A P_n(\ell) \widetilde{\psi}_n(s), \quad \widetilde{\psi}_n(s) = \widetilde{\psi}(s)^n,$$

with unknown amplitude A, and $P_n(\ell)$ the solution to the standard random walk (RW) master equation

$$P_n(\ell) = \sum_{\ell'} p(\ell|\ell') P_{n-1}(\ell'). \qquad (6.1.4)$$

One is often interested in the probability of arriving at the site ℓ at time t irrespective of the number of steps. Therefore, introduce the density

$$R(\ell,t) = \sum_{n \geq 0} R_n(\ell,t), \tag{6.1.5}$$

and note that if the walker starts at the origin, then

$$R_0(\ell,t) = \delta_{\ell,0}\Psi(t), \quad \Psi(t) = \int_t^\infty \psi(t')dt'.$$

Here $\Psi(t)$ is the probability that the walker has not yet taken a first step at time t. Summing both sides of the solution below equation (6.1.3) with respect to n shows that

$$\widetilde{R}(\ell,s) = \sum_{n \geq 0} \widetilde{R}_n(\ell,s) = A \sum_{n \geq 0} P_n(\ell)\widetilde{\psi}(s)^n. \tag{6.1.6}$$

Summing both sides with respect to ℓ and using the normalization $\sum_\ell P_n(\ell) = 1$, we then have

$$\sum_\ell \widetilde{R}(\ell,s) = A \sum_{n \geq 0} \widetilde{\psi}(s)^n = \frac{A}{1 - \widetilde{\psi}(s)}.$$

In addition, the normalization $\sum_\ell R(\ell,t) = 1$ implies $\sum_\ell \widetilde{R}(\ell,s) = 1/s$. Hence, $A = (1 - \widetilde{\psi}(s))/s$ and

$$\widetilde{R}(\ell,s) = \frac{1 - \widetilde{\psi}(s)}{s}\Gamma[\ell, \widetilde{\psi}(s)], \tag{6.1.7}$$

where $\Gamma(\ell,z) = \sum_{n \geq 0} P_n(\ell)z^n$, which is the generating function for the distribution $P_n(\ell)$ and fixed ℓ.

We would like to determine the mean-square displacement for different choices of the waiting time density $\psi(t)$. First, we need to determine the generating function Γ for the underlying RW, which from equation (6.1.4) and $P_0(\ell) = \delta_{\ell,0}$ satisfies

$$\Gamma(\ell,z) = \delta_{\ell,0} + z\sum_{\ell'} p(\ell|\ell')\Gamma(\ell',z).$$

For simplicity, suppose that $p(\ell|\ell') = p(\ell - \ell')$. Taking discrete Fourier transforms and using the convolution theorem on the (infinite) lattice

$$\widehat{\Gamma}(\mathbf{k},z) := \sum_\ell e^{i\mathbf{k}\cdot\ell}\Gamma(\ell,z) = 1 + z\widehat{p}(\mathbf{k})\widehat{\Gamma}(\mathbf{k},z),$$

where

$$\widehat{p}(\mathbf{k}) = \sum_\ell e^{i\mathbf{k}\cdot\ell}p(\ell).$$

Hence

$$\widehat{\Gamma}(\mathbf{k},z) = \frac{1}{1 - z\widehat{p}(\mathbf{k})}.$$

Taking the discrete Fourier transform of (6.1.7) with $\widehat{R}(\mathbf{k},s) := \sum_\ell e^{i\mathbf{k}\cdot\ell}\widetilde{R}(\ell,s)$ thus gives

$$\widehat{R}(\mathbf{k},s) = \frac{1-\widetilde{\psi}(s)}{s}\frac{1}{1-\widetilde{\psi}(s)\widehat{p}(\mathbf{k})}.$$ (6.1.8)

For the sake of illustration, consider a 1D lattice so that $\mathbf{k} \to k$. Using

$$\left(-i\frac{\partial}{\partial k}\right)^n \widehat{R}(k,s)\bigg|_{k=0} = \sum_\ell \ell^n \widetilde{R}(\ell,s),$$

which is the Laplace transform of the nth order moment of ℓ, we obtain the following results:

$$\mathcal{L}(\langle X\rangle)(s) = \sum_\ell \ell R(\ell,s) = \frac{-i\widetilde{\psi}(s)\widehat{p}'(0)}{s(1-\widetilde{\psi}(s))},$$

and

$$\mathcal{L}(\langle X^2\rangle)(s) = \sum_\ell \ell^2 R(\ell,s) = -\frac{\widetilde{\psi}(s)\widehat{p}''(0)}{s(1-\widetilde{\psi}(s))} - \frac{2\widetilde{\psi}(s)^2\widehat{p}'(0)^2}{s(1-\widetilde{\psi}(s))^2}.$$

We are assuming that the distribution $p(\ell)$ of displacements on the lattice has finite moments. In particular, if the CTRW is unbiased with $\widehat{p}'(0) = 0$ and $\widehat{p}''(0) = -\sigma^2$, then

$$\mathcal{L}(\langle X\rangle)(s) = 0, \quad \mathcal{L}(\langle X^2\rangle)(s) = \frac{\widetilde{\psi}(s)\sigma^2}{s(1-\widetilde{\psi}(s))}.$$

Now suppose that the mean waiting time density is also finite:

$$\tau = \int_0^\infty t\psi(t)dt < \infty.$$

We can then Taylor expand the Laplace transform around $s = 0$,

$$\widetilde{\psi}(s) = \int_0^\infty e^{-st}\psi(t)dt \approx \int_0^\infty [1-st]\psi(t)dt = 1-s\tau,$$

so that in the limit $s \to 0$,

$$\mathcal{L}(\langle X^2\rangle)(s) \sim \frac{\sigma^2}{s^2\tau},$$

which implies that

$$\langle X(t)^2\rangle \sim \frac{\sigma^2 t}{\tau}.$$

On the other hand, if

$$\widetilde{\psi}(s) \sim 1 - Bs^\beta$$

as $s \to 0$ with $0 < \beta < 1$ then the waiting time density has infinite mean and

$$\psi(t) \sim t^{-1-\beta}$$

as $t \to \infty$. One now obtains subdiffusive behavior with

$$\langle X(t)^2 \rangle \sim \frac{\sigma^2 t^\beta}{B\Gamma(1+\beta)}. \tag{6.1.9}$$

The relationship between the power laws with respect to t and the Laplace variable s are determined using Tauberian Theorems (see Box 6A).

One possible biological mechanism for generating a heavy-tailed CTRW is a protein transiently binding to a macromolecular complex that grows with time [851]. Assuming a nucleation period t_0, one can postulate a time-dependent unbinding rate $k(t) = \alpha/(t+t_0)$ so that the survival probability of a bound protein evolves as

$$\frac{dS}{dt} = -\frac{\alpha}{t+t_0} S,$$

which has the power law solution $S(t) = t_0^\alpha (t+t_0)^{-\alpha}$. The corresponding waiting time density is

$$\psi(\tau) = -\frac{dS}{d\tau} \sim \tau^{-(1+\alpha)}$$

for large τ. Another possible mechanism is a protein interacting with different complexes, resulting in a distribution of dissociation rates [153]. Suppose that there is a distribution of binding energies E_B (in units of $k_B T$), $p(E_B) = \alpha e^{-\alpha E_B}$. From the Arrhenius rate law, the dissociation constant associated with the complex having binding energy E_B is of the form $k = k_0 e^{-E_B}$. It follows that there is a distribution of dissociation constants

$$p(k) = \int_0^\infty \delta(k - k_0 e^{-E_B}) p(E_B) dE_B = \alpha k_0^{-\alpha} k^{\alpha-1}, \quad 0 < k < k_0.$$

The conditional waiting time density is $\psi(\tau|k) = k e^{-k\tau}$ so that

$$\psi(\tau) = \int_0^{k_0} k e^{-k\tau} p(k) dk = \frac{\alpha}{k_0^\alpha} \gamma(\alpha+1, k_0 t) \tau^{-(1+\alpha)},$$

where $\gamma(s,t)$ is the lower incomplete gamma function,

$$\gamma(\alpha+1, t) = \int_0^t \tau^\alpha e^{-\tau} d\tau.$$

In the long-time limit, $\gamma(\alpha+1, k_0 t) \to \Gamma(\alpha+1)$ and $\psi(\tau) \sim \tau^{-(1+\alpha)}$.

Box 6A. Tauberian theorems [403].

Weak Tauberian Theorem. If $\psi(t) \geq 0$, $0 \leq \rho < \infty$ and L is slowly varying at ∞, then each of the relations

$$\widetilde{\psi}(s) \sim L(1/s) s^{-\rho}$$

as $s \to 0$ and

$$\int_0^t \psi(t')dt' \sim \frac{t^\rho L(t)}{\Gamma(1+\rho)}$$

as $t \to \infty$ implies the other.

Strong Tauberian Theorem. If $\psi(t) \geq 0$, $0 \leq \rho < \infty$, $\psi(t)$ is ultimately monotonic as $t \to \infty$, and L is slowly varying at ∞, then each of the relations

$$\widetilde{\psi}(s) \sim L(1/s)s^{-\rho}$$

as $s \to 0$ and

$$\psi(t) \sim \frac{t^{\rho-1}L(t)}{\Gamma(\rho)}$$

as $t \to \infty$ implies the other.

6.1.3 Molecular crowding and homogenization theory

We now turn to another characteristic feature of the intracellular environment, namely, molecular crowding. One consequence of molecular crowding is that it can drastically alter biochemical reactions in cells [735, 894]. That is, volume or area exclusion effects increase the effective solute concentration, thus increasing the chemical potential of the solute. Here, however, we focus on the role of molecular crowding in hindering diffusion. There is an ongoing debate regarding to what extent the latter results in anomalous diffusion rather than a simple reduction in the normal diffusion coefficient [45, 219, 853]. Consider, for example, the effects of obstacles on protein diffusion [725, 788]. The presence of obstacles reduces the space available for diffusion and consequently decreases the effective diffusion coefficient. (One way to model the partial filling of some space by obstacles is to treat it as a site percolation problem. A brief review of percolation theory is given in Box 6B.) As the volume or area fraction of obstacles ϕ is increased, there is a fragmentation of the available space in the sense that many paths taken by a diffusing protein terminate in a dead end and thus do not contribute to diffusive transport, see Fig. 6.5. The region of free

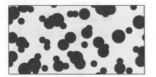 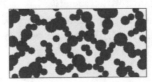

● · obstacle
□ · bulk phase

Fig. 6.5: Filling of bulk cytoplasm by obstacles with low ($\phi < \phi_c$) and high ($\phi > \phi_c$) volume fraction, where ϕ_c is the percolation threshold.

diffusion develops a fractal-like structure resulting in anomalous diffusion at intermediate times, $\langle R^2 \rangle \sim t^\alpha$, $\alpha < 1$. (For sufficiently small times we have $\sqrt{Dt} \ll \xi$, where ξ is the mean distance between obstacles, so that diffusion is normal.) However, assuming that the volume or area fraction is below the percolation threshold [106, 363] (i.e., there is still an unobstructed path across the domain, see Box 6B), diffusion is expected to be normal on sufficiently long time scales, $\langle R^2 \rangle \sim t$. On the other hand, above the percolation threshold, proteins are confined and $\langle R^2 \rangle$ saturates as $t \to \infty$. The time it takes to cross over from anomalous to normal diffusion increases with the volume or area fraction ϕ and diverges at the percolation threshold ϕ_c where $\langle R^2 \rangle \sim t^\alpha$ for all times.

Homogenization theory has been used to develop a fast numerical scheme to calculate the effects of excluded volume due to molecular crowding on diffusion in the cytoplasm [620]. The basic idea is to model the heterogeneous environment in terms of randomly positioned overlapping obstacles. (Note, however, that this is an oversimplification, since single-particle tracking experiments indicate that the cytoplasm is more properly treated as a dynamic, viscoelastic environment.) Although obstacles don't overlap physically, when the finite size of a diffusing molecule (tracer) is taken into account, the effective volume excluded by an obstacle increases so that this can result in at least partially overlapping exclusion domains, see Fig. 6.6(a). In the absence of any restrictions on the degree of overlap, the fraction of inaccessible volume is $\phi = 1 - e^{-V}$, where V is the sum of the individual obstacles per unit volume. A simple argument for this [620] is to consider a set of N identical overlapping objects placed in a box of total volume $|\Omega|$. Let v denote the volume of each obstacle. The probability $P(x)$ that a randomly selected point $x \in \Omega$ is outside any given obstacle is $1 - v/|\Omega|$. Hence, the probability for that point to be outside all obstacles is $P(x)^N = (1 - v/|\Omega|)^N$. The volume fraction of accessible space at fixed number density $n = N/|\Omega|$ is then

$$\lim_{N \to \infty} (1 - v/|\Omega|)^N = \lim_{N \to \infty} (1 - vn/N)^N = e^{-nv} = e^{-V},$$

and the result follows. The mean distance between obstacles can then be determined in terms of ϕ and the geometry of each obstacle.

As previously discussed, three regimes of diffusion are expected below the percolation threshold, $\phi < \phi_c$, as illustrated in Fig. 6.6(b). For sufficiently short times there is unobstructed diffusion, for intermediate times there is anomalous diffusion, and for long times there is normal effective diffusion. Novak *et. al.* [620] used homogenization theory to estimate the effective diffusion coefficient D_{eff} in the last regime. The starting point for their analysis is to consider a periodic arrangement of identical obstacles in a large rectangular box of volume Ω with accessible volume Ω_1 and $\phi = 1 - |\Omega_1|/|\Omega|$. The spatial periods of the arrangement in Cartesian coordinates are $a_j, j = 1, 2, 3$ such that the ratio

$$\varepsilon = \sqrt{a_1^2 + a_2^2 + a_3^2} / \sqrt[3]{|\Omega|} \ll 1. \tag{6.1.10}$$

The heterogeneous diffusion coefficient is

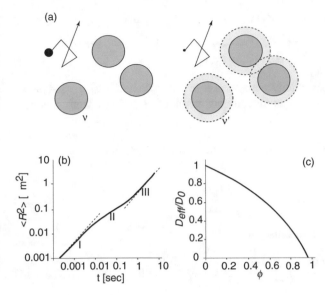

Fig. 6.6: (a) Diffusion of a finite size particle (tracer) between obstacles of volume v (left) can be modeled as diffusion of a point particle between effective obstacles of volume v' (right). Effective obstacles can partially overlap. (b) Sketch of MSD $\langle R^2 \rangle$ against time t: unobstructed diffusion (I), anomalous intermediate diffusion (II), and normal effective diffusion (III). (c) Illustrative plot of the normalized effective diffusion coefficient $D_{\text{eff}}(\phi)/D_0$ for random spheres. The scale of the curves in (b,c) is based on the results of [620].

$$D_\varepsilon(\mathbf{x}) = \begin{cases} D_0 & \text{if } \mathbf{x} \in \Omega_1 \\ 0, & \text{otherwise.} \end{cases} \tag{6.1.11}$$

Inhomogeneous Dirichlet conditions are imposed on the boundaries of the box in order to maintain a steady-state diffusive flux. In the case of a heterogeneous diffusion coefficient, the flux is determined by the steady-state diffusion equation for the tracer distribution $u(\mathbf{x})$:

$$\nabla \cdot (D_\varepsilon(\mathbf{x})\nabla u(\mathbf{x})) = 0. \tag{6.1.12}$$

The basic idea of the homogenization method is to represent the diffusive behavior of a tracer on two different spatial scales [646, 808]: one involving a macroscopic slow variable $\mathbf{x}$ and the other a microscopic fast variable $\mathbf{y} \equiv \mathbf{x}/\varepsilon$ so that u is periodic with respect to $\mathbf{y}$. (Spatial coordinates are non-dimensionalized by taking $\Omega = 1$.) Thus, we write

$$u = u(\mathbf{x},\mathbf{y}), \quad \nabla u = \nabla_{\mathbf{x}} u(\mathbf{x},\mathbf{y}) + \varepsilon^{-1}\nabla_{\mathbf{y}} u(\mathbf{x},\mathbf{y}).$$

Also $D_\varepsilon(\mathbf{x}) \equiv D(\mathbf{x}/\varepsilon) = D(\mathbf{y})$ with D and u having the same periodicity in $\mathbf{y}$.

A solution to equation (6.1.12) is then constructed in terms of the asymptotic expansion

$$u = u_0(\mathbf{x},\mathbf{y}) + \varepsilon u_1(\mathbf{x},\mathbf{y}) + \varepsilon^2 u_2(\mathbf{x},\mathbf{y}) + \dots.$$

Collecting terms of the same order in ε then yields a hierarchy of equations, which up to $O(1)$ are as follows:

$$\nabla_{\mathbf{y}} \cdot [D(\mathbf{y}) \nabla_{\mathbf{y}} u_0(\mathbf{x}, \mathbf{y})] = 0 \tag{6.1.13a}$$

$$\nabla_{\mathbf{y}} \cdot [D(\mathbf{y}) \nabla_{\mathbf{y}} u_1(\mathbf{x}, \mathbf{y})] = -\nabla_{\mathbf{y}} \cdot [D(\mathbf{y}) \nabla_{\mathbf{x}} u_0(\mathbf{x}, \mathbf{y})] - \nabla_{\mathbf{x}} \cdot [D(\mathbf{y}) \nabla_{\mathbf{y}} u_0(\mathbf{x}, \mathbf{y})] \tag{6.1.13b}$$

$$\nabla_{\mathbf{y}} \cdot [D(\mathbf{y}) \nabla_{\mathbf{y}} u_2(\mathbf{x}, \mathbf{y})] = -\nabla_{\mathbf{x}} \cdot [D(\mathbf{y}) \nabla_{\mathbf{x}} u_0(\mathbf{x}, \mathbf{y})] - \nabla_{\mathbf{y}} \cdot [D(\mathbf{y}) \nabla_{\mathbf{x}} u_1(\mathbf{x}, \mathbf{y})]$$
$$- \nabla_{\mathbf{x}} \cdot [D(\mathbf{y}) \nabla_{\mathbf{y}} u_1(\mathbf{x}, \mathbf{y})]. \tag{6.1.13c}$$

Equation (6.1.13a) and periodicity with respect to $\mathbf{y}$ establish that $u_0(\mathbf{x}, \mathbf{y}) \equiv u_0(\mathbf{x})$, that is, u_0 corresponds to an homogenized solution. It follows from equation (6.1.13b) that

$$\nabla_{\mathbf{y}} D(\mathbf{y}) \cdot \nabla_{\mathbf{x}} u_0(\mathbf{x}) + \nabla_{\mathbf{y}} \cdot [D(\mathbf{y}) \nabla_{\mathbf{y}} u_1(\mathbf{x}, \mathbf{y})] = 0,$$

which has the solution

$$u_1(\mathbf{x}, \mathbf{y}) = \sum_{i=1}^{3} \frac{\partial u_0(\mathbf{x})}{\partial x_i} w_i(\mathbf{y}), \tag{6.1.14}$$

with $w_i(\mathbf{y})$ a periodic function satisfying

$$\frac{\partial D(\mathbf{y})}{\partial y_i} + \sum_{j=1}^{3} \frac{\partial}{\partial y_j} \left[D(\mathbf{y}) \frac{\partial w_i(\mathbf{y})}{\partial y_j} \right] = 0. \tag{6.1.15}$$

(Note that one usually has to solve equation (6.1.15) numerically. A particular example where an exact solution can be found is considered in Ex. 6.18.) Finally, averaging both sides of equation (6.1.13c) with respect to $\mathbf{y}$ over a unit volume $|\omega_0|/\varepsilon^3$ of the periodic structure, using the divergence theorem, and expressing u_1 in terms of u_0 yield the homogenized diffusion equation

$$\sum_{i,j=1}^{3} \widetilde{D}_{\text{eff},ij} \frac{\partial^2 u_0(\mathbf{x})}{\partial x_i \partial x_j} = 0,$$

with the anisotropic diffusion tensor

$$\widetilde{D}_{\text{eff},ij} = \frac{\varepsilon^3}{|\omega_0|} \int_{\omega_0} \left(D(\mathbf{y}) \delta_{i,j} + D(\mathbf{y}) \frac{\partial w_i(\mathbf{y})}{\partial y_j} \right) d\mathbf{y}.$$

Finally, rewriting the diffusion tensor in a more symmetric form using equation (6.1.15) and integration by parts give [620]

$$\widetilde{D}_{\text{eff},ij} = \frac{D_0}{|\omega_0|} \int_{\omega_1} \sum_{k=1}^{3} \left(\delta_{i,k} + \frac{\partial \widehat{w}_i(\mathbf{x})}{\partial x_k} \right) \left(\delta_{j,k} + \frac{\partial \widehat{w}_j(\mathbf{x})}{\partial x_k} \right) d\mathbf{x}, \tag{6.1.16}$$

where ω_1 is the accessible region of the fundamental domain ω_0. The function $w(x)$ has been rescaled according to $\widehat{w}(\mathbf{x}) = \varepsilon w(\mathbf{x}/\varepsilon)$ so that

$$\frac{\partial D_\varepsilon(\mathbf{x})}{\partial x_i} + \sum_{j=1}^{3} \frac{\partial}{\partial x_j} \left[D_\varepsilon(\mathbf{x}) \frac{\partial \widehat{w}_i(\mathbf{x})}{\partial x_j} \right] = 0 \tag{6.1.17}$$

over a unit cell with periodic boundary conditions. Note that the concentration $u_0(\mathbf{x})$ is only defined in free space so that the macroscopic concentration is actually $u(\mathbf{x}) = (1-\phi)u_0(\mathbf{x})$ and the macroscopic diffusion tensor is $D_{\mathrm{eff},ij} = \widetilde{D}_{\mathrm{eff},ij}/(1-\phi)$. In the case of isotropic periodic structures $D_{\mathrm{eff},ij} = D_{\mathrm{eff}}\delta_{i,j}$.

Novak *et. al.* [620] numerically extended the homogenization method to a random arrangement of obstacles by approximating the disordered medium with a periodic one, in which the unit cell consists of N randomly placed obstacles. N is taken to be sufficiently large so that for a given density of obstacles, one obtains a statistically stationary D_{eff}. Comparing the homogenized diffusion coefficient with that obtained from Monte Carlo simulations, the numerical homogenization method yields reasonable agreement for $N = O(100)$. One of the interesting results of their study was that the variation of D_{eff} with the excluded volume fraction ϕ can be approximated by the power law

$$D_{\mathrm{eff}}(\phi) = D_0 \frac{(1-\phi/\phi_c)^\mu}{1-\phi}, \tag{6.1.18}$$

where the parameters ϕ_c, μ depend on the geometry of the obstacles. For example, for randomly arranged spheres, $\phi_c \approx 0.96$ and $\mu \approx 1.5$. A typical plot of $D_{\mathrm{eff}}(\phi)$ is shown in Fig. 6.6(c). Previously, the above power law behavior had been predicted close to the percolation threshold [106], but these results suggest it also holds for a wider range of volume fractions.

One-dimensional domain. In the 1D case, we have $u(x,y) = u_0(x) + \varepsilon u_1(x,y) + \dots$, with equations (6.1.14) and (6.1.15) simplifying according to

$$u_1(x,y) = \frac{du_0(x)}{dx} w(y), \tag{6.1.19}$$

with $w(y)$ a periodic function satisfying

$$\frac{dD(y)}{dy} + \frac{d}{dy}\left[D(y)\frac{dw(y)}{dy}\right] = 0. \tag{6.1.20}$$

The latter equation implies that

$$\frac{dw(y)}{dy} = \frac{A}{D(y)} - 1,$$

for some constant A. Integrating both sides over a single period (taken to be unity) shows that

$$A^{-1} = \int_0^1 \frac{dy}{D(y)}.$$

It follows that

$$\widetilde{D}_{\mathrm{eff}} = \varepsilon \int_0^1 \left(D(y) + D(y)\frac{dw(y)}{dy}\right) dy = \left[\int_0^1 \frac{dy}{D(y)}\right]^{-1}. \tag{6.1.21}$$

In other words, to leading order the effective diffusion coefficient is the inverse of the harmonic mean of $D(x/\varepsilon)$. A further example of a 1D homogenization problem is considered in Ex. 6.2.

Box 6B. Percolation theory.

We introduce some of the basic concepts of percolation theory by considering the case of site percolation on a lattice. For a much more detailed survey, see Ref. [781]. Suppose that each site of the lattice is occupied at random with probability p and is thus empty with probability $1 - p$. Define a cluster as a group of nearest neighboring occupied sites. Percolation theory deals with how the numbers and properties of the clusters depend on the site probability. Let $n_s(p)$ denote the number of s-clusters (clusters of size s) per lattice site. For a large lattice of linear size L, the expected number of s-clusters will be $L^d n_s(p)$, where d is the dimension of the lattice. For a finite lattice ($L < \infty$), the probability that there exists a cluster that percolates between two opposite boundaries will be negligible when p is small. On the other hand, such a cluster will almost certainly exist when $p \approx 1$. The percolation threshold p_c is the occupation probability at which an infinite cluster appears for the first time in an infinite lattice ($L \to \infty$). Some well-known site percolation thresholds are as follows: $p_c = 1$ (1D lattice); $p_c \approx 0.696$ (2D square lattice); $p_c \approx 0.593$ (2D hexagonal lattice); $p_c = 1/2$ (2D triangular lattice).

Site percolation has a counterpart known as bond percolation (Fig. 6.7), whereby each bond between neighboring lattice sites can be occupied (open) with probability p and empty (closed) with probability $1 - p$. Now clusters consist of nearest neighboring open bonds. Corresponding percolation thresholds are as follows: $p_c = 1$ (1D lattice); $p_c = 1/2$ (2D square lattice); $p_c = 1 - 2\sin(\pi/18) \approx 0.653$ (2D hexagonal lattice); $p_c = 2\sin(\pi/18) \approx 0.347$ (2D triangular lattice).

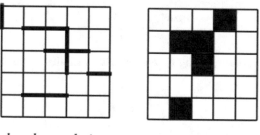

bond percolation   *site percolation*

Fig. 6.7: Comparison of (a) bond and (b) site percolation.

Site percolation in 1D. The site percolation problem can be solved analytically in 1D and shares many of the characteristic features observed in higher dimensions. Since a percolating cluster in 1D spans from $-\infty$ to ∞, it can only exist if every site is occupied, that is, $p = 1$. Hence the percolation threshold is $p_c = 1$. A more precise statement is as follows. Let $\Pi(p,L)$ denote the probability that a lattice of size L percolates at occupation probability p. Then

$$\lim_{L \to \infty} \Pi(p,L) = \begin{cases} 0 & \text{for } p < p_c, \\ 1 & \text{for } p \geq p_c \end{cases}. \tag{6.1.22}$$

In then case of 1D lattice with the sites filled independently, we have $\Pi(p,L) = p^L$ and thus $p_c = 1$. Next we consider properties of the cluster number $n_s(p)$. A cluster of size s is formed whenever s successive sites are occupied and bounded by two empty sites. Ignoring boundary effects for large L,

$$n_s(p) = (1-p)^2 p^s = (1-p)^2 e^{s \ln(p)} = (1-p)^2 e^{-s/s_\xi}, \tag{6.1.23}$$

where $s_\xi = -1/\ln(p)$ is known as the characteristic cluster size. Note that

$$s_\xi = -\frac{1}{\ln[1-(1-p)]} = -\frac{1}{-(1-p)-(1-p)^2/2 - \dots} \to \frac{1}{1-p}$$

as $p \to 1^-$. This is a special case of the following power law, which occurs in higher dimensions where $p_c < 1$:

$$s_\xi \sim |p_c - p|^{-1/\sigma} \text{ for } p \to p_c, \tag{6.1.24}$$

where σ is known as a critical exponent. For a 1D lattice, $p_c = 1$ and $\sigma = 1$.

Mean cluster size. Another quantity of interest for $p < p_c$ is the mean size of a cluster to which a randomly chosen occupied site belongs. First, note that the probability an arbitrary site belongs to an s-cluster is $s n_s(p)$ and the probability of belonging to any finite cluster is simply the occupation probability p. Hence,

$$\sum_{s=1}^{\infty} s n_s(p) = p.$$

Let w_s be the probability that the cluster to which a randomly chosen occupied site belongs has size s:

$$w_s = \frac{s n_s(p)}{\sum_{s=1}^{\infty} s n_s(p)} = \frac{s n_s(p)}{p}.$$

Then the mean cluster size is

$$S(p) = \frac{1}{p} \sum_{s=1}^{\infty} s^2 n_s(p) = \frac{(1-p)^2}{p} \sum_{s=1}^{\infty} s^2 p^s$$

$$= \frac{(1-p)^2}{p} \left(p \frac{d}{dp} \right)^2 \sum_{s=1}^{\infty} p^s = \frac{1+p}{1-p}. \tag{6.1.25}$$

It can be seen that $S(p)$ diverges as a power law as $s \to 1^-$ according to $S(p) \sim (1-p)^{-1}$. An analogous phenomenon occurs in higher dimensions:

$$S(p) \sim |p_c - p|^{-\gamma} \text{ for } p \to p_c. \tag{6.1.26}$$

In the case of 1D percolation, the critical exponent $\gamma = 1$.

Correlation function. A third example of a power law arises when considering the correlation function $g(r)$, which is the probability that a site at a distance r from an occupied site belongs to the same finite cluster. Clearly $g(0) = 1$, since the site is occupied by definition. For $r > 0$ we require that the site at a distance r from the chosen site is occupied, as all are the intermediate sites. Thus

$$g(r) = p^r = e^{r \ln(p)} = e^{-r/\xi}, \quad \xi = -\frac{1}{\ln(p)}, \tag{6.1.27}$$

with ξ known as the correlation length. For 1D percolation we have $\xi = s_\xi$ and thus ξ diverges according to $(1-p)^{-1}$. In higher dimensions $(d > 1)$, we find that $s_\xi \sim \xi^H$ and

$$\xi \sim |p_c - p|^{-\nu} \text{ for } p \to p_c, \tag{6.1.28}$$

with $\nu \neq 1/\sigma$. Here H is known as the fractal dimension (see below). Note that more generally, the correlation length can be defined in terms of the correlation function $g(\mathbf{r})$ according to

$$\xi^2 = \frac{\sum_{\mathbf{r}} r^2 g(\mathbf{r})}{\sum_{\mathbf{r}} g(\mathbf{r})}, \tag{6.1.29}$$

which represents the average distance of two sites belonging to the same cluster.

In summary, we have identified certain quantities, including the characteristic cluster size s_ξ, the mean cluster size $S(p)$, and the correlation length ξ, which diverge at the percolation threshold. Moreover, this divergence can be described by simple power laws of the distance $\Delta p = |p_c - p|$ from the percolation threshold as $\Delta p \to 0$. In 1D the critical exponent of each power law can be obtained analytically, which is not usually possible in higher dimensions. One exception is an infinite Bethe lattice, see Ex. 6.3.

Higher-dimensional lattices. In higher dimensions it is not possible to derive an exact expression for the cluster number $n_s(p)$ as there are a large number of different ways in which clusters can be arranged. Nevertheless, the behavior of the cluster number can be captured by the general scaling form

$$n_s(p) = As^{-\tau} f(B(p - p_c)s^{\sigma}) \text{ for } s \gg 1 \text{ and } p \to p_c. \tag{6.1.30}$$

Here A and B are non-universal constants in the sense that that they depend on the lattice details, whereas the critical exponents τ, σ and the scaling function f are universal. Typically, for $d > 1$ one finds that $f(z)$ is approximately constant when $|z| \ll 1$ ($s \ll s_{\xi}$) and quickly decaying when $|z| \gg 1$ ($s \gg s_{\xi}$). Thus s_{ξ} acts as a cut-off. Another significant differences between $d = 1$ and $d > 1$ is that in the latter case, $p_c < 1$ and there exists a change of behavior when one crosses p_c. This can be explored by defining the percolation strength $P(p)$, which is the probability that an arbitrary site belongs to an infinite cluster. Since all occupied sites must belong to a finite or an infinite cluster, we require

$$P(p) + \sum_{s=1}^{\infty} sn_s(p) = p. \tag{6.1.31}$$

It can be shown that

$$P(p) \sim \begin{cases} (p - p_c)^{\beta} & \text{for } p \to p_c^+ \\ 0 & \text{for } p \le p_c \end{cases}, \tag{6.1.32}$$

with a positive critical exponent β. ($P(p) = 0$ for all $p \in [0, 1)$ in 1D.) The strength $P(p)$ is an example of an order parameter, and the phenomenon whereby an order parameter becomes nonzero for $p > p_c$ is known as a phase transition at the critical point p_c. In addition, the order parameter approaches zero continuously as $p \to p_c^+$, which means that we have a continuous or second-order phase transition. (If the order parameter jumped to zero, then we would have a first-order phase transition.) As is common with other examples of phase transitions, scaling arguments can be used to derive relationships between various critical exponents.

Fractal structures. At $p = p_c$ the percolating cluster is a fractal. In particular, the number of occupied sites (mass) of the largest cluster on a lattice of size L scales as $M(L) \sim L^H$, where H is not an integer (for $D > 1$). On the other hand, when $p \ne p_c$, large clusters appear fractal on length scales up to the correlation length $\xi \sim |p - p_c|^{-\nu}$, whereas they appear homogeneous on much larger length scales. Let $\mathbf{r}_i$ denote the position of the ith occupied site in a cluster of size s. The center of mass and radius of gyration of the cluster are defined as

$$\mathbf{r}_{cm} = \frac{1}{s} \sum_{i=1}^{s} \mathbf{r}_i, \quad R_s^2 = \frac{1}{s} \sum_{i=1}^{s} |\mathbf{r}_i - \mathbf{r}_{cm}|^2. \tag{6.1.33}$$

Note that R_s can be expanded as

$$R_s^2 = \frac{1}{s} \sum_i (\mathbf{r}_i^2 + \mathbf{r}_{cm}^2 - 2\mathbf{r}_i \cdot \mathbf{r}_{cm}) = \frac{1}{s} \sum_i \mathbf{r}_i^2 - \mathbf{r}_{cm}^2 = \frac{1}{2}\frac{1}{s^2} \sum_{i,j} |\mathbf{r}_i - \mathbf{r}_j|^2.$$

Hence, the radius of gyration is equal to half the averaged square distance between two cluster sites. In particular, R_s characterizes the spatial extent of a cluster, which is particularly useful when the cluster is in an irregular shape such as a fractal. (The radius of gyration also plays a role in characterizing the effective size of polymers, see Chap. 12.) The scaling of $M(L)$ then suggests the scaling law $s \sim R_s^H$ for $s \gg 1$ and $p = p_c$. Finally, given that ξ determines the length scale at which there is a cross-over between fractal ($R_s \ll \xi$) and non-fractal ($R_s \gg \xi$) behavior, and s_ξ determines the corresponding cross-over point with respect to the number of occupied sites in a cluster, it follows that $s_\xi \sim \xi^H$.

6.1.4 Diffusion in the plasma membrane

At the simplest level, the plasma membrane can be treated as a 2D sheet of membrane lipids into which proteins are embedded, see Fig. 6.8. Membrane lipids are a group of compounds (structurally similar to fats and oils) which form the double-layered surface of cells. The three major classes of membrane lipids are phospholipids, glycolipids, and cholesterol. Lipids are amphiphilic: they have one end that is soluble in water ("polar") and the other that is soluble in fat ("nonpolar"). By forming a double layer with the polar ends pointing outwards and the nonpolar ends pointing inwards membrane lipids can form a "lipid bilayer" which keeps the watery interior of the cell separate from the watery exterior. The arrangements of lipids and various proteins, acting as receptors and channel pores in the membrane, control the entry and exit of other molecules as part of the cell's metabolism. (The self-assembly of lipid membranes will be considered in Chap. 13.)

In the fluid mosaic model of [762], the membrane lipids are treated as the solvent (water concentrations are very low within the membrane) into which proteins are dissolved. One of the consequences of the fluid mosaic model is that protein clustering, which alters the effective size of a diffusing particle, has only a weak effect on diffusion in the plasma membrane. This follows from the hydrodynamic membrane diffusion model of Saffman and Delbruck [713], which implies that the diffusion coefficient for a cylinder of radius r in a 2D membrane varies as $\ln r$.

Although the diffusion of lipids appears to be Brownian in pure lipid bilayers, single-particle tracking experiments indicate that lipids and proteins undergo anomalous diffusion in the plasma membrane [466, 480, 729]. This has led to a modification of the original fluid mosaic model, whereby lipids and transmembrane

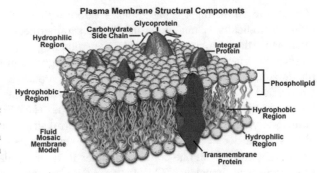

Fig. 6.8: Fluid mosaic model of the plasma membrane. [Public domain figure downloaded from Wikispaces.]

proteins undergo confined diffusion within, and hopping between, membrane microdomains or corrals [480, 481, 827]; the corralling could be due to "fencing" by the actin cytoskeleton or confinement by anchored protein "pickets", see Fig. 6.9. These microdomains could also be associated with lipid rafts [413, 481]. Partitioning the membrane into a set of corrals implies that anomalous diffusion of proteins will be observed on intermediate time scales, due to the combined effects of confinement and binding to the actin cytoskeleton. However, on time scales over which multiple hopping events occur, normal diffusion will be recovered. A rough estimate of the corresponding diffusion coefficient is $D \sim L^2/\tau$, where L is the average size of a microdomain and τ is the mean hopping rate between microdomains. A typical range of values for various types of mammalian cell is $L \sim 30 - 240$ nm and $\tau \sim 1 - 20$ ms.

In the case of confinement by anchored protein pickets, τ can be estimated by treating each corral as a domain with a set of small holes (gaps) between anchored proteins and solving a narrow escape problem [385, 386], see Sect. 6.4. (Another

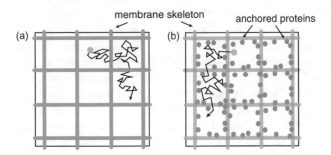

Fig. 6.9: Picket–fence model of membrane diffusion. The plasma membrane is parceled up into compartments whereby both transmembrane proteins and lipids undergo short-term confined diffusion within a compartment and long-term hop diffusion between compartments. This corralling is assumed to occur by two mechanisms. (a) The membrane-cytoskeleton (fence) model: transmembrane proteins are confined within the mesh of the actin-based membrane skeleton. (b) The anchored protein (picket) model: transmembrane proteins, anchored to the actin-based cytoskeleton, effectively act as rows of pickets along the actin fences.

approach to estimating τ is based on a random walker moving on a 1D lattice with either periodically or randomly distributed semipermeable barriers [437].) On the other hand, the membrane cytoskeleton surrounding a corral is usually modeled as an effective energy barrier over which a diffusing protein must escape. For example, Saxton carried out a computational study of a particle diffusing inside a corral surrounded by a static energy barrier [726]. It was assumed that when the particle hits the barrier it had a fixed probability of escape. The MFPT out of the corral was numerically determined for a wide range of corral sizes, shapes, and escape probabilities. In earlier work, Saxton considered a static fence model in which a protein could only move from one corral to another if the particular barrier separating the two corrals was dissociated [724]. In this particular model, large-scale diffusion only occurs if there exists a percolation network of dissociated barriers. However, estimates of the density of the actin cytoskeleton in red blood cells (erythrocytes), for example, suggest that the fraction of dissociated cytoskeleton is below the percolation threshold. Hence, it is necessary to modify the percolation model by considering time-dependent, fluctuating energy barriers. A simplified model of this process is to treat the corral as a well-mixed compartment model, where diffusion within the compartment is relatively fast so that one can ignore spatial effects, and to model the fluctuating barrier as a stochastic gate [144, 504], see Sect. 6.6.1.

6.2 Diffusion–trapping model of protein receptor trafficking in dendrites

Neurons are among the largest and most complex cells in biology. Their intricate geometry presents many challenges for cell function, in particular with regards to the efficient delivery of newly synthesized proteins from the cell body or soma to distant locations on the axon or dendrites. The axon contains ion channels for action potential propagation and presynaptic active zones for neurotransmitter release, whereas each dendrite contains postsynaptic domains (or densities) where receptors that bind neurotransmitter tend to cluster, see Fig. 3.10. At most excitatory synapses in the brain, the postsynaptic density (PSD) is located within a dendritic spine, which is a small, sub-micrometer membranous extrusion that protrudes from a dendrite [779], see Fig. 6.10. Typically spines have a bulbous head that is connected to the parent dendrite through a thin spine neck, and there can exist thousands of spines distributed along a single dendrite. It is widely thought that spines act to compartmentalize chemical signals generated by synaptic activity, thus impeding their diffusion into dendrites [710, 882]. Conversely, in the case of signaling molecules diffusing along the dendrite, the spines act as transient traps, as illustrated in Fig. 6.11(a). Following along similar arguments to the case of diffusion in the presence of obstacles, normal diffusion is expected at short and long times and anomalous subdiffusion at intermediate times. Anomalous subdiffusion was indeed observed experimentally by Santamaria et al [721], such that the mean-square displacement $\langle \mathbf{x}^2(t) \rangle \sim D_0 t^{2/\beta}$ at intermediate times with $\beta > 2$ and D_0 the free diffusion coef-

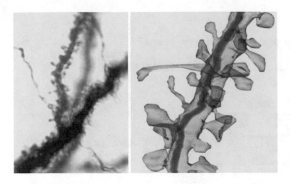

Fig. 6.10: An example of a piece of spine studded dendritic tissue (from rat hippocampal region CA1 stratum radiatum). The dendrite on the right-hand side is $\sim 5\mu$m in length. Taken with permission from SynapseWeb, Kristen M. Harris, PI, http://synapses.clm.utexas.edu/.

ficient. As might be expected, β increases (slower diffusion) with increasing spine density. β also increases when the volume of the spine head is increased relative to the spine neck, reflecting the fact there is an enhanced bottleneck. Note that anomalous diffusion can occur at all times if the reactions within each spine are taken to have a non-exponential waiting time density [268], see also Sect. 6.1.2.

A related problem is the diffusion–trapping of neurotransmitter protein receptors within the plasma membrane of a dendrite, with each spine acting as a transient trap that localizes the receptors at a synapse. The majority of fast excitatory synaptic transmission in the central nervous system is mediated by AMPA (α-amino-3-hydroxy-5-methyl-4-isoxazole-propionic acid) receptors, which respond to the neurotransmitter glutamate. There is now a large body of experimental evidence that the fast trafficking of AMPA receptors into and out of spines is a major contributor to activity-dependent, long-lasting changes in synaptic strength [109, 176, 187, 367, 650]. Single-particle tracking experiments suggest that surface AMPA receptors diffuse freely within the dendritic membrane until they enter a spine, where they are temporarily confined by the geometry of the spine and through interactions with scaffolding proteins such as PSD-95 and cytoskeletal elements [174, 175, 240, 305, 705, 813], see Fig. 6.11(b). (Analogous diffusion–trapping interactions occur between inhibitory glycine receptors (GlyRs) and the scaffolding molecule gephyrin [231, 568].) A surface receptor may also be internalized via endocytosis and stored within an intracellular compartment, where it is either recycled to the surface via recycling endosomes and exocytosis or sorted for degradation by late endosomes and lysosomes [239], see Fig. 6.2. A number of models have explored the combined effects of diffusion, trapping, receptor clustering, and recycling on the number of synaptic AMPA receptors within spines [205, 235, 236, 386, 812]. In such models, the synapse is treated as a self-organizing compartment in which the number of AMPA receptors is a dynamic steady state that determines the strength of the synapse; activity-dependent changes in the strength of the synapse then correspond to shifts in the dynamical set-point. When receptor–receptor interactions are included a synapse can exhibit bistability between a non-clustered and clustered state [753], which can also be understood in terms of a "liquid–vapor" phase transition [152].

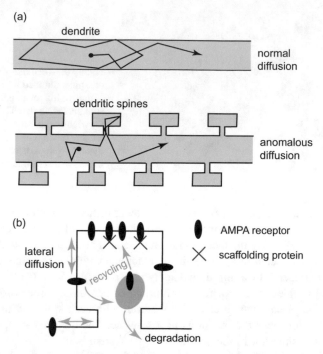

Fig. 6.11: (a) Schematic illustration of the anomalous diffusion model of [721], which was used to carry out detailed 3D simulations of diffusion in a spiny dendrite treated as a system of connected cylinders with the following baseline parameter values: spine neck diameter $0.2\,\mu$m, neck length $0.6\,\mu$m, head length and diameter $0.6\,\mu$m, dendrite diameter $1\,\mu$m, and a spine density of 15 spines/μm. The dendritic spines act as transient traps for a diffusing particle within the dendrite, which leads to anomalous diffusion on intermediate time scales. (b) Schematic illustration of various pathways of AMPA receptor trafficking at a dendritic spine, including lateral diffusion, binding with scaffolding proteins, and recycling between the surface and intracellular compartments.

Most diffusion–trapping models assume that the number of trapping sites or "slots" within a given synapse is fixed. However, it is known that scaffolding proteins and other synaptic components are also transported into and out of a synapse, albeit at slower rates [718]. Several studies have analyzed the joint localization of scaffolding proteins and receptors at synapses [358, 359, 746], showing how stable receptor-scaffold domains could arise via spontaneous pattern formation. The number of receptors or scaffolding proteins within a single synapse ranges from zero to a few hundred, which means that there could be significant fluctuations in the size or strength of an individual synapse over time scales of days and hours. That is, the capacity for a synapse to maintain its particular properties, synaptic tenacity, may be limited [899]. On the other hand, the distribution of properties of a local population of synapses could be relatively stable [751].

6.2.1 One-dimensional cable model

Given the tubular-like structure of a dendrite, it is possible to model the diffusion of proteins in the plasma membrane using a reduced 1D model (analogous to the cable equation for voltage changes along a dendrite [431]). Since the Green's function of the 1D diffusion equation is non-singular (in contrast to higher dimensions, see Box 6C), one can treat the dendritic spines as point-like sources or sinks [111]. This simplification makes it possible to investigate the heterosynaptic effects of diffusion across a population of spines, but at the cost of ignoring the detailed microstructure of the PSD and the geometry of spines. Recent imaging studies of the microstructure of the PSD suggest that there exist PSD microdomains containing higher densities of scaffolding proteins, where most of the receptors are stabilized [175, 439, 543]. Outside of these domains, receptors tend to diffuse more freely, irrespective of whether or not they are bound to scaffolding proteins. However, diffusion is still slower than in the dendritic shaft due to the crowded environment of the PSD and the geometry of a spine. For simplicity, we will ignore the latter feature by treating the dendritic membrane plus spine (excluding the PSD microdomains) as a homogeneous medium, with the PSD microdomains modeled as point-like traps. This is also more appropriate when considering inhibitory synapses located outside spines.[1]

Let $x \in [0, \infty)$ denote the axial coordinate along a semi-infinite dendritic cable with $x = 0$ the location of the somatic end. Consider a population of N spines distributed along a finite section of the cable, $x \in [0, L]$, with $x_j \in [0, L]$, $j = 1, \ldots, N$, the position (axial coordinate) of the jth spine. The positions are ordered such that $x_j < x_{j+1}$ for all $j = 1, \ldots, N - 1$. We will assume that the ith spine has a fixed number of slots S_i, each of which can transiently trap a receptor. (On longer time scales than considered here, the number of slots can also change.) Let $u(x, t)$ denote the concentration (per unit length) of surface receptors within the dendritic membrane at position x at time t. Similarly, let r_i denote the fraction of slots in the ith spine that are transiently occupied by a receptor. For the moment, we will assume that the number of slots is sufficiently large so we can use a simple kinetic equation for r_j. In addition to lateral membrane diffusion, receptors can be recycled between the surface membrane and intracellular pools. There exist sites of exocytosis and endocytosis within the proximity of synapses as well as various extrasynaptic locations along the dendrite. For simplicity, we will model the effects of extrasynaptic sites as a uniform source of receptor insertion at a rate σ and a uniform source of receptor

---

[1] One way to take into account the effects of reduced diffusion in spines would be to treat each spine as a compartment consisting of freely diffusing receptors, which are exchanged with the bulk concentration in the dendrite [236]. The diffusive flux into the jth spine is then $\Omega_j[u_j - R_j]$ with R_j the concentration of free receptors within the spine and Ω_j the strength of diffusive coupling. (The latter depends on the detailed geometry of the spine neck and head and can be estimated by solving a first passage time problem [3, 97, 286, 385, 387, 761].) Receptors within the spine compartment undergo various reactions, such as binding to and unbinding from scaffolding proteins (slots) or being exchanged with intracellular pools.

removal at a rate γ. (The rate of endocytosis $\gamma = \phi \gamma_0$, where γ_0 is the internalization rate of an individual receptor and ϕ is a scale factor that takes into account the fact that sites of endocytosis only cover a fraction of the dendritic surface.) In order to ensure that the total number of receptors is bounded, we assume that the sites of exocytosis are restricted to the domain $[0, L]$. Finally, we associate with each spine an intracellular pool that can exchange receptors with the surface at x_j, $j = 1, \ldots, N$.

The 1D diffusion–trapping model of receptor trafficking is defined according to the following system of equations (see Ex. 6.4 for an alternative model):

$$\frac{\partial u}{\partial t} = D \frac{\partial^2 u}{\partial x^2} + \sigma H(L - x) - \gamma u \tag{6.2.1a}$$

$$+ \sum_{j=1}^{N} S_j [k_- r_j - k_+ u_j (1 - r_j)] \delta(x - x_j) + \sum_{j=1}^{N} \left(\kappa_{\text{exo}} q_j - \kappa_{\text{end}} u_j \right) \delta(x - x_j),$$

$$\frac{dr_j}{dt} = [k_+ u_j (1 - r_j) - k_- r_j], \tag{6.2.1b}$$

$$\frac{dq_j}{dt} = h - \kappa q_j - \kappa_{\text{exo}} q_j + \kappa_{\text{end}} u_j, \tag{6.2.1c}$$

where $H(x)$ is the Heaviside function, $u_j(t) = u(x_j, t)$, D is the diffusivity of receptors in the plasma membrane, k_+ is the rate of binding of a receptor in the dendrite to a slot, and k_- is the corresponding rate of detaching from a slot and returning to the dendritic shaft. (Note that k_+ and k_- have different units.) Moreover, q_j is the number of receptors in the jth intracellular pool, and $\kappa_{\text{exo/end}}$ are the rates of exocytosis and endocytosis. We assume that the pool can be supplied with newly synthesized receptors at a rate h, and that intracellular receptors are sorted for degradation by lysosomes at a rate κ. Equation (6.2.1a) is supplemented by the boundary conditions

$$-D \frac{\partial u}{\partial x} \bigg|_{x=0} = 0, \quad u(x) \to 0 \text{ as } x \to \infty.$$

Integrating equation (6.2.1a) with respect to x and adding to equations (6.2.1b,6.2.1c) leads to the conservation equation

$$\frac{dU}{dt} + \sum_{j=1}^{N} S_j \frac{dr_j}{dt} + \sum_{j=1}^{N} \frac{dq_j}{dt} = \sigma L + Nh - \gamma U - \kappa \sum_{j=1}^{N} q_j, \tag{6.2.2}$$

where $U(t) = \int_0^\infty u(x, t) dx$ is the total number of freely diffusing receptors in the dendritic membrane. Unfortunately, only a few model parameters are known explicitly [812]. These include the unbinding rate $k_- \sim 2.5 \times 10^{-2} \text{ s}^{-1}$ [368], the rate of internalization $\gamma_0 \sim 10^{-3} \text{ s}^{-1}$ [240], and the membrane diffusivity $D \sim 0.1 \, \mu\text{m}^2/\text{s}$ [240]. One general observation, however, is that the basal rates of receptor binding and unbinding are at least an order of magnitude faster than the rates of receptor internalization and externalization. We will exploit this when developing a perturbation analysis of solutions.

6.2.2 Steady-state solution

The steady-state versions of equation (6.2.1b,6.2.1c) yield

$$r_j^* = \frac{k_+ u(x_j)}{k_- + k_+ u(x_j)}, \quad q_j^* = \frac{h + \kappa_{\text{end}} u(x_j)}{\kappa_{\text{exo}} + \kappa}, \tag{6.2.3}$$

with $u(x)$ satisfying the steady-state equation

$$0 = D\frac{d^2 u}{dx^2} + \sigma H(L-x) - \gamma u + \sum_{j=1}^{N} [h - \kappa q_j^*]\delta(x - x_j). \tag{6.2.4}$$

We will assume that $x_j \ll L$ for all $j = 1, \ldots, N$ and take $L \to \infty$. Equation (6.2.4) can then be solved in terms of 1D Neumann Green's function $G(x, \xi)$ on $[0, \infty)$, which is the solution to the equation

$$D\frac{d^2 G(x, \xi)}{dx^2} - \gamma G(x, \xi) = -\delta(x - \xi), \quad \left. \frac{dG(x, \xi)}{dx} \right|_{x=0} = 0. \tag{6.2.5}$$

One finds that (see Box 6C)

$$G(x, \xi) = \frac{1}{2}\sqrt{\frac{1}{D\gamma}} \left[e^{-|x-\xi|\sqrt{\gamma/D}} + e^{-(x+\xi)\sqrt{\gamma/D}} \right]. \tag{6.2.6}$$

It follows that the dendritic surface receptor concentration has an implicit solution of the form

$$u(x) = \sum_{j=1}^{N} [h - \kappa q_j^*]G(x, x_j) + \sigma \int_0^L G(x, \xi)d\xi. \tag{6.2.7}$$

Evaluating the integral on the right-hand side yields

$$\int_0^L G(x, \xi)d\xi = \frac{1}{\gamma}\left[1 - e^{-L\sqrt{\gamma/D}}\sinh(x\sqrt{\gamma/D})\right]$$

for $0 \le x \le L$, and

$$\int_0^L G(x, \xi)d\xi = \frac{1}{\gamma}e^{-x\sqrt{\gamma/D}}\cosh(L\sqrt{\gamma/D})$$

for $x > L$.

We can now generate a closed set of equations for the concentration of dendritic receptors u_i at the ith synapse, $i = 1, \ldots, N$, by setting $x = x_i$ in equation (6.2.7):

$$u_i = \sum_{j=1}^{N} \frac{h\kappa_{\text{exo}} - \kappa\kappa_{\text{end}}u_j}{\kappa_{\text{exo}} + \kappa}G(x_i, x_j) + \frac{\sigma}{\gamma}f(x_i), \tag{6.2.8}$$

with

$$f(x) = \left[1 - e^{-L\sqrt{\gamma/D}}\sinh(x\sqrt{\gamma/D})\right].$$ (6.2.9)

This can be rewritten as the matrix equation

$$\sum_{j=1}^{N} M_{ij}u_j = F_i,$$ (6.2.10a)

with

$$M_{ij} = \delta_{i,j} + \frac{\kappa\kappa_{end}}{\kappa_{exo}+\kappa}G(x_i,x_j), \quad F_i = \frac{\sigma}{\gamma}f(x_i) + \sum_{j=1}^{N}\frac{h\kappa_{exo}}{\kappa_{exo}+\kappa}G(x_i,x_j).$$ (6.2.10b)

One way to invert the matrix is to use perturbation theory. That is, exploiting a separation of time scales, suppose that

$$\frac{\kappa_{end}}{\sqrt{D\gamma}} = \varepsilon, \frac{h}{\sqrt{D\gamma}} = \varepsilon\frac{\sigma}{\gamma},$$ (6.2.11)

with $0 < \varepsilon \ll 1$. If $D = 0.1\mu m^2 \, s^{-1}$, $\gamma_0 = 10^{-3} \, s^{-1}$ and the area fraction $\phi = 10^{-2}$, then $\sqrt{D\gamma} \sim 10^{-3}\mu m/s$. We can now carry out a perturbation expansion of the steady-state solution given by equations (6.2.7) and (6.2.8) with respect to ε. This yields to $O(\varepsilon)$

$$u(x) \approx \frac{\sigma}{\gamma}\left[f(x) + \varepsilon\sum_{j=1}^{N}\left(\frac{\kappa_{exo}-\kappa f(x_j)}{\kappa_{exo}+\kappa}\right)G_j(x)\right]$$ (6.2.12)

for $0 \leq x \leq L$, where we have set $G(x,x_j) = G_j(x)/\sqrt{D\gamma}$. Hence, the steady-state membrane concentration $u(x)$ couples to the spines at $O(\varepsilon)$.

One of the major consequences of the above analysis is that there is a characteristic length scale for receptor trafficking given by $\sqrt{D/\gamma}$. If $D = 0.1 \, \mu m^2/s$, $\gamma_0 = 10^{-3} \, s^{-1}$, and the area fraction $\phi = 10^{-2}$, then $\sqrt{D/\gamma} = 100\mu m$. On the other hand, if $\phi = 10^{-1}$, then $\sqrt{D/\gamma} \approx 30\mu m$. It follows that if the spines are clustered over some interval $[X, X+l]$ with $l \ll \sqrt{D/\gamma}$, then $G(x_i,x_j) \approx G(X,x_j)$ and $u(x_i) \approx u(X)$ for all $x_i \in [X, X+l]$. One thus finds that the fraction of filled slots in each synapse is the same, irrespective of the size of the synapse:

$$r_j = \Gamma := \frac{k_+u(X)}{k_- + k_+u(X)}, \quad j = 1,\ldots,N.$$ (6.2.13)

Suppose that we identify $w_j = S_j\Gamma$ as the strength of the jth synapse. This implies that the ratio of synaptic strengths of any two synapses within the domain is equal to the ratio of their number of slots. Moreover, if Γ is modified due to changes in the influx rate σ, say, then all synaptic strengths are scaled by the same factor (multiplicative scaling). This result was previously obtained in a non-spatial compartmental model of the form [812]

$$\frac{du}{dt} = \sigma - \gamma u + \sum_{j=1}^{N} S_j[k_- r_j - k_+ u(1 - r_j)], \qquad (6.2.14a)$$

$$\frac{dr_j}{dt} = [k_+ u(1 - r_j) - k_- r_j], \qquad (6.2.14b)$$

where u is now the total number of extrasynaptic surface receptors in the compartment and k_+ has been redefined accordingly. In this model a local population of spines compete for receptors supplied at a rate σ and removed at a rate γ. Equation (6.2.13) holds with $u(X) \to \sigma/\gamma$, so that

$$r_j = \frac{1}{(k_- \gamma/k_+ \sigma) + 1}, \qquad j = 1, \dots, N. \qquad (6.2.15)$$

Our more detailed diffusion–trapping model shows that this scaling rule breaks down for a set of spines distributed over a domain of size L with $L \geq \sqrt{D/\gamma}$. In particular, one cannot decouple the steady-state concentration of surface receptors from the bound receptor numbers in the spines, unless additional assumptions are made.

Box 6C. Constructing one-dimensional Green's functions.

Consider a second-order differential equation on the interval $[a, b]$ with either Neumann or Dirichlet boundary conditions at $x = a, b$. We will assume that the differential equation is of the so-called Sturm–Liouville form

$$\mathbb{L}u := (p(x)u')' + q(x)u(x) = f(x), \quad x \in [a, b], \qquad (6.2.16)$$

with p, q, f suitably smooth functions. The classical method for solving such an inhomogeneous equation is in terms of Green's function $G(x, \xi)$, defined according to

$$\mathbb{L}G(x, \xi) = \delta(x - \xi). \qquad (6.2.17)$$

It immediately follows that $G(x, \xi)$ must be continuous at $x = \xi$, otherwise the second-order derivative will generate a derivative of a Dirac delta function. This motivates writing Green's function as

$$G(x, \xi) = \begin{cases} Au_L(x)u_R(\xi), & x < \xi \\ Au_L(\xi)u_R(x), & x > \xi \end{cases}, \qquad (6.2.18)$$

which ensures continuity. We take $u_L(x)$ to be a solution of $\mathbb{L}u = 0$ on $x \in [a, \xi)$ and satisfies the left-hand boundary condition, whereas $\mathbb{L}u_R = 0$ on $(\xi, b]$ and satisfies the right-hand boundary condition. This guarantees that Green's function satisfies the homogeneous equation $\mathbb{L}G(x, \xi) = 0$ for all $x \neq \xi$. In order to generate the Dirac delta function, there must be a discontinuity in the first derivative at $x = \xi$. In order to determine the jump condition, integrate equation (6.2.17) from $\xi - \varepsilon$ to $\xi + \varepsilon$ to give

$$p(\xi) \lim_{\varepsilon \to 0^+} [\partial_x G(\xi + \varepsilon, \xi) - \partial_x G(\xi - \varepsilon, \xi)] = 1.$$

Substituting the product form for G thus yields the jump condition

$$Ap(\xi) \left(u_L(\xi)u_R'(\xi) - u_L'(\xi)u_R(\xi) \right) = 1. \tag{6.2.19}$$

This determines the constant A according to

$$A = \frac{1}{pW}, \quad W(u_L, u_R; \xi) = u_L(\xi)u_R'(\xi) - u_L'(\xi)u_R(\xi),$$

where $W(u_L, u_R; \xi)$ is the so-called Wronskian. One of the nice features of the Sturm–Liouville equations is that the product pW is a constant.

Example (i). Consider the inhomogeneous diffusion equation

$$-\frac{d^2 u}{dx^2} = f(x), \quad u(0) = u(1) = 0.$$

for which $p(x) = -1$. It is straightforward to show that

$$u_L(x) = x, \quad u_R(x) = 1 - x, \quad u_L u_R' - u_L' u_R = -1,$$

and hence

$$G(x, \xi) = \begin{cases} x(1 - \xi), \, x < \xi \\ \xi(1 - x), \, x > \xi \end{cases}. \tag{6.2.20}$$

Example (ii). Consider an equation similar to the cable equation for transport in dendrites, namely,

$$D\frac{d^2 u}{dx^2} - \gamma u(x) = f(x), \quad \left.\frac{du}{dx}\right|_{x=0} = 0, \quad u(x) \to 0, \quad x \to \infty.$$

This is a Sturm–Liouville equation with $p(x) = D$ and $q(x) = -\gamma$. In this case

$$u_L(x) = \left[e^{x\sqrt{\gamma/D}} + e^{-x\sqrt{\gamma/D}} \right], \quad u_R(x) = e^{-x\sqrt{\gamma/D}}.$$

The Wronskian is then

$$W = u_L u_R' - u_L' u_R = -2\sqrt{\frac{\gamma}{D}},$$

and hence

$$G(x, \xi) = \begin{cases} -\frac{1}{2\sqrt{\gamma D}} \left[e^{(x-\xi)\sqrt{\gamma/D}} + e^{-(x+\xi)\sqrt{\gamma/D}} \right], \, x < \xi \\ -\frac{1}{2\sqrt{\gamma D}} \left[e^{(\xi-x)\sqrt{\gamma/D}} + e^{-(x+\xi)\sqrt{\gamma/D}} \right], \, x > \xi \end{cases}. \tag{6.2.21}$$

This is consistent with equation (6.2.6) under $G \to -G$ and $\gamma \to s + \gamma$.

6.2.3 Accumulation time.

The above steady-state analysis does not provide any information regarding the rate at which the system approaches steady state, and how this depends on the location of spines along the dendrite. This is important because one would like to know how the system responds to time-dependent changes in model parameters such as the number of slots in a synapse or the rate of receptor-slot binding. Both of these can be transiently altered during activity-dependent synaptic plasticity. In light of such time-dependent changes, the spines may never reach steady state. An important quantity in characterizing the time-dependent approach to steady state of a diffusion process is the accumulation time. (This is commonly used to estimate the time to form a protein concentration gradient during morphogenesis [79, 80, 332], which has to be consistent with developmental time scales, see also Chap. 11 and Chap. 14.) In order to construct the accumulation time, consider the function

$$R(x,t) = 1 - \frac{u(x,t)}{u(x)}, \qquad (6.2.22)$$

which represents the fractional deviation of the concentration from the steady state. Assuming that there is no overshooting, $1 - R(x,t)$ is the fraction of the steady-state concentration that has accumulated at x by time t. It follows that $-\partial_t R(x,t)dt$ is the fraction accumulated in the interval $[t, t+dt]$. The accumulation time is then defined by analogy to MFPTs [79, 80, 332]

$$\tau(x) = \int_0^\infty t \left(-\frac{\partial R(x,t)}{\partial t} \right) dt = \int_0^\infty R(x,t)dt. \qquad (6.2.23)$$

Note that a finite accumulation time implies that the steady-state $u(x)$ is a stable solution.

We will calculate the accumulation time using Laplace transforms under the assumption that $r_j \ll 1$ for all $j = 1, \ldots, N$. First, setting $\widetilde{F}(x,s) = s\widetilde{u}(x,s)$ and Laplace transforming equation (6.2.22) yields

$$s\widetilde{R}(x,s) = 1 - \frac{\widetilde{F}(x,s)}{\widetilde{F}(x,0)},$$

and

$$\tau(x) = \lim_{s \to 0} \widetilde{R}(x,s) = \lim_{s \to 0} \frac{1}{s} \left[1 - \frac{\widetilde{F}(x,s)}{\widetilde{F}(x,0)} \right] = -\frac{1}{\widetilde{F}(x,0)} \left. \frac{d\widetilde{F}(x,s)}{ds} \right|_{s=0} \qquad (6.2.24)$$

Laplace transforming equations (6.2.1a-6.2.1c) using the initial conditions $u(x,0) = 0$, $r_j(0) = 0$, and $q_j(0) = 0$ gives

$$D\frac{\partial^2 \tilde{u}}{\partial x^2} + \frac{\sigma}{s}H(L-x) - (s+\gamma)\tilde{u} \tag{6.2.25a}$$

$$= \sum_{j=1}^{N} S_j[k_+\tilde{u}_j - k_-\tilde{r}_j]\delta(x-x_j) + \sum_{j=1}^{N} (\kappa_{end}\tilde{u}_j - \kappa_{exo}\tilde{q}_j)\,\delta(x-x_j),$$

$$s\tilde{r}_j = [k_+\tilde{u}_j - k_-\tilde{r}_j], \tag{6.2.25b}$$

$$s\tilde{q}_j = \frac{h}{s} - \kappa\tilde{q}_j - \kappa_{exo}\tilde{q}_j + \kappa_{end}\tilde{u}_j, \tag{6.2.25c}$$

where $\tilde{u}_j(s) = \tilde{u}(x_j,s)$, together with the boundary condition $\partial\tilde{u}/\partial x = 0$ at $x = 0$.

Analogous to equation (6.2.5), we introduce s-dependent Neumann Green's function on $[0,\infty)$ according to

$$D\frac{\partial^2 G(x,y;s)}{\partial x^2} - (s+\gamma)G(x,y;s) = -\delta(x-y), \qquad \left.\frac{\partial G(x,y;s)}{\partial x}\right|_{x=0} = 0. \tag{6.2.26}$$

From equation (6.2.6) we have

$$G(x,y;s) = \frac{1}{2}\sqrt{\frac{1}{D[s+\gamma]}}\left[e^{-|x-y|\sqrt{(s+\gamma)/D}} + e^{-(x+y)\sqrt{(s+\gamma)/D}}\right]. \tag{6.2.27}$$

It then follows that

$$\tilde{u}(x,s) = \frac{\sigma}{s}\int_0^L G(x,y;s)dy - \sum_{j=1}^{N}\frac{sk_+\tilde{u}(x_j,s)}{s+k_-}S_j G(x,x_j;s)$$

$$- \sum_{j=1}^{N}\left[(s+\kappa)\frac{hs^{-1} + \kappa_{end}\tilde{u}(x_j,s)}{s+\kappa_{exo}+\kappa} - \frac{h}{s}\right]G(x,x_j;s). \tag{6.2.28}$$

Multiplying this equation by s and taking the limit $s \to 0$ immediately recover the steady-state solution (6.2.7). In order to simplify the calculation of the accumulation time, we assume that the scalings (6.2.11) hold and carry out a perturbation expansion in ε. Keeping only $O(1)$ terms,

$$\lim_{s\to 0}\frac{d}{ds}\tilde{F}(x,s) = \sigma\lim_{s\to 0}\partial_s\int_0^L G(x,y;s)dy - 2\sum_{j=1}^{N}\frac{k_+u(x_j)}{k_-}S_j G(x,x_j),$$

with $G(x,\xi) = G(x,\xi;0)$. Evaluating the integral with respect to y shows that

$$\int_0^L G(x,y;s)dy = \frac{1}{s+\gamma}\left[1 - e^{-L\sqrt{[\gamma+s]/D}}\sinh(x\sqrt{[s+\gamma]/D})\right]$$

for $0 \le x \le L$. Taking $u(x) = \sigma f(x)/\gamma$ to leading order, we have

$$\tau(x) = \tau_0(x) + 2 \sum_{j=1}^{N} \frac{k_+ f(x_j)}{k_- f(x)} S_j G(x, x_j) \tag{6.2.29}$$

for $0 \leq x \leq L$, where

$$\tau_0(x) = \frac{1}{\gamma} \left(1 - \frac{1}{4f(x)} \left[(L-x) \sqrt{\frac{\gamma}{D}} e^{-(L-x)\sqrt{\gamma/D}} - (L+x) \sqrt{\frac{\gamma}{D}} e^{-(L+x)\sqrt{\gamma/D}} \right] \right) \tag{6.2.30}$$

is the accumulation time without spines. It can be seen that, in contrast to the steady-state solution (6.2.12), the spines have an effect on the accumulation time at $O(1)$.

6.3 Diffusion-limited reactions

Diffusion-limited (or diffusion-controlled) reactions are reactions that occur so quickly that the rate-limiting step is the transport mechanism that brings reactants together. The rate-limiting step is typically governed by diffusion in solution.

6.3.1 Smoluchowski reaction rate theory

We begin by describing the Smoluchowski rate theory for diffusion-limited reactions [188, 434, 674, 696, 771]. The simplest version of the theory concerns the bimolecular reaction $A + B \rightarrow AB$ for which the concentrations evolve according to the following law of mass action:

$$\frac{d[AB]}{dt} = k[A][B].$$

We assume that an A molecule and a B molecule react immediately to form the complex AB when they encounter each other within a reaction radius, so that the speed of reaction k is limited by their encounter rate via diffusion. (Note that k has units of volume s^{-1}.) Following Smoluchowski, the problem is formulated as an idealized FPT process, in which one A molecule, say, is fixed and treated as the center of a spherical target domain of reaction radius a, while the B molecules diffuse and are absorbed if they hit the boundary of the target domain, see Fig. 6.12a. It is assumed that the density of the particles is sufficiently small, so that reactions with other A molecules have a negligible effect on the concentration of B molecules in a neighborhood of the target molecule. The steady-state flux to the target (if it exists) is then identified as the mean reaction rate k across many targets. Let Ω denote the target domain and $\partial\Omega$ its absorbing boundary. We then need to solve the diffusion equation for the concentration $c(\mathbf{x}, t)$ of background molecules exterior to the domain Ω:

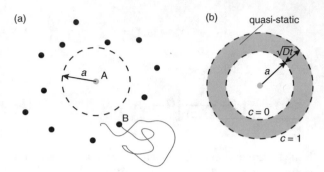

Fig. 6.12: Diffusion-limited reaction rate. (a) Diffusing molecules B in a neighborhood of a fixed target molecule A with reaction radius a. (b) Quasi-static approximation for calculating time-dependent reaction rate.

$$\frac{\partial c(\mathbf{x},t)}{\partial t} = D\nabla^2 c(\mathbf{x},t), \quad c(\mathbf{x} \in \partial\Omega, t) = 0, \, c(\mathbf{x},0) = c_0, \tag{6.3.1}$$

subject to the far-field boundary condition $c(\mathbf{x},t) = c_0$ for $\mathbf{x} \to \infty$. The flux through the target boundary is

$$J = D \int_{\partial\Omega} \nabla c \cdot d\mathbf{S}.$$

Note the sign, which is due to the fact that the flux is from the exterior to the interior of the target.

Let d denote the spatial dimension of the target. For $d > 2$, a diffusing particle is transient, which means that there is a nonzero probability of never reaching the target, see Sect. 2.1. Hence, the loss of reactants by target absorption is balanced by their resupply from infinity. It follows that there exists a steady state in which the reaction rate is finite. On the other hand, for $d \le 2$, reactants are sure to hit the target (recurrent diffusion) and a depletion zone continuously develops around the target so that the flux and reaction rate decay monotonically to zero with respect to time. Although a reaction rate does not strictly exist, it is still useful to consider the time-dependent flux as a time-dependent reaction rate. The 2D case is particularly important when considering interactions of molecules embedded in the plasma membrane of a cell or the lipid bilayer surrounding an intracellular compartment.

First consider the case of a spherical target of radius a ($d = 3$). Exploiting the radial symmetry of the problem, it is possible to set $u(r,t) = rc(r,t)$ such that the 3D diffusion equation for c reduces to a 1D diffusion equation for u [674]:

$$\frac{\partial u(r,t)}{\partial t} = D\frac{\partial^2 u(r,t)}{\partial r^2},$$

with $u(r,0) = rc_0$, $u(a,t) = 0$, and $u(r,t) = rc_0$ as $r \to \infty$. Laplace transforming this equation gives $s\tilde{u}(r,s) - rc_0 = D\tilde{u}''(r,s)$, which has the solution

$$\tilde{u}(r,s) = \frac{c_0}{s}\left[r - ae^{-(r-a)\sqrt{s/D}}\right].$$

Since the inverse Laplace transform of $s^{-1}[1 - e^{-r\sqrt{s/D}}]$ is the error function $\mathrm{erf}(r/\sqrt{4Dt})$, see Table 2.1, where

$$\mathrm{erf}(z) = \frac{2}{\sqrt{\pi}} \int_0^z e^{-r^2} dr,$$

one finds that

$$c(r,t) = c_0 \left(1 - \frac{a}{r}\right) + \frac{ac_0}{r} \mathrm{erf}\left[\frac{r-a}{\sqrt{4Dt}}\right]. \qquad (6.3.2)$$

It follows that the time-dependent flux is

$$J(t) = 4\pi a^2 D \left.\frac{\partial c}{\partial r}\right|_{r=a} = 4\pi a D c_0 \left(1 + \frac{a}{\sqrt{\pi Dt}}\right) \underset{t\to\infty}{\longrightarrow} 4\pi a D c_0. \qquad (6.3.3)$$

Hence, we obtain the Smoluchowski reaction rate $k = 4\pi a D$. As highlighted by Redner [674], it is straightforward to generalize the steady-state result to other 3D targets by making a connection with electrostatics. That is, setting $\phi(\mathbf{x}) = 1 - c(\mathbf{x})/c_0$ in steady state, it follows that ϕ satisfies Laplace's equation with $\phi = 1$ on the target boundary and $\phi = 0$ at infinity, so that ϕ is equivalent to the electrostatic potential generated by a perfectly conducting object Ω held at unit potential. Moreover, the steady-state reaction rate $k = 4\pi D Q$ where Q is the total charge on the surface of the conductor, which for a unit potential is equal to the capacitance, $Q = C$. Thus, determining the reaction rate for a general 3D target is equivalent to finding the capacitance of a perfect conductor with the same shape; see also [169].

Although it is possible to calculate the exact time-dependent flux for $d \leq 2$, a much simpler method is to use a quasi-static approximation [674]. Consider, for example, a target disk of radius $r = a$. The region exterior to the disk is divided into a near zone that extends a distance $\sqrt{Dt}$ from the surface and a complementary far zone, see Fig. 6.12b. In the near zone, it is assumed that diffusing particles have sufficient time to explore the domain before being absorbed by the target so that the concentration in the near zone can be treated as almost steady or quasi-static. Conversely, it is assumed that the probability of a particle being absorbed by the target is negligible in the far zone, since a particle is unlikely to diffuse more than a distance $\sqrt{Dt}$ over a time interval of length t. Thus, $c(r) \approx c_0$ for $r > \sqrt{Dt} + a$. The near zone concentration is taken to be a radially symmetric solution of Laplace's equation, which for $d = 2$ is $c(r) = A + B \ln r$. Matching the solution to the boundary conditions $c(a) = 0$ and $c(a + \sqrt{Dt}) = c_0$ then gives (for $\sqrt{Dt} \gg a$)

$$c(r,t) \approx \frac{c_0 \ln(r/a)}{\ln(\sqrt{Dt}/a)}. \qquad (6.3.4)$$

The corresponding time-dependent flux is

$$J(t) \approx \frac{2\pi D c_0}{\ln(\sqrt{Dt}/a)}. \qquad (6.3.5)$$

Radiation boundary condition. Collins and Kimball [188] generalized the Smoluchowski theory to the case of a radiation or partially absorbing boundary condition

$$-4\pi r^2 J(r,t) = 4\pi r^2 D \frac{dc(r,t)}{dr} = k_{on} c(a,t), \qquad (6.3.6)$$

where k_{on} is the association rate for a single receptor in close proximity to a ligand (intrinsic binding rate). If $k_{on} \to \infty$ then we recover the fully diffusion-limited regime. At steady state, we can integrate the radiation condition to give

$$c(r) - c(a) = \int_a^r \frac{k_{on} c(a)}{4\pi D r^2} dr = \frac{k_{on} c(a)}{4\pi D} \left(\frac{1}{a} - \frac{1}{r} \right).$$

Finally, using the far-field condition $c(\infty) = c_0$, the concentration at the surface is

$$c(a) = \frac{c_0}{1 + k_{on}/(4\pi D a)}.$$

We thus have the following result: the steady-state reaction rate is

$$k = \frac{4\pi D a k_{on}}{4\pi D a + k_{on}}. \qquad (6.3.7)$$

In the limit $k_{on} \to \infty$ one obtains the result for a perfect absorber with $k \to 4\pi D a$. On the other hand, if $k_{on} \ll D a$ then the depletion rate is so slow that $c(a) \approx c_0$, the background concentration, and $k \to k_{on}$.

Chemoreception. One application of diffusion-limited reactions is to chemoreception. In the case of a bacterium such as *E. coli*, the cell surface is covered with receptors that detect signaling molecules in the surrounding environment, see Fig. 6.13. Treating the cell as a perfect absorber assumes that there is a sufficient number of receptors distributed on the cell surface and that binding of a signaling molecule is instantaneous when it hits the surface. There are two major simplifications of such a model—(i) the rate of receptor/ligand binding k_{on} is finite, and (ii) receptors tend to be nonuniformly distributed on the cell surface. The role of receptor clustering in signal amplification will be addressed in Chap. 10, where we discuss the biochemical networks involved in bacterial chemosensing. In the case of a uniform distribution of N receptors, we can simply use equation (6.3.7) under the scaling $k_{on} \to N k_{on}$:

$$k = \frac{4\pi D a N k_{on}}{4\pi D a + N k_{on}}. \qquad (6.3.8)$$

Note that Shoup and Szabo [752] interpreted the factor $\pi = N k_{on}/(4\pi D a + N k_{on})$ as a capture probability of a ligand on to the cell surface membrane and defined an effective dissociation rate according to the product of an intrinsic rate k_{off} and the escape probability $1 - \pi$:

$$k_r = k_{off} \frac{4\pi D a}{4\pi D a + N k_{on}}. \qquad (6.3.9)$$

(a)

(b)

Fig. 6.13: Schematic diagram of (a) a polarized cell such as *E. coli* with a cluster of chemoreceptors and (b) a spherical cell with a uniform distribution of receptors.

In practice, one expects the effective dissociation rate from a population of receptors to depend on the frequency of rebinding, which itself will depend on the diffusive wandering of a recently detached ligand and its probability of reattachment. Such a process is only implicitly taken into account in the formula for k_r. A number of mathematical models of ligand rebinding at cell membrane surfaces have also been developed [307, 328, 329, 488].

Chemoreceptors allow motile *E. coli* to detect changes in concentration of a chemoattractant (food source). *E. coli* propels itself by rotating its flagella. In order to move forward, the flagella rotate together counterclockwise enabling the bacterium to "swim" at low Reynolds number. However, when the flagella rotation abruptly changes to clockwise, the bacterium "tumbles" in place and seems incapable of going anywhere. Then the bacterium begins swimming again in some new, random direction. Swimming is more frequent as the bacterium approaches a chemoattractant (food). Tumbling, hence direction change, is more frequent as the bacterium moves away from the chemoattractant. It is the complex combination of swimming and tumbling that keeps them in areas of higher food concentrations. One important issue is why *E. coli* has to move in order to detect changes in concentration rather than simply comparing differences across its body length. The answer is that there are limitations to the sensitivity of chemoreception due to thermal noise, which means that typical concentration changes along a cell body of size 1 μm are below the signal-to-noise ratio. This observation was first made in a classical paper of Berg and Purcell [83], whose analysis will be presented in Chap. 10. One heuristic way to estimate the sensitivity is to assume that a bacterium integrates signals from chemoreceptors for a mean time τ_{avg}. Assuming a perfect absorber for simplicity, the total number of signaling molecules absorbed is then $N \sim aDc\tau_{avg}$. Based on the law of large numbers, we expect fluctuations in the number of molecules to vary as $\sqrt{N}$. Hence,

$$\frac{\delta c}{c} \sim \frac{\delta N}{N} \sim \frac{1}{\sqrt{Dac\tau_{avg}}}. \tag{6.3.10}$$

Taking $D \sim 10^{-5}$ cm^2/s, $a \sim 1\mu$m, and a typical concentration $c = 6 \times 10^{11}$ molecules per cm^3, we have $Dac \sim 600$ s^{-1}. Assuming that the bacterium integrates for a time $\tau_{avg} \sim 1.5$ s, then $\delta c/c \sim 1/30$. Changes in c across 1 μm are just too small to detect. However, since the speed of motion is $v \sim 10 - 20\,\mu$m/s, it is possible to sample concentration changes of a length scale up to 30 times longer. Note that there is a limit to how large a time τ_{avg} the bacterium can integrate a chemical signal during a run, since rotational diffusion will interfere with the run's direction over longer time scales. (The problem of rotational diffusion was considered in Ex. 2.8.)

6.3.2 Facilitated diffusion and protein search for DNA binding sites

A crucial step in gene regulation is the binding of a protein transcription factor (see Sect. 5.3) to a specific target sequence of base pairs (target site) on a long DNA molecule. The precise mechanism whereby a protein finds its DNA binding site remains unclear. However, it has been observed experimentally that reactions occur at very high rates, of around $k = 10^{10}$ M^{-1} s^{-1} [697, 698]. This is around 100 times faster than the rate based on the Smoluchowski theory of diffusion-limited reaction rates and 1000 times higher than most known protein–protein association rates. This apparent discrepancy in reaction rates suggests that some form of facilitated diffusion occurs. The best-known theoretical model of facilitated diffusion for proteins searching for DNA targets was originally developed by Berg, Winter, and von Hippel (BHW) [87, 88, 862] and subsequently extended by a number of groups [194, 354, 399, 580]. The basic idea of the BHW model is to assume that the protein randomly switches between two distinct phases of motion, 3D diffusion in solution and 1D diffusion along DNA (sliding), see Fig. 6.14a. Such a mechanism is one example of a random intermittent search process (see Sect. 7.3). The BHW model assumes that there are no correlations between the two transport phases, so that the main factor in speeding up the search is an effective reduction in the dimensionality of the protein motion. However, there are a number of discrepancies between the BHW model and experimental data, which has led to several alternative theoretical approaches to facilitated diffusion. We first review the BHW model and then briefly discuss these alternative models.

A simple method for estimating the effective reaction rate of facilitated diffusion in the BHW model is as follows [580]. Consider a single protein searching for a single binding site on a long DNA strand of N base pairs, each of which has length b. Suppose that on a given search, there are R rounds labeled $i = 1, \ldots, R$. In the ith round the protein spends a time $T_{3,i}$ diffusing in the cytosol followed by a period $T_{1,i}$ sliding along the DNA. The total search time is thus $T = \sum_{i=1}^{R}(T_{3,i} + T_{1,i})$, and the mean search time is $\tau = r(\tau_3 + \tau_1)$. Here r is the mean number of rounds and τ_3, τ_1 are the mean durations of each phase of 3D and 1D diffusion. Let n denote the mean number of sites scanned during each sliding phase with $n \ll N$. If the binding site of DNA following a 3D diffusion phase is distributed uniformly along the DNA, then the probability of finding the specific promoter site is $p = n/N$. It follows that the probability of finding the site after R rounds is $(1-p)^{R-1}p$. Hence, the mean number of rounds is $r = 1/p = N/n$. Assuming that 1D sliding occurs via normal diffusion, then $nb = 2\sqrt{D_1 \tau_1}$ where D_1 is the 1D diffusion coefficient, and we have [580]

$$\tau = \frac{N}{n}(\tau_1 + \tau_3), \quad n = \frac{2\sqrt{D\tau_1}}{b}. \tag{6.3.11}$$

Since τ_3 depends primarily on the cellular environment and is thus unlikely to vary significantly between proteins, it is reasonable to minimize the mean search time with respect to τ_1 while τ_3 is kept fixed. Setting $d\tau/d\tau_1 = 0$ implies that the optimal search time occurs when $\tau_1 = \tau_3$ with $\tau_{\text{opt}} = 2N\tau_3/n$. On the other hand, the expected search time for pure 3D diffusion gives $\tau_{3D} = N\tau_3$, which is the approximate time to find one out of N sites by randomly binding to a single site of DNA every τ_3 seconds and no sliding ($\tau_1 = 0$). Thus facilitated diffusion is faster by a factor $n/2$.

Further insights into facilitated diffusion may be obtained by using the Smoluchowski formula for the rate at which a diffusing protein can find any one of N binding sites of size b, namely, $\tau_3^{-1} = 4\pi D_3 N b [DNA]$, where $[DNA]$ is the concentration of DNA. (We are simplifying the problem by not worrying about the 3D geometry of DNA.) Using this to eliminate N shows that the effective reaction rate of facilitated diffusion is [580]

$$k \equiv \frac{1}{\tau[DNA]} = 4\pi D_3 \left(\frac{\tau_3}{\tau_1 + \tau_3}\right) nb.$$

This equation identifies two competing mechanisms in facilitated diffusion. First, sliding diffusion effectively increases the reaction cross section from 1 to n base pairs, thus accelerating the search process compared to standard Smoluchowski theory. This is also known as the antenna effect [399]. However, the search is also slowed down by a factor $\tau_3/(\tau_1 + \tau_3)$, which is the fraction of the time the protein spends in solution. That is, a certain amount of time is lost by binding to nonspecific sites that are far from the target. Note that typical experimental values are

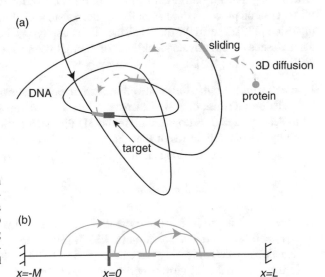

Fig. 6.14: (a) Mechanism of facilitated diffusion involving alternating phases of 3D diffusion and 1 D diffusion (sliding along the DNA). (b) 1D representation of facilitated diffusion.

$D_3 = 10 \, \mu m^2/s$, $b = 0.34$ nm, and $n = 200$, and one has to convert k into units of inverse molars per sec.

There have been a number of extensions of the BHW model that incorporate various biophysical effects. For example, sequence-dependent protein–DNA interactions generate a rugged energy landscape during sliding motion of the protein [580]. This observation then leads to an interesting speed-stability paradox [580, 749]. On the one hand, fast 1D search requires that the variance σ^2 of the protein–DNA binding energy be sufficiently small, that is, $\sigma \sim k_B T$, whereas stability of the protein at the DNA target site requires $\sigma \sim 5k_B T$. One suggested resolution of this paradox is to assume that a protein–DNA complex has two conformational states: a recognition state with large σ and a search state with small σ [580]. If the transitions between the states are sufficiently fast, then target stability and fast search can be reconciled. (For a recent review of the speed-stability paradox and its implications for search mechanisms see [749].) Other effects include changes in the conformational state of DNA and the possibility of correlated association/dissociation of the protein [399], and molecular crowding along DNA [516] or within the cytoplasm [407].

The BHW model and its extensions provide a plausible mechanism for facilitated diffusion that has some support from experimental studies, which demonstrate that proteins do indeed slide along DNA [333, 516, 862]. In particular, recent advances in single-molecule spectroscopy mean that the motion of fluorescently labeled proteins along DNA chains can be quantified with high precision, although it should be noted that most of these studies have been performed *in vitro*. A quantitative comparison of the BHW model with experimental data leads to a number of discrepancies, however. For example, it is usually assumed that $D_1 \approx D_3$ in order to obtain a sufficient level of facilitation. On the other hand, single-molecule measurements indicate that $D_1 \ll D_3$ [872]. Such experiments have also shown that $\tau_1 \gg \tau_3$, which is significantly different from the optimal condition $\tau_1 = \tau_3$. Hence the intermittent search process could actually result in a slowing down compared to pure 3D diffusion [399]. The BHW model also exhibits unphysical behavior in certain limits. These issues have motivated a number of alternative models of facilitated diffusion, as highlighted in [462].

1. *Electrostatic interactions.* One alternative hypothesis is that the observed fast association rates are due to electrostatic interactions between oppositely charged molecules, and thus do not violate the 3D diffusion limit [355]. This is motivated by the theoretical result that the maximal association rate in Smoluchowski theory when there are long-range interactions between the reacting molecules is

$$k = 4\pi Da/\beta, \quad \beta = \int_a^\infty e^{U(r)/k_B T} \frac{dr}{r^2},$$

where $U(r)$ is the interaction potential. The standard result is recovered when $U(r) = 0$ for $r > a$, see equation (6.3.3). It follows that long-range attractive interactions can significantly increase diffusion-limited reaction rates. It has been further argued that in vitro experiments tend to be performed at low salt concentrations so that the effects of screening could be small. However, experimentally based estimates of the Debye length, which specifies the size of the region where electrostatic forces are important, indicate that it is comparable to the size of the target sequence. Hence, electrostatic forces are unlikely to account for facilitated diffusion.

2. *Colocalization.* Another proposed mechanism is based on the observation that in bacteria, genes responsible for producing specific proteins are located close to the binding sites of these proteins. This colocalization of proteins and binding sites could significantly speed up the search process by requiring only a small number of alternating 3D and 1D phases [580]. However, such a mechanism might not be effective in eukaryotic cells, where transcription and translation tend to be spatially and temporally well separated. Moreover, colocalization breaks down in cases where proteins have multiple targets on DNA.

3. *Correlations.* Yet another theoretical mechanism involves taking into account correlations between 1D sliding and 3D bulk diffusion. These correlations reflect the fact that attractive interactions between a protein and nonspecific binding sites mean that there is a tendency for a protein to return back to a neighborhood of the DNA site from which it recently dissociated [167, 893]. Although such interactions tend to slow down proteins moving along DNA, they also increase the local concentration of proteins absorbed to DNA. This suggests that facilitated diffusion can occur at intermediate levels of protein concentration and protein–DNA interactions.

One-dimensional discrete-state model. We now describe a recent modeling approach due to Kolomeisky and collaborators [455, 456, 489, 490, 756–759, 825], which provides an analytically tractable framework for exploring various aspects of facilitated diffusion. We begin by considering the simplest version of the model as illustrated in Fig. 6.15 [825]. The DNA chain is treated as having L discrete binding sites, $n = 1, \ldots, L$, with one at position $n = n_f$ being the target for the protein molecule. Since diffusion in the bulk is typically much faster than sliding motion along DNA, all solution states of the protein are lumped together into a single state $n = M$. It is assumed that the protein in solution can bind to any DNA site with equal probability such that the total association rate is k_{on}. The corresponding dissociation rate is denoted by k_{off}. A protein bound to a nonspecific site can hop to a

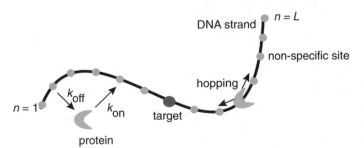

Fig. 6.15: Discrete-state model of facilitated diffusion.

neighboring site at a rate η. Let $P_n(t)$ be the probability that the protein is in state $n \in \{1,\ldots,L,M\}$ at time t given the initial condition $n = n_0$ at $t = 0$. The forward master equation for P_n is then

$$\frac{dP_n}{dt} = \eta[P_{n+1}(t) + P_{n-1}(t) - 2P_n(t)] + \frac{k_{on}}{L}P_M(t) - k_{off}P_n(t), \quad 2 \leq n \leq L-1,$$

$$(6.3.12a)$$

$$\frac{dP_1}{dt} = \eta[P_2(t) - P_1(t)] + \frac{k_{on}}{L}P_M(t) - k_{off}P_1(t), \tag{6.3.12b}$$

$$\frac{dP_L}{dt} = \eta[P_{L-1}(t) - P_L(t)] + \frac{k_{on}}{L}P_M(t) - k_{off}P_L(t), \tag{6.3.12c}$$

$$\frac{dP_M}{dt} = -k_{on}P_M(t) + k_{off}\sum_{n=1}^{L} P_n(t). \tag{6.3.12d}$$

Solution of backwards equation. We are interested in determining the mean first passage time (MFPT) for the protein to reach the target. Therefore, we impose an absorbing boundary condition at $n = n_f$ and introduce the FPT density $F_n(t)$ for first finding the target at time t given that the protein started in state n at time $t = 0$. As shown in Box 5G, $F_n(t)$ satisfies a backward master equation, which is obtained by taking the transpose of the generator of the forward master equation (see Ex. 6.5). It follows that

$$\frac{dF_n}{dt} = \eta[F_{n+1}(t) + F_{n-1}(t) - 2F_n(t)] + k_{off}F_M(t) - k_{off}F_n(t), \quad 2 \leq n \leq L-1,$$

$$(6.3.13a)$$

$$\frac{dF_1}{dt} = \eta[F_2(t) - F_1(t)] + k_{off}F_M(t) - k_{off}F_1(t), \tag{6.3.13b}$$

$$\frac{dF_L}{dt} = \eta[F_{L-1}(t) - F_L(t)] + k_{off}F_M(t) - k_{off}F_L(t), \tag{6.3.13c}$$

$$\frac{dF_M}{dt} = \frac{k_{on}}{L}\sum_{n=1}^{L} F_n(t) - k_{on}F_n(t). \tag{6.3.13d}$$

The simplest method for solving these equations is to use Laplace transforms with $\widetilde{F}_n(s) = \int_0^\infty e^{-st}F_n(t)dt$. One finds that

$$(s + 2\eta + k_{off})\widetilde{F}_n(s) = \eta[\widetilde{F}_{n+1}(s) + \widetilde{F}_{n-1}(s)] + k_{off}\widetilde{F}_M(s), \tag{6.3.14a}$$

$$(s + \eta + k_{off})\widetilde{F}_1(s) = \eta\widetilde{F}_2(s) + k_{off}\widetilde{F}_M(s), \tag{6.3.14b}$$

$$(s + \eta + k_{off})\widetilde{F}_L(s) = \eta\widetilde{F}_{L-1}(s) + k_{off}\widetilde{F}_M(s), \tag{6.3.14c}$$

$$(s + k_{on})\widetilde{F}_M(s) = \frac{k_{on}}{L}\sum_{n=1}^{L} \widetilde{F}_n(s). \tag{6.3.14d}$$

We also have $\widetilde{F}_{n_f}(s) = 1$. Equations (6.3.14a-6.3.14c) can be converted to a system of homogeneous linear equations by setting $\widetilde{F}_n(s) = \widetilde{G}_n(s) + B$ with

$$B = \frac{k_{off}}{k_{off} + s}\widetilde{F}_M(s). \tag{6.3.15}$$

Taking $\widetilde{G}_n(s) = y(s)^n$ we find from equation (6.3.14a) that y satisfies a quadratic equation with a pair of roots $y_\pm = y^{\pm 1}$ and

$$y = \frac{s + 2\eta + k_{\text{off}} - \sqrt{(s + 2\eta + k_{\text{off}})^2 - 4\eta^2}}{2\eta}. \tag{6.3.16}$$

The general solution for $2 \leq n \leq L - 1$ is thus

$$\widetilde{F}_n(s) = A_+ y^n + A_- y^{-n} + B. \tag{6.3.17}$$

The coefficients $A_\pm$ can be determined by imposing the boundary equations (6.3.14b,6.3.14c) and the initial condition $\widetilde{F}_{n_f}(s) = 1$. It is useful to rewrite equations (6.3.14b,6.3.14c) in an identical form to (6.3.14a) by introducing the virtual sites $n = 0, L+1$ and imposing the reflecting boundary conditions

$$\widetilde{F}_0(s) = \widetilde{F}_1(s), \quad \widetilde{F}_L(s) = \widetilde{F}_{L+1}(s).$$

The general solution then holds for $1 \leq n \leq L$. Analogous to Green's function calculations, we solve for $A_\pm$ in the domain $1 \leq n \leq n_f$ by imposing the left-hand boundary condition and solve for $A_\pm$ in the domain $n_f \leq n \leq L$ by imposing the right-hand boundary condition. One finds that (see Ex. 6.5)

$$\widetilde{F}_n(s) = \frac{(1 - B)(y^n + y^{1-n})}{y^{n_f} + y^{1-n_f}} + B, \quad 1 \leq n \leq n_f, \tag{6.3.18}$$

and

$$\widetilde{F}_n(s) = \frac{(1 - B)(y^{n-L} + y^{1-n+L})}{y^{n_f - L} + y^{1-n_f + L}} + B, \quad n_f \leq n \leq L. \tag{6.3.19}$$

$\widetilde{F}_M(s)$ can now be determined by substituting for $\widetilde{F}_n(s)$ into equation (6.3.14d) and summing the various geometric series:

$$(s + k_{\text{on}})\widetilde{F}_M(s) = \frac{k_{\text{on}}}{L}\left[LB + (1 - B)\Gamma(s)\right],$$

with [759]

$$\Gamma(s) = \frac{y(1 + y)(y^{-L} + y^L)}{(1 - y)(y^{1-n_f} + y^{n_f})(y^{n_f - L} + y^{1+L-n_f})}. \tag{6.3.20}$$

Finally, substituting for B using equation (6.3.15) and rearranging yields

$$\widetilde{F}_M(s) = \frac{k_{\text{on}}(k_{\text{off}} + s)\Gamma(s)}{Ls(k_{\text{off}} + k_{\text{on}} + s) + k_{\text{off}}k_{\text{on}}\Gamma(s)}. \tag{6.3.21}$$

Using equation (5.5.23) of Box 5G the MFPT T_m to find the target site having started in state m is

$$T_m = \int_0^\infty F_m(\tau)\tau d\tau = -\left.\frac{d\widetilde{F}_m(s)}{ds}\right|_{s=0}. \tag{6.3.22}$$

For example, if the protein starts in solution, then from equation (6.3.21) we have

$$T_M = \frac{1}{k_{\text{on}}}\frac{L}{\Gamma(0)} + \frac{1}{k_{\text{off}}}\frac{L - \Gamma(0)}{\Gamma(0)}. \tag{6.3.23}$$

Using equation (6.3.14a) it follows that the MFPT to reach the target site having started with equal probability on any DNA site is

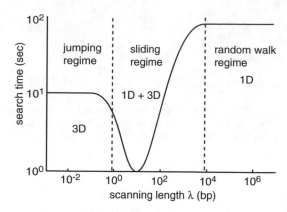

Fig. 6.16: Mean search time as a function of the scaling length $\lambda = \sqrt{\eta/k_{\text{off}}}$. Parameters are $L = 10^3$ bp, $\eta = k_{\text{on}} = 10^5 \text{s}^{-1}$, and $n_f = L/2$. [Redrawn from Shvets et al [759].]

$$\overline{T} = \frac{1}{L}\sum_{n=1}^{L} T_n = -\frac{1}{L}\sum_{n=1}^{L} \left.\frac{d\widetilde{F}_n(s)}{ds}\right|_{s=0} = \frac{(k_{\text{on}}+k_{\text{off}})(L-\Gamma(0))}{k_{\text{on}}k_{\text{off}}\Gamma(0)} = T_M - \frac{1}{k_{\text{on}}}.$$

$$(6.3.24)$$

Equation (6.3.23) has a clear physical interpretation [759]. The quantity $\Gamma(0)$ describes the mean number of distinct sites that the protein scans during each 1D search along the DNA, so that the protein has to make $L/\Gamma(0)$ visits to the DNA on average. Each visit lasts an average time $1/k_{\text{on}}$. Similarly, the protein dissociates back into solution $L/\Gamma(0) - 1$ times, and each period in solution lasts an average time $1/k_{\text{off}}$.

Shvets et al [759] use their analysis to identify three distinct dynamics search regimes, which depend on the particular values of the various kinetic parameters. In Fig. 6.16 we illustrate this by sketching how the mean search time varies with a dimensionless scanning length $\lambda = \sqrt{\eta/k_{\text{off}}}$, which is related to the average distance (number of sites) that the protein travels along the DNA during each search cycle. (This is distinct from $\Gamma(0)$ since the same site may be visited several times during a single search.) The three dynamic regimes are as follows.

(i) Random walk regime (small k_{off}, $\lambda > L$). In this case the protein has a strong nonspecific binding affinity to DNA, so that there is a single search cycle consisting of a simple unbiased 1D random walk. Hence, the mean search time scales as $T_M \sim L^2$ and tends to be slow due to multiple visits to the same sites.

(ii) Jumping regime (large k_{off}, $\lambda < 1$). Although the protein can bind to DNA, it cannot slide as it quickly dissociates back into solution. The search process is effectively 3D and on average the protein has to make L visits to the DNA so that $T_M \sim L$.

(iii) Sliding regime (intermediate values of k_{off}, $1 < \lambda < L$.) One now has facilitated diffusion involving a mixture of 1D and 3D search cycles, and this leads to the smallest search times.

One of the powerful features of the discrete-state model is that it provides a theoretical framework for including a variety of additional biophysical features, as reviewed in [759]. These include multiple targets and traps [489, 490], molecular crowding [757, 758], sequence heterogeneity [756], and conformational changes in the protein bound to DNA [455].

6.3.3 Stochastically gated diffusion-limited reactions

Consider the problem of a single stationary protein surrounded by ligands that can bind to the protein. Furthermore, suppose that the target protein can switch between two conformational states $n = 0, 1$, and is only reactive in the open state $n = 1$. That is, the target can be treated as a stochastic gate. For unbounded domains, this problem was first studied by Szabo *et al* [792], who assumed that it is irrelevant whether it is the target protein or the diffusing ligands that switch between conformational states. Although this symmetry holds for a pair of reacting particles, it breaks down when a single protein is surrounded by many ligands [64, 73, 74, 554, 780, 892]. In particular, one finds that the kinetics is faster when the gating is due to the ligands rather than the protein. In order to understand this result, note that one of the basic simplifying assumptions of Smoluchowski theory is that one can ignore many-particle effects, namely, correlations in the dynamics of the ligands. However, correlations arise when the target protein switches between reactive states, since this is simultaneously experienced by all of the ligands. In other words, all of the ligands diffuse in the same randomly switching environment (see also Sect. 6.6). On the other hand, such correlations wouldn't arise if the ligands are assumed to independently switch between conformational states. For the moment, we will ignore the effects of correlations; a many-particle formulation will be considered later.

Let $N(t) \in \{0, 1\}$ denote the state of the target gate, which evolves according to the two-state Markov process

$$(\text{closed}) \underset{\beta}{\overset{\alpha}{\rightleftharpoons}} (\text{open}). \tag{6.3.25}$$

We will treat the target as a fixed sphere Ω of radius a at the origin in $\mathbb{R}^3$ and denote the concentration of background diffusing molecules exterior to Ω by $c(\mathbf{x},t)$. The associated diffusion equation takes the form

$$\frac{\partial c(\mathbf{x},t)}{\partial t} = D\nabla^2 c(\mathbf{x},t), \quad c(\mathbf{x},0) = 1,$$

subject to the far-field boundary condition $c(\mathbf{x},t) = 1$ for $\mathbf{x} \to \infty$. The boundary conditions are

$$\left.\frac{\partial c}{\partial r}\right|_{r=a} = 0, \quad \text{if } N(t) = 0, \quad 4\pi a^2 D \left.\frac{\partial c}{\partial r}\right|_{r=a} = k_{\text{on}} c(a,t), \quad \text{if } N(t) = 1.$$

Introducing the first moments

$$c_n(\mathbf{x},t) = \mathbb{E}[c(\mathbf{x},t)1_{N(t)=n}], \tag{6.3.26}$$

with respect to realizations of the gate, it can be shown that the moments satisfy the differential Chapman–Kolmogorov (CK) equation (see Sect. 6.6 and Box 6E)

$$\frac{\partial c_0}{\partial t} = D\nabla^2 c_0 - \alpha c_0 + \beta c_1, \tag{6.3.27a}$$

$$\frac{\partial c_1}{\partial t} = D\nabla^2 c_1 + \alpha c_0 - \beta c_1, \tag{6.3.27b}$$

with boundary conditions

$$\left.\frac{\partial c_0}{\partial r}\right|_{r=a} = 0, \quad 4\pi a^2 D \left.\frac{\partial c_1}{\partial r}\right|_{r=a} = k_{\mathrm{on}} c_1(a,t). \tag{6.3.27c}$$

The far-field conditions become

$$c_0(\infty) = \rho_0 \equiv \frac{\beta}{\alpha+\beta}, \quad c_1(\infty) = \rho_1 \equiv \frac{\alpha}{\alpha+\beta}, \tag{6.3.27d}$$

and the initial conditions are $c_n(\mathbf{x},0) = \rho_n$. Given the steady-state solution $(c_0(\mathbf{x}),c_1(\mathbf{x}))$, which exists in 3D, the stochastically gated reaction rate is given by

$$k_{\mathrm{SG}} = 4\pi a^2 D \left.\frac{\partial c_1(\mathbf{x})}{\partial r}\right|_{r=a}. \tag{6.3.28}$$

The calculation of k_{SG} was originally carried out by Szabo et al. [792].

Exploiting the spherical symmetry of the problem, we set $u_n(r) = rc_n(r)$ such that the steady-state version of (6.3.27) becomes

$$0 = D\frac{d^2 u_0}{dr^2} - \alpha u_0 + \beta u_1, \quad 0 = D\frac{d^2 u_1}{dr^2} + \alpha u_0 - \beta u_1, \tag{6.3.29}$$

with $u_n(r) \to \rho_n r$ as $r \to \infty$. Adding the pair of equations in (6.3.29) and setting $u(r) = u_0(r) + u_1(r)$ yields $d^2 u/dr^2 = 0$, which has the solution $u(r) = A + r$ for some unknown constant A. The first equation in (6.3.29) can then be rewritten as

$$D\frac{d^2 u_0}{dr^2} - (\alpha+\beta)u_0 = -\beta(A+r),$$

which has the solution

$$u_0(r) = \Gamma e^{-\lambda r} + \frac{\beta}{\alpha+\beta}(r+A), \quad \lambda = \sqrt{\frac{\alpha+\beta}{D}}.$$

Hence, the steady-state solution takes the form

$$c_0(r) = \frac{\Gamma}{r}e^{-\lambda r} + \frac{\beta}{\alpha+\beta} + \frac{A\beta}{r}, \quad c_1(r) = -\frac{\Gamma}{r}e^{-\lambda r} + \frac{\alpha}{\alpha+\beta} + \frac{A\alpha}{r}, \qquad (6.3.30)$$

with two unknown constants A, Γ. (We have rescaled according to $A/(\alpha+\beta) \to A$.) The constants are determined by imposing the boundary conditions and it leads to the final result

$$k_{SG} = \frac{k_{on}k_D\alpha[1+\lambda a]}{k_D(\alpha+\beta)(1+\lambda a) + k_{on}[\beta + (1+\lambda a)\alpha]}. \qquad (6.3.31)$$

In the limit $\beta \to 0$ (gate always open), we recover the classical result (6.3.7).

Many-particle formulation. Here we describe one approach to understanding the differences between a stochastically gated protein and stochastically gated ligands [72, 74, 554]. Consider a large sphere of volume Ω containing $N = c\Omega$ ligands and one protein at the center of the sphere. For the moment, ignore any gating so that the protein is always available to react with a ligand. Initially each ligand can occupy any position within the sphere, excluding the sphere of radius a centered at the origin. Consider a particular configuration η of initial positions, $\{\mathbf{r}_{j,\eta}, j = 1,\dots,N\}$, where $\mathbf{r}_{j,\eta}$ is the initial position of the jth ligand. Introduce the conditional survival probability $S_\eta(t)$ that no ligand has reacted with the protein up to time t. The fundamental assumption of Smoluchowski rate theory is to treat all of the ligands as independent. This means that the survival probability $S_\eta(t)$ can be written as the product of the survival probabilities of the individual ligands:

$$S_\eta(t) = \prod_{j=1}^{N} s(t, r_{j,\eta}), \qquad (6.3.32)$$

where $s(t,r)$ is the survival probability of a single protein–ligand pair initially separated by the radial distance r. Finally, we define the unconditional survival probability $S(t)$, which is obtained by averaging over all initial configurations: $S(t) = \mathbb{E}[S_\eta(t)]$. Given $S(t)$ the time-dependent reaction rate is defined by the equation

$$\frac{dS(t)}{dt} = -k(t)S(t),$$

which has the solution

$$S(t) = \exp\left(-\int_0^t k(t')dt'\right). \qquad (6.3.33)$$

It remains to perform the configurational averaging, which is carried out in the limit $\Omega \to \infty$. Treating the initial distribution of ligands as uncorrelated,

$$S(t) = \lim_{\Omega \to \infty} \frac{1}{(\Omega - v_a)^N} \int_\Omega \cdots \int_\Omega \prod_{j=1}^{N} s(t,r_j)H(r_j - a)d\mathbf{r}_j \qquad (6.3.34)$$

$$= \lim_{N \to \infty} \left[\frac{1}{N/c - v_a}\int_\Omega s(t,r)H(r-a)d\mathbf{r}\right]^N = \exp\left[-c\int_{r>a} q(t,r)d\mathbf{r}\right],$$

on using the definition

$$e^x = \lim_{N \to \infty} \left(1 + \frac{x}{N}\right)^N.$$

Here $H(x)$ is the Heaviside function, v_a is the volume of a sphere of radius a, and $q(t,r) = 1 - s(t,r)$ is the probability that a reaction occurs by time t given that a protein–ligand pair is initially separated by r. It immediately follows that the reaction rate is

$$k(t) = c \int_{r>a} \frac{\partial q(t,r)}{\partial t} d\mathbf{r}. \tag{6.3.35}$$

Hence, the many-particle problem has been reduced to calculating the survival probability of a single pair of molecules. Let $p(\mathbf{x},t|\mathbf{r},0)$ be the probability that a single ligand is at position $\mathbf{x}$ at time t given that it started at $\mathbf{r}$ and has not been absorbed by the protein. We then have the FP equation

$$\frac{\partial p}{\partial t} = D\nabla^2 p,$$

with initial condition $p(\mathbf{x},0|\mathbf{r},0) = \delta(\mathbf{x} - \mathbf{r})$ and the boundary condition

$$\mathbf{n} \cdot \nabla p(\mathbf{x},t|\mathbf{r},0)) = k_{on} p(\mathbf{x},t), \quad \text{for } |\mathbf{x}| = a.$$

The survival probability is then

$$s(t,r) = \int_{|\mathbf{x}|>a} p(\mathbf{x},t|\mathbf{r},0) d\mathbf{x}.$$

As an example, in the case of a purely absorbing boundary in three dimensions

$$q(t,r) = \frac{a}{r} \operatorname{erfc}\left(\frac{r-a}{2\sqrt{Dt}}\right),$$

and we recover equation (6.3.3).

Following [74], suppose that the protein is now stochastically gated. Consider a particular realization of the gate $\sigma(t) = \{N(\tau), 0 \leq \tau \leq t\}$, and denote the corresponding protein–ligand survival probability by $s(t,r|\sigma(t))$. At the level of a single ligand, the survival probability is the same if the ligand rather than the protein is gated. However, major differences arise at the multi-ligand level. In the case of independently gated ligands (GL) the total survival probability can be expressed as

$$S_{GL}(t) = \exp\left(-c \int_{r>a} \mathbb{E}[q(t,r)] d\mathbf{r}\right), \tag{6.3.36}$$

where expectation is taken with respect to all realizations $\sigma(t)$ of a gate. On the other hand, if the protein is gated (GP), then all ligands are subject to the same gate realization and

$$S_{GP}(t) = \mathbb{E}_\sigma\left[\exp\left(-c \int_{r>a} q(t,r) d\mathbf{r}\right)\right]. \tag{6.3.37}$$

Hence, one can interpret $S_{GL}(t)$ as the mean-field version of $S_{GP}(t)$ in which all correlations are ignored. Using Jensen's inequality, which states that for any convex function f, we have $\mathbb{E}[f(x)] \geq f(\mathbb{E}[x])$, it immediately follows that

$$S_{GP}(t) \geq S_{GL}(t). \tag{6.3.38}$$

In particular, the survival probability for gated ligands is smaller than the survival probability for a gated protein, which implies that the effective reaction rate of the former is faster.

6.3.4 Enzymatic reactions with multiple binding sites

There is an important class of reactions where renormalizing the forward reaction rates according to equation (6.3.7) is not sufficient to take into account the effects of diffusion, even at low concentration levels. Such reactions involve the catalysis of substrate molecules with multiple binding sites. The basic idea is that an enzyme that has just modified (e.g., phosphorylated) one site and dissociated can either (i) diffuse away such that another enzyme modifies the next binding site (distributive catalysis) or (ii) bind to the second site and modify it (processive catalysis). In other words, in the former case the sites are modified by different enzyme molecules, whereas in the latter case they are modified by the same molecule, see Fig. 6.17. Classical chemical kinetics only takes into account the distributive mechanism, since it assumes fast diffusion, whereas both mechanisms are in play when diffusion is finite. For example, based on many-particle simulations, slowing diffusion can have a significant effect on phosphorylation–dephosphorylation cycles, namely, it can speed up the response and lead to the loss of ultrasensitivity and bistability [796]. (See Chap. 10 for a definition and analysis of ultrasensitivity in phosphorylation–dephosphorylation cycles.)

Here we review the analysis of rebinding reactions due to Gopich and Szabo [331]. Using an exactly solvable many-particle reaction–diffusion system as a motivating example, these authors show that in addition to rescaling reaction rates, one also has to introduce new transitions into the reaction scheme. The rates of the new

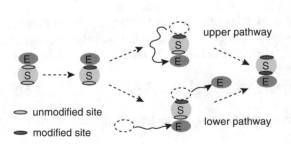

Fig. 6.17: Distributive versus processive catalysis. A substrate S has two sites that can be modified by an enzyme E. After an enzyme has modified one site there are two possible pathways for modifying the second site. Upper pathway: The same enzyme molecule modifies the second site (processive). Lower pathway: a different enzyme modifies the second site (distributive).

reaction pathways depend on the probability that a reactant released from one site rebinds to another rather than diffusing away. (Recently, a rigorous mathematical method for determining these probabilities has been developed, applying the theory of diffusion in randomly switching environments which is presented in Sect. 6.6 [497, 498].)

Exactly solvable model. Consider an immobile substrate with N sites, which is surrounded by M enzymes E diffusing in a cell volume V with the diffusion coefficient D. The enzymes are treated as noninteracting point particles. The substrate can be in states $S_0, S_1, \ldots, S_N$ corresponding to the number n, $0 \leq n \leq N$, of modified sites. It is further assumed that an instantaneous irreversible reaction $S_i \rightarrow S_{i+1}$ occurs at a rate κ_i whenever an enzyme molecule comes within a distance R of the substrate, where R is the reaction radius. This means that we can ignore the complexes ES_i. For the sake of illustration, we will focus on the case of two modifiable sites ($N = 2$), so that the reaction scheme takes the form

$$S_0 + E \xrightarrow{\kappa_0} S_1 + E \xrightarrow{\kappa_1} S_2 + E. \tag{6.3.39}$$

The classical law of mass action leads to the following set of kinetic equations for the substrate concentrations $[S_i](t)$:

$$\frac{d[S_0]}{dt} = -\kappa_0[E][S_0], \tag{6.3.40a}$$

$$\frac{d[S_1]}{dt} = \kappa_0[E][S_0] - \kappa_1[E][S_1], \tag{6.3.40b}$$

$$\frac{d[S_2]}{dt} = \kappa_1[E][S_1]. \tag{6.3.40c}$$

In a well-mixed compartment with a large number of molecules, the enzyme concentration $[E] = M/V$. Let $[\mathbf{S}] = ([S_0], [S_1], [S_2])^\top$ and rewrite these equations in the matrix form

$$\frac{d[\mathbf{S}]}{dt} = -\mathbf{K}[E][\mathbf{S}], \quad \mathbf{K} = \begin{pmatrix} \kappa_0 & 0 & 0 \\ -\kappa_0 & \kappa_1 & 0 \\ 0 & -\kappa_1 & 0 \end{pmatrix}. \tag{6.3.41}$$

The matrix $\mathbf{K}$ has eigenvalues $\lambda_0 = \kappa_0, \lambda_1 = \kappa_1, \lambda_2 = 0$. The zero eigenvalue reflects conservation of the total substrate concentration, $S_{\text{tot}} = \sum_{i=0}^{N} S_i(t)$. Similarly, for $N > 2$ binding sites, the corresponding rate matrix $\mathbf{K}$ has the nonzero elements $K_{ii} = -K_{i+1,j} = \kappa_i$, $i = 0, 1, \ldots, N-1$ and eigenvalues $\kappa_0, \ldots, \kappa_{N-1}, 0$.

In order to take into account the effects of diffusion, let $P_i(r_1, \ldots, r_M, t)$ denote the probability density for the enzymes to be located at radial positions $r_1, \ldots, r_M$ relative to the substrate, and the substrate has i modified sites at time t. It follows that

$$\frac{[S_i](t)}{S_{\text{tot}}} = \int_V P_i d\mathbf{x}_1 \ldots d\mathbf{x}_M, \quad d\mathbf{x}_m = 4\pi^2 r_m^2 dr_m. \tag{6.3.42}$$

Outside the reaction sphere of radius R, each enzyme undergoes diffusion so that

$$\frac{\partial P_i}{\partial t} = D \sum_{m=1}^{M} \nabla_{r_m}^2 P_i, \quad |\mathbf{x}_l| > R \text{ for } l = 1, \dots, M. \tag{6.3.43}$$

The transitions between the substrate states are incorporated as boundary conditions obtained by equating the diffusive and reactive fluxes at the surface of the reaction sphere:

$$4\pi R^2 D \frac{\partial P_i}{\partial r_m} = \kappa_i P_i - \kappa_{i-1} P_{i-1}, \quad r_m = R, \tag{6.3.44}$$

with $\kappa_{-1} \equiv 0$. Finally, the enzymes are taken to be randomly distributed at $t = 0$,

$$P_i(r_1, \dots, r_m, 0) = \frac{1}{V_M} \frac{[S_i(0)]}{S_{\text{tot}}}.$$

It is convenient to rewrite the above equations in matrix form:

$$\frac{\partial \mathbf{P}}{\partial t} = D \sum_{m=1}^{M} \nabla_{r_m}^2 \mathbf{P}, \quad |\mathbf{x}_l| > R \text{ for } l = 1, \dots, M, \tag{6.3.45a}$$

$$4\pi R^2 D \frac{\partial \mathbf{P}}{\partial r_m} = \mathbf{KP}, \quad r_m = R. \tag{6.3.45b}$$

The next step is to use a similarity transform in order to diagonalize the rate matrix $\mathbf{K}$, which is straightforward since it has real, non-degenerate eigenvalues. Let $\mathbf{T}$ denote the matrix whose inverse $\mathbf{T}^{-1}$ has columns given by the eigenvectors of $\mathbf{K}$. Setting $\mathbf{P} = \mathbf{T}\widetilde{\mathbf{P}}$, we have

$$\frac{\partial \widetilde{\mathbf{P}}}{\partial t} = D \sum_{m=1}^{M} \nabla_{r_m}^2 \widetilde{\mathbf{P}}, \quad |\mathbf{x}_l| > R \text{ for } l = 1, \dots, M, \tag{6.3.46a}$$

$$4\pi R^2 D \frac{\partial \widetilde{\mathbf{P}}}{\partial r_m} = \mathbf{K}_D \widetilde{\mathbf{P}}, \quad r_m = R, \tag{6.3.46b}$$

where

$$\mathbf{K}_d = \mathbf{TKT}^{-1} = \text{diag}(\lambda_0, \dots \lambda_N) = \text{diag}(\kappa_0, \dots \kappa_{N-1}, 0),$$

and $\widetilde{\mathbf{P}}(r_1, \dots, r_M, 0) = V^{-M} \mathbf{p}^0$ with $\mathbf{p}^0 = \mathbf{T}^{-1}[\mathbf{S}(0)]/S_{\text{tot}}$. The diagonalized system is now separable and one can proceed along analogous lines to Smoluchowski theory (Sect. 6.3.1). That is,

$$\widetilde{P}_i(r_1, \dots, r_M, t)/p_i^0 = V^{-M} \prod_{m=1}^{M} g_i(r_m, t) \tag{6.3.47}$$

for $i = 0,,\dots,N$, with

$$\frac{\partial g_i(r,t)}{\partial t} = D \nabla_r^2 g_i(r,t), \quad r > R, \tag{6.3.48a}$$

supplemented by the Robin boundary condition

$$4\pi R^2 D \frac{\partial g_i(r,t)}{\partial r}\bigg|_{r=R} = \lambda_i g_i(R,t), \quad i = 0,1,\ldots,N-1, \tag{6.3.48b}$$

and the initial condition $g_i(r,0) = 1$. Note that $g_N(r,t) = 1$ for all $t \geq 0$, since $\lambda_N = 0$. Equation (6.3.48) has the solution

$$g_i(R,t) = \varepsilon_i + (1 - \varepsilon_i)e^{\gamma_i t}\mathrm{erfc}(\sqrt{\gamma_i t}), \quad i = 0,\ldots,N-1, \tag{6.3.49}$$

where

$$\varepsilon_i = \frac{k_D}{\lambda_i + k_D}, \quad \gamma_i = \frac{D}{(\varepsilon_i R)^2}, \tag{6.3.50}$$

and $k_D = 4\pi RD$. Note that $g_i(R,t) \to \varepsilon_i$ as $t \to \infty$.

Substituting $\mathbf{P} = \mathbf{T}\widetilde{\mathbf{P}}$ with $\widetilde{\mathbf{P}}$ given by equation (6.3.47) into the normalization condition (6.3.42) yields

$$[\mathbf{S}(t)] = \mathbf{T}\mathbf{F}(t)\mathbf{T}^{-1}[\mathbf{S}(0)], \tag{6.3.51}$$

where $F_{ij}(t) = F_i(t)\delta_{i,j}$ and

$$F_i(t) = \left(V^{-1}\int_V g_i(r,t)d\mathbf{x}\right)^M, \quad i = 0,\ldots,N-1,$$

and $F_N(t) = 1$. Now note that

$$\frac{dF_i}{dt} = \frac{M}{V}\left(\int_V \frac{\partial g_i(r,t)}{\partial t}d\mathbf{x}\right)\left(V^{-1}\int_V g_i(r,t)d\mathbf{x}\right)^{M-1}.$$

Recall that M and V are assumed to be sufficiently large so that we can make the identification $[E] = M/V$. (More formally, we take the thermodynamic limit $M,V \to \infty$ with $M/V = [E]$ fixed.) This means that $M - 1 \approx M$ and we can write

$$\frac{dF_i}{dt} = [E]\left(\int_V \frac{\partial g_i(r,t)}{\partial t}d\mathbf{x}\right)F_i(t).$$

Applying these results to the time derivative of equation (6.3.51) leads to the modified kinetic equations

$$\frac{d[\mathbf{S}]}{dt} = -\mathbf{K}(t)[E][\mathbf{S}], \quad \mathbf{K}(t) = \mathbf{T}\Lambda(t)\mathbf{T}^{-1}, \tag{6.3.52}$$

where $\Lambda(t) = \mathrm{diag}(k_0(t),\ldots,k_{N-1}(t),0)$ and

$$k_i(t) := -\left(\int_V \frac{\partial g_i(r,t)}{\partial t}d\mathbf{x}\right) = \lambda_i g_i(R,t). \tag{6.3.53}$$

The latter follows from equation (6.3.48) and integration by parts.

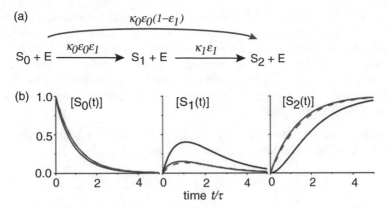

Fig. 6.18: Effect of diffusion on the kinetic scheme of a simple model of an enzymatic reaction with two modifiable sites. (a) In addition to rescaling of the original reaction rates by a corresponding escape probability ε_i, there is a new connection in the reaction diagram indicated in red. (b) Comparison of the exact kinetics (red curves) with the predictions of the classical mass-action scheme (black curves) and the modified scheme (dashed curves). Time is expressed in units of $\tau = (\kappa_0 \varepsilon_0 [E])^{-1}$. Parameters are $\kappa_0 = 4k_D$, $\kappa_1 = 2k_D$, $4\pi R^3 [E]/3 = 0.01$, D=1, R=1, and $[S_{tot}] = 1$. [Adapted from Gopich and Szabo (2013).]

Now suppose $t \gg R^2/D$ so that the rate constants approach their steady-state values $k_i(t) \to \lambda_i \varepsilon_i$ and return to the case $N = 2$. The solution of equation (6.3.52), assuming that initially all substrates are in the completely unmodified state S_0, takes the form

$$\frac{[S_0(t)]}{S_{tot}} = e^{-[E]\kappa_0 \varepsilon_0 t}, \quad \frac{[S_1(t)]}{S_{tot}} = \frac{\kappa_0}{\kappa_0 - \kappa_1} \left(e^{-[E]\kappa_1 \varepsilon_1 t} - e^{-[E]\kappa_0 \varepsilon_0 t} \right), \quad (6.3.54)$$

with $[S_2(t)] = S_{tot} - [S_0(t)] - [S_1(t)]$. The major point is that this solution cannot be obtained from the standard kinetic equation (6.3.40) under the simple replacement $\kappa_i \to \kappa_i \varepsilon_i$, which would correspond to the kinetic scheme

$$S_0 + E \overset{\kappa_0 \varepsilon_0}{\to} S_1 + E \overset{\kappa_1 \varepsilon_1}{\to} S_2 + E. \quad (6.3.55)$$

Instead, equation (6.3.54) solves the modified kinetic equations (see Ex. 6.6),

$$\frac{d[S_0]}{dt} = -\kappa_0 \varepsilon_0 [E][S_0], \quad (6.3.56a)$$

$$\frac{d[S_1]}{dt} = \kappa_0 \varepsilon_0 (1 - \pi_1)[E][S_0] - \kappa_1 \varepsilon_1 [E][S_1], \quad (6.3.56b)$$

$$\frac{d[S_2]}{dt} = \kappa_1 \varepsilon_1 [E][S_1] + \kappa_0 \varepsilon_0 \pi_1 [E][S_0], \quad (6.3.56c)$$

where $\pi_1 = 1 - \varepsilon_1$. Equation (6.3.56) corresponds to a reaction scheme with an additional connection representing a direct transition from S_0 to S_2 [331], see Fig. 6.18. This additional connection has the following physical interpretation. Immediately following the transition $S_0 \to S_1$, the enzyme molecule dissociates and either diffuses away with probability ε_1 or remains close to the substrate and modifies the second site with rebinding probability $1 - \varepsilon_1$. Note that the analysis can be extended to the case of reversible catalysis [331].

Michaelis–Menten model. We now generalize the theory to a substrate with N sites modified via a Michaelis–Menten mechanism (see Sect. 1.4):

$$S_i + E \underset{\beta_i}{\overset{\alpha_i}{\rightleftharpoons}} S_i E \overset{c_i}{\rightarrow} S_{i+1} + E, \quad i = 0, \ldots, N-1. \tag{6.3.57}$$

The conventional mass-action kinetic equations are

$$\frac{d[S_i]}{dt} = -\alpha_i[E][S_i] + \beta_i[S_iE] + c_{i-1}[S_{i-1}E], \tag{6.3.58a}$$

$$\frac{d[S_iE]}{dt} = \alpha_i[E][S_i] - \beta_i[S_iE] - c_i[S_iE], \tag{6.3.58b}$$

$$\frac{d[E]}{dt} = \sum_{i=0}^{N-1} \frac{d[S_i]}{dt}. \tag{6.3.58c}$$

The effect of diffusion is obtained by modifying the biomolecular reaction rate according to

$$\alpha_i[E](t)[S_i](t) \to \alpha_i \rho_i(R,t), \tag{6.3.59}$$

where $\rho_i(r,t)$ is the pair distribution function for molecules E and S_i separated by r, and R is the reaction radius. When $r \to \infty$, the enzyme and substrate are uncorrelated, so $\rho_i(r,t) \to [E][S_i]$. On the other hand, the flux at contact ($r = R$) must balance the forward and backward reactions, that is,

$$4\pi R^2 D \left.\frac{\partial \rho_i}{\partial r}\right|_{r=R} = \alpha_i \rho_i(R,t) - \beta_i[S_iE] - c_{i-1}[S_{i-1}E]. \tag{6.3.60}$$

It remains to determine the equation for $\rho(r,t)$. Unfortunately, this is a complicated many-body problem. The simplest approximation is to assume that spatially [331]

$$D\nabla_r^2 \rho = 0, \tag{6.3.61}$$

with the time-dependence driven by the boundary condition. The solution has the form $\rho_i = [E][S_i] + c_i/r$ with the constant c_i determined by the boundary condition at $r = R$. We thus find that

$$\rho_i(r,t) = [E][S_i] - \frac{R}{r}\frac{\pi_i}{\alpha_i}\left(\alpha_i[E][S_i] - \beta_i[S_iE] - c_{i-1}[S_{i-1}E]\right),$$

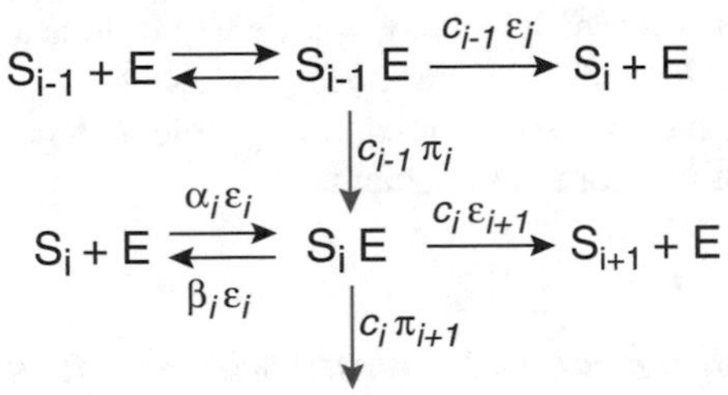

$$S_{i-1} + E \rightleftharpoons S_{i-1}E \xrightarrow{c_{i-1}\,\varepsilon_i} S_i + E$$

$$\Big\downarrow c_{i-1}\,\pi_i$$

$$S_i + E \overset{\alpha_i \varepsilon_i}{\underset{\beta_i \varepsilon_i}{\rightleftharpoons}} S_i E \xrightarrow{c_i\,\varepsilon_{i+1}} S_{i+1} + E$$

$$\Big\downarrow c_i\,\pi_{i+1}$$

Fig. 6.19: Effect of diffusion on the kinetic scheme of a multisite modification based on the Michaelis–Menten kinetics with N modifiable sites. In addition to rescaling of the original reaction rates, there are two new connections in the reaction diagram indicated in red.

where $\pi_i = \alpha_i/(\alpha_i + 4\pi RD)$ is the capture probability, see Sect. 6.3.1. Substituting for ρ_i into equation (6.3.58) after making the modification (6.3.59) yields the modified rate equations

$$\frac{d[S_i]}{dt} = -\alpha_i \varepsilon_i [E][S_i] + \beta_i \varepsilon_i [S_i E] + c_{i-1} \varepsilon_i [S_{i-1}E], \tag{6.3.62a}$$

$$\frac{d[S_i E]}{dt} = \alpha_i \varepsilon_i [E][S_i] - \beta_i \varepsilon_i [S_i E] - c_i [S_i E] + \pi_i c_{i-1} [S_{i-1}E], \tag{6.3.62b}$$

where $\varepsilon_i = 1 - \pi_i$ is the escape probability. If we decompose c_i as $c_i = c_i \varepsilon_{i+1} + c_i \pi_{i+1}$, then all the original terms in the mass-action kinetics are multiplied by an escape probability, as expected from Smoluchowski theory, but there are two extra terms in equation (6.3.62b), namely, $\pi_i c_{i-1}[S_{i-1}E] - \pi_{i+1} c_i [S_i E]$. Diagrammatically, this is equivalent to two additional reaction channels at the ith level, as shown in Fig. 6.19.

6.4 Narrow capture and escape problems, small traps, and singular perturbation methods

There are an increasing number of examples of diffusive transport that involve the presence of small targets or traps within the interior or on the boundary of a sub-cellular domain. In many cases absorption by a target initiates a downstream signaling cascade, as in the immune response induced by a T cell finding its antigen. Alternatively, the traps may act as small absorbing windows on an otherwise reflecting boundary that allow the escape of a freely diffusing molecule from the domain. Examples of the latter include the confinement of neurotransmitter receptors within the synapse of a neuron and the confinement of calcium within intracellular compartments such as dendritic spines. One is often interested in the distribution of escape or capture times, as determined by moments of the conditional FPT densities. The latter satisfy boundary value problems (BVPs) that can be solved in the small target limit using matched asymptotic expansions and Green's functions [68, 111, 119, 168, 170, 189, 193, 385, 389, 390, 474, 522–524, 660, 738, 845]. This is known as the narrow escape or capture problem, depending on whether the targets are exterior or interior to the domain.

6.4.1 Diffusion in a bounded domain with small interior traps

Consider a bounded domain $\mathcal{U} \subset \mathbb{R}^d$ that contains a set of N small interior targets $\mathcal{U}_k$, $k = 1, \ldots, N$, with $\bigcup_{j=1}^{N} \mathcal{U}_k = \mathcal{U}_a \subset \mathcal{U}$, see Fig. 6.20. Let $p(\mathbf{x}, t | \mathbf{x}_0)$ be the probability density that at time t a Brownian particle is at $\mathbf{X}(t) = \mathbf{x}$, having started at position $\mathbf{x}_0$. Then

$$\frac{\partial p(\mathbf{x}, t | \mathbf{x}_0)}{\partial t} = D\nabla^2 p(\mathbf{x}, t | \mathbf{x}_0), \ \mathbf{x} \in \mathcal{U} \backslash \mathcal{U}_a, \quad \nabla p \cdot \mathbf{n} = 0, \ \mathbf{x} \in \partial \mathcal{U}, \quad (6.4.1a)$$

$$p(\mathbf{x}, t | \mathbf{x}_0) = 0, \ \mathbf{x} \in \partial \mathcal{U}_a, \quad (6.4.1b)$$

together with the initial condition $p(\mathbf{x}, t | \mathbf{x}_0) = \delta(\mathbf{x} - \mathbf{x}_0)$. Each target is assumed to have a volume $|\mathcal{U}_j| \sim \varepsilon^d |\mathcal{U}|$ with $\mathcal{U}_j \to \mathbf{x}_j \in \mathcal{U}$ uniformly as $\varepsilon \to 0$, $j = 1, \ldots, N$. The targets are also taken to be well separated in the sense that $|\mathbf{x}_i - \mathbf{x}_j| = O(1)$, $j \neq i$, and $\mathrm{dist}(x_j, \partial \mathcal{U}) = O(1)$. For the sake of illustration, we take each target to be a d-dimensional sphere of radius $\varepsilon \ell$ and fix length scales by setting $\ell = 1$. Thus $\mathcal{U}_i = \{\mathbf{x} \in \mathcal{U}, |\mathbf{x} - \mathbf{x}_i| \leq \varepsilon\}$.

In order to proceed, we first have to generalize the definitions of splitting probabilities and conditional FPTs introduced in Sect. 2.4. Let $\mathcal{T}_k(\mathbf{x}_0)$ denote the FPT that the particle is captured by the kth trap, with $\mathcal{T}_k(\mathbf{x}_0) = \infty$ indicating that it is not captured:

$$\mathcal{T}_k(\mathbf{x}_0) = \inf\{t \geq 0; \mathbf{X}(t) \in \partial \mathcal{U}_k, \mathbf{X}(0) = \mathbf{x}_0\}. \quad (6.4.2)$$

Define $\Pi_k(\mathbf{x}_0, t)$ to be the probability that the particle is captured by the kth trap after time t, given that it started at $\mathbf{x}_0$:

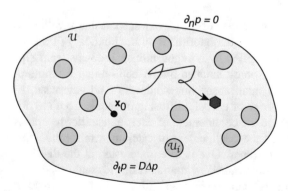

Fig. 6.20: Diffusion of a Brownian particle in a bounded domain $\mathcal{U}_i \subset \mathcal{U} \subset \mathbb{R}^d$ containing N small traps or holes labeled $i = 1, \ldots, N$. The exterior boundary $\partial \mathcal{U}$ is reflecting, whereas the interior boundaries $\partial \mathcal{U}_i$ are absorbing.

$$\Pi_k(\mathbf{x}_0,t) = \mathbb{P}[t < \mathcal{T}_k(\mathbf{x}_0) < \infty] = \int_t^\infty J_k(\mathbf{x}_0,t')dt', \qquad (6.4.3)$$

where

$$J_k(\mathbf{x}_0,t) = -\int_{\partial \mathcal{U}_k} \mathbf{J}(\mathbf{y},t|\mathbf{x}_0) \cdot d\mathbf{y} \qquad (6.4.4)$$

is the probability flux into the trap or, equivalently, the conditional FPT density. The negative sign indicates that the flux is into the domain $\mathcal{U}_k$. Integrating equation (6.4.1) with respect to $\mathbf{x}$ and t implies that the survival probability up to time t is

$$Q(\mathbf{x}_0,t) = \int_{\mathcal{U}\backslash\mathcal{U}_a} p(\mathbf{x},t|\mathbf{x}_0)d\mathbf{x} = \sum_{k=1}^N \Pi_k(\mathbf{x}_0,t). \qquad (6.4.5)$$

The splitting probability $\pi_k(\mathbf{x}_0)$ for being captured by the kth trap is then

$$\pi_k(\mathbf{x}_0) = \Pi_k(\mathbf{x}_0,0) = \int_0^\infty J_k(\mathbf{x}_0,t')dt'. \qquad (6.4.6)$$

We will assume that $\sum_k \pi_k(\mathbf{x}_0) = 1$, which implies that the particle is eventually captured by a trap with probability one. Using the backward diffusion equation, it follows that the splitting probability $\pi_k(\mathbf{x})$ satisfies the BVP

$$\nabla^2 \pi_k(\mathbf{x}) = 0, \ \mathbf{x} \in \mathcal{U}\backslash\mathcal{U}_a; \quad \partial_n \pi_k(\mathbf{x}) = 0, \ \mathbf{x} \in \partial\mathcal{U}, \qquad (6.4.7a)$$

$$\pi_k(\mathbf{x}) = \delta_{j,k}, \ \mathbf{x} \in \partial\mathcal{U}_j. \qquad (6.4.7b)$$

The corresponding conditional MFPT $T_k(\mathbf{x}_0)$ is given by

$$T_k(\mathbf{x}_0) = \mathbb{E}[\mathcal{T}_k|\mathcal{T}_k < \infty] = \frac{1}{\pi_k(\mathbf{x}_0)}\int_0^\infty t J_k(\mathbf{x}_0,t)dt = \frac{1}{\pi_k(\mathbf{x}_0)}\int_0^\infty \Pi_k(\mathbf{x}_0,t)dt. \ (6.4.8)$$

We have used the fact that $J_k = -d\Pi_k/dt$ and integrated by parts. More generally, moments of the conditional FPT density can be generated from the flux J_k:

$$\pi_k(\mathbf{x}_0)\mathbb{E}[(\mathcal{T}_k(\mathbf{x}_0))^n|\mathcal{T}_k < \infty] = \int_0^\infty t^n J_k(\mathbf{x}_0,t)dt = \left(-\frac{d}{ds}\right)^n \int_0^\infty e^{-st} J_k(\mathbf{x}_0,t)dt\Big|_{s=0}$$

$$= -\left(-\frac{d}{ds}\right)^n \int_0^\infty e^{-st}\frac{\partial \Pi_k}{\partial t}dt\Big|_{s=0}$$

$$= \left(-\frac{d}{ds}\right)^n (1 - s\widetilde{\Pi}_k(\mathbf{x}_0,s))\Big|_{s=0}, \qquad (6.4.9)$$

where $\widetilde{\Pi}_k(\mathbf{x}_0,s)$ is the Laplace transform of $\Pi_k(\mathbf{x}_0,t)$. In particular,

$$\pi_k(\mathbf{x}_0)T_k(\mathbf{x}_0) \quad = \mathbb{E}[\mathcal{T}_k(\mathbf{x}_0)1_{\mathcal{T}_k<\infty}] = \widetilde{\Pi}_k(\mathbf{x}_0,0),$$

$$\pi_k(\mathbf{x}_0)T_k^{(2)}(\mathbf{x}_0) = \mathbb{E}[\mathcal{T}_k(\mathbf{x}_0)^2 1_{\mathcal{T}_k<\infty}] = -2\widetilde{\Pi}_k'(\mathbf{x}_0,0), \qquad (6.4.10)$$

where $'$ indicates differentiation with respect to s. BVPs for the moments can now be derived from the backward equation for $\Pi(\mathbf{x}_0, t)$:

$$\frac{\partial \Pi_k(\mathbf{x}_0, t)}{\partial t} = D\nabla^2 \Pi_k(\mathbf{x}_0, t), \quad \mathbf{x}_0 \in \mathcal{U} \backslash \mathcal{U}_a, \tag{6.4.11a}$$

$$\partial_n \Pi_k(\mathbf{x}_0, t) = 0, \ \mathbf{x} \in \partial \mathcal{U}; \ \Pi_k(\mathbf{x}_0, t) = 0, \ \mathbf{x} \in \partial \mathcal{U}_a, \quad t > 0. \tag{6.4.11b}$$

Laplace transforming these equations gives

$$s\widetilde{\Pi}_k(\mathbf{x}_0, s) - \pi_k(\mathbf{x}_0) = D\nabla^2 \widetilde{\Pi}_k(\mathbf{x}_0, s), \ \mathbf{x}_0 \in \mathcal{U} \backslash \mathcal{U}_a, \tag{6.4.12a}$$

$$\partial_n \widetilde{\Pi}_k(\mathbf{x}_0, s) = 0, \ \mathbf{x} \in \partial \mathcal{U}; \ \widetilde{\Pi}_k(\mathbf{x}_0, s) = 0, \ \mathbf{x} \in \partial \mathcal{U}_a. \tag{6.4.12b}$$

It follows from equation (6.4.10) that the first- and second-order moments, respectively, satisfy the BVPs

$$\nabla^2 (\pi_k(\mathbf{x}_0) T_k(\mathbf{x}_0)) = -\frac{\pi_k(\mathbf{x}_0)}{D}, \ \mathbf{x}_0 \in \mathcal{U} \backslash \mathcal{U}_a, \tag{6.4.13a}$$

$$\partial_n (\pi_k(\mathbf{x}_0) T_k(\mathbf{x}_0)) = 0, \ \mathbf{x}_0 \in \partial \mathcal{U}, \tag{6.4.13b}$$

$$\pi_k(\mathbf{x}_0) T_k(\mathbf{x}_0) = 0, \ \mathbf{x}_0 \in \partial \mathcal{U}_a, \tag{6.4.13c}$$

and

$$\nabla^2 (\pi_k(\mathbf{x}_0) T_k^{(2)}(\mathbf{x}_0)) = -\frac{2\pi_k(\mathbf{x}_0) T_k(\mathbf{x}_0)}{D}, \ \mathbf{x}_0 \in \mathcal{U} \backslash \mathcal{U}_a, \tag{6.4.14a}$$

$$\partial_n (\pi_k(\mathbf{x}_0) T_k^{(2)}(\mathbf{x}_0)) = 0, \ \mathbf{x}_0 \in \partial \mathcal{U}, \tag{6.4.14b}$$

$$\pi_k(\mathbf{x}_0) T_k^{(2)}(\mathbf{x}_0) = 0, \ \mathbf{x}_0 \in \partial \mathcal{U}_a. \tag{6.4.14c}$$

Asymptotic analysis in 2D. Each of the above BVPs can be solved in the small target limit using matched asymptotic expansions and Green's functions. The basic idea is to construct an inner or local solution valid in an $O(\varepsilon)$ neighborhood of each trap, and then matching to an outer or global solution that is valid away from each neighborhood. In order to introduce the basic theory, we consider the simpler BVP for the unconditional MFPT $T(\mathbf{x}) = \sum_k \pi_k(\mathbf{x}) T_k(\mathbf{x})$ given by

$$\nabla^2 T(\mathbf{x}) = -\frac{1}{D}, \ \mathbf{x} \in \mathcal{U} \backslash \mathcal{U}_a, \tag{6.4.15a}$$

$$\partial_n T(\mathbf{x}) = 0, \ \mathbf{x} \in \partial \mathcal{U}, \tag{6.4.15b}$$

$$T(\mathbf{x}) = 0, \ \mathbf{x} \in \partial \mathcal{U}_a. \tag{6.4.15c}$$

We have dropped the subscript on the initial position. In the case of a 2D domain the basic steps are as follows.

Inner solution. First, consider the inner solution around the ith trap,

$$\Phi_i(\mathbf{y}) = T(\mathbf{x}_i + \varepsilon \mathbf{y}), \quad \mathbf{y} = \varepsilon^{-1}(\mathbf{x} - \mathbf{x}_i),$$

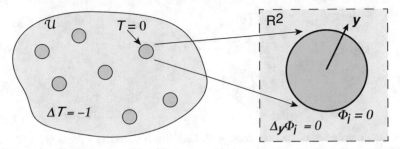

Fig. 6.21: Construction of the inner solution in terms of stretched coordinates $\mathbf{y} = \varepsilon^{-1}(\mathbf{x} - \mathbf{x}_i)$, where $\mathbf{x}_i$ is the center of the ith trap. Rescaled radius is $\rho_i = 1$ and the region outside the trap is taken to be $\mathbb{R}^2$ rather than the bounded domain $\mathcal{U}$.

where we have introduced stretched coordinates and replaced the domain $\mathcal{U}$ by $\mathbb{R}^2$, see Fig. 6.21. It follows that

$$\nabla_{\mathbf{y}}^2 \Phi_i = 0 \text{ for } \mathbf{y} \in \mathbb{R}^2 \backslash \mathcal{U}_i, \; \Phi_i = 0 \text{ on } |\mathbf{y}| = 1,$$

which can be expressed in polar coordinates as

$$\frac{1}{\rho} \frac{d}{d\rho} \rho \frac{d\Phi_i}{d\rho} = 0, \; 1 < \rho < \infty, \; \Phi_i(1) = 0.$$

The solution takes the form

$$\Phi_i(\rho) = \nu A_i(\nu) \ln(\rho/\rho_i), \tag{6.4.16}$$

where

$$\nu = -\frac{1}{\ln \varepsilon}, \tag{6.4.17}$$

and $A_i(\nu)$ is some undetermined function of ν. The corresponding solution in the original coordinates is

$$\Phi_i(\mathbf{x}) = \nu A_i(\nu) \ln(|\mathbf{x} - \mathbf{x}_i|/\varepsilon). \tag{6.4.18}$$

The coefficients $A_i(\nu)$, $i = 1, \ldots, N$, can be determined by matching the inner solutions with the corresponding outer solution (see below). The presence of the small parameter ν rather than ε in the matched asymptotic expansion is a common feature of strongly localized perturbations in 2D domains. It is well-known that $\nu \to 0$ much more slowly than $\varepsilon \to 0$. Hence, if one is interested in obtaining $O(\varepsilon)$ accuracy, then it is necessary to sum over the logarithmic terms non-perturbatively. This can be achieved by matching the inner and outer solutions using Green's functions, which is equivalent to calculating the asymptotic solution for all terms of $O(\nu^k)$ for any k.

Outer solution. The outer solution is obtained by treating each trap as a point source/sink, see Fig. 6.22. The resulting diffusion equation takes the form

$$\nabla^2 T = -\frac{1}{D}, \quad \mathbf{x} \in \mathcal{U} \backslash \{\mathbf{x}_1, \ldots, \mathbf{x}_N\}, \quad \partial_n T = 0, \quad \mathbf{x} \in \partial \mathcal{U}, \tag{6.4.19a}$$

together with the matching condition

$$T(\mathbf{x}) \sim A_j(\nu) + \nu A_j(\nu) \ln |\mathbf{x} - \mathbf{x}_j| \tag{6.4.19b}$$

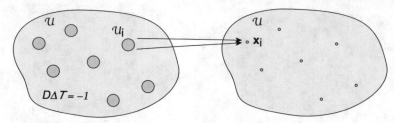

Fig. 6.22: Construction of the outer solution T. Each trap is shrunk to a single point. The outer solution can be expressed in terms of corresponding modified Neumann Green's function and then matched with the inner solution Φ_i around each trap.

as $\mathbf{x} \to \mathbf{x}_j$. The next step is to introduce the 2D Neumann Green's function $G(\mathbf{x},\mathbf{y})$, which is uniquely defined by (see Box 6D)

$$\nabla^2 G = \frac{1}{|\mathcal{U}|} - \delta(\mathbf{x}-\mathbf{y}), \ \mathbf{x} \in \mathcal{U}; \quad \partial_n G = 0 \text{ on } \partial \mathcal{U}, \tag{6.4.20a}$$

$$\int_{\mathcal{U}} G d\mathbf{x} = 0; \quad G(\mathbf{x},\mathbf{y}) = -\frac{\ln|\mathbf{x}-\mathbf{y}|}{2\pi} + R(\mathbf{x},\mathbf{y}) \tag{6.4.20b}$$

for fixed $\mathbf{y}$. Here R is the regular (non-singular) part of the Green's function. We now make the ansatz

$$T(\mathbf{x}) \approx T_\infty - 2\pi v \sum_{i=1}^{N} A_i(v) G(\mathbf{x},\mathbf{x}_i) \tag{6.4.21}$$

for $\mathbf{x} \notin \{\mathbf{x}_j, j = 1,\dots,N\}$ and some constant T_∞. Using the fact that $\int G d\mathbf{x} = 0$, it follows from equation (6.4.21) that

$$T_\infty = |\mathcal{U}|^{-1} \int_{\mathcal{U}} T(\mathbf{x}) d\mathbf{x}.$$

Observe that for $\mathbf{x} \notin \{\mathbf{x}_j, j = 1,\dots,N\}$,

$$\nabla^2 T(\mathbf{x}) \approx -2\pi v \sum_{i=1}^{N} A_i(v) \nabla^2 G(\mathbf{x},\mathbf{x}_i) = -\frac{2\pi v}{|\mathcal{U}|} \sum_{i=1}^{N} A_i(v).$$

Hence, the outer solution satisfies the BVP if and only if

$$\sum_{i=1}^{N} A_i(v) = \frac{|\mathcal{U}|}{2\pi v D}. \tag{6.4.22}$$

Matched asymptotics. As $\mathbf{x} \to \mathbf{x}_j$,

$$T(\mathbf{x}) \to T_\infty + v A_j(v) \ln|\mathbf{x}-\mathbf{x}_j| - 2\pi v A_j(v) R(\mathbf{x}_j,\mathbf{x}_j) - 2\pi v \sum_{i \neq j}^{N} A_i(v) G(\mathbf{x}_j,\mathbf{x}_i).$$

Comparison with the asymptotic limit in equation (6.4.19b) yields the self-consistency conditions

$$-(1 + 2\pi v R(\mathbf{x}_j,\mathbf{x}_j)) A_j(v) - 2\pi v \sum_{i \neq j} A_i(v) G(\mathbf{x}_j,\mathbf{x}_i) = -T_\infty \tag{6.4.23}$$

for $j = 1,\dots,N$. In particular, these can be rewritten as a matrix equation

$$\sum_{i=1}^{N} (\delta_{i,j} + 2\pi v \mathcal{G}_{ji}) A_i(v) = T_\infty, \tag{6.4.24}$$

with

$$\mathcal{G}_{jj} = R(\mathbf{x}_j, \mathbf{x}_j), \quad \mathcal{G}_{ji} = G(\mathbf{x}_j, \mathbf{x}_i), j \neq i. \tag{6.4.25}$$

We thus obtain the solution

$$A_i(v) = T_\infty \sum_{j=1}^{N} [\mathbf{I} + 2\pi v \mathcal{G}]_{ij}^{-1}, \tag{6.4.26}$$

and hence

$$T(\mathbf{x}) = T_\infty \left[1 - 2\pi v \sum_{i,j=1}^{N} [1 + 2\pi v \mathcal{G}]_{ij}^{-1} G(\mathbf{x}, \mathbf{x}_i) \right], \tag{6.4.27}$$

which is clearly non-perturbative with respect to v. It remains to determine the unknown constant T_∞. Imposing the constraint (6.4.22) on equation (6.4.26) implies that

$$T_\infty = \frac{|\mathcal{U}|}{2\pi v D} \left[\sum_{i,j=1}^{N} [1 + v\mathcal{G}]_{ij}^{-1} \right]^{-1}. \tag{6.4.28}$$

In the case of a single target at $\mathbf{x}_1$ ($N = 1$),

$$A_1(v) = \frac{T_\infty}{1 + 2\pi v R(\mathbf{x}_1, \mathbf{x}_1)} = \frac{|\mathcal{U}|}{2\pi v D}. \tag{6.4.29}$$

Hence, we have the explicit solution

$$T(\mathbf{x}) = \frac{|\mathcal{U}|}{2\pi v D} \left[1 + 2\pi v R(\mathbf{x}_1, \mathbf{x}_1) - \frac{|\mathcal{U}|}{D} \sum_{j=1}^{N} G(\mathbf{x}, \mathbf{x}_j) \right]. \tag{6.4.30}$$

For $N > 1$ one either solves the matrix equation numerically or Taylor expands as a power series in v. The latter yields the approximation

$$T(\mathbf{x}) \sim -\frac{|\mathcal{U}|}{DN} \sum_{j=1}^{N} G(\mathbf{x}, \mathbf{x}_j) + \frac{|\mathcal{U}|}{2\pi v D N} \left[1 + \frac{2\pi v}{N} \sum_{i,j=1}^{N} \mathcal{G}_{ij} \right] + O(v). \tag{6.4.31}$$

Finally, note that the same asymptotic method can be used to solve the BVPs for the splitting probabilities $\pi_k(\mathbf{x})$, see Ex. 6.7, and conditional moments $T_k(\mathbf{x}), T^{(2)}(\mathbf{x})$ [139, 474, 523]. A further example can be found in Ex. 6.8.

Asymptotic analysis in 3D. It is useful to briefly explore how the details of the asymptotic analysis change for the narrow capture problem in 3D. The main difference arises from the fact that the Neumann Green's function in 3D has a different singularity structure (see Box 6D)

$$\nabla^2 G(\mathbf{x}; \mathbf{x}') = \frac{1}{|\mathcal{U}|} - \delta(\mathbf{x} - \mathbf{x}'), \mathbf{x} \in \mathcal{U}; \ \partial_n G = 0, \mathbf{x} \in \partial \mathcal{U}, \tag{6.4.32a}$$

$$G(\mathbf{x}, \mathbf{x}') = \frac{1}{4\pi |\mathbf{x} - \mathbf{x}'|} + R(\mathbf{x}, \mathbf{x}'), \ \int_{\mathcal{U}} G(\mathbf{x}, \mathbf{x}') d\mathbf{x} = 0, \tag{6.4.32b}$$

with $R(\mathbf{x}, \mathbf{x}')$ again corresponding to the regular part of the Green's function. We will focus on the BVP (6.4.15) of the unconditional MFPT.

In the outer region, T is expanded as $T = \varepsilon^{-1}T_0 + T_1 + \varepsilon T_2 + \dots$ Here T_0 is an unknown constant, and

$$\nabla^2 T_n(\mathbf{x}) = -\frac{\delta_{n,1}}{D}, \ \mathbf{x} \in \mathcal{U} \backslash \{\mathbf{x}_1, \dots, \mathbf{x}_N\}, \quad \partial_n T_n(\mathbf{x}) = 0, \ \mathbf{x} \in \partial \mathcal{U} \qquad (6.4.33)$$

for $n = 1, 2$, together with certain singularity conditions as $\mathbf{x} \to \mathbf{x}_j$, $j = 1, \dots, N$. The latter are determined by matching to the inner solution. In the inner region around the j-th trap, we introduce the stretched coordinates $\mathbf{y} = \varepsilon^{-1}(\mathbf{x} - \mathbf{x}_j)$ and set $u(\mathbf{y}) = T(\mathbf{x}_j + \varepsilon \mathbf{y})$. Expanding the inner solution as $u = \varepsilon^{-1}u_0 + u_1 + \dots$, we have

$$\nabla_{\mathbf{y}}^2 u_n(\mathbf{y}) = 0, \ \mathbf{y} \notin \mathcal{U}_k; \quad u_n(\mathbf{y}) = 0, \mathbf{y} \in \partial \mathcal{U}_j. \qquad (6.4.34)$$

Finally, the matching condition is that the near-field behavior of the outer solution as $\mathbf{x} \to \mathbf{x}_j$ should agree with the far-field behavior of the inner solution as $|\mathbf{y}| \to \infty$, which is expressed as

$$\varepsilon^{-1}T_0 + T_1 + \varepsilon T_2 + \dots \sim \varepsilon^{-1}u_0 + u_1 + \dots.$$

First $T_0 \sim u_0$ so that we can set

$$u_0(\mathbf{y}) = T_0(1 - w(\mathbf{y})), \qquad (6.4.35)$$

with $w(\mathbf{y})$ satisfying the boundary value problem

$$\nabla_{\mathbf{y}}^2 w(\mathbf{y}) = 0, \ \mathbf{y} \notin \mathcal{U}_k; \quad w(\mathbf{y}) = 1, \ \mathbf{y} \in \partial \mathcal{U}_j; \quad w(\mathbf{y}) \to 0 \quad \text{as } |\mathbf{y}| \to \infty. \qquad (6.4.36)$$

This is a well-known problem in electrostatics and has the far-field behavior

$$w(\mathbf{y}) \sim \frac{C_j}{|\mathbf{y}|} + \frac{\mathbf{P}_j \cdot \mathbf{y}}{|\mathbf{y}|^3} + \dots \text{as } |\mathbf{y}| \to \infty, \qquad (6.4.37)$$

where C_j is the capacitance and $\mathbf{P}_j$ the dipole vector of an equivalent charged conductor with the shape $\mathcal{U}_j$. (Here C_j has the units of length.) Some examples of capacitances for various trap shapes are as follows [170]:

$C_j = a$ (sphere of radius a),

$C_j = 2a(1 - 1/\sqrt{3})$ (hemisphere of radius a),

$C_j = \dfrac{\sqrt{a^2 - b^2}}{\cosh^{-1}(a/b)}$ (prolate spheroid with semi-major and minor axes a, b),

$C_j = \dfrac{\sqrt{a^2 - b^2}}{\cosh^{-1}(b/a)}$ (oblate spheroid with semi-major and minor axes a, b).

It now follows that T_1 is given by equation (6.4.33) together with the singularity condition

$$T_1(\mathbf{x}) \sim -\frac{T_0 C_j}{|\mathbf{x} - \mathbf{x}_j|} \quad \text{as } \mathbf{x} \to \mathbf{x}_j.$$

In other words, T_1 satisfies the inhomogeneous equation

$$\nabla^2 T_1(\mathbf{x}) = -\frac{1}{D} + 4\pi T_0 \sum_{j=1}^{N} C_j \delta(\mathbf{x} - \mathbf{x}_j), \ \mathbf{x} \in \mathcal{U}; \quad \partial_n T_1(\mathbf{x}) = 0, \quad \mathbf{x} \in \partial \mathcal{U}. \qquad (6.4.38)$$

This can be solved in terms of the Neumann Green's function:

$$T_1(\mathbf{x}) = -4\pi T_0 \sum_{j=1}^{N} C_j G(\mathbf{x}, \mathbf{x}_j) + \chi, \tag{6.4.39}$$

with unknown constant

$$\chi = \frac{1}{|\mathcal{U}|} \int_{\mathcal{U}} T_1(\mathbf{x}) d\mathbf{x}. \tag{6.4.40}$$

In order to fully specify the $O(1/\varepsilon)$ and $O(1)$ contributions to the MFPT $T(\mathbf{x})$ we have to determine the constants T_0 and χ. The first follows immediately from integrating equation (6.4.38) over the domain $\mathcal{U}$ and using the divergence theorem. This yields the solvability condition

$$T_0 = \frac{|\mathcal{U}|}{4\pi D N \overline{C}}, \quad \overline{C} = \frac{1}{N} \sum_{j=1}^{N} C_j. \tag{6.4.41}$$

The calculation of χ is more involved, since it is obtained by imposing a solvability condition on the $O(\varepsilon)$ contribution T_2. This requires matching T_2 with the far-field behavior of u_1, which is itself found by matching u_1 with the near-field behavior of T_1. The latter takes the form

$$T_1(\mathbf{x}) \sim -\frac{4\pi T_0 C_j}{|\mathbf{x} - \mathbf{x}_j|} - 4\pi T_0 C_j \mathcal{G}_{jj} - 4\pi T_0 \sum_{k \neq j}^{N} C_k \mathcal{G}_{jk} + \chi.$$

It follows that

$$u_1(\mathbf{y}, s) \rightarrow \Theta_j = -4\pi T_0 \sum_{k=1}^{N} C_k \mathcal{G}_{jk} + \chi \text{ as } |\mathbf{y}| \rightarrow \infty. \tag{6.4.42}$$

The solution for u_1 is thus

$$u_1(\mathbf{y}) = \Theta_j(1 - w(\mathbf{y})), \tag{6.4.43}$$

with $w(\mathbf{y})$ given by equation (6.4.37). It now follows that the outer contribution T_2 satisfies equation (6.4.33) supplemented by the singularity condition

$$T_2(\mathbf{x}) \sim -\frac{\Theta_j C_j}{|\mathbf{x} - \mathbf{x}_j|}, \quad \text{as } \mathbf{x} \rightarrow \mathbf{x}_j.$$

Using the same steps as in the analysis of $T_1(\mathbf{x}, s)$,

$$\nabla^2 T_2(\mathbf{x}) = 4\pi \sum_{j=1}^{N} \Theta_j C_j \delta(\mathbf{x} - \mathbf{x}_j), \ \mathbf{x} \in \mathcal{U}; \quad \partial_n T_2(\mathbf{x}) = 0, \quad \mathbf{x} \in \partial \mathcal{U}. \tag{6.4.44}$$

Integrating equation (6.4.44) and applying the divergence theorem gives the solvability condition $\sum_{j=1}^{N} \Theta_j C_j = 0$. Substituting for Θ_j and rearranging then determines χ:

$$\chi = \frac{4\pi T_0}{N\overline{C}} \sum_{k=1}^{N} C_j C_k \mathcal{G}_{jk}. \tag{6.4.45}$$

The final result is [170]

$$T(\mathbf{x}) \sim \frac{|\mathcal{U}|}{4\pi \varepsilon D N \overline{C}} \left[1 - 4\pi \varepsilon \sum_{j=1}^{N} C_j G(\mathbf{x}, \mathbf{x}_j) + \frac{4\pi \varepsilon}{N\overline{C}} \sum_{i,j=1}^{N} C_i \mathcal{G}_{ij} C_j \right] + o(1). \tag{6.4.46}$$

The analogous asymptotic analysis of splitting probabilities and conditional MFPTs can be found in [170, 189]. The asymptotic expansion of the survival probability is considered in Ex. 6.9.

Box 6D. The 2D and 3D Green function for Laplace's equation.

Consider the steady-state inhomogeneous diffusion equation on a bounded domain $\mathcal{U} \subset \mathbb{R}^d$ with $d = 2,3$ and an inhomogeneous Dirichlet boundary condition on $\partial\mathcal{U}$:

$$\nabla^2 u(\mathbf{x}) = -f(\mathbf{x}), \quad \mathbf{x} \in \mathcal{U} \subset \mathbb{R}^d, \quad u(\mathbf{x}) = g(\mathbf{x}), \mathbf{x} \in \partial\mathcal{U}. \tag{6.4.47}$$

One way to solve the inhomogeneous equation is to use Green's functions. Green's function $G(\mathbf{x},\mathbf{y})$ for the Dirichlet boundary value problem is defined by the equation

$$\nabla_y^2 G(\mathbf{x},\mathbf{y}) = -\delta(\mathbf{x}-\mathbf{y}), \quad \mathbf{x},\mathbf{y} \in \mathcal{U}, \quad G(\mathbf{x},\mathbf{y}) = 0, \quad \mathbf{y} \in \partial\mathcal{U}, \tag{6.4.48}$$

where $\delta(\mathbf{x})$ is the Dirac delta function in $\mathbb{R}^d$. Hence, in Cartesian coordinates for $d = 3$, $\delta(\mathbf{x}) = \delta(x)\delta(y)\delta(z)$. Once one has determined Green's function, the solution of the inhomogeneous boundary value problem can be obtained from the following Green identity:

$$\int_{\mathcal{U}} \left[u(\mathbf{y})\nabla_y^2 G(\mathbf{x},\mathbf{y}) - G(\mathbf{x},\mathbf{y})\nabla_y^2 u(\mathbf{y}) \right] d\mathbf{y}$$
$$= \int_{\mathcal{U}} \nabla_y \cdot [u(\mathbf{y})\nabla_y G(\mathbf{x},\mathbf{y}) - G(\mathbf{x},\mathbf{y})\nabla_y u(\mathbf{y})] d\mathbf{y}.$$

Applying the steady-state equations to both terms on the left-hand side and using the divergence theorem on the right-hand side show that

$$-u(\mathbf{x}) + \int_{\mathcal{U}} G(\mathbf{x},\mathbf{y})f(\mathbf{y})d\mathbf{y} = \int_{\partial\mathcal{U}} [u(\mathbf{y})\nabla_y G(\mathbf{x},\mathbf{y}) - G(\mathbf{x},\mathbf{y})\nabla_y u(\mathbf{y})] \cdot \mathbf{n}d\mathbf{y},$$

where $\mathbf{n}$ is the outward normal along the boundary $\partial\mathcal{U}$. Imposing the boundary conditions on u and G and rearranging yield the solution

$$u(\mathbf{x}) = \int_{\mathcal{U}} G(\mathbf{x},\mathbf{y})f(\mathbf{y})d\mathbf{y} - \int_{\partial\mathcal{U}} \partial_n G(\mathbf{x},\mathbf{y})g(\mathbf{y})d\mathbf{y}, \tag{6.4.49}$$

where $\partial_n G$ denotes the normal derivative of G.

Eigenfunction expansion of Green's function. Using the properties of the spectrum of the negative Laplacian listed in Box 2D, it follows that Green's function has a formal expansion in terms of the complete set of orthonormal eigenfunctions:

$$G(\mathbf{x}, \mathbf{y}) = \sum_{n=1}^{\infty} \frac{\phi_n(\mathbf{x})\phi_n(\mathbf{y})}{\lambda_n}. \tag{6.4.50}$$

This is straightforward to establish, since

$$\nabla_y^2 G(\mathbf{x}, \mathbf{y}) = \nabla_y^2 \left(\sum_{n=1}^{\infty} \frac{\phi_n(\mathbf{x})\phi_n(\mathbf{y})}{\lambda_n} \right) = \sum_{n=1}^{\infty} \frac{\phi_n(\mathbf{x})\nabla_y^2 \phi_n(\mathbf{y})}{\lambda_n}$$

$$= -\sum_{n=1}^{\infty} \frac{\phi_n(\mathbf{x})\lambda_n\phi_n(\mathbf{y})}{\lambda_n} = -\sum_{n=1}^{\infty} \phi_n(\mathbf{x})\phi_n(\mathbf{y}) = -\delta(\mathbf{x} - \mathbf{y}).$$

We have reversed the order of summation and integration and used a completeness relation. Note that the definition of Green's function has to be slightly modified in the case of Neumann boundary conditions, since there exists a zero eigenvalue. The so-called generalized or modified Neumann Green function has the eigenfunction expansion

$$G(\mathbf{x}, \mathbf{y}) = \sum_{n=1}^{\infty} \frac{\phi_n(\mathbf{x})\phi_n(\mathbf{y})}{\lambda_n}. \tag{6.4.51}$$

Since the completeness relation has to be extended to include the constant normalized eigenfunction $\phi_0(\mathbf{x}) = 1/\sqrt{|\mathcal{U}|}$, where $|\mathcal{U}|$ denotes the volume of the bounded domain $\mathcal{U}$, we see that Neumann Green's function satisfies

$$\nabla_y^2 G(\mathbf{x}, \mathbf{y}) = \frac{1}{|\mathcal{U}|} - \delta(\mathbf{x} - \mathbf{y}). \tag{6.4.52}$$

One of the significant features of the Dirichlet or Neumann Green's function $G(\mathbf{x}, \mathbf{y})$ in two and three dimensions is that it is singular in the limit $\mathbf{x} \to \mathbf{y}$. Moreover, these singularities take the specific form

$$G(\mathbf{x}, \mathbf{y}) \sim \ln(|\mathbf{x} - \mathbf{y}|) \text{ (in 2D)}, \quad G(\mathbf{x}, \mathbf{y}) \sim \frac{1}{|\mathbf{x} - \mathbf{y}|} \text{ (in 3D)}. \tag{6.4.53}$$

In order to isolate these singularities, we consider the corresponding fundamental solution of Laplace's equation in R^d.

Fundamental solution of Laplace's equation. First, consider Laplace's equation in $\mathbb{R}^2$:

$$\nabla^2 u(\mathbf{x}) = 0, \quad \mathbf{x} \in \mathbb{R}^2.$$

Since there are no boundaries, this equation is symmetric with respect to rigid body translations and rotations in the plane. This implies that if $u(\mathbf{x})$ is a solution to Laplace's equation then so are $v(\mathbf{x}) = u(\mathbf{x} - \mathbf{a})$ and $w(\mathbf{x}) = u(\mathbf{R}_\theta \mathbf{x})$. Here $\mathbf{a}$ is a constant vector and $\mathbf{R}_\theta$ is the 2×2 rotation matrix about the origin

$$\mathbf{R}_\theta = \begin{pmatrix} \cos(\theta) & -\sin(\theta) \\ \sin(\theta) & \cos(\theta) \end{pmatrix}.$$

This suggests that we look for a radially symmetric solution $u = u(r)$. Introducing polar coordinates, Laplace's equation becomes

$$\frac{d^2u}{dr^2} + \frac{1}{r}\frac{du}{dr} = 0, \quad 0 < r < \infty.$$

The radially symmetric solution is thus of the form

$$u(r) = C_0 \ln(r) + C_1$$

for constants C_0, C_1. Similarly, radially symmetric solutions in $\mathbb{R}^3$ satisfy Laplace's equation

$$\frac{d^2u}{dr^2} + \frac{2}{r}\frac{du}{dr} = 0, \quad 0 < r < \infty,$$

which has the solution

$$u(r) = \frac{C_0}{r} + C_1.$$

For convenience, choosing $C_1 = 0$, $C_0 = 1/4\pi$ (in 3D), and $C_0 = -1/2\pi$ (in 2D), we obtain the fundamental solution of the Laplace equation

$$K(\mathbf{x}) = -\frac{1}{2\pi}\ln|\mathbf{x}| \text{ (in 2D)}, \quad K(\mathbf{x}) = \frac{1}{4\pi|\mathbf{x}|} \text{ (in 3D)}.$$

The fundamental solution satisfies Laplace's equation everywhere except the origin, where it is singular. It turns out that K satisfies the equation

$$\nabla^2 K(\mathbf{x}) = -\delta(\mathbf{x}).$$

We will show this explicitly for the 3D case. Let $f \in L^2(\mathbb{R}^3)$ be a function that vanishes at ∞. Define the function

$$u(\mathbf{x}) = \int_{\mathbb{R}^3} K(\mathbf{x} - \mathbf{y})f(\mathbf{y})d\mathbf{y} = \frac{1}{4\pi}\int_{\mathbb{R}^3} \frac{f(\mathbf{y})}{|\mathbf{x} - \mathbf{y}|}d\mathbf{y}.$$

We will prove that $\nabla^2 u = -f$ and hence $\nabla^2 K = -\delta$. First, it is convenient to rewrite the expression for u as

$$u(\mathbf{x}) = \frac{1}{4\pi}\int_{\mathbb{R}^3} \frac{f(\mathbf{x} - \mathbf{y})}{|\mathbf{y}|}d\mathbf{y}.$$

Since $\nabla_x^2 f(\mathbf{x} - \mathbf{y}) = \nabla_y^2 f(\mathbf{x} - \mathbf{y})$,

$$\nabla^2 u(\mathbf{x}) = \int_{\mathbb{R}^3} \frac{1}{|\mathbf{y}|} \nabla_y^2 f(\mathbf{x} - \mathbf{y}) d\mathbf{y}.$$

We would like to integrate by parts, but since $K(\mathbf{y})$ is singular at $\mathbf{y} = 0$, we first have to isolate the origin by surrounding it with a small sphere $B_r(0)$ of radius r. That is, we write

$$\nabla^2 u(\mathbf{x}) = \left[\int_{B_r(0)} + \int_{\mathbb{R}^3 \backslash B_r(0)} \right] \frac{1}{4\pi|\mathbf{y}|} \nabla_y^2 f(\mathbf{x} - \mathbf{y}) d\mathbf{y} \equiv I_r + J_r.$$

Here $\mathbb{R}^3 \backslash B_r(0)$ denotes $\mathbb{R}^3$ excluding the sphere around the origin. Using spherical polar coordinates,

$$|I_r| \leq \frac{\max |\nabla_y^2 f|}{4\pi} \int_{B_r(0)} \frac{1}{|\mathbf{y}|} d\mathbf{y} = \max |\nabla_y^2 f| \int_0^r \rho d\rho$$

$$= \frac{\max |\nabla_y^2 f|}{2} r^2 \to 0 \text{ as } r \to 0.$$

Recalling that f vanishes at infinity, we can integrate J_r by parts twice. First,

$$J_r = \frac{1}{4\pi} \int_{\mathbb{R}^3 \backslash B_r(0)} \left[\nabla_y \cdot \left(\frac{1}{|\mathbf{y}|} \nabla_y f(\mathbf{x} - \mathbf{y}) \right) - \nabla_y \frac{1}{|\mathbf{y}|} \cdot \nabla_y f(\mathbf{x} - \mathbf{y}) \right] d\mathbf{y}$$

$$= \frac{1}{4\pi} \int_{\mathbb{R}^3 \backslash B_r(0)} \nabla_y \cdot \left(\frac{1}{|\mathbf{y}|} \nabla_y f(\mathbf{x} - \mathbf{y}) - f(\mathbf{x} - \mathbf{y}) \nabla_y \frac{1}{|\mathbf{y}|} \right) d\mathbf{y}$$

$$+ \frac{1}{4\pi} \int_{\mathbb{R}^3 \backslash B_r(0)} f(\mathbf{x} - \mathbf{y}) \nabla_y^2 \frac{1}{|\mathbf{y}|} d\mathbf{y}.$$

Using the fact that $\nabla_y^2(1/|\mathbf{y}|) = 0$ in $\mathbb{R}^3 \backslash B_r(0)$ and applying the divergence theorem, we have

$$J_r = \frac{1}{4\pi} \int_{\partial B_r(0)} \left(\frac{1}{|\mathbf{y}|} \nabla_y f(\mathbf{x} - \mathbf{y}) - f(\mathbf{x} - \mathbf{y}) \nabla_y \frac{1}{|\mathbf{y}|} \right) \cdot \mathbf{n}_y d\mathbf{y}.$$

The first integral vanishes in the limit $r \to 0$, since

$$\frac{1}{4\pi r} \left| \int_{\partial B_r(0)} \nabla_y f(\mathbf{x} - \mathbf{y}) \cdot \mathbf{n}_y d\mathbf{y} \right| \leq r \max |\nabla_y f| \to 0.$$

On the other hand, since $\nabla_y(1/|\mathbf{y}|) = -\mathbf{y}/|\mathbf{y}|^3$ and $\mathbf{n}_y = -\mathbf{y}/r$, the second integral yields

$$\frac{1}{4\pi} \int_{\partial B_r(0)} f(\mathbf{x} - \mathbf{y}) \nabla_y \frac{1}{|\mathbf{y}|} \cdot \mathbf{n}_y d\mathbf{y} = \frac{1}{4\pi r^2} \int_{\partial B_r(0)} f(\mathbf{x} - \mathbf{y}) d\mathbf{y} \to f(\mathbf{x}) \text{ as } r \to 0.$$

We conclude that $I_r \to 0$ and $J_r \to -f(\mathbf{x})$ as $r \to 0$, which implies that $\nabla^2 u = -f$ and $\nabla^2 K = -\delta$. A similar analysis can be carried out in 2D using the logarithmic fundamental solution.

Finally, given the properties of the fundamental solution $K(\mathbf{x})$, we can construct Green's function for a boundary value problem in terms of $K(\mathbf{x})$ and a non-singular or regular part that satisfies Laplace's equation everywhere, that is,

$$G(\mathbf{x}, \mathbf{y}) = K(\mathbf{x} - \mathbf{y}) + R(\mathbf{x}, \mathbf{y}).$$

Examples of Neumann Green's functions in simple geometries. In the main text we consider various boundary value problems that are solved in terms of 2D or 3D Neumann Green's functions. Here we consider a few examples involving simple geometries.

(a) *The disk.* Let $\mathcal{U} \subset \mathbb{R}^2$ be the unit circle centered at the origin. The 2D Neumann Green function is given by

$$G(\mathbf{x}, \boldsymbol{\xi}) = \frac{1}{2\pi} \left[-\ln(|\mathbf{x} - \boldsymbol{\xi}|) - \ln\left(\left| \mathbf{x}|\boldsymbol{\xi}| - \frac{\boldsymbol{\xi}}{|\boldsymbol{\xi}|} \right| \right) \right.$$
$$\left. + \frac{1}{2}(|\mathbf{x}|^2 + |\boldsymbol{\xi}|^2) - \frac{3}{4} \right], \tag{6.4.54}$$

with the regular part obtained by dropping the first logarithmic term.

(b) *The sphere.* Let $\mathcal{U} \subset \mathbb{R}^3$ be the sphere of radius a centered about the origin. The 3D Neumann Green function takes the form [170]

$$G(\mathbf{x}, \boldsymbol{\xi}) = \frac{1}{4\pi|\mathbf{x} - \boldsymbol{\xi}|} + \frac{a}{4\pi|\mathbf{x}|r'} + \frac{1}{4\pi a} \ln\left(\frac{2a^2}{a^2 - |\mathbf{x}||\boldsymbol{\xi}|\cos\theta + |\mathbf{x}|r'} \right)$$
$$+ \frac{1}{6|\mathcal{U}|}(|\mathbf{x}|^2 + |\boldsymbol{\xi}|^2) + B, \tag{6.4.55}$$

where the constant B is chosen so that $\int_{\mathcal{U}} G(\mathbf{x}, \boldsymbol{\xi}) d\mathbf{x} = 0$, and

$$\cos\theta = \frac{\mathbf{x} \cdot \boldsymbol{\xi}}{|\mathbf{x}||\boldsymbol{\xi}|}, \quad \mathbf{x}' = \frac{a^2\mathbf{x}}{|\mathbf{x}|^2}, \quad r' = |\mathbf{x}' - \boldsymbol{\xi}|.$$

It can be shown that B is independent of $\boldsymbol{\xi}$.

(c) *Rectangular domain.* Let $\mathcal{U} \subset \mathbb{R}^2$ be a rectangular domain $[0, L_1] \times [0, L_2]$. The 2D Neumann Green function has the logarithmic expansion [111, 660]

$$G(\mathbf{r}, \mathbf{r}') = \frac{1}{L_1} H_0(y, y') - \frac{1}{2\pi} \sum_{j=0}^{\infty} \sum_{n=\pm} \sum_{m=\pm} \left(\ln|1 - \tau^j z_n \zeta_m| \right.$$
$$\left. + \ln|1 - \tau^j z_n \varsigma_m| \right), \tag{6.4.56}$$

where

$$\tau = e^{-2\pi L_2/L_1}, \quad z_{\pm} = e^{i\pi(x\pm x')/L_1},$$

$$\zeta_{\pm} = e^{-\pi|y\pm y'|/L_1}, \quad \varsigma_{\pm} = e^{-\pi(2L_2-|y\pm y'|)/L_1},$$

and

$$H_0(y,y') = \frac{L_2}{3} + \frac{1}{2L_2}(y^2 + y'^2) - \max\{y,y'\}, \tag{6.4.57}$$

Assuming that $\tau \ll 1$, we have the approximation

$$G(\mathbf{r},\mathbf{r}') = \frac{1}{L_1}H_0(y,y') - \frac{1}{2\pi}\sum_{n=\pm}\sum_{m=\pm}\left(\ln|1-z_n\zeta_m| + \ln|1-z_n\varsigma_m|\right) + O(\tau).$$

The only singularity exhibited by equation (6.4.56) occurs when $\mathbf{r} \to \mathbf{r}'$, $\mathbf{r}' \notin \partial \mathcal{U}$, in which case $z_- = \zeta_- = 1$ and the term $\ln|1-z_-\zeta_-|$ diverges. Writing

$$\ln|1-z_-\zeta_-| = \ln|\mathbf{r}-\mathbf{r}'| + \ln\frac{|1-z_-\zeta_-|}{|\mathbf{r}-\mathbf{r}'|},$$

where the first term on the right-hand side is singular and the second is regular, we find that

$$G(\mathbf{r},\mathbf{r}') = -\frac{1}{2\pi}\ln|\mathbf{r}-\mathbf{r}'| + R(\mathbf{r},\mathbf{r}'), \tag{6.4.58}$$

where R is the regular part of Green's function given by

$$R(\mathbf{r},\mathbf{r}') = -\frac{1}{L_1}H_0(y,y') + \frac{1}{2\pi}\ln\frac{|1-z_-\zeta_-||1-z_-\zeta_+|}{|\mathbf{r}-\mathbf{r}'|}$$

$$+ \frac{1}{2\pi}\ln|1-z_-\varsigma_-||1-z_-\varsigma_+| + \frac{1}{2\pi}\ln|1-z_+\varsigma_-||1-z_+\varsigma_+|$$

$$+ \frac{1}{2\pi}\ln|1-z_+\zeta_-||1-z_+\zeta_+| + O(\tau). \tag{6.4.59}$$

6.4.2 Diffusion-limited reaction for a small target

In the derivation of diffusion-limited reaction rates (Sect. 6.3), it was assumed that diffusion of the background reactants occurs in an unbounded domain with a uniform concentration at infinity. The analysis becomes considerably more involved when reactions occur in a bounded domain. However, asymptotic methods similar to the analysis of the narrow escape and capture problems can be used to determine the reaction rate in the asymptotic limit that the target is much smaller than the domain size [783]. Consider a target disk $\mathcal{U}_\varepsilon$ of radius $\varepsilon \ll 1$ and center $\mathbf{x}_0$ that is located in the interior of a rectangular domain $\mathcal{U}$ of size $O(1)$, see Fig. 6.23. The

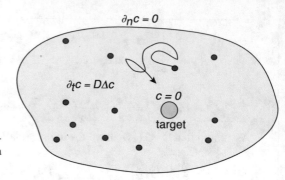

$\partial_n c = 0$

$\partial_t c = D\Delta c$

$c = 0$

target

Fig. 6.23: Diffusion-limited reaction rate for a small target in a
bounded domain.

calculation of the reaction rate can be formulated in terms of the solution to the
following diffusion equation:

$$\frac{\partial c(\mathbf{x},t)}{\partial t} = D\nabla^2 c(\mathbf{x},t), \quad \mathbf{x} \in \mathcal{U}\backslash\mathcal{U}_\varepsilon, \tag{6.4.60}$$

with $\partial_n c = 0$ on the exterior boundary $\partial\mathcal{U}$ and $c = 0$ on the interior boundary $\partial\mathcal{U}_\varepsilon$.
The initial condition is taken to be $c(\mathbf{x},0) = 1$. Following Straube *et al* [783], we
seek a solution in the form of an eigenfunction expansion,

$$c(\mathbf{x},t) = \sum_{j=0}^{\infty} c_j\phi_j(\mathbf{x})e^{-\lambda_j D t}, \tag{6.4.61}$$

where the eigenfunctions $\phi_j(\mathbf{x})$ satisfy the Helmholtz equation

$$\nabla^2\phi_j + \lambda_j\phi_j = 0, \quad \mathbf{x} \in \mathcal{U}\backslash\mathcal{U}_\varepsilon, \tag{6.4.62}$$

subject to the same boundary conditions as $c(\mathbf{r},t)$. The eigenfunctions are orthogo-
nalized as

$$\int_{\mathcal{U}\backslash\mathcal{U}_\varepsilon} \phi_i(\mathbf{x})\phi_j(\mathbf{x})d\mathbf{x} = \delta_{i,j}. \tag{6.4.63}$$

The initial condition then implies that

$$c_j = \int_{\mathcal{U}\backslash\mathcal{U}_\varepsilon} \phi_j(\mathbf{x})d\mathbf{x}. \tag{6.4.64}$$

Taking the limit $\varepsilon \to 0$ results in an eigenvalue problem in a rectangular domain
without a hole. It is well known that the eigenvalues are ordered as $\lambda_0 = 0 < \lambda_1 \leq$
$\lambda_2 \leq \ldots$. This ordering will persist when $0 < \varepsilon \ll 1$ so that in the long-time limit,
the solution will be dominated by the eigenmode with the smallest eigenvalue:

$$c(\mathbf{x},t) \sim c_0\phi_0(\mathbf{x})e^{-\lambda_0 D t}. \tag{6.4.65}$$

The time-dependent flux is then

$$J(t) = Dc_0 e^{-\lambda_0 Dt} \int_0^{2\pi} \left(r \frac{\partial \phi_0}{\partial r} \right) \Big|_{r=\varepsilon} d\theta. \tag{6.4.66}$$

For small ε, the principal eigenvalue λ_0 of the Helmholtz operator has an infinite logarithmic expansion [783, 845]:

$$\lambda_0 = v\Lambda_1 + v^2 \Lambda_2 + \dots, \quad v = -\frac{1}{\ln \varepsilon}. \tag{6.4.67}$$

Moreover, the eigenfunction $\phi_0(\mathbf{x})$ develops a boundary layer in a neighborhood of the target, where it changes rapidly from zero on the boundary $\partial \mathcal{U}_\varepsilon$ to a value of $O(1)$ away from the target. This suggests dividing the domain into inner and outer regions and using matched asymptotics along analogous lines to the study of the narrow escape and narrow capture problems. The logarithmic expansion of λ_0 implies that the right-hand side of the rescaled eigenvalue equation is of $O(\varepsilon^2 v^2) = o(v^k)$ for all $k \geq 0$. Thus to logarithmic accuracy, it follows that the inner problem with stretched coordinates $\mathbf{y} = \mathbf{x}/\varepsilon$ is

$$\nabla^2 \varphi(\mathbf{y}) = 0, \quad \mathbf{y} \in \mathbb{R}^2 \setminus S^1,$$

where S^1 is the unit circle centered about the origin, and $\varphi = 0$ on $|\mathbf{y}| = 1$. Hence, $\varphi(\mathbf{y}) = A \ln |\mathbf{y}|$ and the inner solution has the far-field behavior

$$\varphi \sim A \ln(|\mathbf{x} - \mathbf{x}_0|/\varepsilon). \tag{6.4.68}$$

The outer solution satisfies the equation

$$\nabla^2 \phi_0 + \lambda_0 \phi_0 = 0, \quad \mathbf{x} \in \mathcal{U} \setminus \{\mathbf{x}_0\},$$

$$\phi_0 \sim A \ln(|\mathbf{x} - \mathbf{x}_0|/\varepsilon), \quad \mathbf{x} \to \mathbf{x}_0, \quad \int_{\mathcal{U}} \phi_0^2(\mathbf{x}) d\mathbf{x} = 1.$$

The outer problem can be solved in terms of Neumann Green's function for the Helmholtz equation:

$$\nabla^2 G(\mathbf{x}, \mathbf{x}_0; \lambda_0) + \lambda_0 G(\mathbf{x}, \mathbf{x}_0; \lambda_0) = -\delta(\mathbf{x} - \mathbf{x}_0), \quad \mathbf{x} \in \mathcal{U}, \tag{6.4.69a}$$

$$\partial_n G(\mathbf{x}, \mathbf{x}_0; \lambda_0) = 0, \quad \mathbf{x} \in \partial \mathcal{U}, \tag{6.4.69b}$$

$$G(\mathbf{x}, \mathbf{x}_0; \lambda_0) \sim -\frac{1}{2\pi} \ln |\mathbf{x} - \mathbf{x}_0| + R(\mathbf{x}_0, \mathbf{x}_0; \lambda_0), \quad \mathbf{x} \to \mathbf{x}_0. \tag{6.4.69c}$$

That is,

$$\phi_0(\mathbf{x}) = -2\pi A G(\mathbf{x}, \mathbf{x}_0; \lambda_0). \tag{6.4.70}$$

Matching the near-field behavior of the outer solution with the far-field behavior of the inner solution then yields a transcendental equation for the principal eigenvalue:

$$R(\mathbf{x}_0, \mathbf{x}_0; \lambda_0) = -\frac{1}{2\pi v}. \tag{6.4.71}$$

Finally, the normalization condition for ϕ_0 determines the amplitude A according to

$$4\pi^2 A^2 \int_{\mathcal{U}} G(\mathbf{x}, \mathbf{x}_0; \lambda_0)^2 d\mathbf{x} = 1. \tag{6.4.72}$$

Since $0 < \lambda_0 \ll 1$ for a small target, Green's function has the expansion

$$G(\mathbf{x}, \mathbf{x}_0; \lambda_0) = -\frac{1}{\lambda_0 |\mathcal{U}|} + G_1(\mathbf{x}, \mathbf{x}_0) + \lambda_0 G_2(\mathbf{x}, \mathbf{x}_0) + O(\lambda_0^2),$$

with $\int_{\mathcal{U}} G_j(\mathbf{x}, \mathbf{x}_0) d\mathbf{x} = 0$. Substituting this expansion into equation (6.4.72) shows that to leading order in λ_0,

$$A \approx \frac{\sqrt{|\mathcal{U}|} \lambda_0}{2\pi}. \tag{6.4.73}$$

Similarly, equations (6.4.64) and (6.4.70) imply that

$$c_0 = -2\pi A \int_{\mathcal{U}} G(\mathbf{x}, \mathbf{x}_0; \lambda_0) d\mathbf{x} \approx \frac{2\pi A}{\lambda_0}. \tag{6.4.74}$$

The regular part $R(\mathbf{x}, \mathbf{x}_0; \lambda_0)$ can also be expanded in terms of λ_0. Hence, neglecting terms of $O(\lambda_0)$ and higher, substitute $R(\mathbf{x}, \mathbf{x}_0; \lambda_0) \approx -(\lambda_0 |\mathcal{U}|)^{-1} + R_1(\mathbf{x}, \mathbf{x}_0)$ into (6.4.71). This yields a linear equation for λ_0 such that

$$\lambda_0 \approx \frac{2\pi v}{|\mathcal{U}|} \frac{1}{1 + 2\pi v R_1(\mathbf{x}_0, \mathbf{x}_0)}. \tag{6.4.75}$$

We now have all the components necessary to determine the time-dependent reaction rate. That is, substituting the inner solution $\phi_0(\mathbf{x}) = A \ln(r/\varepsilon)$, $r = |\mathbf{x} - \mathbf{x}_0|$, into (6.4.66) and using (6.4.73) and (6.4.74) yield the result

$$J(t) \approx D|\mathcal{U}|\lambda_0 e^{-\lambda_0 D t}, \quad \lambda_0 = \frac{2\pi v}{\mathcal{U}} + O(v^2). \tag{6.4.76}$$

Finally, note that one can also handle the case of diffusion to a small switching target in a bounded domain by extending the analysis of Sect. 6.3.3 [121].

6.4.3 Narrow escape problem

Another major application of asymptotic methods is analyzing the FPT problem for the escape of a freely diffusing molecule from a 2D or 3D bounded domain through small absorbing windows on an otherwise reflecting boundary [68, 111, 169, 385, 389, 390, 660, 738]. Consider diffusion in a 2D domain $\mathcal{U} \subset \mathbb{R}^2$ whose boundary can be decomposed as $\partial \mathcal{U} = \partial \mathcal{U}_r \cup \partial \mathcal{U}_a$, where $\partial \mathcal{U}_r$ represents the reflecting part of the boundary and $\partial \mathcal{U}_a$ the absorbing part. We then have a narrow escape problem in the limit that the measure of the absorbing set $|\partial \mathcal{U}_a| = O(\varepsilon)$ is asymptotically small,

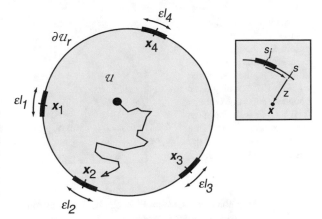

Fig. 6.24: Example trajectory of a Brownian particle moving in a 2D unit disk with small absorbing windows on an otherwise reflecting circular boundary. Inset: a local coordinate system around the jth arc.

that is, $0 < \varepsilon \ll 1$. It follows from the analysis of exit times, see equation (2.4.21), that the MFPT to exit the boundary $\partial \mathcal{U}_a$ satisfies the equation (in the absence of external forces)

$$\nabla^2 \tau(\mathbf{x}) = -\frac{1}{D}, \ \mathbf{x} \in \mathcal{U}; \quad \tau(\mathbf{x}) = 0, \mathbf{x} \in \partial \mathcal{U}_a; \ \partial_n \tau(\mathbf{x}) = 0, \ \mathbf{x} \in \partial \mathcal{U}_r. \quad (6.4.77)$$

The absorbing set is assumed to consist of N small disjoint absorbing windows $\partial \mathcal{U}_j$ centered at $\mathbf{x}_j \in \partial \mathcal{U}$. In the 2D case, each window is a small absorbing arc of length $|\partial \mathcal{U}_j| = \varepsilon l_j$ with $l_j = O(1)$. It is also assumed that the windows are well separated, that is, $|\mathbf{x}_i - \mathbf{x}_j| = O(1)$ for all $i \neq j$. An example of a Brownian particle in a 2D unit disk with small absorbing windows on the circular boundary is illustrated in Fig. 6.24. Since the MFPT diverges as $\varepsilon \to 0$, the calculation of $\tau(\mathbf{x})$ requires solving a singular perturbation problem [68, 169, 385, 660, 738] analogous to the narrow capture problem considered in Sect. 6.4.1. That is, an inner or local solution valid in an $O(\varepsilon)$ neighborhood of each absorbing arc is constructed and then this is matched to an outer or global solution that is valid away from each neighborhood, see Fig. 6.25.

Asymptotic analysis. In order to construct an inner solution near the j-th absorbing arc, equation (6.4.77) is rewritten in terms of a local orthogonal coordinate system (z, s), in which s denotes arc length along $\partial \mathcal{U}$ and z is the minimal distance from $\partial \mathcal{U}$ to an interior point $\mathbf{x} \in \mathcal{U}$, as shown in the inset of Fig. 6.24. Now introduce stretched coordinates $\widehat{z} = z/\varepsilon$ and $\widehat{s} = (s - s_j)/\varepsilon$, and write the solution to the inner problem as $\tau(\mathbf{x}) = w(\widehat{z}, \widehat{s})$. Neglecting terms of $O(\varepsilon)$, it can be shown that w satisfies the homogeneous equation [660]

$$\frac{\partial^2 w}{\partial \widehat{z}^2} + \frac{\partial^2 w}{\partial \widehat{s}^2} = 0, \quad 0 < \widehat{z} < \infty, \quad -\infty < \widehat{s} < \infty, \quad (6.4.78)$$

with the following boundary conditions on $\widehat{z} = 0$:

$$\frac{\partial w}{\partial \widehat{z}} = 0 \text{ for } |\widehat{s}| > l_j/2, \quad w = 0 \text{ for } |\widehat{s}| < l_j/2. \quad (6.4.79)$$

Fig. 6.25: Construction of
the matched asymptotic solution for the narrow es-
cape problem. (a) Inner solution w in the half-plane
$s \in \mathbb{R}, z \in \mathbb{R}^+$ with mixed
boundary conditions on $z =$
0. (b) Outer solution τ in
the disk with a reflecting
boundary condition and the
target treated as a point.

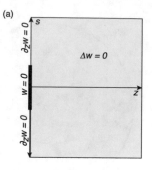
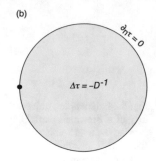

The resulting boundary value problem can be solved by introducing elliptic cylinder coordinates. However, in order to match the outer solution we need only specify the far-field
behavior of the inner solution, which takes the form

$$w(\mathbf{x}) \sim A_j \left[\ln |\mathbf{y}| - \ln d_j + o(1)\right] \text{ as } |\mathbf{y}| \to \infty, \tag{6.4.80}$$

where $d_j = l_j/4$, $|\mathbf{y}| = |\mathbf{x} - \mathbf{x}_j|/\varepsilon = \sqrt{z^2 + s^2}$ and A_j is an unknown constant, which is
determined by matching with the outer solution.

As far as the outer solution is concerned, each absorbing arc shrinks to a point $\mathbf{x}_j \in \partial \mathcal{U}$
as $\varepsilon \to 0$, see Fig. 6.25(b). Each point $\mathbf{x}_j$ effectively acts as a point source that generates a
logarithmic singularity resulting from the asymptotic matching of the outer solution to the
far-field behavior of the inner solution. Thus the outer solution satisfies

$$\nabla^2 \tau(\mathbf{x}) = -\frac{1}{D}, \quad \mathbf{x} \in \mathcal{U}; \quad \partial_n \tau = 0 \text{ for } \mathbf{x} \in \partial \mathcal{U} \backslash \{\mathbf{x}_1, \ldots, \mathbf{x}_N\}, \tag{6.4.81}$$

with

$$\tau(\mathbf{x}) \sim A_j \ln(|\mathbf{x} - \mathbf{x}_j|/\varepsilon d_j) \text{ as } \mathbf{x} \to \mathbf{x}_j, \, j = 1, \ldots, N. \tag{6.4.82}$$

This can be solved in terms of the Neumann Green's function (6.4.20), after noting the
following singular structure at the boundary

$$G(\mathbf{x}, \mathbf{x}_j) \sim -\frac{1}{\pi} \ln |\mathbf{x} - \mathbf{x}_j| + R(\mathbf{x}_j, \mathbf{x}_j) \text{ as } \mathbf{x} \to \mathbf{x}_j \in \partial \mathcal{U}, \tag{6.4.83}$$

where $R(\mathbf{x}, \mathbf{x}')$ is the regular part of $G(\mathbf{x}, \mathbf{x}')$. It follows that the outer solution can be expressed as

$$\tau(\mathbf{x}) = -\pi \sum_{i=1}^{N} G(\mathbf{x}, \mathbf{x}_i) A_i + \chi, \tag{6.4.84}$$

where χ is an unknown constant. Integrating both sides of equation (6.4.84) shows that χ is
the MFPT averaged over all possible starting positions:

$$\chi = \overline{\tau} \equiv \frac{1}{|\mathcal{U}|} \int_{\mathcal{U}} \tau(\mathbf{x}) d\mathbf{x}. \tag{6.4.85}$$

Moreover, differentiating both sides of equation (6.4.84) implies

$$\nabla^2 \tau(\mathbf{x}) = -\pi \sum_{j=1}^{N} A_j \nabla^2 G(\mathbf{x}, \mathbf{x}_j) = -\pi |\mathcal{U}|^{-1} \sum_{j=1}^{N} A_j$$

for $\mathbf{x} \in \mathcal{U}$. Comparison with the outer equation yields the constraint

$$\pi |\mathcal{U}|^{-1} \sum_{j=1}^{N} A_j = \frac{1}{D}.$$ (6.4.86)

The problem has reduced to solving $N+1$ linear equations for $N+1$ unknowns A_i, $i = 1,\dots,N$, and χ. One equation is given by the constraint (6.4.87), whereas the remaining N equations are obtained by matching the near-field behavior of the outer solution as $\mathbf{x} \to \mathbf{x}_j$ with the far-field behavior of the corresponding inner solution (6.4.80). After cancellation of the logarithmic terms, we find that

$$-\pi \sum_{i=1}^{N} \mathcal{G}_{ji} A_i + \chi = \frac{A_j}{v}, \quad v = -\frac{1}{\ln \varepsilon}$$ (6.4.87)

for $j = 1,\dots,N$, where

$$\mathcal{G}_{ji} = G(\mathbf{x}_j, \mathbf{x}_i), \ i \neq j, \quad \mathcal{G}_{jj} = R(\mathbf{x}_j, \mathbf{x}_j) - \frac{\ln d_j}{\pi}.$$

If we now rescale the coefficients according to $A_i \to v A_i$, then the matrix equation (6.4.86) can be rewritten as

$$\sum_{i=1}^{N} (\pi v \mathcal{G}_{ji} + \delta_{i,j}) A_i = \chi,$$

which can be inverted to give

$$A_i = \sum_{j=1}^{N} (\delta_{i,j} + \pi v \mathcal{G}_{ij})^{-1} \chi.$$ (6.4.88)

In the case of a single absorbing window of arc length 2ε ($d = 1/2$), equations (6.4.86) and (6.4.87) are easily solved to give $A_1 = \mathcal{U}|/\pi D$, so that

$$\tau(\mathbf{x}) \sim \frac{|\mathcal{U}|}{D} \left[-\frac{1}{\pi} \ln(\varepsilon/2) + R(\mathbf{x}_1, \mathbf{x}_1) - G(\mathbf{x}, \mathbf{x}_1) \right],$$ (6.4.89)

and

$$\overline{\tau} \sim \frac{|\mathcal{U}|}{D} \left[-\frac{1}{\pi} \ln(\varepsilon/2) + R(\mathbf{x}_1, \mathbf{x}_1) \right].$$ (6.4.90)

All that remains is to calculate the regular part of Neumann Green's function $R(\mathbf{x}, \mathbf{x}_j)$, which will depend on the geometry of the domain $\mathcal{U}$. In certain cases such as the unit disk or a rectangular domain, explicit formulae for R can be obtained, otherwise numerical methods are required [385, 660, 763, 764]. Green's function for a unit disk when the source $\mathbf{x}_j$ is on the unit circle has the well-known formula

$$G(\mathbf{x}, \mathbf{x}_j) = -\frac{1}{\pi} \ln |\mathbf{x} - \mathbf{x}_j| + \frac{|\mathbf{x}|^2}{4\pi} - \frac{1}{8\pi}.$$

It immediately follows that $R(\mathbf{x}_1, \mathbf{x}_1) = 1/8\pi$ (since $|\mathbf{x}_1|^2 = 1$) and

$$\overline{\tau} = \frac{1}{D} [-\ln(\varepsilon) + \ln 2 + 1/8].$$

For a rectangular domain of width L_2 and height L_1, Green's function can be solved using separation of variables and expanding the result in terms of logarithms, see [113, 660]. Finally, note that asymptotic methods can also be used to solve the narrow escape problem in a 3D domain such as the sphere [169]. Now one can use a regular perturbation expansion in ε due to the fact that the 3D Green function has a singularity of the form $1/|\mathbf{x} - \mathbf{x}'|$ rather than $-\ln|\mathbf{x} - \mathbf{x}'|$, see Box 6D and Sect. 6.4.1.

6.5 Extreme statistics and the fastest escape time.

So far we have encountered a number of cellular processes that are triggered when a "searcher" reaches a "target." Examples include calcium-induced calcium release (Sect. 3.5), the triggering of the immune response by T cells binding to antigen presenting cells (Ex. 5.14), the generation of postsynpatic potentials following the binding of diffusing neurotransmitters to synaptic receptors (Sect. 6.2), and gene activation by the arrival of a diffusing transcription factor to a DNA binding site (Sect. 6.3.2). Yet another example is the active motor-driven transport and delivery of vesicular cargo to synaptic targets along the dendrites and axons of neurons,

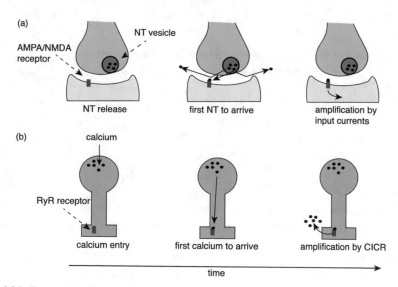

Fig. 6.26: Examples of fast activation via arrival of the first diffusing particles. (a) The fast post-synaptic response to the release of neurotransmitter (NT) is determined by the first NT molecules that bind to the receptor targets to open them. (b) Calcium-induced calcium release (CICR) in a dendritic spine. The first ryanodine receptor (RyR) that opens is triggered by the first calcium ions to reach the base of the spine. An avalanche of calcium release then occurs due to the opening of neighboring receptors. This leads to a rapid amplification at a much shorter time than the MFPT of individual calcium ions. [Adapted from Schuss et al. [740].]

which can be modeled as a random intermittent search process, see Sect. 7.3. In all of these systems, the time scale of activation is set by the FPT of a searcher finding a target. The standard way of estimating the time scale is to calculate the MFPT. However, there are growing number of systems where the arrival of only a few molecules can lead to chemical activation due to signal amplification, as illustrated in Fig. 6.26. The relevant time scale is then not the MFPT of a single searcher, but rather the MFPT of the fastest searcher in a large population of searchers [52–54, 499, 690, 740]. These first arrival times are referred to as order statistics [852, 880, 881] or extreme statistics [348, 776]. The shift in focus regarding FPT problems is captured by the recently formulated "redundancy principle," which suggests that many apparently redundant copies of an object (molecules, proteins, cells, etc.) are not a waste, but rather have the specific function of accelerating search processes [740].

In order to illustrate the basic idea, consider N noninteracting Brownian particles (searchers) diffusing in a bounded domain $\Omega \subset \mathbb{R}^d$ with a reflecting boundary $\partial \Omega$. All particles are taken to have the same diffusivity D and are initially placed at the same location $\mathbf{x}_0 \in \Omega$. Let $\mathbf{X}_n(t) \in \overline{\Omega}$ denote the position of the nth particle at time $t \geq 0$, and define the corresponding FPT to find a target $\partial \Omega_a \subset \partial \Omega$ according to

$$\tau_n = \inf\{t > 0 : \mathbf{X}_n(t) \in \partial \Omega_a\}. \tag{6.5.1}$$

Each individual particle has the same MFPT, $\mathbb{E}[\tau_n] = \mathbb{E}[\tau_1]$ for all n, $1 < n \leq N$. In many biological applications, however, a more relevant time scale is the FPT of the fastest particle,

$$T = \min\{\tau_1, \ldots, \tau_N\}. \tag{6.5.2}$$

In other words, T is the first time that any of the N particles reaches the target. Of particular significance is the observation that in the case of many particles ($N \gg 1$), there is a separation of time scales, $\mathbb{E}[T] \ll \mathbb{E}[\tau_1]$. This reflects the fact that $\mathbb{E}[\tau_1]$ represents a typical particle that wanders around the domain before finding the target, whereas $\mathbb{E}[T]$ depends on extremely rare events in which one of the particles happens to go directly to the target, see Fig. 6.27. Indeed, it has been argued formally that the fastest searcher closely follows the shortest path from the initial point $\mathbf{x}_0$ to the target [740]. This has also been established using more rigorous probabilistic methods [499].

Since the positions of the Brownian particles are independent identically distributed stochastic processes, the statistics of the first particle to arrive at the target can be computed from the single-particle statistics [53, 880, 881]. This is most easily achieved using survival probabilities. Following Sect. 2.4, consider the single-particle survival probability

$$S_1(t) = \mathbb{P}[\tau_1 > t] = \int_\Omega p(\mathbf{x}, t) d\mathbf{x},$$

where $p(\mathbf{x}, t)$ satisfies the following diffusion equation:

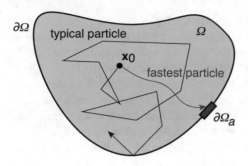

Fig. 6.27: Distinction between a typical path and the fastest path of population of Brownian particles searching for a target $\partial \Omega_T$ on the boundary of a domain Ω.

$$\frac{\partial p(\mathbf{x},t)}{\partial t} = D\nabla^2 p(\mathbf{x},t), \quad p(\mathbf{x},0) = \delta(\mathbf{x}-\mathbf{x}_0) \quad \mathbf{x} \in \Omega, \tag{6.5.3a}$$

$$\frac{\partial p(\mathbf{x},t)}{\partial n} = 0, \quad \mathbf{x} \in \partial\Omega \backslash \partial\Omega_a, \tag{6.5.3b}$$

$$p(\mathbf{x},t) = 0, \quad \mathbf{x} \in \partial\Omega_a. \tag{6.5.3c}$$

For simplicity, we suppress the explicit dependence on the initial position $\mathbf{x}_0$. The single-particle FPT density $f_1(t)$ is then defined according to

$$f_1(t)dt = \mathbb{P}[t < \tau_1 < t + dt],$$

which implies that

$$f_1(t) = -\frac{dS(t)}{dt} = D\int_\Omega \frac{\partial p(\mathbf{x},t)}{\partial t}d\mathbf{x} = -D\int_{\partial\Omega_a} \nabla p(\mathbf{x},t)\cdot \mathbf{n}dS. \tag{6.5.4}$$

Independence of the Brownian particles means that the distribution of the FPT of the fastest particle can be expressed as

$$\Pr[T > t] := S(t) = S_1(t)^N,$$

that is, all particles have survived up to time t. Moreover,

$$\Pr[t < T < t + dt] = NS_1(t)^{N-1}f_1(t)dt,$$

that is, $N-1$ particles survive up to time t but the remaining particle finds the target in the time interval $[t, t+dt]$. The factor of N takes into account the fact that the identity of the successful particle is irrelevant. It follows that the FPT density for the fastest particle is

$$f(t) = -\frac{dS}{dt} = -ND\left(\int_\Omega p(\mathbf{x},t)d\mathbf{x}\right)^{N-1}\int_{\partial\Omega_a} \nabla p(\mathbf{x},t)\cdot \mathbf{n}dS. \tag{6.5.5}$$

Finally, the MFPT for the fastest particle is

$$\overline{T} = \int_0^\infty t f(t)dt = \int_0^\infty S(t)dt = \int_0^\infty S_1(t)^N dt, \qquad (6.5.6)$$

after performing an integration by parts. If the last integral is rewritten as

$$\overline{T} = \int_0^\infty e^{N \ln S_1(t)} dt, \qquad (6.5.7)$$

then an asymptotic expansion can be used to evaluate the integral in the large N limit.

MFPT of the fastest Brownian particle in 1D. The analysis of the MFPT in one spatial dimension was originally carried out in Refs. [852, 880], and subsequently recovered in Ref. [53]. We follow the latter formulation here. For simplicity, take $\Omega = [0,\infty)$ with an absorbing boundary (target) at $x = 0$. Suppose that N Brownian particles are initially at a distance a from the target. The FP equation becomes

$$\frac{\partial p(x,t)}{\partial t} = D \frac{\partial^2 p(x,t)}{\partial x^2}, \qquad (6.5.8a)$$

$$p(x,0) = \delta(x-a), x > 0, t > 0, \quad p(0,t) = 0, \quad t > 0. \qquad (6.5.8b)$$

This has the solution (see Sect. 2.3.3)

$$p(x,t) = \frac{1}{\sqrt{4\pi Dt}} \left[\exp\left\{ -\frac{(x-a)^2}{4Dt} \right\} - \exp\left\{ -\frac{(x-a)^2}{4Dt} \right\} \right].$$

The single-particle survival probability is then

$$S_1(t) = \int_0^\infty p(x,t)dx = 1 - \frac{2}{\sqrt{\pi}} \int_{a/\sqrt{4Dt}}^\infty e^{-y^2} dy.$$

In order to calculate the MFPT $\overline{T}$ given by equation (6.5.7), we note that the major contribution to the integral will occur when $a/\sqrt{4Dt} \gg 1$. Therefore, we use the following asymptotic expansion of the complementary error function:

$$\frac{2}{\sqrt{\pi}} \int_x^\infty e^{-y^2} dy \sim \frac{e^{-x^2}}{x\sqrt{\pi}} \left(1 - \frac{1}{2x^2} + O(x^{-4}) \right) \qquad \text{for } x \gg 1.$$

It follows from equation (6.5.7) that

$$\overline{T} \sim \int_0^\infty \exp\left\{ N \ln\left(1 - \frac{e^{-(a/\sqrt{4Dt})^2}}{(a/\sqrt{4Dt})\sqrt{\pi}} \right) \right\} dt \sim \int_0^\infty \exp\left\{ -N \frac{\sqrt{4Dt}e^{-a^2/4Dt}}{a\sqrt{\pi}} \right\} dt$$

$$\sim \frac{a^2}{4D} \int_0^\infty \exp\left\{ -N \frac{\sqrt{u}e^{-1/u}}{\sqrt{\pi}} \right\} du \quad \text{as } N \to \infty.$$

We have used the additional approximation $\ln(1-r) \approx -r$ for $r \ll 1$. (Here, the notations "$f \sim g$ as $N \to \infty$" means $\lim_{N \to \infty} f/g = 1$.)

The next step is to perform the change of variables

$$w = w(u) = \sqrt{u}e^{-1/u}, \quad w'(u) = \sqrt{u}e^{-1/u}\left(\frac{1}{2u} + \frac{1}{u^2}\right),$$

so that

$$\overline{T} \sim \frac{a^2}{4D} \int_0^\infty \exp\left\{-N\frac{w}{\sqrt{\pi}}\right\} \frac{dw}{f(w)} \quad \text{as } N \to \infty,$$

where $f(w) = w'(u(w))$. Suppose that the integration domain is partitioned according to $\Omega = [0, \delta] \cup (\delta, \infty)$ with $0 < \delta < 1$. It can be shown that the integral over (δ, ∞) is exponentially small compared to the integral over $[0, \delta]$ for large N. Moreover, for small u (and thus small w) we have the approximations $w'(u) \approx w/u^2$ and $\ln w \approx -1/u$, which imply that $f(w) \approx w(\ln w)^2$. Hence,

$$\overline{T} \sim \frac{a^2}{4D} \int_0^\delta \exp\left\{-N\frac{w}{\sqrt{\pi}}\right\} \frac{dw}{w(\ln w)^2}$$

$$\sim \frac{a^2 N}{4D\sqrt{\pi}} \int_0^\delta \exp\left\{-N\frac{w}{\sqrt{\pi}}\right\} \frac{dw}{|\ln w|} + O(e^{-aN}) \quad \text{as } N \to \infty,$$

where the second line follows from integration by parts. Performing the change of variables $v = Nw/\sqrt{\pi}$,

$$\overline{T} \sim \frac{a^2}{4D} \int_0^{N\delta/\sqrt{\pi}} \frac{e^{-v}}{|\ln(\sqrt{\pi}v/N)|} dv$$

$$\sim \frac{a^2}{4D} \int_0^{N\delta/\sqrt{\pi}} \frac{e^{-v}}{\ln(N/\sqrt{\pi})}\left(1 + \frac{|\ln v|}{\ln(N/\sqrt{\pi})}\right) dv \quad \text{as } N \to \infty.$$

Since contributions to the integral in an ε neighborhood of zero are negligible, and assuming $N\delta \gg 1$, we can take the integral domain to be $[\varepsilon, \infty)$, which yields the final asymptotic approximation [53]

$$\overline{T} \sim \frac{a^2}{4D\ln N/\sqrt{\pi}} \quad \text{as } N \to \infty. \tag{6.5.9}$$

This is clearly much smaller than the single-particle MFPT, which is infinite due to the transient nature of 1D random walks (see Ex. 2.13).

A similar analysis to the above can be carried out in the case of a finite interval $[0, L]$ with absorbing boundaries at both ends. The solution of the FP equation is then given by the infinite series (2.3.44). One now finds that [53]

$$\overline{T} \sim \frac{L^2}{16D\ln(2N/\sqrt{\pi})} \quad \text{as } N \to \infty. \tag{6.5.10}$$

Higher dimensions. The calculation of the MFPT for the fastest particle was extended to higher dimensions by Yuste et al. [881], in the special case of escape from a hypersphere of radius a in dimensions $d \geq 1$. That is,

$$\Omega = \{\mathbf{x} \in \mathbb{R}^d : |\mathbf{x}| < a\}, \quad \partial\Omega_a = \partial\Omega = \{\mathbf{x} \in \mathbb{R}^d : |\mathbf{x}| = a\}.$$

In particular, it was found that the asymptotic behavior of the MFPT was independent of the dimension d. More specifically, the mth moment of the fastest FPT satisfies

$$\mathbb{E}[T^m] \sim \left(\frac{a^2}{4D \ln N}\right)^m \quad \text{as } N \to \infty. \tag{6.5.11}$$

More recently, formal asymptotic analysis has been used to argue that the large N behavior of $\mathbb{E}[T]$ in a more general bounded 2D domain with a small target is identical to the 1D case, equation (6.5.9), with a the initial distance of the searchers from the target, but that there is qualitatively different asymptotic behavior in 3D bounded domains [53, 54]. However, a more rigorous probabilistic analysis has shown that, in fact, the asymptotic behavior of the mth moment of the fastest FPT is identical to equation (6.5.11) for a general class of 2D and 3D domains [499]. One of the constraints on such domains is that they contain a straight line path from the initial searcher position to the closest point on the target.

6.6 Diffusion in randomly switching environments

In the analysis of narrow escape problem in Sect. 6.4.3, we assumed that the holes in the boundary of the domain were always open. However, if we view each pore as a stochastically gated ion channel, for example, then one has the problem of analyzing the diffusion equation in a domain with a randomly switching external boundary (or partial boundary). An analogous scenario occurs in the case of cells coupled by stochastically gated gap junctions, which can be formulated as diffusion in a domain with randomly switching interior boundaries, see Sect. 6.7. Mathematically speaking, we can view diffusion in a randomly switching environment as an example of a stochastic hybrid system involving a piecewise deterministic PDE. We will exploit this interpretation in order to develop a method for analyzing the diffusion equation with switching boundary conditions, based on discretizing space and constructing the Chapman–Kolmogorov (CK) equation for the resulting finite-dimensional stochastic hybrid system [119]. We show how the CK equation can be used to determine the moments of the stochastic concentration in the continuum limit. We thus obtain a hierarchy of equations for the rth moments, which take the form of r-dimensional parabolic PDEs that couple to lower-order moments at the boundaries. Although the diffusing particles are noninteracting, statistical correlations arise at the population level due to the fact that they all move in the same randomly switching environment. This is analogous to a stochastically gated protein undergoing diffusion-limited binding with ligands, see Sect. 6.3.3.

6.6.1 Stochastic gating model of confinement

Before proceeding to the analysis of diffusion in randomly switching environments, it is useful first to consider the simpler problem of well-mixed particles confined to a stochastically gated compartment, where spatial effects can be ignored. This provides additional insights into the role of stochastic hybrid systems.

Well-mixed compartment with an open gate. Consider a well-mixed compartment of molecules, where particles can transfer between the exterior and interior of the domain through an open gate, with rates of outflux and influx given by γ and k, respectively. Note that k will depend linearly on the concentration of particles outside the domain, which is assumed to be fixed. Let $N(t)$ be the number of particles in the compartment at time t and set $P_n(t) = \mathbb{P}[N(t) = n | N(0) = n_0]$. The probability distribution $P_n(t)$ evolves according to the birth–death master equation

$$\frac{dP_n}{dt} = kP_{n-1}(t) + \gamma(n+1)\gamma P_{n+1}(t) - (k + \gamma n)P_n(t), \tag{6.6.1}$$

with $P_n(0) = \delta_{n,n_0}$. This is identical in form to the master equation for unregulated mRNA synthesis, see equation (5.2.1). Hence, for $n_0 = 0$ (bounded domain initially empty), we find that $P_n(t)$ is given by a Poisson distribution with a time-dependent rate $\lambda(t)$:

$$P_n(t) = e^{-\lambda(t)}\frac{\lambda(t)^n}{n!}, \quad \lambda(t) = \frac{k}{\gamma}(1 - e^{-\gamma t}). \tag{6.6.2}$$

It immediately follows from properties of a Poisson distribution that

$$\langle N(t) \rangle = \lambda(t), \quad \text{var}[N(t)] = \lambda(t). \tag{6.6.3}$$

In the large time limit $t \to \infty$, we see that $\lambda(t) \to \lambda^* = k/\gamma$ and we have a stationary Poisson distribution.

Well-mixed compartment with a stochastic gate. Now suppose that particles can only be exchanged with the exterior through a two-state stochastically gated ion channel, see Fig. 6.28. Denote the state of the stochastic gate at time t to be the binary random variable $\mu(t)$ with $\mu(t) = 1$ ($\mu(t) = 0$) corresponding to the open (closed) state. The opening and closing of the stochastic gate is governed by the two-state Markov process (see also (3.3.2))

$$\frac{d\mathcal{P}_1}{dt} = -\gamma_- \mathcal{P}_1 + \gamma_+ \mathcal{P}_0,$$

$$\frac{d\mathcal{P}_0}{dt} = \gamma_- \mathcal{P}_1 - \gamma_+ \mathcal{P}_0, \tag{6.6.4}$$

where $\mathcal{P}_1(t)$ and $\mathcal{P}_0(t)$ are the probabilities that the gate is open and closed, respectively, at time t, and $\gamma_\pm$ are the transition rates between the two states.

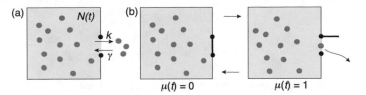

Fig. 6.28: Well-mixed compartment with (a) an open gate (b) a stochastic gate.

The probability distribution $P_n(t)$ now evolves according to the non-autonomous birth–death master equation

$$\frac{dP_n}{dt} = \mu(t)\left[kP_{n-1}(t) + (n+1)\gamma P_{n+1}(t) - (k+\gamma n)P_n(t)\right], \qquad (6.6.5)$$

with $n \geq 0$ and $P_{-1}(t) \equiv 0$. For a given realization of the stochastic gate (a given trajectory $\mu(t)$ through state space), we can repeat the analysis of the autonomous master equation (6.6.1) to obtain the mean and variances (6.6.3) except that

$$e^{-\gamma t} \to X(t) \equiv e^{-\gamma \int_0^t \mu(t')dt'}.$$

Different realizations of $\mu(t)$ will yield different realizations of $X(t)$ and hence different values of the mean and variance. Hence, a more useful characterization of the statistics is obtained by averaging $X(t)$ with respect to all possible stochastic realizations of the gate, which is denoted by $\langle X \rangle_\mu$. The latter can be performed using a method originally developed by Kubo [470] in the study of spectral line broadening in a quantum system, and subsequently extended to chemical rate processes with dynamical disorder by Zwanzig [903].

Following Kubo [470], we note that $X(t)$ is the solution to the stochastic differential equation

$$\frac{dX}{dt} = -\gamma\mu(t)X(t), \quad X(0) = 1, \qquad (6.6.6)$$

where $\mu(t)$ is a discrete random variable that switches between $\mu = 1$ and $\mu = 0$ according to (6.6.4). (This is a simple example of a stochastic hybrid system, see Sect. 5.3.2.) Introduce the probability densities $p_l(x,t)$ with

$$p_l(x,t)dx = \mathbb{P}[\mu(t) = l, x \leq X(t) \leq x+dx], l = 0, 1,$$

and initial conditions $p_l(x,0) = \delta(x-1)\Pi_l$. Here Π_l, $l = 0, 1$, are the stationary probability distributions of the two-state Markov process (6.6.4):

$$\Pi_1 = \frac{\gamma_+}{\gamma_+ + \gamma_-}, \quad \Pi_0 = \frac{\gamma_-}{\gamma_+ + \gamma_-}.$$

These densities evolve according to the Chapman–Kolmogorov equation (see Sect. 5.3.2)

$$\frac{\partial p_0}{\partial t} = \gamma_- p_1 - \gamma_+ p_0, \tag{6.6.7a}$$

$$\frac{\partial p_1}{\partial t} = \gamma \frac{\partial (x p_1)}{\partial x} - \gamma_- p_1 + \gamma_+ p_0. \tag{6.6.7b}$$

We now make the observation that $p(x,t) = p_0(x,t) + p_1(x,t)$ is the probability density for the stochastic process $X(t)$, together with the constraint that the initial state of the gate $\mu(0)$ is a random variable distributed according to the stationary distributions $\Pi_{0,1}$. Hence,

$$\langle X(t) \rangle_\mu = m_0(t) + m_1(t),$$

where

$$m_l(t) = \int_0^\infty x p_l(x,t) dx, \quad l = 0, 1.$$

In order to determine $m_{0,1}(t)$, take first moments of equations (6.6.7a,6.6.7b). That is, multiply both sides by x and integrate by parts. This yields the matrix equation

$$\frac{d}{dt} \begin{pmatrix} m_1(t) \\ m_0(t) \end{pmatrix} = -A \begin{pmatrix} m_1(t) \\ m_0(t) \end{pmatrix}, \quad A = \begin{pmatrix} \gamma_- + \gamma & -\gamma_+ \\ -\gamma_- & \gamma_+ \end{pmatrix}, \tag{6.6.8}$$

which has the solution

$$\begin{pmatrix} m_1(t) \\ m_0(t) \end{pmatrix} = e^{-tA} \begin{pmatrix} \Pi_1 \\ \Pi_0 \end{pmatrix}.$$

One thus finds the μ-averaged mean according to

$$\langle N \rangle_\mu = (n_0 - k/\gamma) \langle X(t) \rangle_\mu + k/\gamma = (n_0 - k/\gamma)[m_0(t) + m_1(t)] + k/\gamma. \tag{6.6.9}$$

A similar analysis can be carried out for second moments:

$$\mathrm{var}_\mu[n] = \langle n \rangle_\mu - n_0 \langle \mathcal{N}^2 \rangle_\mu + (n_0 - k/\gamma)^2 \left(\langle \mathcal{N} \rangle_\mu - \langle \mathcal{N} \rangle_\mu^2 \right), \tag{6.6.10}$$

where

$$\langle \mathcal{N}(t)^q \rangle_\mu = \begin{pmatrix} 1 \\ 1 \end{pmatrix}^T \exp\left[-t \begin{pmatrix} \gamma_- + q\gamma & -\gamma_+ \\ -\gamma_- & \gamma_+ \end{pmatrix} \right] \begin{pmatrix} \Pi_1 \\ \Pi_0 \end{pmatrix} \tag{6.6.11}$$

for $q = 1, 2$. The averages $\langle \mathcal{N}^q \rangle$, $q = 1, 2$, approach zero as time increases, hence the steady-state mean and variance are both equal to k/γ.

6.6.2 1D diffusion with a switching boundary

We now turn to the case of particles diffusing in the finite interval $[0, L]$ with a fixed absorbing boundary at $x = 0$ and a randomly switching gate at $x = L$. (This problem was originally analyzed by Lawley et al. [493] using the theory of random iterative systems.) Let $N(t) \in \{0,1\}$ denote the discrete state of the gate such that it is open when $N(t) = 1$ and is closed when $N(t) = 0$. Assume that $N(t)$ evolves according to a two-state Markov process with switching rates α, β:

$$\text{(closed)} \underset{\beta}{\overset{\alpha}{\rightleftharpoons}} \text{(open)}. \tag{6.6.12}$$

Consider a particular realization $\sigma(T) = \{N(t), 0 \le t \le T\}$ of the gate, and let $u(x,t)$ denote the population density of particles in state x at time t given the realization $\sigma(T)$ up to time T. The population density evolves according to the diffusion equation

$$\frac{\partial u}{\partial t} = D\frac{\partial^2 u}{\partial x^2}, \quad x \in [0,L], \tag{6.6.13a}$$

$$u(0,t) = 0, \quad u(L,t) = \eta \text{ for } N(t) = 1, \quad J(L,t) = 0 \text{ for } N(t) = 0, \tag{6.6.13b}$$

and $J(x,t) = -D\partial_x u(x,t)$. We are assuming that when the gate is open, the system is in contact with a particle bath of density η. Since each realization of the gate will typically generate a different solution $u(x,t)$, it follows that $u(x,t)$ is a random field.

In Ref. [119] a method has been developed for deriving moment equations of the stochastic density $u(x,t)$ in the case of particles diffusing in a domain with randomly switching boundary conditions (see Box 6E). The basic approach is to discretize the piecewise deterministic diffusion equation (6.6.13) with respect to space using a finite-difference scheme, and then to construct the differential CK equation for the resulting finite-dimensional stochastic hybrid system. One of the nice features of finite differences is that one can incorporate the boundary conditions into the resulting discrete linear operators. Since the CK equation is linear in the dependent variables, one can derive a closed set of moment equations for the discretized density. The final step is to retake the continuum limit.

For example, the first-order moments

$$V_n(x,t) = \mathbb{E}[u(x,t)1_{N(t)=n}], \tag{6.6.14}$$

satisfy the system of equations (see Box 6E)

$$\frac{\partial V_0}{\partial t} = D\frac{\partial^2 V_0}{\partial x^2} - \alpha V_0 + \beta V_1, \tag{6.6.15a}$$

$$\frac{\partial V_1}{\partial t} = D\frac{\partial^2 V_1}{\partial x^2} + \alpha V_0 - \beta V_1, \tag{6.6.15b}$$

with

$$V_0(0,t) = V_1(0,t) = 0, \quad V_1(L,t) = \rho_1 \eta > 0, \quad \partial_x V_0(L,t) = 0, \tag{6.6.16}$$

and

$$\rho_0 = \frac{\beta}{\alpha + \beta}, \quad \rho_1 = \frac{\alpha}{\alpha + \beta} \tag{6.6.17}$$

are the components of the stationary distribution of the discrete Markov process. To see why these are the correct boundary conditions, note that if $N(t) = 1$ and $x = L$, then $u(x,t) = \eta$ with probability one, and thus

$$V_1(L,t) = \mathbb{E}[u(L,t)1_{N(t)=1}] = \eta\mathbb{P}(N(t) = 1) = \eta\rho_1.$$

Deriving the other boundary conditions is similar. The steady-state solution of equations (6.6.15a) and (6.6.15b) can be determined explicitly. First, note that

$$\mathbb{E}[u(x,t)] = V_0(x,t) + V_1(x,t). \tag{6.6.18}$$

Since equations equations (6.6.15a) and (6.6.15b) have a globally attracting steady state, it follows that

$$\lim_{t\to\infty} \mathbb{E}[u(x,t)] = V(x) \equiv \sum_{n=0,1} V_n(x), \tag{6.6.19}$$

where $V_n(x) \equiv \lim_{t\to\infty} V_n(x,t)$. Adding equations (6.6.15a) and (6.6.15b) and using the boundary conditions in equation (6.6.16) give

$$\frac{d^2V}{dx^2} = 0, \quad V(0) = 0, \quad V(L) = \rho_1\eta + \kappa, \tag{6.6.20}$$

and $\kappa = V_0(L)$. Hence,

$$V(x) = \frac{x}{L}[\rho_1\eta + \kappa],$$

with

$$D\frac{d^2V_0}{dx^2} - (\alpha+\beta)V_0 = -\frac{\beta}{L}x(\rho_1\eta + \kappa), \tag{6.6.21}$$

and $V_0(0) = 0, \partial_x V_0(L) = 0$. It follows that

$$V_0(x) = ae^{-\xi x} + be^{\xi x} + \frac{\rho_0}{L}(\rho_1\eta + \kappa)x,$$

with $\xi = \sqrt{(\alpha+\beta)/D}$. The boundary conditions imply that

$$a = -b, \quad 2\xi a\cosh(\xi L) = \frac{\rho_0}{L}(\rho_1\eta + \kappa),$$

which yields the solution

$$V_0(x) = \rho_0(\rho_1\eta + \kappa)\left[-\frac{1}{\xi L}\frac{\sinh(\xi x)}{\cosh(\xi L)} + \frac{x}{L}\right]. \tag{6.6.22}$$

Finally, we obtain κ by setting $x = L$:

$$\kappa = \rho_0(\rho_1\eta + \kappa)\left[1 - (\xi L)^{-1}\tanh(\xi L)\right],$$

which can be rearranged to yield

$$\kappa = \rho_0 \rho_1 \eta \frac{1 - (\xi L)^{-1} \tanh(\xi L)}{\rho_1 + \rho_0 (\xi L)^{-1} \tanh(\xi L)},$$

and thus [493]

$$V(x) = \frac{x}{L} \frac{\eta}{1 + (\rho_0/\rho_1)(\xi L)^{-1} \tanh(\xi L)}. \tag{6.6.23}$$

In the limit $\xi \to \infty$ (fast switching),

$$V(x) = \frac{x}{L} \eta.$$

One of the important points to highlight regarding the stochastic diffusion equation (6.6.13) is that it describes a population of particles diffusing in the same random environment. This means that although the particles are noninteracting, statistical correlations arise at the population level. For example,

$$C_n(x,y,t) = \mathbb{E}[u(x,t)u(y,t)1_{N(t)=n}] \neq V_n(x,t)V_n(y,t). \tag{6.6.24}$$

The inequality follows from the observation that the second-order moment equations are non-separable. That is, they take the form (see Box 6E)

$$\frac{\partial C_0}{\partial t} = D \frac{\partial^2 C_0}{\partial x^2} + D \frac{\partial^2 C_0}{\partial y^2} - \alpha C_0 + \beta C_1, \tag{6.6.25a}$$

$$\frac{\partial C_1}{\partial t} = D \frac{\partial^2 C_1}{\partial x^2} + D \frac{\partial^2 C_1}{\partial y^2} + \alpha C_0 - \beta C_1, \tag{6.6.25b}$$

and couple to the first-order moments via the boundary conditions:

$$C_0(0,y,t) = C_0(x,0,t) = C_1(x,0,t) = C_1(0,y,t) = 0, \tag{6.6.26a}$$

and

$$C_1(L,y,t) = \eta V_1(y,t), \, C_1(x,L,t) = \eta V_1(x,t), \, \partial_x C_0(L,y,t) = \partial_y C_0(x,L,t) = 0. \tag{6.6.26b}$$

The second-order moment equations can be solved explicitly using the Fourier transforms. Similarly, higher-order density correlations can be determined from higher-order moments equations. Defining

$$C_n^{(r)}(x,y) = \mathbb{E}[u(x_1,t)u(x_2,t)\ldots u(x_r,t)1_{N(t)=n}], \tag{6.6.27}$$

we have

$$\frac{\partial C_0^{(r)}}{\partial t} = D \sum_{l=1}^{r} \frac{\partial^2 C_0^{(r)}}{\partial x_l^2} - \alpha C_0^{(r)} + \beta C_1^{(r)}, \tag{6.6.28a}$$

$$\frac{\partial C_1^{(r)}}{\partial t} = D \sum_{l=1}^{r} \frac{\partial^2 C_0^{(r)}}{\partial x_l^2} + \alpha C_0^{(r)} - \beta C_1^{(r)}. \tag{6.6.28b}$$

The r-point correlations couple to the $(r-1)$-order moments via the boundary conditions:

$$C_0^{(r)}(x_1,\ldots,x_r,t)\Big|_{x_l=0} = C_1^{(r)}(x_1,\ldots,x_r,t)\Big|_{x_l=0} = \partial_{x_l}C_0^{(r)}(x_1,\ldots,x_r,t)\Big|_{x_l=L} = 0,$$

(6.6.29a)

and

$$C_1^{(r)}(x_1,\ldots,x_r,t)\Big|_{x_l=L} = \eta C_1^{(r-1)}(x_1,\ldots,x_{l-1},x_{l+1}\ldots,x_r,t)$$

(6.6.29b)

for $l = 1,\ldots,r$.

Box 6E. Moment equations for stochastic diffusion equation

Here we indicate how to derive density moment equations for the stochastic diffusion equation (6.6.13).

Finite-difference scheme. The first step is to introduce the lattice spacing a such that $(N+1)a = L$ for integer N, and let $u_j = u(aj)$, $j = 0,\ldots,N+1$. Then we obtain the piecewise deterministic ODE

$$\frac{du_i}{dt} = \sum_{j=1}^{N} \Delta_{ij}^n u_j + \eta_a \delta_{i,M}\delta_{n,1}, \quad i = 1,\ldots,N, \quad \eta_a = \frac{\eta D}{a^2}$$

(6.6.30)

for $n = 0, 1$. Away from the boundaries ($i \neq 1, N$), Δ_{ij}^n is given by the discrete Laplacian

$$\Delta_{ij}^n = \frac{D}{a^2}[\delta_{i,j+1} + \delta_{i,j-1} - 2\delta_{i,j}].$$

(6.6.31a)

On the left-hand absorbing boundary we have $u_0 = 0$, whereas on the right-hand boundary we have

$$u_{N+1} = \eta \text{ for } n = 1, \quad u_{N+1} - u_{N-1} = 0 \text{ for } n = 0.$$

These can be implemented by taking

$$\Delta_{1j}^0 = \frac{D}{a^2}[\delta_{j,2} - 2\delta_{j,1}], \quad \Delta_{Nj}^0 = \frac{2D}{a^2}[\delta_{N-1,j} - \delta_{N,j}],$$

(6.6.31b)

and

$$\Delta_{1j}^1 = \frac{D}{a^2}[\delta_{j,2} - 2\delta_{j,1}], \quad \Delta_{Nj}^1 = \frac{D}{a^2}[\delta_{N-1,j} - 2\delta_{N,j}].$$

(6.6.31c)

The Chapman–Kolomogorov equation. Let $\mathbf{u}(t) = (u_1(t),\ldots,u_N(t))$ and introduce the probability density

$$\text{Prob}\{\mathbf{u}(t) \in (\mathbf{u}, \mathbf{u}+d\mathbf{u}), N(t) = n\} = p_n(\mathbf{u},t)d\mathbf{u},$$

(6.6.32)

where we have dropped the explicit dependence on initial conditions. The probability density evolves according to the following differential CK equation for the stochastic hybrid system (6.6.30), see Sect. 5.3.2,

$$\frac{\partial p_n}{\partial t} = -\sum_{i=1}^{N} \frac{\partial}{\partial u_i} \left[\left(\sum_{j=1}^{N} \Delta_{ij}^n u_j + \eta_a \delta_{i,N} \delta_{n,1} \right) p_n(\mathbf{u},t) \right] + \sum_{m=0,1} A_{nm} p_m(\mathbf{u},t),$$

(6.6.33)

where A is the matrix

$$A = \begin{bmatrix} -\alpha & \beta \\ \alpha & -\beta \end{bmatrix}.$$

(6.6.34)

The left nullspace of the matrix A is spanned by the vector

$$\psi = \begin{pmatrix} 1 \\ 1 \end{pmatrix},$$

(6.6.35)

and the right nullspace is spanned by

$$\rho \equiv \begin{pmatrix} \rho_0 \\ \rho_1 \end{pmatrix} = \frac{1}{\alpha+\beta} \begin{pmatrix} \beta \\ \alpha \end{pmatrix}.$$

(6.6.36)

A simple application of the Perron–Frobenius theorem shows that the two-state Markov process with master equation

$$\frac{dP_n(t)}{\partial t} = \sum_{m=0,1} A_{nm} P_m(t)$$

(6.6.37)

is ergodic with $\lim_{t\to\infty} P_n(t) = \rho_n$. Since the drift terms in the CK equation (6.6.33) are linear in the u_j, it follows that we can obtain a closed set of equations for the moment hierarchy.

Moment equations. Let

$$v_{n,k}(t) = \mathbb{E}[u_k(t) 1_{N(t)=n}] = \int p_n(\mathbf{u},t) u_k(t) d\mathbf{u}.$$

(6.6.38)

Multiplying both sides of the CK equation (6.6.33) by $u_k(t)$ and integrating with respect to $\mathbf{u}$ give (after integrating by parts and using that $p_n(\mathbf{u},t) \to 0$ as $\mathbf{u} \to \infty$ by the maximum principle)

$$\frac{dv_{n,k}}{dt} = \sum_{j=1}^{N} \Delta_{kj}^n v_{n,j} + \eta_a \rho_1 \delta_{k,N} \delta_{n,1} + \sum_{m=0,1} A_{nm} v_{m,k}.$$

(6.6.39)

We have assumed that the initial discrete state is distributed according to the stationary distribution ρ_n so that

$$\int p_n(\mathbf{u},t)d\mathbf{u} = \rho_n.$$

Equations for rth order moments $r \geq 2$ can be obtained in a similar fashion. Let

$$v^{(r)}_{n,k_1\ldots k_r}(t) = \mathbb{E}[u_{k_1}(t)\ldots u_{k_r}(t)1_{N(t)=n}] = \int p_n(\mathbf{u},t)u_{k_1}(t)\ldots u_{k_r}(t)d\mathbf{u}.$$
(6.6.40)

Multiplying both sides of the CK equation (6.6.33) by $u_{k_1}(t)\ldots u_{k_r}(t)$ and integrating with respect to $\mathbf{u}$ give (after integration by parts)

$$\frac{dv^{(r)}_{n,k_1\ldots k_r}}{dt} = \sum_{l=1}^{r}\sum_{j=1}^{N}\Delta^n_{k_l j}v^{(r)}_{n,k_1\ldots k_{l-1}jk_{l+1}\ldots k_r} + \eta_a\delta_{n,1}\sum_{l=1}^{r}v^{(r-1)}_{n,k_1\ldots k_{l-1}k_{l+1}\ldots k_r}\delta_{k_l,N}$$

$$+ \sum_{m=0,1}A_{nm}v^{(r)}_{m,k_1\ldots k_r}.$$
(6.6.41)

Continuum limit. Taking the continuum limit $a \to 0$ in equation (6.6.39) leads to the first-order moment equations (6.6.15a) and (6.6.15b). Similarly, taking $a \to 0$ in equation (6.6.41) generates the higher-order moment equation (6.6.28).

Diffusive flux through a stochastically gated membrane channel. The above model assumes that there is a stochastically gated channel on the boundary of the domain, whose size is negligible compared to the size of the domain, such that when the gate is open the concentration rapidly equilibriates with the reservoir concentration. The natural higher-dimensional analog is the narrow escape problem illustrated in Fig. 6.24, where each $O(\varepsilon)$ absorbing window is now stochastically gated [118, 685]. An alternative scenario is to view the 1D domain itself as the membrane channel. That is, we consider the flux of particles through a stochastically gated membrane channel linking two particle reservoirs [82], as illustrated in Fig. 6.29. The cylindrical channel is taken to be of length L and radius a. There is an absorbing boundary on the left-hand side of the channel ($x = 0$) and a stochastic gate on the right-hand size of the channel ($x = L$). Given the background concentration c_A in the right-hand reservoir A, we would like to determine the flux into reservoir B. One application of such a model is oxygen transport in insect respiration [82, 493]. We will assume that the channel is sufficiently wide so that one does not have to deal with the effects of narrow confinement, see Sect. 6.8.

When the gate is open, diffusing particles freely enter the channel from reservoir A with an influx $J^0_{in} = 4aD_A c_A$ [371], where D_A is the diffusivity within the reservoir. The mean flux entering a stochastically gated channel is then the product of the influx J^0_{in} and the probability of finding the gate open:

$$J_{in} = \rho_1 J_{in}^0 = \frac{\alpha}{\alpha + \beta} 4aD_A c_A. \tag{6.6.42}$$

Once a particle enters the channel, it can either diffuse all the way to reservoir B or return to reservoir A. If P is the probability of the former event then the outflux into reservoir B is

$$J_{out} = PJ_{in} = \frac{\alpha}{\alpha + \beta} 4aPD_A c_A. \tag{6.6.43}$$

The probability P can be calculated by considering a modified, single-particle version of the boundary value problem given by equations (6.6.15a) and (6.6.15b). Let $p_n(x,t)$ be the probability density that a particle entering the channel from reservoir A at time $t = 0$ is at position $X(t) \in (0,L)$ at time t and the gate is in state $N(t) \in \{0,1\}$. The corresponding CK equation for the densities is given by [82]

$$\frac{\partial p_0}{\partial t} = D\frac{\partial^2 p_0}{\partial x^2} - \alpha p_0 + \beta p_1, \tag{6.6.44a}$$

$$\frac{\partial p_1}{\partial t} = D\frac{\partial^2 p_1}{\partial x^2} + \alpha p_0 - \beta p_1, \tag{6.6.44b}$$

where D is the diffusivity within the channel and the boundary conditions are

$$p_0(0,t) = p_1(0,t) = 0, \quad D\partial_x p_0(L,t) = 0, \tag{6.6.44c}$$

and

$$-D\partial_x p_1(L,t) = \kappa_L p_1(L,t), \quad \kappa_L = \frac{4D_A}{\pi a}. \tag{6.6.44d}$$

The initial conditions are $p_0(x,0) = 0$ and $p_1(x,0) = \delta(x-L)$. The partially absorbing boundary condition (6.6.44d) can be derived as follows [82, 92]. First, suppose that the exit to reservoir B is blocked, which means that the channel will reach equilibrium with channel A. At the macroscopic level, the particle concentration (per unit length) within the channel is simply $\pi a^2 c_A$. The outflux into reservoir A will then be of the form $\kappa_L \pi a^2 c_A$ for some constant κ_L. The latter is determined by requiring that the outflux balances the influx J_{in}^0. We can interpret κ_L as the rate at

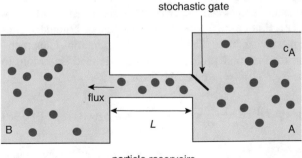

stochastic gate

c_A

flux

L

B

A

particle reservoirs

Fig. 6.29: Membrane channel of length L and radius a linking two particle reservoirs A and B. There is an absorbing boundary at $x = 0$ and a stochastic gate at $x = L$.

which a single particle within the channel can be absorbed at $x = L$ when the gate is open.

The desired probability P is now given by the time-integral of the probability flux at $x = 0$:

$$P = D \int_0^\infty (\partial_x p_0 + \partial_x p_1)|_{x=0} dt = D \frac{d}{dx} \int_0^\infty (p_0(x,t) + p_1(x,t)|_{x=0} dt. \quad (6.6.45)$$

P can be calculated by integrating the CK equation (6.6.44) with respect to time from zero to infinity using the initial conditions. Solving the resulting boundary value problem for P, see Ex. 6.10, finally yields the flux [82]

$$J_{\text{out}} = \frac{4aD_A c_A \alpha}{\alpha + \beta + [\alpha + \beta(\xi L)^{-1} \tanh(\xi L)]4LD_A/\pi aD}, \quad \xi = \sqrt{\frac{\alpha + \beta}{D}}. \quad (6.6.46)$$

Note that when the channel length greatly exceeds its radius, $L \gg a$, the flux reduces to

$$J_{\text{out}} = \frac{\alpha}{\alpha + \beta(\xi L)^{-1} \tanh(\xi L)]} \frac{\pi a^2 c_A D}{L} = D \frac{dV}{dx}, \quad (6.6.47)$$

with $V(x)$ given by equation (6.6.23) and $\eta = \pi a^2 c_A$.

6.6.3 Stochastically gated Brownian motion

So far we have considered the effects of a randomly switching environment on a large population of noninteracting particles evolving according to the diffusion equation. However, it is also possible to consider how such an environment impacts a single Brownian particle evolving according to the SDE

$$dX(t) = -\Phi'(X)dt + \sqrt{2D}dW(t), \quad (6.6.48)$$

for $0 < X(t) < L$, where $\Phi(x)$ is a potential energy function. Suppose that there is an absorbing boundary at $x = 0$ and a stochastic gate at $x = L$ such that the particle is absorbed if $N(t) = 1$ and is reflected if $N(t) = 0$. As in previous examples, we assume that transitions between the two states are given by the two-state Markov process (6.6.12). Equation (6.6.48) with a switching boundary condition can be treated as a piecewise SDE. The associated differential Chapman–Kolmogorov (CK) equation for the joint probability density defined according to $p_n(x,t)dx = \mathbb{P}[x < X(t) < x + dx, N(t) = n]$ takes the form

$$\frac{\partial p_n(x,t)}{\partial t} = \frac{\partial}{\partial x}[\Phi'(x)p_n(x,t)] + D\frac{\partial^2 p_n(x,t)}{\partial x^2} + \sum_{m=0,1} A_{nm} p_m(x,t), \quad (6.6.49)$$

with $\mathbf{A}$ given by equation (6.6.34). Equation (6.6.49) is supplemented by the boundary conditions

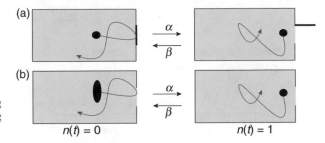

Fig. 6.30: (a) Switching gate and (b) switching conformational state.

$n(t) = 0$ $n(t) = 1$

$$p_0(0,t) = p_1(0,t) = 0, \quad p_1(L,t) = 0, \quad -\Phi'(L)p_0(L,t) - D\left.\frac{\partial p_0(x,t)}{\partial x}\right|_{x=L} = 0,$$

and the initial condition $p_n(x,0) = \delta(x-y)\rho_n$, where ρ_n is the stationary measure of the two-state Markov process generated by the matrix $\mathbf{A}$. (Returning to the stochastic PDE (6.6.13), we note that equation (6.6.49) for a single particle is formally equivalent to the first-order moment equations (6.6.15a) and (6.6.15b) after setting $\Phi = 0$ and $\eta = 0$.)

Equation (6.6.49) and its higher-dimensional analog are the starting point for analyzing the escape of a single particle from a bounded domain with switching gates in the boundary [118, 685]. From this single-particle perspective, one could equally well interpret the source of the switching to be changes in the conformational state of the particle, rather than the gates, such that it can only pass through a gate when in one of the two states, see Fig. 6.30. In the latter case we can also allow the conformational states to have different diffusivities, which is explored further in Sect. 6.6.4. However, it is important to note that the equivalence between switching particles and switching environments breaks down when there are multiple noninteracting particles. As we have already highlighted in the analysis of the diffusion equation, statistical correlations arise when all the particles move in the same randomly switching environment; such correlations would not occur if each particle independently switched between different conformational states.

Recall from Sect. 2.4 that at the single-particle level one is often interested in solving a first passage time problem. Quantities of particular interest are the splitting probability of exiting one end rather than the other, and the associated conditional MFPT. One way to determine these quantities is to consider the corresponding backward CK equation for $q_m(y,t) = p(x,n,t|y,m,0)$ with x,n fixed. For the sake of illustration suppose that the potential $\Phi(x) = 0$ so that

$$\frac{\partial q_0}{\partial t} = D\frac{\partial^2 q_0}{\partial y^2} - \beta[q_0 - q_1], \tag{6.6.50a}$$

$$\frac{\partial q_1}{\partial t} = D\frac{\partial^2 q_1}{\partial y^2} + \alpha[q_0 - q_1]. \tag{6.6.50b}$$

Let $\Pi_m(y,t)$ be the total probability that the particle is absorbed at the end $x = L$, say, after time t given that it started at y in state m. That is,

$$\Pi_m(y,t) = -D \int_t^\infty \frac{\partial p(L,0,t'|y,m,0)}{\partial x} dt'. \tag{6.6.51}$$

Differentiating equations (6.6.50a) and (6.6.50b) with respect to x and integrating with respect to t, we find that

$$\frac{\partial \Pi_0}{\partial t} = D_0 \frac{\partial^2 \Pi_0}{\partial y^2} - \beta[\Pi_0 - \Pi_1], \tag{6.6.52a}$$

$$\frac{\partial \Pi_1}{\partial t} = D_1 \frac{\partial^2 \Pi_1}{\partial y^2} + \alpha[\Pi_0 - \Pi_1]. \tag{6.6.52b}$$

The probability $\Pi_m(y,t)$ can now be used to define the hitting probability,

$$\pi_m(y) = \Pi_m(y,0), \tag{6.6.53}$$

and the conditional mean first passage time $T_m(y)$,

$$T_m(y) = -\int_0^\infty t \frac{\partial_t \Pi_m(y,t)}{\Pi_m(y,0)} dt = \frac{\int_0^\infty \Pi_m(y,t) dt}{\Pi_m(y,0)}, \tag{6.6.54}$$

after integration by parts. Setting $t = 0$ in equations (6.6.52a) and (6.6.52b) and using $\partial_t \Pi_m(y,0) = 0$ for all $y \neq L$ show that

$$D \frac{\partial^2 \pi_0}{\partial y^2} - \beta[\pi_0 - \pi_1] = 0, \tag{6.6.55a}$$

$$D \frac{\partial^2 \pi_1}{\partial y^2} + \alpha[\pi_0 - \pi_1] = 0 \tag{6.6.55b}$$

with boundary conditions

$$\pi_0(0) = \pi_1(0) = 0, \quad \partial_y \pi_0(L) = 0, \quad \pi_1(L) = 1.$$

Similarly, it can be shown that the conditional MFPT satisfies the equations

$$D \frac{\partial^2 \pi_0 T_0}{\partial y^2} - \beta[\pi_0 T_0 - \pi_1 T_1] = 0, \tag{6.6.56a}$$

$$D \frac{\partial^2 \pi_1 T_1}{\partial y^2} + \alpha[\pi_0 T_0 - \pi_1 T_1] = -\pi_1, \tag{6.6.56b}$$

with boundary conditions

$$\pi_0(0)T_0(0) = \pi_1(0)T_1(0) = 0, \quad \pi_0(0)T_0(0) \to 0, \quad \pi_1(L)T_1(L) = 0.$$

In solving for the splitting probabilities, it is convenient to change variables by setting $\pi_n = U_n/\rho_n$:

$$\frac{\partial U_0}{\partial t} = D\frac{\partial^2 U_0}{\partial x^2} - \alpha U_0 + \beta U_1, \tag{6.6.57a}$$

$$\frac{\partial U_1}{\partial t} = D\frac{\partial^2 U_1}{\partial x^2} + \alpha U_0 - \beta U_1, \tag{6.6.57b}$$

with

$$U_0(0,t) = U_1(0,t) = 0, \quad U_1(L,t) = \rho_1 > 0, \quad \partial_x U_0(L,t) = 0. \tag{6.6.58}$$

These are identical to the pair of equations for the first moments of the piecewise deterministic PDE for $\eta = 1$, see equations (6.6.15a) and (6.6.15b). It immediately follows that

$$\pi_0(x) = \frac{V_0(x)}{\rho_0} = \frac{\rho_1}{\rho_1 + \rho_0(\xi L)^{-1}\tanh(\xi L)}\left[-\frac{1}{\xi L}\frac{\sinh(\xi x)}{\cosh(\xi L)} + \frac{x}{L}\right]. \tag{6.6.59}$$

In the limit $\xi \to \infty$ (fast switching), $\pi_0(x) = x/L = \pi_1(x)$. One way to understand the relationship between $\pi_0(x)$ and $V_0(x)$ is to note that

$$V_n(x) = \mathbb{E}[q(x,t)1_{N(0)=n}] = \mathbb{E}[q(x,t)|1_{N(0)=n}]\mathbb{P}[1_{N(0)=n}] = \pi_n(x)\rho_n.$$

6.6.4 Brownian motion with dichotomous fluctuating diffusivity

A number of recent statistical analyzes of single-particle tracking (SPT) experiments [209, 654, 765] suggest that particles within the plasma membrane can switch between different discrete states with different diffusivities. Such switching could be due to interactions between proteins and the actin cytoskeleton [209] or due to protein–lipid interactions [871]. This has motivated several analytical studies of Brownian particles with dichotomously switching diffusivity [8, 128, 129, 315]. For the sake of illustration, consider the following SDE on $\mathbb{R}$:

$$dX(t) = \frac{1}{\kappa}\sqrt{2D_{N(t)}}Y(t)dt, \tag{6.6.60a}$$

$$dY(t) = -\frac{1}{\kappa^2}Y(t)dt + \frac{1}{\kappa}dW(t), \tag{6.6.60b}$$

with $N(t) \in \{0,1\}$ evolving according to the two-state Markov chain (6.6.12) and $D_0 \neq D_1$. The associated differential CK equation for the stochastic hybrid system $(X(t),Y(t),N(t))$ is given by

$$\frac{\partial p_n(x,y,t)}{\partial t} = \frac{1}{\kappa^2}\left(\frac{\partial}{\partial y}y + \frac{1}{2}\frac{\partial^2}{\partial y^2}\right)p_n(x,y,t) - \frac{\sqrt{2D_n}y}{\kappa}\left[\frac{\partial}{\partial x}\right]p_n(x,y,t)$$

$$+ \frac{1}{\varepsilon}\sum_{m=0,1}A_{nm}p_m(x,y,t), \tag{6.6.61}$$

where

$$p_n(x,y,t) = \mathbb{P}[x < X(t) < x+dx, y < Y(t) < y+dy, N(t) = n|X_0, Y_0, n_0]$$

for fixed initial conditions. We have rescaled the transition rates by introducing the positive parameter ε.

As shown elsewhere, the white noise limit $\eta \to 0$ and the adiabatic limit $\varepsilon \to 0$ do not commute [129]. First, applying the projection method to equation (6.6.61) as outlined in Sect. 2.3 shows that in the white noise limit $\kappa \to 0$ with $\varepsilon > 0$, we obtain the reduced CK equation

$$\frac{\partial p_n(x,t)}{\partial t} = D_n \frac{\partial^2 p_n(x,t)}{\partial x^2} + \frac{1}{\varepsilon} \sum_{m=0,1} A_{nm} p_m(x,t), \qquad (6.6.62)$$

corresponding to the SDE

$$dX(t) = \sqrt{2D_{N(t)}}\, dW(t). \qquad (6.6.63)$$

Now taking the limit $\varepsilon \to 0$ yields the SDE

$$dX(t) = \sqrt{2\overline{D}}\, dW(t), \qquad (6.6.64)$$

where

$$\overline{D} = \sum_{n=0,1} \rho_n D_n. \qquad (6.6.65)$$

On the other hand, if we fix $\kappa > 0$ and apply the adiabatic limit $\varepsilon \to 0$ to equation (6.6.61) along the lines of Sect. 5.3.2, we obtain the FP equation corresponding to an effective SDE for $(X(t), Y(t))$ of the form

$$dX(t) = \frac{1}{\kappa} \sqrt{2\widehat{D}} Y(t)\, dt, \qquad (6.6.66a)$$

$$dY(t) = -\frac{1}{\kappa^2} Y(t) dt + \frac{1}{\kappa} dW(t), \qquad (6.6.66b)$$

where

$$\widehat{D} = \left[\sum_{n=0,1} \sqrt{D_n} \rho_n \right]^2. \qquad (6.6.67)$$

Now taking $\kappa \to 0$ along the lines of Sect. 2.3 yields the following SDE for X:

$$dX(t) = \sqrt{2\widehat{D}} dW(t). \qquad (6.6.68)$$

Comparing the formulas for $\overline{D}$ and $\widehat{D}$ in equations (6.6.65) and (6.6.67), we have

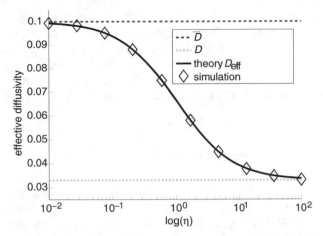

Fig. 6.31: Effective diffusivity of hybrid colored noise process as a function of ratio of correlation time to mean switching time, $\eta = \kappa^2/\varepsilon$, for spatially constant switching rates. The red dashed line is $\overline{D}$ in (6.6.65), the blue dotted line is $\widehat{D}$ in (6.6.67), and the black solid line is the effective diffusivity $D_{\text{eff}} = v_\eta \widehat{D} + (1 - v_\eta)\overline{D}$, with v_η in (6.6.70). The black diamonds are one half of the empirical variance of 10^6 simulations of the hybrid colored noise process until time $t = 1$. The parameters are given by $\varepsilon = 10^{-4}$, $D_0 = 10^{-2}$, $D_1 = 1$, and $\alpha = 1 - \beta = 0.9091$.

$$\widehat{D}/\overline{D} = \frac{4\sqrt{D_0/D_1}}{(\sqrt{D_0/D_1} + 1)^2} < 4\sqrt{D_0/D_1} \qquad (6.6.69)$$

for α, β chosen appropriately. Equation (6.6.69) shows that the possible discrepancy between the effective diffusion coefficients, $\overline{D}$ and $\widehat{D}$, is related to the discrepancy between the two diffusivities, D_0 and D_1. We therefore note that single-particle tracking experiments have found that antigens diffusing on the surface of T cells can switch between two diffusivities D_0 and D_1 with $D_0/D_1 \approx 5 \times 10^{-2}$ (see Table 1 in [209]). Hence, taking $D_0 = 10^{-2} < D_1 = 1$ and $\alpha = 1 - \beta = 0.9091$, we obtain the ratio $\widehat{D}/\overline{D} = 0.3306$.

The above analysis establishes that the order in which the adiabatic and white noise limits are taken has a nontrivial effect on the form of the diffusivity. This can be further illustrated by taking $\varepsilon = \kappa^2/\eta$ for some $\eta > 0$. The previous two cases are then recovered in the limits $\eta \to 0$ (κ approaches zero faster than ε) and $\eta \to \infty$ (ε approaches zero faster than κ). However, for finite η we obtain yet another limit, $\varepsilon, \kappa \to 0$ with ε/κ^2 fixed. The resulting diffusivity interpolates between $\widehat{D}$ and $\overline{D}$ according to [129]

$$v_\eta \widehat{D} + (1 - v_\eta)\overline{D}, \quad v_\eta := \frac{\eta(\alpha + \beta)}{1 + \eta(\alpha + \beta)}. \qquad (6.6.70)$$

In Fig. 6.31 we plot the effective diffusivity of the hybrid colored noise process for a range of ratios of correlation time to mean switching time, $\eta = \kappa^2/\varepsilon$, for spatially constant switching rates. It can be seen that as η varies, there is a sharp transition between the regime in which the effective diffusivity is approximately $\overline{D}$ and the regime in which the effective diffusivity is approximately $\widehat{D}$. Finally, note that if the transition rates α, β and hence ρ_n are x-dependent then, in the double limit $\varepsilon, \kappa \to 0$, we either obtain an Ito SDE with diffusivity $\overline{D}(X) = \sum_n D_n \rho_n(X)$ or a Stratonovich SDE with diffusivity $\widehat{D}(X) = (\sum_n \sqrt{D_n} \rho_n(X))^2$ [129]. This turns out to play an important role in *C. elegans* synaptogenesis [133, 867] (Chap. 11).

6.6.5 Diffusion over a fluctuating barrier

Another interesting example of a diffusion process in a randomly switching environment is overdamped Brownian motion in the presence of a fluctuating barrier [24, 221, 680, 682]. Such systems have arisen in a wide variety of contexts (see the review [681]), including the escape of a ligand molecule from a myoglobin "pocket" after photodissociation [60], ion channel kinetics with fluctuations in the activation energy barriers [292], molecular motor transport [33], and chemical reactions [5, 648, 840]. In the 1990s, fluctuating barriers generated considerable interest following the observation by Doering and Gouda [221] that the mean escape time across a fluctuating barrier may exhibit a non-monotonous dependence on the characteristic time scale of the fluctuations—so-called resonant activation.

Following Refs. [24, 221], consider an overdamped Brownian particle evolving under the influence of a potential that switches between two functions $V_n(x)$, $n = 1, 2$, according to a two-state Markov process $N(t)$:

$$dX(t) = -V_n'(X)dt + \sqrt{2D}dW(t), \quad X \in [-L, L] \tag{6.6.71}$$

for $N(t) = n$, with transition rates $0 \underset{\gamma}{\overset{\gamma}{\rightleftharpoons}} 1$. Equations of the form (6.6.71) have arisen in various contexts in previous chapters, including the flashing ratchet model of a molecular motor (Sect. 4.4) and a two-state gene network with promoter noise (Sect. 5.3). The corresponding CK equation for $p_n(x,t)$ is given by

$$\frac{\partial p_0}{\partial t} = \frac{\partial V_0'(x)p_0}{\partial x} + D\frac{\partial^2 p_0}{\partial x^2} - \gamma p_0 + \gamma p_1, \tag{6.6.72a}$$

$$\frac{\partial p_1}{\partial t} = \frac{\partial V_1'(x)p_1}{\partial x} + D\frac{\partial^2 p_1}{\partial x^2} + \gamma p_0 - \gamma p_1, \tag{6.6.72b}$$

where the friction coefficient has been set to unity. Here $p_n(x,t)dx = \mathbb{P}[x < X(t) < x+dx, N(t) = n | X(0) = x_a, N(0) = m)$ and we have suppressed the initial conditions. The potentials $V_n(x)$ are assumed to be symmetric about $x = 0$, with minima at $x = \pm x_a$ and a maximum at $x = 0$. One is interested in determining the MFPT to cross the barrier at $x = 0$ given that the particle starts in the right-hand well, say. The walls of the domain at $x = \pm L$ are assumed to be reflecting. We thus have the following initial and boundary conditions:

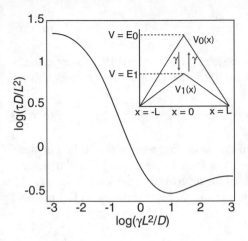

Fig. 6.32: Sketch of MFPT τ as a function of the switching rate γ for $E_0 = -E_1 = 8D$. Inset: Schematic diagram of a fluctuating barrier. The system switches between two piecewise linear potentials $V_n(x)$ with barrier heights E_n, $n = 0,1$. [Adapted from Doering and Gadoua [221].]

$$p_n(x,0) = \frac{1}{2}\delta(x - x_a), \quad -\left(V_n'(x) + D\frac{\partial}{\partial x}\right)p_n(x,t)\bigg|_{x=\pm L} = 0, \quad p_n(0,t) = 0$$

$$(6.6.73)$$

for $n = 0,1$.

The calculation of the MFPT can be developed along analogous lines to the analysis of FPT problems for the Dogterom–Leibler model (Sect. 4.2). Introducing the stopping time

$$T_m(y) = \inf\{t \geq 0; X(t) = 0 | X(0) = x_a, N(0) = m\},$$

the corresponding MFPT can be expressed in terms of the survival probability

$$S_m(y,t) = \int_0^L p(x,t|y,m,0)dx,$$

$$(6.6.74)$$

where $p = p_0 + p_1$. That is,

$$\tau_m(y) := \mathbb{E}[T_m(y)] = \int_0^\infty S_m(y,t)dt.$$

$$(6.6.75)$$

One can either proceed by solving the forward CK equation or solving the corresponding backward equation for τ_m. It is not possible to obtain a solution for general choices of the potentials. However, one special case where an explicit solution is known is for the piecewise potential shown in the inset of Fig. 6.32 [221] for which $x_a = L$. A sketch of the MFPT $\tau = [\tau_0(L) + \tau_1(L)]/2$ as a function of the switching rate is given in Fig. 6.32 for $E_0 = E = -E_1$, illustrating resonant activation. In the limit of slow switching rates, the MFPT converges to the average of the MFPTs for the two different barrier configurations, see also Ex. 2.12:

$$\frac{\tau D}{L^2} \to \frac{D^2}{2E^2}\left[e^{E/D} - 1 - \frac{E}{D}\right] + \frac{D^2}{2E^2}\left[e^{-E/D} - 1 + \frac{E}{D}\right] \tag{6.6.76}$$

as $\gamma \to 0$. On the other hand, in the fast switching limit, the MFPT converges to the value obtained by crossing the barrier with fixed mean height $\overline{E} = (E_0 + E_1)/2$:

$$\frac{\tau D}{L^2} \to \frac{D^2}{2\overline{E}^2}\left[e^{\overline{E}/D} - 1 - \frac{\overline{E}}{D}\right] \tag{6.6.77}$$

as $\gamma \to \infty$. Resonant activation refers to the surprising feature that at intermediate switching rates the MFPT is less than either of these limits, having a local minimum at a rate that is comparable to the inverse of the time to cross the lower barrier [221]. In Ex. 6.11 we consider escape over a fluctuating barrier driven by an OU process, as considered in Refs [680, 682].

6.7 Stochastically gated gap junctions

Gap junctions are arrays of transmembrane channels that connect the cytoplasm of two neighboring cells and thus provide a direct diffusion pathway between the cells. Cells sharing a gap junction channel each provide a hemichannel (also known as a connexon) that connect head-to-head [253, 326, 712], see Fig. 6.33. Each hemichannel is composed of proteins called connexins that exist as various isoforms named Cx23 through Cx62, with Cx43 being the most common. The physiological properties of a gap junction, including its permeability and gating characteristics, are determined by the particular connexins forming the channel. Gap junctions have been found in almost all animal organs and tissues and allow for direct electrical and chemical communication between cells. Electrical coupling is particularly important in cardiac muscle, where the signal to contract is passed efficiently through gap junctions, allowing the heart muscle cells to contract in unison. Gap junctions

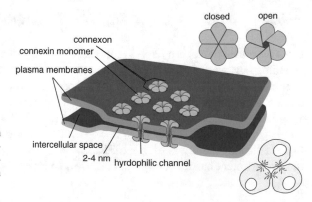

Fig. 6.33: Schematic diagram of gap junction coupling between two cells. [Public domain figure downloaded from Wikimedia Commons.]

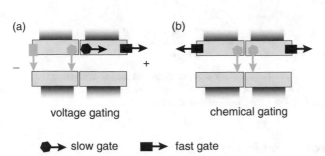

Fig. 6.34: Schematic illustration of a Cx43 gap junction channel containing fast (arrow with square) and slow (arrow with hexagon) gates. (a) Voltage gating is mediated by both fast and slow gating mechanisms. (b) Chemical gating is mediated by the slow gating mechanism in both hemichannels. [Redrawn from [149].]

(or electrical synapses) are also present throughout the central nervous system including the neocortex and hippocampus [190]. Direct chemical communication between cells occurs through the transmission of small second messengers, such as inositol triphosphate (IP3) and calcium (Ca^{2+}). More generally, gap junctions allow small diffusing molecules to undergo cytoplasmic transfer between cells, whereas large biomolecules such as nucleic acids and proteins are blocked. One example of long-range chemical signaling via gap junctions is the propagation of intercellular Ca^{2+} waves (ICWs), which consist of increases in cytoplasmic Ca^{2+} concentration that are communicated between cells and appear as waves that spread out from an initiating or trigger cell [195, 514, 515, 774]. An ICW often propagates at a speed of 10-20 μm/s and lasts for periods of up to tens of seconds, indicating that it can involve the recruitment of hundreds of contiguous cells. Indeed, it has been hypothesized that ICWs in glial cells known as astrocytes could provide a potential mechanism for coordinating and synchronizing the activity of a large group of neurons via the so-called tripartite synapse [514, 774].

Just as with the opening and closing of ion channels (Chap. 3), gap junctions can be gated by both voltage and chemical agents. There appear to be at least two gating mechanisms associated with gap junctions [149], as illustrated in Fig. 6.34. The first involves a fast gate located at the cytoplasmic entrance of a hemichannel, which has transition times of order 1 ms and is voltage controlled. The second is based on a slow gate located toward the center of a cell-to-cell channel (or the extracellular end of a hemichannel), which has transition times of order 10 ms and is both voltage and chemically modulated. Even when a gap junction is open, it tends to restrict the flow of molecules and this is typically modeled by assuming that a gap junction has a certain channel permeability [431]. Given that gap junctions are gated, this suggests that thermal fluctuations could result in the stochastic opening and closing of gap junctions in an analogous fashion to ion channels. There has been relatively little work on the effects of thermal noise on gap junction diffusive coupling, beyond modeling the voltage characteristics of a single stochastically gated gap junction [644].

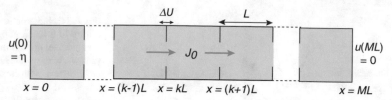

Fig. 6.35: One-dimensional line of cells coupled by gap junctions. At steady state there is a uniform flux J_0 through each cell but a jump discontinuity ΔU in the concentration across each gap junction, where μ is the permeability of each junction. See text for details.

6.7.1 Effective diffusion coefficient for a deterministic 1D model

Following Keener and Sneyd [431], consider a simple 1D model of molecules diffusing along a line of M cells that are connected via gap junctions, see Fig. 6.35. For simplicity, we ignore any nonlinearities arising from chemical reactions. Since gap junctions have relatively high resistance to flow compared to the cytoplasm, we assume that each intercellular membrane junction acts like an effective resistive pore with some permeability μ. Suppose that we label the cells by an integer k, $k = 1,\ldots,M$, and take the length of each cell to be L. Let $u(x,t)$ for $x \in ([k-1]L, kL)$ denote the particle concentration within the interior of the kth cell, and assume that it evolves according to the diffusion equation

$$\frac{\partial u}{\partial t} = D\frac{\partial^2 u}{\partial x^2}, \quad x \in ([k-1]L, kL), t > 0. \tag{6.7.1}$$

However, at each of the intercellular boundaries $x = l_j \equiv jL$, $j = 1,\ldots,M-1$, the concentration is discontinuous due to the permeability of the gap junctions. Conservation of diffusive flux across each boundary then implies that

$$-D\frac{\partial u(l_k^-,t)}{\partial x} = -D\frac{\partial u(l_k^+,t)}{\partial x} = \mu[u(l_k^-,t) - u(l_k^+,t)] \tag{6.7.2}$$

for $k = 1,\ldots,M-1$, where the $+$ and $-$ superscripts indicate that the function values are evaluated as limits from the right and left, respectively. Finally, it is necessary to specify the exterior boundary conditions at $x = 0$ and $x = ML$. We impose the Dirichlet boundary conditions with $u(0,t) = \eta$ and $u(ML,t) = 0$.

In steady state, there is a constant flux $J_0 = -DK$ through the system and the steady-state concentration takes the form

$$u(x) = \begin{cases} Kx + \eta, & x \in [0,L) \\ K(x - [k-1]L) + U_k, & x \in ([k-1]L, kL), k = 2,\ldots,M-1, \\ K(x - ML), & x \in ([M-1]L, ML] \end{cases} \tag{6.7.3}$$

for the $M-1$ unknowns $K, U_k = u((k-1)L)$, $k = 2, \ldots, M-1$. These are determined by imposing the $M-1$ boundary conditions (6.7.2) in steady state:

$$J_0 = \mu[\eta + KL - U_2] = \mu[KL + U_2 - U_3] = \cdots \mu[KL + U_{M-2} - U_{M-1}], \quad (6.7.4a)$$

$$J_0 = \mu[2KL + U_{M-1}]. \tag{6.7.4b}$$

Rearranging equation (6.7.4a) gives

$$U_2 = \eta - \frac{J_0 L}{D} - \frac{J_0}{\mu}, \quad U_k = U_{k-1} - \frac{J_0 L}{D} - \frac{J_0}{\mu}, \quad k = 3, \ldots, M-1, \tag{6.7.5}$$

which can be iterated to give

$$U_{M-1} = \eta - (M-2)J_0 \left[\frac{L}{D} + \frac{1}{\mu} \right].$$

Since we also have

$$U_{M-1} = 2J_0 \left[\frac{L}{D} + \frac{1}{\mu} \right] - \frac{J_0}{\mu},$$

it follows that

$$J_0 = \frac{D\eta}{ML} \left[1 + \frac{D(M-1)}{\mu LM} \right]^{-1}. \tag{6.7.6}$$

Introducing the effective diffusion coefficient D_e according to

$$J_0 = \frac{D_e \eta}{ML}, \tag{6.7.7}$$

we see that for large M

$$\frac{1}{D_e} = \left[\frac{1}{D} + \frac{1}{\mu L} \right]. \tag{6.7.8}$$

6.7.2 Effective permeability for cells coupled by stochastically gated gap junctions

The above deterministic model has recently been extended to incorporate the effects of stochastically gated gap junctions [125]. The resulting model can be analyzed by extending the theory of diffusion in domains with randomly switching exterior boundaries [119, 493], see Sect. 6.6, to the case of switching interior boundaries. Solving the resulting first-order moment equations of the stochastic concentration allows one to calculate the mean steady-state concentration and flux, and thus extract the effective single-gate permeability of the gap junctions.

We start by looking at a pair of stochastically coupled cells of length L, see Fig. 6.36. The basic problem can be formulated as follows: We wish to solve the diffusion equation in the open domain $\mathcal{U} = \mathcal{U}_1 \cup \mathcal{U}_2$ with $\mathcal{U}_1 = (0,L)$ and $\mathcal{U}_2 = (L,2L)$, with the interior boundary between the two subdomains at $x = L$ randomly switching between an open and a closed state. Let $N(t)$ denote the discrete state of the gate at time t with $N(t) = 1$ if the gate is open and $N(t) = 0$ if it is closed. Assume that transitions between the two states $n = 0,1$ are described by the two-state Markov process (6.6.12). The random opening and closing of the gate means that particles diffuse in a random environment according to the piecewise deterministic equation

$$\frac{\partial u}{\partial t} = D\frac{\partial^2 u}{\partial x^2}, \quad x \in [0,L], \tag{6.7.9a}$$

$$u(0,t) = \eta > 0, \quad u(2L,t) = 0, \tag{6.7.9b}$$

$$u(L^-,t) = u(L^+,t), \quad \partial_x u(L^-,t) = \partial_x u(L^+,t) \quad \text{for } N(t) = 1, \tag{6.7.9c}$$

$$\partial_x u(L^-,t) = 0 = \partial_x u(L^+,t) \quad \text{for } N(t) = 0, \tag{6.7.9d}$$

where $L^\pm = \lim_{\varepsilon \to 0^+} L \pm \varepsilon$. That is, u satisfies the Dirichlet boundary conditions on the exterior boundaries of $\mathcal{U}$ and $N(t)$-dependent boundary conditions on the interior boundary at $x = L$: when the gate is open there is continuity of the concentration and the flux across $x = L$, whereas when the gate is closed the right-hand boundary of $\mathcal{U}_1$ and the left-hand boundary of $\mathcal{U}_2$ are reflecting. For simplicity, we assume that the diffusion coefficient is the same in both compartments so that the piecewise nature of the solution is solely due to the switching gate. For the sake of illustration we take the exterior boundary conditions to be Dirichlet, but the analysis is easily modified in the case of a Neumann boundary condition at one of the ends, for example.

First-order moment equations and effective permeability ($M = 2$). In order to determine the effective permeability of a stochastically gated gap junction, we need to calculate the mean of the concentration $u(x,t)$,

$$\mathbb{E}[u(x,t)] = V_0(x,t) + V_1(x,t), \tag{6.7.10}$$

where

$$V_n(x,t) \equiv C_n^{(1)}(x,t) = \mathbb{E}[u(x,t)1_{N(t)=n}]. \tag{6.7.11}$$

The corresponding first-order moment equations for V_n can be derived along similar lines to the case of 1D diffusion in domains with switching exterior boundaries, see Box 6E. One finds that [119]

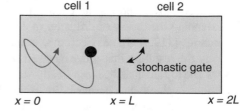

Fig. 6.36: Pair of cells coupled by a stochastically gated gap junction.

$$\frac{\partial V_0}{\partial t} = D\frac{\partial^2 V_0}{\partial x^2} - \alpha V_0 + \beta V_1, \tag{6.7.12a}$$

$$\frac{\partial V_1}{\partial t} = D\frac{\partial^2 V_1}{\partial x^2} + \alpha V_0 - \beta V_1 \tag{6.7.12b}$$

for $x \in \mathcal{U}_1 \cup \mathcal{U}_2$ with exterior boundary conditions

$$V_0(0,t) = \rho_0 \eta, \quad V_1(0,t) = \rho_1 \eta, \quad V_0(2L,t) = V_1(2L,t) = 0, \tag{6.7.13}$$

and interior boundary conditions

$$V_1(L^-,t) = V_1(L^+,t), \partial_x V_1(L^-,t) = \partial_x V_1(L^+,t), \partial_x V_0(L^-,t) = 0 = \partial_x V_0(L^+,t). \tag{6.7.14}$$

(Note that the problem of diffusion in a domain with a switching interface has also been considered by Lawley and Keener [493]. However, they focus on the fast switching limit rather than the steady-state solution for arbitrary switching rates and do not consider the multi-interface case.)

Since equations (6.7.12a) and (6.7.12b) have a globally attracting steady state, it follows that

$$\lim_{t\to\infty} \mathbb{E}[u(x,t)] = V(x) \equiv \sum_{n=0,1} V_n(x), \tag{6.7.15}$$

where $V_n(x) \equiv \lim_{t\to\infty} V_n(x,t)$. From the interior boundary conditions (6.7.14), we set

$$\partial_x V_1(L^-) = \partial_x V_1(L^+) = K_1,$$

with K_1 to be determined later by imposing $V_1(L^-) = V_1(L^+)$. Adding equations (6.7.12a) and (6.7.12b) then gives

$$\frac{d^2 V}{dx^2} = 0, \quad x \in [0,L), \quad V(0) = \eta, \quad \partial_x V(L^-) = K_1, \tag{6.7.16}$$

and

$$\frac{d^2 V}{dx^2} = 0, \quad x \in (L,2L], \quad \partial_x V(L^+) = K_1, \quad V(2L) = 0. \tag{6.7.17}$$

This yields the piecewise linear solution

$$V(x) = \begin{cases} K_1 x + \eta, & x \in [0,L) \\ K_1(x - 2L), & x \in (L, 2L]. \end{cases} \tag{6.7.18}$$

Since $V_1 = V - V_0$, we can rewrite equation (6.7.12a) as

$$D\frac{d^2 V_0}{dx^2} - (\alpha + \beta)V_0(x) = -\beta V(x), \tag{6.7.19}$$

with $V_0(0) = \rho_0 \eta, V_1(2L) = 0$, and $\partial_x V_0(L^-,t) = 0 = \partial_x V_0(L^+,t)$. Substituting for $V(x)$ using equation (6.7.18) we obtain a piecewise solution of the form

$$V_0(x) = B\sinh(\xi x) + \rho_0(K_1 x + \eta), \quad x \in [0, L], \tag{6.7.20a}$$

$$V_0(x) = C\sinh([2L - x]\xi) + \rho_0 K_1(x - 2L), \quad x \in (L, 2L], \tag{6.7.20b}$$

with $\xi = \sqrt{(\alpha + \beta)/D}$. We have imposed the exterior boundary conditions. The interior boundary conditions for V_1 then determine the coefficients B, C in terms of K_1 so that we find

$$V_0(x) = -\frac{\rho_0 K_1}{\xi} \frac{\sinh(\xi x)}{\cosh(\xi L)} + \rho_0(K_1 x + \eta), \quad x \in [0, L], \tag{6.7.21a}$$

$$V_0(x) = \frac{\rho_0 K_1}{\xi} \frac{\sinh(\xi[2L - x])}{\cosh(\xi L)} + \rho_0 K_1(x - 2L), \quad x \in (L, 2L]. \tag{6.7.21b}$$

Finally, we determine the unknown constant K_1 by requiring that $V_1(x)$ is continuous across $x = L$, that is,

$$K_1 L + \eta - V_0(L^-) = -K_1 L - V_0(L^+),$$

which gives

$$\frac{2\rho_0 K_1}{\xi} \tanh(\xi L) = -\rho_1(\eta + 2K_1 L).$$

The latter can be rearranged to yield the following result for the mean flux through the gate, $J_0 = -DK_1$:

$$J_0 = \frac{D\eta}{2L} \frac{1}{1 + (\rho_0/\rho_1)(\xi L)^{-1} \tanh(\xi L)}. \tag{6.7.22}$$

Comparison with equation (6.7.6) for $M = 2$ implies that the stochastically gated gap junction has the effective permeability μ_e with

$$\frac{1}{\mu_e} = \frac{2\rho_0}{\rho_1} \frac{\tanh(\xi L)}{\xi D}. \tag{6.7.23}$$

It is useful to note some asymptotic properties of the solution given by equations (6.7.18) and (6.7.22). First, in the fast switching limit $\xi \to \infty$, we have $J_0 \to \eta D/2L$, $\mu_e \to \infty$ and equation (6.7.18) reduces to the continuous steady-state solution

$$V(x) = \frac{\eta(2L - x)}{2L}, \quad x \in [0, 2L].$$

The mean flux through the gate is the same as the steady-state flux without a gate. On the other hand, for finite switching rates the mean flux J_0 is reduced. In the limit $\alpha \to 0$ (gate always closed), $J_0 \to 0$ so that $V(x) = \eta$ for $x \in [0, L)$ and $V(x) = 0$ for $x \in (L, 2L]$.

Multi-cell model ($M > 2$). Let us return to the general case of a line of M identical cells of length L coupled by $M - 1$ gap junctions at positions $x = l_k = kL$, $1 \leq k \leq M - 1$, see Fig. 6.35. The analysis is considerably more involved if the gap

junctions physically switch because there are significant statistical correlations arising from the fact that all the particles move in the same random environment, which
exists in 2^{M-1} different states if the gates switch independently [125]. Therefore,
we will restrict the analysis to the simpler problem in which individual particles independently switch conformational states along the lines of Fig. 6.30(b). If $V_n(x,t)$
is the concentration of particles in state n, then we have the pair of PDEs given by
equations (6.7.12a) and (6.7.12b) except now the exterior boundary conditions are

$$V_n(0) = \rho_n \eta, \quad V_n(L) = 0, \quad n = 0, 1, \tag{6.7.24}$$

and the interior boundary conditions at the jth gate are

$$[V_1(x)]_{x=l_j^-}^{x=l_j^+} = 0, \quad [\partial_x V_1(x)]_{x=l_j^-}^{x=l_j^+} = 0, \tag{6.7.25a}$$

$$\partial_x V_0(l_j^-) = 0 = \partial_x V_1(l_j^+). \tag{6.7.25b}$$

Adding equations (6.7.12a) and (6.7.12b), we obtain the following equation for
$V(x) = V_0(x) + V_1(x)$:

$$\frac{d^2V}{dx^2} = 0, \quad x \in \bigcup_{k=1}^{M} ((k-1)L, kL), \tag{6.7.26}$$

with exterior boundary conditions

$$V(0) = \eta, \quad V(ML) = 0,$$

and interior boundary conditions

$$\partial_x V(l_j^-) = \partial_x V(l_j^+) = K_1, \quad \forall j = 1, \ldots, M-1.$$

It follows that $V(x)$ has the piecewise solution identical to equation (6.7.3). We thus
have $M - 1$ unknowns $K_0, U_2, \ldots, U_{M-1}$. These can be determined by first solving
equation (6.7.12a) for $V_0(x)$ with boundary conditions (6.7.24) and (6.7.25b), and
then imposing continuity of $V_0(x)$ at each interior boundary (see Ex. 6.12). This
ultimately yields the following expression for the flux J_0:

$$J_0 = \frac{D\eta}{ML} \frac{1}{1 + (\rho_0/\rho_1)(M\xi L)^{-1} \left[2\tanh(\xi L) + (2M-4)\dfrac{\cosh(\xi L) - 1}{\sinh(\xi L)} \right]}. \tag{6.7.27}$$

We deduce that the effective permeability $\mu_e(M)$ in the case of M cells with $M - 1$
independent, stochastically gated gap junctions is

$$\frac{1}{\mu_e(M)} = \frac{\rho_0}{[M-1]\rho_1 \xi D} \left[2\tanh(\xi L) + (2M-4)\frac{\cosh(\xi L) - 1}{\sinh(\xi L)} \right]. \tag{6.7.28}$$

This reduces to equation (6.7.23) when $M = 2$. We conclude that the effective single-gate permeability is M-dependent with

$$\lim_{M \to \infty} \frac{1}{\mu_e(M)} = \frac{2\rho_0}{\rho_1 \xi D} \frac{\cosh(\xi L) - 1}{\sinh(\xi L)}.$$

6.7.3 Splitting probabilities and MFPT

So far we have considered the diffusion of many particles along a 1D line of cells with gap junction coupling, see Fig. 6.35. Here we follow a complementary single-particle perspective and use the probabilistic approach introduced in Sect. 4.2 and Sect.6.6.3 to calculate splitting probabilities and MFPTs [126]. Consider a single particle diffusing in the interval $[0, ML]$ and suppose that it switches conformational state according to a continuous-time Markov jump process $n(t) \in \{0, 1\}$ with fixed transition rates α and β. Suppose that the particle can diffuse freely through a gap junction when $n(t) = 1$, but cannot pass through $x = l_k := kL$ when $n(t) = 0$ for $1 \leq k \leq M - 1$. (We could equivalently have assumed that the gates switch states provided all the gates are perfectly correlated.) We also assume that the particle can be absorbed at $x = 0$ and $x = ML$ only when $n(t) = 1$, otherwise it is reflected.

Splitting probabilities. Let $X(t) \in [0, ML]$ denote the position of the particle at time t and define the stopping time

$$T = \inf\{t \geq 0 : \{X(t) \in \{0, ML\}\} \cap \{n(t) = 1\}\}. \tag{6.7.29}$$

This is the FPT for the particle to reach either external boundary and is in the state $n(t) = 1$, so that it is absorbed by the boundary. Assume $\mathbb{P}(n(0) = 1) = \rho_1$. For $n \in \{0, 1\}$, introduce the splitting probability that the particle is absorbed at the left-hand boundary $x = 0$ given that $X(0) = x$ and $n(0) = n$:

$$\pi_n(x) = \mathbb{P}(X(T) = 0 \mid \{X(0) = x\} \cap \{n(0) = n\}). \tag{6.7.30}$$

From the backward CK equation, one finds that π_n satisfies [119, 493]

$$\begin{pmatrix} 0 \\ 0 \end{pmatrix} = \frac{d^2}{dx^2} \begin{pmatrix} \pi_0 \\ \pi_1 \end{pmatrix} + \begin{pmatrix} -\alpha & \alpha \\ \beta & -\beta \end{pmatrix} \begin{pmatrix} \pi_0 \\ \pi_1 \end{pmatrix}, \tag{6.7.31}$$

with exterior boundary conditions

$$\pi_0'(0) = 0, \quad \pi_0'(ML) = 0, \quad \pi_1(0) = 1, \quad \pi_1(ML) = 0,$$

and the interior boundary conditions at the kth gate, $l_k := kL$, are

$$\pi_0'(l_k-) = \pi_0'(l_k+) = 0, \quad \pi_1(l_k-) = \pi_1(l_k+), \quad \pi_1'(l_k-) = \pi_1'(l_k+).$$

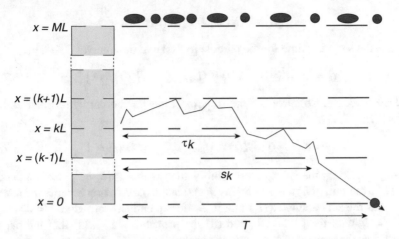

Fig. 6.37: Sample trajectory illustrating the definitions of the conditional MFPTs τ_k and s_k. The red horizontal lines indicate the time intervals during which the particle cannot pass through a gap junction since it is in the state $n = 0$.

A simple rescaling shows that

$$p_n(x) := \rho_n \pi_n(x) = \mathbb{P}(\{X(T) = 0\} \cap \{n(0) = n\} \,|\, X(0) = x) \qquad (6.7.32)$$

satisfies the ODEs

$$\begin{pmatrix} 0 \\ 0 \end{pmatrix} = \frac{d^2}{dx^2} \begin{pmatrix} p_0 \\ p_1 \end{pmatrix} + \begin{pmatrix} -\alpha & \beta \\ \alpha & -\beta \end{pmatrix} \begin{pmatrix} p_0 \\ p_1 \end{pmatrix}, \qquad (6.7.33)$$

with exterior boundary conditions

$$p_0'(0) = 0, \quad p_0'(ML) = 0, \quad p_1(0) = \rho_1, \quad p_1(ML) = 0,$$

and the same interior boundary conditions as π_n.

From the definition of p_n, we have

$$p(x) := p_0(x) + p_1(x) = \mathbb{P}(X(T) = 0 \,|\, X(0) = x). \qquad (6.7.34)$$

By the strong Markov property (see Box 4A) if $0 \le k \le M - 1$ and $x \in (l_k, l_{k+1})$, then

$$p(x) = \frac{1}{\rho_1} \Big(q(s) p_1(l_k) + (1 - q(s)) p_1(l_{k+1}) \Big), \qquad (6.7.35)$$

where $s = x - l_k$, $q(s)$ is the splitting probability that the particle first exits (l_k, l_{k+1}) at the left-hand boundary,

$$q(s) = \mathbb{P}(X(\tau_k) = l_k \,|\, X(0) = s),$$

and τ_k is the stopping time for the particle to first exit the interval (l_k, l_{k+1}):

$$\tau_k = \inf\{t \geq 0 : \{X(t) \notin (l_k, l_{k+1})\} \cap \{n(t) = 1\}\}. \qquad (6.7.36)$$

For later use, we also introduce the stopping time for the particle to first exit the interval (l_{k-1}, l_{k+1}),

$$s_k = \inf\{t \geq 0 : \{X(t) \notin (l_{k-1}, l_{k+1})\} \cap \{n(t) = 1\}\}. \qquad (6.7.37)$$

Roughly speaking, the splitting probability $p(x)$ with $x \in (l_k, l_{k+1})$ can be decomposed into the sum of (i) the probability $q(s)$ that the particle first escapes the interval at l_k times the probability $\pi_1(l_k)$ that it is subsequently absorbed at $x = 0$ starting from $x = l_k$ with the gate open, and (ii) the probability $1 - q(s)$ that the particle first escapes the interval at l_{k+1} times the probability $\pi_1(l_{k+1})$ that it is subsequently absorbed at $x = 0$ starting from $x = l_{k+1}$ with the gate open. A sample trajectory illustrating the definitions of τ_k and s_k is shown in Fig. 6.37.

The calculation of $q(s)$ is equivalent to obtaining the splitting probability of a Brownian particle in $[0, L]$ escaping at the end $x = 0$ first. Let

$$q_n(x) := \mathbb{P}(\{X(\tau) = 0\} \cap \{n(0) = n\} \,|\, X(0) = x).$$

Note that q_n satisfies the same ODE (6.7.33) with the boundary conditions

$$q_0'(0) = 0, \quad q_0'(L) = 0, \quad q_1(0) = \rho_1, \quad q_1(L) = 0.$$

Adding the equations for q_0 and q_1 and setting $q(x) = q_0(x) + q_1(x)$ give

$$\frac{d^2}{dx^2}q(x) = 0, \quad q(0) = \rho_1 + q_0(0), \quad q(L) = q_0(L),$$

with $q_0(x)$ satisfying the equation

$$\frac{d^2}{dx^2}q_0(x) - (\alpha + \beta)q_0(x) = -\beta q(x).$$

It is straightforward to solve this boundary value problem (BVP) and obtain

$$q(x) := q_0(x) + q_1(x) = \frac{\rho_1\xi(L-x) + e^{\xi L}(\rho_1(L\xi - x\xi - 1) + 1) + \rho_1 - 1}{\rho_1(L\xi + 2) + e^{\xi L}(\rho_1(L\xi - 2) + 2) - 2},$$

$$(6.7.38)$$

where $\xi = \sqrt{\alpha + \beta}$.

Since $p_1(0) = \rho_1$, $p_1(ML) = 0$, it follows that $p(x)$ is determined by the remaining $M - 1$ constants $p_1(l_1), \ldots, p_1(l_{M-1})$. As the cells are evenly spaced, we find that each of these constants is the average of its neighbors [126]

$$p_1(l_k) = \frac{1}{2}\left(p_1(l_{k-1}) + p_1(l_{k+1})\right) \tag{6.7.39}$$

for $k = 1, \ldots, M - 1$. Rearranging (6.7.39), we see that the constants satisfy a discretized Laplace's equation

$$p_1(l_{k-1}) - 2p_1(l_k) + p_1(l_{k+1}) = 0 \tag{6.7.40}$$

for $k = 1, \ldots, M - 1$, with boundary conditions $p_1(0) = \rho_1$, $p_1(ML) = 0$. Solving this system of equations, see Ex. 6.13, and applying (6.7.35) and (6.7.38) yield $p(x)$.

Mean first passage times. Calculating the expected absorption time (MFPT) of the particle to either of the switching boundaries,

$$w_n(x) = \mathbb{E}_x[T1_{\{n(0)=n\}}], \tag{6.7.41}$$

proceeds along similar lines to the splitting probability. Again one can show that w_n satisfies the ODEs [119, 493]

$$-\begin{pmatrix} \rho_0 \\ \rho_1 \end{pmatrix} = \frac{d^2}{dx^2}\begin{pmatrix} w_0 \\ w_1 \end{pmatrix} + \begin{pmatrix} -\alpha & \beta \\ \alpha & -\beta \end{pmatrix}\begin{pmatrix} w_0 \\ w_1 \end{pmatrix}, \tag{6.7.42}$$

with exterior boundary conditions

$$w_0'(0) = 0 \quad w_0'(ML) = 0,, \quad w_1(0) = 0 \quad w_1(ML) = 0,$$

and the same interior boundary conditions as π_n. From the definition of w_n, we have that

$$w(x) := w_0(x) + w_1(x) = \mathbb{E}_x[T]. \tag{6.7.43}$$

By the strong Markov property, if $0 \leq k \leq M - 1$ and $x \in (l_k, l_{k+1})$, then

$$w(x) = v(s) + \frac{1}{\rho_1}\left(q(s)w_1(l_k) + (1 - q(s))w_1(l_{k+1})\right), \tag{6.7.44}$$

where $s = x - l_k$, $v(s)$ is the mean exit time to escape the interval (l_k, l_{k+1}) starting at s, and the splitting probability $q(s)$ is given in (6.7.38). We can calculate $v(s)$ by considering the FPT τ for a particle to escape from either end in the interval $[0, L]$. Defining

$$v_n(x) := \mathbb{E}[\tau 1_{\{n(0)=n\}} \mid X_0 = x],$$

we have that v_n satisfies the ODE (6.7.42) with boundary conditions

$$v_0'(0) = 0, \quad v_0'(L) = 0, \quad v_1(0) = 0, \quad v_1(L) = 0.$$

It is straightforward to solve this boundary value problem and obtain

$$v(x) = v_0(x) + v_1(x) = \frac{1}{2}\left(x(L-x) + \frac{L\rho_0 \coth\left(\frac{L\xi}{2}\right)}{\rho_1 \xi}\right). \tag{6.7.45}$$

Since $w_1(0) = w_1(ML) = 0$, it remains to determine the $M-1$ constants $w_1(l_1), \ldots,$ $w_1(l_{M-1})$. For evenly spaced cells, we find that [126]

$$w_1(l_k) = V + \frac{1}{2}\left(w_1(l_{k-1}) + w_1(l_{k+1})\right) \tag{6.7.46}$$

for $k = 1, \ldots, M-1$, where V is the MFPT to escape from an interval of length $2L$ starting at the center of the domain and $n(0) = 1$:

$$V = \frac{L\left(L\rho_1 \xi + 2\rho_0 \tanh\left(\frac{L\xi}{2}\right)\right)}{2\xi}. \tag{6.7.47}$$

Rearranging (6.7.46), we notice that these constants satisfy a discretized Poisson equation

$$w_1(l_{k-1}) - 2w_1(l_k) + w_1(l_{k+1}) = -2V \tag{6.7.48}$$

for $k = 1, \ldots, M-1$, and $w_1(l_0) = w_1(ML) = 0$ can be interpreted as boundary conditions.

Derivation of equations (6.7.39) and (6.7.46). To see why (6.7.39) holds, observe that the splitting probability for the particle to be absorbed at the left-hand boundary $x = 0$ given that $X(0) = l_k$ and $n(0) = 1$ satisfies

$$\begin{aligned}
\pi_1(l_k) &= \mathbb{E}_{l_k,1}[\mathbb{1}_{X(T)=0}] \\
&= \mathbb{E}_{l_k,1}[\mathbb{1}_{X(T)=0}|X(s_k) = l_{k-1}]\mathbb{P}[X(s_k) = l_{k-1}] \\
&\quad + \mathbb{E}_{l_k,1}[\mathbb{1}_{X(T)=0}|X(s_k) = l_{k+1}]\mathbb{P}[X(s_k) = l_{k+1}] \\
&= \mathbb{P}[X(s_k) = l_{k-1}]\pi_1(l_{k-1}) + \mathbb{P}[X(s_k) = l_{k+1}]\pi_1(l_{k+1}).
\end{aligned}$$

The second and third lines express the fact that sample paths that are eventually absorbed at $x = 0$ can be partitioned into those that exit (l_{k-1}, l_{k+1}) at time s_k from the end l_{k-1} and the end l_{k+1}, respectively. The final line exploits the strong Markov property and linearity. By symmetry, $\mathbb{P}[X(s_k) = l_{k-1}] = \mathbb{P}[X(s_k) = l_{k+1}] = 1/2$. Applying the definition of p_1 in (6.7.32) gives (6.7.39). (Implicit in the above derivation is the fact that absorption at the left-hand boundary $x = 0$ is measurable with respect to the filtration $\mathcal{F}_{s_k}$, since the particle has to escape the interval (l_{k-1}, l_{k+1}) first; see Chap. 9 for a definition of filtrations.)

Turning to equation (6.7.46), observe that $w_1(l_k) = \rho_1 \mathbb{E}_{l_k,1}[T]$ and thus

$$\begin{aligned}
\frac{w_1(l_k)}{\rho_1} &= \mathbb{E}_{l_k,1}[s_k] + \mathbb{P}[X(s_k) = l_{k-1}]\mathbb{E}_{l_k,1}[T - s_k|X(s_k) = l_{k-1}] \\
&\quad + \mathbb{P}[X(s_k) = l_{k+1}]\mathbb{E}_{l_k,1}[T - s_k|X(s_k) = l_{k+1}].
\end{aligned}$$

As before, applying the strong Markov property and using linearity gives

$$\begin{aligned}
\frac{w_1(l_k)}{\rho_1} &= \mathbb{E}_{l_k,1}[s_k] + \mathbb{P}[X(s_k) = l_{k-1}]\mathbb{E}_{l_{k-1},1}[T] \\
&\quad + \mathbb{P}[X(s_k) = l_{k+1}]\mathbb{E}_{l_{k+1},1}[T]. \tag{6.7.49}
\end{aligned}$$

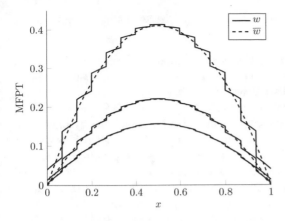

Fig. 6.38: Homogenized diffusion. Plots of MFPT to escape from an interval with many fast switching gates ($w(x)$ in (6.7.43)) and the MFPT to escape an interval with homogenized diffusion coefficient (6.7.51), which is $\overline{w}(x)$ in (6.7.50). The three pairs of curves correspond to $\rho_1 = 1/4$, $1/2$, and $3/4$, with higher curves corresponding to lower values of ρ_1. $w(x)$ has jump discontinuities at gates which we plot as vertical lines. The number of gates is $M = 15$ and $a = 1$.

Again, by symmetry, $\mathbb{P}[X(s_k) = l_{k-1}] = \mathbb{P}[X(s_k) = l_{k+1}] = 1/2$. Moreover, by the reflection principle for Brownian motion, $\rho_1 \mathbb{E}_{l_k,1}[s_k] = V$. Multiplying (6.7.49) by ρ_1 gives (6.7.46).

Limit of many gates and fast switching. Let $L = 1/M \ll 1$ and $\xi = 2M/a$ for some $a > 0$. From (6.7.46), we have

$$M^2\big(w_1(l_{k-1}) - 2w_1(l_k) + w_1(l_{k+1})\big) = -2M^2V = -(\rho_1 + \rho_0 a \tanh(1/a)).$$

Taking the continuum limit of the left-hand side and noting that $w_0 \approx (\rho_0/\rho_1)w_1$ for fast switching, we have $w(x) \approx \overline{w}(x)$ with $\overline{w}(x)$ the solution to the BVP

$$\Delta \overline{w}(x) = -\Big(1 + a\tanh(1/a)\frac{\beta}{\alpha}\Big), \quad \overline{w}(0) = \overline{w}(1) = 0. \qquad (6.7.50)$$

The latter yields the classical MFPT for a diffusing particle with diffusion coefficient

$$D = \Big[1 + a\tanh(1/a)\frac{\beta}{\alpha}\Big]^{-1} \qquad (6.7.51)$$

to escape from the interval $(0,1)$. We illustrate the accuracy of this approximation in Fig. 6.38. The solution for finite M is considered in Ex. 6.13.

6.8 Diffusive transport through nanopores and channels

As we mentioned in the introduction to this chapter, an important cellular transport process is the passage of molecules through membrane channels or pores. Here we will focus on passive aspects of membrane transport, which are characterized by diffusion in confined domains. The details of the process will depend on the relative size of the molecules compared to the channel and any interactions between the diffusing particles and the channel itself. A major consequence of confinement is that it restricts the number of degrees of freedom available to a molecule, which leads to strong entropic effects. We will explore the effects of confinement in three distinct cases.

(i) *Diffusion of ions or lipids through a narrow channel.* Here changes in the motion of a molecule occur mainly in the axial direction along the channel, whereas local equilibrium is rapidly reached in the transverse directions. Thus transport is quasi-one-dimensional and the effects of the boundaries of the channel can be incorporated by introducing an effective (entropic) energy barrier into the dynamics of a Brownian particle, leading to the so-called Fick–Jacobs equation [150, 151, 411, 427, 709, 904], see Sect. 6.8.1. Typically a 3D narrow channel is represented by a cylinder that extends axially in the x-direction and has a periodically varying cross section that is rotationally symmetric about the x-axis, see Fig. 6.39(a). Denoting the space-dependent radius of period L by $w(x)$, with $w(x+L) = w(x)$ for all x, the cross section varies as $A(x) = \pi w(x)^2$. In the case of a corresponding 2D channel, $w(x)$ represents the half-width of the channel.

(ii) *Single-file diffusion.* An extreme version of confined diffusion along a channel is single-file diffusion, in which the channel is so narrow that particles cannot pass each other. In other words, the longitudinal motion of each particle is hindered by the presence of its neighbors, which act as moving obstacles, see Fig. 6.39(b). Hence, interparticle interactions can suppress the Brownian motion and lead to subdiffusive behavior [49, 511, 651, 797], see Sect. 6.8.2.

(iii) *Translocation of polymers through a pore.* Polymer translocation through a membrane pore plays an important role in a number of cellular processes, including the transport of RNA and proteins across nuclear pores (see Sect. 6.8.3), virus infection of cells, and DNA packaging into viral capsids [9]. There are also an increasing number of technological applications, ranging from drug delivery to biochemical sensors, which have been driven by significant progress in experimental studies of translocation at the single-molecule level [600]. Translocation of biopolymers *in vivo* is typically facilitated by interactions with the channel or specialized proteins (chaperones), whereas translocation *in vitro* is driven by the application of external electrical fields. Since the resulting

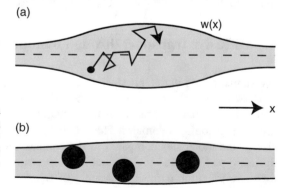

(a)

w(x)

x

Fig. 6.39: Confined diffusion in a narrow cylindrical channel with a periodically modulated boundary $w(x)$ in the axial direction. (a) Small diffusing particle. (b) Single-file diffusion.

(b)

voltage difference across the pore would normally cause the flow of ions, one can measure when a polymer enters the pore since it partially blocks the path of the ions, resulting in a significant decrease in the ionic current. Thus the frequency and duration of translocation events can be recorded. From a theoretical prospective the translocation process also involves diffusion past an entropic barrier, in this case arising from the fact that a free polymer has many more configurational states than one that is threaded through a pore [184, 599, 635, 789].

6.8.1 Confined diffusion and the Fick–Jacobs equation

We begin by deriving the Fick–Jacobs equation for a Brownian particle diffusing in a 2D channel as shown in Fig. 6.39(a). We follow the particular derivation of [904]. It is assumed that the channel walls at $y = \pm w(x)$ confine the motion of the particle but do not exchange energy with it. Thus the probability flux normal to the boundary is zero. This condition can be imposed by introducing a confining potential $U(x,y)$ such that $U(x,y) = 0$ for $|y| < w(x)$ and $U(x,y) = \infty$ for $|y| \geq w(x)$. Let $p(x,y,t)$ denote the probability that the particle is located at position $\mathbf{x} = (x,y)$ at time t with periodic boundary conditions in the longitudinal direction, $p(x+L,y,t) = p(x,y,t)$. For a general potential $U(x,y)$, the 2D FP equation takes the form

$$\frac{\partial p}{\partial t} = -\frac{1}{\gamma} \left[\frac{\partial [F_x p]}{\partial x} + \frac{\partial [F_y p]}{\partial y} \right] + D_0 \left[\frac{\partial^2 p}{\partial x^2} + \frac{\partial^2 p}{\partial y^2} \right],$$

where $F_x = -\partial_x U, F_y = -\partial_y U$. Using the Einstein relations $D_0 \gamma = k_B T = \beta^{-1}$, the FP equation can be rewritten as

$$\frac{\partial p}{\partial t} = D_0 \frac{\partial}{\partial x} e^{-\beta U(x,y)} \frac{\partial}{\partial x} e^{\beta U(x,y)} p(x,y,t)$$
$$+ D_0 \frac{\partial}{\partial y} e^{-\beta U(x,y)} \frac{\partial}{\partial y} e^{\beta U(x,y)} p(x,y,t). \tag{6.8.1}$$

In order to reduce to a 1D equation, first integrate both sides of the FP equation with respect to the transverse coordinate y:

$$\frac{\partial P(x,t)}{\partial t} = D_0 \frac{\partial}{\partial x} \int_{-w(x)}^{w(x)} e^{-\beta U(x,y)} \frac{\partial}{\partial x} e^{\beta U(x,y)} p(x,y,t) dy,$$

where $P(x,t)$ is the reduced probability density

$$P(x,t) = \int_{-w(x)}^{w(x)} p(x,y,t) dy. \tag{6.8.2}$$

The major step in the reduction is to assume that the probability density reaches equilibrium in the transverse direction. That is, $p(x,y,t)$ is assumed to factorize as follows:

$$p(x,y,t) \approx P(x,t)\rho(x,y), \tag{6.8.3}$$

where $\rho(x,y)$ is a normalized Boltzmann–Gibbs probability density (Sect. 1.3):

$$\rho(x,y) = \frac{e^{-\beta U(x,y)}}{A_0 e^{-\beta \mathcal{U}(x)}}, \qquad e^{-\beta \mathcal{U}(x)} = \frac{1}{A_0}\int_{-w(x)}^{w(x)} e^{-\beta U(x,y)}dy, \tag{6.8.4}$$

where $A_0 = 2\int_0^L w(x)dx$ and $\mathcal{U}(x)$ interpreted as an effective x-dependent barrier potential or free energy. Under this factorization the averaged FP equation becomes

$$\frac{\partial P(x,t)}{\partial t} \approx D_0 \frac{\partial}{\partial x} e^{-\beta \mathcal{U}(x)} \frac{\partial}{\partial x} e^{\beta \mathcal{U}(x)} P(x,t). \tag{6.8.5}$$

This holds for a general potential energy function $U(x,y)$ [677]. If U is now taken to be the confining potential of the channel boundary, then $e^{-\beta \mathcal{U}(x)} = 2w(x)/A_0 \equiv \sigma(x)$ and we obtain the Fick–Jacobs equation

$$\frac{\partial P(x,t)}{\partial t} = D_0 \frac{\partial}{\partial x} \sigma(x) \frac{\partial}{\partial x} \frac{P(x,t)}{\sigma(x)}. \tag{6.8.6}$$

The same equation is obtained in 3D with $\sigma(x) = A(x)/A_0$ with $A(x) = \pi w(x)^2$ and $A_0 = \pi \int_0^L w(x)^2 dx$. In the physics literature $\mathcal{U}(x)$ is usually referred to as an entropic barrier, since confinement reduces the volume of the phase space of available states accessible to the particle.[2] The Fick–Jacobs equation is valid provided that $|w'(x)| \ll 1$. However, it has been shown that the introduction of an x-dependent diffusion coefficient into the Fick–Jacobs equation can considerably increase the accuracy of the reduced FP equation and thus extend the domain of validity [427, 677, 904]:

$$D(x) = \frac{D_0}{[1+w'(x)^2]^\alpha}, \tag{6.8.7}$$

with $\alpha = 1/3, 1/2$ for 2D and 3D, respectively.

As it stands, the Fick–Jacobs equation (6.8.6) represents a particle diffusing in a 1D periodic potential $\mathcal{U}(x)$, which means that the mean velocity of the particle is zero. On the other hand, net transport of the particle through the channel does occur in the presence of a constant external force F_0 in the x-direction. Equation (6.8.6) still holds, except that now $\mathcal{U}(x) = -F_0 x - k_B T \ln \sigma(x)$, which yields the classical problem of Brownian motion in a periodic potential with tilt [150, 357, 684, 782], which was analyzed in Sect. 4.3. Given the mean and variance of the particle posi-

---

[2] The effective free energy $\mathcal{U}(x) = -k_B T \ln[A(x)/A_0]$ reflects the existence of an entropic barrier to diffusion [677]. That is, using the standard definition of free energy $\mathcal{U} = E_0 - TS$, where E_0 is internal energy and S is the entropy, it follows that $S(x) \sim \ln A(x)$ where $A(x)$ is the cross-sectional area of the channel at x. This is consistent with the definition of entropy in terms of the logarithm of the number of microstates. That is, in equilibrium there is a uniform probability density ρ_0 in the channel, so that the equilibrium x-dependent density $P_{eq}(x) = \rho_0 A(x)/A_0$ and the number of microstates available to a diffusing particle at location x is proportional to the area of the channel.

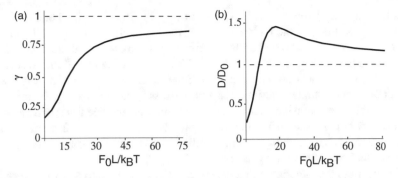

Fig. 6.40: Illustrative sketches of how mobility and diffusivity vary with non-dimensionalized applied force $F_0 L/k_B T$ in the case of a 2D channel with a sinusoidally varying half-width (6.8.10). (a) Effective mobility μ in units of γ. In the limit $F_0 \to \infty$, $\mu \to \gamma^{-1}$. (b) Diffusion coefficient D in units of free diffusivity D_0. In the limit $F_0 \to \infty$, $D \to D_0$. Sketches are based on numerical results of [678] for $a = L/2\pi$ and $b = 1.02$.

tion in the long-time limit, one can define the drift mobility and diffusion coefficient of the particle according to

$$\mu(F_0) \equiv \frac{\langle \dot{X} \rangle}{F_0}, \quad \langle \dot{X} \rangle = \lim_{t \to \infty} \frac{\langle X(t) \rangle}{t}, \tag{6.8.8}$$

and

$$D(F_0) = \lim_{t \to \infty} \frac{\langle X(t)^2 \rangle - \langle X(t) \rangle^2}{2t}. \tag{6.8.9}$$

Note that the relationship between $\langle \dot{X} \rangle$ and the long-time limit of $\langle X(t) \rangle / t$ is a consequence of ergodicity [683], see Sect. 4.3. The force dependence of the mobility and diffusion coefficient have been studied both analytically and numerically in the case of a sinusoidal boundary function [150, 678]

$$w(x) = a[\sin(2\pi x/L) + b], \quad a > 0, b > 1. \tag{6.8.10}$$

The basic results are sketched in Fig. 6.40. A number of interesting observations emerge from this study. First, the mobility only depends on the temperature via the dimensionless parameter $F_0 L/k_B T$. Hence, increasing the temperature reduces the mobility. Second, as the force is increased the effective diffusion coefficient $D(F_0)$ exceeds the free diffusion coefficient D_0. Finally, note that for certain forms of $\sigma(x)$, one can find exact solutions of the Fick–Jacobs equation by a change of variables [703], see Ex. 6.14.

The Fick–Jacobs equation represents diffusion through a narrow channel in terms of a 1D overdamped Brownian particle moving in an effective potential $\mathcal{U}(x)$ that arises from confinement. Such a 1D model has also been the starting point for a series of studies of channel-facilitated membrane transport, where now $\mathcal{U}(x)$ reflects the constructive role of attractive interactions between permeating particles and proteins forming the channel pore. In a series of studies [75–77], mixed boundary conditions are assumed at the ends $x = 0, L$ of the channel: $J(0,t) = -\kappa_0 P(0,t)$ and $J(L,t) = -\kappa_L P(L,t)$. The probability of crossing the channel and the mean time in the channel is then calculated in terms of FPTs and splitting probabilities. It can be shown that there is an optimal form of the interaction potential that maximizes the flux through the channel and involves a play off between increasing the translocation probability through the channel and decreasing the average time particles spend in the channel [77], see also the entropic gate model in Sect. 6.8.2. For a complementary approach to studying channel-facilitated transport that is based on spatially discrete stochastic site-binding models, see [177, 460].

Finally, note that a variety of models have been developed to analyze ion permeation through narrow channels. Broadly speaking, these models can be divided into three classes [373, 512, 708]. (i) Brownian dynamics models based on a Langevin description of the full 3D dynamics of ion motion through channels, which take into account both ion–ion interactions and ion–channel interactions. (ii) Continuum mean-field models based on the Poisson–Nernst–Planck equation. The latter treats a channel as a continuous medium, and the ionic current is determined by coupling the Nernst–Planck electrodiffusion equation for the flux of charged particles in the presence of a concentration gradient and an electric field (see Ex. 3.9) with the Poisson equation describing how the distribution of charges generates an effective mean-field potential [601, 737]. (iii) Barrier models in which ions are localized to specific regions of the channel via local potentials, and the kinetics are represented by the rate constants for hopping between these regions and between the channel and the bulk [373]. A simple barrier model of ion permeation is considered in Ex. 6.15.

6.8.2 Single-file diffusion

When a pore or channel becomes sufficiently narrow, particles are no longer able to pass each other, which imposes strong constraints on the diffusive motion. An idealized model of single-file diffusion considers a 1D collection of diffusing particles with hard-core repulsion. The many-body problem of single-file diffusion was originally tackled by relating the dynamics of the interacting system with the effective motion of a free particle [500, 511, 651]. In particular, in the case of an infinite system and a uniform initial particle density, it was shown that a tagged particle exhibits anomalous subdiffusion on long time scales, $\langle X^2(t) \rangle \sim t^{1/2}$. (On the other hand, the center of mass of the system of particles exhibits normal diffusion). More recently, a variety of complementary approaches to analyzing single-file diffusion have been developed [49, 162, 700, 798]. Here we review the particular formulation of Barkai and Silbey [49], which develops the analysis of a tagged particle in terms of classical reflection and transmission coefficients.

Suppose that the tagged particle is initially at the origin with N particles to its left and N particles to its right, see Fig. 6.41(a). The motion of each particle in the absence of hard core interactions is taken to be overdamped Brownian motion as described by the Langevin equation (2.2.5) or the corresponding FP equation (2.3.1). As a further simplification, the potential energy function $V(x) = \int^x F(x')dx'$ is taken to be symmetric, $V(x) = V(-x)$, as is the initial distribution of particles. That is, if the initial position x_0 of a particle is drawn from $f_R(x_0)$ for $x_0 > 0$ and from $f_L(x_0)$ for $x_0 < 0$, then $f_R(x_0) = f_L(-x_0) \equiv f(x_0)$. This reflection symmetry ensures that $\langle X(t) \rangle = 0$, where $X(t)$ is the stochastic position of the tagged particle at time t. The main underlying idea is to map the many-body problem to a noninteracting one by allowing particles to pass through each other and keeping track of the particle label, see Fig. 6.41(b). That is, assuming that collisions are elastic and neglecting n-body interactions for $n > 2$, it follows that when two particles collide they exchange momenta and this is represented as an exchange of particle labels. The probability density for the tagged particle to be at $X(t) = X_T$ at time t then reduces to the problem of finding the probability that the number of free particle trajectories that started at $x_0 < 0$ and are now to the right of X_T is balanced by the number of free particle trajectories that started at $x_0 > 0$ and are now to the left of X_T.

Thus, let $P_{LL}(x_0^{-j})$ $(P_{LR}(x_0^{-j}))$ denote the probability that the jth free particle trajectory starting from $x_0^{-j} < 0$ at $t = 0$ is to the left (right) of X_T at time t. Similarly, let $P_{RR}(x_0^j)$ $(P_{RL}(x_0^j))$ denote the probability that the jth free particle trajectory starting from $x_0^j > 0$ at $t = 0$ is to the right (left) of X_T at time t. Let α be the net number of free particle trajectories that are on the opposite side of X_T at time t compared to their starting point (with left to right taken as positive). The associated probability distribution for α given $2N$ untagged particles is [49]

$$P_N(\alpha) = \frac{1}{2\pi} \int_{-\pi}^{\pi} \prod_{j=1}^{N} \Gamma(\phi, x_0^{-j}, x_0^j) e^{i\alpha\phi} d\phi, \qquad (6.8.11)$$

where

(a)

(b)

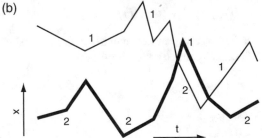

Fig. 6.41: (a) Single-file diffusion of a tagged particle (darker filled circle) surrounded by other impenetrable particles. (b) Equivalent noninteracting picture, in which each trajectory is treated as a noninteracting Brownian particle by keeping track of the exchange of particle label.

$$\Gamma(\phi, x_0^{-j}, x_0^j) = e^{i\phi} P_{LR}(x_0^{-j}) P_{RR}(x_0^j) + P_{LL}(x_0^{-j}) P_{RR}(x_0^j)$$
$$+ P_{LR}(x_0^{-j}) P_{RL}(x_0^j) + e^{-i\phi} P_{LL}(x_0^{-j}) P_{RL}(x_0^j). \tag{6.8.12}$$

The integration with respect to ϕ ensures that the net number of crossings is α, that is, $\int_{-\pi}^{\pi} e^{i\phi n} = \delta_{n,0}$. Since the trajectories are independent and the initial conditions are (iid) random variables, $P_N(\alpha)$ can be averaged with respect to the initial conditions to give

$$\langle P_N(\alpha) \rangle = \frac{1}{2\pi} \int_{-\pi}^{\pi} \langle \Gamma(\phi) \rangle^N e^{i\alpha\phi} d\phi, \tag{6.8.13}$$

where

$$\langle \Gamma(\phi) \rangle = \left(\langle P_{RR} \rangle + e^{-i\phi} \langle P_{RL} \rangle \right) \left(\langle P_{LL} \rangle + e^{i\phi} \langle P_{LR} \rangle \right). \tag{6.8.14}$$

The averages $\langle P_{LR} \rangle$, etc. can be calculated using the fundamental solution $K(x, x_0, t)$ of the corresponding FP equation (2.3.1), that is, the solution with initial condition $K(x, x_0, 0) = \delta(x - x_0)$. For example,

$$\langle P_{LR} \rangle = \int_{-l}^{0} f_L(x_0) \int_{X_T}^{l} K(x, x_0, t) dx\, dx_0, \tag{6.8.15}$$

where $2l$ is the length of the 1D domain. For fixed $x_0 < 0$, $\int_{X_T}^{l} K(x, x_0, t) dx$ is the probability of being to the right of X_T at time t, which is then averaged over all initial conditions to the left of the origin.

Equation (6.8.13) takes the form of the generating function for a discrete random walk of N steps and a net displacement of α, see equation (2.1.4). Hence, for large N, application of the central limit theorem (Sect. 2.1) leads to the Gaussian approximation

$$P_N(0) \sim \frac{1}{\sqrt{2\pi N \sigma^2}} \exp(-N\mu_1^2 / 2\sigma^2), \tag{6.8.16}$$

where $\sigma^2 = \mu_2 - \mu_1^2$ and μ_1, μ_2 are the first two moments of the structure function:

$$\langle \Gamma(\phi) \rangle = 1 + i\mu_1 \phi - \frac{1}{2}\mu_2 \phi^2 + O(\phi^3). \tag{6.8.17}$$

Hence,

$$\mu_1 = \langle P_{LR} \rangle - \langle P_{RL} \rangle, \quad \sigma^2 = \langle P_{RR} \rangle \langle P_{RL} \rangle + \langle P_{LL} \rangle \langle P_{LR} \rangle. \tag{6.8.18}$$

Since $\langle X(t) \rangle = 0$ and N is assumed to be large, μ_1 and σ^2 can be Taylor expanded with respect to X_T about $X_T = 0$. Reflection symmetry then implies that

$$\langle P_{LL} \rangle |_{X_T=0} = \langle P_{RR} \rangle |_{X_T=0} \equiv \mathcal{R},$$

$$\langle P_{LR} \rangle |_{X_T=0} = \langle P_{RL} \rangle |_{X_T=0} \equiv \mathcal{T} = 1 - \mathcal{R},$$

$$\partial_{X_T} \langle P_{LR} \rangle |_{X_T=0} = -\partial_{X_T} \langle P_{RL} \rangle |_{X_T=0} \equiv \mathcal{J}.$$

The time-dependent functions $\mathcal{R}$ and $\mathcal{T}$ may be interpreted as reflection and transmission coefficients determining whether or not a free particle trajectory crosses $X_T = 0$. The resulting mean and variance are

$$\mu_1 = -2\mathcal{J}X_T + O(X_T^2), \quad \sigma^2 = 2\mathcal{R}(1-\mathcal{R}) + O(X_T). \tag{6.8.19}$$

Thus, $\langle P_N(\alpha) \rangle$ for $\alpha = 0$ reduces to a Gaussian distribution for the position $X(t) = X_T$:

$$P(X_T, t) = \frac{1}{\sqrt{2\pi \langle X(t)^2 \rangle}} \exp\left[-\frac{X_T^2}{2\langle X(t)^2 \rangle}\right], \tag{6.8.20}$$

with

$$\langle X(t)^2 \rangle = \frac{\mathcal{R}(1-\mathcal{R})}{2N\mathcal{J}^2}. \tag{6.8.21}$$

Finally, using equation (6.8.15),

$$\mathcal{R} = \int_0^l f(x_0) \int_0^l K(x, x_0, t) dx dx_0, \tag{6.8.22}$$

$$\mathcal{J} = \int_0^l f(x_0) K(0, x_0, t) dx_0. \tag{6.8.23}$$

In the special case of zero external forces and $l \to \infty$, the fundamental solution is

$$K(x, x_0, t) = \frac{1}{\sqrt{4\pi Dt}} e^{-(x-x_0)^2/4Dt}. \tag{6.8.24}$$

That is, $\lim_{t\to 0} K(x, x_0, t) = \delta(x - x_0)$ and $\partial_t K = D\partial_{xx}K$ for $t > 0$, $x \in \mathbb{R}$. Taking a uniform initial distribution $f(x_0) = 1/l$ with $l \to \infty$ and fixed particle density $\rho = N/l$, one finds anomalous subdiffusion for large times t:

$$\langle X(t)^2 \rangle \sim \frac{2}{\sqrt{\pi}} \frac{\sqrt{Dt}}{\rho}. \tag{6.8.25}$$

On the other hand, for particles initially centered at the origin, $f(x_0) = \delta(x_0)$, diffusion is normal

$$\langle X(t)^2 \rangle \sim \frac{\pi Dt}{2N}. \tag{6.8.26}$$

In the case of a bounded domain or a Gaussian initial condition, anomalous diffusion occurs at intermediate times only [49].

6.8.3 Nuclear transport

The nucleus of eukaryotes is surrounded by a protective nuclear envelope (NE) within which are embedded nuclear pore complexes (NPCs), see Fig. 6.42. The NPCs

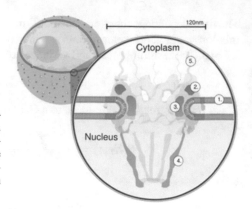

Fig. 6.42: The nuclear pore complex. 1. Nuclear envelope. 2. Outer ring. 3. Spokes. 4. Basket. 5. Filaments. Each of the eight protein sub-units surrounding the actual pore (the outer ring) projects a spoke-shaped protein into the pore channel. (Public domain figure from Wikimedia.)

are the sole mediators of exchange between the nucleus and cytoplasm. In general small molecules of diameter ~ 5 nm can diffuse through the NPCs unhindered, whereas larger molecules up to around 40 nm in diameter are excluded unless they are bound to a family of soluble protein receptors known as karyopherins (kaps), see the reviews [539, 707, 807, 811]. Within the cytoplasm kap receptors bind cargo to be imported via a nuclear localization signal (NLS) that results in the formation of a kap–cargo complex. This complex can then pass through an NPC to enter the nucleus. A small enzyme RanGTP then binds to the kap, causing a conformational change that releases the cargo. The sequence of events underlying the import of cargo is shown in Fig. 6.43(a). In the case of cargo export from the nucleus, kaps bind to cargo with a nuclear export signal (NES) in the presence of RanGTP, and the resulting complex passes through the NPC. Once in the cytoplasm, RanGTP is hydrolyzed by the cytoplasmic factor RanGAP1 to form RanGDP, resulting in the release of the

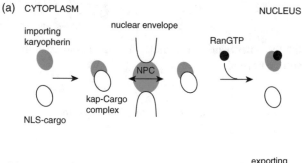

Fig. 6.43: Schematic illustration of the (a) import and (b) export process underlying the karyopherin-mediated transportation of cargo between the nucleus and cytoplasm via a nuclear pore complex (NPC). See text for details.

cargo. The export process is illustrated in Fig. 6.43(b). Finally, RanGDP is recycled to the nucleus by another molecule NFT2 and is reloaded with GTP to begin another import/export cycle. The conversion to RanGTP is mediated by a chromatin-associated guanine exchange factor RanGEF. This cycle allows a single NPC to support a very high rate of transport on the order of 1000 translocations/sec and a typical transit time of ~ 10 ms [692]. Since the transportation cycle is directional and accumulates cargo against a concentration gradient, an energy source combined with a directional cue is required. Both of these are provided by the hydrolysis of RanGTP and the maintenance of a concentration gradient of RanGTP across the NPC. The RanGTP gradient is continuously regenerated by GTP hydrolysis in the cytoplasm, translocation of RanGTD into the nucleus by NFT2, and replacement of GDP by GTP in the nucleus. It is important to note that the energy generated from RanGTP hydrolysis is ultimately used to create a concentration gradient of RanGTP between the nucleus and cytoplasm, so that the actual translocation across the NPC occurs purely via diffusion.

Although the above basic picture is now reasonably well accepted, the detailed mechanism underlying facilitated diffusion of kap–cargo complexes within the NPC is still not understood. The NPC is composed of about 30 distinct proteins known collectively as nucleoporins (nups). It has emerged in recent years that individual nups are directly related to a number of human diseases including influenza and cancers such as leukemia [204], as well as playing an important role in viral infections by providing docking sites for viral capsids [857]. Associated with many of the nups are natively unfolded phenylalanine-glycine (FG) repeats, known collectively as FG-nups [672, 733]. The FG-nups set up a barrier to diffusion for large molecules so that the key ingredient in facilitated diffusion through the NPC is the interaction between kap receptors with the FG-nups. In essence, the major difference between most theoretical models of NPC transport concerns the built-in assumptions regarding the properties and spatial arrangements of FG-nups within the NPC, and the nature of interactions with kaps during translocation through the NPC [59, 807]. One important feature that emerges from the various theoretical studies is that kap complexes should also be mobile in the bound state.

Two complementary approaches to modeling the interior of an NPC are based on polymer brushes and polymer gels, respectively. The former treats the FG-nups as flexible polymer tethers to which kaps can temporarily bind and undergo confined diffusion in the bound state, see Fig. 6.44 [279, 280, 549]. The mechanical prop-

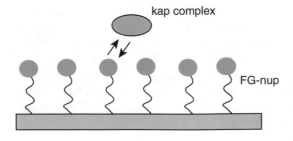

kap complex

FG-nup

Fig. 6.44: Polymer brush model, in which the FG repeats within a NPC are treated as flexible polymer tethers.

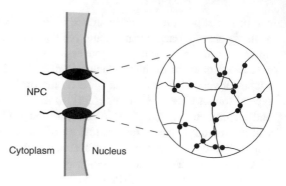

NPC

Fig. 6.45: Selective phase or gel model, in which the FG repeats within a NPC are treated as a reversible polymer gel.

Cytoplasm | Nucleus

erties of the tethers contribute to the effective diffusivity of the complexes in the bound state. It is also possible that complexes undergo interchain transfers without unbinding [672]. The second type of model treats the NPC as a weakly reversible gel [95, 478, 692, 693], see Fig. 6.45. (A gel is a jelly-like material that is mostly liquid by weight, yet behaves like a solid due to a 3D cross-linked polymer network within the liquid, see also Chap. 12. It is the cross-linking within the fluid that gives a gel its solid-like properties such as hardness. A gel is said to be reversible if the cross-linking is reversible.) Particles smaller than the mesh size of the network can diffuse freely through the NPC, whereas non-selective macromolecules larger than the mesh size cannot. On the other hand, kap–cargo complexes can "dissolve" in the gel due to the presence of hydrophobic domains on the surface of the kap receptors, and then diffuse through the pore by breaking the weak bonds of the reversible gel [327, 692, 693]. Alternatively, one can treat the gel as a continuum of binding sites for kap complexes, with the gel properties determining the effective diffusivity in the bound state [873].

Entropic (virtual) gating model. Recall from Sect. 6.8.1 that a macromolecule diffusing in a confined geometry (such as a nuclear pore) experiences an entropic barrier due to excluded volume effects. Within the NPC this would be enhanced by the densely packed FG-nups. One way to counteract the effects of the entropic barrier is for the kaps to have an affinity for and bind to the FG-repeat regions [707, 897], thus lowering the effective free energy of the cargo complex within the NPC. The degree of affinity has to be sufficiently high to overcome the entropic barrier but not too high otherwise the complex can be trapped within the NPC and the rate of translocation would be too small. One possible solution is to have a large number of low-affinity binding sites within the nuclear pore. A mathematical model for the effects of binding on diffusion within the NPC has been developed in Ref. [897], based on diffusion through an effective energy landscape, which approximates the effects of multiple binding sites when the binding/unbinding rates are relatively fast compared to the diffusion rate (see below). The simplest version of the model is illustrated in Fig. 6.46 for the case of nuclear import. The effective potential energy $U(x)$ is taken to be a flat potential well of depth E along an NPC, and zero outside the NPC. Absorbing boundary conditions are placed at the points $x = 0, L$ a distance

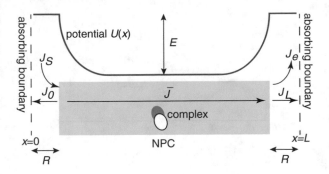

Fig. 6.46: Sketch of model of [897]. Transport of cargo complex through the NPC is modeled as diffusion in an energy landscape. See text for details.

R from either side of the NPC, which has length $L - 2R$. The absorbing boundary conditions represent a cargo complex returning to the cytoplasm at $x = 0$ or entering the nucleus at $x = L$ and diffusing away. Diffusion within the NPC is described by a standard Smoluchowski equation for the density of cargo complexes $\rho(x), x = [0, L]$:

$$\frac{\partial \rho}{\partial t} = -\frac{\partial J}{\partial x}, \quad J = -D\frac{\partial \rho}{\partial x} - D\rho\frac{\partial U}{\partial x}, \quad (6.8.27)$$

with U measured in units of $k_B T$. This equation is supplemented by the absorbing boundary conditions $\rho(0) = \rho(L) = 0$. The diffusion of complexes outside of the NPC is not modeled directly. Instead, two effective fluxes are introduced: J_S is the total flux of complexes injected into the NPC from the cytoplasm, which is proportional to the density of complexes in the cytoplasm, and J_e denotes the flux due to active removal of complexes from the nucleus end of the NPC by RanGTP. The latter depends on the number of complexes at the nuclear exit, $\rho(L-R)R$, and the removal rate J_{ran}: $J_e = J_{ran}\rho(L-R)R$.

The steady-state solution is obtained by assuming that there are constant diffusive fluxes J_0 in $[0, R]$, J_L in $[L-R, L]$, and $\bar{J}$ in $[R, L-R]$ with $J_0 < 0$. These fluxes are related according to the external fluxes according to $J_S = \bar{J} - |J_0|$ and $\bar{J} = J_L + J_e$ with J_S and J_{ran} predetermined. The steady-state rate of transport $\bar{J}$ can now be determined by solving for $\rho(x)$ in terms of $J_0, J_L, \bar{J}$ in each of the three domains and imposing continuity of the density at $x = R$ and $x = R - L$ (see Ex. 6.16). The result is that the fraction of complexes reaching the nucleus is given by [897]

$$P = \frac{\bar{J}}{J_S} = \left[1 + \frac{1}{1+K} + \frac{1}{R}\int_R^{L-R} e^{U(x)} dx\right]^{-1}, \quad (6.8.28)$$

with $K = J_{ran}R^2/D$. It follows that for a sufficiently deep well (large E), where the integral term is negligible, and for sufficiently large K (large J_{ran}), the probability of translocation is $P \approx 1$. On the other hand, if K is small so that RanGTP does not facilitate entry of complexes into the nucleus then $P_{max} = 0.5$. As previously indicated, it is not possible to arbitrarily increase the affinity of binding sites and thus the well depth E, since this will lead to trapping of the complexes so that they accumulate within the NPC, resulting in molecular crowding and an unrealistically long time for

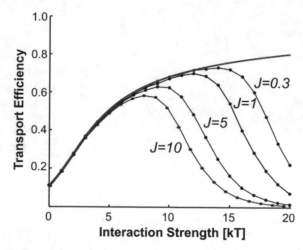

Fig. 6.47: Results of Monte Carlo simulations carried out by Zilman et al. [897] on their entropic gate model. Transport efficiency (probability P to reach the nucleus) is plotted as a function of the NPC interaction strength E. The unimodal curves correspond to four different values of the entrance rate J (in units of $10^{-4}16D/R^2$) and have a peak at a specific value of E, which provides a mechanism of selectivity. The envelope is the theoretical curve calculated from equation (6.8.28). RanGTP activity in the nucleus is fixed by setting $J_{ran}L^2/N^2D = 1.5$. (Adapted from Zilman et al. [897].)

an individual molecule to pass through the NPC. Thus there is some optimal well depth that balances an increase of transport probability P with increased time spent in the NPC [897]. Finally, note that the model is robust with regards to the particular shape of the potential well. For example, one could represent transport through the NPC as diffusion in an array of overlapping potential wells that represent flexible FG-repeat regions. The shape of each well will depend on the number and affinity of binding sites on each FG repeat, and the degree of flexibility of the polymers which will determine the entropic costs of bending and stretching the FG-nups. Results of Monte Carlo simulations are shown in Fig. 6.47. One difference from the analytical model is that only one complex is allowed to occupy a given site in the discretized model, which takes into account molecular crowding. However, this effect is small for small interaction strength E, since the density of complexes in the NPC is low, and there is good agreement between numerics and theory.

It is straightforward to show that for relatively fast binding/unbinding, the multi-well potential can be replaced by a single well along the lines of Fig. 6.46. Suppose that inside the NPC there are N binding sites labeled $i = 1, \ldots, N$ with corresponding potentials $U_i(x)$. Let $U_0(x)$ be the potential of an unbound complex. Denoting the probability density of complexes bound to the jth site by $\rho_i(t)$ and the density of unbound complexes by $\rho_0(t)$, we have the system of equations

$$\frac{\partial \rho_0}{\partial t} = D_0 \frac{\partial}{\partial x} e^{-U_0(x)} \frac{\partial}{\partial x} e^{U_0(x)} \rho_0 + \sum_{i>0} [\omega_{0i}(x)\rho_i(x) - \omega_{i0}(x)\rho_0(x)], \qquad (6.8.29a)$$

$$\frac{\partial \rho_i}{\partial t} = D_i \frac{\partial}{\partial x} e^{-U_i(x)} \frac{\partial}{\partial x} e^{U_i(x)} \rho_i + [\omega_{i0}(x)\rho_0(x) - \omega_{0i}(x)\rho_i(x)]. \qquad (6.8.29b)$$

Here a complex can unbind from the ith site at a rate $\omega_{0i}(x)$ and rebind to the same site or bind to a different site at a rate $\omega_{j0}(x)$. From detailed balance (see Sect. 1.3), the transition rates are related according to

$$\frac{\omega_{i0}(x)}{\omega_{0i}(x)} = e^{-U_i(x)+U_0(x)}.$$

Under the assumption of fast transition rates, the densities in the different states will be at local thermodynamic equilibrium with respect to the internal states. We thus have the Boltzmann–Gibbs distribution

$$\rho_i(x) = \frac{e^{-U_i(x)}}{\sum_{i=0}^{M} e^{-U_i(x)}} \rho(x), \quad \rho(x) = \sum_{i=0}^{N} \rho_i(x).$$

Hence, adding together the $N+1$ differential equations and taking $D_0 = D_i$ for simplicity, we find that

$$\frac{\partial \rho}{\partial t} = D_0 \frac{\partial}{\partial x} e^{-U(x)} \frac{\partial}{\partial x} e^{U(x)} \rho, \qquad (6.8.30)$$

where

$$U(x) = -\ln\left(e^{-U_0(x)} + \sum_{i>0} e^{-U_i(x)} \right). \qquad (6.8.31)$$

Binding-diffusion model. A number of more recent modeling studies have included more details concerning the binding interactions between kap complexes and FG-nups [137, 279, 280, 549, 873]. For the sake of illustration, consider complexes moving through a polymer gel or polymer brush treated as a continuum of binding sites [549, 873]. A complex can either be freely diffusing with concentration $A(\mathbf{r},t)$ or in a bound state with concentration $B(\mathbf{r},t)$. Let $S(\mathbf{r},t)$ be the density of free FG-repeat binding sites such that $S(\mathbf{r},t) + B(\mathbf{r},t) = S_0$, where S_0 is the density of both occupied and unoccupied binding sites, which is taken to be a constant. The resulting two-state diffusion model takes the form

$$\frac{\partial A}{\partial t} = D_A \nabla^2 A + k_{\text{off}} B - k_{\text{on}} A S, \qquad (6.8.32a)$$

$$\frac{\partial B}{\partial t} = D_B \nabla^2 B - k_{\text{off}} B + k_{\text{on}} A S, \qquad (6.8.32b)$$

where k_{on} and k_{off} are the binding and unbinding rates, D_A is the diffusivity in the unbound state, and D_B is the diffusivity in the bound state. The mobility of bound molecules could be due to the dynamics of the polymer network itself or arise from the capacity of molecules to slide across the network in a series of bound states.

The steady-state model can be solved explicitly in 1D under the additional simplifying assumption that the concentration of complexes is sufficiently dilute so that S can be treated as a constant, that is, $S \approx S_0$. This approximation will be valid when the dissociation constant $K_D = k_{off}/k_{on}$ is sufficiently large. Taking the NPC to have width L, the resulting steady-state equations are

$$0 = D_A \frac{d^2A}{dx^2} + k_{off}B - k_{on}S_0A, \tag{6.8.33a}$$

$$0 = D_B \frac{d^2B}{dx^2} - k_{off}B + k_{on}S_0A, \tag{6.8.33b}$$

The corresponding boundary conditions are taken to be

$$A(0) = A_0, \quad A(L) = 0, \quad B'(0) = B'(L) = 0. \tag{6.8.34}$$

That is, there is no flux of bound complexes into or out of the pore. Performing the change of variables $\widehat{B} = B - KS_0A$, with $K = K_D^{-1} = k_{on}/k_{off}$ the binding equilibrium constant, the steady-state equations can be rewritten as

$$0 = D_A \frac{d^2A}{dx^2} + k_{off}\widehat{B}, \tag{6.8.35a}$$

$$0 = D_B \frac{d^2\widehat{B}}{dx^2} - k_{off}\widehat{B} + KS_0D_B \frac{d^2A}{dx^2}. \tag{6.8.35b}$$

Substituting for $k_{off}\widehat{B}$ in equation (6.8.35b) using equation (6.8.35a) leads to the fourth-order equation

$$\frac{d^4A}{dx^4} = \lambda^2 \frac{d^2A}{dx^2}, \quad \lambda^2 = \frac{k_{off}(D_A + KS_0D_B)}{D_AD_B}. \tag{6.8.36}$$

This has a general solution of the form

$$A(x) = a_0 + a_1x + a_+e^{\lambda x} + a_-e^{-\lambda x}, \tag{6.8.37}$$

with the coefficients $a_0, a_1, a_{\pm}$ determined by the boundary conditions, see Ex. 6.17.

Given the solution $A(x)$ one can now determine the selectivity S of the pore by comparing the flux out of the pore at $x = L$, J_L, with the corresponding flux for pure diffusion, which is simply $J_0 = D_AA_0/L$. One finds that [549]

$$\mathcal{S} := \frac{J_L}{J_0} = L(1 + \Gamma_0)\left(L + \frac{2\Gamma_0}{\lambda}\tanh(\lambda L/2)\right)^{-1}, \tag{6.8.38}$$

with

$$\Gamma_0 = \left[\frac{D_A\lambda^2}{k_{off}KS_0} - 1\right]^{-1} = \frac{S_0k_{on}D_B}{k_{off}D_A}. \tag{6.8.39}$$

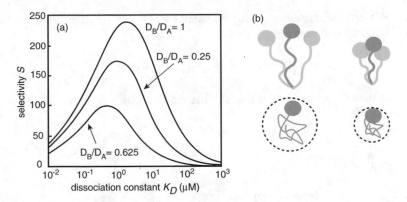

Fig. 6.48: Binding-diffusion model of NPC selectivity. (a) Sketch of selectivity curves as a function of the dissociation constant K_D for different diffusivity ratios D_B/D_A based on numerical simulations of equation (6.8.32). The selectivity $S = 1$ when $D_B = 0$. The linearized model only holds for $K_D > 1$. (b) Schematic illustration of the flexible tether model of bound state diffusion. Shorter tethers more severely constrain the motion of the bound particle [Adapted from Maguire et al. [549].]

Note that in the dilute regime, the selectivity is independent of the exterior concentration A_0. Moreover, in the case of immobile bound complexes, $D_B \to 0$, we have $\Gamma_0 \to 0$ and $S \to 1$. Hence, in order to have facilitated transport, the bound complexes need to be mobile. In Ref. [549], analysis in the linear regime is supplemented by numerical simulations in the full nonlinear regime for which $S \neq S_0$. A sketch of how the selectivity varies with the dissociation constant K_D is shown in Fig. 6.48(a).

The value of the diffusivity D_B of a complex in the bound state will depend on the molecular details of the NPC medium. Here we will follow Ref. [549] by considering the mobility of a complex while bound to a flexible polymer tether. The FG-nups are treated as entropic springs that constrain the motion of the complex, as illustrated in Fig. 6.48(b). More specifically, each bound molecule is modeled as a Brownian particle diffusing in a harmonic potential well:

$$dX = -\frac{kX(t)}{\gamma} + \sqrt{2D_A}dW(t), \tag{6.8.40}$$

where $X(t)$ is the displacement of the complex relative to the center of the well, $W(t)$ is a Wiener process, and γ is a friction coefficient with $D_A\gamma = k_BT$. The spring constant k is estimated by treating the polymer as a worm-like chain (Chap. 12), that is, $k = 3k_BT/2\ell_pL_c$, where ℓ_p is the persistence length of the polymer and L_c is its contour length. Setting $p(x,t)dx = \mathbb{P}[x < X(t) < x+dx]$ and solving the corresponding FP equation with $X(0) = 0$ show that

$$p(x,t) = \frac{1}{\sqrt{2\pi\alpha(t)}}e^{-x^2/2\alpha(t)}, \quad \alpha(t) = \frac{1 - e^{-2kD_At/k_BT}}{k/k_BT}. \tag{6.8.41}$$

It follows that the mean-square displacement (MSD) of a particle that is bound to the FG repeat up to time t is

$$\langle X^2(t) \rangle = \int_{-\infty}^{\infty} p(x,t)x^2 dx = \alpha(t).$$

If we then take into account the fact that the probability density of being bound for a time t is

$$\rho(t) = \tau^{-1}e^{-t/\tau}, \quad \tau = \frac{1}{k_{\text{off}}},$$

the effective MSD is

$$\overline{\langle X^2 \rangle} = \int_0^{\infty} \rho(t')\langle X^2(t') \rangle dt' = \frac{2D_A L_c \ell_p}{L_c \ell_p k_{\text{off}} + 3D_A}. \tag{6.8.42}$$

We have used the explicit formula for the spring constant k. The diffusivity D_B is then estimated by treating the confined diffusion over the mean bound time τ as Fickian, that is,

$$D_B = \frac{\overline{\langle X^2 \rangle}}{2\tau} = \frac{2D_A L_c \ell_p k_{\text{off}}}{L_c \ell_p k_{\text{off}} + 3D_A} = \frac{D_A}{1 + 3D_A/D_p}, \tag{6.8.43}$$

where $D_p = L_c \ell_p k_{\text{off}}$ quantifies how the physical properties of the polymer determine the diffusivity in the bound state. In particular, mobility in the bound state increases with increasing chain length L_c or persistence length ℓ_p and decreases with increasing binding lifetime k_{off}^{-1}. For large D_p it can be seen that D_B approaches the free mobility D_A, whereas $D_B \ll D_A$ when D_p is small.

In addition to tethered diffusion, mobility in the bound state can arise from multivalent interactions that allow transfer of particles between polymer chains while remaining bound [672, 801, 817]. For example, a particle may bind simultaneously to more than one FG repeat, moving hand-over-hand while remaining bound. Thus particles may slide between nearby FG sites rather than fully unbinding and rebinding. The effects of interchain transfer can also be included into the binding-diffusion model [549]. Again consider a particle moving in a harmonic potential according to equation (6.8.40). An interchain transfer probability can be constructed by taking the probability $P(t)\Delta t$ that a transfer occurs in the time interval $[t, t + \Delta]$ satisfies detailed balance. That is,

$$P(t) = r_1 M e^{-\Delta G(t)/2k_B T},$$

where r_1 is a background transfer rate, M is the number of FG repeats in a neighborhood of the bound tether, and $\Delta G(t)$ is the change in free energy associated with the transfer to a new chain:

$$\Delta G(t) = \frac{k}{2}(X(t) - x_{\text{new}}) - \frac{k}{2}(X(t) - x_{\text{cur}}),$$

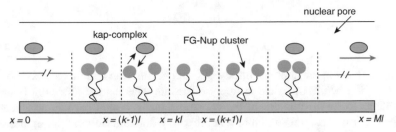

Fig. 6.49: Stochastically gated diffusion model of selective transport in a nuclear pore. Each particle (kap complex) switches between a freely diffusing state ($n = 0$) and a bound state ($n = 1$) in which it is tethered to an FG-nup cluster and is confined to a domain of size L.

where x_{new} and x_{cur} are the locations of the anchor points of the new and current FG-nups. Numerical simulations indicate that the inclusion of interchain transfer can enhance selectivity by up to a factor of six in the case of tight binding and short chains; weaker binding and longer chains leads to a more moderate increase in selectivity [549].

Finally, note that if $D_A = D_B$, then equation (6.8.33) has the same form as the steady-state version of the first moment equations (6.6.15a) and (6.6.15b) for stochastically gated diffusion through gap junctions. However, the boundary conditions are different as is the length scale. Another major difference is that here particles independently switch states rather than the environment, so there are no multi-particle correlations. Nevertheless, this connection has motivated a stochastically gated diffusion model of selective nuclear transport [137], as illustrated in Fig. 6.49. The basic assumption of the model is that the FG-nups are distributed in clusters in such a way that they define confinement domains. The diffusion of a particle when bound to a tether is thus taken to be spatially confined rather than simply reduced (at least in the absence of sliding). The 1D pore of length L is now partitioned into M domains of size $l = L/M$. Let $x \in [(j-1)l, jl]$, $j = 1, \ldots, M$, denote the spatial coordinates of the jth domain. Suppose that a cluster of FG-nups is tethered at the center $x = (j - 1/2)L$ of each domain $j = 1, \ldots, M$, which acts as a transient, partially mobile trap for the diffusing particles. This has two consequences. First, a particle can bind to an FG-nup anywhere in a given domain. For simplicity, the binding rate is taken to be independent of the current location of the head of an FG-nup, which is reasonable if the effective reaction radius is sufficiently large. Thus one can treat the binding sites within a domain as spatially uniform. Second, in the bound state the complex continues to diffuse within the domain, but cannot escape it. That is, each cluster defines a confinement domain of size l. In the dilute concentration regime, the steady-state equation (6.8.33) and exterior boundary conditions still hold with $D_A = D_B$. However, they are now supplemented by the following interior boundary conditions:

$$[A(x)]_{x=l_j^-}^{x=l_j^+} = 0, \quad [\partial_x A(x)]_{x=l_j^-}^{x=l_j^+} = 0, \tag{6.8.44a}$$

$$\partial_x B(l_j^-) = 0 = \partial_x B(l_j^+). \tag{6.8.44b}$$

Equation (6.8.44a) ensures continuity of the concentration and flux across the point $x = l_j = jl$ when the particle is freely diffusing, whereas the reflecting boundary conditions (6.8.44b) implement the constraint that a bound particle attached to a tethered FG-nup cannot switch to a neighboring domain. Solving the model equations in each confinement domain as in the binding-diffusion model and matching boundary conditions, one finds that the selectivity becomes [137]

$$S = \frac{L}{M}(1 + \Gamma_0) \left(\frac{L}{M} + \frac{2\Gamma_0}{\lambda} \tanh(\lambda L/2M) \right)^{-1}, \tag{6.8.45}$$

with Γ_0 given by equation (6.8.39) with $D_A = D_B$. It can thus be established that increasing M has a similar effect to reducing D_B.

6.9 Diffusive transport on a Cayley tree

In Sect. 6.2 we considered the 1D diffusive transport of proteins in the membrane of a dendrite. It is well known that the dendrites of a neuron form an intricate branching structure, as illustrated in Fig. 3.10. This is one biological motivation for considering diffusion on branching structures. Tree-like topologies are also of interest from a mathematical perspective, since they are simpler to analyze compared to a study of the same process defined on a regular lattice. This permits investigations of generic features of interest that can also, in certain cases, be directly relevant to the regular lattice problem in some appropriate limit. For example, it is well known that Cayley trees and Bethe lattices provide insights into the behavior of various processes on both infinite-dimensional lattices and finite-dimensional lattices in the mean-field limit [56, 898]. It turns out that solving the diffusion equation on a tree is much easier in Laplace space, since iterative methods can be used. In this final section, we illustrate the basic method by considering advection–diffusion on a semi-infinite Cayley tree Γ with coordination number z. The example of $z = 3$ is shown in Fig. 6.50(a). We will assume for simplicity that the drift velocity v, diffusivity D, and branch length L are identical throughout the tree so that we can exploit the recursive nature of the infinite tree. Suppose that there is an absorbing boundary at the terminal (primary) node of the tree. We will calculate the Laplace transform of the flux through the target using the iterative method introduced in Ref. [608].

Denote the first branch node opposite to the terminal node by α_0. For every other branching node $\alpha \in \Gamma$ there exists a unique direct path from α_0 to α (one that does not traverse any line segment more than once). We can label each node $\alpha \neq \alpha_0$ uniquely by the index k of the final segment of the direct path from α_0 to α so that the branch node corresponding to a given segment label k can be written $\alpha(k)$.

We denote the other node of segment k by $\alpha'(k)$. Taking the primary branch to be $k = 0$, it follows that $\alpha(0) = \alpha_0$ and $\alpha'(0)$ is the terminal node. We also introduce a direction on each segment of the tree such that every direct path from $\alpha'(0)$ always moves in the positive direction. Consider a single branching node $\alpha \in \Gamma$ and label the set of segments radiating from it by $\mathcal{I}_\alpha$. Let $\bar{\mathcal{I}}_\alpha$ denote the set of $z - 1$ line segments $k \in \mathcal{I}_\alpha$ that radiate from α in a positive direction, see Fig. 6.50(b). Using these various definitions we can introduce the idea of a generation. Take α_0 to be the zeroth generation. The first generation then consists of the set of nodes $\Sigma_1 = \{\alpha(k), k \in \bar{\mathcal{I}}_{\alpha_0}\}$, the second generation is $\Sigma_2 = \{\alpha(l), l \in \bar{\mathcal{I}}_\alpha, \alpha \in \Sigma_1\}$, etc.

Denote the position coordinate along the ith line segment by x, $0 \leq x \leq L$, where L is the length of the segment and $0 \leq i < \infty$. Given the above labeling scheme for nodes, we take $x(\alpha'(i)) = 0$ and $x(\alpha(i)) = L$. Let u_i denote the probability density on the ith segment, which evolves according to the Fokker–Planck equation

$$\frac{\partial p_i}{\partial t} = D\frac{\partial^2 p_i}{\partial x^2} - v\frac{\partial p_i}{\partial x}, \quad 0 < x < L. \tag{6.9.1}$$

As a further simplification, we assume that the searcher initiates its search on the primary branch so that

$$p_i(x,0) = \delta(x - x_0)\delta_{i,0}, \quad 0 < x_0 < L. \tag{6.9.2}$$

(a)

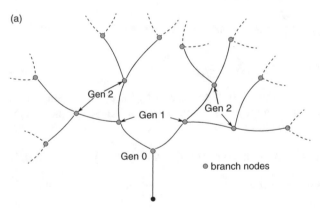

Gen 2

Gen 1

Gen 2

Gen 0

● branch nodes

(b)

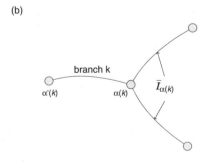

branch k

$\alpha'(k)$ $\alpha(k)$ $\bar{\mathcal{I}}_{\alpha(k)}$

Fig. 6.50: (a) Semi-infinite Cayley tree with coordination number $z = 3$ and an absorbing boundary at the primary node. (b) A branch node $\alpha(k)$ is shown in relation to the neighboring branch node $\alpha'(k)$ closest to the primary node. The branch segments extending out from $\alpha(k)$ in the positive direction together comprise the set $\bar{\mathcal{I}}_{\alpha(k)}$.

This initial condition means that all branches of a given generation are equivalent. Let $\mathcal{J}[p_i]$ denote the corresponding probability current or flux, which is taken to be positive in the direction flowing away from the primary node at the soma:

$$\mathcal{J}[p] \equiv -D\frac{\partial p}{\partial x} + vp. \tag{6.9.3}$$

The open boundary condition on the primary branch is $p_0(0,t) = 0$. At all branch nodes $\alpha \in \Sigma_n$ of the nth generation we impose the continuity conditions

$$p_i(x(\alpha),t) = \Phi_n(t), \quad \text{for all } i \in \mathcal{I}_\alpha, \quad \alpha \in \Sigma_n, \tag{6.9.4}$$

where the $\Phi_n(t)$ are unknown functions, which will ultimately be determined by imposing current conservation at each branch node:

$$\sum_{i \in \mathcal{I}_\alpha} \mathcal{J}[p_i](x(\alpha),t) = 0. \tag{6.9.5}$$

Note that for the upstream segment $j \notin \overline{\mathcal{I}}_\alpha$, $x(\alpha) = L$ and the corresponding flux $\mathcal{J}[p_j](L,t)$ flows into the branch node, whereas for the remaining $z-1$ downstream segments $k \in \overline{\mathcal{I}}_\alpha$ we have $x(\alpha) = 0$ and $\mathcal{J}[p_k](0,t)$ flows out of the branch node.

After Laplace transforming the above system of equations on the Cayley tree, we obtain the following system of equations for any i such that $\alpha(i) \in \Sigma_n$:

$$\left[D\frac{\partial^2}{\partial x^2} - v\frac{\partial}{\partial x} - s\right]\widetilde{p}_i(x,s) = -\delta_{i,0}\delta(x-x_0), \tag{6.9.6a}$$

$$\widetilde{p}_i(0,t) = \widetilde{\Phi}_{n-1}(s), \quad \widetilde{p}_i(L,t) = \widetilde{\Phi}_n(s). \tag{6.9.6b}$$

Note that $\widetilde{\Phi}_{-1} = 0$. The solution in each branch is given by the corresponding finite interval Green function $\mathcal{G}$ with homogeneous boundary conditions supplemented by terms satisfying the boundary conditions. That is,

$$\widetilde{p}_i(x,s) = -\delta_{i,0}\mathcal{G}(x,x_0;s) + \widetilde{\Phi}_{n-1}(s)\widehat{F}(x,s) + \widetilde{\Phi}_n(s)F(x,s) \tag{6.9.7}$$

for $\alpha(i) \in \Sigma_n$. The Green function $\mathcal{G}$ satisfies

$$\left[D\frac{\partial^2}{\partial x^2} - v\frac{\partial}{\partial x} - s\right]\mathcal{G}(x,y;s) = \delta(x-y), \tag{6.9.8}$$

with $\mathcal{G}(0,y;s) = 0 = \mathcal{G}(L,y;s)$. The Green function is given by (see Box 6C)

$$\mathcal{G}(x,y;s) = \begin{cases} \dfrac{\psi(x,s)\psi(y-L,s)}{DW(s)}, & 0 \leq x \leq y \\[4mm] \dfrac{\psi(x-L,s)\psi(y,s)}{DW(s)}, & y \leq x \leq L \end{cases}, \tag{6.9.9}$$

where

$$\psi(x,s) = e^{\mu_+(s)x} - e^{\mu_-(s)x}, \quad \mu_\pm = \frac{v \pm \sqrt{v^2 + 4Ds}}{2D}, \tag{6.9.10}$$

and W is the Wronskian

$$W(s) = \psi(y,s)\psi'(y-L,s) - \psi'(y,s)\psi(y-L,s), \tag{6.9.11}$$

which is independent of y. The functions $F(x,s)$ and $\widehat{F}(x,s)$ satisfy the homogeneous version of Eq. (6.9.6) with boundary conditions $F(0,s) = 0, F(L,s) = 1$ and $\widehat{F}(0,s) = 1, \widehat{F}(L,s) = 0$:

$$F(x,s) = \frac{\psi(x,s)}{\psi(L,s)}, \quad \widehat{F}(x,s) = \frac{\psi(x-L,s)}{\psi(-L,s)}. \tag{6.9.12}$$

In the following we drop the tildes on the Φ_n. The unknown functions Φ_n are determined by imposing the current conservation condition (6.9.5) at each branch node and using the identity $\mathcal{J}[\Phi_n F] = \Phi_n \mathcal{J}[F]$, which follows from the observation that Φ_n is x-independent. At the zeroth generation node α_0, the current conservation equation is given by (suppressing the s variable)

$$\Phi_0 \mathcal{J}[F](L) = (z-1)\Phi_1 \mathcal{J}[F](0) + (z-1)\Phi_0 \mathcal{J}[\widehat{F}](0) + \mathcal{X}_0,$$

where

$$\mathcal{X}_0 \equiv \mathcal{J}[\mathcal{G}](L), \tag{6.9.13}$$

and at all branching nodes $\alpha \in \Sigma_n, 1 \leq n$ we have

$$\Phi_{n-1} \mathcal{J}[\widehat{F}](L) + \Phi_n \mathcal{J}[F](L) = (z-1)\Phi_{n+1} \mathcal{J}[F](0) + (z-1)\Phi_n \mathcal{J}[\widehat{F}](0). \tag{6.9.14}$$

Note that $\mathcal{X}_0$ depends on the source location x_0 through its dependence on finite interval Green's function $\mathcal{G}$; this then generates an x_0 dependence of the functions Φ_α. The four possible contributions to the probability flux at any branch $k \neq 0$ are

$$g(s) \equiv \mathcal{J}[\widehat{F}](L) = \frac{\psi'(0,s)}{\psi(-L,s)} = \frac{D\eta(s)e^{vL/2D}}{\sinh(\eta(s)L)}, \tag{6.9.15a}$$

$$h(s) \equiv \mathcal{J}[F](L) = \frac{\psi'(L,s)}{\psi(L,s)} = -D\eta(s)\coth(\eta(s)L) + \frac{v}{2}, \tag{6.9.15b}$$

$$\bar{g}(s) \equiv \mathcal{J}[F](0) = \frac{\psi'(0,s)}{\psi(L,s)} = -\frac{D\eta(s)e^{-vL/2D}}{\sinh(\eta(s)L)}, \tag{6.9.15c}$$

$$\bar{h}(s) \equiv \mathcal{J}[\widehat{F}](0) = \frac{\psi'(-L,s)}{\psi(-L,s)} = D\eta(s)\coth(\eta(s)L) + \frac{v}{2}, \tag{6.9.15d}$$

where

$$\eta(s) = \frac{\sqrt{v^2 + 4Ds}}{2D}.$$ (6.9.16)

Using these definitions the current conservation equations simplify to

$$-H\Phi_0 + G\Phi_1 = \mathcal{X}_0$$ (6.9.17)

for the first branch, and

$$g\Phi_{n-1} - H\Phi_n + G\Phi_{n+1} = 0$$ (6.9.18)

for $n > 0$, where

$$H = (z-1)\bar{h} - h, \quad G = -\bar{g}(z-1).$$ (6.9.19)

The second-order difference equation can be solved using the ansatz $\Phi_n = \lambda^n \Phi_0$, which yields a quadratic equation for λ:

$$G\lambda^2 - H\lambda + g = 0.$$ (6.9.20)

Taking the smaller root, in order to ensure that the solution is normalizable, we have

$$\lambda = \frac{1}{2G}\left[H - \sqrt{H^2 - 4gG}\right].$$ (6.9.21)

It can be checked that λ is real since

$$H^2 - 4gG > \frac{z^2(D\eta)^2\cosh^2(\eta L)}{\sinh^2(\eta L)} - \frac{4(z-1)(D\eta)^2}{\sinh^2(\eta L)}$$

$$= \frac{(D\eta)^2}{\sinh^2(\eta L)}\left(z^2\cosh^2(\eta L) - 4(z-1)\right) > \frac{(D\eta)^2}{\sinh^2(\eta L)}(z-2)^2 > 0.$$

In addition, $\lambda < 1$ since this inequality is equivalent to the condition

$$H - \sqrt{H^2 - 4gG} < 2G,$$

which can be rewritten as

$$(H - 2G)^2 < H^2 - 4gG \implies g + G < H,$$

and $g + G < H$. We now substitute the solution for Φ_1 into Eq. (6.9.17) and rearrange to obtain the following expression for Φ_0:

$$\Phi_0 = \frac{\mathcal{X}_0}{\lambda G - H}.$$ (6.9.22)

Having determined Φ_0, the Laplace transform of the flux into the terminal node takes the explicit form

$$\tilde{J}(x_0, s) = D\frac{\partial \tilde{p}_0}{\partial x}(0, s|x_0) = -D\frac{\partial \mathcal{G}(x, x_0; s)}{\partial x}\bigg|_{x=0} + D\,\Phi_0(x_0, s)\frac{\partial F(x, s)}{\partial x}\bigg|_{x=0}.$$

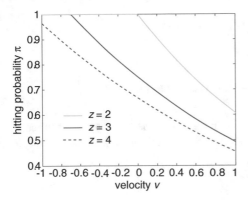

Fig. 6.51: Advection–diffusion on a semi-infinite Cayley tree. Plots of hitting probability π as a function of velocity v for different coordination numbers z. Other parameters are $D = 1$ and $L = 1$.

The function $\widetilde{J}(x_0, s)$ determines the hitting probability that the particle is eventually absorbed at the terminal node, having started at x_0. That is, $\pi = \widetilde{J}(0)$. Note that unbiased one-dimensional diffusion is transient in one dimension ($z = 2$), see Sect. 2.1, and on a Cayley tree. This means that $\pi < 1$ for $v \geq 0$. On the other hand, there exists a critical negative (inward) velocity v_c, $v_c < 0$, such that $\pi = 1$ for $v < v_c$. In other words, there is a critical phase transition from recurrent to transient transport. This transition point is identical to the so-called localization–delocalization threshold [110]. In order to define the latter, suppose that the particle starts at the terminal node, $p_i(x, 0) = \delta_{i,0}\delta(x)$. The initially localized probability density will tend to diffuse away from that origin, but this is counteracted by an inward velocity field on the tree. If, in steady state, the concentration remaining at the origin has not decayed to zero, we say the system is localized, otherwise it is delocalized. By studying the steady-state solution it can be shown that $v_c = -\ln(z - 1)$. Hence $v_c = 0$ for $z = 2$ (a semi-infinite line), $v_c \approx -0.69$ for $z = 3$, and $v_c \approx -1.1$ for $z = 4$. The existence of the recurrent–transient transition for the hitting probability is confirmed in Fig. 6.51. Note that for $v > v_c$, the hitting probability is a decreasing function of both v and z.

6.10 Exercises

Problem 6.1. CTRW and the generalized master equation. Consider the generalized master equation

$$\frac{dQ(\ell, t)}{dt} = \int_0^t \phi(t - t') \sum_{\ell'} [p(\ell|\ell')Q(\ell', t') - p(\ell'|\ell)Q(\ell, t')]dt',$$

where $\phi(t)$ is a memory kernel and $\sum_{\ell'} p(\ell'|\ell) = 1$. Laplace transforming both sides of this equation under the initial condition $Q(\ell, 0) = \delta_{\ell, 0}$ and rearranging, show that

$$\tilde{Q}(\ell,s) = [s + \tilde{\phi}(s)]^{-1} \Gamma(\ell, \xi(s)), \quad \xi(s) = \frac{\tilde{\phi}(s)}{s + \tilde{\phi}(s)},$$

where Γ satisfies the equation

$$\Gamma(\ell, z) = \delta_{\ell,0} + z \sum_{\ell'} p(\ell|\ell') \Gamma(\ell', z).$$

From the analysis of the CTRW (6.1.2), we recognize Γ as the generating function $\Gamma(\ell, z) = \sum_{n \geq 0} z^n P_n(\ell)$ of the solution to the discrete random walk master equation

$$P_n(\ell) = \sum_{\ell'} p(\ell|\ell') P_{n-1}(\ell').$$

Hence, deduce that $\tilde{Q}(\ell, s)$ is identical to the Laplace transform $\tilde{R}(\ell, s)$ of the solution to the CTRW, see equation (6.1.7) provided that the waiting time density $\psi(t)$ and the memory kernel $\phi(t)$ are related according to

$$\tilde{\psi}(s) = \frac{\tilde{\phi}(s)}{s + \tilde{\phi}(s)}.$$

Problem 6.2. 1D homogenization. Following along analogous lines to Sect. 6.1.3, homogenize the steady-state equation

$$\frac{d^2 u}{dx^2} + g(x/\varepsilon) u = 0,$$

with $0 < \varepsilon \ll 1$ and $g(y)$ a periodic function of unit period. That is, introduce the fast spatial variable $y = x/\varepsilon$ and set

$$u = u_0(x) + \varepsilon u_1(x, y) + \varepsilon^2 u_2(x, y) + \ldots, \quad \frac{d}{dx} = \frac{\partial}{\partial x} + \frac{1}{\varepsilon} \frac{\partial}{\partial y},$$

where $u_n(x, y)$ for $n \geq 1$ are periodic functions of y. Substitute into the ODE and collect terms at each order in ε.

(a) By considering the $O(\varepsilon^{-1})$ equation, show that u_1 is independent of y (and can thus be absorbed into u_0). Use the $O(1)$ equation to derive the averaged equation

$$\frac{d^2 u_0}{dx^2} + \bar{g} u_0 = 0, \quad \bar{g} = \int_0^1 g(y) dy.$$

(b) Subtracting the result of part (a) from the $O(1)$ equation, show that u_2 can be decomposed as

$$u_2(x, y) = \chi(y) u_0(x),$$

with

$$\frac{d^2\chi}{dy^2} = \bar{g} - g(y).$$

Integrating the latter equation twice with respect to y and exploiting periodicity, show that the $O(\varepsilon^2)$ contribution can be written in the form

$$u_2(x,y) = \overline{u_2}(x) + \left[\int_0^y H(z)dz - \overline{\int_0^y H(z)dz}\right] u_0(x).$$

with

$$H(y) = \int_0^y [\bar{g} - g(z)]dz - \overline{\int_0^y [\bar{g} - g(z)]dz}.$$

Finally, derive an ODE for $\overline{u_2}(x)$ by averaging the $O(\varepsilon)$ equation with respect to y. The latter determines the $O(\varepsilon^2)$ correction to the homogenization of the distribution $g(x/\varepsilon)$.

Problem 6.3. Percolation theory on a Bethe lattice. Consider site percolation on a Bethe lattice, which is an infinite-dimensional lattice. Each site has z neighboring sites and hence each branch gives rise to $z-1$ other branches, see Fig. 6.52. Choose an arbitrary point on the lattice as the center, and let g denote the generation (distance from the center site).

(a) Show that the total number of sites in the first g generations (including the center) is

$$n_g = \frac{z(z-1)^g - 2}{z-2}.$$

What is the number of surface sites S_g (sites of the gth generation)? Show that

$$\frac{S_g}{n_g} \to \frac{z-2}{z-1} \text{ as } g \to \infty.$$

Contrast this with the surface area to volume ratio for a cube of linear size L as $L \to \infty$.

(b) Starting from the center site and going outwards, determine the probability that at each step there exists an occupied site to continue the cluster. Use this to show that the percolation threshold is

$$p_c = \frac{1}{z-1}.$$

(c) Take $z = 3$. Let Q be the probability that an arbitrary site is not connected to infinity via a given branch originating from that site. Show that the percolation strength can be written as

$$P(p) = p(1 - Q^3).$$

An equation for Q can be derived by conditioning on whether or not the neighboring site on the given branch is occupied:

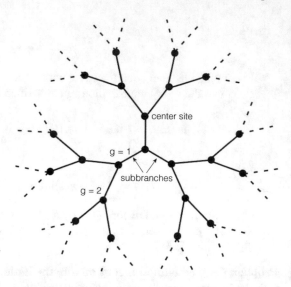

g = 1

subbranches

g = 2

Fig. 6.52: Bethe lattice of coordination number $z = 3$.

$$Q = \mathbb{P}(\text{neighboring site is empty})$$
$$+ \mathbb{P}(\text{neighboring site is occupied})\mathbb{P}(\text{no subbranch leads to infinity}).$$

Use this to derive a quadratic equation for Q and thus show that

$$P(p) = \begin{cases} 0 & \text{for } p < p_c \\ p\left(1 - \left[\frac{1-p}{p}\right]^3\right) & \text{for } p \geq p_c \end{cases}.$$

Finally, Taylor expanding $P(p)$ around $p = p_c = 1/2$, show that $P(p) \sim (p - p_c)$ for $p \to p_c^+$.

Problem 6.4. Steady-state analysis of a diffusion–trapping model. Consider a diffusion–trapping model in a domain of length L, with a set of N traps at locations x_j, $j = 1, \ldots, N$. Let $u(x,t)$ be the concentration of receptors within the domain, which evolves according to the equation

$$\frac{\partial u}{\partial t} = D\frac{\partial^2 u}{\partial x^2} - \beta \sum_{j=1}^{N} u(x_j,t)\delta(x - x_j).$$

Suppose that newly synthesized receptors enter the end $x = 0$ at a rate σ and that the other end is reflecting. The boundary conditions are thus

$$D\frac{\partial u}{\partial x}\bigg|_{x=0} = -\sigma, \quad D\frac{\partial u}{\partial x}\bigg|_{x=L} = 0.$$

(a) Using the boundary conditions, derive the steady-state conservation condition

$$\sigma = \beta \sum_{j=1}^{N} u(x_j).$$

(b) Introducing Neumann Green's function

$$\frac{d^2 G(x,y)}{dy^2} = -\delta(x-y) + L^{-1},$$

with reflecting boundary conditions at the ends $x = 0, L$, derive the following implicit equation

$$u(x) = \chi - \beta \sum_j G(x,x_j)u_j + \frac{\sigma}{D}G(x,0),$$

where $u_j = u(x_j)$. The constant χ is determined from the conservation condition $\sigma = \beta \sum_{j=1}^{N} p_j$.

(c) Using the theory of 1D Green's functions, derive the formula

$$G(x,x') = \frac{L}{12}\left[h([x+x']/L) + h(|x-x'|/L)\right],$$

where $h(x) = 3x^2 - 6|x| + 2$.

(d) By setting $x = x_i$ in the result of part (b), show that

$$u_i = \sum_j M_{ij}^{-1}(\chi + \sigma G(x_j,0)/D), \quad M_{ij} = G(x_i,x_j) + \delta_{ij}.$$

Use this to determine χ.

(e) Consider a set of N identical traps with uniform spacing $d = L/N$ such that $x_j = jd$, $j = 1,\ldots,N$. Using the results of parts (b) and (c), plot $u(x)$ as a function of x for the three cases $D = 0.5, 0.1, 0.05\,\mu m^2/s$, with $L = 200\,\mu m$, $N = 200$, trap spacing $d = 1\,\mu m$, $\sigma = 1\,s^{-1}$, and $\beta = 10^{-4}\,\mu m/s$.

(f) Consider a set of N identical traps with uniform spacing $d = L/N$ such that $x_j = jd$, $j = 1,\ldots,N$, and rewrite the steady-state equation in the form

$$D\frac{d^2 u}{dx^2} - \beta \rho(x)u = 0,$$

where $\rho(x) = \sum_{j=1}^{N} \delta(x - jd)$. In the large N limit, take $\rho = 1/d$ and solve the resulting continuum equation. Compare with the results of part (d). (Note that for finite N, the periodic density of spines $\rho(x)$ can be homogenized using the analysis of Ex. 6.2.)

Problem 6.5. Protein search for a target on DNA. Fill in the details of the calculation of the Laplace transformed FPT density $\widehat{F}_M(s)$ for a protein to find a target site on a DNA chain, given that it started in solution, see Sect. 6.3.2. First, use Box 5G to show that equation (6.3.13) is the appropriate backward equation. Then provide the steps in the derivation of equations (6.3.16), (6.3.18), (6.3.19), and (6.3.21).

Problem 6.6. Effects of ligand rebinding. Extend the analysis of the two-site ir-reversible reaction scheme (6.3.39) to the reversible reaction scheme

$$S_0 + E \underset{\beta_0}{\overset{\kappa_0}{\rightleftharpoons}} S_1 + E \underset{\beta_1}{\overset{\kappa_1}{\rightleftharpoons}} S_2 + E.$$

Assume that initially all substrates are in the completely unmodified state S_0. First write down the classical mass-action kinetics in the matrix form (6.3.41) with the matrix $\mathbf{K}$ modified to include the reverse reactions: $\mathbf{K} \to \mathbf{K}_r$ with

$$\mathbf{K}_r = \begin{pmatrix} \kappa_0 & -\beta_0 & 0 \\ -\kappa_0 & \beta_0 + \kappa_1 & -\beta_1 \\ 0 & -\kappa_1 & \beta_1 \end{pmatrix}.$$

The analysis proceeds as before, leading to equations (6.3.52) and (6.3.53), with the columns of $\mathbf{T}^{-1}$ constructed from the eigenvectors of $\mathbf{K}_r$, and λ_i given by the eigenvalues of $\mathbf{K}_r$. Write down the modified kinetic equations and construct the associated reaction diagram.

Problem 6.7. Asymptotic expansion of splitting probabilities in 2D. Consider the BVP for the splitting probabilities of the 2D narrow capture problem with N small interior traps:

$$\nabla^2 \pi_k(\mathbf{x}) = 0, \ \mathbf{x} \in \mathcal{U} \backslash \mathcal{U}_a; \quad \partial_n \pi_k(\mathbf{x}) = 0, \ \mathbf{x} \in \partial \mathcal{U},$$

with

$$\pi_k(\mathbf{x}) = \delta_{j,k}, \ \mathbf{x} \in \partial \mathcal{U}_j.$$

(a) Introducing the stretched coordinates $\mathbf{y} = \varepsilon^{-1}(\mathbf{x} - \mathbf{x}_j)$ around the jth trap and setting $u_{k,j}(\mathbf{y}) = \pi_k(\mathbf{x}_j + \varepsilon \mathbf{y})$, show that the inner solution takes the form

$$u_{k,j} = \delta_{j,k} + v A_{k,j}(v) \ln|\mathbf{y}| + \dots$$

where $v = -1/\ln \varepsilon$ and $A_{k,j}$ is an unknown constant. (Higher-order terms are assumed to be beyond-all-orders with respect to the logarithmic terms.)

(b) Show that the outer solution can be written in terms of modified Green's function as

$$\pi_k(\mathbf{x}) = -2\pi \sum_{j=1}^{N} v A_{k,j}(v) G(\mathbf{x}, \mathbf{x}_j) + \chi_k,$$

with

$$\pi_k(\mathbf{x}) \sim \delta_{j,k} + v A_{k,j}(v) \ln|\mathbf{x} - \mathbf{x}_j|/d_j \varepsilon, \quad \text{as } \mathbf{x} \to \mathbf{x}_j.$$

Also show that

$$\chi_k = |\mathcal{U}|^{-1} \int_{\mathcal{U}} \pi_k(\mathbf{x}) d\mathbf{x}, \quad \sum_{i=1}^{N} A_i(v) = 0.$$

(c) Matching the far-field behavior of the inner solution $u_{k,j}$ with the near-field behavior of the outer solution π_k in a neighborhood of $\mathcal{U}_j$ for $j = 1, \dots, N$, derive the N conditions

$$-2\pi \sum_{i \neq j} v A_{k,i}(v) G(\mathbf{x}_i, \mathbf{x}_j) - A_{k,j}(v) - v A_{k,j}(v)[2\pi R(\mathbf{x}_j, \mathbf{x}_j) - \ln d_j] + \chi_k = \delta_{j,k}.$$

Solving for $A_{k,j}(v)$ to $o(v)$ and imposing the condition $\sum_j A_j = 0$, show that the outer solution has the asymptotic expansion

$$\pi_k(\mathbf{x}) \sim \frac{1}{N} - 2\pi v \left[\frac{1}{N} \sum_{j=1}^{N} G(\mathbf{x}, \mathbf{x}_j) - G(\mathbf{x}, \mathbf{x}_k) \right] + \frac{2\pi v}{N} \sum_{i,j=1}^{N} (N^{-1} - \delta_{i,k}) \mathcal{G}_{ij} + o(v).$$

Problem 6.8. Diffusion in a bounded 2D domain with N circular traps. Consider a bounded domain $\mathcal{U}$ with N small circular, partially absorbing traps of radius ε within the interior of the domain, see Fig. 6.20. The steady-state diffusion problem is

$$\nabla^2 u(\mathbf{x}) = 0 \quad \mathbf{x} \in \mathcal{U} \setminus \cup_{j=1}^{N} \mathcal{U}_j, \quad \partial_n u = 0 \text{ on } \partial \mathcal{U}, \quad \varepsilon \partial_n u + \alpha_j u = 0 \text{ on } \partial \mathcal{U}_j.$$

Here $\mathcal{U}_j = \{\mathbf{x} \in \mathcal{U}; |\mathbf{x} - \mathbf{x}_j| \leq \varepsilon\}$ where $\mathbf{x}_j$ is the center of the jth hole. Derive an asymptotic expansion for u along the following lines.

(a) Introduce stretched coordinates $\mathbf{y} = \varepsilon^{-1}(\mathbf{x} - \mathbf{x}_j)$ around the jth trap and consider the inner solution $V(\mathbf{y}) = u(\mathbf{x}_j + \varepsilon \mathbf{y})$. Setting $V(\mathbf{y}) = U_j + v A_j(v) V_0(\mathbf{y}) + \cdots$, where U_j is to be determined, show that $V_0(\mathbf{y}) = \ln |\mathbf{y}|$. Use the partially absorbing boundary condition on $\partial \mathcal{U}_j$ to show that $v A_j(v) = \alpha_j U_j$.

(b) The outer solution $U(\mathbf{x})$ satisfies to leading order

$$\nabla_{\mathbf{x}}^2 U = 0 \text{ for } \mathbf{x} \in \mathcal{U} \setminus \{\mathbf{x}_1, \ldots, \mathbf{x}_N\}, \quad \partial_n U = 0 \text{ on } \partial \mathcal{U},$$

and is singular as $\mathbf{x} \to \mathbf{x}_j$. By matching with the far-field behavior of the inner solution using Neumann Green's function show that

$$U(\mathbf{x}) = \chi - 2\pi v \sum_{i=1}^{N} A_i(v) G(\mathbf{x}, \mathbf{x}_i), \quad \mathbf{x} \notin \{\mathbf{x}_1, \ldots, \mathbf{x}_N\},$$

for some constant χ, with the coefficients A_j satisfying the self-consistency conditions

$$-(1 + 2\pi v R(\mathbf{x}_j, \mathbf{x}_j)) A_j - 2\pi v \sum_{i \neq j} A_i G(\mathbf{x}_j, \mathbf{x}_i) = U_j - \chi, \quad j = 1, \ldots, N,$$

and $\sum_{j=1}^{N} A_j(v) = 0$.

(c) Construct formal matrix solutions for the unknown coefficients U_j and χ. Taylor expanding these non-perturbative solutions with respect to v, calculate the leading order expressions for U_j and χ.

Problem 6.9. Asymptotic expansion of the survival probability in 3D. Consider the survival probability $Q(\mathbf{x},t)$ of the 3D narrow capture problem with N small interior traps. Q evolves according to the backward diffusion equation

$$\frac{\partial Q(\mathbf{x},t)}{\partial t} = D\nabla^2 Q(\mathbf{x},t), \mathbf{x} \in \mathcal{U}\backslash\mathcal{U}_a, \quad \mathcal{U}_a = \bigcup_{j=1}^{N} \partial \mathcal{U}_j,$$

with a reflecting boundary condition on the exterior of the domain and absorbing boundary conditions on the target boundaries:

$$\partial_n Q(\mathbf{x},t) = 0, \quad \mathbf{x} \in \partial \mathcal{U}, \, Q(\mathbf{x},t) = 0 \quad \mathbf{x} \in \partial \mathcal{U}_a = \bigcup_{j=1}^{N} \partial \mathcal{U}_j.$$

(For notational convenience, we drop the subscript on the initial position $\mathbf{x}_0$.) The initial condition is $Q(\mathbf{x},0) = 1$. Laplace transforming the diffusion equation gives

$$D\nabla^2 \widetilde{Q}(\mathbf{x},s) - s\widetilde{Q}(\mathbf{x},s) = -1, \mathbf{x} \in \mathcal{U}\backslash\mathcal{U}_a$$

with the same boundary conditions. Consider the following expansion of the outer solution:

$$\widetilde{Q} = \frac{1}{s} + \varepsilon\widetilde{Q}_1 + \varepsilon^2\widetilde{Q}_2 + \dots$$

with (for $n = 0,1$)

$$D\nabla^2\widetilde{Q}_n - s\widetilde{Q}_n = 0, \mathbf{x} \in \mathcal{U}\backslash\{\mathbf{x}_1,\dots,\mathbf{x}_N\}, \, \partial_n\widetilde{Q}_n = 0, \mathbf{x} \in \partial\mathcal{U}.$$

Similarly, in the inner region around the jth target, introduce the stretched coordinates $\mathbf{y} = \varepsilon^{-1}(\mathbf{x} - \mathbf{x}_j)$ and set $q(\mathbf{y},s) = \widetilde{Q}(\mathbf{x}_j + \varepsilon\mathbf{y},s)$. Expand the inner solution as $q = q_0 + \varepsilon q_1 + \dots$ with

$$\nabla_\mathbf{y}^2 q_n = 0, \, \mathbf{y} \in \mathbb{R}^d\backslash\mathcal{U}_j \quad q_n(\mathbf{y},s) = 0, \, \mathbf{y} \in \partial\mathcal{U}_j.$$

Finally, impose the matching condition that the near-field behavior of the outer solution as $\mathbf{x} \to \mathbf{x}_j$ should agree with the far-field behavior of the inner solution as $|\mathbf{y}| \to \infty$, which is expressed as

$$\frac{1}{s} + \varepsilon\widetilde{Q}_1 + \varepsilon^2\widetilde{Q}_2 + \dots \sim q_0 + \varepsilon q_1 + \dots.$$

(a) Show that around the jth target,

$$q_0(\mathbf{y},s) = s^{-1}\left(1 - \frac{C_j}{|\mathbf{y}|}\right),$$

with C_j is the "capacitance" of the target. Hence, write down the singularity condition for $\widetilde{Q}_1$ as $\mathbf{x} \to \mathbf{x}_j$.

(b) Show that the solution for $\widetilde{Q}_1$ can be written as

$$\tilde{Q}_1(\mathbf{x},s) = -\frac{4\pi}{s} \sum_{k=1}^{N} C_k G(\mathbf{x},\mathbf{x}_k;\lambda_s),$$

where G is Green's function of the 3D modified Helmholtz equation,

$$\nabla^2 G(\mathbf{x},\mathbf{x}';\lambda_s) - \lambda_s^2 G(\mathbf{x},\mathbf{x}';\lambda_s) = -\delta(\mathbf{x}-\mathbf{x}'), \ \mathbf{x} \in \mathcal{U}; \ \partial_n G = 0, \ \mathbf{x} \in \partial\mathcal{U},$$

$$\partial_n G = 0, \ \mathbf{x} \in \partial\mathcal{U}, \ \int_\mathcal{U} G(\mathbf{x},\mathbf{x}';\lambda_s)d\mathbf{x} = \lambda_s^{-2}, \ G(\mathbf{x},\mathbf{x}';\lambda_s) = \frac{1}{4\pi|\mathbf{x}-\mathbf{x}'|} + R(\mathbf{x},\mathbf{x}';\lambda_s),$$

with $\lambda_s = \sqrt{s/D}$ and R the regular part of G.

(c) Matching q_1 with the near-field behavior of $\tilde{Q}_1(\mathbf{x},s)$ around the jth target, obtain the solution

$$q_1(\mathbf{y},s) \sim \frac{\chi_j}{s}\left(1 - \frac{C_j}{|\mathbf{y}|}\right),$$

with

$$\chi_j = -4\pi C_j R(\mathbf{x}_j,\mathbf{x}_j;\lambda_s) - 4\pi \sum_{k \neq j}^{N} C_k G(\mathbf{x}_j,\mathbf{x}_k;\lambda_s) = -4\pi \sum_{k=1}^{N} C_k \mathcal{G}_{jk}(s),$$

where $\mathcal{G}_{ij}(s) = G(\mathbf{x}_i,\mathbf{x}_j;\lambda_s)$ for $i \neq j$ and $\mathcal{G}_{ii}(s) = R(\mathbf{x}_i,\mathbf{x}_i;\lambda_s)$. Write down the resulting singularity condition for $\tilde{Q}_2$ and thus show that the outer solution takes the form

$$\tilde{Q}(\mathbf{x},s) = \frac{1}{s}\left[1 - 4\pi\varepsilon\sum_{k=1}^{N} C_k(1+\varepsilon\chi_k)G(\mathbf{x},\mathbf{x}_k;\lambda_s) + o(\varepsilon^2)\right].$$

Problem 6.10. Flux through a stochastically gated membrane channel. Derive the expression (6.6.46) for the steady-state flux through a stochastically gated membrane channel of length L and radius a.

(a) First integrate equation (6.6.44) with respect to $t \in [0,\infty)$ to obtain the following equations for $q_n(x) = \int_0^\infty p_n(x,t)dt$:

$$D\frac{d^2 q_0}{dx^2} - \alpha q_0 + \beta q_1 = 0,$$

$$D\frac{d^2 q_1}{dx^2} + \alpha q_0 - \beta q_1 = -\delta(x-L),$$

with boundary conditions

$$q_0(0) = q_1(0) = 0, \quad Dq_0'(L) = 0, \quad -Dq_1'(L) = \kappa_L q_1(L).$$

(b) Show that the Dirac delta function can be replaced by zero provided that the partially absorbing boundary condition becomes

$$Dq_1'(L) = 1 - \kappa_L q_1(L).$$

[Hint: shift the original partially reflecting boundary condition to $L - \varepsilon$ and integrate both sides of the ODE for q_1 between $L - \varepsilon$ to L.]

(c) Solve the resulting boundary value problem to show that the probability of reaching reservoir B is

$$P = D[q_0'(0) + q_1'(0)] = \frac{\alpha + \beta}{\alpha + \beta + [\alpha + \beta(\xi L)^{-1}\tanh(\xi L)]4LD_A/\pi a D}.$$

Problem 6.11. Fluctuating barrier driven by an OU process. Consider the following SDE for an overdamped Brownian particle:

$$dX = [-U'(X) - V'(X)Y]dt + \sqrt{2D}dW_1(t),$$

where the potential is taken to be of the form $U(x) + V(x)y$, consisting of a static part $U(x)$ and a fluctuating part driven by an OU process:

$$dY = -\frac{Y}{\tau} + \sqrt{2D/\tau}dW_2(t).$$

Here $W_1(t)$ and $W_2(t)$ are independent Wiener processes and τ is the correlation time of the OU process. The latter is assumed to be stationary so that

$$p(y) = \frac{1}{\sqrt{2\pi D}}e^{-y^2/2D}, \quad C(\tau) = De^{-|t|/\tau}.$$

Suppose that $U(x)$ has a minimum at $x = 0$ and a maximum at $x = 1$, and $V(x) \to$ constant as $x \to -\infty$. The 2D deterministic system $(D = 0)$ has a saddle at $(1,0)$ and a stable fixed point at $(0,0)$.

(a) In the white noise limit $\tau \to 0$, to leading order in τ, the system reduces to the Stratonovich SDE

$$dX = -U'(X)dt - \sqrt{2D\tau}V'(X)dW_2(t) + \sqrt{2D}dW_1(t).$$

Write down the corresponding Stratonovich FP equation for the probability density p and determine the corresponding backward equation. Hence, deduce that the MFPT $T_\tau(x)$ to reach $x = 1$ satisfies an ODE of the form

$$-(U(x)' - D_\tau'(x))\frac{\partial T_\tau}{\partial x} + D_\tau(x)\frac{\partial^2 T_\tau}{\partial x^2} = -1, \quad D_\tau(x) = D(1 + \tau V'(x)^2).$$

Solve for $R(x) = T'(x)$ with $R(-\infty) = 0$ and then for $T(x)$ with $T(1) = 0$, to obtain the result

$$T_\tau(x) = \int_x^1 dw \int_{-\infty}^w dz \frac{\exp\left(\int_z^w \frac{U'(y)}{D_\tau(y)}dy\right)}{\sqrt{D_\tau(w)D_\tau(z)}}.$$

(b) In the limit $\tau \to \infty$, y can be treated as a constant. Determine the MFPT for fixed y and integrate the result with respect to y using the Gaussian density. Hence show that the averaged MFPT T_∞ is given by

$$T_\infty(x) = \frac{1}{D} \int_x^0 dw \int_{-\infty}^w \exp(F(w,z)/D),$$

with

$$F(w,z) = U(w) - U(z) + [V(w) - V(z)]^2/2.$$

Since $T_0(0) \leq T_\infty(0)$ it follows that resonant activation will occur if $T_\tau(0)$ is a decreasing function of τ.

Problem 6.12. Effective permeability of N cells coupled by stochastically gated gap junctions. Consider a line of N identical cells of length L coupled by $N-1$ gap junctions at positions $x = l_k = kL$, $1 \leq k \leq N-1$, see Fig. 6.35. If $V_n(x,t)$ is the concentration of particles in state n, then we have the pair of PDEs given by

$$\frac{\partial V_0}{\partial t} = D\frac{\partial^2 V_0}{\partial x^2} - \beta V_0 + \alpha V_1,$$

$$\frac{\partial V_1}{\partial t} = D\frac{\partial^2 V_1}{\partial x^2} + \beta V_0 - \alpha V_1$$

for $x \in \mathcal{U}_1 \cup \mathcal{U}_2$ with exterior boundary conditions

$$V_n(0) = \rho_n \eta, \quad V_n(L) = 0, \quad n = 0,1,$$

and the interior boundary conditions at the jth gate are

$$[V_1(x)]_{x=l_j^-}^{x=l_j^+} = 0, \quad [\partial_x V_1(x)]_{x=l_j^-}^{x=l_j^+} = 0,$$

$$\partial_x V_0(l_j^-) = 0 = \partial_x V_0(l_j^+).$$

Find the steady-state solution of these equations and thus derive the following expression for the effective permeability $\mu_e(N)$:

$$\frac{1}{\mu_e(N)} = \frac{\rho_0}{[N-1]\rho_1 \xi D}\left[2\tanh(\xi L) + (2N-4)\frac{\cosh(\xi L) - 1}{\sinh(\xi L)}\right].$$

Problem 6.13. Splitting probability and MFPT for stochastically gated gap junctions. Consider a 1D array of M cells of size L coupled by stochastically gated gap junctions. In the analysis of Sect. 6.7.3 the strong Markov property was used to calculate the MFPT for a single diffusing particle to reach either end of the array, and the corresponding splitting probability.

(a) Recall the iterative equation (6.7.40) for the rescaled hitting probability p_1:

$$p_1(l_{k-1}) - 2p_1(l_k) + p_1(l_{k+1}) = 0, \quad k = 1,\dots,M-1,$$

with $p_1(0) = \rho_1$ and $p_1(ML) = 0$. Staring at the right-hand side by taking $k+1 = M$, iteratively solve this equation to show that

$$p_1(l_{M-r}) = \frac{r\rho_1}{M}.$$

(b) Recall the iterative equation (6.7.48) for the MFPT w_1:

$$w_1(l_{k-1}) - 2w_1(l_k) + w_1(l_{k+1}) = -2V, \quad = 1, \ldots, M-1,$$

with $w_1(l_0) = w_1(ML) = 0$, and V given by equation (6.7.47). Proceeding along similar lines to part (a), show that

$$w_1(l_{M-r}) = r(M-r)V.$$

Problem 6.14. Exact solutions of the Fick–Jacobs equation. Consider the Fick–Jacobs equation (with $D_0 = 1$)

$$\frac{\partial P(x,t)}{\partial t} = \frac{\partial}{\partial x}\sigma(x)\frac{\partial}{\partial x}\frac{P(x,t)}{\sigma(x)}.$$

(a) Let $f(x) = \frac{1}{2}\ln\sigma(x)$. Performing the change of variables

$$A(x,t) = e^{-f(x)}P(x,t) = \frac{P(x,t)}{\sqrt{\sigma(x)}},$$

show that A satisfies an equation of the form

$$\frac{\partial A(x,t)}{\partial t} = \frac{\partial^2 A}{\partial x^2} - V(x)A(x,t),$$

with

$$V(x) = f'^2(x) + f''(x) = \frac{1}{2\sigma(x)}\frac{d^2\sigma(x)}{dx^2} - \frac{1}{4}\left(\frac{\sigma'(x)}{\sigma(x)}\right)^2.$$

(b) Determine the effective potential $V(x)$ in the two cases (i) $\sigma(x) = \pi(1 + \lambda x)^2$ (conical cross section) and (ii) $\sigma(x) = \sin^2(\gamma x)$. Hence, for each case, express the solution $P(x,t)$ in terms of the initial data $P(x,0)$ and the fundamental solution of the diffusion equation.

Problem 6.15. Multibarrier model of ion channel transport. Suppose that an ion passing through a channel encounters a series of energy barriers and wells as shown in Fig. 6.53. Ion channel movement within the channel can then be modeled in terms of a sequence of hops over the barriers from one site (minimum) to the next. For simplicity, in the absence of a voltage drop across the membrane ($\Delta V = 0$), the barriers are identical, symmetric, and spaced uniformly along the channel. A voltage drop can then be superimposed to bias the jumps in a particular direction. Let α and β denote the right and left hopping rates across each barrier. Assuming that the rates are given by a standard Arrhenius formula (Sect. 2.4), we can write

$$\alpha = \omega e^{-(E-q\Delta V/2n)/k_B T}, \quad \beta = \omega e^{-(E+q\Delta V/2n)/k_B T},$$

where ω is some prefactor that depends on the detailed shape of the channel potential energy around the minimum, and q is the ion charge. Here E is the height of the energy barrier when $\Delta V = 0$ and the symmetric barriers are lowered by an amount $\Delta V/2n$, where n is the number of barriers. Assume the system is in steady state so the flux J through the system is the same everywhere, and let p_i denote the probability that an ion is at the ith site.

(a) Iteratively solve the sequence of flux equations

$$J = \alpha p_i - \beta p_{i+1}, i = 0,\ldots,n-1,$$

with $p_0 = vc_a$ and $p_n = vc_b$. Here c_a and c_b are the bulk concentrations on the left and right sides of the channel, respectively, and v is related to a small volume element at the entry/exit of the channel. In particular, show that

$$J = v\omega e^{-(E-q\Delta V/2n)/k_B T}\frac{(c_a - c_b\phi^n)(1-\phi)}{1-\phi^n}, \qquad \phi = e^{-q\Delta V/nk_B T}.$$

(b) Determine J when $c_a = c_b = c$. Plot the result flux as a function of voltage ΔV measured in units of $q/2k_B T$ for $n = 1,2,3$. Set the scale factor $cv\omega e^{-E/k_B T} = 1$. What happens to the current–voltage curve for large n?

Problem 6.16. Entropic gate model of nuclear pore transport. Consider the steady-state equation for the density $\rho(x)$ of cargo complexes in the nuclear pore complex (NPC):

$$\frac{d}{dx}\left[\frac{d\rho}{dx}+\rho(x)\frac{dU(x)}{dx}\right] = 0,$$

supplemented by the boundary conditions $\rho(0) = \rho(L) = 0$. Here $U(x)$ is given by Fig. 6.46

(a) Solve the steady-state equation using the conditions $J(x) = -|J_0|$ for $0 < x < R$, $J(x) = \bar{J}$ for $R < x < L-R$, and $J(x) = J_L$ for $L-R < x < L$.
(b) Using $\bar{J} = J_L + J_{ran}\rho(L-R)R$ and the solution in the region $L-R < x < L$ show that

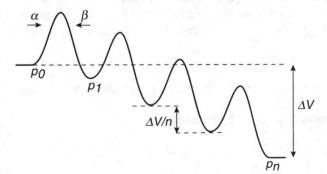

Fig. 6.53: Multibarrier potential of an ion channel.

$$Dp(L-R) = \frac{R\bar{J}}{J_{ran}R^2/D+1}.$$

Now imposing continuity of $\rho(x)$ at $x = L - R$ and setting $\bar{J} = J_S - |J_0|$ derive the result

$$\frac{\bar{J}}{J_S} = \left[1 + \frac{1}{J_{ran}R^2/D+1} + \frac{1}{R}\int_R^{L-R} e^{-U(x')}dx'\right]^{-1}.$$

Problem 6.17. Binding-diffusion model of selective nuclear transport. Recall the equation for the concentration of unbound particles in the binding-diffusion model of Sect. 6.8.3:

$$\frac{d^4A}{dx^4} = \lambda^2\frac{d^2A}{dx^2}, \quad \lambda^2 = \frac{k_{off}(D_A + KS_0D_B)}{D_AD_B},$$

with boundary conditions

$$A(0) = A_0, \quad A(L) = 0, \quad B'(0) = B'(L) = 0,$$

and

$$B(x) = KS_0A(x) - \frac{D_A}{k_{off}}\frac{d^2}{dx^2}$$

for $K = K_D^{-1} = k_{on}/k_{off}$. Given the general solution

$$A(x) = a_0 + a_1x + a_+e^{\lambda x} + a_-e^{-\lambda x},$$

calculate the coefficients $a_0, a_1, a_\pm$ using the boundary conditions. Hence show that the flux at $x = 0$ is given by

$$J_L = A_0D_A(1+\Gamma_0)\left(L + \frac{2\Gamma_0}{\lambda}\tanh(\lambda L/2)\right)^{-1},$$

with

$$\Gamma_0 = \left[\frac{D_A\lambda^2}{k_{off}KS_0} - 1\right]^{-1} = \frac{S_0k_{on}D_B}{k_{off}D_A}.$$

Problem 6.18. Two-dimensional homogenization. Consider a two-dimensional heterogeneous medium with a periodic substructure Ω_p consisting of the rectangular domain $0 \le y_1 \le a$ and $0 \le y_2 \le b$. Take $D(\mathbf{y})$ to have the separable form

$$D(\mathbf{y}) = D_0e^{\alpha(y_1)+\beta(y_2)},$$

where $\alpha(y_1)$ and $\beta(y_2)$ are periodic. The functions $w_i(\mathbf{y})$ in equation (6.1.15) simplify as $w_1 = w_1(y_1), w_2 = w_2(y_2)$.
(a) Show that the solutions of equation (6.1.15) are given by

$$w_1(y_1) = -y_1 + \kappa_1\int_0^{y_1} e^{-\alpha(s)}ds, \quad w_2(y_2) = -y_2 + \kappa_2\int_0^{y_2} e^{-\beta(s)}ds.$$

(b) Determine the integration constants κ_1 and κ_2 by imposing the periodicity requirements $w_1(0) = w_1(a)$ and $w_2(0) = w_2(b)$.

(c) Substituting for $w_i(\mathbf{y})$ in the equation above (6.1.16), show that $D_{12} = D_{21} = 0$ and calculate D_{11}, D_{22}. Hence obtain the homogenized diffusion equation

$$D_0 \left(\lambda_1 \frac{\partial^2 u_0}{\partial x_1^2} + \lambda_2 \frac{\partial^2 u_0}{\partial x_2^2} \right) = 0,$$

where

$$\lambda_1 = \left(\frac{1}{b} \int_0^b e^{\beta(s)} ds \right) \left(\frac{1}{a} \int_0^a e^{-\alpha(s)} ds \right)^{-1},$$

$$\lambda_2 = \left(\frac{1}{a} \int_0^a e^{\alpha(s)} ds \right) \left(\frac{1}{b} \int_0^b e^{-\beta(s)} ds \right)^{-1}.$$

Chapter 7
Active transport

In this chapter we turn to the major mechanism for transporting newly synthesized products from the nucleus to other intracellular compartments and the cell membrane. This is based on the active motor-driven transport of vesicular cargo along cytoskeletal tracks [154]. Active transport is faster and more easily regulated than passive diffusion, but requires a constant supply of energy to do useful work. We begin in Sect. 7.1 by considering an advection-diffusion population model of motor transport in axons and dendrites. This is used to investigate the possible role of reversible vesicular transport in generating a more uniform steady-state distribution of resources along an axon. In Sect. 7.2 we turn to the stochastic dynamics of an individual motor-cargo complex, which is modeled as a velocity jump process. The latter focuses on the transitions between different types of motion rather than the microscopic details of how a motor performs a single step, which were considered in Chap. 4. In the fast switching limit, the associated Chapman–Kolmogorov (CK) equation for the probability density function can be reduced to a Fokker–Planck (FP) equation using a quasi-steady-state reduction, see also Sect. 5.3.2. We then extend the analysis to transport on higher-dimensional cytoskeletal networks, including an application to virus trafficking. In Sect. 7.3 the efficiency of active bidirectional transport in delivering vesicular cargo to subcellular targets is analyzed in terms of a random search-and-capture process. The latter considers a domain with one or more hidden targets that are being searched for by a particle that randomly switches between a slow search phase (e.g., passive diffusion) and a faster non-search phase (e.g., active motor transport). In certain cases, it can be shown that there exists an optimal search strategy, in the sense that the mean search time to find a single hidden target can be minimized by varying the switching rates between different states. In Sect. 7.4 we use queuing theory to analyze the accumulation of resources in a set of targets due to multiple rounds of search-and-capture events. That is, rather than being permanently absorbed or captured by a target, a particle delivers a discrete packet of some resource to the target and then returns to its initial position, where it is reloaded with cargo and another round of search-and-capture is initiated. An alternative type of optimal random search process, which involves so-called stochastic

© Springer Nature Switzerland AG 2021
P. C. Bressloff, *Stochastic Processes in Cell Biology*, Interdisciplinary
Applied Mathematics 41, https://doi.org/10.1007/978-3-030-72515-0_7

resetting, is analyzed in Sect. 7.5. The position of a searcher is now reset to some fixed point $\mathbf{x}_r$ at a random sequence of times that is usually (but not necessarily) generated by a Poisson process with rate r. One often finds that the mean time to find a target may be minimized by varying the resetting rate r. Finally, in Sect. 7.6 we consider the effects of molecular crowding of motors on a filament track, as modeled by so-called asymmetric exclusion processes. We show how, in the mean-field limit, molecular crowding can be treated in terms of quasi-linear PDEs that support shock waves.

7.1 Vesicular transport in axons and dendrites

Axons of neurons can extend up to 1m in large organisms but synthesis of many of their components occur in the cell body. The healthy growth and maintenance of an axon depends on the interplay between the axonal cytoskeleton and the active transport of various organelles and macromolecular proteins along the cytoskeleton [147, 379, 546]. The disruption of axonal transport occurs in many neurodegenerative diseases, including Alzheimer's disease, Parkinson's disease, amyotrophic lateral sclerosis (also known as Lou Gherig's disease), and Huntington's disease [216, 579]. All of these diseases exhibit an aberrant accumulation of certain cellular components and excessive focal swelling of the axon, ultimately leading to axon degeneration.

The axonal cytoskeleton contains microtubules and actin microfilaments, which play a role in axonal transport, and neurofilaments that provide structural support for the axon. The microtubules align axially along the axon, with plus ends pointing away from the cell body. They do not extend over the whole length of an axon, having typical lengths of around $100\,\mu$m, but rather form an overlapping array from the cell body to the axon terminal, see Fig. 7.1. One of their main functions is to act as filament tracks for the long-range transport of membraneous organelles and macromolecular proteins via the action of kinesin and dynein. Actin microfilaments are mainly found beneath the axon membrane, forming evenly spaced ring-like structures that wrap around the circumference of the axon shaft. They are also enriched in growth cones and axon terminals. Actin microfilaments tend to be involved in

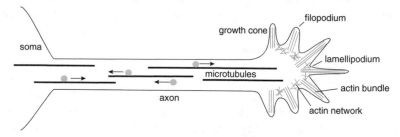

Fig. 7.1: Bidirectional transport of intracellular cargo along an overlapping 1D array of microtubules within an axon.

more short-range transport, such as the transfer of organelles and proteins from microtubules to targets in the membrane via myosin molecular motors.

Axonal transport is typically divided into two main categories based upon the observed speed [146, 147, 869]: fast transport $(1 - 9\,\mu\text{m/s})$ of organelles and vesicles and slow transport $(0.004 - 0.6\,\mu\text{m/s})$ of soluble proteins and cytoskeletal elements. Slow transport is further divided into two groups; actin and actin-bound proteins are transported in slow component A while cytoskeletal polymers such as microtubules and neurofilaments are transported in slow component B. It had originally been assumed that the differences between fast and slow components were due to differences in transport mechanisms, but direct experimental observations now indicate that they all involve fast motors but differ in how the motors are regulated. Membranous organelles such as mitochondria and vesicles, which function primarily to deliver membrane and protein components to sites along the axon and at the axon tip, move rapidly in a unidirectional or bidirectional manner, pausing only briefly. In other words, they have a high duty ratio—the proportion of time a cargo complex is actually moving. On the other hand, cytoskeletal polymers such as neurofilaments move in an intermittent and bidirectional manner, pausing more often and for longer time intervals; such transport has a low duty ratio.

Another example of a transport process in neurons that exhibits bidirectionality is the trafficking of mRNA-containing granules within dendrites. There is increasing experimental evidence that local protein synthesis in the dendrites of neurons plays a crucial role in mediating persistent changes in synaptic structure and function, which are thought to be the cellular substrates of long-term memory [109, 187, 367, 813]. This is consistent with the discovery that various mRNA species, as well as important components of the translational machinery such as ribosomes, are distributed in dendrites. Although many of the details concerning mRNA transport and localization are still unclear, a basic model is emerging. First, newly transcribed mRNA within the nucleus binds to proteins that inhibit translation, thus allowing the mRNA to be sequestered away from the protein-synthetic machinery within the cell body. The repressed mRNAs are then packaged into ribonucleoprotein granules that are subsequently transported into the dendrite via kinesin and dynein motors along microtubules. Finally, the mRNA is localized to an activated synapse by actin-based myosin motor proteins, and local translation is initiated following neutralization of the repressive mRNA–binding protein. Details regarding the motor-driven transport of mRNA granules in dendrites have been obtained by fluorescently labeling either the mRNA or mRNA-binding proteins and using live-cell imaging to track the movement of granules in cultured neurons [234, 454, 704]. It has been found that under basal conditions the majority of granules in dendrites are stationary or exhibit small oscillations around a few synaptic sites. However, other granules exhibit rapid retrograde (toward the cell body) or anterograde (away from the cell body) motion consistent with bidirectional transport along microtubules. These movements can be modified by neuronal activity. In particular, there is an enhancement of dendritically localized mRNA due to a combination of newly transcribed granules being transported into the dendrite, and the conversion of stationary or oscillatory granules already present in the dendrite into anterograde-moving granules.

7.1.1 Slow axonal transport

Radioisotopic pulse labeling experiments provide information about the transport of neurofilaments at the population level, which takes the form of a slowly moving Gaussian-like wave that spreads out as it propagates distally. A number of authors have modeled this slow transport in terms of a system of hyperbolic PDEs [99, 202, 675]. For example, Ref. [99] considered the following system on the semi-infinite domain $0 \leq x < \infty$:

$$\varepsilon \left[\frac{\partial p_1}{\partial t} - v \frac{\partial p_1}{\partial x} \right] = \sum_{j=1}^{n} A_{1j} p_j, \tag{7.1.1a}$$

$$\varepsilon \frac{\partial p_i}{\partial t} = \sum_{j=1}^{n} A_{ij} p_j, \quad 1 < i \leq N, \tag{7.1.1b}$$

where p_1 represents the concentration of moving neurofilament proteins, and $p_i, i > 1$ represent the concentrations in $n - 1$ distinct stationary states. Conservation of mass implies that $A_{jj} = -\sum_{i \neq j} A_{ij}$. The initial condition is $p_i(x, 0) = 0$ for all $1 \leq i \leq n$, $0 < x < \infty$. Moreover $p_1(0, t) = 1$ for $t > 0$. Ref. [675] carried out an asymptotic analysis of equation (7.1.1) that is related to the QSS reduction method described in Box 5D, see also Sect. 7.2. Suppose that u_1 is written in the form

$$p_1(x, t) = Q_\varepsilon \left(\frac{x - ut}{\sqrt{\varepsilon}}, t \right),$$

where u is the effective speed, $u = v\rho_1$, and ρ_1 is the stationary probability of being in the moving state. They then showed that $Q_\varepsilon(s, t) \to Q_0(s, t)$ as $\varepsilon \to 0$, where Q_0 is a solution to the diffusion equation

$$\frac{\partial Q_0}{\partial t} = D \frac{\partial^2 Q_0}{\partial x^2}, \quad Q_0(s, 0) = H(-s),$$

with H the Heaviside function. The diffusivity D can be calculated in terms of v and the transition matrix $\mathbf{A}$. Hence the propagating and spreading waves observed in experiments could be interpreted as solutions to an effective advection–diffusion equation. More recently, a more rigorous analysis of spreading waves has been developed [288, 289].

In contrast to the above population models, direct observations of neurofilaments in axons of cultured neurons using fluorescence microscopy have demonstrated that individual neurofilaments are actually transported by fast motors but in an intermittent fashion [842]. Hence, it has been proposed that the slow rate of movement of a population is an average of rapid bidirectional movements interrupted by prolonged pauses, the so-called stop-and-go hypothesis [145, 422, 518]. Computational simulations of an associated system of PDEs show how fast intermittent transport can account for the slowly spreading wave seen at the population level. One version of the model assumes that the neurofilaments can be in one of six states [145, 518]:

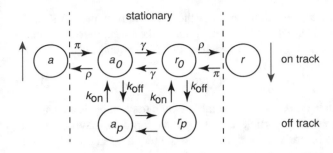

Fig. 7.2: Transition diagram of "stop-and-go" model for the slow axonal transport of neurofilaments. See text for definition of different states.

anterograde moving on track (state a), anterograde pausing on track (a_0 state), anterograde pausing off track (state a_p), retrograde pausing on track (state r_0), retrograde pausing off track (state r_p), and retrograde moving on track (state r). The state transition diagram is shown in Fig. 7.2.

7.1.2 Reversible vesicular transport and synaptic democracy

An important issue regarding the efficacy of active transport is how to ensure a "fair" distribution of resources across a cell. This is a particularly acute challenge in neurons whose distal synaptic sites on axons and dendrites can be significantly far from the major source of newly synthesized proteins within the cell body. A model of the active transport and delivery of vesicles across *en passant* synapses in the axons of neurons has recently been developed [122], based on the following experimental observations in *C. elegans* and *Drosophila* [547, 548, 866]: (i) motor-driven cargo exhibits ballistic anterograde or retrograde motion interspersed with periods of long pauses at presynaptic sites; (ii) the capture of vesicles by synapses during the pauses is reversible, in that vesicular aggregation at a site could be inhibited by signaling molecules resulting in dissociation from the target; (iii) the distribution of resources across synapses is relatively uniform—so-called synaptic democracy. In Ref. [122] the transport and delivery of vesicles to synaptic targets was modeled using a one-dimensional (1D) advection–diffusion equation. It was shown that in the case of irreversible cargo delivery, the steady-state vesicle density decays exponentially from the soma, whereas the steady-state density is relatively uniform in the reversible case. This suggests that reversibility in vesicular delivery plays a crucial role in achieving a "fair" distribution of resources within a cell. Here we review the 1D advection–diffusion model of reversible vesicular transport.

Consider a population of motor-cargo complexes or particles moving on a semi-infinite track, each of which carries a single synaptic vesicle precursor (SVP) to be delivered to a synaptic site. Assume that these particles are injected at the soma ($x = 0$) at a fixed rate J_1 and that the distribution of synaptic sites along the axon is given by

$$\rho(x) = \sum_{j=1} \Theta(a - |x - x_j|), \qquad (7.1.2)$$

where $\Theta(x)$ is the Heaviside function, and a is the half-width of the jth synapse centered at x_j. Neglecting interactions between particles, the dynamics of the motor-cargo complexes can be captured by the advection–diffusion equation [122]

$$\frac{\partial u}{\partial t} = -v\frac{\partial u}{\partial x} + D\frac{\partial^2 u}{\partial x^2} - \kappa\rho(x)u, \qquad x \in (0,\infty), \qquad (7.1.3)$$

where $u(x,t)$ is the particle density along the microtubule track at position x at time t, and κ is the rate at which a particle can deliver its cargo to a synapse. Note that equation (7.1.3) can be derived from more detailed biophysical models of motor transport under the assumption that the rates at which motor-cargo complexes switch between different motile states are relatively fast [122, 609], see Sect. 7.2. In particular, the mean speed will depend on the relative times that the complex spends in different anterograde, stationary, and possibly retrograde states, whereas the diffusivity D reflects the underlying stochasticity of the motion. Equation (7.1.3) is supplemented by the boundary condition at $x = 0$:

$$J(u(0,t)) = J_1, \quad J(u) \equiv vu - D\frac{\partial u}{\partial x}. \qquad (7.1.4)$$

Remark 7.1. Note that in equation (7.1.3), the *en passant* synapses act as partially absorbing targets, in the sense that whenever a particle (searcher) is within the given target domain, it finds (is captured by) the target with a certain probability according to a Poisson rate κ. The capture rate κ will depend on the biophysical mechanisms for transferring a vesicle from a motor complex to a synapse. This is likely to include transport on the local actin cortex via mysosin motors. Thus the actual target is not physically within the search domain. This contrasts with the case where a target lies within the search domain and the boundary of the interior target is partially absorbing, in which case one has a Robin boundary condition.

As a further simplification, suppose that the synapses are sufficiently packed together that we can treat $\rho(x)$ as uniform density ρ_0, which is absorbed into the rate κ. Let $c(x,t)$ denote the concentration of delivered vesicles to the presynaptic sites at x at time t with

$$\frac{\partial c}{\partial t} = \kappa u - \lambda c, \qquad (7.1.5)$$

where λ denotes the degradation rate for vesicles. Note that in the irreversible delivery case, including vesicular degradation is necessary to prevent blowup in the solutions for $c(x,t)$. This consideration is not necessary in the reversible delivery case. The steady-state solution for c is given by

$$c = \frac{\kappa J_1 e^{-\xi x}}{\lambda D\xi + v} \qquad \xi = \frac{-v + \sqrt{v^2 + 4D\kappa}}{2D}, \qquad (7.1.6)$$

which clearly indicates that c decays exponentially with respect to distance from the soma with correlation length ξ^{-1}. Taking the values $D = 1\,\mu\text{m}^2/\text{s}$ for cytoplasmic diffusion and $v = 0.1 - 1\,\mu\text{m/s}$ for motor transport [397], and assuming that $\kappa \ll 1$ s^{-1}, we see that $\overline{\xi} \approx (v/\kappa)\,\mu\text{m}$. Thus, in order to have correlation lengths comparable to axonal lengths of several millimeters, we would require delivery rates of the order $\kappa \sim 10^{-5}\,\text{s}^{-1}$, whereas measured rates tend to be of the order of a few per minute [122, 381, 507]. This simple calculation establishes that injecting motor

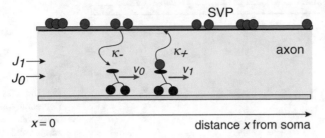

Fig. 7.3: Schematic diagram of the reversible exchange of vesicles between motor-cargo complexes and presynaptic targets. Each motor can carry at most one vesicle (SVP) and the transport of motors with (without) an attached vesicle is modeled using an advection–diffusion equation with average speed v_1 (v_0) and diffusivity D. The corresponding motor fluxes injected at the somatic end ($x = 0$) are J_1 (J_0). Finally, a vesicle can be reversibly exchanged between a motor and synaptic target at the rates $\kappa_\pm$.

complexes from the somatic end of the axon leads to an exponentially decaying distribution of synaptic resources along the axon. We now show, following Ref. [122], that relaxing the irreversible delivery condition in this model allows for a more uniform distribution of vesicles along the axon.

In order to take into account the reversibility of vesicular delivery to synapses, one must consider a generalization of the advection–diffusion model (7.1.3). To that end, let $u_0(x,t)$ and $u_1(x,t)$ denote the density of motor-cargo complexes without and with an attached SVP, respectively, and let κ_+ and κ_- denote the rates at which vesicles are delivered to synaptic sites and recovered by the motors, respectively. Each density evolves according to an advection–diffusion equation combined with transition rates that represent the delivery and recovery of SVPs, see Fig. 7.3:

$$\frac{\partial u_0}{\partial t} = -v_0 \frac{\partial u_0}{\partial x} + D \frac{\partial^2 u_0}{\partial x^2} - \gamma_0 u_0 + \kappa_+ u_1 - \kappa_- c u_0, \qquad (7.1.7a)$$

$$\frac{\partial u_1}{\partial t} = -v_1 \frac{\partial u_1}{\partial x} + D \frac{\partial^2 u_1}{\partial x^2} - \gamma_1 u_1 - \kappa_+ u_1 + \kappa_- c u_0, \qquad (7.1.7b)$$

with $x \in (0, \infty)$. Disparity in the velocities in each state reflects the effect cargo can have on particle motility, while the degradation rates $\gamma_{0,1}$ are included to account for the possibility of particle degradation or recycling. Equations (7.1.7a) and (7.1.7b) are supplemented by the boundary conditions

$$J(u_j(0,t)) = J_j, \qquad j = 0,1, \qquad (7.1.8)$$

where J_j is the constant rate at which particles with or without cargo are injected into the axon from the soma. The dynamics for $c(x,t)$ are now given by

$$\frac{\partial c}{\partial t} = \kappa_+ u_1 - \kappa_- c u_0. \qquad (7.1.9)$$

We need not explicitly include degradation in this case because, provided $J_0 > 0$, $c(x,t)$ will be bounded. The steady-state distribution of vesicles is then

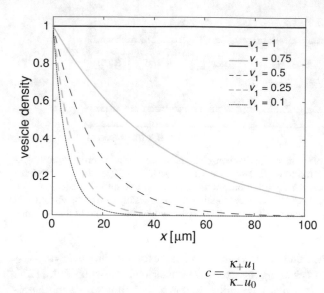

Fig. 7.4: Figure depicting the loss of synaptic democracy as disparity in velocities between free motors and cargo-carrying motors grows normalized so all curves fit in one frame. Parameter values are $D = 1$, $\gamma_{0,1} = 0.1$, $\kappa_{\pm} = 0.1$, $J_{0,1} = 10$, $v_0 = 1$.

$$c = \frac{\kappa_+ u_1}{\kappa_- u_0}.$$

Substitution into the steady-state analogs of equations (7.1.7a) and (7.1.7b) yields

$$u_j(x) = \frac{J_j e^{-\xi_j x}}{D\xi_j + v_j} \qquad \xi_j = \frac{-v_j + \sqrt{v_j^2 + 4D\gamma_j}}{2D}, \qquad (7.1.10)$$

so that

$$c = \frac{\kappa_+}{\kappa_-} \frac{J_1}{J_0} \frac{D\xi_0 + v_0}{D\xi_1 + v_1} e^{-\Gamma x}, \qquad (7.1.11)$$

with $\Gamma \equiv \xi_1 - \xi_0$. It is evident that if $\Gamma = 0$, then c has a spatially uniform distribution.

Suppose that the diffusion and degradation rates of motors do not change when carrying cargo. Then $\Gamma = 0$ would imply that the velocities of the cargo-carrying motors are equal to the velocities of the free motors. However, we would expect $v_1 < v_0$ due to the added load of the cargo on the motor, and that this would lead to a loss of synaptic democracy since $\Gamma > 0$. Indeed, values of v_1 less than v_0 lead to steady-state profiles of vesicle density reminiscent of the exponential decay behavior of the irreversible delivery case; see Fig. 7.4, although the spatial rate of decay is mitigated by the presence of reversible delivery. Hence, attaining synaptic democracy also depends on physical properties of the cargo being carried. Large cargo, for example, may not be uniformly distributed throughout an axon, whereas smaller cargo will.

Finally, note that the important role of reversibility appears to hold under more general conditions. One extension is to consider reversible transport along higher-dimensional microtubular networks [428], see also Sect. 7.2.3. In particular, one can model a cell as a disc or a sphere and assume that the distribution of microtubules is radially symmetric. The source of the motor-cargo complexes is taken to be at the origin of the cell, and the dynamics of the motor densities are represented by advection–diffusion equations transformed into their polar (2D) and spherical (3D) representations. It is also assumed that each motor carries one cargo element and can deliver its cargo at any point within the given domain, that is, that there is a continuum of target sites within the cell. It is straightforward to show that one still obtains exponential-like decay of the steady-state distribution of resources away from the origin in the irreversible case, and a more uniform distribution in the reversible case. A second extension is to remove the restriction that each motor can carry only one vesicle. Using a modified version of the well-known Becker–Döring equations for aggregation–fragmentation phenomena (Chap. 13), the analysis can be extended to the case of motors carrying vesicular aggregates, assuming that only one vesicle can be exchanged with a target at any one time [123]. Finally, in Ref. [124], exclusion effects between motor-cargo complexes were also taken into account (see Sect. 7.6). The axon was treated as a 1D lattice, and the motion of motors was represented by a system of ordinary differential equations for the mean occupation number at each site. Using a combination of mean field and adiabatic approximations, hydrodynamic equations were obtained for the dynamics of motor density in the continuum limit. Again, it was found that synaptic democracy is achieved in the reversible delivery case, provided the cargo-carrying motors' speed is not greatly reduced by their cargo.

Remark 7.2. There is growing experimental evidence that presynaptic, mitochondrial, and ribosomal proteins are locally synthesized in adult axons via the activation of stored mRNAs [443]. Local protein synthesis thus provides an alternative mechanism for allowing distal regions of axons to function and survive. An associated finding is that inhibiting the local translation of key mRNAs results in axon degeneration, with an increasing number of human mutations in RNA metabolic pathways becoming associated with neurodegenerative diseases. The basic idea is that the relevant mRNAs are stored in a translationally repressed state due to the action of RNA-binding proteins, for example. When a particular signal is received, specific mRNAs are released from repression, allowing their translation.

7.2 Intracellular motor transport as a velocity-jump process

When modeling the active transport of intracellular cargo over relatively long distances, it is often convenient to ignore the microscopic details of how a motor performs a single step (as described by the Brownian ratchet models of Sect. 4.4), and to focus instead on the transitions between different velocity states (e.g., anterograde vs. retrograde active transport) as described by a velocity-jump process. The corresponding CK equation takes the form of a system of PDEs, which is the starting point for a variety of mesoscopic models [112, 289, 422, 482, 531, 609, 675, 768].

7.2.1 Active transport along a 1D track

As we described in Sect. 7.1, experimental observations of active transport along axons [147, 548] and dendrites [234, 454, 704] reveal intermittent behavior with constant velocity movement in both directions along an effective 1D array of micro-tubules, interrupted by brief pauses or fast oscillatory movements that may corre-spond to localization at specific targets such as synapses. Motivated by these obser-vations, consider a motor-cargo complex that has N distinct velocity states, labeled $n = 1, \ldots, N$, with corresponding velocities v_n. Take the position $X(t)$ of the complex on a filament track to evolve according to the velocity-jump process

$$\frac{dX}{dt} = v_{N(t)}, \tag{7.2.12}$$

where the discrete random variable $N(t) \in \{1, \ldots, N\}$ indexes the current velocity state $v_{N(t)}$, and transitions between the velocity states are governed by a discrete Markov process with generator $\mathbf{A}$, see Box 3A. (Velocity jump processes are also used to model microtubule catastrophes (Sect. 4.2) and bacterial chemotaxis (Chap. 10).) Define $\mathbb{P}(x, n, t \mid y, m, 0)dx$ as the joint probability that $x \leq X(t) < x + dx$ and $N(t) = n$ given that initially the particle was at position $X(0) = y$ and was in state $N(0) = m$. Setting

$$p_n(x, t) \equiv \sum_m \mathbb{P}(x, n, t | 0, m, 0) \sigma_m, \tag{7.2.13}$$

with initial condition $p_n(x, 0) = \delta(x)\sigma_n$, $\sum_m \sigma_m = 1$, the evolution of the probabili-ty is described by the differential Chapman–Kolmogorov (CK) equation (see Sect. 5.3.2)

$$\frac{\partial p_n}{\partial t} = -v_n \frac{\partial [p_n(x, t)]}{\partial x} + \sum_{n'=1}^{N} A_{nn'} p_{n'}(x, t). \tag{7.2.14}$$

In the case of bidirectional transport, the velocity states can be partitioned such that $v_n > 0$ for $n = 1, \ldots, \mathcal{N}$ and $v_n \leq 0$ for $n = \mathcal{N} + 1, \ldots, N$ with $\mathcal{N} > 0$. Note that one can also construct more complicated versions of the model, in which the transition matrix $\mathbf{A}$ depends on the position x or equation (7.2.12) is replaced by an SDE. In the latter case, an additional diffusion term appears in the CK equation.

Two-state model of unbiased bidirectional transport and the diffusion limit. The simplest example of unbiased bidirectional transport is a two-state model, in which the particle switches equally between a right-moving state ($n = 1$) with ve-locity v and a left-moving state ($n = 2$) with velocity $-v$. Equation (7.2.14) takes the explicit form

$$\frac{\partial p_1}{\partial t} = -v \frac{\partial p_1}{\partial x} - kp_1 + kp_2, \tag{7.2.15a}$$

$$\frac{\partial p_2}{\partial t} = v \frac{\partial p_2}{\partial x} + kp_1 - kp_2, \tag{7.2.15b}$$

where k is the switching rate. This pair of equations is a symmetric version of the Dogterom-Leibler model of microtubule catastrophe [223], see Sect. 4.2, where $x(t)$ is the position of the tip of a microtubule and $\pm v$ represent the rates of growth and shrinkage, respectively. It is convenient to perform the change of variables $p = p_1 + p_2$ and $J = v(p_+ - p_-)$, where p is the marginal density and J is the flux. Adding and subtracting equations (7.2.15a, 7.2.15b) gives

$$\frac{\partial p}{\partial t} = -\frac{\partial J}{\partial x}, \qquad \frac{\partial J}{\partial t} = -v^2 \frac{\partial p}{\partial x} - 2kJ. \qquad (7.2.16)$$

We now notice that in the limit $v^2 \to \infty$ and $k \to \infty$ with $D \equiv v^2/2k$ fixed, we can set the right-hand side of the second equation to zero and thus eliminate J. This then yields the diffusion equation for p [424]

$$\frac{\partial p}{\partial t} = D\frac{\partial^2 p}{\partial x^2}, \qquad D = \frac{v^2}{2k}. \qquad (7.2.17)$$

The diffusion limit could also be obtained by performing the rescaling $t \to \tau = \varepsilon^2 t, x \to X = \varepsilon x$, and carrying out a formal asymptotic expansion in ε [374]. It is equivalent to ignoring short-time transients.

Another way to understand the above limit is to note that by differentiating equations (7.2.15a, 7.2.15b) it can be shown that the marginal probability density $p(x,t)$ satisfies the telegrapher's equation [43, 96, 320]

$$\left[\frac{\partial^2}{\partial t^2} + 2k\frac{\partial}{\partial t} - v^2\frac{\partial^2}{\partial x^2}\right] p(x,t) = 0. \qquad (7.2.18)$$

(The individual densities $p_\pm$ satisfy the same equations.) The telegrapher's equation can be solved explicitly for a variety of initial conditions. More generally, the short-time behavior (for $t \ll \tau_c = 1/2k$) is characterized by wave-like propagation with $\langle x(t) \rangle^2 \sim (vt)^2$, whereas the long-time behavior ($t \gg \tau_c$) is diffusive with $\langle x^2(t) \rangle \sim 2Dt, D = v^2/2k$. As an explicit example, the solution on $\mathbb{R}$ (no boundary conditions) for the initial conditions $p(x,0) = \delta(x)$ and $\partial_t p(x,0) = 0$ is given by

$$p(x,t) = \frac{e^{-kt}}{2}[\delta(x-vt) + \delta(x+vt)]$$
$$+ \frac{ke^{-kt}}{2v}\left[I_0(k\sqrt{t^2 - x^2/v^2}) + \frac{t}{\sqrt{t^2 - x^2/v^2}}I_0(k\sqrt{t^2 - x^2/v^2})\right]$$
$$\times [\Theta(x+vt) - \Theta(x-vt)],$$

where I_n is the modified Bessel function of nth order and Θ is the Heaviside function. The first two terms clearly represent the ballistic propagation of the initial data along characteristics $x = \pm vt$, whereas the Bessel function terms asymptotically approach Gaussians in the large time limit. The steady-state equation for $p(x)$ is simply $p''(x) = 0$, which from integrability means that $p(x) = 0$ point-wise. This is consistent with the observation that the above explicit solution satisfies $p(x,t) \to 0$ as $t \to \infty$.

The two-state velocity-jump process (7.2.15) also occurs in the context of bacterial chemotaxis, where it is known as a run-and-tumble model (see Chap. 10). In the absence of an external chemotactic signal, the motion is unbiased. Performing the rescaling $t \to \tau = \varepsilon^2 t, x \to X = \varepsilon x$, and taking the directional bias in the switching rates to be $O(\varepsilon)$ in the presence of a chemotactic signal, one obtains an advection–diffusion equation in the limit $\varepsilon \to 0$, with the advection term representing cells climbing up a chemotactic concentration gradient [374, 375, 626]. However, biased reaction-transport models of intracellular transport tend to operate in a different scaling regime, where advection dominates diffusion. In this case, the appropriate scaling is $t \to \varepsilon t, x \to \varepsilon x$ (see section 7.2.2).

Tug-of-war model. Recall the tug-of-war model of a motor-cargo complex introduced in Sect. 4.5.2 and Fig. 4.25. There we focused on the various velocity states and the transitions between them, without considering the actual displacement of the motor complex along a filament track. Here we briefly describe how to write down a PDE version of the model based on the CK equation (7.2.14). Consider a motor complex consisting of N_+ anterograde motors and N_- retrograde motors. The internal states of the complex are specified by the number of bound motors (n_+, n_-) and the velocity $v_c(n_+, n_-)$ of a given state is given by equation (4.5.24). Following [609, 610], we introduce the mapping $(n_+, n_-) \to \mathcal{N}(n_+, n_-) \equiv (N_+ + 1)n_- + (n_+ + 1)$ with $0 \le n \le N = (N_+ + 1)(N_- + 1)$. The corresponding probability density $p_n(x,t)$ satisfies equation (7.2.14) with $v_n = v_c(n_+, n_-)$. The components A_{nm}, $n, m = 1, \ldots, N$, of the state transition matrix A are given by the corresponding binding/unbinding rates of equation (4.5.20). That is, the nonzero off-diagonal terms are

$$A_{nm} = \pi_+(n_+ - 1) \text{ for } m = \mathcal{N}(n_+ - 1, n_-),$$
$$A_{nm} = \pi_-(n_- - 1), \text{ for } m = \mathcal{N}(n_+, n_- - 1),$$
$$A_{nm} = \gamma_+(n_+ + 1), \text{ for } m = \mathcal{N}(n_+ + 1, n_-),$$
$$A_{nm} = \gamma_-(n_- + 1), \text{ for } m = \mathcal{N}(n_+, n_- + 1)$$

for $n = \mathcal{N}(n_+, n_-)$ and the diagonal terms are $A_{nn} = -\sum_{m \ne n} A_{mn}$. Given the velocity states and the generator $\mathbf{A}$, we can then describe the stochastic evolution of the complex along a 1D track using the CK equation (7.2.14).

7.2.2 Quasi-steady-state reduction

Suppose that on an appropriate length-scale L, the transition rates are fast compared to v/L where $v = \max_n |v_n|$. (L could be the length of a single filament track or the size of a target for delivery of cargo.) Performing the rescalings $x \to x/L$ and $t \to tv/L$ leads to a non-dimensionalized version of the CK equation

$$\frac{\partial p_n}{\partial t} = -v_n \frac{\partial p_n(x,t)}{\partial x} + \frac{1}{\varepsilon} \sum_{n'=1}^{N} A_{nn'}(x) p_{n'}(x,t), \tag{7.2.19}$$

with $0 < \varepsilon \ll 1$. We are now allowing for the possibility that the generator $\mathbf{A}(x)$ is x-dependent. Suppose that for each x, the matrix $\mathbf{A}(x)$ is irreducible with a unique stationary density (right eigenvector) $\rho_n(x)$. In the limit $\varepsilon \to 0$, $p_n(x,t) \to \rho_n(x)$ and the motor moves deterministically according to the mean-field equation

$$\frac{dx}{dt} = V(x) \equiv \sum_{n=1}^{N} v_n \rho_n(x). \tag{7.2.20}$$

In the regime $0 < \varepsilon \ll 1$, there are typically a large number of transitions between different motor complex states n while the position x hardly changes at all. This suggests that the system rapidly converges to the quasi-steady-state (QSS) $\rho_n(x)$, which will then be perturbed as x slowly evolves. The resulting perturbations can thus be analyzed using the QSS diffusion approximation, in which the CK equation (7.2.19) is approximated by a Fokker–Planck (FP) equation for the total probability density $C(x,t) = \sum_n p_n(x,t)$. The basic steps proceed along the lines of Sect. 5.3.2:

1. Decompose the probability density as

$$p_n(x,t) = C(x,t)\rho_n(x) + \varepsilon w_n(x,t), \qquad (7.2.21)$$

where $\sum_n p_n(x,t) = C(x,t)$ and $\sum_n w_n(x,t) = 0$. Substituting into (7.2.19) yields

$$\frac{\partial C}{\partial t}\rho_n(x) + \varepsilon\frac{\partial w_n(x,t)}{\partial t} = -v_n\frac{\partial[C(x,t)\rho_n(x) + \varepsilon w_n(x,t)]}{\partial x}$$

$$+ \frac{1}{\varepsilon}\sum_{n'=1}^{N} A_{nn'}(x)[C(x,t)\rho_{n'}(x) + \varepsilon w_{n'}(x,t)].$$

Summing both sides with respect to n then gives

$$\frac{\partial C}{\partial t} = -\frac{\partial VC}{\partial x} - \varepsilon\sum_{n=1}^{N} v_n\frac{\partial w_n(x,t)}{\partial x}, \qquad (7.2.22)$$

where $V(x) = \sum_m v_m\rho_m(x)$.

2. Using the equation for C and the fact that $A\rho = 0$, we have

$$\varepsilon\frac{\partial w_n}{\partial t} = \sum_{n'=1}^{N} A_{nn'}(x)w_{n'}(x,t) - v_n\frac{\partial\rho_n(x)C}{\partial x} + \rho_n(x)\frac{\partial V(x)C}{\partial x}$$

$$- \varepsilon\sum_{m=1}^{N}[v_n\delta_{m,n} - \rho_n(x)v_m]\frac{\partial w_m(x,t)}{\partial x}.$$

3. Introduce the asymptotic expansion

$$w_n \sim w_n^{(0)} + \varepsilon w_n^{(1)} + \varepsilon^2 w_n^{(2)} + \dots$$

and collect $O(1)$ terms:

$$\sum_{n'=1}^{N} A_{nn'}(x)w_{n'}^{(0)}(x,t) = v_n\frac{\partial\rho_n(x)C(x,t)}{\partial x} - \rho_n(x)\frac{\partial V(x)C(x,t)}{\partial x}. \qquad (7.2.23)$$

The Fredholm alternative theorem shows that this has a solution, which is unique on imposing the condition $\sum_n w_n^{(0)}(x,t) = 0$.

4. Combining equations (7.2.23) and (7.2.22) shows that C evolves according to the FP equation

$$\frac{\partial C}{\partial t} = -\frac{\partial}{\partial x}(VC) + \varepsilon \frac{\partial}{\partial x}\left(D\frac{\partial C}{\partial x}\right), \tag{7.2.24}$$

with the drift V and diffusion coefficient D given by

$$V(x) = \sum_{m=1}^{N} v_m \rho_m(x), \quad D(x) = \sum_{n=1}^{N} Z_n(x)v_n, \tag{7.2.25}$$

where $Z_n(x)$ is the unique solution to

$$\sum_{m=1}^{N} A_{nm}(x)Z_m(x) = [V(x) - v_n]\rho_n(x), \tag{7.2.26}$$

and $\sum_m Z_m(x) = 0$. We have dropped $O(\varepsilon)$ corrections to the drift term.

Remark 7.3. The FP equation (7.2.24) is often easier to analyze than the full CK equation, particularly when the number of internal states is large or the motor moves along a higher-dimensional microtubular network rather than a 1D track, see Sect. 7.2.3. The drift and diffusion terms also preserve certain details regarding the underlying biophysics of motor transport, due to the dependence of V and D on underlying biophysical parameters. However, as noted more generally in the analysis of stochastic hybrid systems (Box 5D), care has to be taken in bounded domains. That is, the original CK equation is an Nth order hyperbolic PDE, whereas the FP equation is a second-order parabolic PDE. Thus there will generally be a mismatch in the number of boundary conditions at each end of a bounded domain. We thus have a singular perturbation problem, in which the solution to equation (7.2.24) represents an outer solution that is valid in the bulk of the domain, but has to be matched to an inner solution at each boundary [609, 900].

Example 7.1. 3-state model. As an illustration of the QSS reduction, consider a 3-state ($N = 3$) model of a motor consisting of a right-moving state ($n = 1, v_1 = v$), a left-moving state ($n = 2, v_2 = -v$), and a stationary (or possibly slowly diffusing) state ($n = 3$) corresponding to the case that the motor is unbound from the microtubule, say. The corresponding CK equation is

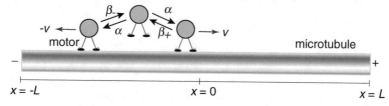

Fig. 7.5: Schematic diagram illustrating a model of a motor-driven particle moving along a one-dimensional track. The particle can transition from a moving state with velocity $\pm v$ to a stationary state at a rate $\beta_\pm$ and back again at a rate α.

$$\frac{\partial p_1}{\partial t} = -v\frac{\partial p_1}{\partial x} - \beta_+ p_1 + \alpha p_3, \tag{7.2.27a}$$

$$\frac{\partial p_2}{\partial t} = v\frac{\partial p_2}{\partial x} - \beta_- p_2 + \alpha p_3, \tag{7.2.27b}$$

$$\frac{\partial p_3}{\partial t} = \beta_+ p_1 + \beta_- p_2 - 2\alpha p_3. \tag{7.2.27c}$$

Here $\alpha, \beta_\pm$ are the transition rates between the stationary and mobile states as indicated in Fig. 7.5. (If the unbound state were to diffuse, then there would be a term of the form $D_3\partial^2 p_3/\partial x^2$ on the right-hand side of equation (7.2.27c).) The transport will be biased in the anterograde direction if $\beta_+ < \beta_-$, which implies that the particle spends more time in the anterograde state than the retrograde state. Unidirectional transport is obtained in the limit $\beta_- \to \infty$.

The generator $\mathbf{A}$ is given by the 3×3 matrix

$$A = \begin{bmatrix} -\beta_+ & 0 & \alpha \\ 0 & -\beta_- & \alpha \\ \beta_+ & \beta_- & -2\alpha \end{bmatrix}, \tag{7.2.28}$$

The left nullspace of the matrix A is spanned by the vector $\psi = (1,1,1)^\top$ and the right nullspace is spanned by

$$\rho = \frac{1}{\gamma}\begin{pmatrix} \frac{1}{\beta_+} \\ \frac{1}{\beta_-} \\ \frac{1}{\alpha} \end{pmatrix}, \quad \gamma = \frac{1}{\beta_+} + \frac{1}{\beta_-} + \frac{1}{\alpha}. \tag{7.2.29}$$

The normalization factor γ is chosen so that $\sum_{n=1}^3 \rho_n = 1$. The QSS reduction yields equation (7.2.24) with x-independent drift and diffusion coefficients:

$$V = v(\rho_1 - \rho_2) = \frac{1}{\gamma}\left(\frac{1}{\beta_+} - \frac{1}{\beta_-}\right), \quad D = v(Z_1 - Z_2), \tag{7.2.30}$$

where Z_1, Z_2 satisfy the equations

$$-\beta_+ Z_1 - \alpha(Z_1 + Z_2) = (V - v)\rho_1,$$
$$-\beta_- Z_2 - \alpha(Z_1 + Z_2) = (V + v)\rho_2.$$

Using the expressions for ρ_n and γ, the solutions can be written as

$$Z_1 = \frac{1}{\gamma^2\beta_+\beta_-\alpha}\left[\frac{(v - V)(\beta_- + \alpha)}{\beta_+} + \frac{(v + V)\alpha}{\beta_-}\right],$$

$$Z_2 = -\frac{1}{\gamma^2\beta_+\beta_-\alpha}\left[\frac{(v + V)(\beta_+ + \alpha)}{\beta_-} + \frac{(v - V)\alpha}{\beta_+}\right].$$

Noting that

$$v - V = \frac{2\alpha + \beta_-}{\alpha\beta_-\gamma}v, \quad v + V = \frac{2\alpha + \beta_+}{\alpha\beta_+\gamma}v,$$

we finally obtain the results

$$D = \frac{(v - V)^2}{\gamma\beta_+^2} + \frac{(v + V)^2}{\gamma\beta_-^2}. \tag{7.2.31}$$

Finally, note that for more complex models such as tug-of-war, one has to solve equation (7.2.26) for Z_n numerically. This will be illustrated in Sect. 7.3.3, where the QSS reduction is used to explore how motor transport is affected by local chemical signaling.

7.2.3 Active transport on microtubular networks

So far we have considered a 1D model of active transport. In the case of axonal transport in neurons, the microtubules tend to be aligned in parallel, effectively forming an overlapping array that extends all the way from the cell body to the axon terminal, see Fig. 7.1. Hence, to a first approximation, one can treat the whole axon as a 1D track even though individual microtubules have lengths of around $100\,\mu$m. On the other hand, intracellular transport within the soma of neurons and most non-polarized animal cells occurs along a microtubular network that projects radially from organizing centers (centrosomes) with outward polarity [154]. This allows the delivery of cargo to and from the nucleus. Moreover, various animal viruses including HIV take advantage of microtubule-based transport in order to reach the nucleus from the cell surface and release their genome through nuclear pores [485]. In contrast, the delivery of cargo from the cell membrane or nucleus to other localized cellular compartments requires a non-radial path involving several tracks. It has also been found that microtubules bend due to large internal stresses, resulting in a locally disordered network. This suggests that *in vivo* transport on relatively short length scales may be similar to transport observed *in vitro*, where microtubular networks are not grown from centrosomes, and thus exhibit orientational and polarity disorder [426, 716]. Another example where a disordered microtubular network exists is within the *Drosophila* oocyte [58]. Kinesin and dynein motor-driven transport along this network is thought to be one of the mechanisms for establishing the asymmetric localization of four maternal mRNAs—*gurken, oskar, bicoid* and *nanos*—which are essential for the development of the embryonic body axes (see Sect. 5.7).

A detailed microscopic model of intracellular transport within the cell would need to specify the spatial distribution of microtubular orientations and polarity, in order to determine which velocity states are available to a motor-cargo complex at a particular spatial location. However, a simplified model can be obtained under the "homogenization" assumption that the network is sufficiently dense so that the set of velocity states (and associated state transitions) available to a motor complex is independent of position. In that case, one can effectively represent the active transport and delivery of cargo to an unknown target within the cell in terms of a two-dimensional (2D) or three-dimensional (3D) model of active transport [67, 70, 113, 531]. For simplicity, consider a disordered 2D microtubular network as illustrated in Fig. 7.6. (The extension to 3D networks is straightforward.) Suppose that after homogenization, a molecular motor at any point $\mathbf{r} = (x, y)$ in the plane can bind to a microtubule with any orientation θ, resulting in ballistic motion with velocity $\mathbf{v}(\theta) = v(\cos\theta, \sin\theta)$ and $\theta \in [0, 2\pi)$. If the motor is unbound then it acts as a Brownian particle with diffusion coefficient D_0. Transitions between the

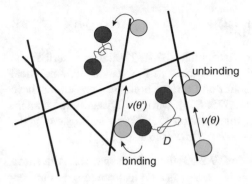

Fig. 7.6: Active 2D transport on a dis-ordered microtubular network. A particle switches between diffusion and ballistic motion in a random direction.

diffusing state and a ballistic state are governed by a discrete Markov process. The transition rate β from a ballistic state with velocity $\mathbf{v}(\theta)$ to the diffusive state is taken to be independent of θ, whereas the reverse transition rate is taken to be of the form $\alpha q(\theta)$ with $\int_0^{2\pi} q(\theta)d\theta = 1$. Suppose that at time t the motor is under-going ballistic motion. Let $(X(t), Y(t))$ be the current position of the motor particle and let $\Theta(t)$ denote the corresponding velocity direction. Introduce the conditional probability density $p(x, y, \theta, t)$ such that $p(x, y, \theta, t)dxdyd\theta$ is the joint probability that $(x, y, \theta) < (X(t), Y(t), \Theta(t)) < (x + dx, y + dy, \theta + d\theta)$, given that the particle is in the ballistic phase. Similarly, take $p_0(x, y, t)$ to be the corresponding condition-al probability density if the particle is in the diffusive phase. (For the moment the initial conditions are left unspecified). The evolution of the probability densities for $t > 0$ can then be described in terms of the following 2D system of PDEs [113]:

$$\frac{\partial p}{\partial t} = -\nabla \cdot (\mathbf{v}(\theta)p) - \frac{\beta}{\varepsilon} p(\mathbf{r}, \theta, t) + \frac{\alpha q(\theta)}{\varepsilon} p_0(\mathbf{r}, t), \tag{7.2.32a}$$

$$\frac{\partial p_0}{\partial t} = D_0 \nabla^2 p_0 + \frac{\beta}{\varepsilon} \int_0^{2\pi} p(\mathbf{r}, \theta', t)d\theta' - \frac{\alpha}{\varepsilon} p_0(\mathbf{r}, t). \tag{7.2.32b}$$

In the case of a uniform density, $q(\theta) = 1/(2\pi)$, equations (7.2.32a) and (7.2.32b) reduce to a 2D model of active transport considered by Benichou *et. al.* [67, 70, 531].

We will consider a simplified version of the model obtained by setting $D_0 = 0$ and $\partial p_0/\partial t = 0$ in equation (7.2.32b) so that $\alpha p_0 = \beta \int_0^{2\pi} p(\mathbf{r}, \theta', t)d\theta'$. This gives

$$\frac{\partial p}{\partial t} = -\mathbf{v}(\theta) \cdot \nabla p(\mathbf{r}, \theta, t) - \frac{\beta}{\varepsilon} p(\mathbf{r}, \theta, t) + \frac{\beta q(\theta)}{\varepsilon} P(\mathbf{r}, t), \tag{7.2.33}$$

with

$$P(\mathbf{r}, t) = \int_0^{2\pi} p(\mathbf{r}, \theta, t)d\theta. \tag{7.2.34}$$

Equation (7.2.33) is a special case of a general class of velocity-jump models given by (see Chap. 10)

$$\frac{\partial A}{\partial t} = -\mathbf{v} \cdot \nabla A(\mathbf{r},\mathbf{v},t) - \frac{\beta}{\varepsilon}A(\mathbf{r},\mathbf{v},t) + \frac{\beta}{\varepsilon}\int_V T(\mathbf{v},\mathbf{v}')A(\mathbf{r},\mathbf{v}',t)d\mathbf{v}', \quad (7.2.35)$$

where $T(\mathbf{v},\mathbf{v}')$ is the so-called turning distribution [374]. The latter describes the probability that a particle with velocity $\mathbf{v}'$ switches to the velocity $\mathbf{v}$. Analytical properties of the associated turning operator and various biological constraints have been discussed in detail elsewhere [374, 375, 626], particularly within the context of animal movement models. We recover equation (7.2.33) by setting $\mathbf{v} = \mathbf{v}(\theta)$, $A(\mathbf{r},\mathbf{v}(\theta),t) \to p(\mathbf{r},\theta,t)$ and $T(\mathbf{v},\mathbf{v}')d\mathbf{v}' \to q(\theta)d\theta'$, see also [375].

Quasi-steady-state reduction. In equation (7.2.33) the units of space and time have been fixed according to $l = 1$ and $l/v = 1$, where l is a typical run length. Furthermore, for the given choice of units, it has been assumed that there exists a small parameter $\varepsilon \ll 1$ such that all transition rates are $O(\varepsilon^{-1})$ and all velocities are $O(1)$. In the limit $\varepsilon \to 0$, the system rapidly converges to the space-clamped (i.e., $\nabla p = 0$) steady-state distribution $p^*(\theta) = q(\theta)$. As in the 1D case, the QSS reduction is based on the assumption that for $0 < \varepsilon \ll 1$, solutions remain close to the steady-state solution. Hence, we set

$$p(\mathbf{r},\theta,t) = P(\mathbf{r},t)q(\theta) + \varepsilon w(\mathbf{r},\theta,t), \tag{7.2.36}$$

where

$$\int_0^{2\pi} w(\mathbf{r},\theta,t)d\theta = 0. \tag{7.2.37}$$

Integrating equation (7.2.33) with respect to θ and using the normalization of $q(\theta)$ yields

$$\frac{\partial P}{\partial t} = -\langle\langle \mathbf{v}(\theta) \cdot \nabla p \rangle\rangle = -\langle \mathbf{v} \cdot \nabla P \rangle - \varepsilon \langle\langle \mathbf{v} \cdot \nabla w \rangle\rangle, \tag{7.2.38}$$

where $\langle f \rangle = \int_0^{2\pi} q(\theta)f(\theta)d\theta$ and $\langle\langle f \rangle\rangle = \int_0^{2\pi} f(\theta)d\theta$ for any function or vector component $f(\theta)$. Next, substituting equation (7.2.36) into equation (7.2.33) gives

$$q(\theta)\frac{\partial P}{\partial t} + \varepsilon\frac{\partial w}{\partial t} = -\mathbf{v}(\theta) \cdot \nabla[q(\theta)P + \varepsilon w] - \beta w. \tag{7.2.39}$$

Now substitute for $\partial P/\partial t$ in equation (7.2.39) using equation (7.2.38), and introduce the asymptotic expansion $w \sim w_0 + \varepsilon w_1 + O(\varepsilon^2)$. Collecting terms to leading order in ε we have

$$\beta w_0(\mathbf{r},\theta,t) \sim q(\theta)(\langle \mathbf{v} \rangle - \mathbf{v}(\theta)) \cdot \nabla P. \tag{7.2.40}$$

Finally, setting $w = w_0$ in equation (7.2.38) yields to $O(\varepsilon)$ the FP equation

$$\frac{\partial P}{\partial t} = -\langle \mathbf{v} \rangle \cdot \nabla P + \varepsilon \nabla \cdot (\mathbf{D}\nabla P). \tag{7.2.41}$$

In components,

$$\nabla \cdot (\mathbf{D}\nabla P) = \sum_{i=1,2} \sum_{j=1,2} D_{ij} \frac{\partial^2 P}{\partial x_i \partial x_j},$$

with the diffusion tensor having the components

$$D_{ij} = \frac{1}{\beta} \int_0^{2\pi} v_i(\theta) q(\theta) [v_j(\theta) - \langle v_j \rangle] \, d\theta. \tag{7.2.42}$$

We have dropped $O(\varepsilon)$ corrections to the drift velocity.

In the case of a uniform direction distribution $q(\theta) = 1/(2\pi)$, the diffusion tensor reduces to a scalar and the turning operator generates a so-called Pearson random walk [375]. This follows from the fact that $v_1 = v\cos\theta, v_2 = v\sin\theta$ so $\langle v_1 \rangle = \langle v_2 \rangle = \langle v_1 v_2 \rangle = 0$ and to leading order

$$D_{11} = \frac{v^2}{2\beta} = D_{22}, \quad D_{12} = 0. \tag{7.2.43}$$

More generally, assuming that $T(\theta)$ is sufficiently smooth, we can expand it as a Fourier series,

$$q(\theta) = \frac{1}{2\pi} + \frac{1}{\pi} \sum_{n=1}^{\infty} (\tau_n \cos(n\theta) + \hat{\tau}_n \sin(n\theta)). \tag{7.2.44}$$

Assume further that $\tau_1 = \hat{\tau}_1 = 0$ so there is no velocity bias i.e., $\langle v_1 \rangle = \langle v_2 \rangle = 0$. Then

$$D_{11} = \frac{v^2}{\beta} \int_0^{2\pi} \cos^2(\theta) q(\theta) d\theta = \frac{v^2}{2\beta} (1 + \tau_2),$$

$$D_{22} = \frac{v^2}{\beta} \int_0^{2\pi} \sin^2(\theta) q(\theta) d\theta = \frac{av^2}{2\beta} (1 - \tau_2), \tag{7.2.45}$$

$$D_{12} = \frac{v^2}{\beta} \int_0^{2\pi} \sin(\theta) \cos(\theta) q(\theta) d\theta = \frac{v^2}{2\beta} \hat{\tau}_2.$$

It follows that only the second terms in the Fourier series expansion contribute to the diffusion tensor. Further aspects of the QSS reduction of 2D models are considered in Ex. 7.1.

Random velocity field model. An alternative formulation of transport on disordered microtubular networks has been developed by Kahana *et. al.* [426] in terms of random velocity fields [673, 902]. In order to describe the basic idea, consider the simplified model analyzed in Ref. [902]. The latter model consists of a set of equally spaced parallel tracks along the x-axis, say, see Fig. 7.7. The tracks are assigned random polarities ± 1 with equal probabilities corresponding to quenched polarity disorder. A particle undergoes a random walk in the y-direction, whereas when a particle attaches to a certain track it moves ballistically with velocity ± 1 according to the track's polarity. It is assumed that when a particle hops to a neighboring track it binds immediately. Let $X(t)$ denote the displacement of a random walker in the longitudinal direction at time t:

$$X(t) = \int_0^t v[y(t')]dt'. \tag{7.2.46}$$

Taking the continuum limit in the y-direction means that

$$p(y,t) = \frac{1}{\sqrt{4\pi Dt}} e^{-y^2/4Dt},$$

where D is the diffusion coefficient, and the velocity field is delta-correlated, that is, $\langle v(y)v(y')\rangle_c = v^2\xi\delta(y-y')$. Here averaging is taken with respect to the quenched polarity disorder and ξ is the infinitesimal spacing between tracks. Now consider the second moment $\langle\langle X^2(t)\rangle\rangle$ of the stochastic process averaged with respect to the quenched disorder and realizations of the random walk:

$$\langle\langle X^2(t)\rangle\rangle = 2\int_0^t dt_1 \int_0^{t_1} dt_2 \langle\langle v[y(t_1)]v[y(t_2)]\rangle\rangle, \tag{7.2.47}$$

where

$$\langle\langle v[y(t_1)]v[y(t_2)]\rangle\rangle = \int_{-\infty}^{\infty} dy_1 \int_{-\infty}^{\infty} dy_2 \langle v(y_1)v(y_2)\rangle_c$$
$$\times p(y_2,t_2)p(y_1-y_2,t_1-t_2). \tag{7.2.48}$$

Using Laplace transforms and the velocity correlation function,

$$\langle\langle\widetilde{X}^2(s)\rangle\rangle = \frac{2v^2\xi}{s}\widetilde{p}(0,s)\int_{-\infty}^{\infty}\widetilde{p}(y,s)dy, \tag{7.2.49}$$

with

$$\widetilde{p}(y,s) = \frac{1}{\sqrt{4Ds}}e^{-|y|\sqrt{s/D}}.$$

Performing the integration with respect to y thus shows that

$$\langle\langle\widetilde{X^2}(s)\rangle\rangle = v^2\xi D^{-1/2}s^{-5/2},$$

Fig. 7.7: Random velocity model of a microtubular network with quenched polarity disorder. Particles move ballistically along parallel tracks in a direction determined by the polarity of the given track. They also hop between tracks according to an unbiased random walk.

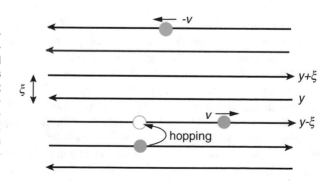

which on inverting the Laplace transform gives

$$\langle\langle X^2(t)\rangle\rangle = \frac{4v^2\xi}{3\sqrt{\pi D}}t^{3/2}. \tag{7.2.50}$$

An equivalent formulation of the problem is to treat $\langle\langle X^2(t)\rangle\rangle$ as the solution to the differential equation [426]

$$\frac{d^2}{dt^2}\langle\langle X^2(t)\rangle\rangle = 2v^2\xi y p(0,t), \tag{7.2.51}$$

where $\xi p(0,t)$ is the probability of y returning to the origin at time t within a single lattice spacing ξ, and $p(0,t) = 1/\sqrt{4\pi Dt}$. In conclusion, the random velocity model supports anomalous superdiffusion in the x-direction.

In Ref. [426] the above construction was extended to 2D (and 3D) disordered networks where there are parallel tracks in the x- and y-directions. The distribution of polarities are unbiased in both directions. A self-consistent description of the dynamics is obtained by taking

$$\frac{d^2}{dt^2}\langle\langle X^2(t)\rangle\rangle = 2v^2\xi p_y(0,t), \quad \frac{d^2}{dt^2}\langle\langle Y^2(t)\rangle\rangle = 2v^2\xi p_x(0,t), \tag{7.2.52}$$

where p_x and p_y are the probability densities of the x and y coordinates. From the symmetry of the network, $p_x(0,t) = p_y(0,t)$. Hence, assuming that $p_x(0,t) = C\langle\langle X^2(t)\rangle\rangle^{-1/2}$ for some constant C, and setting $\phi(t) = \langle\langle X^2(t)\rangle\rangle$ gives

$$\phi^{1/2}\frac{d^2}{dt^2}\phi = 2Cv^2\xi. \tag{7.2.53}$$

It follows that $\phi(t) \sim t^{4/3}$ so that the diffusion is less enhanced than in the case of parallel tracks in one direction. Finally, note that active transport on the randomly oriented network of Fig. 7.6 exhibits normal rather than anomalous diffusion. A major difference from the random velocity model is that the latter has quenched polarity disorder, whereas the former has dynamical polarity disorder.

7.2.4 Virus trafficking

An interesting example of active transport in 2D or 3D is given by virus trafficking. An animal virus typically invades a mammalian cell by first undergoing membrane endocytosis from the exterior to the interior of the cell. It then has to navigate the crowded cytoplasm without being degraded in order to reach a nuclear pore and deliver its DNA to the cell nucleus [206]. Single-particle tracking has established that virus trajectories within the cytoplasm consist of a succession of free or confined diffusion and ballistic periods involving active transport along microtubules or actin networks [107]. A macroscopic computational model of the trafficking of a population of viruses has been developed based on the law of mass action, which takes into account cell geometry but neglects stochastic effects [218]. On the other hand,

Holcman and collaborators [384, 484, 485] have developed a stochastic model of a single virus trafficking inside a cell, which involves reducing an active transport model to an effective Langevin equation, and using the latter to calculate the mean time to reach a nuclear pore based on a narrow escape problem (see Sect. 6.4). A more accurate reduction method has subsequently been developed by Lawley et al. [494], and we will follow their formulation here.

The disc. First, consider a 2D model in which the cell is represented as a radially symmetric disc Ω_2 consisting of an annular region of cytoplasm of outer radius R and inner radius δ, surrounding a central nuclear disc (see Fig. 7.8). N microtubules radiate outwards from the nucleus to the cell membrane, and are assumed to be distributed uniformly so that the angle between two neighboring microtubules is $\Upsilon = 2\pi/N$. (A 2D description of a cell would be reasonable in the case of cultured cells that are flattened due to adhesion to the substrate). The motion of a virus particle alternates between diffusive motion within a wedge region $\widehat{\Omega}$ subtending an angle Υ at the origin, and binding to one of the two microtubules at the boundary of the wedge. Following Lawley et al. [494], we will derive an effective advection–diffusion equation for virus transport by considering the dynamics of a single virus moving within a single slice $\mathcal{U}_2 \equiv [\delta, R] \times [0, \Upsilon] \subset \Omega_2$—restriction to a single slice is allowed because of the symmetric partitioning and the fact that we are only interested in the radial distribution of the virus.

Therefore, consider a single virus originating on the outer boundary $r = R$ and undergoing Brownian motion in the interior of $\mathcal{U}_2$ until it reaches a microtubule, after which it binds to the microtubule and moves ballistically toward the origin for some exponentially distributed amount of time. At this point the motor-cargo complex is reinserted into the slice at the current radius for some randomly selected angle between 0 and Υ. If $X(t)$ represents the motor's radial distance from the origin and $\theta(t)$ represents some angle between $[0, \Upsilon]$, the motor's motion is described by the following system of SDEs [484, 494], see also equation (2.3.28):

$$
dX = \begin{cases} -V dt, & \theta = 0, \Upsilon \\ (D/X)dt + \sqrt{2D}dW_X, & \theta \in (0, \Upsilon) \end{cases},
$$

$$
d\theta = \begin{cases} 0, & \theta = 0, \Upsilon \\ (\sqrt{2D}/X)dW_\theta, & \theta \in (0, \Upsilon) \end{cases}, \tag{7.2.54}
$$

where W_X, W_θ are standard independent Wiener processes, V is the motor-driven velocity of the virus along a microtubule, and D is the diffusion coefficient. In Ref. [494], a coarse graining method is used to derive a single effective SDE describing the overall radial motion of a particle evolving according to equation (7.2.54). They assume there is a continuous-time jump Markov process underlying the particle's switching between diffusive and ballistic dynamics, and that the dynamics of the Markov process are fast relative to all other processes. Suppose that the virus has just been released back into the cytoplasm. Ignoring radial motion, the amount of time it takes the virus to reach a microtubule again is

$$T(r) = \frac{1}{\Upsilon} \int_0^\Upsilon \tau(\theta, r) d\theta, \tag{7.2.55}$$

where τ satisfies the boundary value problem

$$\frac{D}{r^2} \frac{d^2}{d\theta^2} \tau(\theta, r) = -1, \quad \tau(0, r) = 0 = \tau(\Upsilon, r). \tag{7.2.56}$$

For a particle diffusing in the interval $[0, \Upsilon]$ with effective diffusion coefficient D/r^2, the quantity $T(r)$ can be interpreted as the MFPT to reach either 0 or Υ given a uniform initial position. Hence,

$$T(r) = \frac{\Upsilon^2 r^2}{12D}. \tag{7.2.57}$$

Let μ denote the mean for the exponential distribution that dictates the amount of time a particle spends in the ballistic phase. It follows that the mean time on a microtubule is $\mu/(\mu + T(r))$.

Under the above adiabatic (or quasi-steady-state) approximation, one obtains the following coarse-grained effective SDE approximation to equation (7.2.54):

$$dX = \left(\frac{D}{X} \frac{T(X)}{\mu + T(X)} - V \frac{\mu}{\mu + T(X)} \right) dt + \sqrt{2D \frac{T(X)}{\mu + T(X)}} dW, \tag{7.2.58}$$

where $W(t)$ is a standard Wiener process. Let $p(r,t)$ represent the probability that a particle evolving according to equation (7.2.58) is at a distance r from the origin at time t. The corresponding FP equation is

$$\frac{\partial p}{\partial t} = -\frac{\partial}{\partial r} \left(\left[\frac{D}{r} \frac{T(r)}{\mu + T(r)} - V \frac{\mu}{\mu + T(r)} \right] p \right) + \frac{\partial^2}{\partial r^2} \left(D \frac{T(r)}{\mu + T(r)} p \right), \tag{7.2.59}$$

with $\int_a^b p(r,t) dr = \mathbb{P}[a \leq X(t) \leq b]$. (Note that there is no factor of r in the integral measure.) Lawley et al. [494] carried out Monte Carlo simulations of the effective SDE (7.2.58) and showed that the results matched the behavior of the full inter-

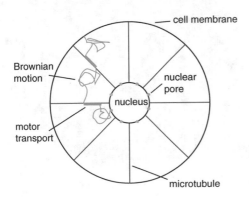

cell membrane

Brownian motion

nuclear pore

nucleus

motor transport

microtubule

Fig. 7.8: Model of Lagache et al [484]. Diagram of a 2D radially symmetric cell with radially equidistant microtubules. A virus trajectory is shown that alternates between ballistic motion along a microtubule and diffusion of the cytoplasm. The trajectory starts at the cell membrane and ends at a nuclear pore.

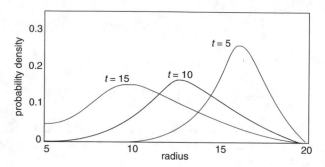

Fig. 7.9: Sketch of distribution of radial position for a 2D cell at times $t = 5, 10, 15$ s based on simulations of the SDE (7.2.58. Each distribution is calculated from 10^5 trials and N = 48. Other parameters are cell radius $R = 20\,\mu$m, nuclear radius $\delta = 5\,\mu$m, cytoplasmic diffusion coefficient $D = 1.3\,\mu$m^2/s, ballistic velocity $V = 0.7\,\mu$m/s, and the average time on a microtubule to be $\mu = 1$. [Redrawn from Lawley et al [494].]

mittent system (7.2.54) very well. Example distributions for the radial position as a function of time are shown in Fig. 7.9.

The sphere. Now suppose that the cell is represented by a sphere Ω_3 as shown in Fig. 7.10. It is natural to partition the sphere according to $\Omega_3 = \omega_1 \cup \omega_2$ with ω_1 defined in the following way. Let N be the number of microtubules emanating out from the small sphere of radius δ enveloping the origin. These can be modeled as cylinders, each of radius ε. Let $\mathbf{c}_i$ for $i = 1...N$ denote a randomly selected fixed position on the δ-sphere. Then each microtubule $\mathcal{M}_i$ is defined as follows:

$$\mathcal{M}_i \equiv \left\{ \mathbf{x} \in \Omega_3 \,\middle|\, \|\mathbf{x} - \rho\mathbf{c}_i\| \leq \varepsilon, \rho \in [\delta, R) \right\}.$$

We take $\omega_1 = \cup_{i=1}^{N} \mathcal{M}_i$ and $\omega_2 = \Omega_3 \setminus \omega_1$. To model the dynamics of motor-cargo complexes in this domain, we must derive PDEs from the SDEs describing the mo-

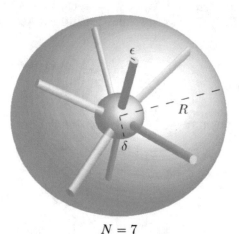

Fig. 7.10: Sketch of Ω_3 showing $N = 6$ microtubules radiating from center.

$N = 7$

tion of a single particle in this domain. We assume a single particle's motion is characterized by standard Brownian motion in ω_2 until it reaches a microtubule, when it undergoes ballistic motion with fixed velocity V toward the origin for some exponentially distributed time. The particle is then released at a random position in Ω_3 with radius equal to how far it reached with ballistic motion. Following along identical lines to the example of a disc, one obtains the following SDE as a coarse-grain approximation to a particle moving through Ω_3 [494]

$$dX = \left(\frac{2D}{X} \frac{T(X)}{\mu + T(X)} - V \frac{\mu}{\mu + T(X)} \right) dt + \sqrt{2D \frac{T(X)}{\mu + T(X)}} dW. \qquad (7.2.60)$$

The calculation of the MFPT $T(r)$ at radial position r is more involved than the 2D case. However, we note that it reduces to a narrow escape problem for diffusion on the surface of a sphere of radius r with N small targets of size ε (see also Sect. 6.4). Coombs, Straube, and Ward [193] provide the following asymptotic approximation for $T(r)$ in the small ε limit:

$$T(r) = \frac{r^2}{D} \left[-\frac{2}{N} \ln\left(\frac{\varepsilon}{r} \right) + \ln 4 - 1 - \frac{4}{N^2} \Psi \right],$$

with

$$\Psi = \sum_{k=1}^{N} \sum_{j=k+1}^{N} \ln \| \mathbf{c}_k - \mathbf{c}_j \|.$$

Let $p(\rho,t)$ represent the probability that a particle is at position ρ at time t. The Fokker–Planck equation associated with equation (7.2.60) is

$$\frac{\partial p}{\partial t} = -\frac{\partial}{\partial \rho} \left[\left(\frac{2D}{\rho} \frac{T(\rho)}{\mu + T(\rho)} - V \frac{\mu}{\mu + T(\rho)} \right) p \right] + D \frac{\partial^2}{\partial \rho^2} \left(\frac{T(\rho)}{\mu + T(\rho)} p \right).$$
$$(7.2.61)$$

Again numerical simulations of the effective SDE compare favorably with the full switching system [494].

7.3 Intracellular transport as a random search-and-capture process

In Sect. 7.2 we focused on the stochastic dynamics of individual motor complexes. However, we ignored the crucial role of these complexes in delivering vesicular cargo to subcellular compartments such as synapses, see Sect. 7.1 and Fig. 7.3. These compartments act as local absorbing traps or targets that are hidden from the motor complex in the absence of long-range signaling. One can thus view the active transport and delivery of vesicles to a subcellular target as a random search-and-capture process. From the perspective of the target, the efficiency of the process will be characterized in terms of the MFPT that a complex is captured by the target. In other words, we have to solve an associated FPT problem. (In the case of multiple competing targets, one has to determine the splitting probabilities and conditional MFPTs.)

The active transport of resources to subcellular targets is one example of a general class of search strategy known as random intermittent search, in which a particle randomly switches between a slow search phase and a faster non-search phase [66, 67, 70, 625]. At the macroscopic scale, intermittent motion has been observed in a variety of animal species during exploratory behavior [62, 830, 831]. One striking example is given by the nematode *C.elegans*, which alternates between a fast displacement along a straight trajectory (roaming) and a much slower displacement along a more sinuous trajectory (dwelling) [659]. During the slow phase, the worm's head, bearing most of its sensory organs, moves and touches the surface nearby suggesting that this is a search phase. As we have already described in Sect. 7.2, motor complexes can also switch between ballistic transport states and stationary or slowly diffusing states. In addition to delivering resources to subcellular compartments, motor complexes can also speed up biochemical reactions by transiently binding diffusing reactant molecules within the cellular environment [401, 531]. The effective reaction rate can then be determined by solving a first passage time (FPT) problem for random intermittent search in 3D or 2D, depending on whether or not the search domain is restricted to the plasma membrane.

A variety of stochastic models of random intermittent search processes have been developed (as reviewed in [70, 531]). In these studies it is typically assumed that (i) a particle (e.g., a motor-cargo complex) is searching for some hidden target (e.g., a subcellular compartment) within a bounded physical domain (e.g., the plasma membrane or cytoplasm), (ii) the motion of the particle is unbiased, (iii) the particle initiates its search at some random location within the domain, and (iv) the probability of eventually finding the target is equal to unity. Under these conditions, it can be shown that there exists an optimal search strategy, in the sense that the mean search time to find a single hidden target can be minimized by varying the switching rates between different states [66–68]. An analogous result holds for protein-DNA interactions (see Sect. 6.3.2). However, for some cellular processes, such as the directed transport of newly synthesized products from the nucleus to targets in the plasma membrane, assumptions (ii)–(iv) no longer hold, since the motion is biased in the anterograde direction and the initial location is always at the nucleus. Moreover, there is now a nonzero probability that the particle does not reach the target due to degradation or absorption by another target [112, 608]. Under these circumstances, an optimal search strategy no longer exists. On the other hand, the failure to find a subcellular target may be mitigated if the target is only partially hidden, in the sense that it emits a local chemical signal that increases the probability of a motor particle stopping in a neighborhood of that target [609].

In this section we will focus on solving the FPT problem for a single search-and-capture event involving a motor complex searching for one or more hidden, partially absorbing targets located along the track. The related FTP problem of a Brownian particle searching for one of N absorbing, small interior targets was analyzed in Sect. 6.4 using asymptotic methods. Another example of a totally absorbing target occurred in the analysis of virus trafficking in Sect. 7.2.4.

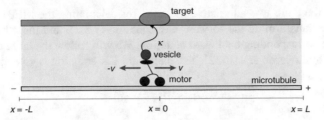

Fig. 7.11: Schematic diagram illustrating a model of a motor-driven particle moving along a 1D track of length $2L$. The particle can transition between two motiles states with speeds $\pm v$ and a stationary state. A hidden target is located at the center of the domain. The particle can be absorbed by the target at a rate κ when in the stationary state and within range of the target.

7.3.1 Optimal unbiased transport

Consider a single motor-driven particle moving along a one-dimensional track of length $2L$ according to the 3-state model of Sect. 7.2 and Fig. 7.5. We now assume that there is a hidden, partially absorbing target of width $2a$ at the center of the domain as shown in Fig. 7.11. If the particle is within a distance a of the target and is in the stationary state, then the particle can detect or, equivalently, be absorbed by the target at a rate κ. We assume throughout that $a \ll L$. Let $X(t)$ and $N(t)$ denote the random position and state of the particle at time t and define $\mathbb{P}(x,t,n \mid y,0,m)dx$ as the joint probability that $x \leq X(t) < x+dx$ and $N(t) = n$ given that initially the particle was at position $X(0) = y$ and was in state $N(0) = m$. Setting $p_n(x,t) \equiv \mathbb{P}(x,t,n|y,0,+)$ with initial condition $p_n(x,0) = \delta(x-y)\delta_{n,+}$, we have the 3-state model (see also Sect. 7.2)

$$\frac{\partial p_+(x,t)}{\partial t} = -v\frac{\partial p_+(x,t)}{\partial x} - \beta p_+(x,t) + \alpha p_0(x,t), \tag{7.3.1a}$$

$$\frac{\partial p_-(x,t)}{\partial t} = v\frac{\partial p_+(x,t)}{\partial x} - \beta p_-(x,t) + \alpha p_0(x,t), \tag{7.3.1b}$$

$$\frac{\partial p_0(x,t)}{\partial t} = \beta p_+(x,t) + \beta p_-(x,t) - 2\alpha p_0(x,t) - \kappa\chi(x)p_0(x,t). \tag{7.3.1c}$$

Here α, β are the transition rates between the stationary and mobile states and χ is the target indicator function

$$\chi(x) = \begin{cases} 1, & \text{if } |x| < a \\ 0, & \text{otherwise.} \end{cases} \tag{7.3.2}$$

Equation (7.3.1) is supplemented by a reflecting boundary condition at each end:

$$p_-(x,t) = p_+(x,t), \quad x = \pm L. \tag{7.3.3}$$

The efficacy of the search process can be characterized in terms of the MFPT to find (be absorbed by) the target. As previously noted for SDEs and discrete Markov processes, there are two alternative methods for calculating the MFPT, one based on Laplace transforming the forward CK equation (7.3.1), and the other based on solving the corresponding backward equation. We will follow the latter approach as developed by Loverdo *et al* [532]. The backward CK equation is given by

$$\frac{\partial q_+}{\partial t} = v\partial_y q_+ - \beta[q_+ - q_0], \tag{7.3.4a}$$

$$\frac{\partial q_-}{\partial t} = -v\partial_y q_- - \beta[q_- - q_0], \tag{7.3.4b}$$

$$\frac{\partial q_0}{\partial t} = \alpha[q_+ + q_- - 2q_0] - \kappa\chi(y)q_0, \tag{7.3.4c}$$

where $q_m(y,t) = \mathbb{P}(x,t,0|y,0,m)$. Let $Q_m(y,t)$ be the survival probability that the particle is absorbed by the target after time t, given that it started at y in state m. That is,

$$Q_m(y,t) = \kappa\int_t^\infty \int_{-a}^a \mathbb{P}(x,t',0|y,0,m)dxdt'. \tag{7.3.5}$$

Integrating equation (7.3.4) with respect to x and t and using

$$\frac{\partial Q_m(y,t)}{\partial t} = -\kappa\int_{-a}^a \mathbb{P}(x,t,0|y,0,m)dx,$$

we find that

$$\frac{\partial Q_+}{\partial t} = v\partial_y Q_+ + \beta(Q_0 - Q_+), \tag{7.3.6a}$$

$$\frac{\partial Q_-}{\partial t} = -v\partial_y Q_- + \beta(Q_0 - Q_-), \tag{7.3.6b}$$

$$\frac{\partial Q_0}{\partial t} = \alpha[Q_+ + Q_- - 2Q_0] - \kappa\chi(y)Q_0. \tag{7.3.6c}$$

Let $T_m(y)$ be the MFPT to find the target, given that the particle is at position y and in state m at $t = 0$. Then

$$T_m(y) = -\int_0^\infty t\partial_t Q_m(y,t)dt = \int_0^\infty Q_m(y,t)dt, \tag{7.3.7}$$

after integration by parts. It follows that T_m evolves according to the equations

$$v\frac{\partial T_+}{\partial y} + \beta(T_0 - T_+) = -1, \tag{7.3.8a}$$

$$-v\frac{\partial T_-}{\partial y} + \beta(T_0 - T_-) = -1, \tag{7.3.8b}$$

$$\alpha(T_+ + T_-) - (2\alpha + \kappa\chi(y))T_0 = -1. \tag{7.3.8c}$$

Solving equation (7.3.8c) for T_0 yields

$$T_0(y) = u(y)(\alpha[T_+(y) + T_-(y)] + 1), \tag{7.3.9}$$

where

$$u(y) = \frac{1}{2\alpha + \kappa\chi(y)}. \tag{7.3.10}$$

Substituting (7.3.9) into (7.3.8a,7.3.8b) gives

$$\frac{\partial T_+}{\partial y} + \frac{\beta}{v}[(\alpha u(y) - 1)T_+(y) + \alpha u(y)T_-(y)] = -\frac{\beta}{v}\left(\frac{1}{\beta} + u(y)\right), \tag{7.3.11a}$$

$$\frac{\partial T_-}{\partial y} - \frac{\beta}{v}[\alpha u(y)T_+(y) + (\alpha u(y) - 1)T_-(y)] = \frac{\beta}{v}\left(\frac{1}{\beta} + u(y)\right). \tag{7.3.11b}$$

It is now necessary to solve for $T_\pm(y)$ in the three regions: $-L < y < -a$, $-a < y < a$, and $a < y < L$. The solution in each of these regions will have two unknown integration constants so that we require six conditions. Two are given by the boundary conditions $T_+(y) = T_-(y)$ for $y = \pm L$, whereas the other four are obtained by requiring continuity in $T_+(y)$ and $T_-(y)$ at $y = \pm a$, see Ex. 7.2. Suppose that the particle starts at a random position inside the domain, and is initially in the $+$ state. We then define the averaged MFPT τ_1 according to

$$\tau_1 = \frac{1}{2L}\int_{-L}^{L} T_+(y)dy = \frac{1}{2L}\int_0^L [T_+(y) + T_-(y)]dy.$$

From the analysis of $T_\pm(y)$, one finds that [68] (see Ex. 7.2)

$$\tau_1 = \frac{1}{L}\left(\frac{1}{\beta} + \frac{1}{2\alpha}\right)\left(\left[\frac{\beta}{v}\right]^2\frac{(L-a)^3}{3} + \frac{\beta}{v}\sqrt{\frac{2\alpha + \kappa}{\kappa}}(L-a)^2\coth(\Lambda a)\right. \tag{7.3.12}$$

$$\left. + \frac{2\alpha + \kappa}{\kappa}(L-a)\right) + \frac{2\alpha + \kappa}{\kappa\beta} + \frac{1}{\kappa}, \tag{7.3.13}$$

where

$$\Lambda = \frac{\beta}{v}\sqrt{\frac{\kappa}{2\alpha + \kappa}}.$$

This expression can be simplified by taking $L \gg a$ and $\Lambda a \ll 1$:

$$\tau_1 \approx \left(\frac{1}{\beta} + \frac{1}{2\alpha}\right)\left(\left[\frac{\beta}{v}\right]^2\frac{L^2}{3} + \frac{2\alpha + \kappa}{\kappa}\frac{L}{a}\right). \tag{7.3.14}$$

It is then straightforward to show that τ_1 has a global minimum as a function of the parameters α, β, which occurs at the values (α^*, β^*) with

$$\frac{1}{\alpha^*} = 2\sqrt{\frac{a}{v\kappa}}\left(\frac{L}{12a}\right)^{1/4}, \quad \frac{1}{\beta^*} = \frac{a}{v}\sqrt{\frac{L}{3a}}.$$

It is possible to extend the above analysis to a modified target detection scheme, in which the particle slowly diffuses in the search phase and is immediately absorbed

(a)   (b)

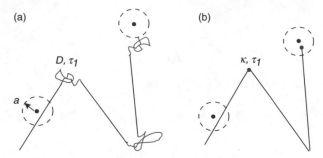

Fig. 7.12: Two models of target detection: the particle alternates between slow reactive phases of mean duration τ_1, and fast non-reactive ballistic phases of mean duration τ_2. (a) The slow reactive phase is diffusive and detection is infinitely efficient. (b) The slow reactive phase is static and detection takes place with finite rate κ.

by the target if it enters the target domain ($\kappa \to \infty$), see Fig. 7.12. In this case, an intermittent search scheme is more efficient than pure diffusion provided that $D/v \ll a$, where D is the diffusivity of the particle in the search phase, v is its speed in a ballistic non-search phase, and a is the size of the target [532]. Interestingly the optimal time spent in the non-search phase, $1/\beta^*$, is independent of the particular target detection mechanism. The existence of a minimum search time for unbiased intermittent search in a bounded domain also extends to higher spatial dimensions [67, 113, 532] and to more detailed molecular motor models such as tug-of-war [610, 611]. However, in these more complicated cases, the calculation of the MFPT becomes considerably more difficult unless some approximation scheme is used such as the quasi-steady-state reduction outlined in Sect. 7.2 [608, 609], see below.

7.3.2 Biased cargo transport

The existence of an optimal search strategy breaks down if one considers a biased search process and allows for the possibility of failure to find the target [112, 608]. In order to illustrate this, we modify the previous 1D model (7.3.1) along the following lines. First the transition rates from the left and right moving states are taken to be different, $\beta \to \beta_\pm$ with $\beta_+ < \beta_-$. The stochastic process is then biased in the anterograde direction since the particle tends to spend more time in the right-moving state. (One could also take the velocities in the two directions to be different). Unidirectional transport is obtained in the limit $\beta_- \to \infty$. Second, the possibility of failure is incorporated into the model by considering a 1D track of length L with a reflecting boundary at $x = 0$ and an absorbing boundary at $x = L$:

$$p_-(0,t) = p_+(0,t), \quad p_-(L,t) = 0. \tag{7.3.15}$$

The absorbing boundary takes into account the fact that a motor particle can be degraded or absorbed by other targets downstream to the given target. Third, the particle always starts from the end $x = 0$ (which could be close to the cell nucleus) and the target is at some unknown location X with $0 < X - a < X + a < L$. In contrast to the unbiased case, the efficacy of the search process is characterized by two quantities. Let $J(t)$ denote the probability flux due to absorption by the target at X, given that the particle started at $x(0) = 0$ in the right-moving state:

$$J(t) = \kappa \int_{X-a}^{X+a} p_0(x,t)dx. \tag{7.3.16}$$

Define the hitting probability Π to be the probability that the particle eventually finds the target, that is, it is absorbed somewhere in the interval $X - a \leq x \leq X + a$ rather than at the end $x = L$:

$$\Pi = \int_0^\infty J(t)dt. \tag{7.3.17}$$

The conditional mean first passage time (MFPT) T is then defined to be the mean time it takes for the particle to find the target given that it is not absorbed at $x = L$ [674]:

$$T = \frac{\int_0^\infty tJ(t)dt}{\int_0^\infty J(t)dt}. \tag{7.3.18}$$

The quantities T and Π determine the efficiency of the stochastic search process.

Clearly it would be advantageous for the particle to minimize the search time T and maximize the hitting probability Π. However, these two requirements compete with each other so that, in contrast to unbiased intermittent search with $\Pi = 1$, there is not a single optimal search strategy. This can be seen heuristically in the case of unidirectional transport where the particle is either stationary or undergoes antero-grade motion. Here the particle can reach the target more quickly by having a higher probability of being in the mobile state. However, this also increases the chance of overshooting the target without detecting it, thus reducing the hitting probability. It could be argued that the only important factor is minimizing the MFPT irrespective of the hitting probability, since active transport typically involves multiple motor-cargo complexes. However, a low hitting probability would require more resources, which costs the cell energy.

In the case of unidirectional transport and a target at $x = X$, equation (7.3.1) reduces to

$$\frac{\partial p_+}{\partial t} = -v\frac{\partial p_+}{\partial x} + \alpha p_0 - \beta p_+, \tag{7.3.19a}$$

$$\frac{\partial p_0}{\partial t} = \beta p_+ - \alpha p_0 - \kappa\chi(x-X)p_0. \tag{7.3.19b}$$

Note that there is no need to introduce any supplementary boundary conditions, since the particle cannot return to the origin nor find the target once it has crossed the point $x = X + a$. Introduce the survival probability $Q_m(y,t)$ that the particle has not yet been absorbed by the target at time t, given that it started at y in state m:

$$Q_m(y,t) = \int_0^\infty [p_0(x,t|y,t,m) + p_+(x,t|y,t,m)]dx. \tag{7.3.20}$$

Note that $Q_m(y,t) = 1$ for $y > X + a$. Setting $m = +$, $y = 0$, and $p_n(x,t) = p_n(x,t|0,0,+)$, we have

$$\frac{\partial Q_+(0,t)}{\partial t} = \int_0^\infty \frac{\partial[p_0(x,t)+p_+(x,t)]}{\partial t}dx$$

$$= \int_0^\infty \left[-v\frac{\partial p_+(x,t)}{\partial x} - \kappa\chi(x-X)p_0(x,t)\right]dx$$

$$= -\kappa\int_{X-a}^{X+a} p_0(x,t)dx = -J(t),$$

where $J(t)$ is the probability flux into the target. It follows that

$$\Pi = 1 - \lim_{t\to\infty} Q_+(0,t), \quad T = \Pi^{-1}\int_0^\infty [Q_+(0,t)-(1-\Pi)]dt. \qquad (7.3.21)$$

The survival probabilities evolve according to the backward CK equations

$$\frac{\partial Q_+}{\partial t} = v\frac{\partial Q_+}{\partial y} - \beta[Q_+ - Q_0], \qquad (7.3.22a)$$

$$\frac{\partial Q_0}{\partial t} = \alpha[Q_+ - Q_0] - \kappa\chi(y-X)Q_0. \qquad (7.3.22b)$$

Using Laplace transforms, one finds that [112, 608]

$$\Pi = 1 - e^{-2\lambda a}, \quad T = (X-a)\mu_1 - \frac{2a\mu_2}{e^{2\lambda a}-1} + \mu_3, \qquad (7.3.23)$$

where

$$\mu_1 = \frac{1}{v}\left(1+\frac{\beta}{\alpha}\right), \mu_2 = \frac{1}{v}\left(1+\frac{\alpha\beta}{(\alpha+\kappa)^2}\right), \mu_3 = \frac{\alpha+\beta+\kappa}{\beta\kappa}, \lambda = \frac{\beta}{v}\frac{\kappa}{\alpha+\kappa}.$$

The calculation of Π and T can also be carried out for the full three-state model, although the analysis is considerably more involved [112]. The results are illustrated in Fig. 7.13. It can be seen that increasing the parameter α, which controls how much time the particle spends in the stationary search mode, decreases both the hitting probability and the conditional MFPT. Similarly, increasing the parameter β_+, which controls how much time the particle spends in the anterograde mobile state, increases both the hitting probability and the MFPT. During unidirectional motion average velocities are found in the range $0.05 - 0.2\,\mu$m/s, whereas the duration of a moving phase tends to be in the range 1-10 s. Dendrites in cultured cells range in length from $10\,to\,100\,\mu$m.

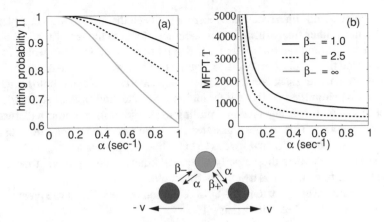

Fig. 7.13: Partially biased anterograde transport. (a) The hitting probability Π and (b) the MFPT T are plotted as functions of the parameter α for fixed $\beta_+ = 1$ s^{-1} and various values of β_- : solid black curve ($\beta_- = 1.5$ s^{-1}), dashed curve ($\beta_- = 2.5$ s^{-1}) and solid gray curve (unidirectional). Other parameter values are $X = 10\,\mu$m, $L = 20\,\mu$m, $a = 1\,\mu$m, $\kappa = 0.05$ s^{-1}, $v_\pm = 0.1\,\mu$m/s . These values are extracted from experimental studies of mRNA transport [234, 454, 704].

7.3.3 Effects of local chemical signaling

Let us now consider a much more general model of motor-driven search by incorporating a hidden target into the PDE model (7.2.14)

$$\frac{\partial p_n}{\partial t} = -v_n \frac{\partial p_n(x,t)}{\partial x} + \sum_{n'=1}^{N} A_{nn'}(x) p_{n'}(x,t) - \kappa_n \chi(x-X), \qquad (7.3.24)$$

where κ_n is the rate of target absorption in the nth internal state. Thus the flux into the target is

$$J(t) = \sum_{n=1}^{N} \kappa_n \int_{X-a}^{X+a} p_n(x,t)dx. \qquad (7.3.25)$$

In general, κ_n will only be nonzero for a subset of states. For example, in the 3-state model the unbound stationary or diffusing state is identified as the search state. In the case of the more biophysically realistic tug-of-war model (Sect. 4.5.2), the identification of the search states is more complicated. The simplest scenario is that the cargo locates its target after it becomes fully detached from the microtubule and diffuses within distance of its target, where it binds to scaffolding proteins and is separated from its molecular motors. However, if many molecular motors are bound to the cargo, the waiting time between diffusive searching events can be too large to reliably deliver the cargo. Moreover, if the cargo is large so that its diffusivity is low or the cargo is moving through a crowded and confined domain, diffusive motion may be restricted, preventing the cargo from reaching the target. Another possibility is that subcellular machinery is present to detach the cargo from its motors or inhibit the activity of the motors so that scaffolding proteins can bind to and sequester the cargo. Delivery then changes from a diffusion-limited reaction to a waiting time that depends on a reaction occurring between the motor-cargo complex and biomolecules (either freely diffusing or anchored) local to the target while the complex is moving along the microtubule. If details of the localization mechanism are unknown then the simplest model is to assume that this waiting time is approximately exponential and to associate a target detection rate κ_n with each motor state. The model can be simplified further by assuming that detection is unlikely while only one species of motors is engaged and pulling the cargo at its maximum (forward or backward) velocity. This suggests assigning a single target detection rate κ to those states that have sufficiently low speeds [611]. Thus, $\kappa_{(n_+,n_-)} = \kappa \Theta(v_h - v(n_+,n_-))$, where $v(n_+,n_-)$ denotes the velocity when n_+ kinesin and n_- dynein motors are attached to the track and v_h is a velocity threshold.

It is straightforward to extend the QSS reduction of Sect. 7.2 in the presence of a target, see Ex. 7.3, and one finds that [609]

$$\frac{\partial C}{\partial t} = -\frac{\partial}{\partial x}(VC) + \frac{\partial}{\partial x}\left(D\frac{\partial C}{\partial x}\right) - \lambda \chi(x - X)C, \qquad (7.3.26)$$

with the drift V and diffusion coefficient D given by equation (7.2.25) and the effective detection rate is

$$\lambda = \sum_{n=1}^{N} \kappa_n \rho_n(x). \qquad (7.3.27)$$

There are now three effective parameters that describe the random search process: the drift V, the diffusivity D, and the target detection rate λ. Each of these parameters are themselves functions of the various cargo velocities, transition rates, and target detection rates contained in the full model. The hitting probability and MFPT are still given by equations (7.3.17) and (7.3.18) except that now the flux is

$$J(t) = \lambda \int_{X-a}^{X+a} C(x,t)dx. \qquad (7.3.28)$$

In general, one finds that there is a playoff between minimizing the MFPT and maximizing the hitting probability [112, 608, 609]. One way to enhance the efficiency of the search process is for the target to generate a local chemical signal that increases the probability of finding the target without a significant increase in the MFPT. This issue has been explored by incorporating a local ATP or tau signal into the tug-of-war model of Sect. 4.5.2 and carrying out a QSS reduction along the lines outlined above [610, 611]. The possible role of ATP is based on the observation that the stall force and other single motor parameters are strongly dependent on the level of [ATP]. Since the ATP concentration is heavily buffered, a small region of intense ATP phosphorylation around a target could create a sharp, localized [ATP] gradient, which would significantly slow down a nearby motor complex, thus increasing the chances of target detection. Here we consider a mechanism based on local tau signaling [610], by including a tau concentration-dependent kinesin binding rate (4.5.28) in the tug-of-war model. Carrying out the QSS reduction of the tug-war-model then leads to the FP equation (7.3.26) with τ-dependent drift V, diffusivity D, and capture rate λ as illustrated in Fig. 7.14. The most significant alteration in the behavior of the motor complex is the change in the drift velocity V as a function

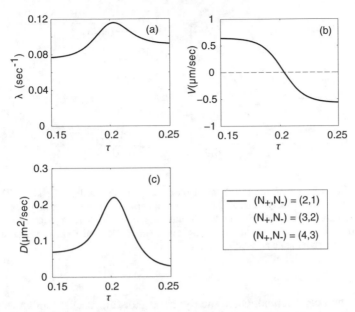

Fig. 7.14: Effects of tau concentration on the tug-of-war model with N_+ kinesin motors and N_- dynein motors. The stall force F_s, forward velocity v_f, and unbinding rate $\bar{\gamma}$ are given by equations (4.5.25)-(4.5.27) with $[ATP] = 10^3\,\mu M$. The other single motor parameters are [596]: $F_d = 3$ pN, $\gamma_0 = 1\,s^{-1}$, $\bar{\pi} = 5\,s^{-1}$, and $v_b = 0.006\,\mu m/s$. The corresponding parameters of the FP equation are obtained using a QSS reduction and plotted as a function of τ. (a) Effective capture rate λ. (b) Drift velocity V. (c) Diffusivity D.

of τ. The drift velocity switches sign when τ is increased past a critical point. That is, by reducing the binding rate of kinesin, the dynein motors become dominant, causing the motor complex to move in the opposite direction. The effects of local changes in τ concentration on the efficiency of random search can now be determined by assuming that within range of the target, $|x-X| < a$, $\tau = \tau_1 > \tau_0$, whereas $\tau = \tau_0$ outside the target, $|x-X| > a$. The FP equation (7.2.24) then has x-dependent drift and diffusivity of the form

$$V(x) = V_0 + \Delta V \chi(x), \quad D(x) = D_0 + \Delta D \chi(x), \tag{7.3.29}$$

where $\chi(x)$ is the indicator function defined in (7.3.2), $V_0 = V(\tau_0), D_0 = D(\tau_0)$, $\Delta V = V(\tau_1) - V_0$, and $\Delta D = D(\tau_1) - D_0$. Solving the piecewise-continuous FP equation then determines the hitting probability Π and MFPT T as functions of τ_1 for fixed τ_0. In Fig. 7.15, the hitting probability Π and the MFPT T are plotted as a function of τ_1. As τ_1 is increased above the critical level $\tau_0 = 0.19$, there is a sharp increase in Π but a relatively small increase in the MFPT, confirming that τ can improve the efficacy of the search process.

One interesting effect of a local increase in MAPs is that it can generate stochastic oscillations in the motion of the motor complex [611], see Fig. 7.16. As a kinesin-driven cargo encounters the MAP-coated trapping region the motors unbind at their

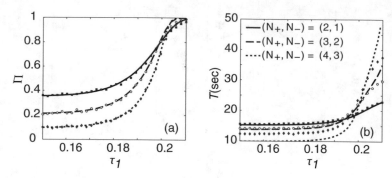

Fig. 7.15: Effect of adding tau to the target on the capture probability Π and MFPT T using parameters from Fig. 7.14. (a) The analytical approximation Π (curves) and results from Monte Carlo simulations (dots). (b) Corresponding results for the MFPT T. The synaptic trap is located at $X = 10\,\mu$m, the trapping region has radius $a = 2\,\mu$m, and the microtubular track has length $L = 20\,\mu$m. The capture rate is taken to be $\kappa_0 = 0.5\,\text{s}^{-1}$.

usual rate and can't rebind. Once the dynein motors are strong enough to pull the remaining kinesin motors off the microtubule, the motor complex quickly transitions to $(-)$ end-directed transport. After the dynein-driven cargo leaves the MAP-coated region, kinesin motors can then re-establish $(+)$ end-directed transport until the motor complex returns to the MAP-coated region. This process repeats until the motor complex is able to move forward past the MAP-coated region. Interestingly, particle tracking experiments have observed oscillatory behavior during mRNA transport in dendrites [234, 704]. In these experiments, motor-driven mRNA granules move rapidly until encountering a fixed location along the dendrite where they slightly overshoot then stop, move backwards, and begin to randomly oscillate back and forth. After a period of time, lasting on the order of minutes, the motor-driven mRNA stops oscillating and resumes fast ballistic motion. Calculating the mean time to escape the target can be formulated as a FPT problem, in which the parti-

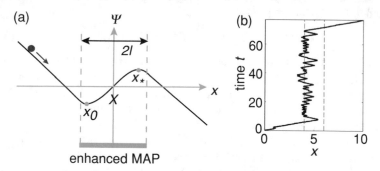

Fig. 7.16: Diagram showing (a) the effective potential well created by a region of tau coating a microtubule, and (b) a representative trajectory showing random oscillations within the well.

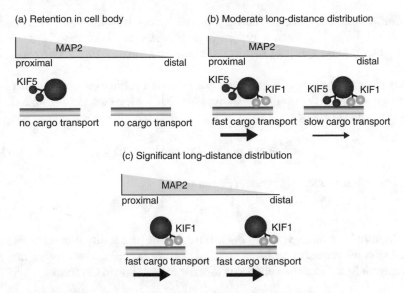

Fig. 7.17: Schematic illustration of how MAP2 regulation of kinesin motor activities leads to cargo sorting and trafficking in axons. [Redrawn from [349].]

cle starts at $x = x_0$ and has to make a rare transition to the unstable fixed point at $x = x_*$. As in the analogous problem of stochastic action potential generation (Sect. 3.4), the QSS diffusion approximation breaks down for small ε, and one has to use singular perturbation methods. The details in the case of the 3-state model can be found elsewhere [612].

Remark 7.4. Interestingly, there is recent evidence that the selective transport of cargo into the axon depends on the localized restriction of MAP2 to the proximal axon [350]. It is known that in both mammalian and Drosophila axons, secretory vesicles are trafficked by the cooperative action of two types of kinesin motors, KIF5 and KIF1 motors. Experimental studies of their motility indicate that MAP2 directly inhibits KIF5 motor activity and that axonal cargo entry and distribution depends on the balanced activities between KIF5 and KIF1 bound to the same cargo. That is, cargoes bound to the dominant motor KIF5 are unable to enter the axon, whereas those bound to motors that are not influenced by MAP2 are able to quickly enter the axon and move to the distal terminals. Moreover, cargoes bound to both KIF1 and KIF5 will enter the axon, but their axonal distribution will be affected by the reactivation of KIF5 past the proximal axon as the inhibition by MAP2 wears off, which slows down the transport, see Fig. 7.17.

7.3.4 Directed search along an array of synaptic targets

Let us now turn to an example of an FPT problem with multiple partially absorbing targets. For the sake of illustration, consider bidirectional transport in a 1D semi-infinite domain with an array of partially absorbing targets and a reflecting boundary at $x = 0$, see Fig. 7.18(a). The kth target has width a and its distal end is at a distance

ka from $x = 0$. The particle can be captured by the kth target at a rate κ when $X(t) \in ((k-1)a, ka)$. Such a system is an idealized representation of a uniform array of en passant synapses distributed along an axon; see Fig. 7.18(b) and Sect. 7.1.2.

Let $p(x,t)$ be the probability density that at time t a particle (searcher) is at $X(t) = x$, having started at the origin $x = 0$. Suppose that p evolves according to a master equation with generator $\mathbb{L}$:

$$\frac{\partial p(x,t)}{\partial t} = \mathbb{L}p(x,t) - \kappa p(x,t), \quad x \in (0, \infty), \tag{7.3.30a}$$

together with the reflecting boundary condition

$$\mathbb{J}[p](0,t) = 0, \tag{7.3.30b}$$

and the initial condition $p(x,t) = \delta(x)$. Here $\mathbb{J}$ is the probability flux operator associated with the generator $\mathbb{L}$, $\mathbb{L}p = -\partial_x \mathbb{J}[p]$. We are assuming that capture rate κ is spatially uniform. This then allows us to write the solution in the form

$$p(x,t) = G(x,t)e^{-\kappa t}, \tag{7.3.31}$$

where $G(x,t)$ is the propagator of the operator $\mathbb{L}$:

$$\frac{\partial G(x,t)}{\partial t} = \mathbb{L}G(x,t), \quad x \in (0, \infty); \quad \mathbb{J}[G](0,t) = 0, \tag{7.3.32}$$

and $G(x,0) = \delta(x)$.

The probability flux into the kth target at time t is

$$J_k(t) = \kappa \int_{(k-1)a}^{ka} G(x,t)e^{-\kappa t}dx, \quad k \geq 1. \tag{7.3.33}$$

Hence, the probability that the particle is captured by the kth target after time t is

$$\Pi_k(t) = \int_t^{\infty} J_k(t')dt', \tag{7.3.34}$$

and the corresponding splitting probability is

$$\pi_k = \Pi_k(0) = \int_0^{\infty} J_k(t')dt'. \tag{7.3.35}$$

It follows that the Laplace transform of $\Pi_k(t)$ is given by

$$s\widetilde{\Pi}_k(s) - \pi_k = -\widetilde{J}_k(s) = -\kappa\Lambda_k(s+\kappa), \tag{7.3.36}$$

where

$$\Lambda_k(s) = \int_{(k-1)a}^{ka} \widetilde{G}(x,s)dx. \tag{7.3.37}$$

The splitting probability is thus

$$\pi_k = \kappa \Lambda_k(\kappa). \tag{7.3.38}$$

Next we introduce the survival probability that the particle hasn't been absorbed by a target in the time interval $[0,t]$, having started at $x = 0$:

$$Q(t) = \int_0^\infty p(x,t)dx = \int_0^\infty G(x,t)e^{-\kappa t}dx. \tag{7.3.39}$$

Integrating the master equation (7.3.30a) with respect to x and using the reflecting boundary condition leads to the condition

$$\frac{\partial Q(t)}{\partial t} = -\kappa Q(t). \tag{7.3.40}$$

Laplace transforming equations (7.3.39) and (7.3.40) imply that

$$\widetilde{Q}(s) = \frac{1}{s+\kappa} = \sum_{k=1}^N \Lambda_k(s+\kappa). \tag{7.3.41}$$

We have used the definition of Λ_k. It immediately follows from equation (6.7.3) that

$$\sum_{k=1}^N \pi_k = 1. \tag{7.3.42}$$

Since the probability of the particle being captured by the kth target is typically less than unity ($\pi_k < 1$), it follows that the corresponding MFPT is infinite unless we condition on the given event. Therefore, let the discrete random variable $K(t) \in \{0,1,\ldots\}$ indicate whether the particle has been captured by the kth target ($K(t) = k \neq 0$) or has not been captured by any target ($K(t) = 0$) in the time interval $[0,t]$. The MFPT $\mathcal{T}_k$ to be captured by the kth target is given by

$$\mathcal{T}_k = \inf\{t > 0; X(t) \in \mathcal{U}_k, \ K(t) = k\}, \tag{7.3.43}$$

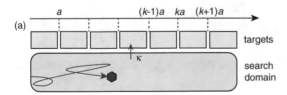

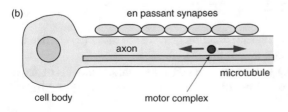

Fig. 7.18: (a) Semi-infinite array of partially absorbing targets. Each target is of width a and the particle can be captured by the kth target cell at a rate κ if $(k-1)a < X(t) < ka$. (b) Motor-driven axonal transport. A motor-cargo complex undergoing bidirectional transport along the axon of a neuron with an array of en passant synaptic targets.

with $\mathcal{T}_k = \infty$ if the particle is captured by another target. Introducing the set of events $\Omega_k = \{\mathcal{T}_k < \infty\}$ we can then define the conditional MFPT

$$T_k = \mathbb{E}[\mathcal{T}_k|\Omega_k] = \mathbb{E}[\mathcal{T}_k 1_{\Omega_k}]/\mathbb{P}[\Omega_k], \tag{7.3.44}$$

where $\pi_k = \mathbb{P}[\Omega_k]$. It follows that

$$\pi_k T_k = \int_0^\infty \tau J_k(\tau) d\tau = -\int_0^\infty \tau \frac{d\Pi(\tau)}{d\tau} d\tau = \widetilde{\Pi}_k(0), \tag{7.3.45}$$

where we have used equation (7.3.17) and integration by parts. Note that the conditional FPT density to be captured by the kth target is $f_k(t) = J_k(t)/\pi_k$.

As a simple illustration of the above, suppose that the particle moves at a constant speed v_+. In this deterministic transport model, $G(t) = \delta(x - v_+ t)$ and

$$\Lambda_k(s) = \frac{1}{v_+} \int_{(k-1)a}^{ka} e^{-sx/v_+} dx = \frac{1}{s} \left[e^{-s(k-1)a/v_+} - e^{-ska/v_+} \right].$$

Substituting into equations (7.3.36) and (7.3.38) implies that

$$\pi_k = e^{-\kappa(k-1)a/v_+} - e^{-\kappa ka/v_+}, \tag{7.3.46}$$

and

$$s\widetilde{\Pi}_k(s) = \pi_k - \frac{\kappa}{\kappa + s} \left[e^{-(s+\kappa)(k-1)a/v_+} - e^{-(s+\kappa)ka/v_+} \right] \equiv f_k(s). \tag{7.3.47}$$

Note that π_k is an exponentially decreasing function of k with space constant v_+/κ, since

$$\pi_{k+1} = e^{-\kappa a/v_+} \pi_k.$$

This is consistent with the analysis of the deterministic population model in Sect. 7.1.2. It then follows from equation (7.3.45) that

$$\pi_k T_k = \lim_{s \to 0} \frac{f_k(s)}{s} = f_k'(0) \tag{7.3.48}$$

$$= \frac{\pi_k}{\kappa} + \frac{a}{v_+} \left[(k-1)e^{-\kappa(k-1)a/v_+} - ke^{-\kappa ka/v_+} \right].$$

Suppose that we set $ka = x$, $T(x) = T_{x/a}$ and $\pi(x) = \pi_{x/a}$ and treat x as a continuous variable. Then

$$T(x) = \frac{1}{\kappa} - \frac{a}{\kappa} \frac{\pi'(x)}{\pi(x)} = \frac{1}{\kappa} - \frac{a}{\kappa} \frac{d}{dx} \ln \pi(x) = \frac{1}{\kappa} + \frac{a}{v_+}.$$

Since $\pi(x)$ is an exponentially decreasing function of x, we see that $T(x)$ is a linearly increasing function of x (in the continuum limit).

The same qualitative behavior holds for more general stochastic models of active transport, except that the space constant differs. For example, in the case of diffusive transport

$$\widetilde{G}(x,s) = \frac{1}{\sqrt{sD}} e^{-\sqrt{s/D}x},$$

and

$$\Lambda_k(s) = \frac{1}{s}\left[e^{-\sqrt{s/D}(k-1)a} - e^{-\sqrt{s/D}ka}\right].$$

Repeating the analysis of the ballistic case yields the same results except that $v_+/\kappa \to \sqrt{D/\kappa}$.

7.4 Multiple search-and-capture events and queuing theory

Now suppose that rather than being permanently absorbed or captured by a target, a particle delivers a discrete packet of some resource to the target and then returns to its initial position x_0, where it is reloaded with cargo and another round of search-and-capture is initiated. We will refer to the delivery of a single packet as a burst event. The sequence of bursts leads to an accumulation of packets within the target, which we assume is counteracted by degradation at some rate γ. This is illustrated in Fig. 7.19 for two targets in a rectangular domain. For each target j, $j = 1,\ldots,N$, a sequence of search-and-capture events can be mapped onto a queuing process as follows (see also Sect. 5.4 and Ref. [136]): individual resource packets are analogous to customers, the delivery of a packet corresponds to a customer arriving at the service station, and the degradation of a resource packet is the analog of a customer exiting the system after being serviced. Finally, assuming that the packets are degraded independently of each other, the effective number of servers in the corresponding queuing model is infinite, that is, the presence of other customers does not affect the service time of an individual customer. It follows that the relevant queuing model is the $G/M/\infty$ system [528, 795]. Recall from Sect. 5.4 that the symbol G denotes a general customer inter-arrival time distribution $F_j(t)$, the symbol M stands for a Markovian or exponential service-time distribution $\Phi(t) = 1 - e^{-\gamma t}$, and "$\infty$" denotes infinite servers.

It remains to determine the distribution $F_j(t)$ for a given target. Let $\mathcal{T}_j$ denote the FPT that the particle finds the jth target during a single search-and-capture event, starting from x_0. Denote the corresponding conditional FPT density by $f_j(t)$ and set

$$\pi_j = \mathbb{P}[\mathcal{T}_j < \infty], \quad T_j = \mathbb{E}[\mathcal{T}_j|\mathcal{T}_j < \infty] = \int_0^\infty t f_j(t)dt.$$

Suppose that the total time for the particle to deliver its cargo, return to x_0 and start a new search process is given by the random variable $\widehat{\tau}$, which for simplicity is taken to be independent of the location of the target. (This is reasonable if the sum of the mean times to load and unload cargo is much larger than a typical return time.) Let $n \geq 1$ label the nth burst event and denote the target that receives the nth packet by j_n. If τ_n is the time of the nth burst, then

$$\tau_n = \widehat{\tau}_n + \mathcal{T}_{j_n} + \tau_{n-1}, \quad n \geq 1. \tag{7.4.1}$$

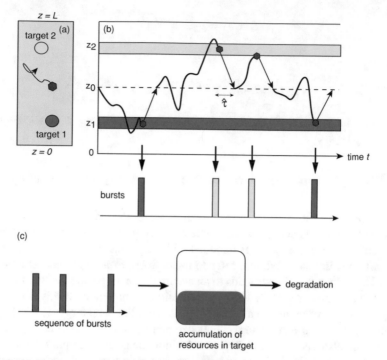

Fig. 7.19: Multiple search-and-capture events. (a) Particle searching in a rectangular domain with two targets. Each time the particle reaches a target it delivers a discrete packet of resources (burst event) and then returns to $\mathbf{x}_0$ where it is loaded with another packet and the process repeats. (b) Sample trajectory projected onto the z-coordinate showing a sequence of burst events. The delay time $\widehat{\tau}$ between a burst event and initiation of a new search is generated from a waiting time density $\rho(\widehat{\tau})$. (The lines with arrows do not represent actual trajectories.) (c) The burst sequence results in an accumulation of resources within each target, which is counteracted by degradation at a rate γ.

The corresponding inter-arrival times are

$$\Delta_n = \tau_n - \tau_{n-1} = \widehat{\tau}_n + T_{j_n}, \quad n \geq 1.$$

Finally, given an inter-arrival time Δ, we denote the identity of the target that captures the particle by $\mathcal{K}(\Delta)$. We can then write for each target k,

$$F_k(t) = \mathbb{P}[\Delta < t, \mathcal{K}(\Delta) = k] = \mathbb{P}[\Delta < t | \mathcal{K}(\Delta) = k]\mathbb{P}[\mathcal{K}(\Delta) = k]$$
$$= \pi_k \int_0^t \mathcal{F}_k(\Delta)d\Delta, \tag{7.4.2}$$

where $\mathcal{F}_k(\Delta)$ is the conditional inter-arrival time density for the kth target. Let $\rho(\widehat{\tau})$ denote the waiting time density of the delays $\widehat{\tau}_n$ with finite mean τ_{cap}. Then

$$\mathcal{F}_k(\Delta) = \int_0^\Delta dt \int_0^\Delta d\widehat{\tau}\, \delta(\Delta - t - \widehat{\tau})f_k(t)\rho(\widehat{\tau}) = \int_0^\Delta f_k(t)\rho(\Delta - t)dt.$$

Laplace transforming the convolution equation then yields

$$\widetilde{\mathcal{F}}_k(s) = \widetilde{f}_k(s)\widetilde{\rho}(s). \tag{7.4.3}$$

7.4.1 Analysis in terms of a $G/M/\infty$ queue

We now focus on a particular labeled target k and use classical queuing theory to determine the steady-state statistics of resource accumulation within the target. For further details see [528, 795]. Let $M_k(t)$ be the number of resource packets in the target that have not yet degraded. In terms of the sequence of arrival times τ_n, we can write

$$M_k(t) = \sum_{n,0\leq\tau_n\leq t} I(t-\tau_n,S_n)\delta_{j_n,k}, \tag{7.4.4}$$

where

$$I(t-\tau_n,S_n) = \begin{cases} 1 \text{ if } t-\tau_n \leq S_n \\ 0 \text{ if } t-\tau_n > S_n \end{cases}. \tag{7.4.5}$$

Here S_n is the degradation time of the nth packet. Introduce the generating function

$$G_k(z,t) = \sum_{l=0}^{\infty} z^l \mathbb{P}[M_k(t) = l], \tag{7.4.6}$$

and the binomial moments

$$B_{m,k}(t) = \sum_{l=m}^{\infty} \frac{l!}{(l-m)!m!} \mathbb{P}[M(t) = l], \quad m = 1,2,\cdots. \tag{7.4.7}$$

Suppose that the targets have no resources at time $t = 0$. We will derive an integral equation for the generating function $G(z,t)$. Conditioning the first arrival time by setting $T_1 = y$, we have

$$M_k(t) = \begin{cases} I(t-y,S_1)\delta_{j_1,k} + M_k^*(t-y) \text{ if } y \leq t \\ 0 \quad\quad\quad\quad\quad\quad\quad \text{ if } y > t \end{cases},$$

where $M_k^*(t)$ has the same distribution as $M_k(t)$. Note that $I(t-y,S_1)\delta_{j_1,k}$ and $M^*(t-y)$ are independent. Moreover,

$$\mathbb{P}[I(t-y,S_1) = a] = [1 - \Phi(t-y)]\delta_{a,1} + \Phi(t-y)\delta_{a,0},$$

so it follows that

$$\sum_{a=0,1} z^j \mathbb{P}[I(t-y,S_1) = a] = z + (1-z)\Phi(t-y).$$

The total expectation theorem then yields

$$\mathbb{E}[z^{I(T_1-y,S_1)\delta_{j_1,k}}] = \mathbb{E}\left[\mathbb{E}[z^{I(T_1-y,S_1)}|T_1=y,j_1=k]\right]$$

$$= \int_0^\infty [z+(1-z)\Phi(t-y)]dF_k(y).$$

Another application of the total expectation theorem gives

$$G_k(z,t) = \mathbb{E}[z^{M_k(t)}] = \mathbb{E}\left[\mathbb{E}[z^{M_k(t)}|T_1=y,j_1=k]\right]$$

$$= \sum_{j=1}^N \int_t^\infty dF_j(y) + \int_0^t [z+(1-z)\Phi(t-y)]G_k(z,t-y)dF_k(y)$$

$$+ \sum_{j\neq k}\int_0^t G_k(z,t-y)dF_j(y). \qquad (7.4.8)$$

Differentiating equation (7.4.8) with respect to z and using

$$B_{m,k}(t) = \frac{1}{m!}\frac{d^m G_k(z,t)}{dz^m}\bigg|_{z=1},$$

we obtain an iterative integral equation for the Binomial moments:

$$B_{m,k}(t) = \sum_{j=1}^N \int_0^t B_{m,k}(t-y)dF_j(y) + \int_0^t B_{m-1,k}(t-y)[1-\Phi(t-y)]dF_k(y).$$

$$(7.4.9)$$

In order to obtain the steady-state binomial moments, we Laplace transform equation (7.4.9) after making the substitutions $dF_j(y) = \pi_j \mathcal{F}_j(y)dy$ and $1-\Phi(t) = e^{-\gamma t}$:

$$\widetilde{B}_{m,k}(s) = \widetilde{B}_{m,k}(s)\sum_{j=1}^N \pi_j\widetilde{\mathcal{F}}_j(s) + \pi_k\widetilde{\mathcal{F}}_k(s)\widetilde{B}_{m-1,k}(s+\gamma),$$

which can be rearranged to give

$$\widetilde{B}_{m,k}(s) = \left[\frac{\pi_k\widetilde{\mathcal{F}}_k(s)}{1-\sum_{j=1}^N \pi_j\widetilde{\mathcal{F}}_j(s)}\right]\widetilde{B}_{m-1,k}(s+\gamma). \qquad (7.4.10)$$

Multiplying both sides by s and taking the limit $s \to 0^+$ yields

$$B_{m,k}^* := \lim_{t\to\infty} B_{m,k}(t) = \lim_{s\to 0^+} s\widetilde{B}_{m,k}(s) = \lambda_k\widetilde{B}_{m-1,k}(\gamma),$$

where

$$\lambda_k := \lim_{s\to 0^+}\frac{s\pi_k\widetilde{\mathcal{F}}_k(s)}{1-\sum_{j=1}^N \pi_j\widetilde{\mathcal{F}}_j(s)} = \frac{\pi_k}{\sum_{j=1}^N \pi_j(T_j+\tau_{\text{cap}})} = \frac{\pi_k}{T+\tau_{\text{cap}}}, \qquad (7.4.11)$$

where $T = \sum_{j=1}^N \pi_j T_j$ is the unconditional MFPT. We have used L'Hopital's rule and equation (7.4.3), together with the following properties:

$$\widetilde{f}_k(0) = 1 = \widetilde{\rho}(0), \quad \frac{d\widetilde{f}_k}{ds}\bigg|_{s=0} = -T_k, \quad \frac{d\widetilde{\rho}}{ds}\bigg|_{s=0} = -\tau_{\text{cap}}.$$

Equations (7.4.10) and (7.4.11) completely determine the steady-state binomial moments. In particular, since $B_0(t) = 1$ and $\widetilde{B}_0(s) = 1/s$, the mean number of packets in the kth target $j = k$ is given by $B_{1,k}^*$ so that

$$\overline{M}_k = \lim_{t\to\infty} \langle M_k(t) \rangle = \frac{\pi_k}{\gamma(T + \tau_{\text{cap}})}. \tag{7.4.12}$$

We can interpret λ_k as the mean rate at which a packet is delivered to the kth target. This is consistent with the observation that $T + \tau_{\text{cap}}$ is the mean time for one successful delivery of a packet to any one of the targets and initiation of a new round of search-and-capture. Hence, its inverse is the mean rate of resource bursts and π_k is the fraction that are delivered to the kth target (over many trials). (Note that equation (7.4.12) is known as Little's law in the queuing theory literature [527] and applies more generally.) The dependence of the mean $\overline{M}_k$ on the target label k specifies the steady-state allocation of resources across the set of targets. It will depend on the details of the particular search process, the geometry of the search domain, the initial position $\mathbf{x}_0$, and the rate of degradation γ. Similarly, the second-order Binomial moment is given by

$$
\begin{aligned}
B_{2,k}^* &\equiv \lim_{t\to\infty} \frac{1}{2}(\langle M_k(t)^2 \rangle - \langle M_k(t) \rangle) = \lambda_k \widetilde{B}_{1,k}(\gamma) \\
&= \frac{\lambda_k}{2\gamma} \frac{\pi_k \widetilde{\mathcal{F}}_k(\gamma)}{1 - \sum_{j=1}^N \pi_j \widetilde{\mathcal{F}}_j(\gamma)}.
\end{aligned}
\tag{7.4.13}
$$

The variance of the number of resource packets is $\text{Var}[M_k] = 2B_{2,k}^* + B_{1,k}^*(1 - B_{1,k}^*)$ so

$$\text{Var}[M_k] = \overline{M}_k \left[\frac{\pi_k \widetilde{\mathcal{F}}_k(\gamma)}{1 - \sum_{k=1}^N \pi_k \widetilde{\mathcal{F}}_k(\gamma)} + 1 - \overline{M}_k \right]. \tag{7.4.14}$$

Eq. (7.4.3) implies that

$$\widetilde{\mathcal{F}}_j(\gamma) = \frac{\widetilde{J}_j(\gamma)}{\pi_j} \widetilde{\rho}(s) = 1 - \gamma(T_j + \tau_{\text{cap}}) + O(\gamma^2). \tag{7.4.15}$$

Defining the coefficient of variation (CV) and Fano factor (FF) for the kth target according to

$$CV_k^2 = \frac{\text{Var}[M_k]}{\overline{M}_k^2}, \quad FF_k = \frac{\text{Var}[M_k]}{\overline{M}_k}, \tag{7.4.16}$$

we see that

$$\lim_{\gamma\to 0} CV_k^2(\gamma) = 0, \quad \lim_{\gamma\to 0} FF_k(\gamma) = 1. \tag{7.4.17}$$

Moreover, using $\widetilde{\mathcal{F}}_k(\gamma) \to 0$ and $\gamma \widetilde{\mathcal{F}}_k(\gamma) \to \mathcal{F}_k(0)$ as $\gamma \to \infty$, it follows that

$$\lim_{\gamma \to \infty} CV_k^2(\gamma) = \infty, \quad \lim_{\gamma \to \infty} FF_k(\gamma) = 1. \tag{7.4.18}$$

Example 7.2. Distribution of resources under diffusive search between concentric spheres.
We now illustrate the above theory by considering the classical problem of diffusive search between two concentric d-dimensional spheres of radii R_1 and R_2, respectively, with $R_2 > R_1$, see Fig. 7.20. In the 1D case this reduces to the problem of diffusive search in a finite interval of length $L = R_2 - R_1$ with absorbing boundaries at the ends $x = 0, L$. (Note that diffusive search could approximate unbiased active transport on a microtubular network, see Sect. 7.2.3.) Using radial symmetry, the diffusion equation reduces to the 1D equation

$$\frac{\partial p(x,t|x_0)}{\partial t} = D \left[\frac{\partial^2 p(x,t|x_0)}{\partial x^2} + \frac{d-1}{x} \frac{\partial p(x,t|x_0)}{\partial x} \right], \tag{7.4.19a}$$

where x represents the radial coordinate. (We keep the symbol r as the resetting rate.) This is supplemented by the boundary conditions

$$p(R_1,t|x_0) = 0 = p(R_2,t|x_0), \tag{7.4.19b}$$

and the initial condition

$$p(x,0|x_0) = \frac{\delta(x-x_0)}{\Omega_d x_0^{d-1}},$$

with Ω_d the surface area of the d-dimensional unit sphere. That is, the initial condition is uniformly distributed around the spherical surface of radius x_0. Laplace transforming the radial diffusion equation gives

$$D \left[\frac{\partial^2 \widetilde{p}(x,s|x_0)}{\partial x^2} + \frac{d-1}{x} \frac{\partial \widetilde{p}(x,s|x_0)}{\partial x} \right] - s\widetilde{p}(x,s|x_0) = -\frac{\delta(x-x_0)}{\Omega_d x_0^{d-1}}, \tag{7.4.20a}$$

supplemented by the boundary conditions

$$\widetilde{p}(R_1,s|x_0) = 0 = \widetilde{p}(R_2,s|x_0). \tag{7.4.20b}$$

The solution of Eq. (7.4.20) is carried out in Ref. [674]. We sketch the basic steps here. In each subdomain $x < x_0$ and $x > x_0$, the general solution of Eq. (7.4.20a) is a linear combination of $x^\nu I_\nu(\sqrt{s/D}x)$ and $x^\nu K_\nu(\sqrt{s/D}x)$, where I_ν and K_ν are the modified Bessel functions of the first and second kind, respectively, and $\nu = 1 - d/2$. Imposing the absorbing boundary conditions implies that

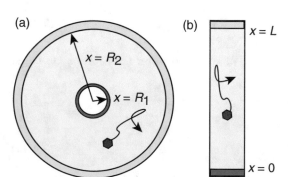

(a)

$x = R_2$

$x = R_1$

(b)

$x = L$

$x = 0$

Fig. 7.20: (a) Search domain consisting of the region between a pair of concentric d-dimensional spheres of radii R_1, R_2. (b) In 1D the search domain reduces to a finite interval of length $L = R_2 - R_1$ with absorbing boundaries at $x = 0, L$.

$$\widetilde{p}(x,s|x_0) = Ax^\nu C_\nu(x,R_1;s), \quad x < x_0,$$
$$\widetilde{p}(x,s|x_0) = Bx^\nu C_\nu(x,R_2;s), \quad x > x_0,$$

where

$$C_\nu(a,b;s) = I_\nu(\sqrt{s/D}a)K_\nu(\sqrt{s/D}b) - I_\nu(\sqrt{s/D}b)K_\nu(\sqrt{s/D}a).$$

Note that $C_\nu(a,b) = -C_\nu(b,a)$ and $C_\nu(a,b) > 0$ for $a > b$. The coefficients A and B are determined by imposing continuity of the modified Helmholtz Green's function at $x = x_0$ together with the discontinuity condition

$$\lim_{x \to x_0^+} \frac{\partial \widetilde{p}(x,s|x_0)}{\partial x} - \lim_{x \to x_0^-} \frac{\partial \widetilde{p}(x,s|x_0)}{\partial x} = -\frac{1}{D\Omega_d x_0^{d-1}}.$$

The final result is [674]

$$\widetilde{p}(x,s|x_0) = \frac{(xx_0)^\nu}{D\Omega_d} \frac{C_\nu(x,R_1;s)C_\nu(x_0,R_2;s)}{C_\nu(R_1,R_2;s)}, \quad x < x_0,$$
$$\widetilde{p}(x,s|x_0) = \frac{(xx_0)^\nu}{D\Omega_d} \frac{C_\nu(x_0,R_1;s)C_\nu(x,R_2;s)}{C_\nu(R_1,R_2;s)}, \quad x > x_0. \qquad (7.4.21)$$

The FPT properties without resetting can now be obtained by integrating the probability flux over the surface boundary of each sphere and taking Laplace transforms. Hence, for the inner target of radius R_1,

$$\widetilde{J}_1(x_0,s) = D\Omega_d R_1^{d-1} \left.\frac{\partial \widetilde{p}(x,s|x_0)}{\partial x}\right|_{x=R_1} = \left(\frac{x_0}{R_1}\right)^\nu \frac{C_\nu(x_0,R_2)}{C_\nu(R_1,R_2)}. \qquad (7.4.22a)$$

Similarly, for the outer target of radius R_2,

$$\widetilde{J}_2(x_0,s) = -D\Omega_d R_2^{d-1} \left.\frac{\partial \widetilde{p}(x,s|x_0)}{\partial x}\right|_{x=R_2} = -\left(\frac{x_0}{R_2}\right)^\nu \frac{C_\nu(x_0,R_1)}{C_\nu(R_1,R_2)}. \qquad (7.4.22b)$$

We have used the Bessel function identities

$$I_\nu'(x) = -\frac{\nu}{x}I_\nu(x) + I_{\nu-1}(x), \quad K_\nu'(x) = -\frac{\nu}{x}K_\nu(x) - K_{\nu-1}(x),$$

which imply that

$$\left.\frac{\partial C_\nu(x,y;s)}{\partial x}\right|_{y=x} = I_{\nu-1}(x)K_\nu(x) + I_\nu(x)K_{\nu-1}(x) = \frac{1}{x}.$$

Splitting probabilities and conditional FPT densities. We now use the above solutions to solve the FPT problem for a single search-and-capture event. Let $\mathcal{T}_k(x_0)$ denote the FPT that the particle is captured by the inner sphere ($k = 1$) or outer sphere ($k = 2$), with $\mathcal{T}_k(x_0) = \infty$ indicating that it is not captured. That is, set

$$\mathcal{T}_k(x_0) = \inf\{t \geq 0; X(t) = R_k, \, X(0) = x_0\}.$$

Define $\Pi_k(x_0,t)$ to be the probability that the particle is captured by the kth target after time t, given that it started at x_0:

$$\Pi_k(x_0,t) = \mathbb{P}[t < \mathcal{T}_k(x_0) < \infty] = \int_t^\infty J_k(x_0,t')dt'. \qquad (7.4.23)$$

Laplace transforming gives

$$s\widetilde{\Pi}_k(x_0,s) - \pi_k(x_0) = -\widetilde{J}_k(x_0,s). \qquad (7.4.24)$$

The splitting probability $\pi_k(x_0)$ for being captured by the kth target is then

$$\pi_k(x_0) = \Pi_k(\mathbf{x}_0,0) = \int_0^\infty J_k(x_0,t')dt' = \lim_{s\to 0} \widetilde{J}_k(x_0,s). \qquad (7.4.25)$$

As $s \to 0$ for fixed v, one has the asymptotic limits

$$I_v(x) \sim (x/2)^v/\Gamma(v+1),$$

and

$$K_v(x) \sim \frac{1}{2}(x/2)^{-v}/\Gamma(v), v \neq 0, \quad K_0(x) \sim -\ln x.$$

One thus finds that [674]

$$\pi_2(x_0) = \begin{cases} \dfrac{1 - (R_1/x_0)^{d-2}}{1 - (R_1/R_2)^{d-2}}, & d \neq 2 \\[2ex] \dfrac{\ln(R_2/x_0)}{\ln(R_2/R_1)}, & d = 2 \end{cases}, \qquad (7.4.26)$$

with $\pi_1(x_0) = 1 - \pi_2(x_0)$. The Laplace transform of the survival probability is

$$\widetilde{Q}(x_0,s) = \widetilde{\Pi}_1(x_0,s) + \widetilde{\Pi}_2(x_0,s) = \frac{1}{s}\left[1 - \widetilde{J}_1(x_0,s) - \widetilde{J}_2(x_0,s)\right]$$

$$= \frac{1}{s}\left[1 + \left(\frac{x_0}{R_2}\right)^v \frac{C_v(x_0,R_1)}{C_v(R_1,R_2)} - \left(\frac{x_0}{R_1}\right)^v \frac{C_v(x_0,R_2)}{C_v(R_1,R_2)}\right],$$

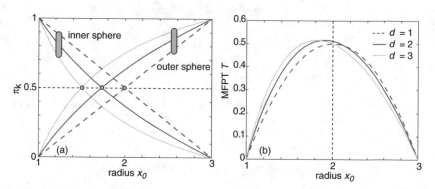

Fig. 7.21: Plots of (a) splitting probabilities π_k of inner ($k=1$) and outer ($k=2$) spheres and (b) unconditional MFPT T as a function of resetting radius x_r for $d=1,2,3$ and $r=0.1,10$. Other parameters are $D=1, R_1=1, R_3=3$. Green dots in (a) indicate the initial location where the splitting probabilities in the two targets are balanced.

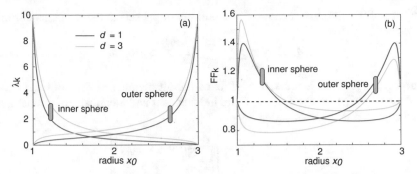

Fig. 7.22: Plot of steady-state (a) accumulation rate λ_k and (b) Fano factor FF_k in the kth sphere as a function of the initial radius x_0 for dimensions $d = 1, 3$ and $\tau_{cap} = 0.1$. Other parameters are as in Fig. 7.21.

and the conditional FPTs are

$$\widetilde{f}_k(x_0, s) = \frac{\widetilde{J}_k(x_0, s)}{\pi_k(x_0)}. \tag{7.4.27}$$

In Fig. 7.21 we plot the splitting probabilities π_k, $k = 1, 2$, and the unconditional MFPT T as a function of the initial position x_0 for $d = 1, 2, 3$. As expected, $\pi_2(R_1) = 0$, $\pi_2(R_2) = 1$ and $T(R_{1,2}) = 0$. The 1D case has a reflection symmetry about the midpoint $(R_1 + R_2)/2$, that is, $\pi_k(x_0) = \pi_k(L - x_0)$ for $L = R_2 - R_1$ and similarly for T. On the other hand, in higher dimensions the curves are skewed toward the inner sphere, since it has a smaller surface area and is thus a less effective trap compared to the outer sphere.

Distribution of resources. The final step of our analysis is to use equations (7.4.22), (7.4.26), and (7.4.27) to calculate the steady-state mean and variance of the resources distributed between the two targets, which are given by equations (7.4.12) and (7.4.14), respectively. For the sake of illustration, we take

$$\widetilde{\rho}(s) = \frac{1}{1 + s\tau_{cap}}.$$

Since T vanishes at the boundaries, it follows from Eq. (7.4.12) that the means $\overline{M}_k$ are singular at the boundaries in the absence of a capture delay. This reflects the fact that without any delays, the frequency at which the particle loads and unloads resources becomes unbounded, which is physically unrealistic. In order to remove this singular behavior, we will take $\tau_{cap} > 0$. In Fig. 7.22 we show sample plots of the steady-state accumulation rate $\lambda_k = \gamma \overline{M}_k$ and the Fano factor FF_k as a function of the radius x_0 for $d = 1, 3$. We find that the accumulation rate increases with the dimension d and is a monotonic function of x_0. On the other hand, the Fano factor is a non-monotonic function of the initial location. One also finds that the Fano factor approaches unity as τ_{cap} becomes large (Poisson-like).

7.4.2 Multiple packets and searchers

The previous example exhibited significant fluctuations. One way to reduce fluctuations is to assume that during each burst event, the particle delivers C packets that degrade independently. In order to establish this result, it is necessary to modify the analysis of the binomial moments. The number of packets at the kth target at time t is now

$$M_k(t) = \sum_{n,0 \leq \tau_n \leq t} \sum_{i=1}^{C} I(t - \tau_n, S_{ni}) \delta_{j_n,k}, \qquad (7.4.28)$$

where S_{ni}, $i = 1, \ldots$, is the service time of the ith member of the nth burst event. Since $I(t - y, S_{1i})$ for $i = 1, 2, \cdots, C$ are independent and identically distributed, the total expectation theorem yields

$$\mathbb{E}[z^{\sum_{i=1}^{C} I(t-\tau_n,S_{ni})\delta_{j_1,k}}] = \mathbb{E}\left[\prod_{i=1}^{C} \mathbb{E}[z^{I(t-y,S_{1i})\delta_{j_1,k}}]\right] \qquad (7.4.29)$$

$$= \int_0^{\infty} [z + (1-z)\Phi(t-y)]^C dF_k(y).$$

Another application of the total expectation theorem gives equation (7.4.8) with the factor $[z + (1-z)\Phi(t-y)]$ replaced by $[z + (1-z)\Phi(t-y)]^C$. Differentiating with respect to z using

$$\frac{d^m}{dz^m}[z + (1-z)\Phi(t-y)]^C \bigg|_{z=1} \qquad (7.4.30)$$

$$= \begin{cases} \dfrac{C!}{(C-m)!}[1 - \Phi(t-y)]^m & \text{if } C \geq m \\ 0 & \text{if } C < m \end{cases},$$

leads to the following integral equation for the binomial moments:

$$B_{m,k}(t) = \sum_{j=1}^{N} \int_0^t B_{m,k}(t-y)dF_j(y) + \int_0^t \mathcal{H}_{m,k}(t-y)dF_k(y), \qquad (7.4.31)$$

where

$$\mathcal{H}_{m,k}(t) = \sum_{l=1}^{\min\{m,C\}} \binom{C}{l} B_{m-l,k}(t)e^{-l\gamma t}. \qquad (7.4.32)$$

Laplace transforming the renewal equation with $dF_k(y) = \pi_k \mathcal{F}_k(y)dy$ and rearranging gives

$$\widetilde{B}_{m,k}(s) = \frac{\widetilde{\mathcal{H}}_{m,k}(s)\pi_k \widetilde{\mathcal{F}}_k(s)}{1 - \pi_k \widetilde{\mathcal{F}}_k(s)} = \frac{\pi_k \widetilde{\mathcal{F}}_k(s)}{1 - \pi_k \widetilde{\mathcal{F}}_k(s)} \sum_{l=1}^{\min\{m,C\}} \binom{C}{l} \widetilde{B}_{m-l,k}(s+l\gamma).$$

$$(7.4.33)$$

Multiplying both sides by s and taking the limit $s \to 0^+$ along identical lines to the analysis of equation (7.4.10) yields

$$B_m^* = \frac{\lambda_k}{\gamma} \sum_{l=1}^{\min\{m,C\}} \binom{C}{l} \widetilde{B}_{m-l,k}(l\gamma). \qquad (7.4.34)$$

In particular, the mean is

$$B^*_{1,k} \equiv \overline{M}_k = \frac{C\lambda_k}{\gamma},$$

(7.4.35)

with λ_k given by equation (7.4.11). As expected, $\overline{M}_k$ scales with the number of packets C per delivery. Similarly, defining $H_{m,C} = 1$ if $m \leq C$ and zero otherwise,

$$\begin{aligned}
B^*_{2,k} &= \lambda_k \left(\widetilde{B}_{1,k}(\gamma)C + \frac{H_{2,C}C(C-1)}{4\gamma} \right) \\
&= \frac{C^2\lambda_k}{4\gamma} \left(\frac{2\widetilde{\mathcal{F}}_k(\gamma)}{1 - \widetilde{\mathcal{F}}_k(\gamma)} + H_{2,C} \right) - \frac{C\lambda_k H_{2,C}}{4\gamma}.
\end{aligned}$$

(7.4.36)

Hence, the variance is

$$\mathrm{Var}[M_k] = a_k C + b_k C^2,$$

(7.4.37)

where

$$a_k = \frac{\lambda_k}{\gamma} \left(1 - \frac{H_{2,C}}{2} \right), \quad b_k = \frac{\lambda_k}{2\gamma} \left(\frac{2\widetilde{\mathcal{F}}_k(\gamma)}{1 - \widetilde{\mathcal{F}}_k(\gamma)} + H_{2,C} \right) - \frac{\lambda_k^2}{\gamma^2}.$$

(7.4.38)

For $C \geq 2$ the coefficients are independent of C so that the C-dependence of the corresponding coefficient of variation (CV) can be expressed as

$$\mathrm{CV}_k = \frac{\gamma}{\lambda_k} \sqrt{b_k + \frac{a_k}{C}}.$$

(7.4.39)

This establishes that increasing C reduces the CV.

A more effective way to reduce fluctuations is to have $\mathcal{M}$ independent, parallel searchers. Statistical independence implies that the steady-state mean and variance become

$$\overline{M}_k = \frac{\mathcal{M}C\lambda_k}{\gamma}, \quad \mathrm{Var}[M_k] = \mathcal{M}C(b_k C + a_k).$$

(7.4.40)

The CV thus scales as

$$\mathrm{CV}_k = \frac{\gamma}{\lambda_k \sqrt{\mathcal{M}C}} \sqrt{b_k C + a_k}.$$

(7.4.41)

Note that the steady-state mean $\overline{M}_k$ depends on the product $\mathcal{M}C$. Hence, for a given mean, one can reduce the CV by decreasing C and increasing $\mathcal{M}$ such that $\mathcal{M}C$ is fixed. That is, a larger number of searchers carrying a smaller number of packets results in smaller fluctuations. (An analogous observation was made in a study of cytoneme-based transport [132].)

Another possible modification of the basic queuing model is to assume that particles are inserted into the search domain from some intracellular pool at a rate $v = 1/\Delta\tau$ [904]; after delivering resources to a target each particle is either degraded or recycled to the particle pool. One possible application is to axonal trans-

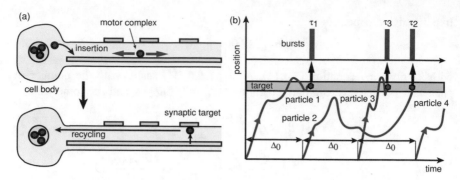

Fig. 7.23: Multiparticle search-and-capture in an axon. (a) Particles from a compartmental pool are inserted periodically into the axon at a rate $\nu = 1/\Delta\tau$. Each particle undergoes bidirectional transport along the axon until it is captured by a target, and secretes a discrete packet of resources (burst event). Following target capture, the particle is either sorted for degradation or recycled to the particle pool. (b) Sample particle trajectories. The jth particle starts its search at time $t_j = (j-1)\Delta\tau$ and is captured by the target at time $\tau_j = t_j + \mathcal{T}_j(\mathbf{x}_0)$.

port as illustrated in Fig. 7.23(a). Let t_j denote the time of the jth insertion event, $j = 1, 2, \ldots$, with $t_1 = 0$ and $\Delta\tau = t_{j+1} - t_j$. We will assume that a single particle is inserted each time, see Fig. 7.23(b). Note that this rule could be generalized in several ways. First, the number of particles injected at time t_j could be a random variable M_j and the inter-insertion times $= t_{j+1} - t_j$ could be randomly generated from some waiting time density such as an exponential. Second, the total number of particles in the compartmental pool could be bounded (finite capacity pool). The latter would significantly complicate the analysis, since one would need to keep track of the total number of particles that have been inserted up to time t, including any particles that have been recycled to the pool.

Denote the target that receives the jth vesicle by k_j and define τ_j to be the time at which the jth particle is captured by the target and delivers its cargo (jth burst event). It follows that

$$\tau_j = t_j + \mathcal{T}_{j,k_j} \quad j \geq 1, \tag{7.4.42}$$

where $\mathcal{T}_{j,k_j}$ is the FPT for the jth particle to find the target k_j. It is important to note that although the insertion times are ordered, $t_j < t_{j+1}$ for $j \geq 1$, there is no guarantee that the burst times are also ordered. That is, the condition $\tau_i < \tau_j$ for $i < j$ need not hold. For example, in Fig. 7.23(b) we see that $\tau_3 < \tau_2$. Suppose that a vesicle is delivered to a given target k at the sequence of times $\tau_{j_1,k}$, $\tau_{j_2,k}$ etc. That is, the nth vesicle is delivered to the given target by the particle labeled j_n. Consider the difference equation

$$\tau_{j_{n+1},k} - \tau_{j_n,k} = t_{j_{n+1}} - t_{j_n} + \mathcal{T}_{j_{n+1},k} - \mathcal{T}_{j_n,k}. \tag{7.4.43}$$

Taking expectations of both sides shows that

$$\mathbb{E}[\tau_{j_{n+1},k}] - \mathbb{E}[\tau_{j_n,k}] = \mathbb{E}[t_{j_{n+1}}] - \mathbb{E}[t_{j_n}] + \mathbb{E}[\mathcal{T}_{j_{n+1},k}] - \mathbb{E}[\mathcal{T}_{j_n,k}]$$
$$= \mathbb{E}[t_{j_{n+1}}] - \mathbb{E}[t_{j_n}]. \tag{7.4.44}$$

We have used the fact that the search particles are independent and identical so $\mathbb{E}[\mathcal{T}_{j,k}] = T_k$ independently of j. It follows that the mean inter-burst interval Δ_k to a given target k is independent of the MFPT T_k. On the other hand, it does depend on the splitting probability π_k, since

$$\Delta_k \equiv \lim_{N \to \infty} \frac{1}{N} \sum_{n=1}^{N} [\mathbb{E}[\tau_{j_{n+1},k}] - \mathbb{E}[\tau_{j_n,k}]] = \frac{\Delta\tau}{\pi_k}. \tag{7.4.45}$$

That is, the mean time between particle injections is $\Delta\tau$ and only a fraction π_k deliver to the kth target so that $\Delta\tau/\pi_k$ is the expected time separating the injection of particles j_n and j_{n+1}.

The accumulation of resources can be analyzed along similar lines to the search scenario shown in Fig. 7.19 using a G/M/∞ queue, although the resulting renewal equation for the Binomial moments differs significantly [904]. see Ex. 7.4. The expressions for the steady-state mean and variance also differ. The former can be written down immediately using Little's law:

$$\overline{M}_k = \frac{\pi_k}{\gamma\Delta\tau}. \tag{7.4.46}$$

Note that $\overline{M}_k$ does not depend on the FPT statistics of the search-and-capture process, which is a major difference from sequential search-and-capture, see equation (7.4.12).

7.5 Search processes with stochastic resetting

A topic of increasing interest is the theory of stochastic processes under resetting [261, 590]. The simplest example of such a process is a Brownian particle whose position is reset randomly in time at a constant rate r (Poissonian resetting) to some fixed point x_r, which could be its initial position [255–257]. Even this simple system exhibits the major features observed in more complex models: (i) convergence to a nontrivial nonequilibrium stationary state; (ii) the mean time for a Brownian particle to find a hidden target is finite and has an optimal value as a function of the resetting rate r. (One of the limitations of a purely diffusive process as a mechanism for the stochastic search for some hidden target in an unbounded domain is that the MFPT for target detection is infinite.) These observations have motivated numerous studies of more general stochastic processes with resetting, including non-diffusive processes such as Levy flights [477] and active run and tumble particles [258], resetting in bounded domains [631], resetting followed by a refractory period [260, 570] (which also arises in certain Michaelis-Menten reaction schemes [688, 689, 706]), and resetting with finite return times [102, 569, 632, 633]. (For further extensions and applications see the review [261] and references therein.) To what extent stochastic resetting plays a role in cell biology is not yet clear, and a particular challenge is identifying plausible active biophysical mechanisms for reset. Nevertheless, potential applications include DNA elongation and backtracking [701] (Sect. 5.5), axonal motor-driven transport, cytoneme-based morphogenesis [134] (Chap. 14), and the binding of focal adhesions during cell motility [138] (Chap. 14).

A common thread through most analytical studies of search processes with stochastic resetting is renewal theory, which exploits the fact that once a particle has returned to its resetting state, it has lost all memory of previous search phases. Often the survival probability with resetting is expressed in terms of the survival probability without resetting using an integral renewal equation, which can then be solved using Laplace transforms [261]. An alternative approach is to decompose the various contributions to the FPT by conditioning on whether or not the searcher resets at least once [61, 135, 630, 633, 689]. This latter approach does not require Laplace transforms and is particularly useful when incorporating features such as finite return times, search failures, and multiple targets. The renewal method based on conditional expectations was also applied in Sect. 6.6.3 to the case of a stochastically gated Brownian particle.

7.5.1 Brownian particle with Poissonian resetting

Consider a single Brownian particle diffusing in 1D with initial position $X(0) = x_0$ and resetting at a rate r to a fixed position x_r [255]. In an infinitesimal time interval dt,

$$X(t+dt) = \begin{cases} x_r & \text{with probability } rdt \\ X(t) + \sqrt{2D}dW(t) & \text{with probability } (1-rdt) \end{cases}, \tag{7.5.1}$$

where $W(t)$ is a Wiener process. Let $p(x,t|x_0,x_r)$ be the probability density for the particle to be at position x at time t. It follows that

$$p(x,t+dt) = \delta(x-x_r)rdt + (1-rdt) \int_{-\infty}^{\infty} p(x-\sqrt{2Ddt}\xi,t) \frac{1}{\sqrt{2\pi}} e^{-\xi^2/2} d\xi.$$

We have dropped the explicit dependence of p on x_0 and x_r. Taylor expanding in dt gives

$$p(x,t+dt) = \delta(x-x_r)rdt$$
$$+ (1-rdt) \int_{-\infty}^{\infty} \left[p(x,t) - \frac{\partial p(x,t)}{\partial x} \sqrt{2Ddt}\xi + \frac{\xi^2}{2}(2Ddt) \frac{\partial^2 p(x,t)}{\partial x^2} + \ldots \right]$$
$$\times \frac{1}{\sqrt{2\pi}} e^{-\xi^2/2} d\xi.$$

Performing the Gaussian integrals, rearranging, dividing through by dt, and taking the limit $dt \to 0$ yields the forward CK equation

$$\frac{\partial p(x,t)}{\partial t} = D \frac{\partial^2 p(x,t)}{\partial x^2} - rp(x,t) + r\delta(x-x_r), \tag{7.5.2}$$

with the initial condition $p(x,0) = \delta(x-x_0)$.

It turns out that is usually more convenient to work with a renewal equation approach [257, 259]. In the absence of resetting ($r = 0$), equation (7.5.2) reduces to the 1D diffusion equation so that $p(x,t|x_0)$ is given by the fundamental solution or propagator $G_0(x,t|x_0)$, see Sect. 2.3:

$$G_0(x,t|x_0) = \frac{1}{\sqrt{4\pi Dt}}e^{-|x-x_0|^2/4Dt}. \tag{7.5.3}$$

When resetting is included, the probability density $p(x,t|x_0)$ has two distinct types of contribution; see Fig. 7.24: paths where no resetting events have occurred up to time t, and paths where the last resetting event occurred at time $\tau_l = t - \tau$ for some $\tau \in (0,t)$. In the case of Poissonian resetting with rate r (see Sect. 3.6.2), the probability density of no resetting events up to time t is e^{-rt}. Similarly, the probability density that the last resetting event occurred at time $t - \tau$ (with no subsequent resetting events) is $re^{-r\tau}$. Using the fact that in the latter case $X(t-\tau) = x_r$ and one has pure diffusion over the time interval $(t-\tau,t)$, the full time-dependent solution to the CK equation (7.5.2) satisfies the so-called last renewal equation

$$p(x,t|x_0) = e^{-rt}G_0(x,t|x_0) + r\int_0^t G_0(x,\tau|x_r)e^{-r\tau}d\tau. \tag{7.5.4}$$

One of the useful features of this particular formulation is that it holds for more general stochastic processes by taking G_0 to be the appropriate propagator. A complementary version is the so-called first renewal equation

$$p(x,t|x_0) = e^{-rt}G_0(x,t|x_0) + r\int_0^t p(x,t-\tau|x_r)e^{-r\tau}d\tau. \tag{7.5.5}$$

The second term now integrates over all trajectories for which a first reset to x_r occurred between time τ and $\tau + d\tau$ followed by possibly multiple further resetting events in the remaining time $t - \tau$.

The stationary state $p^*(x)$ is obtained by taking the limit $t \to \infty$ in equation (7.5.4):

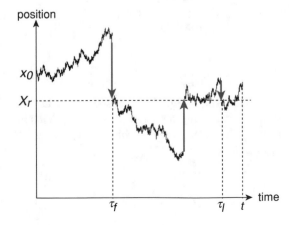

Fig. 7.24: Sample trajectory of 1D Brownian motion with resetting. The last (first) resetting event in the interval $[0,t]$ is τ_l (τ_f).

$$p^*(x) = r \int_0^\infty G_0(x, \tau | x_r) e^{-r\tau} d\tau. \tag{7.5.6}$$

That is, $p^*(x)$ is determined by the r-Laplace transform of the propagator. Using the integral identity

$$\int_0^\infty t^{\nu-1} e^{-\beta/t - \gamma t} dt = 2 \left(\frac{\beta}{\gamma} \right)^{\nu/2} K_\nu(2\sqrt{\beta\gamma}), \tag{7.5.7}$$

where K_ν is the modified Bessel function of the second kind of order ν, and given the formula

$$K_{1/2}(y) = \sqrt{\frac{\pi}{2y}} e^{-y},$$

one finds that the stationary density takes the form of a decaying exponential centered about x_r [255]:

$$p^*(x) = \frac{\alpha_0}{2} e^{-\alpha_0 |x - x_r|}, \quad \alpha_0 = \sqrt{r/D}. \tag{7.5.8}$$

It follows that there is a cusp singularity at $x = x_r$, since the first derivative is discontinuous. Note that $p^*(x)$ represents a nonequilibrium stationary state (NESS) because there exist nonzero probability fluxes (see also Sect. 4.4). That is the point x_r acts as a probability source, whereas all positions $x \neq x_r$ are potential probability sinks.

Relaxation to the stationary-state. The relaxation to the stationary state can be analyzed by carrying out an asymptotic expansion of the exact solution (7.5.4) for large t [553]. It is convenient to choose $x_0 = x_r$. Substituting for G_0 in equation (7.5.4) then gives

$$p(x, t | x_r) = \frac{1}{\sqrt{4\pi Dt}} e^{-rt - |x - x_r|^2 / 4Dt} + r \int_0^t \frac{1}{\sqrt{4\pi D\tau}} e^{-r\tau - |x - x_r|^2 / 4D\tau} d\tau.$$

Performing the change of variables $\tau = wt$ allows us to rewrite the above equation in the form

$$p(x, t | x_r) = \frac{1}{\sqrt{4\pi Dt}} e^{-t\Phi(1, (x - x_r)/t)} + r \sqrt{\frac{t}{4\pi D}} \int_0^1 e^{-t\Phi(w, (x - x_r)/t)} \frac{dw}{\sqrt{w}},$$

with

$$\Phi(w, y) = rw + \frac{y^2}{4Dw}. \tag{7.5.9}$$

For large t with $y = (x - x_r)/t$ kept fixed, the integral with respect to w can be analyzed using steepest descents or the saddle point method (see Box 11H). That is, for large t the integrand is dominated by the value of w for which $\Phi(w, y)$ takes its smallest value. Since

$$\partial_w \Phi(w, y) = r - \frac{y^2}{4Dw^2},$$

it follows that there is a minimum at $w^* = |y|/\sqrt{4Dr}$ with $\Phi(w^*, y) = \alpha_0|y|$. If $w^* < 1$ then the minimum lies within the integration domain and

$$p(x,t) \sim e^{-t\Phi(w^*,(x-x_r)/t)} = e^{-\alpha_0|x-x_r|} = p^*(x).$$

That is, the probability density has converged toward the stationary state. On the other hand, if $w^* > 1$, then the function $\Phi(w,y)$ attains its lowest value at $w = 1$. The integrand is thus dominated by the region in a neighborhood of $w = 1$, and is comparable to the first exponential. Thus

$$p(x,t) \sim e^{-t\Phi(1,(x-x_r)/t)} = e^{-rt+|x-x_r|^2/4Dt}.$$

This represents trajectories that have undergone very few or zero resettings up to time t and are thus far from the stationary state. Note that there is a phase transition at $w^* = 1$, which can be interpreted as a traveling front

$$|x - x_r| = \sqrt{4Drt},$$

which separates the spatial regions for which the probability density has relaxed to the NESS from those where it has not, see Fig. 7.25. Since the trajectories contributing to the transient region are rare events, one can view the above asymptotic analysis as a large deviation principle (Chap. 8). It should also be noted that occurrence of sharp boundaries between transient and NESS regions is a typical feature of stochastic processes with resetting [259]. For an extension of the above analysis to Brownian motion with stochastic resetting in the presence of a potential see Ref. [631].

Higher spatial dimensions. It is straightforward to extend the analysis of Brownian motion with resetting to higher spatial dimensions [257], $\mathbf{X}(t) \in \mathbb{R}^d$. The analogs of the CK equation (7.5.2) and the last renewal equation (7.5.4) are

$$\frac{\partial p(\mathbf{x},t)}{\partial t} = D\nabla^2 p(\mathbf{x},t) - rp(\mathbf{x},t) + r\delta^d(\mathbf{x} - \mathbf{x}_r), \qquad (7.5.10)$$

and

$$p(\mathbf{x},t|\mathbf{x}_0) = e^{-rt} G_0(\mathbf{x},t|\mathbf{x}_0) + r\int_0^t G_0(\mathbf{x},\tau|\mathbf{x}_r)e^{-r\tau}d\tau. \qquad (7.5.11)$$

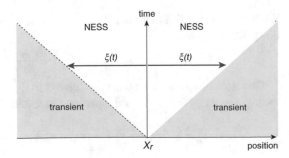

Fig. 7.25: Expanding front $|x - x_r| = \xi(t)$ separating transient and NESS regions.

The higher-dimensional propagator takes the form

$$G_0(\mathbf{x},t|\mathbf{x}_0) = \frac{1}{(4\pi Dt)^{d/2}} e^{-|\mathbf{x}-\mathbf{x}_0|^2/4Dt}.$$

The integral identity (7.5.7) with $v = 1 - d/2$ can again be used to determine the NESS:

$$p^*(\mathbf{x}) = \left(\frac{\alpha_0^2}{2\pi}\right)^{1-v} (\alpha_0|\mathbf{x}-\mathbf{x}_r|)^v K_v(\alpha_0|\mathbf{x}-\mathbf{x}_r|), \qquad (7.5.12)$$

with $\alpha_0 = \sqrt{r/D}$ as before. One finds that in a neighborhood of $\mathbf{x}_r$, $p^*(\mathbf{x}) \sim \ln|\mathbf{x} - \mathbf{x}_r|$ for $d = 2$ and $p^*(\mathbf{x}) \sim 1/|\mathbf{x}-\mathbf{x}_r|$ for $d = 3$.

7.5.2 Single search-and-capture event and renewal theory

Let us now return to the problem of a random search process, see Sect. 7.3. Consider a particle (searcher) subject to stochastic motion in $\mathcal{U} \subseteq \mathbb{R}^d$, and resetting to a fixed point $\mathbf{x}_r = \mathbf{x}_0$ at a rate $\mathcal{T}$. Suppose that there exists some target $\mathcal{U}_0 \subset \mathbb{R}^d$ whose boundary $\partial\mathcal{U}_0$ is absorbing and $\mathbf{x}_r \notin \mathcal{U}_0$. The probability density $p_r(\mathbf{x},t)$ for the particle to be at position $\mathbf{x}$ at time t evolves according to the master equation

$$\frac{\partial p_r(\mathbf{x},t|\mathbf{x}_r)}{\partial t} = \mathbb{L}p_r(\mathbf{x},t|\mathbf{x}_r) - rp_r(\mathbf{x},t|\mathbf{x}_r) + r\delta(\mathbf{x}-\mathbf{x}_r), \qquad (7.5.13)$$

where $\mathbb{L}$ is the infinitesimal generator of the stochastic process without resetting. This is supplemented by the absorbing boundary condition $p_r(\mathbf{x},t|\mathbf{x}_r) = 0$ for all $\mathbf{x} \in \partial\mathcal{U}_0$ and the reflecting boundary condition $J(\mathbf{x},t|\mathbf{x}_r) = 0$ for all $\mathbf{x} \in \partial\mathcal{U}$. Here $J(\mathbf{x},t|\mathbf{x}_r)$ denotes the probability flux. Let $Q_r(\mathbf{x}_r,t)$ be the survival probability of the particle that started at $\mathbf{x}_r$:

$$Q_r(\mathbf{x}_r,t) = \int_{\mathcal{U}\backslash\mathcal{U}_0} p_r(\mathbf{x},t)d\mathbf{x}. \qquad (7.5.14)$$

The MFPT can be expressed in terms of Q_r according to

$$T_r(\mathbf{x}_r) = -\int_0^\infty t\frac{dQ_r(\mathbf{x}_r,t)}{dt}d\tau = \int_0^\infty Q_r(\mathbf{x}_r,t)dt, \qquad (7.5.15)$$

We have used the fact that the FPT density $f_r(\mathbf{x}_r,t)$ is related to the survival probability according to

$$f_r(\mathbf{x}_r,t) = -\frac{dQ_r(\mathbf{x}_r,t)}{dt}. \qquad (7.5.16)$$

One now observes that Q_r can be related to the survival probability without resetting, Q_0, using a last renewal equation [255, 256, 259]:

$$Q_r(\mathbf{x}_r,t) = e^{-rt}Q_0(\mathbf{x}_r,t) + r\int_0^t Q_0(\mathbf{x}_r,\tau)Q_r(\mathbf{x}_r,t-\tau)e^{-r\tau}d\tau. \qquad (7.5.17)$$

The first term on the right-hand side represents trajectories with no resettings. The integrand in the second term is the contribution from trajectories that last reset at time $\tau \in (0,t)$, and consists of the product of the survival probability starting from $\mathbf{x}_r$ with resetting up to time $t - \tau$ and the survival probability starting from $\mathbf{x}_r$ without any resetting for the time interval τ. Since we have a convolution, it is natural to introduce the Laplace transform

$$\widetilde{Q}_r(\mathbf{x}_r, s) = \int_0^\infty Q_r(\mathbf{x}_r, t) e^{-st} dt.$$

Laplace transforming the last renewal equation and rearranging shows that

$$\widetilde{Q}_r(\mathbf{x}_r, s) = \frac{\widetilde{Q}_0(\mathbf{x}_r, r+s)}{1 - r\widetilde{Q}_0(\mathbf{x}_r, r+s)}. \tag{7.5.18}$$

The MFPT to reach the target is then given by

$$T_r(\mathbf{x}_r) = \widetilde{Q}_r(\mathbf{x}_r, 0) = \frac{\widetilde{Q}_0(\mathbf{x}_r, r)}{1 - r\widetilde{Q}_0(\mathbf{x}_r, r)}. \tag{7.5.19}$$

The corresponding FPT density in Laplace space is

$$\widetilde{f}_r(\mathbf{x}_r, s) = \frac{1 - (r+s)\widetilde{Q}_0(\mathbf{x}_r, r+s)}{1 - r\widetilde{Q}_0(\mathbf{x}_r, r+s)}. \tag{7.5.20}$$

Example 7.3. Diffusion on a finite interval. Consider a diffusing particle on the interval $[0, L]$ with an absorbing target at $x = 0$ and a reflecting boundary at $x = L$ [631]. In the absence of resetting the Laplace transformed survival probability $\widetilde{Q}_0(x, s)$ satisfies the equation

$$D\frac{d^2\widetilde{Q}_0}{dx^2} - s\widetilde{Q}_0 = -1, \quad x \in (0, L), \tag{7.5.21}$$

together with the boundary conditions

$$\widetilde{Q}_0(0, s) = 0,, \quad \partial_x \widetilde{Q}_0(L, s) = 0. \tag{7.5.22}$$

The solution takes the form

$$\widetilde{Q}_0(x, s) = \frac{1}{s}\left(1 - \frac{\cosh(\sqrt{s/D}[L-x])}{\cosh(\sqrt{s/D}L)}\right). \tag{7.5.23}$$

Equation (7.5.19) then implies that

$$T_r(x_r) = \frac{\cosh(\sqrt{r/D}L) - \cosh(\sqrt{r/D}[L - x_r])}{r\cosh(\sqrt{r/D}[L - x_r])}. \tag{7.5.24}$$

In the limit $L \to \infty$, we have

$$Q_0(x_r, t) = \mathrm{erf}(x_r/2\sqrt{Dt}), \quad \widetilde{Q}_0(x_r, s) = \frac{1 - e^{-\sqrt{s/D}|x_r|}}{s}. \tag{7.5.25}$$

which is the Laplace transform of the survival probability on the half-line. The latter was calculated in Ex. 2.12 and takes the form of an error function. Equation (7.5.19) then implies that

$$T_r = \frac{1}{r}\left(e^{\sqrt{r/D}x_r} - 1\right), \tag{7.5.26}$$

Note that in the limit $r \to 0$, the MFPT diverges as $T_r \sim 1/\sqrt{r}$, which recovers the result that the MFPT of a Brownian particle without resetting to return to the origin is infinite. One also finds that T_r diverges in the limit $r \to \infty$, since the particle resets to X_r so often that it never has the chance to reach the origin. It turns out that the MFPT has a finite and unique minimum at an intermediate value of the resetting rate r [256, 257]. Similar behavior is found in a wide range of search processes with resetting, another example of which is considered in Ex. 7.5.

The situation is more complicated in the case of the finite interval. One finds that the MFPT $T_r(x_r)$ is a unimodal function of r for reset locations close to the target whereas it is a monotonically increasing function of r at more distal locations, as illustrated in Fig. 7.26; in the former case there exists an optimal resetting rate that minimizes T_r. One way to investigate whether or not the MFPT has at least one turning point is to calculate the sign of the derivative dT_r/dr at $r = 0$ [61, 630, 633, 689], see also Ex. 7.6. If this derivative is

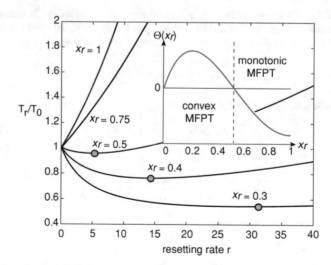

Fig. 7.26: Plot of $T_r(x_r)/T_0(x_r)$ as a function of the resetting rate r for various reset positions x_r. Other parameters are $L = 1 = D$. The filled green dots indicate the optimal resetting rate for a given x_r. Inset: Plot of $\Theta(x_r) := T_0^{(2)}(x_r) - 2T_0(x_r)^2$ as a function of the reset location x_r. The sign of Θ determines whether or not the corresponding MFPT with resetting is initially a decreasing function of the resetting rate r. Green dots indicate minima of the MFPT curves (when they exist).

negative then resetting reduces the MFPT in the small-r regime. Equation (7.5.19) implies that

$$T_r'(\mathbf{x}_r) = \widetilde{Q}_0'(\mathbf{x}_r,0) + \widetilde{Q}_0(\mathbf{x}_r,0)^2 = T_0(\mathbf{x}_r)^2 - \frac{T_0^{(2)}(\mathbf{x}_r)}{2}.$$

Taking the limit $s \to 0$ in equation (7.5.23) and using L'Hopital's rule, yields the classical results

$$T_0(x) = -\frac{x^2}{2D} + \frac{xL}{D}, \; T_0^{(2)}(x) = \frac{1}{12D^2}(x^4 - 4x^3L + 8xL^3).$$

Introducing the variance $\sigma_0^2(\mathbf{x}_r) = T_0^{(2)}(\mathbf{x}_r) - T_0(\mathbf{x}_r)^2$, it follows that adding a small rate of resetting reduces the MFPT for a given $\mathbf{x}_r$ if and only if

$$\frac{\sigma_0(\mathbf{x}_r)}{T_0(\mathbf{x}_r)} > 1. \tag{7.5.27}$$

In the inset of Fig. 7.26 we plot $\Theta(x_r) := T_0^{(2)}(x_r) - 2T_0(x_r)^2$ as a function of x_r. Applying the condition (7.5.27) implies that the sign of Θ determines whether or not the corresponding MFPT T_r with resetting is initially a decreasing function of the resetting rate r. It can be seen that $\Theta(x_r)$ is negative in the case of proximal positions ($x_r < x_c \approx 0.55$) but switches to positive values in the case of distal locations ($x_r > x_c$).

Example 7.4. Diffusive search for a d-dimensional spherical target. As a second example, consider the diffusive search for a d-dimensional spherical target of radius a and center at the origin, see also the example in Sect. 7.4. Exploiting spherical symmetry, the Laplace transformed survival probability $\widetilde{Q}_0(r,s)$, $x = |\mathbf{x}|$, satisfies the backward equation

$$\frac{d^2\widetilde{Q}_0}{dx^2} + \frac{d-1}{x}\frac{d\widetilde{Q}_0}{dx} - s\widetilde{Q}_0 = -1, \quad a < x < \infty, \tag{7.5.28}$$

together with the boundary condition $\widetilde{Q}_0(a,s) = 0$. Following [257], the solution of equation (7.5.28) takes the form

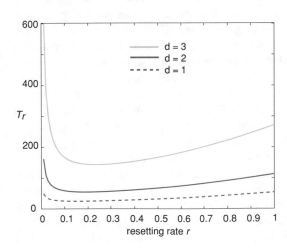

Fig. 7.27: Effect of spatial dimension d on MFPT for a spherical target. Plot of MFPT $T_r(x_r)$ as a function of the resetting rate r for $x_r = 5$ and $d = 1,2,3$. Other parameters are $a = 1 = D$.

$$\widetilde{Q}(r,s) = \frac{1}{r} - A(s)x^{\nu}K_{\nu}(\eta_s x), \tag{7.5.29}$$

where $\nu = 1 - d/2$ and $\eta_s = \sqrt{s/D}$, and K_{ν} is the modified Bessel function of the second kind of order ν. The boundary condition $\widetilde{Q}(a,s) = 0$ then determines $A(s)$ according to

$$A(s) = \frac{1}{sa^{\nu}K_{\nu}(\eta_s a)}. \tag{7.5.30}$$

Substituting equation (7.5.29) into (7.5.19) with $x = x_r$ leads to the following expression for the MFPT T_r in the presence of resetting:

$$T_r(x_r) = \frac{r^{-1} - A(r)x_r^{\nu}K_{\nu}(\sqrt{r/D}x_r)}{rA(r)x_r^{\nu}K_{\nu}(\sqrt{r/D}x_r)} = \frac{a^{\nu}K_{\nu}(\sqrt{r/D}a) - x_r^{\nu}K_{\nu}(\sqrt{r/D}x_r)}{rx_r^{\nu}K_{\nu}(\sqrt{r/D}x_r)}. \tag{7.5.31}$$

Example plots for $d = 1, 2, 3$ are shown in Fig. 7.27.

7.5.3 Finite return times and refractory periods

Most studies of stochastic resetting assume that the particle instantaneously re-turns to x_r and then immediately re-enters the search state [261]. However, both of these assumptions are unrealistic for many physical applications, which moti-vates the incorporation of finite return times [102, 569, 632, 633] and refractory periods [260, 570]. Therefore, suppose that, rather than instantaneously returning to x_r following reset, the particle switches to a ballistic state in which it returns to x_r at a constant speed V. (For simplicity, the particle cannot be absorbed by the target when it is in the return phase. One could also consider a more general dynamical model for the return phase as in Ref. [102, 633].) In addition, whenever the parti-cle returns to x_r, it is subject to a refractory period before the search begins again. The refractory period is itself a random variable with a corresponding waiting time density ϕ, which is taken to have a finite mean τ_{ref}. These features are illustrated in Fig. 7.28 for a random search process on the half-line with an absorbing target at the origin. In order to keep the analysis general, we will take resetting to occur at a random sequence of times generated by a probability density $\psi(\tau)$. It follows that $\Psi(\tau) = 1 - \int_0^{\tau} \psi(s)ds$ is the probability that no resetting has occurred up to time τ.

Finite return times and refractory periods modify the calculation of the FPT as follows. First, since the particle returns to x_r at a constant speed V, there is an ad-ditional contribution to the FPT given by $|X(\sigma) - x_r|/V$ where σ is either the first resetting time or the time since the last reset, and $X(\sigma)$ is the position just pri-or to switching to the return phase. In general, $X(\sigma)$ will be a doubly stochastic variable, since both σ and $X(t)$ are stochastic processes. Each resetting event also adds an additional contribution to the FPT due to refractoriness. Clearly both fea-tures increase the MFPT. Again, renewal theory can be used to calculate the MFPT. However, rather than constructing a renewal integral equation, it is more convenient to decompose the various contributions to the FPT by conditioning on whether or not the particle resets at least once [61, 630, 633, 689]. This exploits the fact that once the particle has returned to $\mathbf{x}_r$ it has lost all memory of previous search phases. In other words, the stochastic process satisfies the strong Markov property. Indeed, one can formulate the analysis along similar lines to previous FPT problems such as microtubule dynamics with sticky boundaries (Sect. 4.2) and stochastically gated Brownian motion (Sect. 6.7) [135].

Let $\mathcal{I}(t)$ denote the number of resetting events up to time t. Consider the following set of FPTs, analogous to the decompositions shown in Fig. 7.28:

$$\mathcal{T} = \inf\{t > 0; \mathbf{X}(t) \in \partial \mathcal{U}_0, \ \mathcal{I}(t) \geq 0\},$$
$$\mathcal{S} = \inf\{t > 0; \mathbf{X}(t) = \mathbf{x}_r, \ \mathcal{I}(t) = 1\}, \tag{7.5.32}$$
$$\mathcal{R} = \inf\{t > 0; \mathbf{X}(t + \mathcal{S} + \mathcal{N}) \in \partial \mathcal{U}_0, \ \mathcal{I}(t + \mathcal{S} + \mathcal{N}) \geq 1\}.$$

Here $\mathcal{T}$ is the FPT for finding the target irrespective of the number of resettings, $\mathcal{S}$ is the FPT for the first resetting and return to the reset point $\mathbf{x}_r$ given that the particle is still free, $\mathcal{N}$ is the first refractory time, and $\mathcal{R}$ is the FPT for finding the target given that at least one resetting has occurred. Next we introduce the sets $\Omega = \{\mathcal{T} < \infty\}$ and $\Gamma = \{\mathcal{S} < \mathcal{T} < \infty\} \subset \Omega$. Here Ω is the set of all events for which the particle is eventually absorbed by the target (which has measure one), and Γ is the subset of events in Ω for which the particle resets at least once. It immediately follows that $\Omega \backslash \Gamma = \{\mathcal{T} < \mathcal{S} = \infty\}$, that is, $\Omega \backslash \Gamma$ is the set of all events for which the particle is captured by the target without any resetting. The Laplace transform of the FPT density with resetting is

$$\tilde{f}_r(\mathbf{x}_r, s) = \mathbb{E}[e^{-s\mathcal{T}} 1_\Omega]. \tag{7.5.33}$$

Following along the lines of Refs. [633], consider the decomposition

$$\mathbb{E}[e^{-s\mathcal{T}} 1_\Omega] = \mathbb{E}[e^{-s\mathcal{T}} 1_{\Omega \backslash \Gamma}] + \mathbb{E}[e^{-s\mathcal{T}} 1_\Gamma]. \tag{7.5.34}$$

The first expectation can be evaluated by noting that it is the FPT density for capture by the target without any resetting, and the probability density for such an event is $-\Psi(t)\partial_t Q_0(\mathbf{x}_r, t)d\tau$, where Q_0 is the survival probability (7.5.14). Hence,

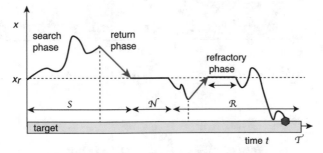

Fig. 7.28: Particle searching for a target at $x = 0$ on the half-line with an absorbing target at the origin, which is found after two resettings. After each resetting event, the particle returns to the point x_r at a constant speed V, after which it remains at x_r for a refractory period before re-entering the search phase. Also shown is the decomposition of the FPT, $\mathcal{T} = \mathcal{S} + \mathcal{N} + \mathcal{R}$.

$$\mathbb{E}[e^{-s\mathcal{T}}1_{\Omega\backslash\Gamma}] = -\int_0^\infty e^{-s\tau}\Psi(\tau)\frac{\partial Q_0(\mathbf{x}_r,\tau)}{\partial\tau}d\tau = 1 - \int_0^\infty e^{-s\tau}\psi(\tau)Q_0(\mathbf{x}_r,\tau)d\tau$$
$$- s\int_0^\infty e^{-s\tau}\Psi(\tau)Q_0(\mathbf{x}_r,\tau)d\tau. \tag{7.5.35}$$

The second expectation can be written as

$$\mathbb{E}[e^{-s\mathcal{T}}1_\Gamma] = \mathbb{E}[e^{-s[\mathcal{S}+\mathcal{N}+\mathcal{R}]}1_\Gamma]$$
$$= \int_0^\infty \psi(\tau_1)\left[\int_{\mathcal{U}\backslash\mathcal{U}_0} e^{-s(\tau_1+|\mathbf{x}-\mathbf{x}_r|/V)}p(\mathbf{x},\tau_1|\mathbf{x}_r)d\mathbf{x}\right]d\tau_1$$
$$\times \left[\int_0^\infty e^{-s\tau_2}\phi(\tau_2)d\tau_2\right]\widetilde{f}_r(\mathbf{x}_r,s). \tag{7.5.36}$$

We have used the fact that the probability that the first return is initiated in the interval $[\tau_1,\tau_1+d\tau_1]$, given that $\mathbf{X}(\tau_1)=\mathbf{x}$ and the particle has not been captured by the target, is $\psi(\tau_1)p(\mathbf{x},\tau_1|\mathbf{x}_r)d\tau_1$. The particle then takes an additional time $|\mathbf{x}-\mathbf{x}_r|/V$ to return to $\mathbf{x}_r$, after which it spends a time τ_2 in the refractory state with waiting time density $\phi(\tau_2)$. The remaining time to find the target has the same FPT density as $\mathcal{T}$.

Combining Eqs. (7.5.34)–(7.5.36) and rearranging yields the general result [633]

$$\widetilde{f}_r(\mathbf{x}_r,s) = \frac{1 - \int_0^\infty e^{-s\tau}\psi(\tau)Q_0(\mathbf{x}_r,\tau)d\tau - s\int_0^\infty e^{-s\tau}\Psi(\tau)Q_0(\mathbf{x}_r,\tau)d\tau}{1 - \widetilde{\phi}_{\text{ref}}(s)\int_0^\infty \psi(\tau_1)\left[\int_{\mathcal{U}\backslash\mathcal{U}_0} e^{-s(\tau_1+|\mathbf{x}-\mathbf{x}_r|/V)}p(\mathbf{x},\tau_1|\mathbf{x}_r)d\mathbf{x}\right]d\tau_1}.$$

The Laplace transform of the FPT density is the moment generator of the FPT $\mathcal{T}$:

$$T_r^{(n)} = \mathbb{E}[\mathcal{T}^n 1_\Omega] = \left(-\frac{d}{ds}\right)^n \mathbb{E}[e^{-s\mathcal{T}}1_\Omega]\bigg|_{s=0}. \tag{7.5.37}$$

For example, the MFPT $T_r = T_r^{(1)}$ is

$$T_r(\mathbf{x}_r) = \frac{\langle Q_0(\mathbf{x}_r,\tau)\rangle_\psi - \langle \tau Q_0(\mathbf{x}_r,\tau)\rangle_\psi}{1 - \langle Q_0(\mathbf{x}_r,\tau)\rangle_\psi}$$
$$+ \frac{-\widetilde{\phi}'_{\text{ref}}(0)\langle Q_0(\mathbf{x}_r,\tau)\rangle_\psi + \langle \tau Q_0(\mathbf{x}_r,\tau)\rangle_\psi + V^{-1}\langle F(\mathbf{x}_r,\tau)\rangle_\psi}{1 - \langle Q_0(\mathbf{x}_r,\tau)\rangle_\psi}$$
$$= \frac{\langle Q_0(\mathbf{x}_r,\tau)\rangle_\psi + \tau_{\text{ref}}\langle Q_0(\mathbf{x}_r,\tau)\rangle_\psi + V^{-1}\langle F(\mathbf{x}_r,\tau)\rangle_\psi}{1 - \langle Q_0(\mathbf{x}_r,\tau)\rangle_\psi}, \tag{7.5.38}$$

where we have set

$$\langle Q_0(\mathbf{x}_r,\tau)\rangle_\psi = \int_0^\infty \psi(\tau)Q_0(\mathbf{x}_r,\tau)d\tau \tag{7.5.39}$$

etc., and

$$F(\mathbf{x}_r,t) = \int_{\mathcal{U}\backslash\mathcal{U}_0} |\mathbf{x}-\mathbf{x}_r| p(\mathbf{x},t|\mathbf{x}_r) d\mathbf{x}. \qquad (7.5.40)$$

We have also used the result

$$-\widetilde{\phi}_{\text{ref}}'(0) = \tau_{\text{ref}} = \int_0^\infty \tau\phi(\tau)d\tau.$$

In the case of exponential resetting,

$$\psi(t) = re^{-r\tau}, \quad \Psi(\tau) = e^{-rt},$$

everything can be expressed in terms of Laplace transforms. For example, the MFPT reduces to

$$T_r(\mathbf{x}_r) = \frac{\widetilde{Q}_0(\mathbf{x}_r,r) + r\tau_{\text{ref}}\widetilde{Q}_0(\mathbf{x}_r,r) + r\widetilde{F}(\mathbf{x}_r,r)/V}{1 - r\widetilde{Q}_0(\mathbf{x}_r,r)}, \qquad (7.5.41)$$

where $\widetilde{F}(\mathbf{x}_r,r)$ is the Laplace transform of $F(\mathbf{x}_r,\tau)$. Similarly, the Laplace transform of the FPT density becomes

$$\widetilde{f}_r(\mathbf{x}_r,s) = \frac{1 - (r+s)\widetilde{Q}_0(\mathbf{x}_r,r+s)}{1 - r\widetilde{\phi}(s)\int_{\mathcal{U}\backslash\mathcal{U}_0} e^{-s|\mathbf{x}-\mathbf{x}_r|/V}\widetilde{p}(\mathbf{x},r+s|\mathbf{x}_r)d\mathbf{x}}. \qquad (7.5.42)$$

Example 7.5. As an illustration of the above result, consider again a diffusing particle on the semi-infinite interval with a target at the origin. Now suppose that resetting is non-instantaneous, with the particle returning to x_r at a constant speed V. It follows from Laplace transforming equation (7.5.40) that $\widetilde{F}(r) = \int_0^\infty |x-x_r|\widetilde{G}(x,r|x_r)dx$, where

$$\widetilde{G}(x,r|x_r) = \frac{1}{2\sqrt{rD}}\left(e^{-\sqrt{r/D}|x-x_r|} - e^{-\sqrt{r/D}(x+x_r)}\right)$$

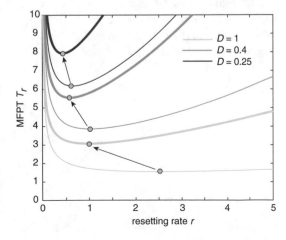

Fig. 7.29: Effect of finite return time on MFPT T_r for diffusion with resetting. Plot of T_r as a function of resetting rate r for various diffusivities D. Thick curves are for $V = 1$ and thin curves are for $V = \infty$. The reset point is $x_r = 1$ and the target is at $x = 0$. The green dots indicate the optimal resetting rate for each curve, and the arrows indicate the shift due to a finite return time.

is the Laplace transform of Green's function for diffusion on the half-line with an absorbing boundary at $x = 0$. Evaluating the integral yields the result

$$\widetilde{F}(r) = \frac{1}{r}\sqrt{\frac{D}{r}}\left[1 - \sqrt{\frac{r}{D}}x_r e^{-\sqrt{r/D}x_r} - e^{-2\sqrt{r/D}x_r}\right]. \qquad (7.5.43)$$

Substituting equations (7.5.25) and (7.5.43) into (7.5.41) with $\tau_{\text{ref}} = 0$ thus determines the MFPT T_r. In Fig. 7.29 we plot T_r as a function of r for various diffusivities D, and compare the two cases $V = 1$ and $V = \infty$ (instantaneous resetting). We see that there still exists a unique minimum of the MFPT curve when $V = 1$ (indicated by green dots in Fig. 7.29), but the minimum is shifted upwards and to the left. In particular, the value of the optimal resetting rate r_{min} is reduced in the case of a finite return time, with the degree of shift a monotonically decreasing function of V. (A reduction in r_{min} also occurs for refractory periods.)

7.5.4 Splitting probabilities and conditional MFPTs for multiple targets

Now suppose that there are N interior targets $\mathcal{U}_i$, $i = 1, \ldots, N$, as shown in Fig. 7.30. First consider the case without resetting ($r = 0$). The FPT problem is identical to that considered in Sect. 6.4.1. Let $\mathbf{J}(\mathbf{x},t|\mathbf{x}_r)$ denote the probability flux of the stochastic search process without resetting such that equation (7.5.13) becomes

$$\frac{\partial p(\mathbf{x},t|\mathbf{x}_r)}{\partial t} = \mathbb{L}p(\mathbf{x},t|\mathbf{x}_r) = -\nabla \cdot \mathbf{J}(\mathbf{x},t|\mathbf{x}_r), \qquad (7.5.44)$$

supplemented by the absorbing boundary conditions

$$p(\mathbf{x},t|\mathbf{x}_r) = 0, \quad \mathbf{x} \in \partial\mathcal{U}_a = \bigcup_{j=1}^{N}\partial\mathcal{U}_j, \qquad (7.5.45)$$

and the initial condition $p(\mathbf{x},0|\mathbf{x}_r) = \delta(\mathbf{x} - \mathbf{x}_r)$. Let $\mathcal{T}_k(\mathbf{x}_r)$ denote the FPT that the particle is captured by the kth target, with $\mathcal{T}_k(\mathbf{x}_r) = \infty$ indicating that it is not captured. Define $\Pi_k(\mathbf{x}_r,t)$ to be the probability that the particle is captured by the kth target after time t, given that it started at $\mathbf{x}_r$:

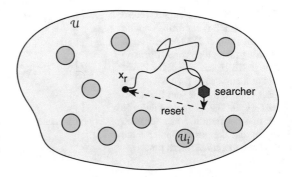

Fig. 7.30: Particle searching for multiple targets $\mathcal{U}_i \subset \mathbb{R}^d$, $i = 1,\ldots,N$, and subject to resetting to position $\mathbf{x}_r$ at a rate r.

$$\Pi_k(\mathbf{x}_r,t) = \mathbb{P}[t < \mathcal{T}_k(\mathbf{x}_r) < \infty] = \int_t^\infty J_k(\mathbf{x}_r,t)dt', \qquad (7.5.46)$$

where

$$J_k(\mathbf{x}_r,t) = -\int_{\partial \mathcal{U}_k} \mathbf{J}(\mathbf{y},t|\mathbf{x}_r) \cdot d\mathbf{y}. \qquad (7.5.47)$$

The negative sign indicates that the flux is into the domain $\mathcal{U}_k$.

The splitting probability $\pi_k(\mathbf{x}_r)$ and conditional MFPT $T_k(\mathbf{x}_r)$ to be captured by the kth target are then

$$\pi_k(\mathbf{x}_r) = \Pi_k(\mathbf{x}_r,0) = \int_0^\infty J_k(\mathbf{x}_r,t)dt, \qquad (7.5.48)$$

and

$$T_k(\mathbf{x}_r) = \mathbb{E}[\mathcal{T}_k|\mathcal{T}_k < \infty] = \frac{1}{\pi_k(\mathbf{x}_r)} \int_0^\infty \Pi_k(\mathbf{x}_r,t)dt. \qquad (7.5.49)$$

Also note that differentiating equation (7.5.46) with respect to t and Laplace transforming implies

$$s\widetilde{\Pi}_k(x_r,s) - \pi_k(\mathbf{x}_r) = -\widetilde{J}_k(\mathbf{x}_r,s), \quad \widetilde{J}_k(\mathbf{x}_r,s) = \pi_k(\mathbf{x}_r)\widetilde{f}_k(\mathbf{x}_r,s), \qquad (7.5.50)$$

where $\widetilde{f}_k(\mathbf{x}_r,s)$ is the Laplace transform of the conditional FPT density. The total probability of being captured by one of the targets, $\pi_{\text{tot}}(\mathbf{x}_r)$, is

$$\pi_{\text{tot}}(\mathbf{x}_r) := \sum_{k=1}^N \pi_k(\mathbf{x}_r).$$

We will allow for the possibility that $\pi_{\text{tot}}(\mathbf{x}_r) < 1$, which holds in the case of diffusion in an unbounded d-dimensional domain for $d > 2$. This should be distinguished from the notion of failure to find a particular target due to absorption by another target. The former type of failure disappears when resetting is included, whereas the latter persists. Finally, note that integrating equation (7.5.44) with respect to $\mathbf{x}$ and t implies that the survival probability up to time t is

$$Q_0(\mathbf{x}_r,t) = \int_{\mathbb{R}^d \backslash \mathcal{U}_a} p(\mathbf{x},t|\mathbf{x}_r)d\mathbf{x} = 1 - \sum_{k=1}^N \Lambda_k(\mathbf{x}_r,t) = 1 - \pi_{\text{tot}}(\mathbf{x}_r) + \sum_{k=1}^N \Pi_k(\mathbf{x}_r,t).$$

$$(7.5.51)$$

We now extend the renewal method introduced in Sect. 7.5.2 to calculate the splitting probability $\pi_{r,k}(\mathbf{x}_r)$ and conditional MFPT $T_{r,k}(\mathbf{x}_r)$ to be captured by the kth target in the presence of resetting. Consider the following set of first passage times, which are the multitarget analogs of equation (7.5.32):

$$\mathcal{T}_k = \inf\{t \geq 0; \mathbf{X}(t) \in \partial \mathcal{U}_k, \mathcal{I}(t) \geq 0\},$$
$$\mathcal{S} = \inf\{t \geq 0; \mathbf{X}(t) = \mathbf{x}_r, \mathcal{I}(t) = 1\},$$
$$\mathcal{R}_k = \inf\{t \geq 0; X(t+\mathcal{S}+\mathcal{N}) \in \partial \mathcal{U}_k, \mathcal{I}(t+\mathcal{S}+\mathcal{N}) \geq 1\}.$$

Here $\mathcal{T}_k$ is the FPT for finding the kth target irrespective of the number of resettings, $\mathcal{S}$ is the FPT for the first resetting and return to $\mathbf{x}_r$ without being captured by any target, $\mathcal{N}$ is the first refractory time, and $\mathcal{R}_k$ is the FPT for finding the kth target given that at least one resetting has occurred. Next we define the sets

$$\Omega_k = \{\mathcal{T}_k < \infty\}, \quad \Gamma_k = \{\mathcal{S} < \mathcal{T}_k < \infty\} \subset \Omega_k,$$

where Ω_k is the set of all events for which the particle is eventually absorbed by the kth target without being absorbed by any other target, and Γ_k is the subset of events in Ω_k for which the particle resets at least once. It immediately follows that

$$\Omega_k \backslash \Gamma_k = \{\mathcal{T}_k < \mathcal{S} = \infty\},$$

where $\Omega_k \backslash \Gamma_k$ is the set of all events for which the particle is captured by the kth target without any resetting.

The splitting probability $\pi_{r,k}$ can be decomposed as

$$\pi_{r,k}(\mathbf{x}_r) := \mathbb{P}[\Omega_k] = \mathbb{P}[\Omega_k \backslash \Gamma_k] + \mathbb{P}[\Gamma_k]. \tag{7.5.52}$$

We note that the probability that the particle is captured by the kth target in the interval $[\tau, \tau + d\tau]$ without any returns to $\mathbf{x}_r$ is $\Psi(\tau)J_k(\mathbf{x}_r, \tau)d\tau$ with $J_k(\tau)$ given by equation (7.5.47). Hence,

$$\mathbb{P}[\Omega_k \backslash \Gamma_k] = \int_0^\infty \Psi(\tau)J_k(\mathbf{x}_r, \tau)d\tau = -\int_0^\infty \Psi(\tau)\frac{d\Pi_k(\mathbf{x}_r, \tau)}{d\tau}d\tau$$

$$= \pi_k(\mathbf{x}_r) - \int_0^\infty \psi(\tau)\Pi_k(\mathbf{x}_r, \tau)d\tau, \tag{7.5.53}$$

after integrating by parts. Next, from the definitions of the first passage times, we have

$$\mathbb{P}[\Gamma_k] = \mathbb{P}[\mathcal{S} < \infty]\mathbb{P}[\mathcal{R}_k < \infty], \tag{7.5.54}$$

and memoryless return to $\mathbf{x}_r$ implies that $\mathbb{P}[\mathcal{R}_k < \infty] = \pi_{r,k}(\mathbf{x}_r)$. In addition

$$\mathbb{P}[\mathcal{S} < \infty] = \int_0^\infty \psi(\tau)Q_0(\mathbf{x}_r, \tau)d\tau = 1 - \pi_{\text{tot}}(\mathbf{x}_r) + \sum_{k=1}^N \int_0^\infty \psi(\tau)\Pi_k(\mathbf{x}_r, \tau)d\tau.$$

$$\tag{7.5.55}$$

We have used the fact that the probability of resetting in the time interval $[\tau, \tau + d\tau]$ is equal to the product of the reset probability $\psi(\tau)d\tau$ and the survival probability $Q_0(\mathbf{x}_r, t)$ that the particle hasn't been captured by a target up to time τ. Hence, equation (7.5.54) becomes

$$\mathbb{P}[\Gamma_k] = \pi_{r,k}(\mathbf{x}_r)\langle Q_0(\mathbf{x}_r, \tau)\rangle_\psi. \tag{7.5.56}$$

Combining equations (7.5.53) and (7.5.56) yields the implicit equation

$$\pi_{r,k}(\mathbf{x}_r) = \pi_k(\mathbf{x}_r) - \langle \Pi_k(\mathbf{x}_r, \tau)\rangle_\psi + \langle Q_0(\mathbf{x}_r, \tau)\rangle_\psi \pi_{r,k}(\mathbf{x}_r),$$

which on rearranging leads to the following result:

$$\pi_{r,k}(\mathbf{x}_r) = \frac{\pi_k(\mathbf{x}_r) - \langle \Pi_k(\mathbf{x}_r, \tau) \rangle_\psi}{\langle Q_0(\mathbf{x}_r, \tau) \rangle_\psi}. \tag{7.5.57}$$

Summing both sides of equation (7.5.57) implies that $\sum_{k=1}^N \pi_{r,k}(\mathbf{x}_r) = 1$. In other words, in the presence of reset, the particle is captured by one of the targets with probability one. Note that the splitting probability $\pi_{r,k}(\mathbf{x}_r)$ is independent of the refractory periods and finite return times. However, implicit in the calculation of $\pi_{r,k}(\mathbf{x}_r)$ is the assumption that the particle returns to $\mathbf{x}_r$ and then escapes from the refractory state in a finite time. In the particular case of exponential resetting, equation (7.5.57) becomes

$$\pi_{r,k}(\mathbf{x}_r) = \frac{\pi_k(\mathbf{x}_r) - r\widetilde{\Pi}_k(\mathbf{x}_r, r)}{1 - r\widetilde{Q}_0(\mathbf{x}_r, r)}. \tag{7.5.58}$$

The conditional MFPT $\mathbb{E}[\mathcal{T}_k 1_{\Omega_k}] = \pi_{r,k}T_{r,k}$ can be analyzed along similar lines to the splitting probability by introducing the decomposition

$$\mathbb{E}[\mathcal{T}_k 1_{\Omega_k}] = \mathbb{E}[\mathcal{T}_k 1_{\Omega_k \backslash \Gamma_k}] + \mathbb{E}[\mathcal{T}_k 1_{\Gamma_k}]. \tag{7.5.59}$$

The first expectation can be evaluated by noting that it is the MFPT for capture by the kth target without any resetting, and the probability density for such an event is $\Psi(\tau)J_k(\mathbf{x}_r, \tau)d\tau$. Hence,

$$\mathbb{E}[\mathcal{T}_k 1_{\Omega_k \backslash \Gamma_k}] = \int_0^\infty \tau\Psi(\tau)J_k(\mathbf{x}_r, \tau)d\tau = -\int_0^\infty \tau\Psi(\tau)\frac{d\Pi_k(\mathbf{x}_r, \tau)}{d\tau}d\tau$$

$$= -\int_0^\infty \tau\psi(\tau)\Pi_k(\mathbf{x}_r, \tau)d\tau + \int_0^\infty \Psi(\tau)\Pi_k(\mathbf{x}_r, \tau)d\tau. \tag{7.5.60}$$

The second expectation can be further decomposed as

$$\mathbb{E}[\mathcal{T}_k 1_{\Gamma_k}] = \mathbb{E}[(\mathcal{S} + \mathcal{N} + \mathcal{R}_k)1_{\Gamma_k}] = \mathbb{E}[\mathcal{S}1_{\Gamma_k}] + \tau_{\text{ref}}\mathbb{P}[\Gamma_k] + \mathbb{E}[\mathcal{R}_k 1_{\Gamma_k}]$$

$$= \mathbb{E}[\mathcal{S}1_{\Gamma_k}] + (\tau_{\text{ref}} + T_{r,k})\mathbb{P}[\Gamma_k], \tag{7.5.61}$$

with $\mathbb{P}[\Gamma_k]$ given by equation (7.5.56). Again $\mathcal{N}$ denotes the random time spent in the refractory state at $\mathbf{x}_r$ before switching back to the search phase, with $\mathbb{E}[\mathcal{N}] = \tau_{\text{ref}}$, and we have used the result $\mathbb{E}[\mathcal{R}_k 1_{\Gamma_k}] = T_{r,k}\mathbb{P}[\Gamma_k]$. The latter follows from the fact that return to $\mathbf{x}_r$ restarts the stochastic process without any memory. In order to calculate $\mathbb{E}[\mathcal{S}1_{\Gamma_k}]$, it is necessary to incorporate the time of return following the first resetting event along the lines of Sect. 7.5.2 The first return is initiated before being captured by a target with probability $\psi(\tau)Q_0(\mathbf{x}_r, \tau)d\tau$ in the interval $[\tau, \tau + d\tau]$. At time τ the particle is at position $\mathbf{X}(\tau)$ and thus takes an additional time $|\mathbf{X}(\tau) - \mathbf{x}_r|/V$ to return to $\mathbf{x}_r$. Using the fact that $\mathbb{P}[\mathcal{R}_k < \infty] = \pi_{r,k}$, we have

$$\mathbb{E}[\mathcal{S}1_{\Gamma_k}] = \pi_{r,k}(\mathbf{x}_r)\int_0^\infty \psi(\tau)\left(\tau + \frac{\langle|\mathbf{X}(\tau) - \mathbf{x}_r|\rangle}{V}\right)Q_0(\mathbf{x}_r, \tau)d\tau,$$

where $\langle \cdot \rangle$ denotes expectation with respect to the probability density $p(\mathbf{x}, t|\mathbf{x}_r)$ evolving according to equation (7.5.44) without resetting, and conditioned on survival up to time τ. Hence, we have

$$\mathbb{E}[\mathcal{S}1_\Gamma] = \pi_{r,k}(\mathbf{x}_r) \int_0^\infty \psi(\tau) \left(\tau Q_0(\mathbf{x}_r, \tau) + \frac{F(\mathbf{x}_r, \tau)}{V} \right) d\tau, \qquad (7.5.62)$$

where $F(\mathbf{x}_r, \tau)$ is given by equation (7.5.40) for p evolving according to (7.5.44). Combining equations (7.5.59)–(7.5.62) yields an implicit equation of the form

$$\pi_{r,k}(\mathbf{x}_r)T_{r,k}(\mathbf{x}_r) = -\langle \tau \Pi_k(\mathbf{x}_r, \tau) \rangle_\psi + \pi_{r,k}(\mathbf{x}_r) \left(\langle \tau Q_0(\mathbf{x}_r, \tau) \rangle_\psi + V^{-1} \langle F(\mathbf{x}_r, \tau) \rangle_\psi \right)$$
$$+ (\tau_{\text{ref}} + T_{r,k}(\mathbf{x}_r)) \pi_{r,k}(\mathbf{x}_r) \langle Q_0(\mathbf{x}_r, \tau) \rangle_\psi.$$

Rearranging then yields the conditional MFPT

$$\pi_{r,k}(\mathbf{x}_r)T_{r,k}(\mathbf{x}_r) = \frac{-\langle \tau \Pi_k(\mathbf{x}_r, \tau) \rangle_\psi + \langle \Pi_k(\mathbf{x}_r, \tau) \rangle_\psi}{1 - \langle Q_0(\mathbf{x}_r, \tau) \rangle_\psi} \qquad (7.5.63)$$
$$+ \frac{\pi_{r,k}(\mathbf{x}_r) \left(\langle [\tau + \tau_{\text{ref}}] Q_0(\mathbf{x}_r, \tau) \rangle_\psi + V^{-1} \langle F(\mathbf{x}_r, \tau) \rangle_\psi \right)}{1 - \langle Q_0(\mathbf{x}_r, \tau) \rangle_\psi},$$

with

$$\langle Q_0(\mathbf{x}_r, \tau) \rangle_\psi = 1 - \pi_{\text{tot}}(\mathbf{x}_r) + \sum_{k=1}^N \langle \Pi_k(\mathbf{x}_r, \tau) \rangle_\psi.$$

In the particular case of exponential resetting, equation (7.5.63) can be expressed in terms of Laplace transforms according to

$$\pi_{r,k}(\mathbf{x}_r)T_{r,k}(\mathbf{x}_r) = \frac{\widetilde{\Pi}_k(\mathbf{x}_r, r) + r\widetilde{\Pi}_k'(\mathbf{x}_r, r)}{1 - r\widetilde{Q}_0(\mathbf{x}_r, r)} \qquad (7.5.64)$$
$$+ \frac{\pi_{r,k}(\mathbf{x}_r)}{1 - r\widetilde{Q}_0(\mathbf{x}_r, r)} \left[\frac{1 - \pi_{\text{tot}}(\mathbf{x}_r)}{r} - r \sum_{k=1}^N \frac{d\widetilde{\Pi}_k(\mathbf{x}_r, r)}{dr} \right.$$
$$+ r\frac{\widetilde{F}(\mathbf{x}_r, r)}{V} + \tau_{\text{ref}} r \widetilde{Q}_0(\mathbf{x}_r, r) \bigg],$$

where $'$ denotes differentiation with respect to r, and

$$r\widetilde{Q}_0(\mathbf{x}_r, r) = 1 - \pi_{\text{tot}}(\mathbf{x}_r) + r \sum_{k=1}^N \widetilde{\Pi}_k(\mathbf{x}_r, r). \qquad (7.5.65)$$

Finally, summing both sides of equation (7.5.64) with respect to k yields the unconditional MFPT $\sum_{k=1}^N \pi_{r,k}(\mathbf{x}_r)T_{r,k}(\mathbf{x}_r) = T_r(\mathbf{x}_r)$, with $T_r(\mathbf{x}_r)$ given by equation (7.5.41). A specific example of a search process with multiple targets and delays is considered in Ex. 7.7, where a stochastic resetting protocol is added to the directed search process of Sect. 7.3.4. One could also use the asymptotic results for the narrow capture problem presented in Sect. 6.4 to analyze the effects of stochastic resetting in the case of small targets [139]. Finally, note that the analysis of Sect. 7.4 and Sect. 7.5 can be combined to analyze the accumulation of resources under multiple rounds of search-an-capture with resetting. This is explored further for a single target in Ex. 7.8.

7.6 Exclusion processes

So far we have considered PDE models of active transport in which individual molecular motors move independently of one another. However, if the density of motors is sufficiently high, then there is a non-negligible chance that motors could interact with each other and exhibit some form of collective behavior. This has motivated a number of studies that model the movement of multiple motor particles as an asymmetric exclusion process (ASEP) [4, 254, 451, 457, 621, 639, 640, 661, 666]. In the simplest version of such models, each particle hops unidirectionally at a uniform rate along a 1D lattice; the only interaction between particles is a hard-core repulsion that prevents more than one particle occupying the same lattice site at the same time. This so-called totally asymmetric exclusion process (TASEP) is combined with absorption/desorption kinetics, in which individual particles can bind to or unbind from the track, see Fig. 7.31. The TASEP has become the paradigmatic model of nonequilibrium stochastic processes, and a variety of analytical methods have been developed to generate exact solutions for the stationary state, see [100, 180, 730] and references therein. However, when chemical kinetic or other biologically motivated extensions of TASEP are included, it is usually no longer possible to obtain exact solutions so that some form of mean-field approximation is required.

7.6.1 Asymmetric exclusion process and the hydrodynamic limit

Let us consider in more detail the system shown in Fig. 7.31, which consists of a finite 1D lattice of N sites labeled $i = 1, \ldots, N$. The microscopic state of the system is given by the configuration $\mathcal{C}$ that specifies the distribution of identical particles on the lattice. That is, $\mathcal{C} = \{n_1, \ldots, n_N\}$ where each occupation number $n_i = 1$ if the ith site is occupied by a single particle and $n_i = 0$ if the site is vacant. Exclusion effects preclude more than one particle at any site. Thus, the state space consists of 2^N configurations. Let $\mathcal{P}(\mathcal{C}, t)$ denote the probability of finding a particular configuration $\mathcal{C}$ at time t. The evolution of this probability distribution is described by a master equation:

$$\frac{d\mathcal{P}(\mathcal{C}, t)}{dt} = \sum_{\mathcal{C}' \neq \mathcal{C}} \left[\mathcal{W}_{\mathcal{C}' \to \mathcal{C}} \mathcal{P}(\mathcal{C}', t) - \mathcal{W}_{\mathcal{C} \to \mathcal{C}'} \mathcal{P}(\mathcal{C}, t) \right]. \qquad (7.6.1)$$

The transition rate $\mathcal{W}_{\mathcal{C} \to \mathcal{C}'}$ from configuration $\mathcal{C}$ to $\mathcal{C}'$ is determined from the following set of rules [639]:

(a) at sites $i = 1, \ldots, N-1$, a particle can jump to site $i+1$ at a unit rate if the latter is unoccupied;

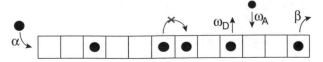

Fig. 7.31: Schematic diagram of TASEP with absorption/desorption kinetics, in which particles can spontaneously detach and attach at rates ω_D and ω_A, respectively.

(b) at site $i = 1$ ($i = N$) a particle can enter (exit) the lattice at a rate α (β) provided that the site is unoccupied (occupied);

(c) in the bulk of the lattice, a particle can detach from a site at a rate ω_D and attach to an unoccupied site at a rate ω_A.

Rules (a) and (b) constitute a TASEP with open boundary conditions, whereas rule (c) describes absorption/desorption kinetics. It follows that the evolution of the particle densities $\langle n_i \rangle$ away from the boundaries is given by the exact equation

$$\frac{d\langle n_i \rangle}{dt} = \langle n_{i-1}(1 - n_i) \rangle - \langle n_i(1 - n_{i+1}) \rangle + \omega_A \langle 1 - n_i \rangle - \omega_D \langle n_i \rangle. \tag{7.6.2}$$

Here $\langle n_i(t) \rangle = \sum_{\mathcal{C}} n_i \mathcal{P}(\mathcal{C}, t)$ etc. Similarly, at the boundaries

$$\frac{d\langle n_1 \rangle}{dt} = -\langle n_1(1 - n_2) \rangle + \alpha \langle 1 - n_1 \rangle - \omega_D \langle n_1 \rangle, \tag{7.6.3a}$$

$$\frac{d\langle n_N \rangle}{dt} = \langle n_{N-1}(1 - n_N) \rangle + \omega_A \langle 1 - n_N \rangle - \beta \langle n_N \rangle. \tag{7.6.3b}$$

Note that in the absence of any exclusion constraints, equation (7.6.2) reduces to a spatially discrete version of a PDE model, with $p_+(n_i \Delta x, t) = \langle n_i \rangle$, $\beta_+ = \omega_D$, and $p_0 \alpha = \omega_A$ and $v_+/\Delta x = 1$. The goal is to find a nonequilibrium stationary state for which the current flux along the lattice is a constant J_0. It then follows that J_0 has the exact form

$$J_0 = \alpha \langle 1 - n_1 \rangle = \langle n_i(1 - n_{i+1}) \rangle = \beta \langle n_N \rangle, \quad i = 1, N - 1.$$

Equations (7.6.2) and (7.6.3) constitute a nontrivial many-body problem, since in order to calculate the time evolution of $\langle n_i \rangle$ it is necessary to know the two-point correlations $\langle n_{i-1}(1 - n_i) \rangle$. The latter obey dynamical equations involving three-point and four-point correlations. Thus, there is an infinite hierarchy of equations of motion. However, progress can be made by using a mean-field approximation and a continuum limit in order to derive a PDE for the density of particles [254, 640]. The mean-field approximation consists of replacing two-point correlations by products of single-site averages:

$$\langle n_i n_j \rangle = \langle n_i \rangle \langle n_j \rangle.$$

(It turns out that for pure TASEP (no binding or unbinding) this yields an accurate phase diagram, see below.) Next introduce the infinitesimal lattice spacing ε and set $x = k\varepsilon$, $\rho(x, t) = \rho_k(t) \equiv \langle n_k(t) \rangle$. The continuum limit is then defined according to $N \to \infty$ and $\varepsilon \to 0$ such that the length of the track $L = N\varepsilon$ is fixed. (Fix length scales by setting $L = 1$). Expanding $\rho_{k\pm1}(t) = \rho(x \pm \varepsilon, t)$ in powers of ε gives

$$\rho(x \pm \varepsilon, t) = \rho(x) \pm \varepsilon \partial_x \rho(x, t) + \frac{1}{2} \varepsilon^2 \partial_{xx} \rho(x, t) + O(\varepsilon^3).$$

Finally, rescaling the absorption/desorption rates according to $\omega_A = \Omega_A \varepsilon$, $\omega_D = \Omega_D \varepsilon$, and rescaling time $\tau = \varepsilon t$, equation (7.6.2) becomes to $O(\varepsilon)$

$$\frac{\partial \rho}{\partial \tau} = \frac{\varepsilon}{2} \frac{\partial^2 \rho}{\partial x^2} - (1 - 2\rho) \frac{\partial \rho}{\partial x} + \Omega_A(1 - \rho) - \Omega_D \rho. \tag{7.6.4}$$

Similarly, equation (7.6.3) reduces to the boundary conditions

$$J(0,t) = \alpha(1 - \rho(0,t)), \quad J(L,t) = \beta\rho(L,t).$$

where the continuum flux is

$$J(x,t) = -\frac{\varepsilon}{2}\frac{\partial\rho}{\partial x} + \rho(1-\rho). \tag{7.6.5}$$

In the following, we describe methods for analyzing the mean-field model for a pure TASEP by setting $\Omega_A = \Omega_D = 0$. For extensions to the full model, we refer the reader to Refs. [254, 639, 640].

7.6.2 Steady-state analysis

In order to develop the basic theory, we will focus on pure TASEP by setting $\Omega_A = \Omega_D = 0$ in equation (7.6.4). We proceed by finding a stationary nonequilibrium state for which the current $J(x,t) = J_0$ is constant, and determining the corresponding stationary density profile. This then generates a phase diagram with respect to the parameters α, β, which can be calculated explicitly [100, 469]. The steady-state current equation takes the form

$$\varepsilon\frac{d\rho}{dx} = \rho(1-\rho) - J_0.$$

Setting $q = \rho - 1/2$, this becomes

$$\varepsilon\frac{dq}{dx} = v^2 - q^2, \quad v^2 = \frac{1}{4} - J_0.$$

It follows that for $v^2 > 0$

$$\varepsilon\int\frac{dq}{(v-q)(v+q)} = x - x_0,$$

where x_0 is an integration constant. Using partial fractions, we find that

$$\frac{v+q}{v-q} = e^{2v(x-x_0)/\varepsilon},$$

which on rearranging yields the density profile

$$\rho(x) = \frac{1}{2} + v\tanh(v(x-x_0)/\varepsilon), \tag{7.6.6}$$

with $v \geq 0$. On the other hand, if $v^2 < 0$ then we have

$$\varepsilon\int\frac{dq}{|v^2| + q^2} = x - x_0.$$

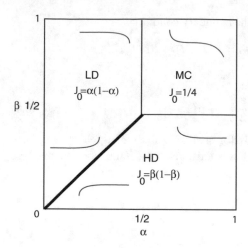

Fig. 7.32: Mean-field phase diagram for the TASEP showing the regions of α, β parameter space where the low-density (LD), high-density (HD), and maximal-current (MC) phases exist. Schematic illustrations of the density profiles in the various regions are shown in red.

Under the change of variables $q = \cot an(u)$, we can evaluate the integral and find that

$$\rho(x) = 0.5 + |v|\cotan(|v|(x - x_0)/\varepsilon). \tag{7.6.7}$$

The two unknown parameters J_0, x_0 can be determined in terms of α, β by imposing the boundary conditions at $x = 0, L$. The resulting phase diagram in the limit of large L and small ε is shown in Fig. 7.32. Three distinct phases can be identified:

1. A *low density* phase in which the bulk density is smaller than $1/2$, $x_0 = O(L)$ and $v^2 > 0$. Since $\varepsilon \ll 1$, we see from equation (7.6.6) that $\rho(x) \approx 0.5 - v$ for all $x < x_0$. In particular, at the left-hand boundary $\alpha(0.5 + v) = J_0$, which can be rewritten as $v = J_0/\alpha - 0.5$. Squaring both sides and using the definition of v gives, to lowest order in ε,

$$\rho(0) = \alpha, \quad J_0 = \alpha(1 - \alpha), \quad \alpha < 1/2.$$

The other boundary condition becomes

$$\beta = \frac{J_0}{0.5 + v \tanh(v(L - x_0)/\varepsilon)} > \frac{J_0}{0.5 + v} = \alpha.$$

In order to satisfy this boundary condition, there is an ε-wide boundary layer at $x = L$ with $L - x_0 = O(\varepsilon)$.

2. A *high density* phase in which the bulk density is larger than $1/2$ and $x_0 \approx 0$. Hence, $\rho(x) \approx 0.5 + v$ in the bulk of the domain and at the right-hand boundary we have $\beta(0.5 + v) = J_0$. Following along similar lines to the low density case, we deduce that

$$\rho(L) = 1 - \beta, \quad J_0 = \beta(1 - \beta), \quad \beta < 1/2,$$

and $\beta < \alpha$. There is now a boundary layer around $x = 0$ in order to match the rate α. The two phases coexist along the line $\alpha = \beta < 1/2$.

3. A *maximal current phase*. In the region $\alpha > 1/2, \beta > 1/2$, we require $J_0 > 1/4$ so that $v^2 < 0$. It turns out that the current takes the form $J_0 = 0.25 + O(\varepsilon^2/L^2)$, that is, it is very close to the maximal value of function $\rho(1 - \rho)$. This follows from the observation that the solution (7.6.7) will blow up unless $0 < |v|(x-x_0)/\varepsilon < \pi$ for all $x \in [0,L]$. This implies that $x_0 = -O(\varepsilon)$ and $|v| < \pi\varepsilon/L$. Under these conditions, equation (7.6.7) ensures that $\rho(x) \approx 0.5$ in the bulk of the domain. The precise values of v and x_0 are then adjusted so that the boundary conditions at $x = 0, L$ are satisfied: $\rho(0) = 1 - 1/(4\alpha) > 0.5$ and $\rho(L) = 1/(4\beta) < 0.5$. Also note away from the left-hand boundary, we have $\cot(|v|(x-x_0)/\varepsilon) \approx \varepsilon/(|v|x)$ so that $\rho(x) \sim 0.5 + \varepsilon/x$.

7.6.3 *Method of characteristics and shocks*

Equation (7.6.4) is mathematically similar in form to the viscous Burger's equation with additional source terms [717]. Thus, one expects singularities such as shocks in the density ρ to develop in the inviscid or non-dissipative limit $\varepsilon \to 0^+$. One can view the formation and propagation of shocks as a way of understanding how the system evolves to the final steady-state solution [254, 458]. Again, we will illustrate this by considering a pure TASEP. Setting $\Omega_A = \Omega_D = 0$ and $\varepsilon = 0$ in equation (7.6.4), yields a kinematic wave equation of the quasi-linear form

$$\frac{\partial \rho}{\partial \tau} + \frac{\partial J(\rho)}{\partial x} = 0, \quad J(\rho) = \rho(1 - \rho). \tag{7.6.8}$$

Equation (7.6.8) is a particular example of a quasi-linear PDE, and can be analyzed using the method of characteristics introduced in Box 2C. Thus one looks for characteristic curves $x = x(\tau)$ along which $\rho(\tau) \equiv \rho(x(\tau), \tau)$ satisfies

$$\frac{d\rho}{d\tau} = \frac{\partial \rho}{\partial \tau} + \frac{dx}{d\tau}\frac{\partial \rho}{\partial x}.$$

Comparison with (7.6.8) leads to the characteristic equations

$$\frac{dx}{d\tau} = J'(\rho) = 1 - 2\rho, \quad \frac{d\rho}{d\tau} = 0. \tag{7.6.9}$$

It can be seen that the characteristics are straight lines along which ρ is constant. Suppose that $x(0) = x_0$ and the corresponding initial density is $\rho(x_0, 0) = \rho_0(x_0)$. For simplicity, we ignore the boundary conditions by taking $x \in \mathbb{R}$. The corresponding characteristic solutions (parameterized by x_0) are then

$$x(\tau) = [1 - 2\rho_0(x_0)]t + x_0, \quad \rho(\tau) = \rho(x(\tau), \tau) = \rho_0(x_0).$$

In other words, the density profile at time t is determined by the propagation of the initial density $\rho(x_0, 0)$ along the straight line characteristics.

For the given kinetic wave equation, one finds that an initial density profile can sharpen up to form a discontinuity, which then propagates as a shock wave. This

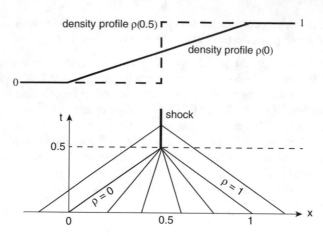

Fig. 7.33: Formation of a shock for equation (7.6.8). The characteristics are straight lines of speed $1 - 2\rho$ with ρ constant along a characteristic. The initial density profile evolves into a stationary shock solution.

is illustrated in Fig. 7.33 for an initial density given by the piecewise linear function $\rho_0(x) = 0$ for $x < 0$, $\rho_0(x) = x$ for $0 \leq x \leq 1$ and $\rho_0(x) = 1$ for $x > 1$. Since higher densities propagate more slowly than lower densities, an initial linear density profile steepens until a shock is formed at the points of intersection where pairs of characteristics meet. In general, a shock propagates with a speed v_S determined by the so-called Rankine–Hugonoit condition [717]:

$$v_S = \frac{J(\rho_2) - J(\rho_1)}{\rho_2 - \rho_1} = 1 - \rho_1 - \rho_2, \tag{7.6.10}$$

where ρ_1, ρ_2 are the densities on either side of the shock. For the particular initial density profile shown in Fig. 7.33, $\rho_1 = 0$ and $\rho_2 = 1$ so that the shock is stationary ($v_S = 0$). The possibility of stationary shocks reflects the fact that the current $J(\rho) = \rho(1 - \rho)$ has a maximum, which means that two different densities can have the same current on either side of the shock. The Rankine–Hugonoit condition is usually derived by considering weak solutions of the kinematic equation, see Box 7A. However, it can also be understood in terms of a traveling wave solution of the corresponding PDE with weak diffusion, see Ex. 7.9. A further example illustrating the method of characteristics and shocks is given in Ex. 7.10.

The method of characteristics and kinematic wave theory yields insights into the dynamics underlying the formation of the various stationary phases shown in Fig. 7.32 [100, 458, 469]. The basic idea is to consider kinematic waves propagating from the left-hand and right-hand boundaries, respectively, by considering an initial density profile such that $\rho(0,0) = \alpha$ and $\rho(L,0) = 1 - \beta$ with L large.

1. If $\alpha, \beta < 1/2$ then a kinematic wave propagates from the left-hand and right-hand boundaries with speeds $1 - 2\alpha > 0$ and $(2\beta - 1) < 0$, respectively. These waves thus propagate into the interior of the domain and meet somewhere in the middle to form a shock that propagates with speed $v_S = \beta - \alpha$. If $\beta > \alpha$ then

the shock moves to the right-hand boundary and the bulk of the domain is in a low-density (LD) state with $\rho \approx \alpha < 1/2$. On the other hand, If $\beta < \alpha$ then the shock moves to the left-hand boundary and the bulk of the domain is in a high-density (HD) state with $\rho \approx 1 - \beta > 1/2$. For weak dissipation the sharp drop in the density at one end is smoothed to form a boundary layer.

2. In the special case $\alpha = \beta < 1/2$ the LD and HD phases coexist. The solution consists of a low-density region separated from a high-density region by a shock. Once higher order dissipative effects are included, this shock diffuses freely between the ends of the domain, so that the average density profile is linear.

3. If both $\alpha > 1/2$ and $\beta > 1/2$ then the steady-state bulk solution has the maximal current density $J = 1/4$. In order to show this, and to determine how bulk solutions match the boundary conditions, it is necessary to include dissipation effects as in the previous section.

The above analysis based on the theory of shocks can be extended to the full molecular motor model that combines TASEP with binding/unbinding kinetics [254, 639, 640]. When $\Omega_A, \Omega_D \neq 0$ the characteristic equation (7.6.9) becomes

$$\frac{dx}{d\tau} = 1 - 2\rho, \quad \frac{d\rho}{d\tau} = \Omega_A(1 - \rho) - \Omega_D\rho. \tag{7.6.11}$$

It follows that the characteristics are now curves in the $x - t$ plane. For example, consider the propagation of density fluctuations along a characteristic starting at the left boundary with $\rho = \alpha < 1/2$ and $\alpha < K/(K+1)$, where $K = \Omega_A/\Omega_D$. It follows from equation (7.6.11) that initially the fluctuation propagates along the characteristic with decreasing speed and increasing density. If $K/(1 + K) < 1/2$ then ρ will approach the constant value $\rho = K/(K+1)$ and the speed approaches a constant value. However, if $K/(1 + K) > 1/2$ then after a finite time the density reaches $\rho = 1/2$ and propagation ceases. A similar analysis holds for characteristics propagating from the right boundary. Furthermore, characteristics propagating from opposite boundaries can again intersect, implying multivalued densities and the formation of shocks. The resulting shock has the same wave speed as pure TASEP. Of particular interest are stationary solutions for which the current $J = \rho(1 - \rho)$ is constant so that any shock solution is stationary ($v_S = 0$). To a first approximation, these can be obtained by finding steady-state solutions of the mean-field equation

$$(1 - 2\rho)\frac{\partial \rho}{\partial x} - \Omega_D[K - (1 + K)\rho] = 0. \tag{7.6.12}$$

The occurrence of stationary shocks is consistent with the observation that this is a first-order ODE but there are two boundary conditions. One thus proceeds by integrating from the left boundary where $\rho(0) = \alpha$ to obtain a density profile $\rho_L(x)$ and then integrating from the right boundary where $\rho(L) = 1 - \beta$ to obtain a second density profile $\rho_R(x)$. The full solution is constructed by matching the two profiles at a shock whose position also has to be determined. If the shock lies outside the interval $[0, L]$, then it is necessary to include at least one boundary layer. A detailed

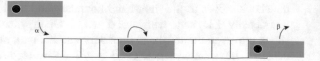

Fig. 7.34: A TASEP with
extended particles of size
$l = 3$.

analysis of the steady-state solutions with coexisting low and high-density phases,
and the corresponding phase diagram with respect to the parameters $(\alpha, \beta, \Omega_D, \Omega_A)$
can be found in [254, 640]. If the effects of dissipation are also taken into account
then the sharp interfaces and boundary layers become smooth fronts of size $O(1/\varepsilon)$.

One of the first examples of a TASEP model in biology was proposed by Gibbs
and collaborators in their study of the translation of messenger RNA (mRNA) by
ribosomes during protein synthesis [540, 541], see also Sect. 5.5. However, it is
necessary to modify pure TASEP to include multisite particles, since ribosomes are
large molecules which extend over several codons or lattice sites (around $l = 12$).
In the case of multisite particles, one has to specify the rules for entry and exit of
a ribosome [180, 896]. One possibility is "complete entry, incremental exit," which
assumes that a ribosome enters completely provided the first l lattice sites are vacant, whereas it exits one step at a time [178], see Fig. 7.34. Inclusion of extended
objects considerably complicates the analysis even though the basic structure of
the phase diagram is preserved [178, 748]. In contrast to pure TASEP, there does
not currently exist an exact solution, although mean-field approximations do provide useful insights. A second biologically motivated modification of TASEP is to
include site-dependent hopping rates [179, 227, 284, 457]. This is motivated by
the fact that the local hopping rate depends on the relative abundance of specific
amino-acid-carrying tRNA. Using a combination of Monte Carlo simulations and
mean-field theory it can be shown, for example, that two defects (regions of slow
hopping rates) decrease the steady-state current more when they are close to each
other. Finally, note that more complex models take into account intermediate steps
in the translocation of a ribosome along the mRNA, including the binding of tRNA
to the ribosome and hydrolysis [55, 185, 298, 679].

Box 7A. Weak formulation of shocks and the Rankine-Hugonoit condition.

In order to deal with discontinuities and shocks, it is necessary to introduce a
more flexible notion of a solution to the kinematic wave equation, in which
derivatives of the solution are not directly involved [717]. Let $\phi(x,t)$ be a
smooth function in $\mathbb{R} \times [0, \infty)$ with compact support, that is, it vanishes outside
a bounded domain. If ρ is a smooth solution of the kinematic wave equation
$\partial_t \rho + \partial_x J(\rho) = 0$ with initial condition $\rho(x,0) = p(x)$, then

$$\int_0^\infty \int_{\mathbb{R}} [\partial_t \rho + \partial_x J(\rho)] \phi \, dx \, dt = 0.$$

We now carry out an integration by parts of the first term with respect to t and the second term with respect to x:

$$\int_0^\infty \partial_t \rho \phi \, dx \, dt = -\int_0^\infty \rho \partial_t \phi \, dx \, dt - \int_{\mathbb{R}} p(x)\phi(x,0) \, dx,$$

and, since $\phi(\pm\infty) = 0$,

$$\int_0^\infty \partial_x J(\rho)\phi \, dx \, dt = -\int_0^\infty J(\rho)\partial_x \phi \, dx \, dt.$$

We thus obtain the integral equation

$$\int_0^\infty \int_{\mathbb{R}} [\rho \partial_t \phi + J(\rho)\partial_x \phi] \, dx \, dt + \int_{\mathbb{R}} p(x)\phi(x,0) \, dx = 0. \qquad (7.6.13)$$

It can be seen that no derivative of ρ appears. We define a weak solution of the kinematic wave equation to be one that satisfies the integral equation (7.6.13) for every test function ϕ in $\mathbb{R} \times [0,\infty)$ with compact support. Note that if ρ is also smooth then we can reverse the integration by parts to recover the PDE.

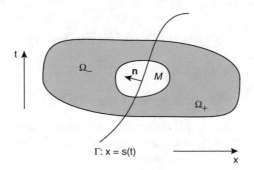

Fig. 7.35 Construction of a weak solution in a domain partitioned by a shock Γ.

The weak formulation can now be used to derive the Rankine–Hugonoit condition for the speed of a shock. Suppose that an open set $\Omega \in \mathbb{R} \times [0,\infty)$ is partitioned into two disjoint domains $\Omega_\pm$ by a smooth shock curve Γ satisfying $x = s(t)$. Suppose that ρ is a weak solution in Ω that is a continuously differentiable function ρ in the closed sets $\overline{\Omega_+}$ and $\overline{\Omega_-}$. That is ρ is a smooth solution of $\partial_t \rho + \partial_x J(\rho) = 0$ in Ω_+ and Ω_- such that ρ and its first derivatives extend continuously up to Γ from either side, see Fig. 7.35. Choose a test function ϕ with support in a compact set $M \subset \Omega$ such that $M \cap \Omega$ is not empty, and take $\phi(x,0) = 0$. The integral equation for the weak solution takes the form

$$0 = \int_0^\infty \int_{\mathbb{R}} [\rho \partial_t \phi + J(\rho)\partial_x \phi] \, dx \, dt$$

$$= \int_{\Omega_+} [\rho \partial_t \phi + J(\rho)\partial_x \phi] \, dx \, dt + \int_{\Omega_-} [\rho \partial_t \phi + J(\rho)\partial_x \phi] \, dx \, dt.$$

Integrating by parts the integral over Γ_+ using Stoke's Theorem, noting that $\phi = 0$ on $\partial\Omega_+/\Gamma$ (the boundary of Ω_+ excluding the curve Γ), we have

$$\int_{\Omega_+}[\rho\partial_t\phi + J(\rho)\partial_x\phi]dx\,dt = -\int_{\Omega_+}[\partial_t\rho + \partial_x J(\rho)]\phi\,dx\,dt$$

$$+ \int_{\Gamma}[\rho_+ n_2 + J(\rho_+)n_1]\phi\,dl$$

$$= \int_{\Gamma}[\rho_+ n_2 + J(\rho_+)n_1]\phi\,dl,$$

where ρ_+ denotes the value of ρ on Γ from the Ω_+ side, $\mathbf{n} = (n_1, n_2)$ is the outward normal vector on the boundary $\partial\Omega_+$, and dl denotes arc length along Γ. Similarly,

$$\int_{\Omega_-}[\rho\partial_t\phi + J(\rho)\partial_x\phi]dx\,dt = -\int_{\Gamma}[\rho_- n_2 + J(\rho_-)n_1]\phi\,dl,$$

where ρ_- denotes the value of ρ on Γ from the Ω_- side.
The above analysis shows that

$$\int_{\Gamma}[(J(\rho_+) - J(\rho_-))n_1 + (\rho_+ - \rho_-)n_2]\,\phi\,dl = 0.$$

The arbitrariness of ϕ means that

$$(J(\rho_+) - J(\rho_-))n_1 + (\rho_+ - \rho_-)n_2 = 0. \tag{7.6.14}$$

If ρ were continuous across Γ then this equation would be an identity. Therefore, suppose that $\rho_+ \neq \rho_-$. The shock curve is given by $x = s(t)$, which implies that

$$\mathbf{n} = (n_1, n_2) = \frac{1}{\sqrt{1 + \dot{s}(t)^2}}(-1, \dot{s}(t)).$$

Substituting for $\mathbf{n}$ in equation (7.6.14) thus gives the Rankine–Hugonoit condition for the speed $\dot{s}(t)$ of the shock:

$$\dot{s} = \frac{J(\rho_+(s,t)) - J(\rho_-(s,t))}{\rho_+(s,t) - \rho_-(s,t)}. \tag{7.6.15}$$

7.7 Exercises

Problem 7.1. 2D transport model. Consider the full intracellular transport model given by

$$\frac{\partial p}{\partial t} = -\nabla \cdot (\mathbf{v}(\theta)p) - \frac{\beta}{\varepsilon}p(\mathbf{r},\theta,t) + \frac{\alpha q(\theta)}{\varepsilon}p_0(\mathbf{r},t),$$

$$\frac{\partial p_0}{\partial t} = \varepsilon D_0 \nabla^2 p_0 + \frac{\beta}{\varepsilon}\int_0^{2\pi} p(\mathbf{r},\theta',t)d\theta' - \frac{\alpha}{\varepsilon}p_0(\mathbf{r},t).$$

Introduce the decompositions

$$p(\mathbf{r},\theta,t) = u(\mathbf{r},t)p^{ss}(\theta) + \varepsilon w(\mathbf{r},\theta,t), \quad p_0(\mathbf{r},t) = u(\mathbf{r},t)p_0^{ss} + \varepsilon w_0(\mathbf{r},t),$$

with

$$p_0^{ss} = \frac{\beta}{\alpha+\beta} \equiv b, \quad p^{ss}(\theta) = \frac{\alpha q(\theta)}{\alpha+\beta} \equiv aq(\theta),$$

and

$$u(\mathbf{r},t) \equiv \int_0^{2\pi} p(\mathbf{r},\theta,t)d\theta + p_0(\mathbf{r},t), \quad \int_0^{2\pi} w(\mathbf{r},\theta,t)d\theta + w_0(\mathbf{r},t) = 0.$$

(a) By carrying out a QSS reduction along the lines outlined in Sect. 7.2, derive the advection–diffusion equation

$$\frac{\partial u}{\partial t} = -\nabla \cdot (\mathbf{V}u) + \varepsilon b D_0 \nabla^2 u + \varepsilon \nabla \cdot (\mathbf{D}\nabla u).$$

where the diffusion tensor $\mathbf{D}$ has components

$$D_{kl} \sim \frac{a}{\beta}\left(\langle v_k v_l \rangle - \langle v_k \rangle \langle v_l \rangle + b^2 \langle v_k \rangle \langle v_l \rangle\right),$$

and the effective drift velocity is given by $\mathbf{V} \sim a\langle \mathbf{v} \rangle$. Here $\langle v_n \rangle = \int_0^{2\pi} v_n(\theta)q(\theta)d\theta$.

(b) Determine the diffusion tensor when

$$q(\theta) = \frac{1}{2\pi} + \frac{1}{\pi}(\hat{\omega}_2 \cos(2\theta) + \hat{\omega}_2 \sin(2\theta)).$$

Problem 7.2. Calculation of MFPT for unbiased random intermittent search.
Consider the unbiased random search process given by equation (7.3.1). The MFPTs $T_\pm(y)$ to find the target, given that the particle starts at position y and state $\pm$ at time t, satisfy the pair of equations (see (7.3.8a,7.3.8b))

$$\frac{\partial T_+(y)}{\partial y} + \frac{\beta}{v}[(\alpha u(y)-1)T_+(y) + \alpha u(y)T_-(y)] = -\frac{\beta}{v}\left(\frac{1}{\beta}+u(y)\right),$$

$$\frac{\partial T_-(y)}{\partial y} - \frac{\beta}{v}[\alpha u(y)T_+(y) + (\alpha u(y)-1)T_-(y)] = \frac{\beta}{v}\left(\frac{1}{\beta}+u(y)\right).$$

(a) Transform the equations using the new variables

$$S_1(y) = \frac{T_+(y)+T_-(y)}{2}, \quad S_2(y) = \frac{T_+(y)-T_-(y)}{2}.$$

(b) Solve the equations for $S_{1,2}(y)$ in each of the following regions: (I) $-L < y < -a$, (II) $-a < y < a$, and (III) $a < y < L$. It is necessary to use the boundary conditions $S_2(\pm L) = 0$ and to impose continuity at $y = \pm a$.

(c) Hence, evaluate the average MFPT defined according to

$$\tau_1 = \frac{1}{L} \int_0^L S_1(y)\,dy,$$

and thus obtain the result

$$\tau_1 = \frac{1}{L}\left(\frac{1}{\beta} + \frac{1}{2\alpha}\right)\left(\left[\frac{\beta}{v}\right]^2 \frac{(L-a)^3}{3} + \frac{\beta}{v}\sqrt{\frac{2\alpha+k}{k}}(L-a)^2\coth(\Lambda a)\right.$$
$$\left. + \frac{2\alpha+k}{k}(L-a)\right) + \frac{2\alpha+k}{k\beta} + \frac{1}{k},$$

where k is the target absorption rate and

$$\Lambda = \frac{\beta}{v}\sqrt{\frac{k}{2\alpha+k}}.$$

Problem 7.3. QSS reduction of random search model. Consider the PDE model of random intermittent search for a hidden target given by equation (7.3.24):

$$\frac{\partial p_n}{\partial t} = -v_n\frac{\partial p_n(x,t)}{\partial x} + \sum_{n'=1}^N A_{nn'}(x)p_{n'}(x,t) - k_n\chi(x-X),$$

where k_n is the rate of target absorption in the nth internal state, and $\chi(x-X) = 1$ if $|x-X| \le a$ and is zero otherwise. Use the QSS reduction method to derive the FP equation (7.3.26) for the total probability density $C(x,t) = \sum_{n=1}^N p_n(x,t)$:

$$\frac{\partial C}{\partial t} = -\frac{\partial}{\partial x}(VC) + \frac{\partial}{\partial x}\left(D\frac{\partial C}{\partial x}\right) - \lambda\chi(x-X)C,$$

with the drift V and diffusion coefficient D given by equation (7.2.25) and the effective detection rate is $\lambda = \sum_{n=1}^N k_n p_n(x)$.

Problem 7.4. Renewal equation for Binomial moments. Consider the multiparticle search-and-capture process introduced at the end of Sect. 7.4.2, see also Fig. 7.23. Suppose that a particle is injected periodically at times $n\Delta\tau$ for integer n. Let $M_k(t)$ be the number of resources within the kth target at time t that have not yet degraded. In terms of the sequence of capture times τ_i, we can write

$$M_k(t) = \sum_{j\ge 1}\chi(t-\tau_j)\delta_{k_j,k}$$

with $\chi(t)$ defined in equation (7.4.5). Introduce the generating function $G_k(x,t)$ and Binomial moments $B_{m,k}(t)$ according to equations (7.4.6) and (7.4.7), respectively. Conditioning on the first capture time τ_1, we have

$$M_k(t) = \chi(t-\tau_1)\delta_{k_1,k} + \Theta(t-\Delta\tau)M_k^*(t-\Delta\tau),$$

where Θ is a Heaviside function, τ_1 is the capture time of the first particle injected at $t_1 = 0$, and $M_k^*(t)$ is the accumulation of resources due to all particles but the first. Note that $M_k^*(t)$ has the same probability distribution as $M_k(t)$, and $\chi(t - \tau_1)$ and $\Theta(t - \Delta\tau)M^*(t - \Delta\tau)$ are statistically independent.

(a) Using the total expectation theorem show that

$$G_k(z,t) = \mathbb{E}[z^{N_k(t)}] = \mathbb{E}\left[\mathbb{E}[z^{\Theta(t-y')N_k^*(t-y')}|t_2 = y']\right] \cdot \mathbb{E}\left[\mathbb{E}[z^{\chi(t-y)\delta_{k_1,k}}|\tau_1 = y, k_1 = k]\right]$$

$$= [\Theta(t - \Delta\tau)G_k(z, t - \Delta\tau) + \Theta(\Delta\tau - t)]\overline{G}_k(z,t),$$

where

$$\overline{G}_k(z,t) = \int_0^t [z + (1-z)H(t-y)]dF_k(y) + \int_t^\infty dF_k(y) + \sum_{k' \neq k} \pi_{k'}.$$

(b) Differentiating the renewal equation of part (a) with respect to z, derive the integral equation

$$B_{r,k}(t) = \delta_{r,1}\mathcal{H}_k(t) + B_{r,k}(t - \Delta\tau) + B_{r-1,k}(t - \Delta\tau)\mathcal{H}_k(t)$$

where

$$\mathcal{H}_k(t) = \int_0^t e^{-\gamma(t-y)}dF_k(y)$$

and $dF_k(t) = \pi_k f_k(t)$.

(c) Using Laplace transforms show that

$$\widetilde{B}_{1,k}(s) = \frac{1}{1 - e^{-s\Delta\tau}}\frac{\widetilde{\mathcal{F}}_k(s)}{\gamma + s}.$$

and hence determine the steady state B_1^*.

(d) Set $r = 2$ in part (b) and square both sides to show that

$$2\mathcal{H}_k(t)B_{1,k}(t - \Delta\tau) = B_{1,k}(t)^2 - \mathcal{H}_k(t)^2 - B_{1,k}(t - \Delta\tau)^2.$$

Setting

$$\mathcal{R}_{2,k}(t) = B_{2,k}(t) - \frac{1}{2}B_{1,k}(t)^2$$

derive the iterative equation

$$\mathcal{R}_{2,k}(t) - \mathcal{R}_{2,k}(t - \Delta\tau) = -\frac{1}{2}\mathcal{H}_k(t)^2.$$

Solving this equation using Laplace transforms, show that

$$B_{2,k}^* = \frac{B_{1,k}^{*\,2}}{2} - \frac{\widetilde{\mathcal{H}_k^2}(0)}{2\Delta\tau},$$

with

$$\widetilde{\mathcal{H}_k^2}(0) = \frac{\pi_k^2}{\gamma}\int_0^\infty e^{-\gamma y'}\int_0^\infty f_k(y)f_k(y+y')dy dy'.$$

Problem 7.5. Velocity jump process with stochastic resetting. Consider the symmetric velocity-jump process given by the CK equation

$$\frac{\partial p_+(x,t)}{\partial t} = -v\frac{\partial p_+(x,t)}{\partial x} - \alpha p_+(x,t) + \alpha p_-(x,t),$$

$$\frac{\partial p_-(x,t)}{\partial t} = v\frac{\partial p_-(x,t)}{\partial x} - \alpha p_-(x,t) + \alpha p_+(x,t).$$

Suppose that there is an absorbing boundary at $x = 0$ and take the initial conditions to be $x(0) = x_0 > 0$ with the initial velocities $\pm v$ equally likely. The system is supplemented by the following stochastic resetting condition: with rate r the particle resets its initial position to $x = X_r > 0$ and the velocity is again chosen to be $\pm v$ with equal probability $1/2$. Since resetting preserves the initial conditions, the renewal equation (7.5.17) holds. The remaining step is to determine the Laplace transform of the survival probability in the absence of reset. Let $Q_0^\pm(x_0,t)$ denote the survival probability without resetting for a particle having started at $x = x_0$ with initial velocity $\pm v$. The total survival probability appearing in equation (7.5.17) is then

$$Q_0(x_0,t) = \frac{1}{2}[Q_0^+(x_0,t) + Q_0^-(x_0,t)].$$

The survival probabilities $Q_0^\pm$ satisfy the backward CK equation

$$\frac{\partial Q_0^+(x_0,t)}{\partial t} = v\frac{\partial Q_0^+(x_0,t)}{\partial x_0} - \alpha Q_0^+(x_0,t) + \alpha Q_0^-(x_0,t),$$

$$\frac{\partial Q_0^-(x_0,t)}{\partial t} = -v\frac{\partial Q_0^-(x_0,t)}{\partial x_0} - \alpha Q_0^-(x_0,t) + \alpha Q_0^+(x_0,t)$$

for $x_0 > 0$ The initial conditions are $Q_0^\pm(x_0,0) = 1$, and there is an absorbing boundary condition at $x = 0$, $Q_0^-(0,t) = 0$.

(a) Laplace transform the backward equations and solve for $\widetilde{Q}_0^\pm(x_0,t)$. Hence show that the Laplace transform of the total survival probability without resetting is

$$\widetilde{Q}_0(x_0,s) = \frac{1}{s} + \frac{1}{2\alpha s}[v\lambda - (s+2\alpha)]e^{-\lambda x_0}, \quad \lambda = \sqrt{\frac{s(s+2\alpha)}{v^2}}.$$

(b) Substitute the result of part (a) into equation (7.5.18) and determine the Laplace transform of the total survival probability with resetting, $\widetilde{Q}_r(x_0,s)$. Setting $x_0 = X_r$ and using the fact that the MFPT with resetting satisfies $T_r = \widetilde{Q}_r(X_r,0)$, show that

$$T_r = -\frac{1}{r} + \frac{2\alpha}{r}\left[\frac{e^{\lambda_r X_r}}{r+2\alpha - \sqrt{r(r+2\alpha)}}\right].$$

(c) Show that in the diffusion limit $\alpha, v \to \infty$ with $D = v^2/2\alpha$ fixed, the expression for the MFPT reduces to equation (7.5.26) for a Brownian particle.

(d) Plot T_r as a function of $R = r/2\alpha$ in the case $\alpha = 1/2$ and $X_r/v = 1$. Hence show that the MFPT has a unique minimum at a finite resetting rate.

Problem 7.6. Small r-expansion for search processes with stochastic resetting.
Consider a search process in a domain with N targets and subject to instantaneous resetting to the initial position $\mathbf{x}_0$ at a rate r. In the absence of resetting, let $J_k(\mathbf{x}_0,t)$ denote the flux into the kth target, which is related to the conditional FPT density according to $f_k(\mathbf{x}_0,t) = J_k(\mathbf{x}_0,t)/\pi_k(\mathbf{x}_0)$, where $\pi_k(\mathbf{x}_0)$ is the splitting probability.

(a) Taylor expand the Laplace transform $\widetilde{J}_k(\mathbf{x}_0,s)$ of the flux $J_k(\mathbf{x}_0,t)$ with respect to s and show that

$$\widetilde{J}_k(\mathbf{x}_0,s) = \pi_k(\mathbf{x}_0) - s\pi_k(\mathbf{x}_0)T_k(\mathbf{x}_0) + \frac{s^2}{2}\pi_k T_k^{(2)}(\mathbf{x}_0) + o(s^2),$$

assuming the moments of f_k exist.

(b) Introduce the survival probability

$$Q_0(\mathbf{x}_0,t) = \sum_{k=1}^{N} \Pi_k(\mathbf{x}_0,t), \quad \Pi(\mathbf{x}_0,t) = \int_t^{\infty} J_k(\mathbf{x}_0,t).$$

Show that

$$s\widetilde{\Pi}_k(\mathbf{x}_0,s) = \pi_k(\mathbf{x}_0) - \widetilde{J}_k(\mathbf{x}_0,s) = \pi_k(\mathbf{x}_0)T_k(\mathbf{x}_0) - \frac{s}{2}\pi_k(\mathbf{x}_0)T_k^{(2)}(\mathbf{x}_0) + o(s).$$

Hence, summing both sides with respect to k, obtain the result

$$\widetilde{Q}_0(\mathbf{x}_0,s) = T(\mathbf{x}_0) - \frac{s}{2}(\sigma_T^2(\mathbf{x}_0) + T^2(\mathbf{x}_0)) + o(s),$$

where $T = \sum_{k=1}^{N} \pi_k T_k$ and $\sigma_T^2 = \sum_{k=1}^{N} \pi_k T_k^{(2)} - T^2$ is the unconditional variance without resetting.

(c) Using the formula for the splitting probability with resetting,

$$\pi_{r,k}(\mathbf{x}_0) = \frac{\pi_k(\mathbf{x}_0) - r\widetilde{\Pi}_k(\mathbf{x}_0,r)}{1 - r\widetilde{Q}_0(\mathbf{x}_0,r)},$$

show that

$$\pi_{r,k}(\mathbf{x}_0) = \pi_k(\mathbf{x}_0) + r\pi_k(\mathbf{x}_0)\left(\sum_{l=1}^{N} \pi_l(\mathbf{x}_0)T_l(\mathbf{x}_0) - T_k(\mathbf{x}_0)\right) + o(r).$$

Similarly, given the formula for the unconditional MFPT with instantaneous resetting,

$$T_r(\mathbf{x}_0) = \frac{\widetilde{Q}_0(\mathbf{x}_0,r)}{1 - r\widetilde{Q}_0(\mathbf{x}_0,r)},$$

show that

$$T_r(\mathbf{x}_0) = T(\mathbf{x}_0) + \frac{r}{2}(T^2(\mathbf{x}_0) - \sigma_T^2(\mathbf{x}_0)) + o(r),$$

(d) Using the analysis of finite return times and refractory periods, see equation (7.5.41), derive the more general result

$$T_r(\mathbf{x}_0) = T(\mathbf{x}_0) + \frac{r}{2}\left(T^2(\mathbf{x}_0) - \sigma_T^2(\mathbf{x}_0) + \frac{2\widetilde{F}(0)}{V} + 2\tau_{\mathrm{ref}}T(\mathbf{x}_0)\right) + o(r).$$

Hence, write down a condition for resetting to reduce the unconditional MFPT in the limit $V \to \infty$. Interpret this result in terms of the mean number of resources in steady-state under multiple rounds of search-and-capture.

Problem 7.7. Directed search with stochastic resetting. Suppose that a stochastic resetting protocol is added to the directed search process of Sect. 7.3.4, as illustrated in Fig. 7.36. Prior to finding one of the synaptic targets, the motor may randomly switch to a retrograde state (resetting), returning to the origin at a speed V. After a refractory period, a new motor is inserted and the process repeats. The waiting time density of the refractory period is taken to be $\phi(t) = \bar{\tau}^{-1}e^{-t/\bar{\tau}}$. Let $p_n(x,t)$ be the probability density that at time t the particle is at $X(t) = x$ and in either the search state ($n = +$) or the return state ($n = -$). Similarly, let $P_0(t)$ denote the probability that the particle is in the refractory state at time t. The corresponding master equation with resetting takes the form

$$\frac{\partial p_+}{\partial t} = -v_+\frac{\partial p_+}{\partial x} - rp_+ - \kappa p_+, \quad x \in (0,\infty),$$

$$\frac{\partial p_-}{\partial t} = v_-\frac{\partial p_-}{\partial x} + rp_+, \quad x \in (0,\infty),$$

$$\frac{dP_0}{dt} = v_-p_-(0,t) - \eta P_0(t),$$

together with the boundary condition

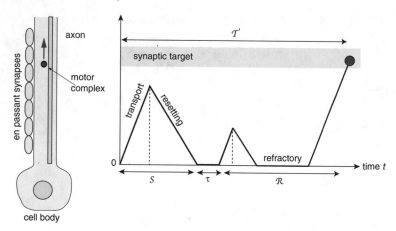

Fig. 7.36 Directed search with resetting. A motor-cargo complex moves along the axon at a constant speed v_+. Prior to finding a synaptic target, the complex may randomly switch to a retrograde state (resetting) and return to the origin at a speed V. After a refractory period, a new motor is inserted and the process repeats. The FPT can be decomposed as $\mathcal{T} = \mathcal{S} + \tau + \mathcal{R}$, where $\mathcal{S}$ is the FPT to return to the start position, τ is a refractory period, and $\mathcal{R}$ is the FPT conditioned on at least one restart. Both $\mathcal{T}$ and $\mathcal{R}$ have the same probability distributions.

$$v_+ p_+(0,t) = \eta P_0(t).$$

(Here we drop the explicit dependence on $x_0 = x_r = 0$.) The case without resetting ($r = 0$) was analyzed in Sect. 7.3.4, where it was shown that

$$\pi_k = e^{-\kappa(k-1)a/v_+} - e^{-\kappa ka/v_+},$$

and

$$s\widetilde{\Pi}_k(s) = \pi_k - \frac{\kappa}{\kappa+s}\left[e^{-(s+\kappa)(k-1)a/v_+} - e^{-(s+\kappa)ka/v_+}\right].$$

(a) Using equations (7.5.58), (7.5.65) and (7.5.41) show that

$$\pi_{r,k} = e^{-(r+\kappa)(k-1)a/v_+} - e^{-(r+\kappa)ka/v_+},$$

and

$$T_r = \frac{\widetilde{Q}_0(r) + r\bar{\tau}\widetilde{Q}_0(r) - r\frac{v_+}{v_-}\widetilde{Q}_0'(r)}{1 - r\widetilde{Q}_0(r)}, \quad \widetilde{Q}_0(r) = \frac{1}{r+\kappa}.$$

(b) Consider the renewal equation for the survival probability with resetting and a refractory period:

$$Q_r(t) = e^{-rt}Q_0(t) + r\int_0^t (1 - \Psi(\sigma))e^{-r(t-\sigma)}Q_0(t-\sigma)d\sigma$$

$$+ r\int_0^t Q_0(t')e^{-rt'}\left[\int_0^{t-t'} \psi(\tau)Q_r(t-t'-\tau)d\tau\right]dt'.$$

The first term on the right-hand side represents trajectories with no resettings, and the second term sums over all trajectories that first reset at some time $t - \sigma, 0 \le \sigma \le t$ and are still in the refractory state at time t. The probability of remaining refractory for a period t is $1 - \Psi(t)$ with $\Psi(t) = \int_0^t \psi(\sigma)d\sigma$. The third term is the contribution from trajectories that last reset at time $t - t' - \sigma$, spent a time σ in the refractory state, and then exited the refractory state at time $t - t'$ without any further resettings. Laplace transforming the renewal equation, and rearranging show that

$$\widetilde{Q}_r(s) = \frac{\widetilde{Q}_0(r+s)\left[1 + \frac{r(1-\widetilde{\psi}(s))}{s}\right]}{1 - r\widetilde{Q}_0(r+s)\widetilde{\psi}(s)}.$$

Taking the limit $s \to 0$, recover the formula

$$T_r = \frac{\widetilde{Q}_0(r)[1+r\bar{\tau}]}{1 - r\widetilde{Q}_0(r)}.$$

(c) Now consider the renewal equation for the survival probability with resetting and a finite return time:

$$Q_r(t) = e^{-rt}Q_0(t) + r\int_0^{t/\xi_+} e^{-r(t-\sigma)}Q_0(t-\sigma)d\sigma$$

$$+ \int_0^t Q_0(t')re^{-rt'}\left[\int_0^{(t-t')/\xi_-} Q_r(t-t'-\sigma\xi_-)re^{-r\sigma}d\sigma\right]dt',$$

where

$$\xi_+ = \frac{v_+ + v_-}{v_+}, \quad \xi_- = \frac{v_+}{v_-}.$$

The first term on the right-hand side represents trajectories with no resettings. The second term sums over all trajectories that first reset at some time $t - \sigma$, $0 \leq \sigma \leq t/\xi_+$ and are still in the process of returning to the origin. Since the particle has been in the ballistic state for time $t - \sigma$, it has traveled a distance $v_+(t - \sigma)$, which means that $\sigma \leq v_+(t - \sigma).v_-$. Rearranging this equation yields the constraint that $\sigma \leq t/\xi_+$. The third term is the contribution from trajectories whose last reset occurred at time $t - t' - v_+\sigma/V$, where σ is the time spent in the ballistic state prior to resetting, after which the particle takes a time $v_+\sigma/v_-$ to return to the origin, and then an additional time t' in the ballistic state without any further resettings. We also require $\sigma \frac{v_+}{v_-} \leq (t - t')$, which yields the constraint $\sigma \leq (t - t')/\xi_-$. Laplace transforming the renewal equation and rearranging show that

$$\widetilde{Q}_r(s) = \frac{\widetilde{Q}_0(r+s) + r\left[\widetilde{Q}_0(r+s) - \widetilde{Q}_0(r+\xi s)\right]/s}{1 - r^2\widetilde{Q}_0(r+s)/(r+s\xi_-)}. \tag{7.7.16}$$

Finally, taking the limit $s \to 0$, recover the formula

$$T_r = \frac{\widetilde{Q}_0(r) - r\frac{v_+}{v_-}\widetilde{Q}_0'(r)}{1 - r\widetilde{Q}_0(r)}.$$

Problem 7.8. Queuing model of target resource accumulation with resetting. Consider search-and-capture under instantaneous resetting for a single target. From equation (7.5.20), the Laplace transform of the FPT density is

$$\widetilde{f}_r(\mathbf{x}_r, s) = \frac{1 - (r+s)\widetilde{Q}_0(\mathbf{x}_r, r+s)}{1 - r\widetilde{Q}_0(\mathbf{x}_r, r+s)},$$

and the MFPT is

$$T_r(\mathbf{x}_r) = \frac{\widetilde{Q}_0(\mathbf{x}_r, r)}{1 - r\widetilde{Q}_0(\mathbf{x}_r, r)},$$

In terms of the FPT density f_0 without resetting, whose Laplace transform is

$$\widetilde{f}_0(\mathbf{x}_r, s) = 1 - s\widetilde{Q}_0(\mathbf{x}_r, s),$$

we can write

$$\widetilde{f}_r(\mathbf{x}_r, s) = \frac{(r+s)\widetilde{f}_0(\mathbf{x}_r, r+s)}{s + r\widetilde{f}_0(\mathbf{x}_r, r+s)}, \quad T_r(\mathbf{x}_r) = \frac{1 - \widetilde{f}_0(\mathbf{x}_r, r)}{r\widetilde{f}_0(\mathbf{x}_r, r)}.$$

(a) Now consider multiple rounds of search-and-capture. The statistics of resource accumulation in the target can be obtained from the Laplace transformed Binomial moments, which satisfy the iterative equation (in the absence of delays)

$$\widetilde{B}_m(s) = \frac{\widetilde{f}_r(\mathbf{x}_r,s)}{1 - \widetilde{f}_r(\mathbf{x}_r,s)} \widetilde{B}_{m-1}(s+\gamma),$$

where γ is the degradation rate. Show that the corresponding steady-state moments satisfy the iterative equations

$$B_m^* = \frac{\widetilde{B}_{m-1}(\gamma)}{T_r(\mathbf{x}_r)}, \quad \widetilde{B}_m(\gamma) = \frac{1}{\gamma T_{r+\gamma}(\mathbf{x}_r)} \widetilde{B}_{m-1}(2\gamma).$$

Hence, the statistics of resource accumulation in the presence of exponential resetting is determined completely in terms of the MFPT at two different resetting rates: r and $r+\gamma$, where γ is the degradation rate.

(b) Show that the steady-state mean and variance of the number of packets in the target are

$$\overline{N} = \frac{1}{\gamma T_r(\mathbf{x}_r)}, \quad \mathrm{Var}[N] = \frac{1}{\gamma^2 T_r(\mathbf{x}_r)} \left[\gamma + \frac{1}{T_{r+\gamma}(\mathbf{x}_r)} - \frac{1}{T_r(\mathbf{x}_r)} \right].$$

(c) Define the Fano factor (FF) by $FF = \mathrm{Var}[N]/\overline{N}$. Obtain the following limits for fixed r:

$$\lim_{\gamma\to\infty} FF = 1, \quad \lim_{\gamma\to 0} FF = 1 + \frac{d}{dr}\left(\frac{1}{T_r}\right).$$

Now suppose that γ is fixed and $r \to 0$. Show that if $\lim_{r\to 0} T_r(\mathbf{x}_r) = \infty$, then the mean and variance both vanish, and

$$\lim_{r\to 0} FF = 1 + \frac{1}{\gamma T_\gamma(\mathbf{x}_r)}.$$

On the other hand, show that if $T_0(\mathbf{x}_r)$ is finite then both $\overline{N}$ and $\mathrm{Var}[N]$ have finite nonzero values such that

$$\lim_{r\to 0} FF = 1 + \frac{1}{\gamma T_\gamma(\mathbf{x}_r)} - \frac{1}{\gamma T_0(\mathbf{x}_r)}.$$

Problem 7.9. Traveling wave approximation of a shock. Construct a traveling wave solution $\rho(x,t) = U(z)$ with $z = (x - ct)/\varepsilon$ of the equation

$$\frac{\partial \rho}{\partial t} + \frac{\partial}{\partial x}(\rho(1-\rho)) = \varepsilon \frac{\partial^2 \rho}{\partial x^2}$$

for $-\infty < x < \infty$ and wavespeed c. Assume that $U(z) \to U_{\pm\infty}$ as $z \to \pm\infty$. Show that

$$\frac{dU}{dz} = U(1-U) - cU + \mathrm{constant},$$

and deduce that

$$c = \frac{U(1-U)]_{-\infty}^{\infty}}{[U]_{-\infty}^{\infty}}.$$

Use phase-plane analysis to that U can only tend to $U(\pm\infty)$ as $z \to \pm\infty$ if $dU/dz < 0$.
Sketch a traveling wave solution and discuss how it relates to a shock solution of
the kinematic wave equation obtained by setting $\varepsilon = 0$.

Problem 7.10. Method of characteristics and shocks. The kinematic wave equation arises in a wide range of transport applications including vehicular traffic. Suppose that $\rho(x,t)$ is the number density of cars on a single-lane road, evolving according to the equation

$$\frac{\partial \rho}{\partial t} + \frac{\partial(\rho V(\rho))}{\partial x} = 0.$$

with $V(\rho)$ the density-dependent car speed. Take

$$V(\rho) = v_m(1 - \rho/\rho_m),$$

where v_m is the speed limit and ρ_m is the maximum (bumper-to-bumper) car density.
Suppose that the initial density profile is

$$\rho(x,0) = \begin{cases} \rho_m/8, & x < 0 \\ \rho_m & x > 0. \end{cases}$$

This represents cars on the left moving with speed $V = 7v_m/8$ encountering a traffic
jam at $x = 0$. Use the method of characteristics to determine the density profile as a
function of time. In particular, show that there is a shock that propagates from $x = 0$
at speed $\dot{s} = -v_m/8$. This represents a back-propagating shock that represents the
slowing down of cars in response to the traffic jam ahead. Sketch the characteristics
and shock in the $t - x$ plane.

Problem 7.11. Computer simulations: random intermittent searcher in 1D.
(a) Write a computer program to simulate an unbiased 3-state random intermittent
searcher moving along a 1D track of length $L = 20\,\mu\text{m}$. Take the velocity to be
$v = 0.1\,\mu\text{m/s}$, and the transition rates $\beta = 1\,\text{s}^{-1}$, $\alpha = 0.5\,\text{s}^{-1}$. Assume that there is a
target of width $a = 1\,\mu\text{m}$ at position $X = 10\,\mu\text{m}$, and the particle finds (is absorbed
by) the target at a rate $k = 0.05\,\text{s}^{-1}$. (Hint: Use the Gillespie algorithm to determine the random transition times and the sequence of states, while keeping track of
changes in position. Hence, if the particle enters the $+ (-)$ state at $t = t_0$ and makes
the next transition at time $t_0 + \tau$, then the position is shifted by an amount $\Delta x = v\tau$
($\Delta x = -v\tau$). There is no shift if the particle is in the stationary state.)

(b) Plot sample trajectories up to the time T the target is found. By averaging T over
many trials determine the MFPT to find the target, starting from a random initial
position. Compare with the analytical expression obtained in part (c) of Ex. [7.9].

(c) Modify the program so that it now simulates a particle executing an unbiased
1D random walk along the track for a random time τ_{1D}, after which it is removed
from the track and placed randomly at a new location on the track after a random
time τ_{3D}. Assume that the times τ_j, $j = 1D, 3D$, are exponentially distributed with
means $\bar{\tau}_j$, and select the new position using the uniform distribution on $[0,L]$. Take
$\bar{\tau}_{3D} = \bar{\tau}_{1D} = 10^{-3}\,\text{s}$ and 1D diffusivity $D_1 = 10\,\mu\text{m}^2/\text{s}$.

Chapter 8
The WKB method, path integrals, and large deviations

In previous chapters, we have encountered a number of examples of noise-induced escape over a potential barrier, including stochastic action potentials (Sect. 3.4), calcium puffs and sparks (Sect. 3.5), genetic switches (Sect. 5.3), epigenetic landscapes (Sect. 5.7), and diffusion over a fluctuating barrier (Sect. 6.5). Noise-induced escape will also arise when we consider bacterial population extinction (Chap. 15). In the absence of noise, the particular state to which a system converges is determined by the initial conditions. On the other hand, when weak noise is included, fluctuations can induce transitions between the metastable states (the states that are stable fixed points in the deterministic limit). Since the noise tends to be weak, transitions are rare events involving large fluctuations that are in the tails of the underlying probability density function. This means that estimates of mean transition times and other statistical quantities can be sensitive to any approximations, including the Gaussian approximation based on a system-size expansion or quasi-steady-state reduction, and can lead to exponentially large errors.

The analysis of metastability has a long history [357], particularly within the context of stochastic differential equations (SDEs) with weak noise. The underlying idea is that the mean rate to transition from a metastable state in the weak noise limit can be identified with the principal eigenvalue of the generator of the underlying stochastic process, which is a second-order differential operator in the case of a Fokker-Planck (FP) equation. Calculating the eigenvalue typically involves obtaining a Wentzel-Kramers-Brillouin (WKB) approximation of a quasi-stationary solution and then using singular perturbation theory to match the solution to an absorbing boundary condition [356, 551, 563, 602, 739]. The latter is defined on the boundary that marks the region beyond which the system rapidly relaxes to another metastable state, becomes extinct, or escapes to infinity. In one-dimensional (1D) systems ($d = 1$), this boundary is simply an unstable fixed point, whereas in higher dimensions ($d > 1$), it is generically a $(d − 1)$-submanifold. As we have already mentioned, in the weak noise limit, the most likely paths of escape through an absorbing boundary are rare events. From a mathematical perspective, the analysis of rare events is known as large deviation theory [213, 274, 287, 809], which provides a

© Springer Nature Switzerland AG 2021
P. C. Bressloff, *Stochastic Processes in Cell Biology*, Interdisciplinary
Applied Mathematics 41, https://doi.org/10.1007/978-3-030-72515-0_8

rigorous probabilistic framework for interpreting the WKB solution in terms of optimal fluctuational paths. The analysis of metastability in chemical master equations has been developed along analogous lines to SDEs, combining WKB methods and large deviation principles [222, 232, 243, 251, 274, 296, 356, 378, 453, 702] with path integral or operator methods [224, 225, 649, 723, 849]. The study of metastability in stochastic hybrid systems is more recent [114–116, 433, 612, 613, 617]. Again there is a strong connection between WKB methods, large deviation principles [127, 264, 440], and formal path integral methods [117, 120], although the connection is now more subtle.

In order to give a heuristic definition of a large deviation principle [287, 739, 809], consider some random dynamical system in $\mathbb{R}^n$ for which there exists a well-defined probability density functional $P_\varepsilon[x]$ over the different sample trajectories $\{x(t)\}_0^T$ in a given time interval $[0, T]$ with $x(0) = x_0$ and $x(T) = x_T$ fixed. Here ε determines the noise amplitude. A large deviation principle (LDP) for the random paths is that

$$P_\varepsilon[x] \sim e^{-S[x]/\varepsilon}, \quad \varepsilon \to 0,$$

where $S[x]$ is known as an action functional or the rate function of the LDP. (The term "functional", or "function of a function," refers to the fact that $S[x]$ depends on the continuous set of values of the function $x(t)$ over the time interval $t \in [0, T]$, see Box 8A and Chap. 12.) Solving the first passage time (FPT) problem for escape from a fixed-point attractor of the underlying deterministic system involves finding the most probable paths of escape, which minimize the action functional with respect to the set of all trajectories emanating from the fixed point (under certain additional constraints). Evaluating the action functional along a most probable path from the fixed point to another point x generates a corresponding quasipotential $\Phi(x)$. Consider for example a single-variable stochastic process that exhibits bistability in the deterministic limit, that is, there exists a pair of stable fixed points at $x = x_\pm$ separated by an unstable fixed point at $x = x_0$. Given a quasipotential Φ, the mean FPT (MFPT) τ to escape from the fixed point at x_- takes the general Arrhenius form (see also Sect. 2.4)

$$\tau \sim \frac{\Gamma(x_0, x_-)}{\sqrt{|\Phi''(x_0)||\Phi''(x_-)}} e^{[\Phi(x_0) - \Phi(x_-)]/\varepsilon},$$

where Γ is an appropriate prefactor. Moreover, $\Phi(x_0) - \Phi(x_-)$ is the value of the action along the optimal path from x_- to x_0. In the weak noise limit $\varepsilon \to 0$, any errors in the form of the quasipotential can generate exponentially large errors in the MFPT to escape from a metastable state. One method for estimating the quasipotential is to use the WKB approximation. In many cases, one can interpret the WKB equation for the quasipotential in terms of a Hamilton-Jacobi equation, whose corresponding Hamiltonian $H(x, q)$ with "momentum" variable q is related to the action of large deviation theory according to

$$S[x] = \int_0^T [q\dot{x} - H(x, q)]dt.$$

Thus, large deviation theory provides a rigorous foundation for the application and interpretation of WKB methods, in particular, ensuring that the solutions to Hamilton's equations correspond to optimal paths of the full stochastic system. However, large deviation theory does not determine the prefactor.

We begin by applying the WKB approximation and asymptotic methods to the analysis of noise-induced escape in an SDE with weak noise (Sect. 8.1). We also show how WKB theory can be interpreted in terms of optimal paths and large deviation theory by constructing a path integral representation of the SDE. We then carry out an analogous set of analyses for birth–death processes (Sect. 8.2) and stochastic hybrid systems (Sect. 8.3). The theory of noise-induced escape and metastability is illustrated by considering applications to an autoregulatory gene network (Sect. 8.4) and to a conductance-based model of a neuron (Sect. 8.5). In Sect. 8.6, we use the path integral representation of an SDE to derive the Feynman-Kac formula for Brownian functionals. The latter are random variables defined by some integral measure of a Brownian path. Finally, a brief review of large deviation theory is presented in Sect. 8.7, including a discussion of generalized central limit theorems and Lévy stable distributions.

8.1 Metastability analysis of SDEs with weak noise

8.1.1 The WKB method for one-dimensional SDEs

In order to develop the theory of metastability for SDEs with weak noise, we begin by considering the simple example of a 1D system, which was previously analyzed using more direct methods in Sect. 2.4.2. The advantage of the WKB and asymptotic methods introduced here is that they can be extended to higher-dimensional SDEs, as well as discrete Markov processes and stochastic hybrid systems. Therefore, consider the non-dimensionalized scalar SDE

$$dX(t) = F(X)dt + \sqrt{\varepsilon}dW(t), \tag{8.1.1}$$

in the weak noise regime $0 < \varepsilon \ll 1$. The corresponding FP equation is[1].

$$\frac{\partial P}{\partial t} = -\frac{\partial J(x,t)}{\partial x}, \quad J(x,t) = -\frac{\varepsilon}{2}\frac{\partial P(x,t)}{\partial x} + F(x)P(x,t). \tag{8.1.2}$$

Suppose that the deterministic equation $\dot{x} = F(x)$ has a stable fixed point x_-, $F(x_-) = 0$, and a basin of attraction given by the interval $\Omega = (0, x_0)$; the point x_0 corresponds to an unstable fixed point. For small but finite ε, fluctuations about the steady state can induce rare transitions out of the basin of attraction due to a metastable trajectory crossing the point x_0. Assume that the stochastic system is ini-

---

[1] In order to distinguish between probability distributions and "momentum" variables, the former will be denoted by P and the latter by p throughout this chapter.

tially at x_- so that $p(x,0) = \delta(x-x_-)$. In order to solve the FPT problem for escape from the basin of attraction of x_-, we impose an absorbing boundary condition at x_0, $p(x_0,t) = 0$, and a reflecting boundary condition at $x = 0$. Let T denote the (stochastic) FPT for which the system first reaches x_0, given that it started at x_-. Recall from Sect. 2.4.2 that the distribution of FPTs is related to the survival probability that the system hasn't yet reached x_0

$$S(t) \equiv \int_\Omega P(x,t)dx. \tag{8.1.3}$$

That is, $\mathbb{P}\{T > t\} = S(t)$ and the FPT density is

$$f(t) = -\frac{dS}{dt} = -\int_\Omega \frac{\partial P}{\partial t}(x,t)dx. \tag{8.1.4}$$

Substituting for $\partial p/\partial t$ using the FP equation (8.1.2) shows that

$$f(t) = \int_\Omega \frac{\partial J(x,t)}{\partial x}dx = J(x_0,t) = -\frac{\varepsilon}{2}\frac{\partial P(x_0,t)}{\partial x}. \tag{8.1.5}$$

We have used $J(0,t) = 0$ and $P(x_0,t) = 0$. The FPT density can thus be interpreted as the probability flux $J(x_0,t)$ at the absorbing boundary.

The FPT problem in the weak noise limit ($\varepsilon \ll 1$) has been well studied in the case of FP equations, see for example [551, 563, 602, 739]. One of the characteristic features of the weak noise limit is that the flux through the absorbing boundary is exponentially small. Let $\langle T \rangle = \int_0^\infty f(t)t\,dt$ denote the mean first passage time (MFPT) to reach the absorbing boundary. Then $\lambda = 1/\langle T \rangle \sim e^{-C/\varepsilon}$ for some constant C, which reflects the existence of an underlying LDP. In order to make this connection more explicit, we consider the eigenfunction expansion

$$P(x,t) = \sum_r c_r e^{-\lambda_r t}\phi_r(x), \tag{8.1.6}$$

where $(-\lambda_r, \phi_r)$ is an eigenpair of the linear operator appearing on the right-hand side of (8.1.2). That is,

$$\mathbb{L}\phi_r(x) = -\lambda_r\phi_r(x), \quad \mathbb{L} = -\frac{\partial}{\partial x}F(x) + \frac{\varepsilon}{2}\frac{\partial^2}{\partial x^2}, \tag{8.1.7}$$

together with the absorbing boundary conditions $\phi_r(x_0) = 0$. (For ease of notation we drop the explicit dependence on ε.) We also assume that the eigenvalues λ_r all have positive definite real parts and the smallest eigenvalue λ_0 is real and simple, so that we can introduce the ordering $0 < \lambda_0 < \text{Re}[\lambda_1] \leq \text{Re}[\lambda_2] \leq \dots$. The exponentially slow rate of escape through x_0 in the weak-noise limit means that λ_0 is exponentially small, $\lambda_0 \sim e^{-C/\varepsilon}$, whereas $\text{Re}[\lambda_r]$ is only weakly dependent on ε for $r \geq 1$. Under the above assumptions, we have the quasi-stationary approximation for large t

$$P(x,t) \sim c_0 e^{-\lambda_0 t}\phi_0(x), \tag{8.1.8}$$

and from equation (8.1.4), the FPT density takes the form

$$f(t) \sim \lambda_0 e^{-\lambda_0 t} c_0 \int_\Omega \phi_0(x) dx.$$

Requiring $\int_0^\infty f(t) dt = 1$ determines c_0 and establishes that $1/\lambda_0$ corresponds to the MFPT. Since we also have

$$f(t) \sim c_0 J_0(x_0) e^{-\lambda_0 t}, \quad J_0(x) = -\frac{\varepsilon}{2} \frac{\partial \phi_0}{\partial x},$$

it follows that

$$\lambda_0 = \frac{J_0(x_0)}{\int_\Omega \phi_0(x) dx}. \tag{8.1.9}$$

The calculation of the principle eigenvalue λ_0 consists of two major components [551, 563, 602, 739]: (i) a WKB approximation of the quasi-stationary state, and (ii) the use of matched asymptotics in order to match the outer quasi-stationary solution with an inner solution within a boundary layer around x_0 so that the absorbing boundary condition is satisfied.

WKB approximation of the quasi-stationary state. The first step involves seeking a quasi-stationary solution of the WKB form

$$\phi_0(x) \sim K(x; \varepsilon) e^{-\Phi(x)/\varepsilon}, \tag{8.1.10}$$

with $K(x; \varepsilon) \sim \sum_{m=0}^\infty \varepsilon^m K_m(x)$ and $\Phi(x)$ the so-called quasipotential. Substitute equation (8.1.10) into the eigenvalue equation $\mathbb{L}\phi_0(x) = -\lambda_0 \phi_0(x)$ and Taylor expand with respect to ε using the fact that λ_0 is exponentially small. Collecting the $O(1/\varepsilon)$ terms gives

$$\frac{1}{2} \left(\frac{\partial \Phi(x)}{\partial x} \right)^2 + F(x) \frac{\partial \Phi(x)}{\partial x} = 0. \tag{8.1.11}$$

This is a quadratic in $\Phi'(x)$ with roots

$$\Phi'(x) = 0, \quad \Phi'(x) = -2F(x). \tag{8.1.12}$$

Similarly, collecting $O(1)$ terms yields the following equation for the leading contribution K_0 to the prefactor

$$\left[\frac{\partial \Phi}{\partial x} + F(x) \right] \frac{\partial K_0}{\partial x} = -\left[F'(x) + \frac{1}{2} \frac{\partial^2 \Phi(x)}{\partial x^2} \right] K_0(x). \tag{8.1.13}$$

The latter either has the pair of solutions

$$K_0(x) = \frac{1}{F(x)} \text{ for } \Phi'(x) = 0, \quad K_0(x) = 1 \text{ for } \Phi'(x) = -2F(x). \tag{8.1.14}$$

(The amplitude of $K_0(x)$ can be absorbed into the constant c_0 of equation (8.1.8).)

It remains to understand the meaning of the two possible solutions for $\Phi(x)$. We proceed by noting that equation (8.1.11) has the form of a Hamilton-Jacobi (HJ) equation for a classical Newtonian particle

$$H(x, \Phi'(x)) = 0, \quad H(x, p) = \frac{p^2}{2} + F(x)p, \tag{8.1.15}$$

where H is an effective time-independent Hamiltonian. The HJ structure suggests a classical interpretation, in which the Hamiltonian H describes the motion of a fictitious "particle" with position x and conjugate momentum p evolving according to Hamilton's equations (see Box 8A)

$$\dot{x} = \frac{\partial H}{\partial p} = p + F(x), \tag{8.1.16a}$$

$$\dot{p} = -\frac{\partial H}{\partial x} = -pF'(x). \tag{8.1.16b}$$

There are then two types of zero energy solutions satisfying $H(x, p) = 0$

$$\dot{x} = F(x), \quad p = 0, \tag{8.1.17a}$$
$$\dot{x} = -F(x), \quad p = -2F(x). \tag{8.1.17b}$$

The first recovers the deterministic dynamics, suggesting that $\Phi'(x) = 0$ represents deterministic solutions that flow into the fixed point at $x = x_-$. Hence, the quasipotential associated with the quasi-stationary density (8.1.10) is given by the equation $\Phi'(x) = -2F(x)$, that is

$$\Phi(x) = -2 \int_{x_-}^{x} F(y) dy. \tag{8.1.18}$$

As shown in Box 8A, the Hamiltonian (8.1.15) is related to a classical Lagrangian of the form

$$L(x, \dot{x}) = \frac{1}{2}(\dot{x} - F(x))^2 \tag{8.1.19}$$

according to the Legendre transformation

$$H(x, p) = p\dot{x} - L(x, \dot{x}), \quad p = \frac{\partial L}{\partial \dot{x}}. \tag{8.1.20}$$

Introducing the classical action

$$S[x] = \int_0^{\tau} L(x, \dot{x}) dt, \tag{8.1.21}$$

Hamilton's equations of motion (8.1.16) are equivalent to the so-called Euler-Lagrange equation

$$\frac{\partial L(x, \dot{x})}{\partial x} = \frac{d}{dt} \frac{\partial L(x, \dot{x})}{\partial \dot{x}}, \tag{8.1.22}$$

which is obtained by minimizing the action with respect to possible paths (the least action principle). This yields another representation of the quasipotential, namely, as the value of the classical action along the optimal path of escape from x_- to x_0

$$\Phi(x) = \inf_{\{x(t):x(-\infty)=x_-,x(0)=x_0\}} S[x]. \tag{8.1.23}$$

At first sight, the existence of an effective classical dynamical system in the analysis of noise-induced escape may seem rather puzzling. However, it emerges naturally from a path integral representation of an SDE as shown in Sect. 8.1.3, and can be derived rigorously using large deviation theory [287].

Asymptotic expansion. Returning to the FPT problem, we now need to match the outer quasi-stationary solution with an appropriate inner solution $\Pi(x)$ in a neighborhood of the point $x = x_0$. This is necessary since the quasi-stationary solution does not satisfy the absorbing boundary condition at x_0. There are a number of different ways of carrying out the matched asymptotics [357, 551, 563, 602, 739]. Here we will proceed by fixing the probability flux $J_0(x_0)$ and carrying out a diffusion approximation of the inner solution. Introducing stretched coordinates $y = (x - x_0)/\sqrt{\varepsilon}$, the inner solution $\Pi(y)$ satisfies the stationary FP equation for constant flux, that is

$$J_0(x_0) = F(x_0 + \sqrt{\varepsilon}y)\Pi(y) - \frac{\sqrt{\varepsilon}}{2}\frac{\partial \Pi(y)}{\partial y}, \tag{8.1.24}$$

which yields

$$\Pi(y) = \frac{2J_0}{\sqrt{\varepsilon}}e^{-\Phi(x_0+\sqrt{\varepsilon}y)/\varepsilon}\int_y^\infty e^{\Phi(x_0+\sqrt{\varepsilon}y')/\varepsilon}dy',$$

with $2F(x) = -\Phi'(x)$. Taylor expanding the potential to second order in $\varepsilon^{1/2}y$ and reintroducing unstretched coordinates gives the inner solution

$$\phi_i(x) = \varepsilon^{-1/2}\Pi(x/\sqrt{\varepsilon}) = \frac{2J_0(x_0)}{\varepsilon}e^{(x-x_0)^2/\sigma^2}\int_x^\infty e^{-(x'-x_0)^2/\sigma^2}dx', \tag{8.1.25}$$

where $\sigma = \sqrt{2\varepsilon/|\Phi''(x_0)|}$ determines the size of the boundary layer around x_0. In order to match with the outer solution, we note that for $x_0 - x \gg \sigma$ we can take the lower limit in the integral to be $-\infty$ and evaluate the resulting Gaussian integral

$$\phi_i(x) = \frac{2J_0(x_0)\sigma\sqrt{\pi}}{\varepsilon}e^{(x-x_0)^2/\sigma^2}. \tag{8.1.26}$$

Next, Taylor expanding the outer solution $\phi_0(x) = e^{-\Phi(x)/\varepsilon}$ to second order in $x - x_0$ and matching with the inner solution shows that

$$J_0(x_0) = \frac{1}{2}\sqrt{\frac{\varepsilon|\Phi''(x_0)|}{2\pi}}e^{-\Phi(x_0)/\varepsilon}. \tag{8.1.27}$$

Finally, using steepest descents (See Sect. 2.4.2 and Box 11H), we evaluate the normalization of the outer solution

$$\int_{\Omega} \phi_{\varepsilon}^{(0)}(x)dx \approx \int_{-\infty}^{\infty} e^{-\varepsilon^{-1}[\Phi(x_-)+\Phi''(x_-)(x-x_-)^2/2]} = \sqrt{\frac{2\pi\varepsilon}{\Phi''(x_-)}}. \qquad (8.1.28)$$

From equation (8.1.9), the mean transition rate is then

$$\lambda_{\varepsilon}^{(0)} = \frac{1}{4\pi}\sqrt{|\Phi''(x_0)|\Phi''(x_-)}\, e^{-(\Phi(x_0)-\Phi(x_-))/\varepsilon}. \qquad (8.1.29)$$

This recovers the Kramers rate formula of Sect. 2.4, on setting $D = \varepsilon/2$, $k_B T = D$ and $\Phi = 2U$.

Box 8A. Least action principle of classical mechanics.

One way to formulate the dynamics of a classical point particle with spatial coordinate $x(t)$ is in terms of the Lagrangian $L(x,\dot{x})$ and action [418]

$$S[x] = \int_0^t L(x(s),\dot{x}(s))ds \Big|_{x(0)=x_0}^{x(t)=x}. \qquad (8.1.30)$$

(For simplicity, we consider an autonomous system for which L is not explicitly dependent on time t.) If the particle has kinetic energy $m\dot{x}^2/2$ and potential energy $U(x)$ then

$$L(x,\dot{x}) = \frac{m}{2}\dot{x}^2 - U(x).$$

One can think of $S[x]$ as a functional (see also Box 12C), since it depends on the continuous set of values $x(t)$ over some time interval $t \in [0,T]$ with the end points fixed; the functional dependence is usually indicated by square brackets. The least action principle states that the trajectories actually realized by the particle are those that extremize the action, $\delta S[x] = 0$. This means that, for any smooth function $y(t)$ with $y(0) = y(T) = 0$

$$\lim_{\varepsilon \to 0} \frac{1}{\varepsilon}(S[x+\varepsilon y] - S[x]) = 0.$$

If we apply this definition to the action given by the time-integral of a Lagrangian, we have

$$\delta S[x] = \int (L(x + \varepsilon y, \dot{x} + \varepsilon \dot{y}) - L(x, \dot{x})) dt$$

$$= \varepsilon \int \left(\frac{\partial L(x, \dot{x})}{\partial x} y + \frac{\partial L(x, \dot{x})}{\partial \dot{x}} \dot{y} \right) dt + O(\varepsilon^2)$$

$$= \varepsilon \int \left(\frac{\partial L(x, \dot{x})}{\partial x} - \frac{d}{dt} \frac{\partial L(x, \dot{x})}{\partial \dot{x}} \right) y dt + O(\varepsilon^2).$$

The last step is obtained by performing an integration by parts and using the boundary conditions on y. Since $y(t)$ is arbitrary, the least action principle generates the Euler-Lagrange equation

$$\frac{\partial L(x, \dot{x})}{\partial x} = \frac{d}{dt} \frac{\partial L(x, \dot{x})}{\partial \dot{x}}.$$

Finally, substituting for L gives

$$m\ddot{x} = -\frac{dU(x)}{dx} \equiv F(x),$$

where $F(x)$ is the force generated by the potential $U(x)$. Thus the least action principle is equivalent to Newton's law of motion.

Hamiltonian. Given the Lagrangian L, the corresponding classical Hamiltonian H is defined by the Legendre transformation

$$H(x, p) = p\dot{x} - L(x, \dot{x}), \quad p := \frac{\partial L}{\partial \dot{x}}. \tag{8.1.31}$$

Taking the differential of $H = H(x, p)$ we have

$$dH = \frac{\partial H}{\partial x} dx + \frac{\partial H}{\partial p} dp.$$

Similarly, taking differentials of the Legendre transformation,

$$dH = \dot{x} dp + p d\dot{x} - \frac{\partial L}{\partial x} dx - \frac{\partial L}{\partial \dot{x}} d\dot{x}.$$

Using the definition of p to eliminate the $d\dot{x}$ terms and applying the Euler-Lagrange equation, we obtain Hamilton's equations

$$\dot{x} = \frac{\partial H}{\partial p}, \quad \dot{p} = -\frac{\partial H}{\partial x}. \tag{8.1.32}$$

For the specific Lagrangian, $p = m\dot{x}$ (the momentum) and

$$H(x, p) = \frac{p^2}{2m} + U(x),$$

which is the total energy of the particle. Note that for autonomous systems (L and H not explicitly dependent on time), H is a constant. This follows from Hamilton's equations:

$$\frac{dH}{dt} = \frac{\partial H}{\partial x}\dot{x} + \frac{\partial H}{\partial p}\dot{p} = \frac{\partial H}{\partial x}\frac{\partial H}{\partial p} - \frac{\partial H}{\partial p}\frac{\partial H}{\partial x} = 0.$$

Finally, the classical action can be expressed in terms of the Hamiltonian according to

$$S = \int_0^t [p\dot{x} - H(x,p)]d\tau,$$

and Hamilton's equations can be derived using the least action principle.

Hamilton-Jacobi equation. Allowing the end-point x of the action to vary yields a function $\Phi(x)$ that satisfies a Hamilton-Jacobi equation. First, calculating δS with respect to variation of the end-point coordinate

$$
\begin{aligned}
\delta S &= \int_0^t L(x(s)+\delta x(s), \dot{x}(s)+\delta\dot{x}(s))ds \Big|_{\substack{x(t)+\delta x(t)=x+\delta x \\ x(0)=x_0}} \\
&\quad - \int_0^t L(x(s),\dot{x}(s))ds \Big|_{\substack{x(t)=x \\ x(0)=x_0}} \\
&= \int_0^t \left(\frac{\partial L(x,\dot{x})}{\partial x}\delta x(s) + \frac{\partial L(x,\dot{x})}{\partial \dot{x}}\delta\dot{x}(s)\right)ds \Big|_{\substack{x(t)=x,\delta x(t)=\delta x \\ x(0)=x_0}}.
\end{aligned}
$$

Using the Euler-Lagrange equation

$$
\begin{aligned}
\delta S &= \int_0^t \left(\frac{d}{ds}\frac{\partial L(x,\dot{x})}{\partial \dot{x}}\delta x(s) + \frac{\partial L(x,\dot{x})}{\partial \dot{x}}\frac{d}{ds}\delta x(s)\right)ds \Big|_{\substack{x(t)=x,\delta x(t)=\delta x \\ x(0)=x_0}} \\
&= \int_0^t \frac{d}{ds}\left(\frac{\partial L(x,\dot{x})}{\partial \dot{x}}\delta x(s)\right)ds \Big|_{\substack{x(t)=x,\delta x(t)=\delta x \\ x(0)=x_0}} \\
&= p(t)\delta x.
\end{aligned}
$$

Hence, $\partial S/\partial x = p$. Similarly, calculating the variation of the action when the final time is changed

$$
\begin{aligned}
\delta S &= \int_0^{t+\delta t} L(x(s),\dot{x}(s))ds \Big|_{\substack{x(t+\delta t)=x \\ x(0)=x_0}} - \int_0^t L(x(s),\dot{x}(s))ds \Big|_{\substack{x(t)=x \\ x(0)=x_0}} \\
&= \int_0^{t+\delta t} L(x(s),\dot{x}(s))ds \Big|_{\substack{x(t+\delta t)=x \\ x(0)=x_0}} - \int_0^t L(x(s),\dot{x}(s))ds \Big|_{\substack{x(t)=x-\dot{x}\delta t \\ x(0)=x_0}}
\end{aligned}
$$

$$+ \int_0^t L(x(s),\dot{x}(s))ds \Big|_{x(0)=x_0}^{x(t)=x-\dot{x}\delta t} - \int_0^t L(x(s),\dot{x}(s))ds \Big|_{x(0)=x_0}^{x(t)=x}$$

$$= L(x(t),\dot{x}(t))\delta t - \frac{\partial L}{\partial \dot{x}}\dot{x}\delta t = [L(x,\dot{x}) - p\dot{x}]\delta t = -H(x,p)\delta t.$$

We thus obtain the Hamilton-Jacobi equation for the action along a classical trajectory

$$\frac{\partial S}{\partial t} + H(x, \partial S/\partial x) = 0. \tag{8.1.33}$$

In the case of an energy conserving system, this reduces to

$$H(x, \partial S/\partial x) = E, \quad \frac{\partial S}{\partial t} = -E, \tag{8.1.34}$$

where E is a constant.

Remark. It is important to emphasize that the WKB method for a stochastic system works backwards. The WKB solution of the quasi-stationary density yields a quasipotential Φ that satisfies a Hamilton-Jacobi equation. From a mathematical perspective, there is an underlying Hamiltonian structure and an associated variational principle. However, the physical interpretation of the latter is not given *a priori*. Instead, one needs to apply some version of large deviation theory to identify solutions of the variational problem with most likely or optimal paths of the underlying stochastic process. This can be established formally using path integrals, see Sect. 8.1.3.

Momentum representation. It is also possible to express Lagrangian dynamics in momentum space. Let

$$L^* = L(x,\dot{x}) - (x\dot{p} + \dot{x}p).$$

Then

$$dL^* = dL - (\dot{p}dx + xd\dot{p} + \dot{x}dp + pd\dot{x}).$$

From the Euler-Lagrange equations,

$$dL = \frac{\partial L}{\partial x}dx + \frac{\partial L}{\partial \dot{x}}d\dot{x} = \dot{p}dx + pd\dot{x},$$

which implies that $L^* = L^*(p,\dot{p})$ with

$$dL^* = \frac{\partial L^*}{\partial p}dp + \frac{\partial L^*}{\partial \dot{p}}d\dot{p} = -xd\dot{p} - \dot{x}dp.$$

Comparison of the two differential expressions for dL^* yields

$$x = -\frac{\partial L^*}{\partial \dot{p}}, \quad \dot{x} = -\frac{\partial L^*}{\partial p}.$$

We thus obtain the momentum space Euler-Lagrange equations

$$\frac{d}{dt}\frac{\partial L^*}{\partial \dot{p}} = \frac{\partial L^*}{\partial p}. \tag{8.1.35}$$

The action can now be expressed as

$$S[p] = \int_0^t L^*(p(\tau), \dot{p}(\tau))d\tau \Big|_{x(0)=x_0}^{p(t)=p}, \tag{8.1.36}$$

and the Hamiltonian can be expressed as

$$-H(x,p) = L^*(p,\dot{p}) + x\dot{p}.$$

Analyzing the variation of S with respect to the endpoint q and the final time t yields a momentum space version of the Hamilton-Jacobi equation

$$\frac{\partial S}{\partial t} + H(-\partial S/\partial p, p) = 0. \tag{8.1.37}$$

Higher dimensions. Finally, note that all of the above analysis extends to higher-dimensional systems, consisting of the canonical coordinates x_i and conjugate momenta p_i, $i = 1, \ldots, N$. For example, given the Lagrangian $L = L(\mathbf{x}, \dot{\mathbf{x}})$ with $\mathbf{x} = (x_1, \ldots, x_N)$, the conjugate momenta are $p_i = \partial L/\partial \dot{x}_i$. The Hamiltonian is

$$H(\mathbf{x}, \mathbf{p}) = \sum_{i=1}^{N} p_i \dot{x}_i - L(\mathbf{x}, \dot{\mathbf{x}}),$$

with $\mathbf{p} = (p_1, \ldots, p_N)$, and Hamilton's equations become

$$\dot{x}_j = \frac{\partial H}{\partial p_j}, \quad \dot{p}_j = -\frac{\partial H}{\partial x_j}. \tag{8.1.38}$$

The associated Hamilton-Jacobi equations for the action S is

$$\frac{\partial S}{\partial t} + H(\mathbf{x}, \nabla S) = 0. \tag{8.1.39}$$

Some examples of classical Hamiltonian dynamics are considered in Ex. 8.1–8.3.

8.1.2 Higher-dimensional SDEs.

In the case of a scalar SDE, one can solve the escape problem directly along the lines of Sect. 2.4. The power of WKB and asymptotic methods, and the underlying large deviation theory, is that one can extend the definition of the quasipotential to multivariate SDEs and to nonlinear systems having multiple attractors [91, 287, 551, 739]. For example, consider the multivariate SDE

$$dX_i(t) = F_i(\mathbf{X})dt + \sqrt{\varepsilon}\sum_j B_{ij}(\mathbf{X})dW_i(t), \tag{8.1.40}$$

for $i = 1,\dots,d$ with $W_i(t)$ a set of independent Wiener processes. The simplest case to analyze is when the drift term is given by the gradient of a potential so that

$$F_i(\mathbf{X}) = -\nabla_i V(\mathbf{X}), \quad i = 1,\dots,d,$$

and there is additive noise, $B_{ij} = \delta_{ij}$. Following Ref. [91], suppose that the potential has two local minima at $\mathbf{x}^*$ and $\mathbf{y}^*$, respectively, as illustrated in Fig. 8.1 for $d = 2$. Let $B_\delta(\mathbf{y}^*)$ denote a ball of radius δ centered at $\mathbf{y}^*$. We can define an escape problem in terms of the first-hitting time of this ball, starting at $\mathbf{x}^*$

$$\tau(\mathbf{x}^*) = \inf\{t > 0; \mathbf{X}(t) \in B_\delta(\mathbf{y}^*)\}.$$

Since $\mathbf{x}^*$ and $\mathbf{y}^*$ are both local minima, it follows that the potential along any continuous path γ from $\mathbf{x}^*$ to $\mathbf{y}^*$ must first increase and then decrease at least once. In order to determine an effective barrier height $\Delta V(\mathbf{x}^*, \mathbf{y}^*)$, we can determine the maximal value of V along the path γ, and then minimize this value with respect to all continuous paths:

$$\Delta V(\mathbf{x}^*, \mathbf{y}^*) = \inf_{\gamma:\mathbf{x}^*\to\mathbf{y}^*}\left(\sup_{\mathbf{z}\in\gamma} V(\mathbf{z})\right).$$

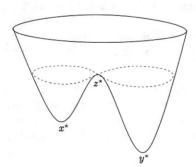

Fig. 8.1: Two-dimensional potential with two local minima $\mathbf{x}^*$ and $\mathbf{y}^*$ separated by a unique saddle at $\mathbf{z}^*$.

Generically, one finds that effective barrier height is reached at a unique point $\mathbf{z}^*$, known as the relevant saddle between $\mathbf{x}^*$ and $\mathbf{y}^*$, so that $\Delta V(\mathbf{x}^*, \mathbf{y}^*) = V(\mathbf{z}^*)$, see Fig. 8.1. Moreover, $\mathbf{z}^*$ is generically a critical point of index 1, that is, in a local neighborhood of $\mathbf{z}^*$ the potential decreases in one direction and increases in the other $d-1$ directions. Mathematically speaking, $\nabla V(\mathbf{z}^*) = 0$ and the Hessian $\nabla^2 V(\mathbf{z}^*)$ has exactly one strictly negative and $d-1$ strictly positive eigenvalues. Under these various assumptions, one can derive the classical Kramers-Eyring formula

$$\lambda_\varepsilon^{(0)} = \frac{|\mu_1(\mathbf{z}^*)|}{2\pi} \sqrt{\frac{\det(\mathbf{Z}(\mathbf{x}^*))}{|\det(\mathbf{Z}(\mathbf{z}^*))|}} e^{-2(V(\mathbf{z}^*) - V(\mathbf{x}^*))/\varepsilon}, \qquad (8.1.41)$$

where $\mu_1(\mathbf{z}^*)$ is the single negative eigenvalue of $\mathbf{Z}(\mathbf{z}^*)$. Here $\mathbf{Z}(\mathbf{x})$ denotes the Hessian matrix with components $Z_{ij} = \partial_i \partial_j V$. A rigorous derivation of equation (8.1.41) can be obtained using potential theory, where the FPT problem is mathematically reformulated as a problem in electrostatics [91].

One of the major mathematical challenges is deriving a version of the Kramers-Eyring formula under more general conditions, including boundaries with multiple saddles and unstable fixed points, non-smooth gradient systems, and non-gradient systems. Maier and Stein [551] have used WKB methods and matched asymptotics to investigate two-dimensional non-gradient systems with multiplicative noise. Now the physical potential V is replaced by a quasipotential $\Phi/2$ satisfying an HJ equation $H(\mathbf{x}, \nabla \Phi) = 0$ with Hamiltonian

$$H(\mathbf{x}, \mathbf{p}) = \frac{1}{2} \sum_{i,j} D_{ij}(\mathbf{x}) p_i p_j + \sum_{i=1}^n F_i(\mathbf{x}) p_i, \quad p_i = \frac{\partial \Phi}{\partial x_i}. \qquad (8.1.42)$$

They show that the Eyring formula breaks down unless the drift is locally given by some gradient potential around $\mathbf{x}_H$ and $|\mu_s(\mathbf{x}_H)|/|\mu_u(\mathbf{x}_H)| > 1$. (If these conditions do not hold then the Hessian matrix of the quasipotential develops a discontinuity at $\mathbf{x}_H$ so standard Gaussian approximations of the inner solution break down). Here μ_s and μ_u are the negative and positive eigenvalues of the 2D Hessian matrix at $\mathbf{x}_H$; the only modification is that one has to determine the prefactor $K_0(\mathbf{x}_H)$, since it is no longer a constant. The equation for the prefactor K_0 now takes the form

$$\frac{dK_0}{dt} \equiv \sum_{i=1}^n \frac{\partial H}{\partial p_i} \frac{\partial K_0}{\partial x_i} = -\left[\sum_i \frac{\partial^2 H}{\partial p_i \partial x_i} + \frac{1}{2} \sum_{i,j} \frac{\partial^2 \Phi}{\partial x_i \partial x_j} \frac{\partial^2 H}{\partial p_i \partial p_j} \right] K_0. \qquad (8.1.43)$$

We have used the fact that along a trajectory $\mathbf{x}(t)$, $\dot{K}_0(\mathbf{x}(t)) = \dot{\mathbf{x}} \cdot \nabla K_0$. Hence, K_0 can be determined numerically by integrating along trajectories originating from a neighborhood of the fixed point $\mathbf{x}_S$, provided that the Hessian $\mathbf{Z}$ of Φ is known. It turns out that the Hessian also satisfies an evolution equation. That is, differentiating the HJ equation twice using

$$\left(\frac{\partial}{\partial x_i} + \sum_k \frac{\partial p_k}{\partial x_i} \frac{\partial}{\partial p_k} \right) \left(\frac{\partial}{\partial x_j} + \sum_k \frac{\partial p_k}{\partial x_j} \frac{\partial}{\partial p_k} \right) H = 0,$$

one finds that [551]

$$\dot{Z}_{ij} = -\sum_{k,l} \frac{\partial^2 H}{\partial p_k \partial p_l} Z_{ik} Z_{jl} - \sum_k \frac{\partial^2 H}{\partial x_j \partial p_k} Z_{ik} - \sum_k \frac{\partial^2 H}{\partial x_i \partial p_k} Z_{jk} - \frac{\partial^2 H}{\partial x_i \partial x_j}. \quad (8.1.44)$$

Thus, one can proceed numerically by simultaneously integrating Hamilton's equations together with equations (8.1.43) and (8.1.44).

8.1.3 Path integral representation of an SDE

One way to understand the occurrence of an effective classical system when determining the quasi-stationary density (8.1.10) is to construct the so-called Martin-Siggia-Rose-Janssen-de Dominicis (MSRJD) path integral representation of the probability density functional for sample paths $\{X(t), 0 \leq t \leq T, X(0) = x_0\}$ of the underlying SDE (8.1.1). Discretizing time by dividing the interval $[0, T]$ into N equal subintervals of size Δt such that $T = N \Delta t$ and setting $X_n = X(n \Delta t)$, we have

$$X_{n+1} - X_n = F(X_n) \Delta t + \sqrt{\varepsilon} \Delta W_n,$$

with $n = 0, 1, \ldots, N-1$, $\Delta W_n = W((n+1)\Delta t) - W(n \Delta t)$

$$\langle \Delta W_n \rangle = 0, \quad \langle \Delta W_m \Delta W_n \rangle = \Delta t \delta_{m,n}.$$

Let $\mathbf{X}$ and $\mathbf{W}$ denote the vectors with components X_n and W_n, respectively.

Formally, the conditional probability density function for $\mathbf{X} = \mathbf{x}$ given a particular realization $\mathbf{w}$ of the stochastic process $\mathbf{W}$ (and initial condition x_0) is

$$P(\mathbf{x}|\mathbf{w}) = \prod_{n=0}^{N-1} \delta\left(x_{n+1} - x_n - F(x_n)\Delta t - \sqrt{\varepsilon} \Delta w_n \right).$$

(Note that in general a multivariate Dirac delta function of the form $\delta(\mathbf{f}(\mathbf{x}))$ would be multiplied by the determinant of the Jacobian matrix with components $\partial f_n / \partial x_m$. However, in the case of additive noise and multiplicative noise of the Ito form, this determinant is unity.) It follows that the conditional expectation of a variable $O(\mathbf{x})$ defined on realizations of the stochastic process can be written as

$$\mathbb{E}[O(\mathbf{x})|\mathbf{w}] = \int O(\mathbf{x}) P(\mathbf{x}|\mathbf{w}) D\mathbf{x}$$

$$= \int O(\mathbf{x}) \left[\prod_{n=0}^{N-1} \delta\left(x_{n+1} - x_n - F(x_n)\Delta t - \sqrt{\varepsilon} \Delta w_n \right) \right] D\mathbf{x},$$

where $Dx = \prod_{n=1}^{N} dx_n$. Now inserting the Fourier representation of the Dirac delta function

$$\delta(x_{m+1} - z_m) = \frac{1}{2\pi} \int_{-\infty}^{\infty} e^{ip_m(x_{m+1} - z_m)} dp_m, \tag{8.1.45}$$

gives

$$P(\mathbf{x}|\mathbf{w}) = \prod_{m=0}^{N-1} \left[\int_{-\infty}^{\infty} e^{ip_m \left(x_{m+1} - x_m - F(x_m)\Delta t - \sqrt{\varepsilon}\Delta w_m \right)} \frac{dp_m}{2\pi} \right].$$

The Gaussian random variable ΔW_n has the probability density function

$$P(\Delta w_n) = \frac{1}{\sqrt{2\pi\Delta t}} e^{-\Delta w_n^2/2\Delta t}.$$

Hence, setting

$$P(\mathbf{x}) = \int P[\mathbf{x}|\mathbf{w}] \prod_{n=0}^{N-1} P(\Delta w_n) d\Delta w_n,$$

and performing the integration with respect to Δw_n by completing the square, we obtain the result

$$P(\mathbf{x}) = \prod_{m=0}^{N-1} \left[\int_{-\infty}^{\infty} e^{ip_m(x_{m+1} - x_m - F(x_m)\Delta t)} e^{-\varepsilon p_m^2 \Delta t/2} \frac{dp_m}{2\pi} \right]. \tag{8.1.46}$$

It immediately follows that

$$\mathbb{E}[O(\mathbf{x})] = \int O(\mathbf{x}) P(\mathbf{x}) Dx$$

$$= \int O(\mathbf{x}) \exp\left\{ ip_m \left(\frac{x_{m+1} - x_m}{\Delta t} - F(x_m) \right) \Delta t - \varepsilon p_m^2 \Delta t/2 \right\} DxDp,$$

where

$$DxDp = \frac{dx_N dp_0}{\pi} \prod_{m=1}^{N-1} \frac{dx_m dp_m}{\pi}.$$

The final step in the derivation of the path integral is to take the continuum limit $\Delta t \to 0, N \to \infty$ with $N\Delta t = T$ fixed. We thus obtain the expectation of a functional $O[x]$ over the different paths $\{x(t)\}_0^T$ realized by the original SDE (8.1.1) with $X(0) = x_0$ such that

$$\mathbb{E}[O(\mathbf{x})] = \int O[x] \exp\left\{ \int_0^T \left[(ip(\dot{x} - F(x)) - \varepsilon p^2/2 \right] dt \right\} \mathcal{D}[x]\mathcal{D}[p],$$

where $\mathcal{D}[x]\mathcal{D}[p]$ is an appropriately defined functional measure. In particular, taking $O[x] = \delta(x(T) - x)$ we obtain the path integral representation of the conditional probability density

$$P(x,T|x_0,0) = \int_{x(0)=x_0}^{x(T)=x} \exp\left\{\int_0^T \left[(ip(\dot{x}-F(x))-\varepsilon p^2/2\right] dt\right\} \mathcal{D}[x]\mathcal{D}[p].$$

$$(8.1.47)$$

It can be shown that the continuum limit and associated integral measure are mathematically well defined. However, in terms of practical calculations, the precise interpretation of the measure is not required. One can relate the path integral action to the one obtained using WKB analysis by performing the change of variables $p \to ip/\varepsilon$

$$P(x,T|x_0,0) = \int_{x(0)=x_0}^{x(T)=x} \exp\left\{-\frac{1}{\varepsilon}\int_0^T [p\dot{x}-H(x,p)]dt\right\} \mathcal{D}[x]\mathcal{D}[p], \qquad (8.1.48)$$

with

$$H(x,p) = pF(x) + \frac{1}{2}p^2, \qquad (8.1.49)$$

which is identical to the WKB Hamiltonian (8.1.15).

Onsager-Machlup path integral. Let us return to the discretized path integral (8.1.46). Since the exponential is quadratic in p_m, we can perform the Gaussian integration with respect to p_m. This gives

$$P(\mathbf{x}) = \prod_{m=0}^{N-1} \frac{1}{\sqrt{2\pi\varepsilon\Delta t}} e^{-(x_{m+1}-x_m-F(x_m)\Delta t)^2/(2\varepsilon\Delta t)}$$

$$= \mathcal{N} \exp\left[-\sum_{m=0}^{N-1} (x_{m+1}-x_m-F(x_m)\Delta t)^2/(2\varepsilon\Delta t)\right]$$

$$= \mathcal{N} \exp\left[-\frac{1}{2\varepsilon}\sum_{m=0}^{N-1}\left(\frac{x_{m+1}-x_m}{\Delta t}-F(x_m)\right)^2 \Delta t\right], \qquad (8.1.50)$$

with

$$\mathcal{N} = \frac{1}{(2\pi\varepsilon\Delta t)^{N/2}}.$$

Note that $P(\mathbf{x})dx_1\ldots dx_N$ is the probability that a given realization $X(t)$ of the discretized stochastic process lies within an infinitesimal domain given by $x_n < X(n\Delta t) < x_n + dx_n$ for $n = 1,\ldots N$ and with initial condition $X(0) = x_0$. Taking the continuum limit then yields the so-called Onsager-Machlup (OM) path integral for the expectation of a functional $O[x]$ [336, 337]

$$\mathbb{E}[O[x]] = \int O[x]\exp\left[-\frac{1}{2\varepsilon}\int_0^T (\dot{x}-F(x))^2 dt\right] \mathcal{D}[x], \qquad (8.1.51)$$

where the probability measure $\mathcal{D}[x]$ on the space of trajectories is known as the Wiener measure. The OM representation of the conditional probability density is then

$$P(x,T|x_0) = \int_{x(0)=x_0}^{x(\tau)=x} \exp\left[-\frac{1}{2\varepsilon}\int_0^T (\dot{x}-F(x))^2 dt\right] \mathcal{D}[x]. \qquad (8.1.52)$$

In the limit $\varepsilon \to 0$, we can use the method of steepest descents (see Box 11H) to obtain the approximation

$$P(x,T|x_0) \sim \exp\left[-\frac{\Phi(x,T|x_0)}{\varepsilon}\right], \qquad (8.1.53)$$

where Φ is the stochastic or quasipotential

$$\Phi(x,T|x_0) = \inf_{x(0)=x_0,x(T)=x} S[x], \qquad (8.1.54)$$

with

$$S[x] = \int_0^T L(x,\dot{x})dt, \quad L(x,\dot{x}) = \frac{1}{2}(\dot{x}-F(x))^2. \qquad (8.1.55)$$

We thus recover the least action principle of Sect. 8.1.1

We now briefly highlight a subtle feature of the above derivation, namely, there is a problem with interpreting the meaning of the term $\dot{x}$ in the OM path integral, since the majority of paths in a diffusion process are non-differentiable. It turns out that within the context of large deviation theory in the limit $\varepsilon \to 0$, the optimal paths are smooth so one can use the classical action (8.1.55). Nevertheless, it is important to clarify the meaning of the OM path integral. Here we follow the particular formulation of Adib [2], see also [262, 901]. The basic idea is to expand the quadratic term in equation (8.1.50) before taking the continuum limit:

$$\left(\frac{x_{m+1}-x_m}{\Delta t}-F(x_m)\right)^2 = \left(\frac{x_{m+1}-x_m}{\Delta t}\right)^2 - 2\frac{x_{m+1}-x_m}{\Delta t}F(x_m) + F(x_m)^2.$$

The term quadratic in Δt^{-1} can still be written formally as $\dot{x}^2$ in the continuum limit, but it is really shorthand for the purely diffusive contribution to the path integral, which determines the Wiener measure. On the other hand, the cross-term becomes an Ito integral

$$\lim_{N\to\infty}\lim_{\Delta t\to 0}\sum_{m=0}^{N-1}(x_{m+1}-x_m)F(x_m) = \int_{x_0}^{x}F(x)dx.$$

Recall from Chap. 2 that the usual rules of calculus do not apply. In particular, suppose that we set $F(x) = -V'(x)$. Then $\Delta V(x) = V(x) = V(x_0)$ becomes

$$\Delta V(x) = \lim_{N\to\infty}\lim_{\Delta t\to 0}\sum_{m=0}^{N-1}V(x_{m+1})-V(x_m) = \lim_{N\to\infty}\lim_{\Delta t\to 0}\sum_{m=0}^{N-1}V(x_m+\Delta x_m)-V(x_m)$$

$$= \lim_{N\to\infty}\lim_{\Delta t\to 0}\sum_{m=0}^{N-1}V'(x_m)\left[\Delta x_m+\frac{1}{2}V''(x_m)\Delta x_m^2\right]$$

$$= \lim_{N\to\infty}\lim_{\Delta t\to 0}\sum_{m=0}^{N-1}V'(x_m)\left[\Delta x_m+\frac{\varepsilon}{2}V''(x_m)\Delta t\right] = \int_{x_0}^{x}V'(x)dx+\frac{\varepsilon}{2}\int_0^t V''(x(t))dt.$$

Therefore,

$$\int_{x_0}^{x} F(x)dx = -\Delta V - \frac{\varepsilon}{2} \int_0^t F'(x(t))dt,$$

so that we can rewrite the action as

$$S[x] = \Delta V + \frac{1}{2} \int_0^\tau \left[\dot{x}^2 + F(x)^2 + \varepsilon F'(x) \right] dt. \qquad (8.1.56)$$

Except for the formal $\dot{x}$ term in the above action, the remaining terms involve ordinary integrals only (assuming that F is smooth), and are thus insensitive to the particular choice of discretization.

8.2 Metastability analysis of birth–death processes

The analysis of metastability in chemical master equations has been developed along analogous lines to SDEs, combining WKB methods [232, 243, 251, 356, 378, 453, 702], large deviation principles [274], and path integral or operator methods [224, 225, 649, 723, 837]. Given that one can often approximate a chemical master equation using an FP equation (see Sect. 5.2), one might be inclined to estimate the MFPT to escape from a metastable state using the results outlined above. However, it is clear from the Arrhenius-like formula (8.1.29) that the escape rate is sensitive to the precise form of the quasipotential Φ. That is, since a diffusion approximation of a master equation leads to an approximation of the corresponding quasipotential, this can lead to exponentially large errors in estimates of the escape rate. Hence, it is often necessary to apply WKB methods and large deviation principles directly to the underlying master equation. (Similar comments apply to the CK equation of stochastic hybrid systems, see Sect. 8.3.)

8.2.1 The WKB method for a birth–death process

For the sake of illustration, consider the birth–death master equation for the discrete stochastic process $N(t) \in \mathbf{Z}^+$ (see Sect. 3.3.2)

$$\frac{d}{dt}P(n,t) = \sum_m A_{nm}P(m,t) \equiv \omega_+(n-1)P(n-1,t) + \omega_-(n+1)P(n+1,t) \quad (8.2.1)$$

$$- [\omega_+(n) + \omega_-(n)]P(n,t),$$

with a reflecting boundary condition at $n = 0$. We assume the scaling $\omega_\pm = N\Omega_\pm$, where N is the system size. Suppose that in the deterministic limit $N \to \infty$, the rate equation for $x = n/N$

$$\frac{dx}{dt} = \Omega_+(x) - \Omega_-(x)$$

exhibits bistability, with stable fixed points at $x = x_\pm$ and an unstable fixed point at $x = x_0$. (Throughout the analysis we will switch between n/N and x, with x treated as a continuous variable for large N.) Since we are interested in calculating the MFPT to escape from n_-, say, we impose an absorbing boundary condition at $n = n_0$, that is, $P(n_0, t) = 0$ and take $0 \leq n \leq n_0$. Let T denote the (stochastic) FPT for which the birth–death process first reaches n_0, given that it started at n_-. The distribution of FPTs is related to the survival probability that the system hasn't yet reached n_0

$$S(t) = \sum_{n < n_0} P(n, t).$$ (8.2.2)

That is, $\mathbb{P}\{T > t\} = S(t)$ and the FPT density is

$$f(t) = -\frac{dS}{dt} = -\sum_{n < n_0} \frac{dP(n, t)}{dt}.$$ (8.2.3)

We now note that equation (8.2.1) can be written in the form

$$\frac{dP(n, t)}{dt} = J(n, t) - J(n + 1, t),$$ (8.2.4)

with

$$J(n, t) = \omega_-(n)P(n, t) - \omega_+(n - 1)P(n - 1, t).$$

Using the reflecting boundary condition $J(0, t) = 0$, it follows that $f(t) = J(n_0, t)$. The FPT density can thus be interpreted as the probability flux $J(n_0, t)$ at the absorbing boundary.

It is convenient to rewrite the master equation (8.2.1) for $n = 0, \ldots, n_0$ as the linear system

$$\frac{dP(n, t)}{dt} = \sum_{n < n_0} A_{nm}P(m, t),$$ (8.2.5)

with $\mathbf{A}$ the generator of the continuous-time Markov chain (see Box 3A). Suppose that the Markov chain is irreducible, that is, any state can be reached from any other state through a sequence of transitions. If the absorbing boundary condition at $n = n_0$ was replaced by a reflecting boundary condition, then $\mathbf{A}$ would have a simple zero eigenvalue with corresponding left eigenvector $\mathbf{1}$, whose components are all unity, that is, $\sum_n A_{nm} = 0$ for all m. The latter follows immediately from conservation of probability in the case of reflecting boundaries. The Perron-Frobenius theorem (see Box 3A) then ensures that all other eigenvalues of $-\mathbf{A}$ have positive real part and that equation (8.2.5) has a globally attracting steady-state ρ_n such that $\sum_m A_{nm}\rho_m = 0$ and $P(n, t) \to \rho_n$ as $t \to \infty$. On the other hand, in the case of an absorbing boundary, probability is no longer conserved since there is an exponentially small but nonzero flux at $n = n_0$ (for large N). The eigenvalues of $-\mathbf{A}$ can now be ordered according to $0 < \lambda_0 \leq \text{Re}[\lambda_1] \leq \text{Re}[\lambda_2] \leq \ldots$ with $\lambda_0 \sim e^{-\eta N}$ for $\eta = O(1)$, whereas λ_r for $r > 0$ are only weakly dependent on N. The exponentially small principal eigenvalue reflects the fact that the flux through the absorbing boundary is exponentially small and in the limit $N \to 0$ reduces to the principal or Perron eigenvalue.

Now consider the eigenfunction expansion

$$P(n,t) = \sum_{r=0}^{n_0} c_r e^{-\lambda_r t} \phi_r(n), \qquad (8.2.6)$$

where ϕ_r is the eigenvector corresponding to λ_r. It follows from the ordering of the eigenvalues that all eigenmodes $\phi_r(n)$, $r > 0$, decay to zero much faster than the perturbed stationary density $\phi_0(n)$. Thus at large times, we have the quasi-stationary approximation

$$P(n,t) \sim c_0 e^{-\lambda_0 t} \phi_0(n). \qquad (8.2.7)$$

Substituting into equation (8.2.5) and summing over n, $0 \le n \le n_0$, shows that the FPT density takes the form $f(t) \sim \lambda_0 e^{-\lambda_0 t}$ with λ_0 identified as the mean transition rate (inverse MFPT) and

$$\lambda_0 = \frac{J_0(n_0)}{\sum_{n \le n_0} \phi_0(n)}, \quad J_0(n) = \omega_-(n)\phi_0(n) - \omega_+(n-1)\phi_0(n-1). \qquad (8.2.8)$$

As in the case of an SDE (Sect. 8.1), the calculation of the principle eigenvalue λ_0 consists of two major components, namely, a WKB approximation of the principal eigenfunction in the bulk of the domain (outer solution), and the asymptotic matching of the outer solution with an inner solution around x_0, which satisfies the absorbing boundary condition.

WKB approximation of the quasi-stationary state. Dropping exponentially small terms, and writing $\phi_0(n) = \phi^\varepsilon(x)$ with x treated as a continuous variable and $\varepsilon = N^{-1}$, we have

$$0 = \Omega_+(x - 1/N)\phi^\varepsilon(x - 1/N)$$
$$+ \Omega_-(x + 1/N)\phi^\varepsilon(x + 1/N) - (\Omega_+(x) + \Omega_-(x))\phi^\varepsilon(x). \qquad (8.2.9)$$

We seek a WKB solution of the form

$$\phi^\varepsilon(x) \sim K(x;\varepsilon)e^{-\Phi(x)/\varepsilon}, \qquad (8.2.10)$$

with $K(x;\varepsilon) \sim \sum_{m=0}^{\infty} \varepsilon^m K_m(x)$. Substituting equation (8.2.10) into equation (8.2.9), Taylor expanding with respect to ε, and collecting the $O(1)$ terms gives

$$\Omega_+(x)(e^{\Phi'(x)} - 1) + \Omega_-(x)(e^{-\Phi'(x)} - 1) = 0, \qquad (8.2.11)$$

where $\Phi' = d\Phi/dx$. Solving this quadratic equation in $e^{\Phi'}$ shows that either $\Phi =$ constant or

$$\Phi = \int^x \ln \frac{\Omega_-(y)}{\Omega_+(y)} dy. \qquad (8.2.12)$$

Proceeding to the next level, equating terms at $O(\varepsilon)$ gives

$$\Omega_+ e^{\Phi'}\left(-\frac{K_0'}{K_0}+\frac{\phi''}{2}\right)+\Omega_- e^{-\Phi'}\left(\frac{K_0'}{K_0}+\frac{\phi''}{2}\right)-\Omega_+' e^{\Phi'}+\Omega_- e^{-\Phi'}=0. \quad (8.2.13)$$

Substituting for Φ using (8.2.11) and solving for K_0 yields the following leading order forms for ϕ^ε

$$\phi^\varepsilon(x)=\frac{A}{\sqrt{\Omega_+(x)\Omega_-(x)}}e^{-N\Phi(x)}, \quad (8.2.14)$$

with Φ given by (8.2.12), which is sometimes called the activation solution, and

$$\phi^\varepsilon(x)=\frac{B}{\Omega_+(x)-\Omega_-(x)}, \quad (8.2.15)$$

which is sometimes called the relaxation solution. The constants A, B are determined by matching solutions around x_0 (see below). Clearly, (8.2.15) is singular at any fixed point x_j, where $\Omega_+(x_j)=\Omega_-(x_j)$, so is not a valid solution for the required quasi-stationary density.

In an analogous fashion to SDEs, equation (8.2.11) has the form of a stationary Hamilton–Jacobi equation for Φ [418]

$$H(x,\Phi'(x))=0, \quad H(x,p)=\sum_{r=\pm 1}\Omega_r(x)\left[e^{rp}-1\right]. \quad (8.2.16)$$

Hamilton's equations take the form

$$\dot{x}=\frac{\partial H}{\partial p}=\sum_{r=\pm 1}r\Omega_r(x)e^{rp}, \quad (8.2.17a)$$

$$\dot{p}=-\frac{\partial H}{\partial x}=\sum_{r=\pm 1}\frac{\partial\Omega_r}{\partial x}(x)\left[1-e^{rp}\right]. \quad (8.2.17b)$$

Analogous to SDEs, the quasipotential $\Phi(x)$ can be expressed as the infimum of a classical action with respect to trajectories from x_- to x, which is a reflection of an underlying large deviation principle [274]. From a dynamical systems perspective, the most probable fluctuational path of large deviation theory is the unstable manifold of the fixed point $(x_-,0)$ in the Hamiltonian phase space $[0,x_0]\times\mathbb{R}$. Since $p=\Phi'(x)$ along this trajectory, it follows that the quasipotential can be computed from

$$\Phi(x)=\int_{x_-}^x p(x)dx. \quad (8.2.18)$$

The corresponding stable manifold is obtained by setting $p=0$ in Hamilton's equations, which recovers the deterministic solution of $\dot{x}=\Omega_+(x)-\Omega_-(x)$ with $\lim_{t\to\infty}x(t)=x_-$. The other solution for Φ is (8.2.12), which can be interpreted as the action along a non-deterministic path that represents the most probable path of escape from x_- to x_0 [232, 243, 251]. (An alternative WKB method for obtaining the quasipotential, based on generating functions is considered in Ex. 8.4 and Ex. 8.5.)

Asymptotic expansion around x_0. Given the quasi-stationary approximation, the rate of escape from the metastable state centered about $x = x_-$ can be calculated by matching it with an appropriate inner solution in a neighborhood of the point $x = x_0$ [232, 243, 251, 356, 378]. This is necessary since the quasi-stationary solution (8.2.14) does not satisfy the absorbing boundary condition at the point x_0 separating the two metastable states. There are a number of different ways of carrying out the matched asymptotics, see for example [378]. Analogous to the analysis of the SDE in Sect. 8.1.1, we will fix the probability flux J_0 through x_0 and then match the activation solution for $x < x_0$ with the relaxation solution for $x > x_0$ using a diffusion approximation of the full master equation (8.2.1) in the vicinity of x_0 [243, 251, 356]. The latter yields an FP equation, which can be written in the form of a conservation equation

$$\frac{\partial}{\partial t}P(x,t) = -\frac{\partial}{\partial x}J(x,t),\qquad(8.2.19)$$

with

$$J(x,t) = (\Omega_+(x) - \Omega_-(x))P(x,t) - \frac{1}{2N}\frac{\partial}{\partial x}[(\Omega_+(x) + \Omega_-(x))P(x,t)].$$

Substituting the quasi-stationary solution $P(x,t) = c_0 e^{-\lambda_0 t}\Pi(x)$ into equation (8.2.19) and using the fact that λ_0 is exponentially small, gives $J(x,t) = c_0 J_0 e^{-\lambda_0 t}$

$$J_0 = (\Omega_+(x) - \Omega_-(x))\Pi(x) - \frac{1}{2N}\frac{\partial}{\partial x}[(\Omega_+(x) + \Omega_-(x))\Pi(x)],$$

where J_0 is the constant flux through x_0. In a neighborhood of x_0, this equation can be Taylor expanded to leading order in $x - x_0$ and integrated to obtain the solution

$$\Pi(x) = \frac{J_0 N}{\Omega_+(x_0)}e^{(x-x_0)^2/\sigma^2}\int_x^\infty e^{-(y-x_0)^2/\sigma^2}dy,\qquad(8.2.20)$$

where

$$\sigma = \sqrt{\frac{2\Omega_+(x_0)}{N[\Omega'_+(x_0) - \Omega'_-(x_0)]}}\qquad(8.2.21)$$

determines the size of the boundary layer around x_0.

In order to match the activation and relaxation solutions, the following asymptotic behavior of the inner solution (8.2.20) is used:

$$\Pi(x) = \begin{cases} \dfrac{N J_0 \sigma^2}{(x - x_0)\Omega_+(x_0)}, & x - x_0 \gg \sigma \\[2ex] \dfrac{N J_0 \sigma \sqrt{\pi}}{\Omega_+(x_0)}e^{(x-x_0)^2/\sigma^2}, & x_0 - x \gg \sigma. \end{cases}\qquad(8.2.22)$$

The solution to the right of the saddle matches the relaxation solution (8.2.15) since $\Omega_+(x) - \Omega_-(x) \approx (x - x_0)[\Omega'_+(x_0) - \Omega'_-(x_0)]$ for $x \approx x_0$ such that $B = J_0$. In order to match the solution on the left–hand side of x_0 with the activation solution (8.2.14), Taylor expand $\Phi(x)$ about x_0 using $\Phi'(x_0) = 0$ and $\Phi''(x_0) = 2/N\sigma^2$. It follows that

$$J_0 = \frac{A\Omega_+(x_0)}{\sqrt{\Omega_+(x_0)\Omega_-(x_0)}}\sqrt{\frac{|\Phi''(x_0)|}{2\pi N}}e^{-N\Phi(x_0)}. \qquad (8.2.23)$$

The final step in the analysis is to link the flux J_0 with the escape rate λ_0. This is achieved by substituting the quasi-stationary solution into the continuity equation (8.2.19), and integrating over the interval $x \in [0, x_0]$ with a reflecting boundary condition at $x = 0$

$$\frac{1}{\lambda_0} = \frac{1}{J_0}\int_0^{x_0}\phi^\varepsilon(y)dy. \qquad (8.2.24)$$

Since the activation solution is strongly peaked around the fixed point x_-, a Gaussian approximation of $\phi^\varepsilon(x)$ around x_- shows that the mean time τ_- to escape the metastable state x_- is

$$\tau_- = \lambda_0^{-1} = \frac{2\pi}{\Omega_+(x_-)}\frac{1}{\sqrt{|\Phi''(x_0)|\Phi''(x_-)}}e^{N[\Phi(x_0)-\Phi(x_-)]}, \qquad (8.2.25)$$

with

$$\Phi''(x) = \frac{d}{dx}\ln\left(\frac{\Omega_-(x)}{\Omega_+(x)}\right) = \frac{\Omega_-'(x)}{\Omega_-(x)} - \frac{\Omega_+'(x)}{\Omega_+(x)}.$$

(We have used the fact that $\Omega_+(x) = \Omega_-(x)$ at the fixed points x_0, x_-.) Finally, we note that certain care has to be taken if the fixed point x_- is too close to the boundary at $x = 0$, as highlighted in Ref. [378].

8.2.2 Path integral representation of a birth–death process

The connection between WKB methods for solving FPT problems and large deviation variational principles also extends to chemical master equations in the large N limit (with N^{-1} playing the role of a weak noise parameter ε). Once again this connection can be established using path integral methods, which were first developed for master equations and reaction-diffusion systems by Doi and Peliti [224, 225, 649]. More recently, an alternative path integral framework has been introduced that avoids the use of operator methods [849]. The latter considers PDEs for the generating function and marginalized probability distribution, which are solved using integral representations of the Dirac delta function along analogous lines to MSRJD. Here we will focus on the original Doi-Peliti construction.

Doi-Peliti operator formalism. For the sake of illustration, consider the birth–death master equation (8.2.1). (The theory can also be applied to more general master equations, with multiple rather than single step reactions.) The starting point of the operator formalism developed by Doi and Peliti is to introduce an abstract "bosonic" vector space (also known as a Fock space) with elements $|n\rangle$ representing the discrete states, together with a pair of creation–annihilation linear operators that satisfy the commutation rule

$$[a, a^\dagger] \equiv aa^\dagger - a^\dagger a = 1. \tag{8.2.26a}$$

These operators generate the full vector space by acting on the "vacuum" state $|0\rangle$, with $a|0\rangle = 0$. The state $|n\rangle$ is then generated according to $|n\rangle = a^{\dagger n}|0\rangle$. Inner products in this state space are defined by $\langle 0|0\rangle = 1$ and the commutation relation. It follows that the dual of the vector $a^\dagger|0\rangle$ is $\langle 0|a$ and $\langle n|m\rangle = \delta_{n,m}n!$. Other standard operator equations are (see Ex. 8.6)

$$a|n\rangle = n|n-1\rangle, \quad a^\dagger|n\rangle = |n+1\rangle, \quad a^\dagger a|n\rangle = n|n\rangle. \tag{8.2.26b}$$

In addition, the basis vectors $|n\rangle$ satisfy the completeness relation (or resolution of the identity)

$$\sum_{n\geq 0} \frac{1}{n!} |n\rangle\langle n| = 1. \tag{8.2.27}$$

That is, for an arbitrary vector $|c\rangle = \sum_{m\geq 0} c_m|m\rangle$,

$$\sum_{n\geq 0} \frac{1}{n!} |n\rangle\langle n|c\rangle = \sum_{n,m\geq 0} \frac{1}{n!} |n\rangle c_m\langle n|m\rangle = \sum_{n,m\geq 0} \frac{1}{n!} |n\rangle c_m m! \delta_{n,m} = \sum_{m\geq 0} c_m|m\rangle = |c\rangle.$$

The next step is to construct an operator representation of the master equation (8.2.1). Given the probability distribution $P(n,t)$, we define the corresponding state vector by

$$|\psi(t)\rangle = \sum_{n\geq 0} P(n,t)a^{\dagger n}|0\rangle = \sum_{n\geq 0} P(n,t)|n\rangle. \tag{8.2.28}$$

Introducing the projection state

$$|\emptyset\rangle = \exp\left(a^\dagger\right)|0\rangle = \sum_{n=0}^{\infty} \frac{1}{n!} |n\rangle, \tag{8.2.29}$$

with $a|\emptyset\rangle = |\emptyset\rangle$ and $\langle\emptyset|m\rangle = 1$, we can then express expectation values in terms of inner products. For example

$$\langle\emptyset|a^\dagger a|\psi(t)\rangle = \sum_{n=0}^{\infty} nP(n,t) = \langle N(t)\rangle. \tag{8.2.30}$$

Differentiating the state vector $|\psi(t)\rangle$ with respect to t and using the master equation (8.2.1) yields the operator equation

$$\frac{d}{dt}|\psi(t)\rangle = \hat{H}|\psi(t)\rangle, \quad \hat{H} = (a - a^\dagger a)\overline{\omega}_-(a^\dagger a) + (a^\dagger - 1)\omega_+(a^\dagger a), \tag{8.2.31}$$

with $n\overline{\omega}_-(n) = \omega_-(n)$. Formally speaking, the solution to the operator version of the master equation can be written as

$$|\psi(t)\rangle = e^{\hat{H}t}|\psi(0)\rangle, \tag{8.2.32}$$

and the expectation value of some physical quantity such as the number $N(t)$ expressed as

$$\langle N(t)\rangle = \langle 0|a^\dagger a e^{\hat{H}t}|\psi(0)\rangle. \tag{8.2.33}$$

The final ingredient of the bra-ket operator formalism is the choice of basis vectors. For example, the construction of the Doi-Peliti integral works with the coherent-state representation

$$|\varphi\rangle = \exp\left(-\frac{1}{2}|\varphi|^2\right)\exp\left(\varphi a^\dagger\right)|0\rangle, \tag{8.2.34}$$

where φ is the complex-valued eigenvalue of the annihilation operator a, with complex conjugate φ^*. Coherent states satisfy the completeness relation

$$\int \frac{d\varphi d\varphi^*}{\pi}|\varphi\rangle\langle\varphi| = 1. \tag{8.2.35}$$

In order to determine the action of $\hat{H}$ on $|\varphi\rangle$, it is first necessary to normal-order $\hat{H}$ by moving all creation operators to the left of all annihilation operators using the commutation relation. For example, if $w_+(n) = n^2$, then

$$\omega_+(a^\dagger a) = a^\dagger a a^\dagger a = a^\dagger[a,a^\dagger]a + a^\dagger a^\dagger a a = a^\dagger a + a^\dagger a^\dagger a a = [\omega]_+(a,a^\dagger),$$

where $[\omega]$ denotes the transition rate after normal ordering. It follows that

$$\langle\varphi|\hat{H}|\varphi\rangle = H(\varphi,\varphi^*) \equiv (\varphi - \varphi^*\varphi)[\overline{\omega}]_-(\varphi,\varphi^*) + (\varphi^* - 1)[\omega]_+(\varphi,\varphi^*). \tag{8.2.36}$$

Another basis can be constructed when the birth–death operator $\hat{H}$ has a discrete spectrum, whose corresponding eigenfunctions form a complete orthogonal basis set for the underlying Fock space. This is particularly useful when the discrete state space is finite, rather than unbounded. (It will also play a major role in the construction of the path integral for a stochastic hybrid system, see Sect. 8.3.2). Let λ_μ denote the μ-th eigenvalue with associated eigenvector $|r_\mu\rangle$

$$\hat{H}|r_\mu\rangle = \lambda_\mu|r_\mu\rangle, \quad |r_\mu\rangle = \sum_{n\geq 0} r_\mu(n)|n\rangle. \tag{8.2.37}$$

It is important to note that the operator $\hat{H}$ is non-Hermitian, which means its right and left (or dual) eigenvectors are not simply adjoints of each other. More specifically,

$$\langle\bar{r}_\mu|\hat{H} = \lambda_\mu\langle\bar{r}_\nu|, \quad \langle\bar{r}_\mu| = \sum_{n\geq 0} \frac{\bar{r}_\mu(n)}{n!}\langle n|, \tag{8.2.38}$$

such that

$$\langle\bar{r}_\mu|r_\nu\rangle = \sum_{n\geq 0} \bar{r}_\mu(n)r_\nu(n) = \delta_{\mu,\nu}. \tag{8.2.39}$$

One also has the completeness relation

$$\sum_{\mu \geq 0} |r_\mu\rangle\langle \bar{r}_\mu| = 1. \tag{8.2.40}$$

This follows from the corresponding completeness relation

$$\sum_{\mu \geq 0} \bar{r}_\mu(m) r_\mu(n) = \delta_{m,n}, \tag{8.2.41}$$

which gives

$$\sum_{\mu \geq 0} |r\rangle\langle \bar{r}| = \sum_{\mu \geq 0} \sum_{m,n \geq 0} \frac{1}{m!} \bar{r}_\mu(m) r_\mu(n) |n\rangle\langle m| = \sum_{m,n \geq 0} \frac{1}{m!} |n\rangle\langle m| \sum_{\mu \geq 0} \bar{r}_\mu(m) r_\mu(n)$$

$$= \sum_{\mu \geq 0} \sum_{m,n \geq 0} \frac{1}{m!} |n\rangle\langle m| \delta_{m,n} = \sum_{n \geq 0} \frac{1}{n!} |n\rangle\langle n| = 1.$$

(We are assuming that if the number of discrete states is infinite, then it is possible to change the summation order.) Moreover, in the original basis

$$\hat{H}|\mu\rangle = \sum_{n \geq 0} r_\mu(n) \hat{H}|n\rangle = \sum_{n \geq 0} \left[\omega_+(n-1) r_\mu(n-1) + \omega_-(n+1) r_\mu(n+1) \right.$$

$$\left. - [\omega_+(n) + \omega_-(n)] r_\mu(n) \right] |n\rangle.$$

Taking the inner product of the operator eigenvalue equation with respect to $\langle m|$ then yields the matrix eigenvalue equation

$$\sum_{m \geq 0} Q_{nm} r_\mu(m) = \lambda_\mu r_\mu(n). \tag{8.2.42}$$

That is, the vector with components $r_\mu(n)$ is a (right) eigenvector of the birth–death matrix generator.

Alternative operator pair. In the case of single step reaction schemes such as birth–death processes, a simpler operator formalism can be developed as follows [36]. First note that the birth–death master equation can be rewritten in the form

$$\frac{dP(x,t)}{dt} = [(\mathbb{E}_1 - 1)\omega_-(Nx) + (\mathbb{E}_{-1} - 1)\omega_+(Nx)] P(x,t), \tag{8.2.43}$$

where $\mathbb{E}_{\pm 1} f(x) = f(x \pm 1/N)$ and N is the system size. As in the system-size expansion of master equations, we treat $x = n/N$ as a continuous variable $x \in \mathbb{R}^+$ and consider a Hilbert space spanned by the vectors $|x\rangle$ with inner product and completeness relation

$$\langle x'|x\rangle = \delta(x - x'), \qquad \int_0^\infty dx |x\rangle\langle x| = 1. \tag{8.2.44}$$

(If $N = 1$ then the above construction is still valid, since the operators $\mathbb{E}_{\pm 1}$ shift x by ± 1. Hence, if x is initially an integer, then it remains an integer.) Introduce the conjugate pair of operators $\hat{x}, \hat{p}$ such that

$$\hat{x}|x\rangle = Nx|x\rangle, \quad \hat{p} = -\frac{1}{N}\frac{\overleftarrow{\partial}}{\partial x}|x\rangle, \quad [\hat{x},\hat{p}] = 1. \tag{8.2.45}$$

The arrow on the differential operator indicates that it operates to the left. This is equivalent to defining the action of $\hat{p}$ on a general state vector $|\phi\rangle = \int_0^\infty dx \phi(x)|x\rangle$

$$\hat{q}|\phi\rangle = \int_0^\infty dx \phi(x)|x\rangle = -\frac{1}{N}\int_0^\infty dx \phi(x)\left[\frac{\overleftarrow{\partial}}{\partial x}\right]|x\rangle = -\frac{1}{N}\int_0^\infty dx \phi'(x)]|x\rangle.$$

That is, $\langle x|\hat{p}|\phi\rangle = -\phi'(x)/N$. Introducing the state vector $|\psi(t)\rangle = \int_0^\infty P(x,t)|x\rangle$, the master equation can be rewritten in the operator form

$$\frac{d}{dt}|\psi(t)\rangle = \hat{H}'|\psi(t)\rangle, \quad \hat{H}' = (e^{-\hat{p}} - 1)\omega_-(\hat{x}) + (e^{\hat{p}} - 1)\omega_+(\hat{x}), \tag{8.2.46}$$

which again has the formal solution $|\psi(t)\rangle = e^{\hat{H}'t}|\psi(0)\rangle$. Using the fact that the "momentum" vector

$$|p\rangle = \int_0^\infty e^{-px}|x\rangle \tag{8.2.47}$$

is an eigenvector of $\hat{p}$ with eigenvalue p, we have

$$\langle p|\hat{H}'|x\rangle = H'(x,p) \equiv (e^{-p} - 1)\omega_-(Nx) + (e^p - 1)\omega_+(Nx). \tag{8.2.48}$$

This recovers the Hamiltonian (8.2.16) of the WKB approximation. Analogous to taking an inverse Laplace transform using a Bromwich integral, the corresponding completeness relation for $|p\rangle$ is [824]

$$\int_{-i\infty}^{i\infty} \frac{dp}{2\pi}|p\rangle\langle p| = 1. \tag{8.2.49}$$

Finally, note that $\hat{H}'$ is equivalent to $\hat{H}$ of equation (8.2.31) under the mapping

$$a = e^{-\hat{p}}\hat{x}, \quad a^\dagger = e^{\hat{p}}, \tag{8.2.50}$$

which is a form of Cole-Hopf transformation [36].

Coherent states and construction of the path integral. In order to convert the operator form of the expectation value into a path integral, we divide the time interval $[0,t]$ into $\mathcal{N}$ intervals of length $\Delta t = t/\mathcal{N}$ and set $t_r = r\Delta t$, $r = 0,1,\ldots,\mathcal{N}$. Equation (8.2.32) becomes

$$|\psi(t)\rangle = e^{\hat{H}\Delta t}e^{\hat{H}\Delta t}\cdots e^{\hat{H}\Delta t}|\psi(0)\rangle. \tag{8.2.51}$$

We then insert multiple copies of the completeness relation (8.2.35), so that

$$|\psi(t)\rangle = \left\{ \prod_{j=0}^{N} \int \frac{d\varphi_j d\varphi_j^*}{\pi} \right\} |\varphi_N\rangle\langle\varphi_N|e^{\hat{H}\Delta t}|\varphi_{N-1}\rangle\langle\varphi_{N-1}|e^{\hat{H}\Delta t}|\varphi_{N-2}\rangle$$

$$\dots \langle\varphi_1|e^{\hat{H}\Delta t}|\varphi_0\rangle\langle\varphi_0|\phi(0)\rangle. \tag{8.2.52}$$

In the limit $N \to \infty$ and $\Delta t \to 0$ with $N\Delta t = t$ fixed, we can make the approximation

$$\langle\varphi_{j+1}|e^{\hat{H}\Delta t}|\varphi_j\rangle \approx (1 + H(\varphi_j, \varphi_j^*)\Delta t)\langle\varphi_{j+1}|\varphi_j\rangle,$$

where $H(\varphi, \varphi^*)$ is given by equation (8.2.36). In addition

$$\langle\varphi_{j+1}|\varphi_j\rangle = \exp\left(-\frac{1}{2}|\varphi_{j+1}|^2 - \frac{1}{2}|\varphi_j|^2\right)\langle 0|e^{\varphi_{j+1}^* a}e^{\varphi_j a^\dagger}|0\rangle$$

$$= \exp\left(-\frac{1}{2}|\varphi_{j+1}|^2 - \frac{1}{2}|\varphi_j|^2\right)e^{\varphi_{j+1}^* \varphi_j}$$

$$= \exp\left(-\frac{1}{2}|\varphi_{j+1}|^2 + \frac{1}{2}|\varphi_j|^2\right)e^{-\varphi_{j+1}^*[\varphi_{j+1}-\varphi_j]}.$$

Combining a product of these terms for increasing j shows that the first exponential term cancels except at the initial and final times, whereas the second exponential yields a factor

$$\exp\left(-\varphi_{j+1}^* \frac{d\varphi_{j+1}}{dt}\Delta t + O(\Delta t^2)\right).$$

Finally, taking the limits $N \to \infty$ and $\Delta t \to 0$, after noting that

$$\langle 0|a^\dagger a|\varphi_N\rangle = e^{\varphi_N - |\varphi_N|^2/2}e^{\varphi_N},$$

we obtain the following path integral representation of the expectation value:

$$\langle N(t)\rangle = \int \mathcal{D}\varphi\mathcal{D}\varphi^* \, \varphi(t)e^{-S[\varphi,\varphi^*]}, \tag{8.2.53}$$

where S is given by the action

$$S[\varphi, \varphi^*] = \int_0^t \varphi^*[\partial_\tau \varphi - H(\varphi, \varphi^*)] \, d\tau, \tag{8.2.54}$$

and we have dropped terms dependent on initial and final times. It turns out if the initial probability density is taken to be a Poisson distribution with mean $\bar{n}$, then the integration with respect to $\varphi(0)$ and $\varphi^*(0)$ simply enforces the initial condition $\varphi(0) = \bar{n}$. As with the path integral representation for SDEs, we do not dwell on the precise mathematical formulation of the path integral measure $\mathcal{D}\varphi\mathcal{D}\varphi^*$.

In anticipation of deriving a large deviation principle for large N, we now make explicit the N dependence of the transition rates,

$$\omega_\pm(n) = N\Omega_\pm(n/N).$$

It follows that $\overline{\omega}_-(n) = \overline{\Omega}_-(n/N)$ with $x\overline{\Omega}_-(x) = \Omega_-(x)$. In order to incorporate this scaling into the path integral, we rescale the variable φ according to $\varphi \to \varphi/N$. It follows that $H \to NH$ with the rescaled Hamiltonian

$$H(\varphi, \varphi^*) = (\varphi^* - 1)\Omega_+(\varphi^*\varphi) - (\varphi^* - 1)\varphi\overline{\Omega}_-(\varphi^*\varphi). \tag{8.2.55}$$

(In the large N limit we can ignore the effects of normal ordering.) We then have for $x(t) = n(t)/N$

$$\langle x(t) \rangle = \int \mathcal{D}\varphi \int \mathcal{D}\varphi^* \; \varphi(t) e^{-NS[\varphi, \varphi^*]}. \tag{8.2.56}$$

It follows from the saddle point method or steepest descents (see Box 11H) that in the large N limit, the path integral is dominated by the classical solution, which is obtained by minimizing the action (8.2.54) with respect to time-dependent trajectories in the phase space $(\varphi(t), \varphi^*(t))$ with the initial condition $\varphi(0) = \bar{n}$ and final condition $\varphi(t) = \varphi$. Denoting the minimal action by $\Phi(\varphi, t | \bar{n})$, we see that

$$\langle n(t) \rangle \sim \int \varphi e^{-N\Phi(\varphi, t | \bar{n})} d\varphi,$$

which has the form of a large deviation principle.

As in the case of the FP equation with weak noise, we have an effective Hamiltonian system in which the minimal action is evaluated along the trajectory (most probable path) given by the solution to Hamilton's equations

$$\dot{\varphi} = \frac{\partial H}{\partial \varphi^*} = \Omega_+(\varphi^*\varphi) - \varphi\overline{\Omega}_-(\varphi^*\varphi) + (\varphi^* - 1)\varphi\Omega_+'(\varphi^*\varphi)$$
$$- (\varphi^* - 1)\varphi^2\overline{\Omega}_-'(\varphi^*\varphi)$$

$$\dot{\varphi}^* = -\frac{\partial H}{\partial \varphi} = (\varphi^* - 1)\overline{\Omega}_-(\varphi^*\varphi) + (\varphi^* - 1)\varphi^*\varphi\overline{\Omega}_-'(\varphi^*\varphi)$$
$$- (\varphi^* - 1)\varphi^*\Omega_+'(\varphi^*\varphi).$$

We now make the observation that $\varphi^* = 1$ is an invariant submanifold of the dynamics for which $\dot{\varphi}^* = 0$ and

$$\dot{\varphi} = \Omega_+(\varphi) - \varphi\overline{\Omega}_-(\varphi) = \Omega_+(\varphi) - \Omega_-(\varphi),$$

which we recognize as the kinetic equation obtained in the deterministic limit $N \to \infty$. However, when $\varphi^* \neq 1$, it is clear   that we cannot identify φ as the physical variable $n(t)/N$, since the transition rates depend on the product $\varphi^*\varphi$. (Moreover, in order to impose the initial condition $\varphi(0) = \bar{n}$ we had to take the initial distribution to be Poisson rather than $\delta_{n,\bar{n}}$.) Nevertheless, one can perform a canonical change of variables that allows us to identify the position variable as n/N, namely, the analog of the Cole-Hopf transformation (8.2.50)

$$x = \varphi^*\varphi, \quad p = \ln(\varphi^*). \tag{8.2.57}$$

The corresponding Hamiltonian becomes identical to equation (8.2.48), which is itself identical to the Hamiltonian (8.2.16) derived using WKB methods in Sect. 8.2, thus establishing the connection between WKB methods and variational principles for a birth–death master equation.

8.3 Metastability analysis of stochastic hybrid systems

The study of metastability in stochastic hybrid systems (SHS) is more recent, and much of the theory has been developed in a series of papers on stochastic ion channels [117, 612, 614, 616], gene networks [613, 615, 617] and stochastic neural networks [115]. Again there is a strong connection between WKB methods, large deviation principles [263, 264, 440], and formal path integral methods [117, 120], although the connection is now more subtle. Examples of stochastic hybrid systems encountered so far in this book include membrane voltage fluctuations (Sect. 3.4), microtubule catastrophes (Sect. 4.2), gene networks with promoter noise (Sect. 5.3), and motor-driven intracellular transport (Sect. 7.2). For concreteness, suppose that there is a single continuous variable evolving according to the piecewise deterministic ODE

$$\frac{dX}{dt} = F(n, X), \tag{8.3.1}$$

for $N(t) = n$, where $X(t) \in \mathbb{R}$ and the discrete variables $N(t) \in \mathbb{Z}^+$ evolves according to the birth–death master equation (8.2.1). The corresponding CK equation takes the form (assuming relatively fast switching)

$$\frac{\partial P(n, x, t)}{\partial t} = -\frac{\partial [F(n, x)P(n, x, t)]}{\partial x} + \frac{1}{\varepsilon} \sum_{m \geq 0} A_{nm}(x)P(m, x, t). \tag{8.3.2}$$

In the limit $\varepsilon \to 0$, equation (8.3.1) reduces to the deterministic or mean-field equation

$$\frac{dx}{dt} = \overline{F}(x) \equiv \sum_{n \geq 0} F(n, x)\rho_n(x), \tag{8.3.3}$$

where $\rho_n(x)$ is the unique steady-state density satisfying $\sum_{m \geq 0} A_{nm}(x)\rho_m(x) = 0$. (As in the case of the birth–death master equation (8.2.1), we are assuming that for fixed x, the matrix $A_{nm}(x)$ is the generator of an irreducible Markov chain.)

8.3.1 The WKB method for a stochastic hybrid system

Following along similar lines to the analysis of the birth–death master equation (8.2.1), suppose that the mean-field equation (8.3.3) is bistable with a pair of stable fixed points $x_\pm$ separated by an unstable fixed point x_0. Assume that the stochastic system is initially at x_-. On short time scales ($t \ll 1/\varepsilon$) the system rapidly converges to a quasi-stationary solution within the basin of attraction of x_-, which can be approximated by a Gaussian solution of the reduced FP equation obtained using a quasi-steady-state (QSS) reduction of the CK equation (8.3.2), see Sect. 5.4. However, on longer time scales, the survival probability slowly decreases due to rare transitions across x_0 at exponentially small rates, which cannot be calculated accurately using the QSS diffusion approximation. One thus has to work with the full CK equation (8.3.2) supplemented by an absorbing boundary conditions at x_0. The initial condition is taken to be

$$P(n,x,0) = \delta(x-x_-)\delta_{n,n_0}. \tag{8.3.4}$$

Let T denote the (stochastic) FPT for which the system first reaches x_0, given that it started at x_-. The distribution of FPTs is related to the survival probability that the system hasn't yet reached x_0, that is

$$\mathbb{P}\{T > t\} = S(t) \equiv \int_{-\infty}^{x_0} \sum_{n \geq 0} P(n,x,t)dx.$$

The FPT density is then

$$f(t) = -\frac{dS}{dt} = -\int_{-\infty}^{x_0} \sum_{n \geq 0} \frac{\partial P(x,n,t)}{\partial t}dx. \tag{8.3.5}$$

Substituting for $\partial P/\partial t$ using the CK equation (8.3.2) shows that

$$f(t) = \int_{-\infty}^{x_0} \left[\sum_{n \geq 0} \frac{\partial [F(n,x)P(n,x,t)]}{\partial x} \right] dx = \sum_{n \geq 0} P(n,x_0,t)F(n,x_0). \tag{8.3.6}$$

We have used $\sum_n A_{nm}(x) = 0$ and $\lim_{x \to -\infty} F(n,x)P(n,x,t) = 0$. The FPT density can thus be interpreted as the probability flux $J(x,t)$ at the absorbing boundary, since we have the conservation law

$$\sum_{n \geq 0} \frac{\partial P(n,x,t)}{\partial t} = -\frac{\partial J(x,t)}{\partial x}, \quad J(x,t) = \sum_{n \geq 0} F(n,x)P(n,x,t). \tag{8.3.7}$$

As in the case of SDEs and birth–death processes, we identify the mean transition rate with the principal eigenvalue λ_0 of the CK operator $-\mathbb{L}$ in equation (8.3.2), assuming λ_0 exists and is exponentially small. We can then make the quasi-stationary approximation

$$P(n,x,t) \sim C_0 e^{-\lambda_0 t} \phi_0(n,x). \tag{8.3.8}$$

Substituting such an approximation into equation (8.3.6) gives

$$f(t) \sim C_0 e^{-\lambda_0 t} \sum_{n \geq 0} F(n,x_0)\phi_0(n,x_0), \tag{8.3.9}$$

and thus

$$\lambda_0 = \frac{\sum_{n\geq 0} F(n,x_0)\phi_0(n,x_0)}{\sum_{n\geq 0}\int_{\Omega}\phi_0(n,x)dx}. \tag{8.3.10}$$

WKB approximation of the quasi-stationary solution. Proceeding along analogous lines to the birth–death master equation, we seek a WKB approximation of the quasi-stationary solution of the form

$$\phi^{\varepsilon}(n,x) \sim Z(n,x)\exp\left(-\frac{\Phi(x)}{\varepsilon}\right), \tag{8.3.11}$$

where $\Phi(x)$ is the WKB quasipotential. Substituting into the time-independent version of equation (8.3.2) yields

$$\sum_{m\geq 0}\left(A_{nm}(x) + \Phi'(x)\delta_{n,m}F(m,x)\right)Z(m,x) = \varepsilon\frac{dF(n,x)Z(n,x)}{dx}, \tag{8.3.12}$$

where $\Phi' = d\Phi/dx$. Introducing the asymptotic expansions $\Phi \sim \Phi_0 + \varepsilon\Phi_1$ and $Z \sim Z_0 + \varepsilon Z_1$, the leading order equation is

$$\sum_{m\in\Gamma} A_{nm}(x)Z_0(m,x) + \Phi_0'(x)F(n,x)Z_0(n,x) = 0. \tag{8.3.13}$$

(Since the prefactor is a component of a vector, we separately expand Φ and Z_0.) Positivity of the quasi-stationary density ϕ^{ε} requires positivity of the corresponding solution Z_0. One positive solution is the trivial solution $Z_0(n,x) = \rho_n(x)$ for all x, where $\rho(x)$ is the unique right eigenvector of $\mathbf{A}(x)$, for which $\Phi_0' = 0$. Establishing the existence of a nontrivial positive solution requires more work, and is related to the fact that the connection of the WKB solution to optimal fluctuational paths and large deviation principles is less direct. We will address this issue using path integrals in the next section. For the moment, we simply assume such a solution exists.

Calculation of principal eigenvalue. In order to calculate the principal eigenvalue, we have to determine the first-order correction Φ_1 to the quasipotential of the WKB solution (8.3.11). Proceeding to the next order in the asymptotic expansion of equation (8.3.12), we have

$$\sum_m \left(A_{nm}(x) + \Phi_0'(x)\delta_{n,m}F(n,x)\right)Z_1(m,x)$$
$$= \frac{dF(n,x)Z_0(n,x)}{dx} - \Phi_1'(x)F(n,x)Z_0(n,x). \tag{8.3.14}$$

For fixed x and WKB potential Φ_0, the matrix operator

$$\bar{A}_{nm}(x) = A_{nm}(x) + \Phi_0'(x)\delta_{n,m}F(m,x)$$

on the left-hand side of this equation has a one-dimensional null space spanned by the positive WKB solution $Z_0(x)$. The Fredholm alternative theorem (see Box 2E) then implies that the right-hand side of (8.3.14) is orthogonal to the left null vector S_0 of $\bar{A}$. That is, we have the solvability condition

$$\sum_{n\geq 0} S_0(n,x) \left[\frac{dF(n,x)Z_0(n,x)}{dx} - \Phi_1'(x)F(n,x)Z_0(n,x) \right] = 0,$$

with S_0 satisfying

$$\sum_{n\geq 0} S_0(n,x)\left(A_{nm}(x) + \Phi_0'(x)\delta_{n,m}F(m,x)\right) = 0. \tag{8.3.15}$$

Given $\mathbf{Z}_0, \mathbf{S}_0$ and Φ_0, the solvability condition yields the following equation for Φ_1

$$\Phi_1'(x) = \frac{\sum_{n\geq 0} S_0(n,x)[F(n,x)Z_0(n,x)]'}{\sum_{n\geq 0} S_0(n,x)F(n,x)Z_0(n,x)}. \tag{8.3.16}$$

Combining the various results, and defining

$$k(x) = \exp\left(-\Phi_1(x)\right), \tag{8.3.17}$$

gives to leading order in ε

$$\phi_\varepsilon^{(0)}(n,x) \sim \mathcal{N}k(x)\exp\left(-\frac{\Phi_0(x)}{\varepsilon}\right)Z_0(n,x), \tag{8.3.18}$$

where we choose $\sum_n Z_0(n,x) = 1$ for all x and $\mathcal{N}$ is the normalization factor

$$\mathcal{N} = \left[\int_\Sigma k(x)\exp\left(-\frac{\Phi_0(x)}{\varepsilon}\right)\right]^{-1}.$$

The latter can be approximated using Laplace's method to give

$$\mathcal{N} \sim \frac{1}{k(x_-)}\sqrt{\frac{|\Phi_0''(x_-)|}{2\pi\varepsilon}}\exp\left(\frac{\Phi_0(x_-)}{\varepsilon}\right). \tag{8.3.19}$$

The final step is to use singular perturbation theory to match the outer quasi-stationary solution to the absorbing boundary condition at x_0. The analysis is quite involved, see [433, 613, 615], so here we simply quote the result for the 1D model: the mean time to escape from the metastable state x_- is

$$\tau_- \sim \frac{\pi k(x_-)}{k(x_0)D(x_0)}\frac{1}{\sqrt{\Phi_0''(x_-)|\Phi_0''(x_0)|}}\exp\left(\frac{\Phi_0(x_0) - \Phi_0(x_-)}{\varepsilon}\right). \tag{8.3.20}$$

with $D(x)$ the effective   diffusion coefficient obtained using a QSS reduction (see Box 5D).

8.3.2 Path integral representation of a stochastic hybrid system

We have found that the WKB analysis of the stochastic hybrid system given by equation (8.3.2) differs from the WKB analysis of master equations (and FP equations) in the weak noise limit, since there is no obvious Hamiltonian structure. That is, the quasipotential satisfies a matrix equation (8.3.13) rather than a scalar Hamilton-Jacobi equation. A path integral formulation of a stochastic hybrid system has been developed that provides a variational principle for the correct Hamiltonian [117]. We describe this construction for the one-dimensional case, although the extension to higher dimensions is straightforward. For the sake of generality, we also include Gaussian noise in the dynamics of the continuous variable. Let the state of the system at time t consist of the pair $(X(t), N(t))$, where $X(t) \in \mathbb{R}$ and $N(t) \in \mathbb{Z}^+$. Suppose that the discrete process evolves according to a birth–death process of the form (8.2.1), except now the transition rates may depend on the continuous state variable $X(t)$. That is, $\omega_\pm = \omega_\pm(n,x)$ for $X(t) = x$ and $N(t) = n$. In between jumps in the discrete variable, $X(t)$ evolves according to the Ito SDE

$$dX = F(n,x)dt + \sqrt{2D(n,x)}dW \qquad (8.3.21)$$

for $N(t) = n$, where $W(t)$ is a Wiener process. The probability density now evolves according to the modified CK equation

$$\frac{\partial P(n,x,t)}{\partial t} = -\frac{\partial F(n,x)P(n,x,t)}{\partial x} + \frac{\partial^2 D(n,x)P(n,x,t)}{\partial x^2} + \sum_{m \geq 0} A_{nm}(x)P(m,x,t),$$

$$(8.3.22a)$$

$$\sum_{m \geq 0} A_{nm}(x)P(m,x,t) = \omega_+(n,x-1)P(n,x-1,t)$$

$$+ \omega_-(n,x+1)P(n,x+1,t) - [\omega_+(n,x) + \omega_-(n,x)]P(n,x,t).$$

$$(8.3.22b)$$

Operator formalism. The hybrid path integral was originally derived using integral representations of the Dirac delta function, along analogous lines to MSRJD [117]. Here we present an alternative derivation based on the use of operators and bra-kets, which is similar in spirit to the Doi-Peliti formalism. The operator approach provides a more efficient and flexible framework for constructing hybrid path integrals, and eliminates certain ad hoc steps of Ref. [117]. The first step is to introduce an appropriate bra-ket representation of a Hilbert or Fock space. This is taken to be the tensor product $|n,x\rangle = |n\rangle \otimes |x\rangle$. The discrete vector $|n\rangle$ is identical to the one introduced by Doi-Peliti with corresponding annihilation and creation operators $a, a^\dagger$ defined according to equation (8.2.26a). The Hilbert space spanned by the continuous vectors $|x\rangle$ with inner product $\langle y|x\rangle = \delta(x-y)$, is analogous to the Hilbert space (8.2.44) introduced for a birth-death process (see also Ref. [824]). That is, we introduce a second conjugate pair of operators, $\hat{x}, \hat{p}$, such that

$$\hat{x}|x\rangle = x|x\rangle, \quad \hat{p}|x\rangle = -i\frac{\overleftarrow{d}}{dx}|x\rangle. \tag{8.3.23}$$

Again the arrow on the differential operator indicates that it operates to the left. (Alternatively, given a state vector $|\phi\rangle = \int_{-\infty}^{\infty} dx\,\phi(x)|x\rangle$, we have $\langle x|\hat{p}|\phi\rangle = -i\phi'(x)$.) The operators satisfy the commutation relation

$$[\hat{x}, \hat{p}] = i, \tag{8.3.24}$$

since

$$[\hat{x}, \hat{p}]|x\rangle = \hat{x}\hat{p}|x\rangle - \hat{p}\hat{x}|x\rangle = \hat{x}\left(-i\frac{\overleftarrow{d}}{dx}\right)|x\rangle - \hat{p}x|x\rangle$$

$$= \left(-i\frac{\overleftarrow{d}}{dx}\right)x|x\rangle - x\left(-i\frac{\overleftarrow{d}}{dx}\right)|x\rangle = i|x\rangle.$$

The Hilbert space also has the completion relation

$$\int_{-\infty}^{\infty} dx\,|x\rangle\langle x| = 1. \tag{8.3.25}$$

That is, for an arbitrary vector $|c\rangle = \int_{-\infty}^{\infty} dx\,c(x)|x\rangle$,

$$\int_{-\infty}^{\infty} dx\,|x\rangle\langle x|c\rangle = \int_{-\infty}^{\infty} dx \int_{-\infty}^{\infty} dy\,|x\rangle c(y)\langle x|y\rangle$$

$$= \int_{-\infty}^{\infty} dx \int_{-\infty}^{\infty} dy\,|x\rangle c(y)\delta(x-y) = \int_{-\infty}^{\infty} dy\,c(y)|y\rangle = |c\rangle.$$

Another useful choice of basis vectors is the momentum representation (analogous to taking Fourier transforms),

$$|p\rangle = \int_{-\infty}^{\infty} dx\,e^{ipx}|x\rangle. \tag{8.3.26}$$

It immediately follows that $|p\rangle$ is an eigenvector of the momentum operator $\hat{p}$, since

$$\hat{p}|p\rangle = \int_{-\infty}^{\infty} dx\,e^{ipx}\left(-i\frac{\overleftarrow{d}}{dx}\right)|x\rangle = \int_{-\infty}^{\infty} dx\,p\,e^{ipx}|x\rangle = p|p\rangle. \tag{8.3.27}$$

Using the inverse Fourier transform, we also have

$$|x\rangle = \int_{-\infty}^{\infty} \frac{dp}{2\pi}\,e^{-ipx}|p\rangle, \tag{8.3.28}$$

and the completeness relation

$$\int_{-\infty}^{\infty} \frac{dp}{2\pi} |p\rangle\langle p| = 1. \qquad (8.3.29)$$

We carry over the definitions of the operator pairs $a, a^\dagger$ and $\hat{x}, \hat{p}$ on the tensor product space as follows:

$$a|n,x\rangle = n|n-1,x\rangle, \quad a^\dagger|n,x\rangle = |n+1,x\rangle, \qquad (8.3.30a)$$

$$\hat{x}|n,x\rangle = x|n,x\rangle, \quad \hat{p}|n,x\rangle = -i\frac{\overleftarrow{d}}{dx}|n,x\rangle. \qquad (8.3.30b)$$

These preserve the commutation relations (8.2.26a) and (8.3.24). We also have the completeness relation

$$\sum_{n\geq 0}\int_{-\infty}^{\infty} dx\,|n,x\rangle\langle n,x| = 1. \qquad (8.3.31)$$

Introduce the state vector

$$|\psi(t)\rangle = \sum_{n\geq 0}\int_{-\infty}^{\infty} dx\,P(n,x,t)|n,x\rangle. \qquad (8.3.32)$$

Differentiating both sides with respect to time gives

$$\frac{d}{dt}|\psi(t)\rangle = \sum_{n\geq 0}\int_{-\infty}^{\infty} dx\left[-\frac{\partial F(n,x)P(n,x,t)}{\partial x} + \frac{\partial^2 D(n,x)P(n,x,t)}{\partial x^2}\right]|n,x\rangle$$

$$+ \sum_{n,m\geq 0}\int_{-\infty}^{\infty} dx\,Q_{nm}(x)P(m,x,t)|n,x\rangle$$

$$= \int_{-\infty}^{\infty} dx\left[-F(n,x)P(n,x,t)\frac{\overleftarrow{\partial}}{\partial x} + D(n,x)P(n,x,t)\frac{\overleftarrow{\partial^2}}{\partial x^2}\right]|n,x\rangle$$

$$+ \hat{H}_{bd}\sum_{n\geq 0}\int_{-\infty}^{\infty} dx\,P(n,x,t)|n,x\rangle$$

$$= \left[-i\hat{p}A(a^\dagger a,\hat{x}) - \hat{p}^2 D(a^\dagger a,\hat{x}) + \hat{H}_{bd}\right]\sum_{n\geq 0}\int_{-\infty}^{\infty} dx\,P(n,x,t)|n,x\rangle,$$

where $\hat{H}_{bd}$ is an extended version of equation (8.2.31):

$$\hat{H}_{bd} = (a - a^\dagger a)\omega_-(a^\dagger a,\hat{x}) + (a^\dagger - 1)\omega_+(a^\dagger a,\hat{x}). \qquad (8.3.33)$$

Hence, the CK equation can be written in the operator form

$$\frac{d}{dt}|\psi(t)\rangle = \hat{H}|\psi(t)\rangle, \quad \hat{H} = -i\hat{p}A(a^\dagger a,\hat{x}) - \hat{p}^2 D(a^\dagger a,\hat{x}) + \hat{H}_{bd}. \qquad (8.3.34)$$

The formal solution of the FP equation is then

$$|\psi(t)\rangle = e^{\hat{H}t}|\psi(0)\rangle, \qquad (8.3.35)$$

and we can define expectations of the continuous and discrete variables via

$$\langle X(t) \rangle = \sum_{n \geq 0} \frac{1}{n!} \int_{-\infty}^{\infty} dx \, \langle n, x | \hat{x} e^{\hat{H}t} | \psi(0) \rangle, \tag{8.3.36a}$$

$$\langle N(t) \rangle = \sum_{n \geq 0} \frac{1}{n!} \int_{-\infty}^{\infty} dx \, \langle n, x | a^{\dagger} a e^{\hat{H}t} | \psi(0) \rangle. \tag{8.3.36b}$$

Construction of stochastic hybrid path integral. As with the previous stochastic processes, the first step in constructing a path integral is to divide the time interval $[0, t]$ into $\mathcal{N}$ subintervals of size $\Delta t = t / \mathcal{N}$ and rewrite the formal solution (8.3.35) as

$$|\psi(t)\rangle = e^{\hat{H}\Delta t} e^{\hat{H}\Delta t} \cdots e^{\hat{H}\Delta t} |\psi(0)\rangle. \tag{8.3.37}$$

with $\hat{H}$ given by equation (8.3.34). We then insert multiple copies of the completeness relation for the basis vectors $|n, x\rangle$ so that

$$|\psi(t)\rangle = \sum_{n_0 \geq 0} \frac{1}{n_0!} \cdots \sum_{n_{\mathcal{N}} \geq 0} \frac{1}{n_{\mathcal{N}}!} \int_{-\infty}^{\infty} dx_0 \cdots \int_{-\infty}^{\infty} dx_{\mathcal{N}} |n_{\mathcal{N}}, x_{\mathcal{N}}\rangle$$

$$\times \langle n_{\mathcal{N}}, x_{\mathcal{N}} | e^{\hat{H}\Delta t} | n_{\mathcal{N}-1}, x_{\mathcal{N}-1} \rangle \langle n_{\mathcal{N}-1}, x_{\mathcal{N}-1} | e^{\hat{H}\Delta t} | n_{\mathcal{N}-2}, x_{\mathcal{N}-2} \rangle$$

$$\cdots \times \langle n_1, x_1 | e^{\hat{H}\Delta t} | n_0, x_0 \rangle \langle n_0, x_0 | \psi(0) \rangle, \tag{8.3.38}$$

with $\langle n_0, x_0 | \psi(0) \rangle = n_0! P(n_0, x_0, 0)$. In the limit $\mathcal{N} \to \infty$ and $\Delta t \to 0$ with $\mathcal{N} \Delta t = t$ fixed, we make the approximation

$$\langle n_{j+1}, x_{j+1} | e^{\hat{H}\Delta t} | n_j, x_j \rangle \approx \langle n_{j+1}, x_{j+1} | 1 + \hat{H} \Delta t | n_j, x_j \rangle = \delta(x_{j+1} - x_j) n_j! \delta_{n_{j+1}, n_j}$$

$$+ \Delta t \langle n_{j+1}, x_{j+1} | \left(-i\hat{p}F(n_j, x_j) - \hat{p}^2 D(n_j, x_j) \right) \delta_{n_j, m_j} + \sum_{m_j \geq 0} Q_{m_j n_j}(x_j) | m_j, x_j \rangle.$$

It follows that we can express each small-time propagator in the matrix form

$$\langle n_{j+1}, x_{j+1} | e^{\hat{H}\Delta t} | n_j, x_j \rangle = \langle n_{j+1}, x_{j+1} | 1 + \Sigma_{m_j} K_{m_j n_j}(x_j, \hat{p}) | m_j, x_j \rangle \Delta t + O(\Delta t^2), \tag{8.3.39}$$

where

$$K_{nm}(x, p) = [-ipF(n, x) - p^2 D(n, x)] \delta_{m,n} + A_{nm}(x). \tag{8.3.40}$$

Suppose that for fixed x, p, there exists a complete orthonormal set of right and left eigenvectors $R_{\mu}(n, x, p)$ and $\overline{R}_{\mu}(n, x, p)$ such that

$$\sum_{m \geq 0} K_{nm}(x, p) R_{\mu}(m, x, p) = \Lambda_{\mu}(x, p) R_{\mu}(n, x, p), \tag{8.3.41a}$$

$$\sum_{m \geq 0} K_{mn}(x, p) \overline{R}_{\mu}(m, x, p) = \Lambda_{\mu}(x, p) \overline{R}_{\mu}(n, x, p), \tag{8.3.41b}$$

with

$$\sum_{m\geq 0} R_\mu(m,x,p)\overline{R}_\nu(m,x,p) = \delta_{\mu,\nu}, \quad \sum_{\mu\geq 0} R_\mu(m,x,p)\overline{R}_\mu(n,x,p) = \delta_{m,n}. \quad (8.3.41c)$$

For a given x and p, we introduce the vectors

$$|R_\mu(x,p)\rangle = \sum_{n\geq 0} R_\mu(n,x,p)|n,p\rangle. \quad (8.3.42)$$

and their duals

$$\langle \overline{R}_\mu(x,p)| = \sum_{n\geq 0} \frac{1}{n!}\overline{R}_\mu(n,x,p)\langle n,p|. \quad (8.3.43)$$

(We are assuming all functions are real. In the case of complex functions, the dual is defined in terms of the complex conjugate $\overline{R}_\mu^*$.) It immediately follows that

$$\begin{aligned}
\langle \overline{R}_\nu(x,q)|R_\mu(x,p)\rangle &= \sum_{n,m\geq 0} \frac{1}{n!}\overline{R}_\nu(n,x,q)R_\mu(m,x,p)\langle n,q|m,p\rangle \\
&= 2\pi \sum_{n,m\geq 0} \overline{R}_\nu(n,x,p)R_\mu(m,x,p)\delta(p-q)\delta_{n,m} \\
&= 2\pi\delta(p-q)\sum_{n\geq 0}\overline{R}_\nu(n,x,p)R_\mu(n,x,p) = 2\pi\delta(p-q)\delta_{\mu,\nu}.
\end{aligned}$$

The associated completeness relation is

$$\sum_{\mu\geq 0}\int_{-\infty}^{\infty}\frac{dp}{2\pi}|R_\mu(x,p)\rangle\langle\overline{R}_\mu(x,p)| = 1. \quad (8.3.44)$$

Note that the given set of basis vectors is a natural generalization of the eigenvectors of the matrix generator for a birth–death process; see equation (8.2.37).

Substituting the completeness relation (8.3.44) into the small-time propagator (8.3.39) gives

$$\begin{aligned}
\langle n_{j+1},x_{j+1}|e^{\hat{H}\Delta t}|n_j,x_j\rangle &= \sum_{\mu_j\geq 0}\int_{-\infty}^{\infty}\frac{dp_j}{2\pi}\langle n_{j+1},x_{j+1}|R_{\mu_j}(x_j,p_j)\rangle\langle\overline{R}_{\mu_j}(x_j,p_j) \\
&\times |1+\sum_{m_j}K_{m_jn_j}(x_j,p_j)\Delta t|m_j,x_j\rangle + O(\Delta t^2).
\end{aligned}$$

Note that

$$\begin{aligned}
L_{\mu_j} &= \sum_{n,m_j\geq 0}\frac{1}{n!}\overline{R}_{\mu_j}(n,x_j,p_j)\langle n,p_j|1+K_{m_jn_j}(x_j,p_j)\Delta t|m_j,x_j\rangle \\
&= e^{-ip_jx_j}\sum_{m_j}\overline{R}_{\mu_j}(m_j,x_j,p_j)(1+K_{m_jn_j}(x_j,p_j)\Delta t) \\
&= e^{-ip_jx_j}(1+\Lambda_\mu(x_j,p_j)\Delta t)\overline{R}_{\mu_j}(n_j,x_j,p_j),
\end{aligned}$$

where $L_{\mu_j}\equiv\langle\overline{R}_{\mu_j}(x_j,p_j)|1+\sum_{m_j}K_{m_jn_j}(x_j,p_j)\Delta t|m_j,x_j\rangle$. We have used equation (8.3.41b) and the definition (8.3.43). Moreover, from equation (8.3.42)

$$\langle n_{j+1}, x_{j+1} | R_{\mu_j}(x_j, p_j) \rangle = \langle n_{j+1}, x_{j+1} | \sum_{n \geq 0} R_{\mu_j}(n, x_j, p_j) | n, p_j \rangle$$

$$= (n_{j+1}!) e^{i p_j x_{j+1}} R_{\mu_j}(n_{j+1}, x_j, p_j).$$

Therefore,

$$\langle n_{j+1}, x_{j+1} | e^{\hat{H}\Delta t} | n_j, x_j \rangle \approx (n_{j+1}!) \sum_{\mu_j \geq 0} \int_{-\infty}^{\infty} \frac{dp_j}{2\pi}$$

$$\times e^{i p_j (x_{j+1} - x_j)} e^{\Lambda_{\mu_j}(x_j, p_j)\Delta t} R_{\mu_j}(n_{j+1}, x_j, p_j) \overline{R}_{\mu_j}(n_j, x_j, p_j) + O(\Delta t^2).$$

Substituting the expression for the propagator back into equation (8.3.38) yields

$$P(n, x, t) = \frac{1}{n!} \langle n, x | \psi(t) \rangle \approx \sum_{n_0 \geq 0} \cdots \sum_{n_{\mathcal{N}} \geq 0} \sum_{\mu_0 \geq 0} \cdots \sum_{\mu_{\mathcal{N}-1} \geq 0}$$

$$\times \int_{-\infty}^{\infty} \int_{-\infty}^{\infty} \frac{dx_0 dp_0}{2\pi} \cdots \int_{-\infty}^{\infty} \int_{-\infty}^{\infty} \frac{dx_{\mathcal{N}} dp_{\mathcal{N}}}{2\pi} \delta(x - x_{\mathcal{N}}) \qquad (8.3.45)$$

$$\times \prod_{j=0}^{\mathcal{N}-1} \left[e^{i p_j (x_{j+1} - x_j)} e^{\Lambda_{\mu_j}(x_j, p_j)\Delta t} R_{\mu_j}(n_{j+1}, x_j, p_j) \overline{R}_{\mu_j}(n_j, x_j, p_j) \right] P(n_0, x_0, 0).$$

Finally, we take the continuum limit $\mathcal{N} \to \infty, \Delta t \to 0$ with $\mathcal{N} \Delta t = t$ fixed, $x_j = x(j\Delta t)$ and $p_j = p(j\Delta t)$. Exploiting the fact that the resulting path integral sums over continuous paths, we can take

$$\sum_{n_{j+1} \geq 0} R_{\mu_j}(n_{j+1}, x_j, p_j) \overline{R}_{\mu_{j+1}}(n_{j+1}, x_{j+1}, p_{j+1})$$

$$\xrightarrow[\Delta t \to 0]{} \sum_{n_{j+1} \geq 0} R_{\mu_j}(n_{j+1}, x_j, p_j) \overline{R}_{\mu_{j+1}}(n_{j+1}, x_j, p_j) = \delta_{\mu_{j+1}, \mu_j}$$

for $j = 0, 1, \ldots \mathcal{N} - 1$. This enforces the condition $\mu_j = \mu$ for all j in the product of exponentials so that in the limit $\Delta t \to 0$

$$\prod_{j=0}^{\mathcal{N}-1} \left[e^{i p_j (x_{j+1} - x_j)} e^{\Lambda_\mu(x_j, p_j)\Delta t} \right] = \exp\left(\sum_{j=0}^{\mathcal{N}-1} i p_j (x_{j+1} - x_j) + \Lambda_\mu(x_j, p_j)\Delta t \right)$$

$$\to \exp\left(\int_0^t [i p \dot{x} + \Lambda_\mu(x, p)] d\tau \right).$$

We thus obtain the path integral representation

$$P(n, x, t) = \sum_{n_0 \geq 0} \int_{-\infty}^{\infty} dx_0 \, G(n, x, t | n_0, x_0) P(n_0, x_0, 0), \qquad (8.3.46a)$$

with

$$G(n,x,t|n_0,x_0) = \int\limits_{x(0)=x_0}^{x(t)=x} \mathcal{D}[p]\mathcal{D}[x]$$

$$\times \sum_{\mu\geq0} R_\mu(n,x,p(t))\exp\left(-\int_0^t [p\dot{x} - \Lambda_\mu(x,p)]d\tau\right)\overline{R}_\mu(n_0,x_0,p(0)). \quad (8.3.46b)$$

We have also performed a Wick rotation of the momentum variable, $p \to -ip$, so that the eigenvalue equations (8.3.41a) and (8.3.41b) become

$$\sum_{m\geq0}\left\{[pF(n,x)+p^2D(n,x)]\delta_{m,n}+A_{nm}(x)]\right\}R_\mu(m,x,p)$$

$$\equiv \sum_{m\geq0} Q_{nm}(x)R_\mu(n,x,p) = \Lambda_\mu(x,p)R_\mu(n,x,p), \quad (8.3.47a)$$

$$\sum_{m\geq0}\overline{R}_\mu(m,x,p)\left\{[pF(n,x)+p^2D(n,x)]\delta_{m,n}+A_{mn}(x)\right\}$$

$$\equiv \sum_{m\geq0}\overline{R}_\mu(m,x,p)Q_{mn}(x) = \Lambda_\mu(x,p)\overline{R}_\mu(n,x,p), \quad (8.3.47b)$$

Remark 8.1. If A and D are independent of the discrete state n, then the eigenvalue equation (8.3.47a) becomes

$$\sum_{m\geq0} A_{nm}(x)R_\mu(m,x,p) = [\Lambda_\mu(x,p) - pA(x) - p^2D(x)]R_\mu(n,x,p). \quad (8.3.48)$$

Let $r_\mu(m,x)$ be an eigenvector of the matrix generator $A(x)$ for fixed x; see equation (8.2.42). That is $\sum_{m\geq0}A_{nm}(x)r_\mu(m,x) = \lambda_\mu(x)r_\mu(m,x)$. Then $R_\mu(m,x,p) = c(p)r_\mu(m,x)$ for arbitrary $c(p)$ and

$$\Lambda_\mu(x,p) = pA(x) + p^2D(x) + \lambda_\mu(x). \quad (8.3.49)$$

The path integral (8.3.46) reduces to the form

$$G(n,x,t|n_0,x_0) = \int\limits_{x(0)=x_0}^{x(t)=x} \mathcal{D}[p]\mathcal{D}[x]\exp\left(-\int_0^t [p\dot{x} - pA(x) - p^2D(x)]d\tau\right) \quad (8.3.50)$$

$$\times \sum_{\mu\geq0} r_\mu(n,x)\overline{r}_\mu(n_0,x_0)\exp\left(\int_0^t \lambda_\mu(x)d\tau\right).$$

That is, the path integral of the continuous stochastic process decouples from the discrete process and we recover the standard action of a one-dimensional Ito SDE [212, 415, 559, 824]:

$$S[x,p] = \int_0^t [p\dot{x} - pA(x) - p^2D(x)]dt.$$

It also follows that if the initial state of the system is $P(n_0,x_0,0) = \rho_{n_0}(\bar{x})\delta(x_0 - \bar{x})$, say, then

$$\sum_{n_0\geq0}\int_{-\infty}^{\infty} dx_0\overline{r}_\mu(n_0,x_0)P(n_0,x_0,0) = \delta_{\mu,0},$$

and the sum over μ in the path integral (8.3.46b) is restricted to $\mu = 0$. On the other hand, when the continuous and discrete processes are mutually coupled, the resulting dynamics mixes the eigenstates of the Markov chain. In that case, a further approximation is needed in order to restrict the sum over μ.

Remark 8.2. The derivation carries over to higher-dimensional stochastic hybrid systems with M continuous variables x_α, $\alpha = 1, \ldots M$. The Ito SDE becomes

$$dX_\alpha = F_\alpha(n, \mathbf{x})dt + \sqrt{2D_\alpha(n, \mathbf{x})}dW_\alpha \tag{8.3.51}$$

for $N(t) = n$ and the multivariate CK equation takes the form

$$\frac{\partial P}{\partial t} = \sum_{\alpha=1}^{M} \left[-\frac{\partial}{\partial x_\alpha}(F_\alpha(n,\mathbf{x})P(n,\mathbf{x},t)) + \frac{\partial^2}{\partial x_\alpha^2}(D_\alpha(n,\mathbf{x})P(n,\mathbf{x},t)) \right]$$
$$+ \sum_m A_{nm}(\mathbf{x})P(m,\mathbf{x},t). \tag{8.3.52}$$

Following along identical lines to the one-dimensional case, we can derive a path integral representation of the solution to equation (8.3.52)

$$P(n,\mathbf{x},t) = \sum_{n_0 \geq 0} \int_{-\infty}^{\infty} d\mathbf{x}_0 \, G(n,\mathbf{x},t|n_0,\mathbf{x}_0)P(n_0,\mathbf{x}_0,0), \tag{8.3.53a}$$

$$G(n,\mathbf{x},t|n_0,\mathbf{x}_0)$$

$$= \iint_{\substack{\mathbf{x}(t)=\mathbf{x} \\ \mathbf{x}(0)=\mathbf{x}_0}} \mathcal{D}[\mathbf{p}]\mathcal{D}[\mathbf{x}] \sum_{\mu \geq 0} R_\mu(n,\mathbf{x},\mathbf{p}(t)) \exp\left(-S_\mu[\mathbf{x},\mathbf{p}]\right) \overline{R}_\mu(n_0,\mathbf{x}_0,\mathbf{p}(0)), \tag{8.3.53b}$$

with

$$S_\mu[\mathbf{x},\mathbf{p}] = \int_0^t \left[\sum_{\alpha=1}^{M} p_\alpha \dot{x}_\alpha - \Lambda_\mu(\mathbf{x},\mathbf{p}) \right] d\tau. \tag{8.3.54}$$

Here Λ_μ is an eigenvalue of the linear equation

$$\left[\sum_m Q_{nm}(\mathbf{x}) + \sum_{\alpha=1}^{M} (p_\alpha A_\alpha(n,\mathbf{x}) + p_\alpha^2 D_\alpha(n,\mathbf{x}))\delta_{n,m} \right] R_\mu(m,\mathbf{x},\mathbf{p})$$
$$= \Lambda_\mu(\mathbf{x},\mathbf{p})R_\mu(n,\mathbf{x},\mathbf{p}). \tag{8.3.55}$$

8.3.3 Finite discrete systems and the Perron-Frobenius theorem

Although we took the discrete part to evolve according to a birth–death process with associated operator $\hat{H}_{bd}$, the above derivation holds for any master equation with matrix generator $\mathbf{A}$, provided that there exists a complete orthonormal set of right and left eigenvectors $R_\mu(x,p,n)$ and $\overline{R}_\mu(x,p,n)$ satisfying equations (8.3.47a) and (8.3.47b). This may not hold if the discrete system is infinite-dimensional. On the other hand, if the underlying discrete space is finite-dimensional, then one can make stronger statements about the spectrum using the Perron-Frobenius theorem. We first recall some results for finite continuous-time Markov chains [343], see Sect. 3.6. Let $N(t)$, $0 \leq N(t) \leq N_{max}$, be a discrete random variable whose Markov chain is irreducible. Under such a condition, one can apply the Perron-Frobenius theorem. In particular, there exists a unique zero eigenvalue of $\mathbf{A}$ with positive left and right eigenvectors $\psi = (1, 1, \ldots, 1)$ and ρ such that $\sum_n Q_{nm} = 0$ and $\sum_m Q_{nm}\rho_m = 0$. The distribution ρ is the unique stationary density. Moreover, the Perron Frobenius theorem ensures that all other eigenvalues have negative real point, ensuring that the distribution $P(m,t) \to \rho(m)$ as $t \to \infty$. In the particular case of a birth–death process, the stationary density (if it exists) is given by

$$\rho(n) = \rho(0) \prod_{m=1}^{n} \frac{\omega_+(m-1)}{m\omega_-(m)}, \quad \rho(0) = \left(1 + \sum_{n=1}^{N_{\max}} \prod_{m=1}^{n} \frac{\omega_+(m-1)}{\omega_-(m)}\right)^{-1}. \quad (8.3.56)$$

For finite $N_{\max}$, ρ exists provided that all birth/death rates are positive definite.

Returning to the stochastic hybrid system (8.3.1), suppose that the discrete component is finite-dimensional and $A_{nm}(x)$ for any fixed x is the generator of an irreducible Markov chain. For fixed (x,p), introduce the modified matrix

$$\widehat{Q}_{nm}(x,p) = Q_{nm}(x,p) + \Gamma \delta_{n,m}, \quad (8.3.57)$$

with

$$\Gamma = \max_n \left\{ [pF(n,x) + p^2 D(n,x)]\delta_{m,n} + A_{nn}(x). \right\}. \quad (8.3.58)$$

$\widehat{Q}(x,p)$ is then an irreducible matrix and we can apply the Perron-Frobenius theorem. This establishes that $\mathbf{Q}(x,p)$ for fixed (x,p) has a unique positive right eigenvector, which we identify with $\mu = 0$, and Λ_0 is the corresponding principal eigenvalue with $\Lambda_0 > \mathrm{Re}(\Lambda_\mu)$ for all $\mu = 1,\ldots,N_{\max}$. This plays a crucial role in analyzing the noise-induced escape from a metastable state in the weak noise limit. The latter is obtained by introducing the scalings $\mathbf{A} \to \mathbf{A}/\varepsilon$ and $D \to \varepsilon D$. The former represents fast switching between the discrete states (adiabatic limit), whereas the latter represents weak Gaussian noise. If one also scales the momentum variable according to $p \to p/\varepsilon$, then the matrix $\mathbf{Q} \to \mathbf{Q}/\varepsilon$ and $\Lambda_\mu \to \Lambda_\mu/\varepsilon$. The resulting spectral gap means that when considering the "least action" path, we can restrict the summation over μ in equation (8.3.46) to $\mu = 0$

$$G(n,x,t|n_0,x_0) = \int\limits_{x(0)=x_0}^{x(t)=x} \mathcal{D}[p]\mathcal{D}[x] \quad (8.3.59)$$

$$R_0(n,x,p(t)) \exp\left(-\frac{1}{\varepsilon} \int_0^t [p\dot{x} - \Lambda_0(x,p)]d\tau\right) \overline{R}_0(n_0,x_0,p(0)).$$

Finding the "least action" path then reduces to a classical variational problem with action

$$S[x,p] = \int_0^t [p\dot{x} - \Lambda_0(x,p)]d\tau. \quad (8.3.60)$$

and Hamiltonian given by the principal eigenvalue $\Lambda_0(x,p)$. In particular, the least action path is the solution to Hamilton's equations

$$\frac{dx}{dt} = \frac{\partial \Lambda_0}{\partial p}, \quad \frac{dp}{dt} = -\frac{\partial \Lambda_0}{\partial x}. \quad (8.3.61)$$

Although the above derivation uses formal methods, it generates the same action principle as obtained rigorously using large deviation theory, as detailed in the monograph by Kifer [440].

We can now link the least action principle obtained using path integrals to the previous WKB approximation. Comparison of equation (8.3.13) with equation (8.3.62), where Λ_0 is the principal eigenvalue of the linear equation (8.3.47a) with zero Gaussian noise, shows that there exists a nontrivial positive solution of equation (8.3.13) given by $Z_0(n,x) = R_0(n,x,p)$ with $p = \Phi_0'(x)$ and Φ_0 satisfying the corresponding HJ equation

$$\Lambda_0(x, \Phi_0'(x)) = 0. \tag{8.3.62}$$

The interesting feature of a stochastic hybrid system is that the WKB method does not directly generate the Hamiltonian of large deviation theory except along least action paths where the Hamiltonian is zero. In summary, we have established that solving an escape problem in the weak noise limit only requires determining the principal eigenvalue Λ_0 and the associated left and right eigenvectors $R_0, \overline{R}_0$. In the case of small N_{max} such as a two-state Markov chain, these can be calculated by brute force. However, there are only a few examples where it has been possible to calculate Λ_0 explicitly for arbitrarily large N_{max} [117]. One example occurs in a model of stochastic ion channels, where the steady-state distribution of the underlying birth–death process is a binomial distribution, see Sect. 8.5. Here we consider an example for which a positive eigenfunction can be constructed even though the Perron-Frobenius theorem does not strictly apply.

Example 8.1. Consider the discrete process $N(t) \in \mathbf{Z}^+$ with transition rates

$$\omega_+(n,X) = f(X), \quad \omega_-(n,X) = n, \tag{8.3.63}$$

where f is a bounded positive function, and suppose that the continuous variable evolves according to the piecewise SDE

$$dX = -(X+n)dt + \sqrt{\varepsilon n D_0} dW \tag{8.3.64}$$

for $N(t) = n$. One application of this type of PDMP is to a one-population neural network model [115, 119]. If $X(t) = x$ is fixed for all t, then the steady-state distribution of the birth–death process is determined from equation (8.3.56), and is given by a Poisson distribution

$$\rho_n(x) = \frac{f(x)^n}{n!} e^{-f(x)}. \tag{8.3.65}$$

The eigenvalue equation (8.3.47a) becomes

$$[p(-x+n) + p^2 D_0]R_\mu(n,x,p) + f(x)R_\mu(n-1,x,p) + (n+1)R_\mu(n+1,x,p)$$
$$- (f(x)+n)R_\mu(n,x,p) = \Lambda_\mu(x,p)R_\mu(n,x,p). \tag{8.3.66}$$

In terms of a candidate principal eigenvalue and associated positive eigenvector, we can formally solve this equation using the trial positive solution

$$R_0(n,x,p) = \frac{\Gamma^n(x,p)}{n!}. \tag{8.3.67}$$

This yields the following equation relating Λ_0 and Γ:

$$\left[\frac{f(x)}{\Gamma} - 1\right] n + \Gamma - f(x) + p(-x+n) + p^2 D_0 = \Lambda_0.$$

Collecting terms independent of n and linear in n, respectively, we find that

$$\Gamma = \frac{f(x)}{1-p-p^2 D_0}, \quad \Lambda_0 = \frac{pf(x)}{1-p-p^2 D_0} - px. \tag{8.3.68}$$

The candidate principal eigenvalue has a pair of real singularities at

$$p = p_\pm \equiv \frac{1}{2D_0}\left(-1 \pm \sqrt{1+4D_0}\right), \quad D_0 > 0. \tag{8.3.69}$$

Moreover, $\Gamma(x,p) < 0$ for $p > p_+$ and $p < p_-$, contradicting the requirement that the eigenfunction R_0 is positive. In the special case $D_0 = 0$, there is a single pole at $p = 1$ and $\Gamma(x,p) < 0$ for $p > 1$. The origin of the singularity can be understood by considering a large but finite population size N_{max}. The Perron-Frobenius theorem then holds but the solution of the eigenvalue equation becomes nontrivial. The basic difficulty arises because the above ansatz for R_0 does not satisfy the boundary condition at $n = N_{max}$. For the sake of illustration suppose that $D_0 = 0$. Setting $n = N_{max}$, $\mu = 0$ and $R_0(N_{max}+1, x, p) = 0$ in equation (8.3.66), we have

$$p(-x+N_{max})R_0(N_{max},x,p) + f(x)R_0(N_{max}-1,x,p) - (f(x)+N_{max})R_0(N_{max},x,p)$$
$$= \Lambda_0(x,p)R_0(N_{max},x,p),$$

which can be rearranged to give

$$R_0(N_{max},x,p) = \frac{f(x)R_0(N_{max}-1,x,p)}{\Lambda_0(x,p)+f(x)+px+N_{max}(1-p)}.$$

This clearly does not satisfy the ansatz (8.3.67). However, if $p < 1$ then in the large-N_{max} limit, we set that

$$R_0(N_{max},x,p) \to \frac{f(x)}{N_{max}(1-p)}R_0(N_{max}-1,x,p) = \frac{\Gamma(x,p)}{N_{max}}R_0(N_{max}-1,x,p).$$

This shows that the given ansatz is a good approximation to the eigenvector for large N_{max} and $p < 1$. Keeping N_{max} large but finite means that we can ensure a spectral gap using Perron-Frobenius.

Finally, note that the path integral (8.3.59), or its higher-dimensional analog, provides an alternative method for carrying out a diffusion approximation in the fast switching limit, see Ex. 8.7. That is, a small p-expansion of the action leads to a Gaussian in p that can be integrated to yield an Onsager-Machlup path integral. One finds that the corresponding SDE is identical to the one obtained   by performing a quasi-steady-state approximation of the underlying stochastic hybrid system (see Box 5D).

8.4 Noise-induced transitions in an autoregulatory gene network

As our first application of metastability analysis to discrete Markov processes and stochastic hybrid systems, we consider a stochastic version of the autoregulatory network introduced in Sect. 5.1.1, in which the dynamics of mRNA is ignored. Let $M(t)$ denote the current state of the gene with $M(t) = 1$ (active) or $M(t) = 0$ (inactive), and let $N(t)$ be the number of protein molecules. The master equation for the joint probability distribution $P(m,n,t) = \mathbb{P}[M(t) = m, N(t) = n]$, where $m \in \{0,1\}$ and $n \geq 0$, is a generalization of equation (5.3.1):

$$\frac{dP(m,n,t)}{dt} = [k_+ f(n/N)m + k_-(1-m)]P(1-m,n,t)$$
$$+ \kappa_p N m P(m,n-1,t) + \gamma_p(n+1)P(m,n+1,t) \qquad (8.4.1)$$
$$- [k_+ f(n/N)(1-m) + k_- m + \kappa_p N m + \gamma_p n] P(m,n,t).$$

The function f incorporates the effects of positive feedback, and for the sake of illustration is taken to be $f(x) = e^{\eta(x-\theta)}$. We will consider separately the two limiting cases $k_\pm \to 0$ (adiabatic or fast switching limit) and $N \to \infty$ (thermodynamic limit). Note, however, that one could analyze the more general model by combining the two cases [615].

Adiabatic limit. In the limit $k_\pm \to 0$ for fixed N and $k_+/k_- = 1$, we have $P(m,n,t) \to \rho_m(n/M)P(n,t)$, where

$$\rho_0(x) = \frac{1}{1+f(x)}, \quad \rho_1(x) := F(x) = \frac{f(x)}{1+f(x)} = \frac{1}{1+e^{-\eta(x-\theta)}},$$

and $P(n,t)$ evolves according to the birth–death master equation (8.2.1) with transition rates

$$\omega_+(n) = N\kappa_p F(n/N), \quad \omega_-(n) = \gamma_p n. \qquad (8.4.2)$$

If we now take the thermodynamic limit $N \to \infty$, we obtain the deterministic equation

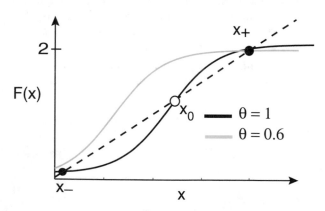

Fig. 8.2: Fixed points of 1D equation $\dot{x} = -x + F(x)$ with $F(x) = 2[1 + e^{-\eta(x-\theta)}]^{-1}$. If $\eta > 1$ then the network can exhibit bistability for a range of thresholds θ.

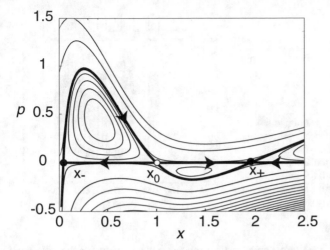

Fig. 8.3: Phase portrait of Hamiltonian equations of motion for $\Omega_+(x) = \kappa_p/(1 + e^{-\eta(x-\theta)})$ and $\Omega_-(x) = \gamma_p x$ with $\eta = 4, \theta = 1.0, \kappa_p = 2$ and $\gamma_p = 1$. The zero energy solutions are shown as thicker curves.

$$\dot{x} = -\gamma_p x + \kappa_p F(x). \tag{8.4.3}$$

We will assume that the deterministic system (8.4.3) is bistable with stable fixed points $x_\pm$ separated by an unstable fixed point x_0, see Fig. 8.2. Carrying out the WKB analysis of Sect. 8.2, we obtain a Hamiltonian system of the form (8.2.16) with

$$\Omega_+(x) = \frac{\kappa_p}{1 + e^{-\kappa(x-\theta))}}, \quad \Omega_-(x) = \gamma_p x. \tag{8.4.4}$$

In Fig. 8.3 we illustrate the corresponding Hamiltonian phase space for these transition rates. The constant energy solutions of the Hamiltonian are shown, with the zero energy activation and relaxation trajectories through the fixed points of the deterministic system highlighted as thicker curves. Note that $N\Phi(x_0)$, where $\Phi(x_0)$ is the area enclosed by the heteroclinic connection from x_- to x_0, gives the leading order contribution to $\log \tau_-$, where τ_- is the mean escape time from x_-. Similarly, the area under the heteroclinic connection from x_+ to x_0 gives the corresponding leading contribution to $\log \tau_+$, where τ_+ is the mean escape time from x_+. We immediately deduce that for the chosen parameter values $\tau_+ < \tau_-$.

Note that the long–term behavior of the birth–death process operating in a bistable regime can be approximated by a two–state Markov process that only keeps track of which metastable state the system is close to [826]:

$$\frac{d}{dt}\begin{pmatrix} P_- \\ P_+ \end{pmatrix} = \widehat{\mathbf{Q}}\begin{pmatrix} P_- \\ P_+ \end{pmatrix}, \quad \widehat{\mathbf{Q}} = \begin{pmatrix} -r_- & r_+ \\ r_- & -r_+ \end{pmatrix}, \tag{8.4.5}$$

where $P_\pm$ are the probabilities of being in a neighborhood of $x_\pm$, and $r_\pm = \tau_\pm^{-1}$. The matrix $\widehat{\mathbf{Q}}$ has eigenvectors $\widehat{\lambda}_0 = 0$ and $\widehat{\lambda}_1 = -(r_+ + r_-)$ and corresponding eigenvectors

$$\widehat{\mathbf{v}}_0 = \begin{pmatrix} r_+/(r_+ + r_-) \\ r_-/(r_+ + r_-) \end{pmatrix}, \quad \widehat{\mathbf{v}}_1 = \begin{pmatrix} 1/2 \\ -1/2 \end{pmatrix}. \tag{8.4.6}$$

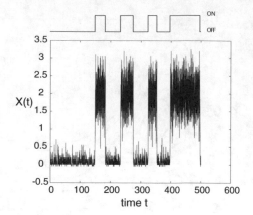

Fig. 8.4: Time series showing a single realization of a discrete Markov process
evolving according to the birth–death master equation (8.2.1) with transition rates
(8.4.2). Parameters are $\theta = 0.86$, $\eta = 4.0$,
$\kappa_p = 2$, $\gamma_p = 1$, and $N = 20$. The bistable
nature of the process is clearly seen.

Switches of the two-state system generated by the master equation for the autoregulatory network is shown in Fig. 8.4. Note that the two-state model generates an
exponential density for the residence times within a given metastable state. This
captures the behavior of the full master equation at large residence times but fails
to capture the short-term dynamics associated with relaxation trajectories within a
neighborhood of the metastable state.

Metastability in the thermodynamic limit. Now suppose that we set $k_\pm = 1/\varepsilon$ and
take the thermodynamic limit $N \to \infty$ for fixed ε. In this case, we obtain a stochastic
hybrid system evolving according to a CK equation of the form (8.3.2), where m
represents the state of the gene, and

$$\mathbf{A}(x) = \begin{pmatrix} -f(x) & 1 \\ f(x) & -1 \end{pmatrix}, \quad F(m,x) = F_m(x) \equiv m\kappa_p - \gamma_p x.$$

Note that if $x(0) \in \Sigma = [0, \kappa_p/\gamma_p]$ then $x(t) \in \Sigma$ for all $t > 0$. If we take the fast
switching limit $\varepsilon \to 0$, then we recover the deterministic system (8.4.3). Again assume that the latter is bistable. Carrying out the metastability analysis of Sect. 8.3
generates a Hamiltonian system with Hamiltonian given by the principal eigenvalue Λ_0 of the linear equation (8.3.47a). For the autoregulatory network, we have the
two-dimensional linear system

$$\begin{pmatrix} -f(x)+pF_0(x) & 1 \\ f(x) & -1+pF_1(x) \end{pmatrix} \begin{pmatrix} R_0 \\ R_1 \end{pmatrix} = \Lambda \begin{pmatrix} R_0 \\ R_1 \end{pmatrix}. \tag{8.4.7}$$

The corresponding characteristic equation is

$$0 = \Lambda^2 + \Lambda[1+f(x)-p(F_0(x)+F_1(x))] + (pF_1(x)-1)(pF_0(x)-f(x)) - f(x).$$

It follows that the principal eigenvalue is given by

$$\Lambda_0(x,p) = \frac{1}{2}\left[\Sigma(x,p) + \sqrt{\Sigma(x,p)^2 - 4h(x,p)}\right], \tag{8.4.8}$$

where

$$\Sigma(x,p) = p(F_0(x) + F_1(x)) - [1 + f(x)],$$

and

$$h(x,p) = p^2 F_1(x) F_0(x) - p[F_0(x) + f(x)F_1(x)].$$

A little algebra shows that

$$\mathcal{D}(x,p) \equiv \Sigma(x,p)^2 - 4h(x,p) = [p(F_0 - F_1) + 1 - f(x)]^2 + 4f(x) > 0,$$

so that Λ_0 is real. The quasipotential $\Phi_0(x)$ satisfies the HJ equation $\Lambda_0(x,p) = 0$ with $p = \Phi_0'(x)$, which reduces to the condition

$$h(x, \Phi_0'(x)) = 0, \quad \Sigma(x, \Phi_0'(x)) < 0. \tag{8.4.9}$$

This has two solutions: the classical deterministic solution $q = 0$ with $\Phi_0'(x) = 0$ and a nontrivial solution whose quasipotential satisfies

$$\Phi_0'(x) = \frac{1}{F_1(x)} + \frac{f(x)}{F_0(x)}. \tag{8.4.10}$$

Note that $F_n(x)$ does not vanish anywhere and $F_0(x)F_1(x) < 0$. The quasipotential can be determined by numerically integrating with respect to x. As shown in [615], the resulting quasipotential differs significantly from the one obtained by carrying out a QSS diffusion approximation of the stochastic hybrid system along the lines outlined in Box 5D.

For this simple model, it is also straightforward to determine the various prefactors in equation (8.3.20). For example, the normalized positive eigenvector $\mathbf{Z}^{(0)}$ has components

$$Z_0^{(0)} = \frac{F_1(x)}{F_1(x) - F_0(x)}, \quad Z_1^{(0)} = \frac{-F_0(x)}{F_1(x) - F_0(x)}.$$

Since $F_0(x) < 0$ and $F_1(x) > 0$ for $0 < x < \kappa_p/\gamma_p$, it follows from equation (8.4.10) that $Z_0^{(0)}$ is positive. The components of the adjoint eigenvector $\mathbf{S}$ satisfy

$$\frac{S_1}{S_0} = \frac{-f(x) + \Phi_0'(x)F_0(x)}{f(x)} = -1 + \Phi_0'(x)F_1(x).$$

It then follows from equation (8.3.16) that the first correction to the quasipotential satisfies

$$\Phi_1'(x) = \frac{1}{F_0(x)F_1(x)} \frac{d}{dx}(F_0(x)F_1(x)). \tag{8.4.11}$$

Hence

$$k(x) \equiv e^{-\Phi_1(x)} = \frac{1}{|F_0(x)|F_1(x)}. \tag{8.4.12}$$

Finally, $\mathcal{D}(x_0)$ is given by

$$D(x_0) = \frac{|F_0(x_0)F_1(x_0)|}{\alpha(x_0) + \beta(x_0)}.$$

For extensions to the two-dimensional gene regulatory networks considered in Sect. 5.3, namely, the mutual repressor network and the autoregulatory network with both mRNA and protein dynamics included, see [438, 613, 617].

8.5 Noise-induced transitions in the stochastic Morris-Lecar model

In Sect. 3.4, we considered a stochastic version of the Morris-Lecar model in order to investigate the effects of stochastic ion channels on membrane voltage fluctuations in a neuron. Following most previous approaches [181, 285, 323], we analyzed noise-induced action potentials by carrying out a system-size expansion with respect to the number of ion channels, which reduced the dynamics to an effective SDE. However, such a reduction can break down if the number of ion channels is small. Moreover, it can lead to exponentially large errors when estimating the rate of spontaneous action potential generation. More accurate results can be achieved by applying the metastability analysis of stochastic hybrid systems [433, 612, 614, 616].

8.5.1 Bistability in a reduced Morris-Lecar model

Following Keener and Newby [433, 612], consider the simpler problem of how ion channel fluctuations affect the initiation of an action potential due to the opening of a finite number of Ca^{2+} channels. (In Ref. [433, 612] the fast ion channels were taken to be Na^+ rather then Ca^{2+}. Since they act in a similar fashion at this level of modeling, we will still refer to them as Ca^{2+} channels.) The slow K^+ channels are assumed to be frozen, so that they effectively act as a leak current, and each calcium channel is treated as a simple two-state system. The Morris-Lecar equations thus reduce to a piecewise deterministic equation for the membrane voltage $X(t)$

$$\frac{dx}{dt} = F(n,x) \equiv \frac{n}{N_{max}}f(x) - g(x), \tag{8.5.1}$$

where $f(x) = g_{Ca}(V_{Ca} - x)$ represents the gated calcium current, $g(x) = -g_{eff}[V_{eff} - x] - I_{ext}$ represents the sum of effective leakage currents and external inputs, and n is the number of open calcium channels. Note that the right-hand side of (8.5.1) is negative for large x and positive for small x. This implies that the stochastic voltage x is confined to some bounded domain $[x_1, x_2]$.

    The opening and closing of the ion channels is described by a birth–death process of the form (8.2.1) with x-dependent transition rates

$$\omega_+(n,x) = \alpha(x)(N_{\max} - n), \quad \omega_-(n) = \beta n, \quad \alpha(x) = \beta \exp\left(\frac{2(x - v_1)}{v_2}\right)$$

for constants β, v_1, v_2. The associated CK equation is thus given by equation (8.3.22a). The equation for the steady–state distribution $\rho_n(x)$ of the discrete Markov process for fixed x can be obtained from the zero current condition

$$J(n,x) \equiv \omega_-(n)\rho_n(x) - \omega_+(n - 1,x)\rho_{n-1}(x) = 0, \quad n \geq 0.$$

Solving this equation iteratively using the unit normalization condition on $\sum_n \rho_n = 1$ gives the binomial distribution

$$\rho_n(x) = \frac{N_{\max}!}{(N_{\max} - n)!n!} a(x)^n (1 - a(x))^{N_{\max}-n}, \quad a(x) = \frac{\alpha(x)}{\alpha(x) + \beta}. \tag{8.5.2}$$

The mean number of open channels is then

$$\langle n \rangle = \sum_{n=1}^{N_{\max}} n\rho_n(x) = Na(x). \tag{8.5.3}$$

It follows that the mean-field equation obtained in the $\varepsilon \to 0$ limit is

$$\frac{dx}{dt} = \sum_{n=0}^{N_{\max}} \rho_n(x)F(n,x) = \overline{F}(x) = a(x)f(x) - g(x). \tag{8.5.4}$$

It is straightforward to find physiologically reasonable parameter values under which the deterministic equation exhibits bistability. Defining the effective potential $\Psi(x)$ according to $\Psi'(x) = -\overline{F}(x)$, we thus obtain a double well potential with x_- representing a stable resting state and x_+ a stable active state.

The spontaneous initiation of an action potential now reduces to a first passage time problem of noise-induced escape from the resting metastable state x_- to the active state x_+ by crossing the barrier at x_0. (Once this transition has occurred, the opening of the K^+ channels has to be taken into account in order that the system eventually returns to the rest state.) The calculation proceeds along analogous lines lines to the autoregulatory network in the thermodynamic limit, see Sect. 8.4. In particular, the mean time of escape τ_- is given by equation (8.3.20). Hence, in order to determine τ_- for the ion channel model, one has to calculate the quasipotential $\Phi_0(x)$, the prefactor $k(x)$ and the effective diffusivity $D(x)$. First note that the generator of the discrete Markov process is now a tridiagonal matrix with

$$A_{n,n-1}(x) = \omega_+(n - 1,x), A_{n,n}(x) = -\omega_+(n,x) - \omega_-(n), A_{n,n+1}(x) = \omega_-(n+1)$$

for $n = 0, 1, \ldots, N_{\max}$. In the case of the stochastic ion channel model, the eigenvalue equation (8.3.47a) takes the explicit form

$$p\left(\frac{n}{N_{\max}}f - g\right)R_0(n,x,p) + (N_{\max} - n + 1)\alpha R_0(n-1,x,p) \qquad (8.5.5)$$

$$- [n\beta + (N_{\max} - n)\alpha]R_0(n,x,p) + (n+1)\beta R_0(n+1,x,p) = \Lambda_0 R_0(n,x,p).$$

Consider the trial Binomial solution

$$R_0(n,x,p) = \frac{\Gamma(x,p)^n}{(N_{\max} - n)!n!}, \qquad (8.5.6)$$

which yields the following equation relating Γ and Λ_0

$$\frac{n\alpha}{\Gamma} + \Gamma\beta(N_{\max} - n) - \Lambda_0 - n\beta - (N_{\max} - n)\alpha = -p\left(\frac{n}{N_{\max}}f - g\right).$$

Collecting terms independent of n and terms linear in n yields the pair of equations

$$p = -\frac{N_{\max}}{f(x)}\left(\frac{1}{\Gamma(x,p)} + 1\right)(\alpha(x) - \beta(x)\Gamma(x,p)), \qquad (8.5.7)$$

and

$$\Lambda_0(x,p) = -N(\alpha(x) - \Gamma(x,p)\beta(x)) - pg(x). \qquad (8.5.8)$$

Eliminating Γ from these equation leads to a quadratic equation for Λ_0 of the form

$$\Lambda_0^2 + \sigma(x,p)\Lambda_0 - h(x,p) = 0, \qquad (8.5.9)$$

with

$$\sigma(x,p) = p(2g(x) - f(x)) + N_{\max}(\alpha(x) + \beta(x)),$$
$$h(x,p) = p[-N_{\max}\beta(x)g(x) + (N_{\max}\alpha(x) + pg(x))(f(x) - g(x))].$$

Along the zero energy surface $\Lambda_0(x,p) = 0$, we have $h(x,p) = 0$ which yields the pair of solutions

$$p = 0 \text{ and } p = p(x) \equiv N_{\max}\frac{\alpha(x)f(x) - (\alpha(x) + \beta)g(x)}{g(x)(f(x) - g(x))}. \qquad (8.5.10)$$

It follows that the nontrivial WKB quasipotential (or action) is given by

$$\Phi_0(x) = \int_{x_-}^{x} p(y)dy. \qquad (8.5.11)$$

In Fig. 8.5, we show solutions to Hamilton's equations in the (x,p)-plane, highlighting the zero energy maximum likelihood curve linking x_- and x_0. Note that $N\Phi(x_0)$, where $\Phi(x_0)$ is the area enclosed by the heteroclinic connection from x_- to x_0, gives the leading order contribution to $\log \tau_-$.

Keener and Newby [433] also calculate the subleading-order contributions to the mean escape time. First, the null eigenfunction $\eta_n(x) = S_0(n,x)$ of equation (8.3.15), which becomes

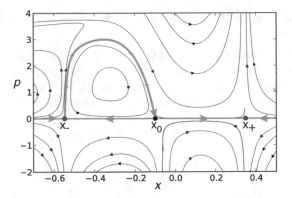

Fig. 8.5: Phase portrait of Hamilton's equations of motion for the ion channel model with Hamiltonian given by the Perron eigenvalue (8.5.8). (x and p are taken to be dimensionless.) The zero energy solution representing the maximum likelihood path of escape from x_- is shown as the gray curve. (The corresponding path from x_+ is not shown.) Parameter values are $V_{Ca} = 120$ mV, $V_L = -62.3$ mV, $g_{Ca} = 4.4$ mS/cm^2, $g_L = 2.2$mS/cm^2, and $\alpha(x) = \beta \exp[(x - v_1)/v_2]$ with $\beta = 0.8$ s^{-1}, $v_1 = -1.2$ mV, $v_2 = 18$ mV.

$$(N_{\max} - m)\alpha\eta_{m+1} - [(N_{\max} - m)\alpha + m\beta]\eta_m + m\beta\eta_{m-1}$$
$$= \Phi_0'(x)\left(\frac{m}{N_{\max}}f(x) - g(x)\right)\eta_m.$$

Trying a solution of the form $\eta_m(x) = \Gamma(x)^m$ yields

$$(N_{\max} - m)\alpha\Gamma - ((N_{\max} - m)\alpha + m\beta) + m\beta\Gamma^{-1} = \Phi_0'(x)\left(\frac{m}{N_{\max}}f(x) - g(x)\right).$$

Γ is then determined by canceling terms independent of m, which gives

$$\eta_n(x) = \left(\frac{(1 - a(x))g(x)}{a(x)(f(x) - g(x)))}\right)^n. \tag{8.5.12}$$

The prefactor $k(x)$ may now be determined using equations (8.3.16) and (8.3.17). Finally, the effective diffusion coefficient $D(x)$ is given by [433]

$$D(x_0) = \frac{f(x_0)^2\alpha(x_0)\beta}{N_{\max}(\alpha(x_0) + \beta)^3}. \tag{8.5.13}$$

Keener and Newby show that the resulting expression for the mean escape time agrees very well with Monte Carlo simulations of the stochastic model [433]. On the other hand, estimating the mean escape time using a QSS reduction leads to exponentially large errors when the resting state of the neuron is well below the firing threshold (small external input).

8.5.2 Excitability in the full stochastic Morris-Lecar model

There have been various extensions of the above analysis to more complicated ion channel models, including a stochastic version of the full Morris-Lecar model [614, 616] and a model of dendritic action potentials [117]. We will consider the former here. The deterministic model is given by the planar dynamical system

$$\frac{dx}{dt} = a(x)f_{Ca}(x) + wf_K(x) + f_L(x), \tag{8.5.14a}$$

$$\frac{dw}{dt} = \frac{w_\infty(x) - w}{\tau_w(x)}, \tag{8.5.14b}$$

where the membrane voltage is denoted by x rather than V, and

$$\tau_w(x) = \frac{1}{\alpha_w(x) + \beta_w(x)}, \qquad w_\infty(x) = \alpha_w(x)\tau_w(x).$$

As we showed in Sect. 3.4, the generation of action potentials in the deterministic model can be analyzed using a slow/fast analysis, since the dynamics of the recovery variable w (representing the fraction of open K^+ ion channels) is slow relative to that of the membrane voltage x. In the analysis of membrane voltage fluctuations, it was assumed that the potassium channel dynamics could be ignored during initiation of a spontaneous action potential (SAP). This corresponds to keeping the recovery variable w fixed. The resulting stochastic bistable model supported the generation of SAPs due to fluctuations in the opening and closing of fast Ca^{2+} channels. However, it turns out that this slow/fast analysis breaks down when the effects of K^+ channel noise are included [614]. That is, it is possible to generate a SAP due to fluctuations causing several K^+ channels to close simultaneously, effectively decreasing w, and thereby causing v to rise. This can be confirmed by numerically solving the full stochastic model. It follows that keeping w fixed in the stochastic model excludes the latter mechanism, and thus the resulting MFPT calculation underestimates the spontaneous rate of action potentials. In order to investigate the above phenomenon, it is necessary to consider the full stochastic ML model given by

$$C\frac{dV}{dt} = F(V,m,n) \equiv \frac{n}{N_0}f_{Ca}(V) + \frac{m}{M_0}f_K(V) + f_L(V) + I_{app}, \tag{8.5.15}$$

and each ion channel opens and closes according to the two-state Markov process

$$C \underset{\beta_i(V)}{\overset{\alpha_i(V)}{\rightleftharpoons}} O, \quad i = Ca,\ K \tag{8.5.16}$$

for N calcium channels and M potassium channels. An additional complication is that the full model is an excitable rather than a bistable system, so it is not straightforward to relate the generation of SAPs with a noise-induced escape problem. Nevertheless, Newby et al. [614] used WKB methods to identify the most probable paths

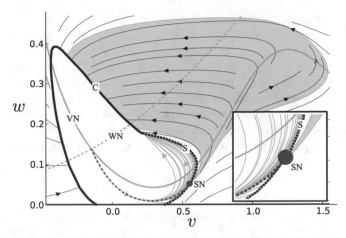

Fig. 8.6: Most probable paths of escape from the resting state of the stochastic ML model calculated using a WKB approximation [614]. All paths of escape that enter the framed blue region
represent large excursions in state space and coincide with SAPs. All of the SAP trajectories are
initially bunched together (red dashed curve) until they cross the bottleneck or metastable saddle
node (SN). Curves that don't pass through SN are bounded by a curve (S) that acts like a stochastic
separatirix. Also shown are a caustic (C) where paths of escape intersect, the v nullcline (VN), and
the w nullcline (WN). The resting state is surrounded by an effective "basin of attraction" bounded
by C and S. Here $N = M = 40$ and $\varepsilon = 0.1$. Other parameter values can be found in [614].

of escape from the resting state and obtained the following results, which are illustrated in Fig. 8.6:

(i) Most probable paths of escape dip significantly below the resting value for w,
indicating a breakdown of the deterministic slow/fast decomposition.

(ii) Escape trajectories   all pass through a narrow region of state space (bottleneck
or stochastic saddle node) so that although there is no well-defined separatrix for an
excitable system, suggesting that it is possible to formulate an escape problem by
determining the mean FPT to reach the bottleneck from the resting state.

8.6 Brownian functionals

An important quantity in the mathematical theory of stochastic processes is the occupation time [513], which is the time spent by a Brownian motion above the origin
within a time window of size t. That is, given the Brownian motion $X(t) \in \mathbb{R}$, the
occupation time $\mathcal{T}$ is

$$\mathcal{T} = \int_0^t \Theta(X(\tau))d\tau, \tag{8.6.1}$$

where $\Theta(X)$ denotes the Heaviside function. In addition to being a fundamental quantity in the mathematical theory of random walks, occupation times have figured prominently in a variety of physical applications under the alternative name of residence times. One relevant example in molecular biology is fluorescent imaging [7]. This involves a single fluorescent particle diffusing under the objective of a confocal microscope. Every time it enters the focus of the laser beam, it is excited and emits photons, so that the total number of emitted photons is proportional to the mean residence time of the molecule in the laser beam's cross-section. If V denotes the volume occupied by the beam, then the residence time is defined according to

$$\mathcal{T} = \int_0^t I_V(X(\tau)))d\tau, \tag{8.6.2}$$

where $X(t) \in \mathbb{R}^3$ is now three-dimensional Brownian motion, $I_V(x)$ denotes the indicator function of the set $V \subset \mathbb{R}^3$, that is, $I_V(x) = 1$ if $x \in V$ and is zero otherwise. (Note that for one-dimensional (1D) motion, $\Theta(x) = I_{\mathbb{R}^+}(x)$.)

A related quantity is the local time [513], which characterizes the amount of time that a diffusion process such as Brownian motion spends in the neighborhood of a point in space. Given the Brownian motion $X(t) \in \mathbb{R}$, let $\mathcal{T}(A,t)$ denote the occupation time of the set $A \subset \mathbb{R}$ during the time interval $[0,t]$

$$\mathcal{T}(A,t) = \int_0^t I_A(X(\tau))d\tau. \tag{8.6.3}$$

From this definition, the local time density $\mathcal{T}(a,t)$ at a point $a \in \mathbb{R}$ is defined by setting $A = [a - \varepsilon, a + \varepsilon]$ and taking

$$\mathcal{T}(a,t) = \lim_{\varepsilon \to 0^+} \frac{1}{2\varepsilon} \int_0^t I_{[a-\varepsilon,a+\varepsilon]}(X(s))ds. \tag{8.6.4}$$

We thus have the following formal representation of the local time density:

$$\mathcal{T}(a,t) = \int_0^t \delta(X(\tau) - a)d\tau, \tag{8.6.5}$$

where $\mathcal{T}(a,t)da$ is the amount of time the Brownian particle spends in the infinitesimal interval $[a, a + da]$. Note, in particular, that

$$\int_{-\infty}^{\infty} \mathcal{T}(a,t)da = \int_{-\infty}^{\infty} \int_0^t \delta(X(\tau) - a)d\tau da = \int_0^t d\tau = t.$$

In probability theory, local time plays an important role in the path-wise formulation of reflected Brownian motion [565]. For the sake of illustration, consider a Wiener process confined to the interval $[0,L]$ with reflecting boundaries at $x = 0, L$. Sample paths are generated from the SDE

$$dX(t) = \sigma dW(t) + \frac{\sigma^2}{2}d\mathcal{T}(0,t) - \frac{\sigma^2}{2}d\mathcal{T}(L,t), \tag{8.6.6}$$

where $\mathcal{T}(x,t)$ is given by equation (8.6.5) so that, formally speaking,

$$dT(0,t) = \delta(X(t))dt, \quad dT(L,t) = \delta(X(t)-L)dt.$$

In other words, each time the Brownian particle hits the end at $x = 0$ ($x = L$) it is given an impulsive kick to the right (left).

Occupation and local times are two examples of a Brownian functional. More generally, suppose that $X(t)$ is a solution to the Ito SDE

$$dX = A(X)dt + \sigma dW(t), \tag{8.6.7}$$

where $W(t)$ is a Wiener process. A Brownian functional over a fixed time interval $[0,T]$ is defined as a random variable $\mathcal{U}_T$ given by [552, 711]

$$\mathcal{U}_T = \int_0^T U(X(\tau))d\tau, \tag{8.6.8}$$

where $U(x)$ is some prescribed function or distribution such that $\mathcal{U}_T$ has positive support and $X(0) = x_0$ is fixed. Since $X(\tau)$ is a continuous stochastic process, it follows that each realization of a Brownian path will typically yield a different value of $\mathcal{U}_T$, which means that $\mathcal{U}_T$ will be distributed according to some probability density function (pdf) $\mathcal{P}(u,T|x_0,0))$. The statistical properties of a Brownian functional can be analyzed using path integrals, and leads to the well-known Feynman-Kac formula [423]. Brownian functionals are finding increasing applications in probability theory, finance, data analysis, and the theory of disordered systems. It is also possible to extend the notion of a Brownian functional to include stochastic resetting (defined in Sect. 7.5) [214, 576].

Feynman-Kac formula. Since $\mathcal{U}_T \geq 0$, we can introduce the following Laplace transform of $\mathcal{P}(u,t|x_0,0)$:

$$Q_s(x_0,t|0) = \mathbb{E}[e^{-s\mathcal{U}_t}] = \int_0^\infty e^{-su}\mathcal{P}(u,t|x_0,0)du. \tag{8.6.9}$$

The latter can be evaluated by using a path integral representation

$$\mathcal{P}(u,t|x_0,0) = \int_{-\infty}^\infty \left[\int_{x(0)=x_0}^{x(t)=x} \delta\left(u - \int_0^t U(x(\tau))d\tau\right) P[x]\mathcal{D}[x] \right] dx,$$

where

$$P[x] = \exp\left[-\int_0^\tau \frac{(\dot{x}-A(x))^2}{2\sigma^2} dt \right]. \tag{8.6.10}$$

Since only the initial point x_0 is fixed, we are also integrating with respect to the final position x. That is

$$Q_s(x_0, t|0) = \int_0^\infty e^{-su} \int_{-\infty}^\infty \left[\int_{x(0)=x_0}^{x(t)=x} \delta\left(u - \int_0^t U(x(\tau))d\tau\right) P[x]\mathcal{D}[x] \right] dx\,du$$

$$= \int_{-\infty}^\infty \left[\int_{x(0)=x_0}^{x(t)=x} \exp\left(-s\int_0^t U(x(\tau))d\tau\right) P[x]\mathcal{D}[x] \right] dx. \qquad (8.6.11)$$

In order to derive a Feynman-Kac equation for Q_s we take the initial time to be $t-\tau$, and consider how Q_s varies under the shift $\tau \to \tau + \Delta\tau$. That is

$$Q_s(x_0, t|t-\tau-\Delta\tau) = \int_{\mathbb{R}} dx \left\langle \exp\left(-s\int_{t-\tau-\Delta\tau}^t U(x(s))ds\right)\right\rangle_{x(t-\tau-\Delta\tau)=x_0}^{x(t)=x}$$

$$\approx e^{-sU(x_0)\Delta\tau} \int p(\Delta W) \left\langle \exp\left(-s\int_{t-\tau}^t U(x(s))ds\right)\right\rangle_{x(t-\tau)=x_0+\Delta x_0}^{x(t)=x} d\Delta W.$$

We have split the time interval $[t-\tau-\Delta\tau, t]$ into two parts $[t-\tau, t]$ and $[t-\tau-\Delta\tau, t]$ and introduced the intermediate state $x(t-\tau) = x_0 + \Delta x_0$ with Δx_0 determined by

$$\Delta x_0 = A(x_0)\Delta t + \sigma \Delta W(t).$$

It follows that

$$Q_s(x_0, t|t-\tau-\Delta\tau) = \exp(-sU(x_0)\Delta t) \int p(\Delta W) Q_s(x_0 + \Delta x_0, t|t-\tau)d\Delta W$$

$$= e^{-sU(x_0)\Delta t} \left(Q_s(x_0, t|t-\tau) + \langle\Delta x_0\rangle \frac{\partial}{\partial x_0} Q_s(x_0, t|t-\tau) \right.$$

$$\left. + \frac{\langle\Delta x_0^2\rangle}{2} \frac{\partial^2}{\partial x^2} Q_s(x_0, t|t-\tau) + \dots \right).$$

Using the fact that

$$\lim_{\Delta t\to 0} \frac{\langle\Delta x_0\rangle}{\Delta t} = A(x_0), \quad \lim_{\Delta t\to 0} \frac{\langle(\Delta X_0)^2\rangle}{\Delta t} = \sigma^2, \qquad (8.6.12)$$

taking the limit $\Delta t \to 0$ and setting $\tau = t$, we obtain the Feynman-Kac formula in the form of a modified backward Fokker-Planck equation:

$$\frac{\partial Q_s}{\partial t} = \frac{\sigma^2}{2} \frac{\partial^2 Q_s}{\partial x_0^2} + A(x_0)\frac{\partial Q_s}{\partial x_0} - sU(x_0)Q_s. \qquad (8.6.13)$$

This is supplemented by the initial condition $Q_s(x_0, 0) = 1$. (The s-dependent differential operator on the right-hand side is also known as a tilted generator.)

Calculation of occupation time. Consider pure brownian motion with $\sigma = 1$ and $A(x) = 0$. In the case of the occupation time in $\mathbb{R}^+$, we have $U(x) = \Theta(x)$ and equation (8.6.13) reduces to

$$\frac{\partial Q_s}{\partial t} = \frac{1}{2}\frac{\partial^2 Q_s}{\partial x^2} - s\Theta(x)Q_s. \qquad (8.6.14)$$

We have dropped the subscript on x_0. The occupation time for pure Brownian motion was originally analyzed by Levy [513] and has also been extended to Brownian motion in the presence of ordered and spatially disordered potentials [711] and randomly switching environments [130]. Laplace transforming the backward FPE (8.6.14) with respect to time t yields the ODE

$$z\widetilde{Q}(x;s,z) - 1 = \frac{1}{2}\widetilde{Q}''(x;s,z) - s\Theta(x)\widetilde{Q}(x;s,z), \qquad (8.6.15)$$

with $\widetilde{Q}' = d\widetilde{Q}/dx$ and

$$\widetilde{Q}(x;s,z) = \int_0^\infty e^{-z\tau}Q_s(x,\tau)d\tau = \int_0^\infty \int_0^\infty e^{-z\tau-su}\mathcal{P}(u,\tau|x,0)dud\tau. \qquad (8.6.16)$$

Following the standard analysis of 1D Green's functions (see Box 6C), we have to solve equation (8.6.15) separately for $\widetilde{Q} = q_+(x;s,z)$ in $x \in (0,\infty)$ and $\widetilde{Q} = q_-(x;s,z)$ in $x \in (-\infty,0)$, and then match the solutions at $x = 0$. That is, $q_\pm$ satisfy the equations

$$\frac{1}{2}q_+'' - (z+s)q_+ = -1, \; x > 0; \quad \frac{1}{2}q_-'' - zq_- = -1, \; x < 0. \qquad (8.6.17)$$

The matching conditions are obtained by (i) imposing continuity of the solution at $x = 0$ and (ii) integrating equation (8.6.15) across $x = 0$:

$$q_+(0;s,z) = q_-(0;s,z) = U(s,z), \quad q_+'(0;s,z) = q_-'(0;s,z)$$

for $U(s,z) = \widetilde{Q}(0;s,z)$. In order to determine the far-field boundary conditions for $x \to \pm\infty$, we note that if a particle starts at $x = \pm\infty$ then it will never cross the origin a finite time τ in the future, that is,

$$\mathcal{P}(u,\tau|\infty,0) = \delta(t-u), \quad \mathcal{P}(u,\tau|-\infty,0) = \delta(u).$$

Substituting this into the definition of $\widetilde{Q}$ shows that

$$q_+(\infty;s,z) = \frac{1}{z+s}, \quad q_-(-\infty;s,z) = \frac{1}{z}. \qquad (8.6.18)$$

If we now perform the shifts

$$q_+(x;s,z) = \frac{1}{z+s} + B_+y_+(x;s,z), \quad q_-(x;s,z) = \frac{1}{z} + B_-y_-(x;s,z), \qquad (8.6.19)$$

then we obtain the homogeneous equations

$$\frac{1}{2}y''_+ - (z+s)y_+ = 0, \ x > 0; \quad \frac{1}{2}y''_- - zy_- = 0, \ x < 0, \tag{8.6.20a}$$

with $y_+(\infty;s,z) = 0$ and $y_-(-\infty;s,z) = 0$. The constants $B_\pm$ can be found by imposing the matching conditions, which take the explicit form

$$\frac{1}{z+s} + B_+ y_+(0;s,z) = \frac{1}{z} + B_- y_-(0;s,z) = U(s,z), \tag{8.6.21a}$$

$$B_+ y'_+(0;s,z) = B_- y'_-(0;s,z). \tag{8.6.21b}$$

Setting

$$\lambda_\pm(s,z) = \frac{y'_\pm(0;s,z)}{y_\pm(0;s,z)},$$

equation (8.6.21) can be used to express $U(s,z)$ as

$$U(s,z) = \frac{L_1(s,z)}{z} + \frac{L_2(s,z)}{z+s}, \tag{8.6.22}$$

where (see Ex. 8.9)

$$L_1(s,z) = -\frac{\lambda_-(s,z)}{\lambda_+(s,z) - \lambda_-(s,z)}, \quad L_2(s,z) = \frac{\lambda_+(s,z)}{\lambda_+(s,z) - \lambda_-(s,z)}. \tag{8.6.23}$$

Note that $L_1(s,z) + L_2(s,z) = 1$. Finally, explicitly solving equation (8.6.20) yields

$$\lambda_+(s,z) = -\sqrt{2(z+s)}, \quad \lambda_-(s,z) = \sqrt{2z}, \tag{8.6.24}$$

so that combining equations (8.6.22) and (8.6.23), and using the definition (8.6.16), we have

$$U(s,z) \equiv \int_0^\infty \int_0^\infty e^{-z\tau - su} \mathcal{P}(u,\tau|0,0)du\,d\tau = \frac{1}{\sqrt{z(s+z)}}. \tag{8.6.25}$$

Inverting the double Laplace transform with respect to s and then z recovers the well known "arcsine" law [513] for the probability density of the occupation time for pure Brownian motion starting at the origin

$$\mathcal{P}(u,\tau|0,0) = \frac{1}{\pi\sqrt{u(\tau-u)}}, \quad 0 < u < \tau. \tag{8.6.26}$$

Calculation of local time. We now repeat the above analysis for $U(x) = \delta(x)$ so that $\mathcal{T}$ is the local time density at the origin. Laplace transforming the backward FP equation (8.6.13) with respect to time t yields the ODE

$$z\tilde{Q}(x;s,z) - 1 = \frac{1}{2}\tilde{Q}''(x;s,z) - s\delta(x)\tilde{Q}(x;s,z). \tag{8.6.27}$$

If a particle starts at $x = \pm\infty$ then it will never cross the origin a finite time τ in the future, that is, $P(u, \tau | \pm\infty, 0) = \delta(u)$. Substituting this into the definition of $\widetilde{Q}$ shows that $\widetilde{Q}(\pm\infty; s, z) = z^{-1}$. Following the standard analysis of 1D Green's functions, we have to solve equation (8.6.27) separately for $\widetilde{Q} = q_+(x; s, z) + z^{-1}$ in $x \in (0, \infty)$ and $\widetilde{Q} = q_-(x; s, z) + z^{-1}$ in $x \in (-\infty, 0)$, and then match the solutions at $x = 0$. (We have performed a uniform shift of the solutions for convenience.) That is, $q_\pm$ satisfy the equations

$$\frac{1}{2}q_\pm''(x; s, z) - z q_\pm(x; s, z) = 0, \tag{8.6.28}$$

with corresponding boundary conditions $q_\pm(\pm\infty; s, z) = 0$. Note that these equations are independent of the Laplace variable s. The s-dependence emerges from the matching conditions, which are obtained by (i) imposing continuity of the solution at $x = 0$ and (ii) integrating equation (8.6.27) across $x = 0$

$$q_+(0; s, z) = q_-(0; s, z) = U(s, z) - z^{-1},$$
$$q_+'(0; s, z) - q_-'(0; s, z) = 2sU(s, z)$$

for some unknown U. Rearranging this pair of equations shows that

$$U = U(s, z) = \frac{\lambda(z)}{z(s + \lambda(z))}, \tag{8.6.29}$$

where

$$\lambda(z) = \frac{q_-'(0; s, z)/q_-(0; s, z) - q_+'(0; s, z)/q_+(0; s, z)}{2}. \tag{8.6.30}$$

Using the fact that $\lambda(z)$ is independent of s and $\widetilde{Q}(0; s, z) = U(s, z)$, we can perform the inverse Laplace transform with respect to s to give

$$F(u, z) \equiv \int_0^\infty e^{-z\tau} P(u, \tau | 0, 0) d\tau = \frac{\lambda(z)}{z} e^{-\lambda(z)u}. \tag{8.6.31}$$

The function $F(u, z)$ can be calculated explicitly in the case of pure Brownian motion. In particular, equation (8.6.28) has the solution

$$q_\pm(x; s, z) = -\frac{sU(s, z)}{\sqrt{2z}} e^{\mp\sqrt{2z}x}. \tag{8.6.32}$$

This implies that $\lambda(z) = \sqrt{2z}$ and

$$F(u, z) = \sqrt{\frac{2}{z}} e^{-\sqrt{2z}u}.$$

Inverting the inverse Laplace transform with respect to z then shows that the distribution of local times around the origin is a Gaussian

$$P(u,t|0,t-\tau) = \sqrt{\frac{2}{\pi\tau}}e^{-u^2/2\tau}. \tag{8.6.33}$$

Hence, the first and second moments of the local time density (starting at $x = 0$) are

$$\langle T(\tau)\rangle = \int_0^\infty uP(u,\tau|0,0)du = \sqrt{\frac{2\tau}{\pi}}, \tag{8.6.34a}$$

$$\langle T^2(\tau)\rangle = \int_0^\infty u^2 P(u,\tau|0,0)du = \tau. \tag{8.6.34b}$$

8.7 Large deviation theory

In this final section, we present a basic introduction to a mathematical approach to rare events, namely, large deviation theory. We follow closely the notes of Touchette [810], which are a shorter version of the review [809].

8.7.1 Sums of random variables and generalized central limit theorems

Let us begin with a simple motivating example of large deviations. Consider the sum (sample mean)

$$S_n = \frac{1}{n}\sum_{i=1}^n X_i,$$

where $X_1, X_2, \ldots$ is an i.i.d. sequence of random variables generating from the probability density $p(X)$ with $X \in \mathbb{R}$. The joint probability density function (pdf) is

$$p(X_1,\ldots,X_n) = \prod_{j=1}^n p(X_j).$$

The corresponding pdf of the sum S_n is obtained as follows:

$$p_n(s) = \mathbb{P}[S_n = s] = \int_{-\infty}^\infty dx_1 \ldots \int_{-\infty}^\infty dx_n \delta(\sum_{i=1}^n x_i - ns)p(x_1,\ldots,x_n). \tag{8.7.1}$$

Introducing the Fourier representation of the Dirac delta function, which is $\delta(x) = \int_{-\infty}^\infty e^{ikx}dk/2\pi$, and using the product decomposition of the joint pdf

$$p_n(s) = \int_{-\infty}^\infty \frac{dk}{2\pi}e^{-ikns}\prod_{j=1}^n\left(\int_{-\infty}^\infty e^{ikx_j}p(x_j)dx_j\right) = \int_{-\infty}^\infty \widehat{p}(k)^n e^{-ikns}\frac{dk}{2\pi}. \tag{8.7.2}$$

In the case of a Gaussian pdf with mean μ and variance σ^2,

$$\widehat{p}(k) = e^{ik\mu}e^{-\sigma^2 k^2/2},$$

and in the case of an exponential pdf on $[0,\infty)$ with mean μ,

$$\widehat{p}(k) = \frac{\mu}{1 - i\mu k}.$$

Substituting into the integral expression for $p_n(s)$ then gives

$$p_n(s) = \int_{-\infty}^{\infty} e^{ikn(\mu-s)}e^{-n\sigma^2 k^2/2}\frac{dk}{2\pi} = \frac{1}{\sqrt{2\pi n\sigma^2}}e^{-n(\mu-s)^2/2\sigma^2}$$

for the Gaussian, and

$$p_n(s) = \int_{-\infty}^{\infty} e^{-ikns}\left(\frac{\mu}{1-i\mu k}\right)^n \frac{dk}{2\pi} = \frac{\mu^n}{\mu}\oint \frac{e^{-izns/\mu}}{[1-iz]^n}\frac{dz}{2\pi}$$

$$= \frac{\mu^{n-1}}{(n-1)!}\left(\frac{-ns}{\mu}\right)^{n-1}e^{-ns/\mu} = \frac{n^{n-1}}{(n-1)!}\left(\frac{s}{\mu}\right)^n e^{-ns/\mu}$$

for the exponential. We have closed the contour in the lower-half complex z-plane and used the following residue theorem for an analytic function $f(z)$:

$$f^{(n-1)}(z_0) = \frac{1}{(n-1)!}\oint \frac{f(z)}{(z-z_0)^n}\frac{dz}{2\pi i}.$$

We now note that in both cases the leading order behavior in n for large n can be expressed as

$$p_n(s) \approx e^{-nI(s)}, \quad I(s) = \frac{(s-\mu)^2}{2\sigma^2}, \quad s \in \mathbb{R} \tag{8.7.3}$$

for the Gaussian pdf and

$$I(s) = -\frac{1}{n}\{(n-1)\ln n - \ln[(n-1)!] - n\ln(s/\mu) - ns/\mu\}$$

$$= \frac{s}{\mu} - 1 - \ln(s/\mu), \quad s \geq 0 \tag{8.7.4}$$

for the exponential pdf (after applying Stirling's formula). In both cases $I(s) \geq 0$ and $I(s) = 0$ only when $s = \mu = \mathbb{E}[X]$. Since the pdf of S_n is normalized, it becomes more and more concentrated around $s = \mu$ as $n \to \infty$, that is, $p_n(s) \to \delta(s-\mu)$ in the large-n limit. The leading order exponential form $e^{-nI(s)}$ found for the Gaussian and exponential pdfs is the fundamental property of large deviation theory, which is known as the large deviation principle (see Sect. 8.7.2).

Generalized central limit theorems (CLTs). It is clear that both of the above examples satisfy the central limit theorem, since in the large-n limit we simply Taylor expand $I(s)$ to second order in $s - \mu$ using the fact that $I(\mu) = 0$ and $I'(\mu) = 0$. In the case of the exponential distribution, we find that

$$p_n(s) \sim e^{-n(s-\mu)^2/2\mu^2},$$

so that $(s_n - \mu)/\sqrt{n}$ converges in distribution to a Gaussian variable with zero mean and variance μ^2. Recall that the CLT applies to any distribution whose tail decays sufficiently fast so that the mean and variance are finite. It turns out that one can obtain a generalized CLT for certain distributions with large tails (infinite means) using a modified scaling rule. We will develop this idea following closely Ch. 6 of the book by Amir [13].

Suppose that n i.i.d. variables are drawn from a probability distribution $p(x)$ with positive support ($x \geq 0$) whose tails fall off as a power law, that is, beyond a point x^*, $p(x) = A/x^{1+\mu}$ with $\mu > 0$ (for normalizability). If $\mu > 2$ then the classical CLT holds and the sum converges to a Gaussian. Here we focus on the case $0 < \mu < 1$, for which neither the mean nor variance are finite. We proceed by investigating the Laplace transform of $p(x)$, which we write as

$$\widetilde{p}(z) = \int_0^\infty p(x) \left[1 + e^{-zx} - 1\right] dx = 1 + \int_0^\infty p(x) \left[e^{-zx} - 1\right] dx. \tag{8.7.5}$$

Exploring the tail of the distribution $p(x)$ corresponds to considering the small-z limit of the Laplace transform. The integral I is separated into two parts $I = I_1 + I_2$, with

$$I_1 = \int_0^{x^*} p(x) \left[e^{-zx} - 1\right] dx, \quad I_2 = \int_{x^*}^\infty p(x) \left[e^{-zx} - 1\right] dx. \tag{8.7.6}$$

Using the inequality $|e^{-zx} - 1| \leq zx$, it follows that

$$|I_1| \leq \int_0^{x^*} p(x) zx dx \leq zx^*.$$

Substituting the power law for $p(x)$ in the second integral gives

$$I_2 = \int_{x^*}^\infty \frac{A}{x^{1+\mu}} \left[e^{-zx} - 1\right] dx = -\frac{A}{\mu} x^{-\mu} \left[e^{-zx} - 1\right]_{x^*}^\infty - \int_{x^*}^\infty \frac{A}{\mu} x^{-\mu} z e^{-zx} dx,$$

after integration by parts. The first term on the right-hand side has the upper bound $A(x^*)^{-\mu} zx^*/\mu$, which is at most linear in z. The second term can be approximated by a gamma function for small z. That is, under the change of variables $zx = y$, we have

$$\frac{Az^\mu}{\mu} \int_{x^*z}^\infty y^{-\mu} e^{-y} dy \approx \frac{Az^\mu}{\mu} \Gamma(1-\mu)$$

as $x^* z \to 0$. Combining the various results implies that for $\mu < 1$,

$$\widetilde{p}(z) \approx 1 - Cz^\mu + O(z), \quad C = \frac{A}{\mu}\Gamma(1-\mu). \tag{8.7.7}$$

We see that the power-law behavior in the tail of $p(x)$ determines the behavior of $\widetilde{p}(z)$ for small z. Moreover, if we rescale the random variable by setting $y = ax$ for some positive constant a, then $\rho(y) = p(y/a)/a$ and

$$\widetilde{\rho}(z) = \int_0^\infty \frac{1}{a} p(y/a) e^{-zy} dy = \int_0^\infty p(x) e^{-zax} = \widetilde{p}(az).$$

Hence, a broader distribution in real space (for $a > 1$) corresponds to a narrower distribution in Laplace space. We now use this to explore the effects of rescaling on the Laplace transform of the sum of N i.i.d. variables drawn from $p(x)$. First note that

$$\widetilde{p}_N(z) = \int_0^\infty dX e^{-zX} \left(\prod_{j=1}^N \int_0^\infty dx_j p(x_j) \right) \delta(X - \sum_{i=1}^N x_i) = \widetilde{p}(z)^N. \tag{8.7.8}$$

Suppose that the sum X is scaled by $a = N^{-\alpha}$ and denote the distribution of Y by $P_N(y)$. It follows that

$$\widetilde{P}_N(z) = \widetilde{p}(z/N^\alpha)^N \approx \left[1 - \frac{Cz^\mu}{N^{\alpha\mu}} \right]^N. \tag{8.7.9}$$

Hence, choosing $\alpha = 1/\mu$ and using the definition of the exponential, we see that the Laplace transformed distribution of the rescaled sum converges to the limit

$$\widetilde{P}_\infty(z) = e^{-Cz^\mu}. \tag{8.7.10}$$

This is an example of a generalized CLT, whereby the rescaled sum of i.i.d. random variables

$$X = \frac{X_1 + \ldots + X_N}{N^{1/\mu}},$$

converges in distribution in the limit $N \to \infty$. The limiting distribution is given by the inverse transform of equation (8.7.10) in the case of a heavy-tail given by a power law. (For the classical CLT the scaling is given by $\mu = 2$.)

Lévy stable distributions. One noticeable feature of the limiting distribution given by the inverse Laplace transform of equation (8.7.10) is that when two variables are independently drawn from this distribution, the Laplace transform of the distribution of the sum is given by e^{-2Cz^μ}. That is, the resulting distribution is identical up to a rescaling. This is an example of a so-called Lévy stable distribution. More generally, if the sum of a large number N of i.i.d. variables converges in distribution to a Lévy stable distribution, then the sum of $2N$ such variables converges to the same distribution up to a rescaling (or possibly a shift).

Although the probability distribution for a general stable distribution cannot be constructed analytically, a general formula for the corresponding characteristic function does exist. That is, a random variable X is called stable if its characteristic function can be written as

$$\widehat{p}(\omega) = \frac{1}{2\pi} \int_{-\infty}^{\infty} e^{i\omega x} p(x) dx = \exp\left(i\omega\mu - |c\omega|^\alpha (1 - i\beta \operatorname{sgn}(\omega)\Phi)\right), \quad (8.7.11)$$

where

$$\Phi = \begin{cases} \tan(\pi\alpha/2) & \text{for } \alpha \neq 1 \\ -\frac{2}{\pi}\log|\omega| & \text{for } \alpha = 1 \end{cases}. \quad (8.7.12)$$

Here $\mu \in \mathbb{R}$ is a shift parameter and $c \in \mathbb{R}^+$ is a scale parameter that is a measure of the width of the distribution. On the other hand, $\beta \in [-1, 1]$ is a measure of asymmetry and α is the index or exponent of the distribution that specifies the asymptotic behavior. The characteristic function (8.7.11) gives a stable distribution, since the sum of two random variables equals the product of the two corresponding characteristic functions, which leads to the same values of the shape parameters α, β, but possibly different values of μ and c. A well-known example of a stable distribution is the Lévy distribution:

$$p(x) = \sqrt{\frac{C}{2\pi}} x^{-3/2} e^{-C/2x}, \quad x \geq 0. \quad (8.7.13)$$

The corresponding characteristic function or Fourier transform is

$$\widehat{p}(\omega) = e^{-\sqrt{-2iC\omega}}, \quad (8.7.14)$$

which is a special case of (8.7.11) with $\alpha = 1/2$ and $\beta = 1$.

8.7.2 Large deviation principle

LDPs arise over a much wider range of stochastic process then sums of i.i.d. random variables. Following Touchette [810], we will avoid the technical aspects of the rigorous formulation of the large deviation principle. For our purposes, a random variable S_n or its pdf $p(S_n)$ satisfies a large deviation principle (LDP) if the following limit exists:

$$\lim_{n \to \infty} -\frac{1}{n} \ln[p_n(s)] = I(s), \quad (8.7.15)$$

with $I(s)$ the so-called rate function. A more rigorous definition involves probability measures on sets rather than in terms of pdfs, and gives lower and upper bounds on these probabilities rather than a simple limit [809, 823]. One of the main goals of large deviation theory is to identify stochastic processes that satisfy an LDP, and to develop analytical (and numerical) methods for determining the associated rate function. Touchette [810] identifies three approaches to establishing an LDP for a random variable S_n:

1. Direct method: Derive an explicit expression for $p(S_n)$ and show that it has the form of an LDP. (This was the method used to derive the LDP for Gaussian and exponential sample means.)
2. Indirect method: Calculate certain functions of S_n that can be used to infer that S_n satisfies an LDP. One example is to use a generating function and apply the Gartner-Ellis Theorem (see below).
3. Contraction method: Relate S_n to another random variable A_n, say, that is known to satisfy an LDP and use this to derive an LDP for S_n.

Varadhan and Gartner-Ellis theorems. In order to discuss in more detail the indirect method, we need to introduce two theorems. Again, we will sacrifice mathematical rigor for ease of presentation. The first theorem is due to Varadhan, and is concerned with the calculation of the functional expectation

$$W_n[f] = \mathbb{E}[e^{nf(S_n)}] = \int_{\mathbb{R}} p_n(s) e^{nf(s)} ds. \tag{8.7.16}$$

If S_n satisfies an LDP with rate function $I(s)$, then for large n

$$W_n[f] \approx \int_{\mathbb{R}} e^{n[f(s)-I(s)]} ds \approx e^{n \sup_s [f(s)-I(s)]}, \tag{8.7.17}$$

after making a saddle-point approximation. We have exploited the fact that the remaining corrections are sub-exponential in n. We now introduce a functional $\lambda[f]$ such that

$$\lambda[f] \equiv \lim_{n \to \infty} \frac{1}{n} \ln W_n[f] = \sup_{s \in \mathbb{R}} \{f(s) - I(s)\}. \tag{8.7.18}$$

This is a statement of the Varadhan Theorem [823], which was originally applied to bounded functions f, but also holds for unbounded functions.

Consider the special case $f(s) = ks$ with $k \in \mathbb{R}$. The functional $W_n[f]$ reduces to a scaled generating function of S_n, that is,

$$W[k_n] = \mathbb{E}[e^{knS_n}],$$

and $\lambda[f]$ reduces to the scaled cumulant generating function for S_n, $\lambda(k)$, with

$$\lambda(k) \equiv \lim_{n \to \infty} \frac{1}{n} \ln \mathbb{E}[e^{nkS_n}] = \sup_{s \in \mathbb{R}} \{ks - I(s)\}. \tag{8.7.19}$$

Hence, the Varadhan Theorem implies that if S_n satisfies an LDP with rate function $I(s)$, then the scaled cumulant generating function $\lambda(k)$ of S_n is the Legendre-Fenchel transform of $I(s)$. It turns out that this result can be inverted, which is the basis of the Gartner-Ellis Theorem [245]: if $\lambda(k)$ is differentiable, then S_n satisfies an LDP and the corresponding rate function $I(s)$ is given by the Legendre-Fenchel transform of $\lambda(k)$

$$I(s) = \sup_{k \in \mathbb{R}} \{ks - \lambda(k)\}. \tag{8.7.20}$$

The usefulness of this result is that one can often calculate $\lambda(k)$ without knowing the full pdf $p(S_n)$. However, it is important to note that not all rate functions can be calculated using the Gartner-Ellis Theorem.

Some properties of $I(s)$ and $\lambda(k)$. We first discuss some properties of $\lambda(k)$. By definition, $\lambda(0) = 0$, since a pdf is normalized to unity. Moreover

$$\lambda'(0) = \lim_{n\to\infty} \frac{\mathbb{E}[S_n e^{nkS_n}]}{\mathbb{E}[e^{nkS_n}]}\bigg|_{k=0} = \lim_{n\to\infty} \mathbb{E}[S_n], \qquad (8.7.21)$$

assuming $\lambda'(0)$ exists. Similarly

$$\lambda''(0) = \lim_{n\to\infty} n \operatorname{Var} S_n. \qquad (8.7.22)$$

Another important feature of $\lambda(k)$ is that it is always convex. This follows from an integral version of Holder's inequality:

$$\sum_i |y_i z_i| \le \left(\sum_j |y_j|^{1/p}\right)^p \left(\sum_j |z_j|^{1/q}\right)^q, \qquad 0 \le p, q \le 1, \quad p + q = 1.$$

Setting $y_a = e^{n\alpha k_1 a}, z_a = e^{n(1-\alpha)k_2 a}$ and $p = \alpha$, we have

$$\left\langle e^{nk_1 S_n}\right\rangle^\alpha + \left\langle e^{nk_2 S_n}\right\rangle^{1-\alpha} \ge \left\langle e^{n(\alpha k_1 + (1-\alpha)k_2)S_n}\right\rangle.$$

Taking the logarithm of both sides gives

$$\alpha \ln \left\langle e^{nk_1 S_n}\right\rangle + (1-\alpha)\ln \left\langle e^{nk_2 S_n}\right\rangle \ge \ln \left\langle e^{n(\alpha k_1 + (1-\alpha)k_2)S_n}\right\rangle.$$

Finally, multiplying both sides by $1/n$ and taking the limit $n \to \infty$ shows that

$$\alpha \lambda(k_1) + (1-\alpha)\lambda(k_2) \ge \lambda(\alpha k_1 + (1-\alpha)k_2), \qquad (8.7.23)$$

which establishes that $\lambda(k)$ is convex. It can also be shown from properties of the Legendre-Fenchel transform that rate functions obtained from the Gartner-Ellis Theorem are strictly convex. We thus infer one limitation of the Gartner-Ellis Theorem, namely, it cannot generate non-convex rate functions with more than one minimum.

Suppose that a rate function can be determined from the Legendre-Fenchel transform of $\lambda(k)$ according to the Gartner-Ellis theorem. Differentiability and convexity of $\lambda(k)$ then imply that the Legendre-Fenchel transform reduces to a standard Legendre transform, that is

$$I(s) = k(s)s - \lambda(k(s)), \qquad (8.7.24)$$

with $k(s)$ the unique root of $\lambda'(k) = s$. Differentiating the Legendre transform equation with respect to s gives

$$I'(s) = k'(s)s + k(s) - \lambda'(k(s))k'(s) = k(s), \quad I''(s) = k'(s) = \frac{1}{\lambda''(k)}.$$

It follows that if $\lambda(k)$ is strictly convex, $\lambda''(k) > 0$, then so is $I(s)$. Finally, note that any rate function obtained from the Gartner-Ellis Theorem has a unique global minimum whose root s^* satisfies $I'(s^*) = k(s^*) = 0$. This implies that

$$s^* = \lambda'(k(s^*)) = \lambda'(0) = \lim_{n \to \infty} \mathbb{E}[S_n],$$

and $I(s^*) = 0$. Thus, in the large-n limit the pdf concentrates around the mean, which is an expression of the law of large numbers. This suggests that one interpretation of the large deviation principle is that it quantifies the likelihood of a large deviation from the mean.

The contraction principle. Let A_n be a random variable that has an LDP with rate function $I_A(a)$. Consider a second random variable $B_n = f(A_n)$. Does this also satisfy an LDP and, if so, what is its rate function? In order to address this issue, first write the pdf of B_n in terms of the pdf of A_n:

$$p_{B_n}(b) = \int_{\{a: f(a) = b\}} p_{A_n}(a) da.$$

Using the LDP for A_n and the saddle-point method gives

$$p_{B_n}(b) \approx \int_{\{a: f(a) = b\}} e^{-n I_A(a)} da \approx \exp\left(-n \inf_{\{a: f(a) = b\}} I_A(a)\right).$$

Hence, $p(B_n)$ also satisfies an LDP with rate function

$$I_B(b) = \inf_{\{a: f(a) = b\}} I_A(a). \tag{8.7.25}$$

The latter formula is known as the contraction principle, since f may be many-to-one, that is, there may be several a's for which $b = f(a)$, in which case information regarding the rate function of A is "contracted" down to B_n.

8.7.3 Large deviation principle for Brownian functionals

We now illustrate the indirect method for deriving an LDP by considering the rescaled Brownian functional

$$\mathcal{U}_T = \frac{1}{T} \int_0^T U(X(t)) dt, \tag{8.7.26}$$

where U is a real function (not necessarily positive). Denote the corresponding probability density for $\mathcal{U}_T$ (assuming it exists) by $\mathcal{P}(a, T)$. Suppose that in the large-T

limit, the probability density has an LDP of the form [213, 244, 391, 809]

$$\mathcal{P}(a,T) = \mathrm{e}^{-TI(a)+o(T)}, \tag{8.7.27}$$

with $I(a)$ the so-called rate function. This implies that the probability of observing fluctuations in $\mathcal{U}_T$ at large times is exponentially small. Let us introduce the scaled cumulant function of $\mathcal{U}_T$, which is defined as

$$\lambda(k) = \lim_{T \to \infty} \frac{1}{T} \ln \mathbb{E}_\sigma[\mathrm{e}^{kT\mathcal{F}_T}], \tag{8.7.28}$$

where $k \in \mathbb{R}$ and $\mathbb{E}_X[\cdot]$ denotes the expectation with respect to different realizations of the stochastic process $X(t)$, given that $X(0) = x_0$. If $\lambda(k)$ can be obtained and is differentiable with respect to k, then one can use the Gartner-Ellis theorem, which ensures that $\mathcal{U}_T$ satisfies an LDP with a rate function given by the Legendre-Fenchel transform of $\lambda(k)$

$$I(a) = \sup_k \{ka - \lambda(k)\}. \tag{8.7.29}$$

In the case of SDEs, $\lambda(k)$ can be identified with the principal eigenvalue of the corresponding backward Feynman-Kac operator for the moment generator

$$\mathcal{Q}_k(x_0,t) = \mathbb{E}_\sigma[\mathrm{e}^{k\int_0^t U(X(\tau))d\tau}], \tag{8.7.30}$$

which takes the form, see also equation (8.6.13),

$$\frac{\partial \mathcal{Q}_k}{\partial t} = \frac{\sigma^2}{2} \frac{\partial^2 \mathcal{Q}_k}{\partial x_0^2} + A(x_0) \frac{\partial \mathcal{Q}_k}{\partial x_0} + kU(x_0)\mathcal{Q}_k. \tag{8.7.31}$$

Suppose that the solution to this equation has the spectral decomposition

$$\mathcal{Q}_k(x_0,t) = \sum_{l=0}^{\infty} \mathrm{e}^{\lambda_{0,l}(k)t} \psi_{l,n}(k,x_0), \tag{8.7.32}$$

where $\lambda_{0,l}(k)$ are the eigenvalues of the operator $Ł_k^\dagger$ and $\psi_{l,n}$ are the corresponding eigenfunctions. We can then identify $\lambda(k)$ with the largest eigenvalue, since this will be the dominant mode in the large T limit

$$\lambda(k) = \max_l \lambda_{0,l}(k). \tag{8.7.33}$$

We will illustrate the theory by considering an OU process

$$dX = -\gamma X dt + \sigma dW(t), \tag{8.7.34}$$

and the functional

$$\mathcal{U}_T = \frac{1}{T} \int_0^T X(t)dt. \tag{8.7.35}$$

The corresponding Feynman-Kac equation is (on dropping the subscript on x_0)

$$\frac{\partial Q_k}{\partial t} = Ł_k Q_k := \frac{\sigma^2}{2} \frac{\partial^2 Q_k}{\partial x^2} - \gamma x \frac{\partial Q_k}{\partial x} + kx Q_k. \tag{8.7.36}$$

The differential operator $Ł_k$ is nonself-adjoint with respect to the $L^2(\mathbb{R})$ norm. However, it can be transformed into a self-adjoint form along the lines of Sect. 2.3.4. That is, setting $Q_k(x,t) = e^{\lambda t} e^{-\gamma x^2/2\sigma^2} \psi(x)$, we find that ψ satisfies an eigenvalue equation with a self-adjoint operator

$$-\lambda \psi = -\frac{\sigma^2}{2} \frac{d^2 \psi}{dx^2} + \left(\frac{\gamma^2 x^2}{2\sigma^2} - \frac{\gamma}{2} - kx \right) \psi. \tag{8.7.37}$$

Mapping this ODE onto the time-independent Schrodinger equation for a harmonic oscillator in an electric field, we can identify $-\lambda$ with the energy eigenvalues of the quantum system. In particular, the maximal eigenvalue $\lambda(k)$ corresponds to the ground state energy. Rearranging and completing the square shows that

$$\left(\frac{\gamma}{2} + \frac{\sigma^2 k^2}{2\gamma^2} - \lambda \right) \psi = -\frac{\sigma^2}{2} \frac{d^2 \psi}{dx^2} + \frac{\gamma^2}{2\sigma^2} \left(x - \frac{\sigma^2 k}{\gamma^2} \right)^2 \psi.$$

The shift in x does not affect the eigenvalues so using the spectrum of the harmonic oscillator, we deduce that the eigenvalues λ_n are

$$\frac{\gamma}{2} + \frac{\sigma^2 k^2}{2\gamma^2} - \lambda_n = \left(n + \frac{1}{2} \right) \gamma,$$

and thus

$$\lambda(k) = \frac{\sigma^2 k^2}{2\gamma^2}. \tag{8.7.38}$$

Finally, by performing a Legendre transform, the corresponding rate function is

$$I(a) = \frac{\gamma^2 a^2}{2\sigma^2}. \tag{8.7.39}$$

We deduce that the fluctuations of U_T about the expected value $U_T = 0$ are Gaussian, reflecting the fact that linear integrals of Gaussian processes are also Gaussian.

8.8 Exercises

Problem 8.1. Hamiltonian dynamics in spherical polar coordinates. Consider the Lagrangian of a particle moving according to Newton's law in 3D, expressed in spherical polar coordinates

$$L = \frac{m}{2} (\dot{r}^2 + r^2 \dot{\theta}^2 + r^2 \sin^2 \theta \dot{\phi}^2) - U(r, \theta, \phi),$$

where U is the potential energy. Construct the Hamiltonian $H(r,\theta,\phi,q_r,q_\theta,q_\phi)$ and derive Hamilton's equations in spherical polar coordinates. Show that if $U = U(r)$, radial symmetry, then the momentum conjugate to ϕ is conserved.

Problem 8.2. Canonical transformations. A canonical transformation of a Hamiltonian dynamical system is one that takes the coordinates (x,q) and Hamiltonian $H(x,q)$ to a new coordinate system (X,Q) with Hamiltonian $K(X,Q)$ such that

$$\dot{x} = \frac{\partial H}{\partial q}, \quad \dot{p} = -\frac{\partial H}{\partial x}, \quad \dot{X} = \frac{\partial K}{\partial Q}, \quad \dot{Q} = -\frac{\partial K}{\partial X}.$$

Since Hamilton's equations minimize the same action, it follows that

$$q\dot{x} - H = Q\dot{X} - K + \frac{dF}{dt},$$

where F is an arbitrary function of coordinates.

(a) Suppose that $F = F(x(t),X(t))$. Using the chain rule, show that the above relationship between H and K holds if

$$q = \frac{\partial F}{\partial x}, \quad Q = -\frac{\partial F}{\partial X}, \quad K = H.$$

Hence, show that the function F generates a canonical transformation.

(b) Show that the canonical transformation $F = xX$ swaps the original coordinate and momentum variables.

(c) Show that the action S acts as a generator of a canonical transformation to the simplest possible Hamiltonian system, in which X,Q are both constants of the motion.

Problem 8.3. Harmonic oscillator. Here we consider some indirect methods for solving the equations of motion of a harmonic oscillator.

(a) Show that the Hamiltonian of a particle of mass m attached to a Hookean spring with spring constant k, is

$$H = \frac{1}{2m}(q^2 + m^2\omega^2 x^2), \quad \omega = \sqrt{k/m}.$$

(b) Introducing the generating function

$$F = \frac{m\omega}{2}x^2\cot X,$$

obtain equations for $x = x(X,Q)$ and $q = q(x,X)$. Hence, show that

$$K(X,Q) = \omega Q.$$

(c) Using part (b) and Hamilton's equations in the new coordinate system obtain the solution

$$x = \sqrt{\frac{2E}{m\omega^2}} \sin(\omega t + \beta),$$

where β is an integration constant.

(d) Write down the Hamiltonian-Jacobi equation for the harmonic oscillator. Show that it has the solution

$$S(x,E,t) = \sqrt{2m} \int^x \sqrt{E - U(x')} dx' - Et,$$

where E is the conserved energy. Using the fact that $\partial S/\partial E = \Gamma$ where Γ is a constant, obtain a solution for $q(t)$ by quadrature.

Problem 8.4. Analysis of metastability in a birth–death process using generating functions. In this problem, we consider an alternative approach to WKB analysis based on generating functions, which is often used to study rate events and extinction in population dynamics [29–31, 243]. Consider the birth–death process master equation (8.2.1) with transition rates having the scaling

$$\omega_\pm(n) = N\Omega_\pm(n/N),$$

where N is the system size and $\omega_-(0) = 0$. Define the generating function according to

$$G(z,t) = \sum_{n=0}^{\infty} z^n P(n,t).$$

(a) Show that the generator satisfies a nonlinear PDE of the form

$$\frac{\partial G}{\partial t} = (z-1)\omega_+(\partial/\partial z)G(z,t) + (z^{-1}-1)\omega_-(\partial/\partial z)G(z,t).$$

Note: if $\omega_-(n) = e^{an} = 1 + an + a^2 n^2 + \ldots$, for example, then $\omega_+(\partial/\partial z)$ is the differential operator

$$\omega_+(\partial/\partial z) = e^{a\partial/\partial z} = 1 + a\frac{\partial}{\partial z} + a^2\frac{\partial^2}{\partial z^2} + \ldots$$

A similar interpretation holds for other functions.

(b) Setting $\omega_\pm = N\Omega_\pm(N^{-1}\partial/\partial z)$ and making the WKB ansatz $G(z,t) \sim e^{-NS(z,t)}$, show that to leading order S satisfies the nonlinear first-order PDE

$$\frac{\partial S}{\partial t} = -(z-1)\Omega_+(-\partial_z S) - (z^{-1}-1)\Omega_-(-\partial_z S).$$

(Hint: Suppose that $\Omega_\pm$ are quadratic functions and collect terms with equal powers in N. Then note that the same result holds for arbitrary order polynomials.) The PDE is a time-dependent Hamilton-Jacobi equation for a Hamiltonian dynamical system in the momentum representation (see Box 8A), with z interpreted as a momentum variable and $x = -\partial_z S$ interpreted as the coordinate variable:

$$H(x,z) = (z-1)\Omega_+(x) + (z^{-1}-1)\Omega_-(x).$$

Note that the Hamiltonian is equivalent to the one derived in Sect. 8.2 under the mapping $z \to e^p$, and S is the associated action.

Problem 8.5. Branching and annihilation reaction. Consider the reaction scheme

$$A + A \xrightarrow{\lambda} \emptyset, \quad A \xrightarrow{\sigma} 2A.$$

The law of mass action yields the following kinetic equation for the concentration $a = [A]$:

$$\frac{da}{dt} = \sigma a(t) - \lambda a^2(t).$$

This has a stable fixed point $a = n_s = \sigma/\lambda$ and an unstable fixed point $a = 0$. Hence, the deterministic system avoids the zero state if $a(0) > 0$. In this problem, we explore the fact that the full stochastic system decays to the zero state in finite time due to a noise-induced transition. We shall identify n_s as the system size with $n_s \gg 1$, and consider the corresponding master equation

$$\frac{dP_n}{dt} = \frac{\lambda}{2}[(n+2)(n+1)P_{n+2}(t) - n(n-1)P_n(t)] + \sigma[(n-1)P_{n-1}(t) - nP_n(t)].$$

(a) Determine the PDE for the generating function $G(z,t) = \sum_{n\geq 0} z^n P_n(t)$.

(b) Introduce the WKB ansatz $G(z,t) \sim e^{-S(z,t)}$. Assuming that higher-order derivative terms can be neglected, show that Ψ is the solution to the Hamilton-Jacobi equation

$$\frac{\partial S}{\partial t} + H(-\partial_z S, z) = 0,$$

with Hamiltonian

$$H(x,z) = \frac{\lambda}{2}(1-z^2)x^2 - \sigma(1-z)zx.$$

Note: It can be shown *a posteriori* that $S(x,t) = n_s s(z,t)$ for $s = O(1)$, which justifies neglecting higher-order derivatives in S.

(c) Write down Hamilton's equations and show that the deterministic dynamics is recovered when $z = 1$, with $\dot{z} = 0$. Also establish that there are three lines of zero energy, $H(x,z) = 0$ in the (x,z) plane: $z = 1, x = 0$ and $x = 2n_2 z/(1+z)$.

(d) Construct a phase portrait of solutions to Hamilton's equations in the x,z plane, including the zero energy curves. Hence, establish that the curve $x = x(z) = 2n_2 z/(1+z)$ is a trajectory linking $(n_s, 1)$ to $(0,0)$, This represents the most likely stochastic path to reach the zero state starting from a quasi-stationary state. Determine the mean exit time $\tau \sim e^{S_0}$, where

$$S_0 = -\int_{\infty}^{\infty} x\dot{z} dt = -\int_1^0 x(z) dz.$$

Problem 8.6. Operator representation of the birth–death master equation.
Consider the annihilation and creation operators acting on the occupation state $|n>$
according to

$$a|0\rangle = 0, \quad a^{\dagger n}|0\rangle = |n\rangle, \quad aa^{\dagger} - a^{\dagger}a = 1.$$

(a) Show that $\langle n|m\rangle = n!\delta_{n,m}$.

(b) Prove that

$$a|n\rangle = n|n-1\rangle, \quad a^{\dagger}|n\rangle = |n+1\rangle.$$

(c) Introducing the projection state

$$|\emptyset\rangle = \exp\left(a^{\dagger}\right)|0\rangle = \sum_{n=0}^{\infty} \frac{1}{n!}|n\rangle,$$

show that

$$\langle\emptyset|a^{\dagger}a|\phi(t)\rangle = \sum_{n\geq 0} nP(n,t) = \langle n(t)\rangle.$$

(d) Show that the birth–death master equation

$$\frac{d}{dt}P(n,t) = \omega_+(n-1)P(n-1,t) + \omega_-(n+1)P(n+1,t) - [\omega_+(n) + \omega_-(n)]P(n,t),$$

can be rewritten in the operator form

$$\partial_t|\phi(t)\rangle = \hat{H}|\phi(t)\rangle,$$

with $|\phi(t)\rangle = \sum_{n\geq 0} P(n,t)|n\rangle$,

$$\hat{H} = (a - a^{\dagger}a)\overline{\omega}_-(a^{\dagger}a) + (a^{\dagger} - 1)\omega_+(a^{\dagger}a),$$

and $\overline{\omega}_-(n) = \omega_-(n)/n$. (Assume that $\omega_{\pm}(n)$ can be expanded as a Taylor series in
n.)

Problem 8.7. Gaussian approximation of stochastic hybrid path integral. Consider the principal eigenvalue equation for a stochastic hybrid system

$$\sum_{m\geq 0} [A_{nm}(x) + p\delta_{n,m}F(m,x)] R(m,x,p) = \Lambda(x,p)R(n,x,p).$$

(a) Carry out a perturbation expansion in the small p limit by performing the rescaling $p \to \varepsilon p$ and introducing the series expansions

$$R(m,x,\varepsilon p) = R_0(m,x) + \varepsilon p R_1(m,x) + \varepsilon^2 p^2 R_2(m,x) + \ldots,$$

and

$$\Lambda(x,\varepsilon p) = \Lambda_0(x) + \varepsilon p \Lambda_1(x) + \varepsilon^2 p^2 \Lambda_2(x) + \ldots,$$

with

$$\sum_{m \geq 0} R_k(m,x) = \delta_{k,0}.$$

Substitute these expansions into the eigenvalue equation and collect terms with equal powers of ε. From the $O(1)$ equation show that

$$\Lambda_0 = 0, \quad R_0(m,x) = \rho_m(x).$$

Applying the Fredholm alternative to the $O(\varepsilon)$ and $O(\varepsilon^2)$, obtain the results

$$\Lambda_1(x) = \sum_n F(n,x)\rho_n(x) = \overline{F}(x),$$

$$\Lambda_2(x) = \sum_n \left[F(n,x) - \overline{F}(x)\right] R_1(n,x) = \sum_n F(n,x)R_1(n,x).$$

(b) Substitute the Gaussian approximation

$$\Lambda(x,\varepsilon q) \approx \varepsilon q \overline{F}(x) + \varepsilon^2 q^2 \sum_n F(n,x)R_1(n,x),$$

into the path integral

$$P(x,t|x_0,0) = \int_{x(0)=x_0}^{x(\tau)=x} D[x]D[q]e^{-S[x,q]/\varepsilon},$$

with

$$S[x,q] = \int_0^\tau \left[\varepsilon p \dot{x} - \Lambda(x,\varepsilon p)\right] dt.$$

Performing the Gaussian integral with respect to q after rotating in the complex plane $(q \to iq)$, derive the Onsager-Machlup path integral

$$P(x,t) = \int_{x(0)=x_0}^{x(\tau)=x} D[x] \exp\left(-\int_0^\tau \frac{[\dot{x} - \overline{F}(x)]^2}{4\varepsilon D(x)} dt\right),$$

with

$$D(x) = \sum_n F(n,x)R_1(n,x).$$

Show how this recovers the QSS approximation of the stochastic hybrid system (see Box 5D).

Problem 8.8. 1D stochastic hybrid system. Consider a 1D stochastic hybrid system with piecewise dynamics

$$\frac{dx}{dt} = F_n(x) \equiv -x + \gamma n$$

for $n \geq 0$ with the discrete process evolving according to a birth-death process with x-dependent transition rates

$$\omega_+(n;x) = f(x), \quad \omega_-(n) = n.$$

Here $f(x)$ is a sigmoidal function – a bounded, monotonically increasing function of x with $f(x) \to f_0$ as $x \to \infty$ and $f(x) \to 0$ as $x \to -\infty$. The corresponding generator is given by the tridiagonal matrix

$$A_{n,n-1}(x) = f(x), \quad A_{nn}(x) = -f(x) - n, \quad A_{n,n+1}(x) = n+1,$$

(a) Show that the steady-state distribution of the birth-death process for fixed x is given by the Poisson distribution

$$p_n(x) = \frac{[f(x)]^n e^{-f(x)}}{n!}, \tag{8.8.40}$$

After rescaling the transition matrix by $1/\varepsilon$, show that the deterministic equation obtained in the $\varepsilon \to 0$ limit is

$$\frac{dx}{dt} = \sum_{n=0}^{\infty} F_n(x) p_n(x) = -x + \gamma f(x).$$

(b) Consider the corresponding eigenvalue equation

$$f(x) R_{n-1}(x,q) - (f(x)+n) R_n(x,q) + (n+1) R_{n+1}(x,q)$$
$$= [\Lambda(x,q) - q(-x+\gamma n)] R_n(x,q).$$

Show by direct substitution that

$$R_n(x,q) = \frac{\Gamma(x,q)^n}{n!},$$

with

$$\Gamma = \frac{f(x)}{1-\gamma q}, \quad \Lambda = \frac{\gamma q f(x)}{1-\gamma q} - qx.$$

(c) Using the solution of part (b), determine the zero energy solutions $q = q(x)$ with $\Lambda(x,q(x)) = 0$.

Remark 8.3. Note that certain care must be taken in the above example, since the discrete-process is infinite. This means that the Perron-Frobenius theorem does not strictly apply.

Problem 8.9. Occupation time for a Brownian particle. Fill in the details of the calculation of the occupation time for a Brownian particle, see Sect. 8.6. In particular, show that

$$U(s,z) \equiv \tilde{Q}(0;s,z) = \frac{1}{\sqrt{z(s+z)}},$$

by solving the system of equations

$$\frac{1}{2}y''_+ - (z+s)y_+ = 0, \quad x > 0,$$

$$\frac{1}{2}y''_- - zy_- = 0, \quad x < 0,$$

with $y_+(\infty; s, z) = 0$, $y_-(-\infty; s, z) = 0$, and

$$\frac{1}{z+s} + B_+ y_+(0; s, z) = \frac{1}{z} + B_- y_-(0; s, z) = U(s, z),$$

$$B_+ y'_+(0; s, z) = B_- y'_-(0; s, z)$$

for constants $B_\pm$.

Problem 8.10. Stochastic hybrid path integrals and spinors. Consider the hybrid SDE

$$dX = F_n(X)dt + \sqrt{D_n}dW$$

for $N(t) = n$, where $X(t) \in \mathbb{R}$, $N(t) \in \{1, \ldots, M\}$, and $W(t)$ is a Wiener process such that

$$\langle W(t) \rangle = 0, \quad \langle W(t)W(t') \rangle = \min\{t, t'\}.$$

The corresponding CK equation is

$$\frac{\partial P_n(x,t)}{\partial t} = -\frac{\partial F_n(x)P_n(x,t)}{\partial x} + D_n \frac{\partial^2 P_n(x,t)}{\partial x^2} + \sum_{m=1}^{M} Q_{nm}P_m(x,t),$$

with $Q_{nm} = T_{nm} - \delta_{n,m}\sum_{l=1}^{M} T_{lm}$ and $\mathbf{T}$ the transition matrix.

(a) Consider the state vectors

$$|\psi_n(t)\rangle = \int_{-\infty}^{\infty} dx\, P_n(x,t)|x\rangle, \quad n = 1, \ldots, M.$$

Differentiating both sides with respect to t, show that the CK equation can be rewritten in the operator form

$$\frac{d}{dt}|\psi(t)\rangle = \widehat{\mathbf{H}}|\psi(t)\rangle, \quad |\psi(t)\rangle = (|\psi_1(t)\rangle, \ldots, |\psi_M(t)\rangle)^\top,$$

where

$$\widehat{\mathbf{H}} = -i\hat{p}\,\mathrm{diag}(F_1(\hat{x}), \ldots, F_n(\hat{x})) - \hat{p}^2\mathrm{diag}(D_1(\hat{x}), \ldots, D_n(\hat{x})) + \mathbf{Q},$$

and the operators $\hat{x}, \hat{p}$ are given by equation (8.3.23).

(b) Introduce the multicomponent spinors

$$|s\rangle = \begin{pmatrix} z_1 e^{i\phi_1/2} \\ \vdots \\ z_M e^{i\phi_M/2} \end{pmatrix}, \quad \langle s| = \begin{pmatrix} e^{-i\phi_1/2} & \cdots & e^{-i\phi_M/2} \end{pmatrix},$$

with $z_j \in [0,1]$, $\phi_j \in [0,2\pi)$, and the normalization condition $\sum_{m=1}^{M} z_m = 1$. Show that

$$\langle s + \Delta s | s \rangle = 1 - \frac{1}{2} i \sum_{m=1}^{M} z_m \Delta \phi_m + O(\Delta \phi^2).$$

and

$$\langle s | \hat{\mathbf{H}} | s \rangle = H(\mathbf{z}, \phi, \hat{x}, \hat{p}) = -i\hat{p} \sum_{m=1}^{M} [-i\hat{p} F_m(\hat{x}) - \hat{p}^2 D_m] z_m + Q(\mathbf{z}, \phi),$$

where

$$Q(\mathbf{z}, \phi) := \langle s | \mathbf{Q} | s \rangle = \sum_{\substack{n,m=1 \\ n \neq m}}^{M} T_{nm} \left[e^{i(\phi_m - \phi_n)/2} - 1 \right] z_m.$$

(c) Formally integrating the evolution equation of part (b) yields the solution

$$|\psi(t)\rangle = e^{\hat{\mathbf{H}} t} |\psi(0)\rangle$$

Dividing the time interval $[0,t]$ into N subintervals of size $\Delta t = t/N$ then gives

$$|\psi(t)\rangle = e^{\hat{\mathbf{H}} \Delta t} e^{\hat{\mathbf{H}} \Delta t} \cdots e^{\hat{\mathbf{H}} \Delta t} |\psi(0)\rangle$$

Consider the product Hilbert space $|s,x\rangle = |s\rangle \otimes |x\rangle$ and the associated completeness relations

$$\int_{-\infty}^{\infty} dx \, |x\rangle\langle x| = 1, \quad \prod_{m=1}^{M-1} \left\{ \int_0^1 \frac{dz_m}{2} \int_0^{4\pi} \frac{d\phi_m}{4\pi} \right\} |s\rangle\langle s| = 1.$$

Inserting multiple copies of the completeness relations along analogous lines to (8.3.38) leads to a product of small time propagators. Inserting the momentum completeness relation (8.3.29) into the j-th propagator and using the inner products calculated in part (b), show that

$$\langle s_{j+1}, x_{j+1} | e^{\hat{\mathbf{H}} \Delta t} | s_j, x_j \rangle$$
$$\approx \int_{-\infty}^{\infty} \frac{dp_j}{2\pi} \langle x_{j+1} | p_j \rangle \langle p_j | x_j \rangle \exp\left(\left[H(\mathbf{z}_j, \phi_j, x_j, p_j) - \frac{i}{2} \sum_{m=1}^{M} \frac{d\phi_{j,m}}{dt} z_{j,m} \right] \Delta t \right).$$

Hence, obtain the discretized path integral

$$|\psi(t)\rangle = \int_{\Omega} ds_0 \cdots \int_{\Omega} ds_N \int_{-\infty}^{\infty} \int_{-\infty}^{\infty} \frac{dx_0 dp_0}{2\pi} \cdots \int_{-\infty}^{\infty} \int_{-\infty}^{\infty} \frac{dx_N dp_N}{2\pi} |s_N, x_N\rangle$$
$$\times \prod_{j=0}^{N-1} \exp\left(\left[H(\mathbf{z}_j, \phi_j, x_j, p_j) - \frac{i}{2} \sum_{m=1}^{M} \frac{d\phi_{j,m}}{dt} z_{j,m} + i p_j \frac{dx_j}{dt} \right] \Delta t \right) \langle s_0, x_0 | \psi(0) \rangle.$$

where

$$\int_\Omega ds_j = \prod_{m=1}^{M-1} \int_0^1 \frac{dz_{j,m}}{2} \int_0^{4\pi} \frac{d\phi_{j,m}}{4\pi}.$$

(d) Taking the continuum limit, setting $p \to -ip$, and performing an integration by parts gives the hybrid path integral

$$P_n(x,t|x_0,0) = \int_{x(0)=x_0}^{x(t)=x} \mathcal{D}[\phi]\mathcal{D}[\mathbf{z}]\mathcal{D}[p]\mathcal{D}[x] \exp\left(-\int_0^t \left[p\frac{dx}{d\tau} - \frac{i}{2}\sum_{m=1}^M \phi_m \frac{dz_m}{d\tau} - \mathcal{H}\right] d\tau\right)$$

where $\mathcal{H}$ is the effective Hamiltonian

$$\mathcal{H} = \sum_{m=1}^M [pF_m(x) + p^2 D_m]z_m + \sum_{\substack{n,m=1 \\ n\neq m}}^M T_{nm}\left[e^{i(\phi_m-\phi_n)/2} - 1\right]z_m.$$

Chapter 9
Probability theory and martingales

In the bulk of this book, we have avoided the rigorous formulation of stochastic processes used by probabilists. We have sacrificed the level of rigor in order to make the material accessible to applied mathematicians and biological physicists who tend not to have a background in advanced probability theory. However, it is useful to have some exposure to the concepts and notation used by probabilists. Therefore, in this chapter we give a very brief introduction to probability theory with an emphasis on martingales. There are a number of excellent textbooks on modern probability theory, see for example [343, 859]. Both of these books tend to focus on discrete-time processes. However, the notation and concepts can be extended to continuous-time processes as detailed in [624].

9.1 Probability spaces, random variables and conditional expectations

Consider a set of possible outcomes (of an experiment), which is denoted by the sample space Ω. An event is defined to be a subset A of Ω, which is some collection of single outcomes or elementary events $\omega \in \Omega$. In general not all subsets of Ω can be treated as events so that the set of events forms a subcollection $\mathcal{F}$ of all subsets. Within a probabilistic setting, this subcollection is required to be a so-called σ-algebra with the following properties:

1. $\emptyset \in \mathcal{F}$,
2. if $A_1, A_2, \ldots \in \mathcal{F}$ then $\cup_{i=1}^{\infty} A_i \in \mathcal{F}$,
3. if $A \in \mathcal{F}$ then $\Omega \backslash A \in \mathcal{F}$.

It can be shown that σ-algebras are closed under the operation of taking countable intersections. A probability measure $\mathbb{P}$ on $(\Omega, \mathcal{F})$ is a function $\mathbb{P} : \mathcal{F} \to [0, 1]$ with

1. $\mathbb{P}(\emptyset) = 0, \mathbb{P}(\Omega) = 1.$

© Springer Nature Switzerland AG 2021
P. C. Bressloff, *Stochastic Processes in Cell Biology*, Interdisciplinary
Applied Mathematics 41, https://doi.org/10.1007/978-3-030-72515-0_9

2. if $A_i, A_j, \ldots \in \mathcal{F}$ with $A_i \cap A_j = \emptyset$, $i \neq j$, then

$$\mathbb{P}(\cup_{i=1}^{\infty} A_i) = \sum_{i=1}^{\infty} \mathbb{P}(A_i).$$

The triple $(\Omega, \mathcal{F}, \mathbb{P})$ is called a probability space.

Given a function f on the sample space Ω, we can use the probability measure $\mathbb{P}$ to define the integral of this function over a set $A \in \mathcal{F}$ according to

$$f(A) = \int_A f(\omega) d\mathbb{P}(\omega).$$

If $f(\omega) = 1$ for all $\omega \in \Omega$, then $f(A) = \mathbb{P}(A)$. Note that for certain choices of σ-algebra, it is necessary to consider measures other than the standard Lebesgue measure. However, we will not consider this technicality here. A random variable is a function $X : \Omega \to \mathbb{R}$ such that

$$\{\omega \in \Omega : X(\omega) \leq x\} \in \mathcal{F}, \quad \forall x \in \mathbb{R}.$$

If this condition holds, then X is said to be $\mathcal{F}$-measurable. If $X \in \mathbb{R}$ then we have a continuous random variable, whereas if X belongs to a countable set then it is said to be a discrete random variable. The distribution function of a random variable X is the function $F : \mathbb{R} \to [0, 1]$ given by

$$F(x) = \text{Prob}(X \leq x) = \mathbb{P}(X^{-1}(-\infty, x)),$$

where $X^{-1}(-\infty, x)$ is the set of events ω for which $X \leq x$.

In the case of two discrete random variables X, Y on the same probability space $(\Omega, \mathcal{F}, \mathbb{P})$, one can construct the conditional expectation of a random variable Y with respect to a sub-σ-algebra $\mathcal{G} \subset \mathcal{F}$. First note that the conditional expectation $\widehat{Y} \equiv \mathbb{E}(Y|X)$ with respect to the random variable X can be interpreted as a random variable satisfying

$$\int_{A(x)} \widehat{Y}(\omega) d\mathbb{P}(\omega) = \int_{A(x)} Y(\omega) d\mathbb{P}(\omega) \quad \forall \quad x \in \mathbb{R},$$

where

$$A(x) = \{\omega \in \Omega : X(\omega) \leq x\} \subset \mathcal{F}.$$

The conditional expectation $\widehat{Y} = \mathbb{E}(Y|\mathcal{G})$ for a general sub-σ-algebra $\mathcal{G}$ is then defined to be a random variable satisfying

$$\int_A \widehat{Y}(\omega) d\mathbb{P}(\omega) = \int_A Y(\omega) d\mathbb{P}(\omega) \quad \forall \quad A \in \mathcal{G}.$$

It immediately follows from taking $A = \Omega$ that

$$\mathbb{E}[\mathbb{E}[Y|\mathcal{G}]] = \mathbb{E}[Y].$$

Given two sub-σ algebras $\mathcal{G}, \mathcal{H}$ with $\mathcal{G} \subset \mathcal{H} \subset \mathcal{F}$, one also has the tower property

$$\mathbb{E}[\mathbb{E}[Y|\mathcal{H}]|\mathcal{G}] = \mathbb{E}[Y|\mathcal{G}].$$

9.2 Discrete-time stochastic processes and martingales

A stochastic process involves one or more random variables evolving in time. Each random variable will have an additional time label: $X \to X_n, n \in \mathbf{Z}^+$ for discrete-time processes and $X \to X(t), t \in \mathbb{R}^+$ for continuous-time processes. Roughly speaking, one can treat n (or t) as a parameter so that for fixed n, X_n is a random variable in the above sense. In this book, we mainly focus on continuous-time processes. One exception is the analysis of discrete-time random walks in Sect. 2.1. Discrete-time processes also naturally arise within the context of cell division and branching processes (Chap. 15). Here we introduce a few ideas from the theory of stochastic processes that are more easily understood using discrete time, although continuous-time analogs can be developed [624].

9.2.1 Filtrations and martingales

Much of applied probability concerns establishing various limit theorems. Often this is achieved by showing that a particular sequence of random variables is a martingale. Such a sequence could be generated by a discrete-time stochastic process. (For the extension of martingales to continuous-time stochastic processes see [624]). In the following, we fix a probability space $(\Omega, \mathcal{F}, \mathbb{P})$ and assume that there exists a given sequence $\mathcal{F}_n$, $n = 1, 2, \ldots$, of σ-algebras $\mathcal{F}_n \subseteq \mathcal{F}$.

Definition 9.1. A filtration on $(\Omega, \mathcal{F})$ is an increasing sequence

$$\mathcal{F}_1 \subseteq \mathcal{F}_1 \subseteq \mathcal{F}_2 \subseteq \cdots$$

of σ-algebras. A (discrete-time) stochastic process $(X_n)_{n \geq 1}$ is adapted to a filtration $(\mathcal{F}_n)_{n \geq 1}$ if and only if each X_n is $\mathcal{F}_n$-measurable, that is,

$$\{\omega \in \Omega : X_n(\omega) \leq x\} \in \mathcal{F}_n, \quad \forall x \in \mathbb{R}.$$

Example 9.1. The canonical or natural filtration generated by a stochastic process (X_n) is given by

$$\mathcal{F}_n = \sigma(X_1, X_2, \ldots, X_n),$$

that is, the minimal event specifying the values of X_j, $j = 1, \ldots, n$. If the filtration is not specified explicitly, it will be assumed to be the canonical filtration. In general (X_n) is adapted to a filtration $(\mathcal{F}_n)$ if and only if $\sigma(X_1, X_2, \ldots, X_n) \subseteq \mathcal{F}_n$. Roughly speaking, as n increases, the statistical information about a larger class of random variables is included within the σ-algebra $\mathcal{F}_n$, as one might expect from the evolution of a discrete-time stochastic process.

Definition 9.2. A sequence of random variables $\{Y_n; n \geq 1\}$ on the probability space $(\Omega, \mathcal{F}, \mathbb{P})$ is called a martingale with respect to the filtration $(\mathcal{F}_n)$ if and only if

(i) (Y_n) is adapted to $(\mathcal{F}_n)$.
(ii) $\mathbb{E}[|Y_n|] < \infty$.
(iii) $\mathbb{E}(Y_{n+1}|\mathcal{F}_n) = Y_n$.

If $\mathcal{F}_n = \sigma(X_1, \ldots, X_n)$ for some other random sequence (X_n) then (Y_n) is said to be a martingale with respect to X. If (i) and (ii) hold but instead of (iii) we have

$$\mathbb{E}(Y_{n+1}|\mathcal{F}_n) \leq Y_n,$$

then (Y_n) is said to be a supermartingale. Similarly, if

$$\mathbb{E}(Y_{n+1}|\mathcal{F}_n) \geq Y_n,$$

then we have a submartingale.

Intuitively, the random sequence (Y_n) models the outcomes of a random process in time, whereas the filtration $(\mathcal{F}_n)$ specifies what is known at each step. Hence, if (Y_n) is a martingale then we have at least partial information about the values $Y_1, \ldots, Y_n$, i.e., their probability distributions, and conditional on this information, the expected value of Y_{n+1} is equal (almost surely) to the observed value Y_n. Given the definition of a martingale, the following results hold:

(a) $\mathbb{E}(Y_n|\mathcal{F}_n) = Y_n$, which reflects the fact that Y_n is $\mathcal{F}_n$-measurable. It immediately follows that

$$\mathbb{E}(Y_{n+1} - Y_n|\mathcal{F}_n) = 0.$$

(b)

$$\mathbb{E}(Y_{n+k}|\mathcal{F}_n) = Y_n, \quad \forall \quad n \geq 1, k \geq 0,$$

which can be shown using induction. The case $k = 0$ holds from (a). Moreover the assertion for $k - 1$ implies that

$$
\begin{aligned}
\mathbb{E}(Y_{n+k}|\mathcal{F}_n) &= \mathbb{E}(\mathbb{E}(Y_{n+k}|\mathcal{F}_{n+k-1})|\mathcal{F}_n) \quad \text{from the tower property} \\
&= \mathbb{E}(Y_{n+k-1}|\mathcal{F}_n) \quad (Y_n) \text{ is a martingale} \\
&= Y_n.
\end{aligned}
$$

(c) Setting $n = 1$ in (b) shows that

$$\mathbb{E}(Y_n) = \mathbb{E}(Y_1) \quad \forall \quad n \geq 1,$$

that is, the expected outcome is independent of time.

Example 9.2. One simple example of a martingale is a 1D random walk, see also Sect. 2.1, in which S_n is the position of the walker at the nth time step and $S_n = \sum_{j=1}^n X_j$ where X_j is the jth independent increment, with $X_j = 1$ (probability p) or $X_j = -1$ (probability $q = 1 - p$), and $S_0 = 0$. In this case, we take

$$\mathcal{F}_n = \sigma(X_1, \ldots, X_n) = \sigma(S_1, \ldots, S_n).$$

The result follows from

$$\mathbb{E}(S_{n+1}|X_1, X_2, \ldots, X_n) = \mathbb{E}(S_n + X_{n+1}|X_1, X_2, \ldots, X_n)$$
$$= \mathbb{E}(S_n|X_1, X_2, \ldots, X_n) + \mathbb{E}(X_{n+1}|X_1, X_2, \ldots, X_n)$$
$$= S_n + p - q.$$

Thus, $Y_n = S_n - n(p-q)$ defines a martingale with respect to X. For an unbiased random walk ($p = q$, S_n is itself a martingale

Example 9.3. Let $X_1, X_2, \ldots$ be a discrete time Markov chain taking discrete values in some countable space Γ, see also Chap. 3, with conditional probabilities

$$p_{ij} = \mathbb{P}(X_{n+1} = j | X_n = i).$$

Suppose that $\psi : \Gamma \to \mathbb{R}$ is a bounded function satisfying

$$\sum_{j \in \Gamma} p_{ij} \psi(j) = \psi(i), \quad \forall i \in \Gamma.$$

Then $Y_n = \psi(X_n)$ constitutes a martingale (with respect to X), since

$$\mathbb{E}(Y_{n+1}|X_1, \ldots X_n) = \mathbb{E}(\psi(X_{n+1})|X_1, \ldots X_n)$$
$$= \mathbb{E}(\psi(X_{n+1})|X_n) \quad \text{Markov property}$$
$$= \sum_{j \in \Gamma} p_{X_n, j} \psi(j) = \psi(X_n) = Y_n.$$

Example 9.4. Let X be a random variable on $(\Omega, \mathcal{F}, \mathbb{P})$. One can then define a sequence of conditional expectations with respect to a filtration $(\mathcal{F}_n)$ in $\mathcal{F}$ given by

$$X_n = \mathbb{E}(X|\mathcal{F}_n).$$

Note that X_n is itself a function of all the random variables contained within $\mathcal{F}_n$. From the tower property of conditional expectations and the fact that $\mathcal{F}_n \subseteq \mathcal{F}_{n+1}$, we have

$$\mathbb{E}(X_{n+1}|\mathcal{F}_n) = \mathbb{E}(\mathbb{E}(X|\mathcal{F}_{n+1})|\mathcal{F}_n) = \mathbb{E}(X|\mathcal{F}_n) = X_n.$$

9.2.2 Stopping times

A stopping time is based on the idea that the decision to stop a stochastic process at time m can be modeled as a random variable that is measurable with respect to $\mathcal{F}_m$, which represents the available information at time m.

Definition 9.3. A random variable $T : \Omega \to \{1, 2, \ldots\} \cup \{\infty\}$ is called a stopping time with respect to the filtration $(\mathcal{F}_n)$ if and only if

$$\{T = n\} \in \mathcal{F}_n \quad \text{for any } n \geq 1.$$

Clearly $\{T \leq n\} \in \mathcal{F}_n$, since we have a filtration.

Example 9.5. A common example is the first passage time for a discrete stochastic process (X_n) adapted to a filtration $(\mathcal{F}_n)$:

$$T_A = \min\{n \geq 0 : X_n \in A\}$$

for some measurable subset A of the state-space. This follows from the observation that

$$\{T_A = m\} = \{X_1 \notin A\} \cap \ldots \cap \{X_{m-1} \notin A\} \cap \{X_m \in A\} \in \sigma(X_1, \ldots, X_n).$$

Note that $T_A = \infty$ if $X_n \notin A$ for all $n \geq 1$.

Let (X_n) be a sequence of random variables adapted to the filtration $(\mathcal{F}_n)$ in $\mathcal{F}$. Suppose that the sequence (X_n) is stopped according to a stopping time T with respect to $(\mathcal{F}_n)$. The sequence of random variables that will actually occur is (X_{T_n}) where

$$T_m = \min\{T, m\}.$$

In other words given a particular realization ω of the stochastic process

$$X_{T_m}(\omega) = \begin{cases} X_m(\omega) & \text{if } m \leq T(\omega) \\ X_T(\omega), & \text{if } m > T(\omega) \end{cases}.$$

It can be shown that if (X_n) is a martingale with respect to the filtration $(\mathcal{F}_n)$ then so is the sequence (Y_n), where $Y_n = X_{T_n}$. If a martingale sequence (Y_n) is stopped at a fixed time n, the mean value satisfies $\mathbb{E}(Y_n) = \mathbb{E}(Y_1)$. An important issue within the context of first passage properties of martingales is whether or not the expectation remains constant when the martingale is stopped after a random time T, that is, $\mathbb{E}(Y_T) = \mathbb{E}(Y_1)$. It turns out that such a result holds provided that T is a stopping time and there are a few additional constraints.

Theorem 9.1. (Optional Stopping theorem). *Let (X_n) be a martingale with respect to the filtration $(\mathcal{F}_n)$ and let T be a stopping time. Then $\mathbb{E}(X_T) = \mathbb{E}(X_1)$ if:*

(a) $\mathbb{P}(T < \infty) = 1$.
(b) $\mathbb{E}(X_T) < \infty$.
(c) $\mathbb{E}(X_n I_{\{T>n\}}) \to 0$ *as $n \to \infty$.*

Here I_A for $A \in \mathcal{F}$ is an indicator function with $I_A(\omega) = 1$ if $\omega \in A$ and $I_A(\omega) = 0$ if $\omega \notin A$.

Proof. First note that the difference between X_T and X_{T_n} is zero if $T < n$. Therefore, we have the decomposition

$$X_T - X_{T_n} = (X_T - X_n)I_{\{T>n\}}.$$

Taking the expectation of both sides,

$$\mathbb{E}(X_T) - \mathbb{E}(X_{T_n}) = \mathbb{E}(X_T I_{\{T>n\}}) - \mathbb{E}(X_n I_{\{T>n\}}).$$

Since (X_{T_n}) is a martingale, it follows that $\mathbb{E}(X_{T_n}) = \mathbb{E}(X_1)$. Moreover, in the limit $n \to \infty$ the last term on the right-hand side vanishes due to condition (c). Also,

$$\mathbb{E}(X_T I_{\{T>n\}}) = \sum_{k=n+1}^{\infty} \mathbb{E}(X_T I_{\{T=k\}}),$$

which is the tail of the convergent series

$$\mathbb{E}(X_T) = \sum_{k=1}^{\infty} \mathbb{E}(X_T I_{\{T=k\}}) < \infty.$$

Thus, taking the limit $n \to \infty$ establishes that $\mathbb{E}(X_T) = \mathbb{E}(X_1)$.

Example 9.6. Let us return to the random walk of example 1.2 for $p = q = 1/2$, in which case (S_n) is a martingale. Let a, b be positive integers. Introduce the stopping time

$$T = \min\{n : S_n = -a \text{ or } S_n = b\},$$

which is the first time that the walker visits the endpoints $-a, b$ of the interval $[-a, b]$ with $-a < 0 < b$. It can be shown that T satisfies the conditions of the optional stopping theorem. Let p_a be the splitting probability that the walker reaches $-a$ before it reaches b. By the optional stopping theorem,

$$\mathbb{E}(S_T) = -ap_a + b(1 - p_a) = \mathbb{E}(S_0) = 0.$$

This then determines p_a:

$$p_a = \frac{b}{a+b}.$$

Using the fact that $Y_n = S_n^2 - n$ also generates a martingale sequence, the optional stopping theorem implies

$$0 = \mathbb{E}(S_T^2 - T) = a^2 p_a + b^2(1 - p_a) - \mathbb{E}(T),$$

that is,

$$\mathbb{E}(T) = a^2 p_a + b^2(1 - p_a) = \frac{a^2 b}{a+b} + \frac{ab^2}{a+b} = ab.$$

Martingale convergence theorems. Let $X_1, X_2, \ldots, X$ be random variables on some probability space $(\Omega, \mathcal{F}, \mathbb{P})$. There are four basic ways of interpreting the meaning of $X_n \to X$.

1. $X_n \xrightarrow{\text{a.s.}} X$ almost surely if

$$\{\omega \in \Omega : X_n(\omega) \to X(\omega) \text{ as } n \to \infty\}$$

is an event of probability one.

2. $X_n \xrightarrow{r} X$ in rth mean, $r \geq 1$, if

$$\mathbb{E}(|X_n - X|^r) \to 0 \text{ as } n \to \infty.$$

The case $r = 2$ is often called mean-square or L^2 convergence.

3. $X_n \xrightarrow{\text{P}} X$ in probability if

$$\mathbb{P}(|X_n - X| > \varepsilon) \to 0 \text{ as } n \to \infty \quad \forall \varepsilon > 0.$$

4. $X_n \overset{D}{\to} X$ in distribution if

$$\mathbb{P}(X_n \leq x) \to \mathbb{P}(X \leq x) \text{ as } n \to \infty$$

for all $x \in \mathbb{R}$ satisfying $\mathbb{P}(X = x) = 0$, where $\mathbb{P}(X \leq x)$ is continuous.

Given these definitions, the following implications hold

$$X_n \overset{a.s.}{\to} X \text{ or } X_n \overset{r}{\to} X \implies X_n \overset{P}{\to} X \implies X_n \overset{D}{\to} X.$$

One of the useful properties of martingales is that one can prove various limit theorems depending on the particular choice of convergence. Here we consider mean-square or L^2 convergence, which is the easiest to establish using standard analysis. Let (X_n) be a martingale with respect to the filtration $(\mathcal{F}_n)$ and assume square-integrability $\mathbb{E}(X_n^2) < \infty$ for all $n \geq 1$ and set $X_0 = 0$. One useful property of square-integrable martingales is that the increments $Y_n = X_n - X_{n-1}$ have zero mean and are uncorrelated in L^2. That is, since (X_n) is a martingale

$$\mathbb{E}[Y_n|\mathcal{F}_{n-1}] = \mathbb{E}[X_n|\mathcal{F}_{n-1}] - \mathbb{E}[X_{n-1}|\mathcal{F}_{n-1}] = X_{n-1} - X_{n-1} = 0$$

for all $n \geq 1$. Hence, from the properties of conditional expectations, $\mathbb{E}[Y_n] = \mathbb{E}[\mathbb{E}[Y_n|\mathcal{F}_{n-1}]] = 0$, and

$$\mathbb{E}[Y_m Y_n] = \mathbb{E}[Y_m \mathbb{E}[Y_n|\mathcal{F}_{n-1}]] = 0, \quad \text{for } m < n.$$

Theorem 9.2. (L^2 **martingale convergence theorem**). *If (X_n) is a nonnegative martingale, such that $\mathbb{E}[X_n] < \infty$ for all n, then there exists a random variable X with $\mathbb{E}[X] < \infty$ and satisfying the following limits:*

1. *$\lim_{n \to \infty} X_n = X$ almost surely*
2. *If $\mathbb{E}[X_n^2] < M < \infty$ for some M and all n then X_n also converges to X in the L^2 sense. In particular, $var(X) = \lim_{n \to \infty} var(X_n)$.*

9.3 The Galton–Watson branching process

The Galton–Watson branching process, which is often used to model cell proliferation, see Chap. 3 of Kimmel and Axelrod [445]. The process starts with a single ancestor who produces a random number of progeny according to a given probability distribution. Each member of the first generation behaves independently and produces second-generation progeny in an identical fashion to the ancestor. Iterating this procedure leads to the Galton–Walton branching process. Let Z_n denote the number of members (events) of the nth generation (time step). Each member of the nth generation gives birth to a family, which could be empty, of members of the $(n+1)$th generation with the following assumption: the family sizes of the individuals of the branching process are independent identically distributed random variables. An example realization of a branching process is shown in Fig. 9.1.

9.3.1 Generating functions and basic properties

A useful method for analyzing a branching process is to use generating functions. Let $G_n(s) = \mathbb{E}(s^{Z_n})$ be the generating function of the random variable Z_n with probability $P_n(m) = \text{Prob}(Z_n = m)$. Each member of the $(n+1)$th generation has a unique ancestor in the nth generation such that

$$Z_{n+1} = X_1 + X_2 + \ldots + X_{Z_n},$$

where X_i is the size of the family produced by the ith member of the nth generation. It follows that

$$G_{n+1}(s) = \mathbb{E}(s^{Z_{n+1}}) = \mathbb{E}(s^{X_1 + \ldots + X_{Z_n}})$$

$$= \sum_{m=0}^{\infty} \mathbb{E}(s^{X_1 + \ldots + X_m} | Z_n = m) P_n(m)$$

$$= \sum_{m=0}^{\infty} \mathbb{E}(s^{X_1} s^{X_2} \ldots^{X_m} | Z_n = m) P_n(m)$$

$$= \sum_{m=0}^{\infty} \left[\prod_{j=1}^{m} \mathbb{E}(s^{X_j}) \right] P_n(m)$$

$$= \sum_{m=0}^{\infty} [G_1(s)]^m P_n(m) = G_n(G_1(s)).$$

Iterating this resulting and dropping the subscript on G_1, we have the recurrence relation

$$G_n(s) = G_{n-1}(G(s)) = G(G(\ldots(G(s))\ldots)). \tag{9.3.1}$$

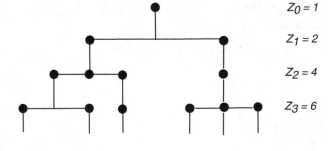

$Z_0 = 1$

$Z_1 = 2$

$Z_2 = 4$

$Z_3 = 6$

Fig. 9.1: Illustration of a branching process with three generations.

Let $\mu = \mathbb{E}(Z_1)$ and $\sigma^2 = \text{var}(Z_1)$. In order to determine the mean and variance of Z_n, we use the recursive structure of the generating functions. First,

$$\mathbb{E}(Z_n) = G_n'(1) = \frac{d}{ds}G(G_{n-1}(s))\Big|_{s=1} = G'(1)G_{n-1}'(s) = \mu G_{n-1}'(s).$$

Iterating this result shows that

$$\mathbb{E}(Z_n) = \mu^n. \tag{9.3.2}$$

Similarly,

$$\mathbb{E}(Z_n(Z_n - 1)) = G_n''(1) = G''(1)G_{n-1}'(1)^2 + G'(1)G_{n-1}''(1).$$

This gives the iterative result

$$\text{var}(Z_n) = \sigma^2 \mu^{2n-2} + \mu \text{var}(Z_{n-1}),$$

from which one finds that

$$\text{var}(Z_n) = \begin{cases} n\sigma^2 & \text{if } \mu = 1 \\ \dfrac{\sigma^2(\mu^n - 1)\mu^{n-1}}{\mu - 1} & \text{if } \mu \neq 1. \end{cases} \tag{9.3.3}$$

Let T_n be the total number of individuals up to and including the nth generation. Then

$$\begin{aligned} \mathbb{E}(T_n) &= \mathbb{E}(Z_0 + Z_1 + Z_2 + \ldots + Z_n) \\ &= 1 + \mathbb{E}(Z_1) + \mathbb{E}(Z_2) + \ldots + \mathbb{E}(Z_n) \\ &= 1 + \mu + \mu^2 + \ldots + \mu^n \\ &= \begin{cases} \frac{\mu^{n+1}-1}{\mu-1}, & \mu \neq 1, \\ n+1, & \mu = 1. \end{cases} \end{aligned}$$

It follows that

$$\lim_{n \to \infty} \mathbb{E}(T_n) = \begin{cases} \infty, & \mu \geq 1, \\ \frac{1}{1-\mu}, & \mu < 1. \end{cases}$$

Let $H_n(s) = \mathbb{E}(s^{T_n})$ be the generating function for the random variable T_n. The generating functions satisfy the recurrence relation

$$H_{n+1}(s) = sG(H_n(s)). \tag{9.3.4}$$

9.3.2 Extinction and criticality

An important property of a branching process is whether or not it eventually becomes extinct, that is, $Z_n = 0$ for some finite n. This motivates the classification of a branching process is in terms of the asymptotic properties of the mean number of progeny $\mathbb{E}(Z_n)$. Since $\mathbb{E}(Z_n) = \mu^n$ for a Galton–Watson process, we see that it grows geometrically if $\mu > 1$, stays constant if $\mu = 1$ and decays geometrically if $\mu < 1$. These three cases are labeled supercritical, critical, and subcritical, respectively:

$$\mu > 1 \text{ (supercritical)} \implies \lim_{n \to \infty} \mathbb{E}(Z_n) = \infty,$$
$$\mu = 1 \text{ (critical)} \implies \mathbb{E}(Z_n) = 1,$$
$$\mu < 1 \text{ (subcritical)} \implies \lim_{n \to \infty} \mathbb{E}(Z_n) = 0.$$

The relationship between criticality and the probability of extinction η, where

$$\eta = \lim_{n \to \infty} \text{Prob}(Z_n = 0), \qquad (9.3.5)$$

is quite subtle. First note that

$$\eta_n \equiv \text{Prob}(Z_n = 0) = G_n(0) = G(G_{n-1}(0)) = G(\eta_{n-1}).$$

Taking the limit $n \to \infty$ shows that η is a root of the equation $\eta = G(\eta)$. Moreover, if ψ is any nonnegative root of $s = G(s)$ then $\eta \leq \psi$. This follows from the fact that G is a non-decreasing function on $[0, 1]$ so

$$\eta_1 = G(0) \leq G(\psi) = \psi, \eta_2 = G(\eta_1) \leq G(\psi) = \psi, \ldots,$$

so by induction $\eta_n \leq \psi$ for all $n \geq 0$. Hence η is the smallest nonnegative root of the equation $s = G(s)$. The value of η can now be determined graphically by noting that $G(1) = 1$ and $G(s)$ is a convex function of s, see Fig. (9.2). The latter is a

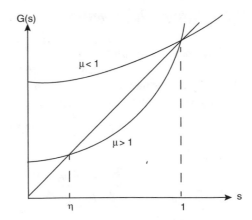

Fig. 9.2: Graphical construction of the probability of extinction η. The convex function $G(s)$, $s \geq 0$, intersects the diagonal twice for $\mu > 1$ but only once when $\mu \leq 1$.

consequence of the result

$$G''(s) = \mathbb{E}(Z_1(Z_1 - 1)s^{Z_1 - 2}) \geq 0, \quad \forall \quad s \geq 0.$$

Hence, if $\mu = G'(1) > 1$ then there exists a root $\eta = G(\eta)$ with $0 \leq \eta < 1$, whereas if $\mu = G'(1) \leq 1$ then $\eta = 1$ (the two roots coincide). We see that the critical process is counterintuitive, since the process becomes extinct almost surely although the mean $\mathbb{E}(Z_n) = 1$ for all $n \geq 0$.

Example 9.7. For the sake of illustration, consider a geometric branching process where the distribution of family sizes is given by $\text{Prob}(Z_1 = k) \equiv f(k) = qp^k$ with $q = 1 - p$. In this case, one can calculate the generating function and other quantities explicitly:

$$G(s) = q(1 - ps)^{-1}, \quad \mu = \frac{p}{q}, \quad \sigma^2 = \frac{p^2}{q^2} + \frac{p}{q}. \tag{9.3.6}$$

Moreover, it can be shown by induction that

$$G_n(s) = \begin{cases} \dfrac{n - (n-1)s}{n + 1 - ns} & \text{if } p = q = \frac{1}{2}, \\[3mm] \dfrac{q[p^n - q^n - ps(p^{n-1} - q^{n-1})]}{p^{n+1} - q^{n+1} - ps(p^n - q^n)} & \text{if } p \neq q. \end{cases} \tag{9.3.7}$$

It follows that

$$\text{Prob}(Z_n = 0) = G_n(0) = \begin{cases} \dfrac{n}{n+1} & \text{if } p = q, \\[3mm] \dfrac{q(p^n - q^n)}{p^{n+1} - q^{n+1}} & \text{if } p \neq q, \end{cases} \tag{9.3.8}$$

and, hence, $\eta = 1$ if $p \leq q$ and $\eta = q/p$ if $p > q$. We conclude that for a geometric branching process, extinction occurs almost surely if $\mathbb{E}(Z_1) = \mu = p/q \leq 1$, otherwise there is a finite probability of persistent growth. We can identify the regime $p < q$ as subcritical, the regime $p > q$ as supercritical and the point $p = q = 1/2$ as critical.

9.3.3 Asymptotic properties

The asymptotic properties of the Galton–Watson process can be analyzed using the convergence properties of martingales, as summarized in Theorem 13.2. The applicability of this theorem to the Galton–Watson process follows from the fact that $W_n = Z_n/\mu^n$ is a martingale. That is, since the branching process is a Markov chain,

$$\mathbb{E}(Z_{n+1}|Z_n, \ldots, Z_0) = \mathbb{E}(Z_{n+1}|Z_n),$$

and thus

$$\mathbb{E}(W_{n+1}|W_n, \ldots, W_0) = \mathbb{E}(W_{n+1}|W_n).$$

Moreover, since Z_{n+1} is the sum of Z_n independent families, we have

$$\mathbb{E}(Z_{n+1}|Z_n) = \mu Z_n.$$

Similarly,

$$\mathbb{E}(W_{n+1}|W_n) = \mu^{-(n+1)}\mathbb{E}(Z_{n+1}|Z_n) = \frac{Z_n}{\mu^n} = W_n.$$

We deduce that (W_n) is a martingale:

$$\mathbb{E}(W_{n+1}|W_n,\ldots,W_0) = W_n.$$

The convergence theorem thus establishes that there exists a random variable W for which

$$\lim_{n\to\infty} W_n = W \quad \text{almost surely.}$$

Since $\eta = 1$ in the critical and subcritical cases, it follows that $W \equiv 0$. Hence, we assume the process is supercritical. Now introduce the discrete Laplace transform of W_n given by $\phi_n(s) = \mathbb{E}(e^{-sW_n})$. We have

$$\begin{aligned}
\phi_n(s) = \mathbb{E}(e^{-sW_n}) &= \mathbb{E}[(e^{-s/\mu^n})^{Z_n}] \\
&= G_n(e^{-s/\mu^n}) = G(G_{n-1}(e^{-s/\mu^n})) \\
&= G(\phi_{n-1}(s/\mu)).
\end{aligned}$$

Since $W_n \to W$ almost surely implies $W_n \to W$ in distribution, taking the limit $n \to \infty$ shows that $\phi_n(s) \to \phi(s)$, with ϕ satisfying the so-called Abel's equation

$$\phi(s) = G[\phi(s/\mu)]. \tag{9.3.9}$$

There are also nontrivial limit theorems for subcritical and critical process conditions on non-extinction. We simply quote the theorems here (see Athreya and Ney [34] for proofs).

Theorem 9.3. *If $\mu < 1$ (subcritical) then* $\text{Prob}(Z_n|Z_n > 0)$ *converges as $n \to \infty$ to a probability distribution whose generating function $B(s)$ satisfies*

$$B[G(s)] = \mu B(s) + 1 - \mu,$$

and the probability of non extinction has the asymptotic form

$$1 - \eta_n \sim \frac{\mu^n}{B'(1)}, \quad n \to \infty.$$

Theorem 9.4. *If $\mu = 1$ (critical) and $\sigma^2 = var(Z_1) < \infty$, then*

$$\lim_{n\to\infty} \text{Prob}(Z_n/n > z|Z_n > 0) = e^{-2z/\sigma^2}, \quad z \geq 0.$$

9.3.4 Application to gene amplification

There are many applications of the Galton–Watson process to models of cell proliferation and population genetics, as reviewed in [445]. Here we illustrate the theory
by considering the particular example of gene amplification [444, 445]. This refers
to the increase in the number of copies of a gene through successive cell generations. One important example is the amplification of genes coding for the enzyme
dihydrofolate reductase (DHFR), which has been associated with cellular resistance
to the anticancer drug methotrexate (MTX). A Galton–Watson process can be used
to model the number Z_n of DHFR genes in a cell randomly selected from cell progeny in the nth generation of repeated cycles of cell replication and division, see Fig.
9.3. During the lifetime of a cell, each DHFR gene is either replicated with probability a or not replicated with probability $1-a$. In the former case, at the time of cell
division, DHFR is assigned to one of the two daughter cells with probability $1/2$.
On the other hand, if replication occurs, then during cell division both copies are
assigned to one daughter cell with probability $\alpha/2$ or each daughter cell receives
one copy with probability $1-\alpha$. It follows that for a randomly selected cell of the
first generation, given $Z_0 = 1$, we have

$$\text{Prob}(Z_1 = 0) = (1-a)/2 + a\alpha/2 \equiv A_0, \quad \text{Prob}(Z_1 = 1) = 1 - A_0 - A_2,$$
$$\text{Prob}(Z_1 = 2) = a\alpha/2 = A_2.$$

Hence

$$G(s) = \mathbb{E}(s^{Z_1}) = A_0 + (1 - A_0 - A_2)s + A_2 s^2.$$

It is assumed that in the absence of selection, DHFR gradually disappears from the
cell population so that the branching process is subcritical. This means that

$$\mu \equiv G'(1) = 1 - A_0 + A_2 < 1 \implies A_2 < A_0.$$

A cell is said to be resistant if it contains at least one copy of the DHFR gene,
otherwise it is called sensitive. Suppose that cells are initially cultured in a medium
rich in MTX so that they develop drug resistance in the sense that all cells have at
least one copy of DHFR. After N generations the distribution of copy numbers is
given by $\text{Prob}(Z_N | Z_N > 0)$. Moreover, since the probability of nonextinction scales
as μ^n for large n, it follows that the probability of a daughter cell be resistant is μ so
that the number of resistant cells grows on average by 2μ per generation (assuming
$1/2 < \mu < 1$). Now suppose that at the Nth generation the drug-resistant cells are
placed in a drug-free medium so sensitive cells also proliferate. Let $R(n)$ and $S(n)$
denote the number of resistant and sensitive cells after n cycles within the drug-free
medium with $S(0) = 0$. Using the fact that $R(n) = (2\mu)^n R(0)$ and $S(n) + R(n) = 2^n R(0)$, we see that the fraction of drug-resistant cells scales as

$$\frac{r(n)}{r(0)} = \mu^n \to 0 \text{ as } n \to \infty, \quad r(n) = \frac{R(n)}{R(n) + S(n)}.$$

Meanwhile the distribution of copy numbers among the drug-resistant population is preserved after normalization, consistent with experimental observations.

9.4 Modeling chemical reaction networks as counting processes

As our final example illustrating the application of martingales, we turn to a stochastic formulation of biochemical reaction networks based on counting processes, which provides a framework for applying rigorous probabilistic methods such as the theory of continuous-time martingales [17, 476]. The reaction network is treated as a continuous-time Markov chain with reactions corresponding to transitions along the chain. The number of occurrences of each reaction is modeled as a counting process, which is itself represented in terms of a scaled Poisson process. This so-called *random time-change representation* yields a stochastic equation for the Markov chain. Our presentation follows closely the introductory review by Anderson and Kurtz [17].

9.4.1 Poisson processes and counting processes

Let us first briefly recall properties of a Poisson process $Y(t)$, which represents the number of events or observations up to time t with the following properties:

1. Observations occur one at a time.

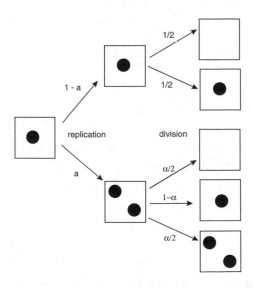

Fig. 9.3: Schematic illustration of a branching process model of gene amplification. See text for details.

2. The number of observations in disjoint time intervals are independent random variables. That is, if $t_0 < t_1 < \cdots < t_m$ then $Y(t_k) - Y(t_{k-1})$, $k = 1, \ldots, m$ are independent random variables.
3. The distribution of $Y(t+a) - Y(t)$ is independent of t.

Under the above assumptions it can be shown that there exists a constant $\lambda > 0$ such that, for $t < s$, $Y(s) - Y(t)$ is Poisson distributed with rate λ:

$$\mathbb{P}[Y(s) - Y(t) = k] = \frac{\lambda(s-t)^k}{k!} e^{-\lambda(s-t)}.$$

In the following, we will take $Y(t)$ to be a unit rate Poisson process ($\lambda = 1$) and denote a Poisson process with rate λ by the time-change representation $Y(\lambda t)$. Suppose that $\mathcal{F}_t$ represents the information obtained by observing $Y(\lambda s)$ for $s \leq t$. It follows that

$$\mathbb{P}[Y(\lambda(t+\Delta t)) - Y(\lambda t) = 1|\mathcal{F}_t] = \mathbb{P}[Y(\lambda(t+\Delta t)) - Y(\lambda t) = 1] = \lambda \Delta t e^{-\lambda \Delta t}.$$

Taking the limit $\Delta t \to 0$, we have the formal limit

$$\mathbb{P}[dY(\lambda t) = 1|\mathcal{F}_t] \equiv \mathbb{E}[dY(\lambda t)|\mathcal{F}_t] = \lambda dt. \tag{9.4.1}$$

Thus one can interpret the rate λ as the expected number of jumps per unit time (transition intensity or propensity). The usefulness of this definition of λ is that it is intuitive and can easily be generalized to a large class of counting processes[1], including those that keep track of the number of single-step chemical reactions (see below). Moreover, it forms the starting point of a mathematically rigorous theory of counting process that involves the theory of continuous-time martingales. Indeed, a more precise version of the formal relation (9.4.1) is that in terms of the filtration $\{\mathcal{F}_t, t \geq 0\}$, the stochastic process $M(t) = Y(\lambda t) - \lambda t$ is a martingale. This follows from

$$\mathbb{E}[Y(\lambda[t+s]) - Y(\lambda t)|\mathcal{F}_t)] = \lambda s \implies \mathbb{E}[Y(\lambda[t+s])|\mathcal{F}_t)] = Y(\lambda t) - \lambda t + \lambda(t-s),$$

that is,

$$\mathbb{E}[M(t+s)|\mathcal{F}_t] = M(t)$$

for all $s, t \geq 0$.

The above results generalize to a Poisson process with a time-dependent rate function $\lambda(t)$ so that the counting process $N(t) = Y(\lambda(t))$:

1. The random variable $Y(\lambda(t)) - Y(\lambda(s))$ is independent of $\mathcal{F}_s$ for $t > s$.
2. The conditional distribution of the increments is given by

---

[1] A counting process is a stochastic process $\{N(t), t \geq 0\}$ satisfying the following properties: $N(t) \geq 0$; $N(t)$ is an integer; If $s \leq t$ then $N(s) \leq N(t)$ (a positive, increasing integer).

$$\mathbb{P}[Y(\lambda(t)) - Y(\lambda(s)) = k|\mathcal{F}_s] = \frac{\Lambda_{s,t}^k}{k!}e^{-\Lambda_{s,t}}, \quad \Lambda_{s,t} = \int_s^t \lambda(u)du.$$

3. The stochastic process

$$M(t) = Y(\lambda(t)) - \int_0^t \lambda(u)du$$

is a martingale with respect to the filtration $\{\mathcal{F}_t\}$.

The converse relation also holds, namely, if $M(t)$ is a martingale with respect to $\mathcal{F}_t$ then $N(t) = Y(\lambda(t))$, that is, $N(t)$ is a Poisson process (Watanabe Theorem).

In order to develop a mathematical theory of chemical reaction networks, it is first necessary to consider a more general counting process $N(t)$ in which the intensity $\lambda(t;N)$ is itself a stochastic process adapted to the filtration $\{\mathcal{F}_t\}$. In other words, if $\mathcal{F}_t$ represents all information about the counting process up to time t then $\lambda(t;N)$ is specified, that is, it is non-anticipating. For concreteness, take $\lambda(t;N) = \lambda(N(t))$ and consider the counting process given by the solution to the stochastic equation

$$N(t) = Y\left(\int_0^t \lambda(N(s))ds\right), \tag{9.4.2}$$

with $\int_0^t \lambda(N(s))ds < \infty$ for all $t \geq 0$. Note that in the infinitesimal interval $(t, t+\Delta t]$,

$$\mathbb{P}[N(t+\Delta t) > N(t)|\mathcal{F}_t] = 1 - \mathbb{P}[N(t+\Delta t) = N(t)|\mathcal{F}_t]$$

$$= 1 - \mathbb{P}\left[Y\left(\int_0^{t+\Delta t} \lambda(N(s))ds\right) - Y\left(\int_0^t \lambda(N(s))ds\right) = 0\right]$$

$$= 1 - e^{-\lambda(N(t))\Delta t} \approx \lambda(N(t))\Delta t.$$

Thus $\lambda(N(t))$ can still be identified as a transition intensity. The relationship to martingales is now a little more involved. First, define the stochastic jump times

$$\tau_k = \inf\{t, t \geq 0 | N(t) \geq k\},$$

with τ_k the time of the kth jump, and introduce the notation $\tau_k \wedge t = \min(\tau_k, t)$. Define the following set of stochastic processes

$$M_k(t) = N(t \wedge \tau_k) - \int_0^{t \wedge \tau_k} \lambda(N(s))ds. \tag{9.4.3}$$

Given the counting process $N(t)$ satisfying equation (9.4.2), $M_k(t)$ is a martingale for all $k \geq 0$, see Box 9A, that is

$$\mathbb{E}[M_k(t+s)|\mathcal{F}_t] = M_k(t), \quad t, s \geq 0.$$

Moreover, if $\mathbb{E}[N(t)] < \infty$ for all $t \geq 0$, then $\lim_{k \to \infty} \tau_k \equiv \tau_\infty = \infty$ and

$$M(t) \equiv \lim_{k \to \infty} M_k(t) = N(t) - \int_0^t \lambda(N(u))du$$

is a martingale. Since $M(0) = 0$ (assuming $N(0) = 0$), it follows that $\mathbb{E}[M(t)] = 0$ for all $t \geq 0$, that is,

$$\mathbb{E}\left[Y\left(\int_0^t \lambda(N(s))ds\right)\right] = \mathbb{E}\left[\int_0^t \lambda(N(u))du\right].$$

Again, the converse relation holds: suppose that $N(t)$ is a counting process with transition intensity $\lambda(N(t))$ such that for each k, $M_k(t)$ is a martingale. Then $N(t)$ is the solution to the stochastic equation (9.4.2).

Box 9A: Martingales of a counting process.

We sketch a proof that $M_k(t)$ is a martingale for the counting process $N(t)$ satisfying the stochastic equation (9.4.2). Suppose that $\tau_{k-1} < t < \tau_k$. Since τ_{k-1} is adapted to $\mathcal{F}_t$, it follows that τ_{k-1} is known when conditioning expectations with respect to $\mathcal{F}_t$. On the other hand, τ_k is a random variable. It is clear from the definition of $M_k(t)$ that for all $j \leq k - 1$

$$\mathbb{E}[M_j(t+s)|\mathcal{F}_t] = N(\tau_j) - \int_0^{\tau_j} \lambda(N(u))du = M_j(t).$$

In the case $j = k$, we have to consider two cases: (i) $\tau_k < t + s$, which means that at least one jump occurs in the interval $(t, t+s]$ and $N(t+s) - N(t) > 0$; (ii) $\tau_k \geq t + s$, which means that no jump occurs in $(t, t+s)$ and $N(t+s) = N(t)$. Therefore,

$$\mathbb{E}[M_k(t+s)|\mathcal{F}_t] = \int_{t+s}^{\infty} \mathbb{P}(\tau_k|\mathcal{F}_t)\left[N(t+s) - \int_0^{t+s} \lambda(N(u))du\right]d\tau_k$$

$$+ \int_t^{t+s} \mathbb{P}(\tau_k|\mathcal{F}_t)\left[N(\tau_k) - \int_0^{\tau_k} \lambda(N(u))du\right]d\tau_k$$

$$= \int_{t+s}^{\infty} \mathbb{P}(\tau_k|\mathcal{F}_t)\left[N(t) - \int_0^t \lambda(N(u))du - \int_t^{t+s} \lambda(N(u))du\right]d\tau_k$$

$$+ \int_t^{t+s} \mathbb{P}(\tau_k|\mathcal{F}_t)\left[N(t) + 1 - \int_0^t \lambda(N(u))du - \int_t^{\tau_k} \lambda(N(u))du\right]d\tau_k$$

$$= \int_{t+s}^{\infty} \mathbb{P}(\tau_k|\mathcal{F}_t)\left[M_k(t) - \int_t^{t+s} \lambda(N(u))du\right]d\tau_k$$

$$+ \int_t^{s+t} \mathbb{P}(\tau_k|\mathcal{F}_t)\left[M_k(t) + 1 - \int_t^{\tau_k} \lambda(N(u))du\right]d\tau_k.$$

Hence,

$$\mathbb{E}[M_k(t+s)|\mathcal{F}_t] - M_k(t) = -\int_{s+t}^{\infty} \mathbb{P}(\tau_k|\mathcal{F}_t)d\tau_k \left[\int_t^{t+s} \lambda(N(u))du \right]$$
$$+ \int_t^{s+t} \mathbb{P}(\tau_k|\mathcal{F}_t) \left[1 - \int_t^{\tau_k} \lambda(N(u))du \right] d\tau_k,$$

where

$$\mathbb{P}(\tau_k|\mathcal{F}_t) = C\exp\left(-\int_t^{\tau_k} \lambda(N(u))du \right),$$

and C is a normalization factor. Conditioning in $\mathcal{F}_t$ and the absence of any subsequent jumps means that $\lambda(N(u))$ can be treated as a constant λ_0. Then

$$\mathbb{E}[M_k(t+s)|\mathcal{F}_t] - M_k(t) = -Cs\lambda_0 \int_{s+t}^{\infty} e^{-\lambda_0(\tau_k-t)}d\tau_k$$
$$+ C\int_t^{s+t} e^{-\lambda_0(\tau_k-t)} \left[1 - (\tau_k - t)\lambda_0 \right] d\tau_k$$
$$= -Cse^{-\lambda_0 s} + C\frac{1-e^{-\lambda_0 s}}{\lambda_0} + C\lambda_0 \frac{d}{d\lambda_0} \frac{1-e^{-\lambda_0 s}}{\lambda_0} = 0.$$

Thus

$$\mathbb{E}[M_k(t+s)|\mathcal{F}_t] = M_k(t) \quad \text{for all } 0 \leq t < \tau_k, \quad s \geq 0.$$

Since $\mathbb{E}[M_k(t+s)|\mathcal{F}_t] = M_k(t)$ for all $t > \tau_k$, it follows that $M_k(t)$ is a martingale. This result holds for all $k \geq 0$.

The above martingale formulation can be used to derive the master equation for the counting process. Introduce an arbitrary, bounded function f on $\mathbb{Z}$. Suppose that $N(t) = n$ and τ_k is the nth jump time. Then

$$f(N(t)) = f(0) + \sum_{k=1}^{n} [f(k) - f(k-1)] = f(0) + \sum_{k=1}^{n} [f(N(\tau_k)) - f(N(\tau_{k-1}))]$$
$$= f(0) + \sum_{k=1}^{n} [f(N(\tau_{k-1}) + 1) - f(N(\tau_{k-1}))]$$
$$= f(0) + \sum_{k=1}^{\infty} \int_0^t [f(N(s) + 1) - f(N(s))]\delta(s - \tau_{k-1})ds,$$
$$= f(0) + \sum_{k=1}^{\infty} \int_0^t [f(N(s-) + 1) - f(N(s-))]\delta(s - \tau_k)ds,$$

which is independent of the particular value n. Formally speaking,

$$\sum_{k=1}^{\infty} \delta(s - \tau_k) = dN(t),$$

so that

$$f(N(t)) = f(0) + \int_0^t [f(N(s-)+1) - f(N(s-))]dN(s). \qquad (9.4.4)$$

Setting $N(t) = M(t) + \int_0^t \lambda(N(u))du$ with $M(t)$ a martingale, we have

$$f(N(t)) - f(0) - \int_0^t \lambda(N(s))[f(N(s)+1) - f(N(s))]ds$$

$$= \int_0^t [f(N(s-)+1) - f(N(s-))]dM(s).$$

Introducing the generator

$$\mathbb{A}f(n) = \lambda(n)[f(n+1) - f(n)],$$

and using the fact that $M(s)$ is a martingale, it follows that

$$f(N(t)) - f(N(0)) - \int_0^t \mathbb{A}f(N(s))ds \qquad (9.4.5)$$

is a martingale and, in particular

$$\mathbb{E}[f(N(t))] = \mathbb{E}[f(N(0))] + \int_0^t \mathbb{E}[\mathbb{A}f(N(s))]ds. \qquad (9.4.6)$$

Consider the index function $f(N) = \chi_n(N) = 1$ if $N = n$ and zero otherwise. Then $\mathbb{E}[f(N(t))] = \mathbb{P}[N(t) = n]$ and

$$\mathbb{P}[N(t) = n] = \mathbb{P}[N(0) = n] + \int_0^t [\lambda(n-1)\mathbb{P}[N(s) = n-1] - \lambda(n)\mathbb{P}[N(s) = n]].$$

Differentiating both sides with respect to t and setting $p(n,t) = \mathbb{P}[N(t) = n]$ yields the forward master equation for the counting process

$$\frac{dp(n,t)}{dt} = [\lambda(n-1)p(n-1,t) - \lambda(n)p(n,t)]. \qquad (9.4.7)$$

9.4.2 Chemical reactions and counting processes

Now consider the simple single-step reaction

$$A + B \xrightarrow{\kappa} C.$$

Let $\mathbf{X}(t) = (X_A(t), X_B(t), X_C(t))$ be the state of the stochastic process at time t with $X_i(t)$ the number of molecules of chemical species i at time t. From simple bookkeeping, we can write

$$\mathbf{X}(t) = \mathbf{X}(0) + R(t) \begin{pmatrix} -1 \\ -1 \\ 1 \end{pmatrix}, \tag{9.4.8}$$

where $R(t)$ is the number of reactions that has occurred by time t, $\mathbf{X}(0)$ is the initial state, and the constant vector specifies the stoichiometric coefficients. We will assume that the probability of a reaction in an infinitesimal interval $(t, t + \Delta t]$ is

$$\mathbb{P}[\text{reaction occurs in } (t, t + \Delta t] | \mathcal{F}_t] \approx \kappa X_A(t) X_B(t) \Delta t, \tag{9.4.9}$$

where $\mathcal{F}_t$ is the information obtained by observing the stochastic process up to time t. Equation (9.4.9) is consistent with the law of mass action in the deterministic limit. The basic idea of the time-change representation is that the number of reactions can be expressed in terms of a unit rate Poisson process according to

$$R(t) = Y \left(\int_0^t \kappa X_A(s) X_B(s) ds \right). \tag{9.4.10}$$

In order to establish that this is consistent with equation (9.4.9), set $\lambda(\mathbf{X}(t)) = \kappa X_A(t) X_B(t)$ and note that the probability a reaction occurs in a small time interval $(t, t + \Delta t]$ is

$$\mathbb{P}[R(t + \Delta t) > R(t) | \mathcal{F}_t] = 1 - \mathbb{P}[R(t + \Delta t) = R(t) | \mathcal{F}_t]$$
$$= 1 - \mathbb{P}\left[Y \left(\int_0^{t+\Delta t} \lambda(\mathbf{X}(s)) ds \right) - Y \left(\int_0^t \lambda(\mathbf{X}(s)) ds \right) = 0 \right]$$
$$= 1 - e^{-\lambda(\mathbf{X}(t))\Delta t} \approx \lambda(\mathbf{X}(t)) \Delta t.$$

We have used the fact that $Y \left(\int_0^t \lambda(\mathbf{X}(s)) ds \right)$ and $\mathbf{X}(t)$ are part of the information in $\mathcal{F}_t$, that is, they are $\mathcal{F}_t$-measurable. Similarly, for a general set of chemical reactions involving N chemical species $i = 1, \dots, N$ and K single-step reactions with stoichiometric vectors $\mathbf{S}_a \in \mathbb{Z}^N$ and propensity functions λ_a, $a = 1, \dots, K$ (Chap. 5), we have

$$\mathbf{X}(t) = \mathbf{X}(0) + \sum_a \mathbf{S}_a R_a(t), \tag{9.4.11}$$

where $\mathbf{X}(t) = (X_1(t), \dots, X_N(t))$ and $R_a(t)$ is the number of occurrences of reaction a up to time t. Moreover,

$$R_a(t) = Y_a \left(\int_0^t \lambda_a(\mathbf{X}(s)) ds \right), \tag{9.4.12}$$

where the Y_a are independent unit rate Poisson processes. Following along similar lines to our analysis of a simple counting process, it can be shown that the counting processes $(R_1, \dots, R_K)$ are solutions of the stochastic process defined by equations (9.4.11) and (9.4.12) if and only if they are solutions of a corresponding martingale problem with respect to the intensities λ_a: there exists a filtration $\{\mathcal{F}_t\}$ to which the R_a are adapted and

$$M_{a,k}(t) \equiv R_a(t \wedge \tau_k) - \int_0^{t \wedge \tau_k} \lambda_a(\mathbf{X}(s))ds$$

is a $\{\mathcal{F}_t\}$-martingale.

The martingale problem can be used to derive the master equation for a chemical reaction network. For simplicity, suppose that $\tau_\infty = \infty$ and $\mathbb{E}[R_a(t)] < \infty$. (This condition can be relaxed [17]). It follows that

$$M_a(t) = R_a(t) - \int_0^t \lambda_a(\mathbf{X}(s))ds$$

is a martingale so that if $R_a(0) = 0$ for all a, then

$$\mathbb{E}[R_a(t)] = \mathbb{E}\left[\int_0^t \lambda_a(\mathbf{X}(s))ds\right].$$

Introduce an arbitrary, bounded function f on $\mathbb{Z}^N$. Generalizing equation (9.4.4), we have

$$f(\mathbf{X}(t)) = f(\mathbf{X}(0)) + \sum_a \int_0^t [f(\mathbf{X}(s-) + \mathbf{S}_a) - f(\mathbf{X}(s-))] dR_a(s),$$

which can be rewritten as

$$f(\mathbf{X}(t)) - f(\mathbf{X}(0)) - \int_0^t \sum_a \lambda_a(\mathbf{X}(s))[f(\mathbf{X}(s) + \mathbf{S}_a) - f(\mathbf{X}(s))]ds$$

$$= \sum_a \int_0^t [f(\mathbf{X}(s-) + \mathbf{S}_a) - f(\mathbf{X}(s-))] dM_a(s).$$

Introducing the generator (of the master equation)

$$\mathbb{A}f(x) = \sum_a \lambda_a(\mathbf{x})[f(\mathbf{x} + \mathbf{S}_a) - f(\mathbf{x})],$$

and using the fact that $M_a(s)$ is a martingale, it follows that

$$f(\mathbf{X}(t)) - f(\mathbf{X}(0)) - \int_0^t \mathbb{A}f(\mathbf{X}(s))ds \qquad (9.4.13)$$

is a martingale and, in particular

$$\mathbb{E}[f(\mathbf{X}(t))] = \mathbb{E}[f(\mathbf{X}(0))] + \int_0^t \mathbb{E}[\mathbb{A}f(\mathbf{X}(s))]ds. \qquad (9.4.14)$$

Consider the index function $f(\mathbf{x}) = \chi_{\mathbf{y}}(\mathbf{x}) = 1$ if $\mathbf{y} = \mathbf{x}$ and zero otherwise. Then $\mathbb{E}[f(\mathbf{X}(t))] = \mathbb{P}[\mathbf{X}(t) = \mathbf{y}]$ and

$$\mathbb{P}[\mathbf{X}(t) = \mathbf{y}] = \mathbb{P}[\mathbf{X}(0) = \mathbf{y}]$$
$$+ \int_0^t \sum_a (\lambda_a(\mathbf{y} - \mathbf{S}_a)\mathbb{P}[\mathbf{X}(s) = \mathbf{y} - \mathbf{S}_a] - \lambda_a(\mathbf{y})\mathbb{P}[\mathbf{X}(s) = \mathbf{y}]) \, ds.$$

Differentiating both sides with respect to t and setting $p(\mathbf{y}, t) = \mathbb{P}[\mathbf{X}(t) = \mathbf{y}]$ yields the forward master equation

$$\frac{dp(\mathbf{y}, t)}{dt} = \sum_a [\lambda_a(\mathbf{y} - \mathbf{S}_a)p(\mathbf{y} - \mathbf{S}_a, t) - \lambda_a(\mathbf{y})p(\mathbf{y}, t)]. \tag{9.4.15}$$

Having shown how to reformulate chemical reaction networks in terms of counting processes and martingales, we briefly highlight some of the applications of the latter [17].

1. The martingale properties of the counting process R_a provide an effective method for evaluating moments of the chemical processes $X_i(t)$.
2. One can obtain rigorous asymptotic estimates for convergence to the deterministic rate equations in the large N limit, where N is the system size (number of molecules). One can also carry out a rigorous slow/fast decomposition in the case of multiple time scales.
3. The concept of the generator $\mathbb{A}$ and its associated martingales can be extended to the case of continuous stochastic processes based on a Langevin equation. Indeed, carrying out a system-size expansion of the master equation generator leads to the second-order differential operator of the Fokker–Planck equation
4. Equations (9.4.11) and (9.4.12) provide the basis for stochastic simulation algorithms such as Gillespie's direct method [310] (Chap. 5).

References

1. Acar, M., Mettetal, J.T., van Oudenaarden, A.: Stochastic switching as a survival strategy in fluctuating environments. Nat. Gen. **40**, 471–475 (2008)
2. Adib, A.: Stochastic actions for diffusive dynamics? Reweighting, sampling, and minimization. J. Phys. Chem. B **112**, 5910–5916 (2008)
3. Adrian, M., Kusters, R., Storm, C., Hoogenraad, C.C., Kapitein, L.C.: Probing the interplay between dendritic spine morphology and membrane-bound diffusion. Biophys. J. **113**, 2261–2270 (2017)
4. Aghababaie, Y., Menon, G.I., Plischke, M.: Universal properties of interacting Brownian motors. Phys. Rev. E **59**, 2578–2586 (1999)
5. Agmon, N., Hopfield, J.J.: Transient kinetics of chemical reactions with bounded diffusion perpendicular to the reaction coordinate: intramolecular processes with slow conformational changes. J. Chem. Phys. **80**, 592 (1984)
6. Agmon, N., Szabo, A.: Theory of reversible diffusion-influenced reactions. J. Chem. Phys. **92**, 5270–5284 (1990)
7. Agmon, N.: Single molecule diffusion and the solution of the spherically symmetric residence time equation. J. Phys. Chem. **115**, 5838–5846 (2011)
8. Akimoto, T., Yamamoto, E.: Distributional behaviors of time-averaged observables in the Langevin equation with fluctuating diffusivity: normal diffusion but anomalous fluctuations. Phys. Rev. E **93**, 062109 (2016)
9. Alberts, B., Johnson, A., Lewis, J., Raff, M., Walter, K.R.: Molecular Biology of the Cell, 5th edn. Garland, New York (2008)
10. Allen, L.J.S.: An Introduction to Stochastic Processes with Applications to Biology, 2nd edn. Chapman and Hall/CRC (2010)
11. Alon, U.: An Introduction to Systems Biology: Design Principles of Biological Circuits. Chapman and Hall/CRC (2007)
12. Alsing, A.K., Sneppen, K.: Differentiation of developing olfactory neurons analysed in terms of coupled epigenetic landscapes. Nucl. Acids Res. **41**, 4755–4764 (2013)
13. Amir, A.: Thinking Probabilistically. Cambridge University Press, Cambridge (2021)
14. Anderson, D.F.: A modified next reaction method for simulating chemical systems with time dependent propensities and delays. J. Chem. Phys. **127**, 214107 (2007)
15. Anderson, D.F., Craciun, G., Kurtz, T.G.: Product-form stationary distributions for deficiency zero chemical reaction networks. Bull. Math. Biol. **72**, 1947–1970 (2010)

© Springer Nature Switzerland AG 2021

P. C. Bressloff, *Stochastic Processes in Cell Biology*, Interdisciplinary
Applied Mathematics 41, https://doi.org/10.1007/978-3-030-72515-0

16. Anderson, D.F., Shiu, A.: The dynamics of weakly reversible population processes near facets. SIAM J. Appl. Math. **70**, 1840–1858 (2010)

17. Anderson, D.F., Kurtz, T.G.: Continuous time Markov chain models for chemical reaction networks. In: Design and Analysis of Biomolecular Circuits, pp. 3–42 (2011)

18. Anderson, D.F., Ermentrout, G.B., Thomas, P.J.: Stochastic representations of ion channel kinetics and exact stochastic simulation of neuronal dynamics. J. Comp Neurosci. **38**, 67–82 (2015)

19. Andrews, S., Bray, D.: Stochastic simulation of chemical reactions with spatial resolution and single molecule detail. Phys. Biol. **1**, 137–151 (2004)

20. Andrews, S.S.: Serial rebinding of ligands to clustered receptors as exemplified by bacterial chemotaxis. Phys. Biol. **2**, 111–122 (2005)

21. Angelani, L.: Run-and-tumble particles, telegrapher's equation and absorption problems with partially reflecting boundaries. J. Phys. A **48**, 495003 (2015)

22. Angelani, L.: Confined run-and-tumble swimmers in one dimension. J. Phys. A **50**, 325601 (2017)

23. Angeli, D., De Leenheer, P., Sontag, E.D.: A petri net approach to the study of persistence in chemical reaction networks. Math. Biosci. **210**, 598–618 (2007)

24. Ankerhold, J., Pechukas, P.: Mathematical aspects of the fluctuating barrier problem. Explicit equilibrium and relaxation solutions. Physica A **261**, 458–470 (1998)

25. Antal, T., Krapivsky, P. L., Redner, S., Mailman, M., Chakraborty.: Dynamics of an idealized model of microtubule growth and catastrophe. Phys. Rev. E **76**, 041907 (2007)

26. Arkin, A., Ross, J., McAdams, H.H.: Stochastic kinetic analysis of developmental pathway bifurcation in phage infected escherichia coli cells. Genetics **149**, 1633–1648 (1998)

27. Ashall, L., Horton, C.A., Nelson, D.E., Paszek, P., Harper, C.V., et al.: Pulsatile stimulation determines timing and specificity of NF-kappaB-dependent transcription. Science **324**, 242–246 (2009)

28. Ashe, H.L., Briscoe, J.: The interpretation of morphogen gradients. Development **133**, 385–394 (2006)

29. Assaf, M., Meerson, B.: Spectral theory of metastability and extinction in birth-death systems. Phys. Rev. Lett. **97**, 200602 (2006)

30. Assaf, M., Kamenev, A., Meerson, B.: Population extinction risk in the aftermath of a catastrophic event. Phys. Rev. E **79**, 011127 (2009)

31. Assaf, M.: Meerson, B: Extinction of metastable stochastic populations. Phys. Rev. E **81**, 021116 (2010)

32. Assmann, M.-A., Lenz, P.: Characterization of bidirectional molecular motor-assisted transport models. Phys. Biol. **10**, 016003 (2013)

33. Astumian, R.D., Bier, M.: Mechanochemical coupling of the motion of molecular motors to ATP hydrolysis. Biophys. J. **70**, 637–653 (1996)

34. Athreya, K.B., Ney, P.E.: Branching Processes. Springer, Berlin (1972)

35. Atkinson, M.R., Savageau, M.A., Myers, J.T., Ninfa, A.J.: Development of genetic circuitry exhibiting toggle switch or oscillatory behavior in Escherichia coli. Cell **113**, 597 (2003)

36. Altland, A., Simons, B.D.: Condensed Matter Field Theory, 2nd edn. Cambridge University Press (2010)

37. Aurell, E., Sneppen, K.: Epigenetics as a first exit problem. Phys. Rev. Lett. **88**, 048101 (2002)

38. Aurell, E., Sneppen, K.: Stability puzzles in phage λ. Phys. Rev. E **65**, 051914 (2002)

39. Axelrod, D., Koppel, D.E., Schlessinger, J., Elson, E., Webb, W.W.: Mobility measurement by analysis of fluorescence photobleaching recovery kinetics. Biophys. J. **16**, 1055–1069 (1976)

40. Baas, P.W., Deitch, J.S., Black, M.M., Banker, G.A.: Polarity orientation of microtubules in hippocampal neurons: Uniformity in the axon and nonuniformity in the dendrite. Proc. Natl. Acad. Sci. USA **85**, 8335–8339 (1988)

41. Bai, L., Santangelo, T.J., Wang, M.D.: Single-molecule analysis of RNA polymerase transcription. Annu. Rev. Biophys. Biomol. Struct. **35**, 343–360 (2006)

42. Bakhtin, Y., Hurth, T., Lawley, S.D., Mattingly, J.C.: Smooth: invariant densities for random switching on the torus. arXiv:1708.01390 (2017)
43. Balakrishnan, V., Chaturvedi, S.: Persistent diffusion on a line. Physica A **148**, 581–596 (1988)
44. Banjade, S., Rosen, M.K.: Phase transitions of multivalent proteins can promote clustering of membrane receptors. eLife **3**, e04123 (2014)
45. Banks, D.S., Fradin, C.: Anomalous diffusion of proteins due to molecular crowding. Biophys. J. **89**, 2960–2971 (2005)
46. Bannai, H., Inoue, T., Nakayama, T., Hattori, M., Mikoshiba, K.: Kinesin dependent, rapid, bi-directional transport of ER sub-compartment in dendrites of hippocampal neurons. J. Cell Sci. **117**, 163–175 (2004)
47. Baras, F., Mansour, M., Malek, M., Pearson, J.E.: Microscopic simulation of chemical bistability in homogeneous systems. J. Chem. Phys. **105**, 8257–8261 (1996)
48. Barkai, N., Leibler, S.: Circadian clocks limited by noise. Nature **403**, 267–268 (2000)
49. Barkai, E., Silbey, R.: Theory of single file diffusion in a force field. Phy. Rev. Lett. **102**, 050602 (2009)
50. Barnhart, E.L., Lee, K.C., Keren, K., Mogilner, A., Theriot, J.A.: An adhesion-dependent switch between mechanisms that determine motile cell shape. PLoS Biol. **9**, e1001059 (2011)
51. Barthelemy, M., Barrat, A., Pastor-Satorras, R., Vespignani, A.: Velocity and hierarchical spread of epidemic outbreaks in scale-free networks. Phys. Rev. Lett. **92**, 178701 (2004)
52. Basnayake, K., Hubl, A., Schuss, Z., Holcman, D.: Shortest paths for the first particles among N to reach a target window. Phys. Lett. A **382**, 3449–3454 (2018)
53. Basnayake, K., Schuss, Z., Holcman, D.: Asymptotic formulas for extreme statistics of escape times in 1, 2 and 3-dimensions. J. Nonlin. Sci. **29**, 461–499 (2019)
54. Basnayake, K., Mazoud, D., Bemelmans, A., Rouach, N., Korkotian, E., Holcman, D.: Fast calcium transients in dendritic spines driven by extreme statistics. PLOS Biol. **17**, e2006202 (2019)
55. Basu, A., Chowdhury, D.: Traffic of interacting ribosomes: effects of single-machine mechanochemistry on protein synthesis. Phys. Rev. E **75**, 021902 (2007)
56. Baxter, R.J.: Exactly Solved Models in Statistical Mechanics. Academic Press, London (1982)
57. Beard, D.A., Qian, H.: Chemical Biophysics: Quantitative Analysis of Cellular Systems. Cambridge University Press, Cambridge, UK (2008)
58. Becalska, A.N., Gavis, E.R.: Lighting up mRNA localization in Drosophila oogenesis. Development **136**, 2493–2503 (2009)
59. Becskei, A., Mattaj, I.W.: Quantitative models of nuclear transport. Curr. Opin. Cell Biol. **17**, 27–34 (2005)
60. Beece, D., Eisenstein, L., Frauenfelder, H., Good, D., Marden, M.C., Reinisch, L., Reynolds, A.H., Sorensen, L.B., Yue, K.T.: Solvent viscosity and protein dynamics. Biochemistry **19**, 5147–5157 (1980)
61. Belan, S.: Restart could optimize the probability of success in a Bernouilli trial. Phys. Rev. Lett. **120**, 080601 (2018)
62. Bell, J.W.: Searching Behavior, the Behavioral Ecology of Finding Resources. Chapman and Hall, London (1991)
63. Bena, I.: Dichotomous Markov noise: exact results for out-of-equilibrium systems. Int. J. Mod. Phys. B **20**, 2825 (2006)
64. Benichou, O., Moreau, M., Oshanin, G.: Kinetics of stochastically gated diffusion-limited reactions and geometry of random walk trajectories. Phys. Rev. E **61**, 3388–3406 (2000)
65. Benishou, O., Coppey, M., Moreau, M., Oshanin, G.: Kinetics of diffusion-limited catalytically activated reactions: an extension of the Wilemski-Fixman approach. J. Chem. Phys. **123**, 194506 (2005)
66. Benichou, O., Coppey, M., Moreau, M., Suet, P., Voituriez, R.: A stochastic model for intermittent search strategies. J. Phys. Cond. Matter **17**, S4275-4286 (2005)

67. Benichou, O., Loverdo, C., Moreau, M., Voituriez, R.: A minimal model of intermittent search in dimension two. J. Phys. A **19**, 065141 (2007)

68. Benichou, O., Voituriez, R.: Narrow escape time problem: time needed for a particle to exit a confining domain through a small window. Phys. Rev. Lett. **100**, 168105 (2008)

69. Benichou, O., Chevalier, C., Klafter, J., Meyer, B., Voituriez, R.: Geometry-controlled kinetics. Nat. Chem. **2**, 472–477 (2010)

70. Benichou, O., Loverdo, C., Moreau, M., Voituriez, R.: Intermittent search strategies. Rev. Mod. Phys. **83**, 81–129 (2011)

71. Ben-Naim, A.: Cooperativity and Regulation in Biochemical Processes. Kluwer Academic, New York (2010)

72. Berezhkovskii, A.M., Makhnovskii, Yu.A., Suris, R.A.: Kinetics of diffusion-controlled reactions. Chem. Phys. **137**, 41–49 (1989)

73. Berezhkovskii, A.M., Yang, D.-Y., Sheu, S.-Y., Lin, S.H.: Stochastic gating in diffusion-influenced ligand binding to proteins: gated protein versus gated ligands. Phys. Rev. E **54**, 4462–4464 (1996)

74. Berezhkovskii, A.M., Yang, D.-Y., Lin, S.H., Makhnovskii, Yu.A., Sheu, S.-Y.: Smoluchowski-type theory of stochastically gated diffusion-influenced reactions. J. Chem. Phys. **106**, 6985 (1997)

75. Berezhkovskii, A.M., Pusovoit, M.A., Bezrukov, S.M.: Channel-facilitated membrane transport: transit probability and interaction with the channel. J. Chem. Phys. **116**, 9952–9956 (2002)

76. Berezhkovskii, A.M., Pusovoit, M.A., Bezrukov, S.M.: Channel-facilitated membrane transport: average lifetimes in the channel. J. Chem. Phys. **119**, 3943–3951 (2003)

77. Berezhkovskii, A.M., Bezrukov, S.M.: Channel-facilitated membrane transport: constructive role of the particle attraction to the channel pore. Chem. Phys. **319**, 342–349 (2005)

78. Berezhkovskii, A.M., Coppey, M., Shvartsman, S.Y.: Signaling gradients in cascades of two-state reaction-diffusion systems. Proc. Natl. Acad. Sci. **106**, 1087–1092 (2009)

79. Berezhkovskii, A.M., Sample, C., Shvartsman, S.Y.: How long does it take to establish a morphogen gradient? Biophys. J. **99**, L59–L61 (2010)

80. Berezhkovskii, A.M., Sample, C., Shvartsman, S.Y.: Formation of morphogen gradients: local accumulation time. Phys. Rev. E **83**, 051906 (2011)

81. Berezhkovskii, A.M., Szabo, A.: Effect of ligand diffusion on occupancy fluctuations of cell-surface receptors. J. Chem. Phys. **139**, 121910 (2013)

82. Berezhkovskii, A.M., Shvartsman, S.Y.: Diffusive flux in a model of stochastically gated oxygen transport in insect respiration. J. Chem. Phys. **144**, 204101 (2016)

83. Berg, H.C., Purcell, E.M.: Physics of chemoreception. Biophys. J. **20**, 93–219 (1977)

84. Berg, H.C.: Random Walks in Biology. University Press, Princeton (1983)

85. Berg, H.C.: Motile behavior of bacteria. Phys. Today **53**, 24–28 (2000)

86. Berg, O.G.: A model for the statistical fluctuations of protein numbers in a microbial population. J. Theor. Biol. **4**, 587–603 (1978)

87. Berg, O.G., Winter, R.B., von Hippel, P.H.: Diffusion-driven mechanisms of protein translocation on nucleic acids. 1. models and theory. Biochem. **20**, 6929–6948 (1981)

88. Berg, O.G., von Hippel, P.H.: Diffusion-controlled macromolecular interactions. Ann. Rev. Biophys. Biophys. Chem. **14**, 131–160 (1985)

89. Berg, O.G., Paulsson, J., Ehrenberg, M.: Fluctuations and quality of control in biological cells: zero-order ultrasensitivity reinvestigated. Biophys. J. **79**, 1228–1236 (2000)

90. Berger, S.L.: The complex language of chromatin regulation during transcription. Nature **447**, 407–412 (2007)

91. Berglund, N.: Kramers' law: validity, derivations and generalizations. Markov Process. Relat. Fields **19**, 459–490 (2011)

92. Bezrukov, S.M., Berezhkovskii, A.M., Pusovoit, M.A., Szabo, A.: Particle number fluctuations in a membrane channel. J. Chem. Phys. **113**, 8206–8211 (2000)

93. Bhalla, U.S.: Signaling in small subcellular volumes. I. Stochastic and diffusion effects on individual pathways. Biophys. J. **87**, 733–744 (2004)

94. Bialek, W.: Biophysics. Princeton University Press, Princeton (2012)
95. Bickel, T., Bruinsma, R.: The nuclear pore complex mystery and anomalous diffusion in reversible gels. Biophys. J. **83**, 3079–3987 (2002)
96. Bicout, D.J.: Green's functions and first passage time distributions for dynamic instability of microtubules. Phys. Rev. E **56**, 6656–6667 (1997)
97. Biess, A., Korkotian, E., Holcman, D.: Diffusion in a dendritic spine: the role of geometry. Phys. Rev. E **76**, 021922 (2007)
98. Bird, A.: DNA methylation patterns and epigenetic memory. Genes Dev. **16**, 6–21 (1998)
99. Blum, J., Reed, M.C.: A model for slow axonal transport and its application to neurofilamentous neuropathies. Cell Motil. Cytoskeleton. **12**, 53–65 (1989)
100. Blythe, R.A., Evans, M.R.: Nonequilibrium steady states of matrix-product form: a solver's guide. J. Phys. A **40**, R333–R441 (2007)
101. Boal, D.: Mechanics of the Cell, 2nd edn. Cambridge University Press, Cambridge (2010)
102. Bodrova, A.S., Sokolov, I.M.: Resetting processes with noninstantaneous return. Phys. Rev. E **101**, 052130 (2020)
103. Boeckh, J., Kaissling, K.E., Schneider, D.: Insect olfactory receptors. Cold Spring Harbor Symp. Quant. Biol. **30**, 1263–1280 (1965)
104. Boland, R.P., Galla, T., McKane, A.J.: How limit cycles and quasi-cycles are related in systems with intrinsic noise. J. Stat. Mech.: Theory Exp. **P09001**, 1–27 (2008)
105. Boland, R.P., Galla, T., McKane, A.J.: Limit cycles, complex Floquet multipliers, and intrinsic noise. Phys. Rev. E **79**, 051131 (2009)
106. Bouchaud, J.P., Georges, A.: Anomalous diffusion in disordered media—statistical mechanisms, models and physical applications. Phys. Rep. **195**, 127–293 (1990)
107. Brandenburg, B.: Virus trafficking-learning from single-virus tracking. Nat. Rev. Microbiol. **5**, 197–208 (2007)
108. Bray, D., Levin, M.D., Morton-Firth, C.J.: Receptor clustering as a cellular mechanism to control sensitivity. Nature **393**, 85–88 (1998)
109. Bredt, D.S., Nicoll, R.A.: AMPA receptor trafficking at excitatory synapses. Neuron **40**, 361–379 (2003)
110. Bressloff, P.C., Dwyer, V.M., Kearney, M.J.: Classical localization and percolation in random environments on trees. Phys. Rev. E **55**, 6765–6775 (1997)
111. Bressloff, P.C., Earnshaw, B.A., Ward, M.J.: Diffusion of protein receptors on a cylindrical dendritic membrane with partially absorbing traps. SIAM J. Appl. Math. **68**, 1223–1246 (2008)
112. Bressloff, P.C., Newby, J.M.: Directed intermittent search for hidden targets. New J. Phys. **11**, 023,033 (2009)
113. Bressloff, P.C., Newby, J.M.: Quasi-steady state analysis of motor-driven transport on a two-dimensional microtubular network. Phys. Rev. E. **83**, 061139 (2011)
114. Bressloff, P.C., Newby, J.M.: Stochastic models of intracellular transport. Rev. Mod. Phys. **85**, 135–196 (2013)
115. Bressloff, P.C.: Waves in Neural Media: From Single Neurons to Neural Fields. Springer, New York (2014)
116. Bressloff, P.C., Newby, J.M.: Stochastic hybrid model of spontaneous dendritic NMDA spikes. Phys. Biol. **11**, 016006 (2014)
117. Bressloff, P.C., Newby, J.M.: Path-integrals and large deviations in stochastic hybrid systems. Phys. Rev. E. **89**, 042701 (2014)
118. Bressloff, P.C., Lawley, S.D.: Escape from subcellular domains with randomly switching boundaries. Multiscale Model. Simul. **13**, 1420–1445 (2015)
119. Bressloff, P.C., Lawley, S.D.: Moment equations for a piecewise deterministic PDE. J. Phys. A **48**, 105001 (2015)
120. Bressloff, P.C.: Path-integral methods for analyzing the effects of fluctuations in stochastic hybrid neural networks. J. Math. Neurosci. **5**, 33pp (2015)
121. Bressloff, P.C., Lawley, S.D.: Stochastically-gated diffusion-limited reactions for a small target in a bounded domain. Phys. Rev. E **92**, 062117 (2015)

122. Bressloff, P.C., Levien, E.: Synaptic democracy and active intracellular transport in axons. Phys. Rev. Lett. **114**, 168101 (2015)

123. Bressloff, P.C.: Aggregation-fragmentation model of vesicular transport in neurons. J. Phys. A **49**, 145601 (2016)

124. Bressloff, P.C., Karamched, B.R.: Model of reversible vesicular transport with exclusion. J. Phys. A **49**, 345602 (2016)

125. Bressloff, P.C.: Diffusion in cells with stochastically-gated gap junctions. SIAM J. Appl. Math. **76**, 1658–1682 (2016)

126. Bressloff, P.C., Lawley, S.D.: Diffusion on a tree with stochastically-gated nodes. J. Phys. A **49**, 245601 (2016)

127. Bressloff, P.C., Faugeras, O.: On the Hamiltonian structure of large deviations in stochastic hybrid systems. J. Stat. Mech. 033206 (2017)

128. Bressloff, P.C., Lawley, S.D.: Temporal disorder as a mechanism for spatially heterogeneous diffusion. Phys. Rev. E **95**, 060101(R) (2017)

129. Bressloff, P.C., Lawley, S.D.: Hybrid colored noise process with space-dependent switching rates. Phys. Rev. E **96**, 012129 (2017)

130. Bressloff, P.C.: Stochastically-gated local and occupation times of a Brownian particle. Phys. Rev. E **95**, 012130 (2017)

131. Bressloff, P.C.: Stochastic switching in biology: from genotype to phenotype (Topical Review). J. Phys. A **50**, 133001 (2017)

132. Bressloff, P.C., Kim, H.: Search-and-capture model of cytoneme-mediated morphogen gradient formation. Phys. Rev. E **99**, 052401 (2019)

133. Bressloff, P.C., Lawley, S.D., Murphy, P.: Protein concentration gradients and switching diffusions. Phys. Rev. E **99**, 032409 (2019)

134. Bressloff, P.C.: Directed intermittent search with stochastic resetting. J. Phys. A **53**, 105001 (2020)

135. Bressloff, P.C.: Search processes with stochastic resetting and multiple targets. Phys. Rev. E **102**, 022115 (2020)

136. Bressloff, P.C.: Queuing theory of search processes with stochastic resetting. Phys. Rev. E **102**, 032109 (2020)

137. Bressloff, P.C.: Stochastically-gated diffusion model of selective nuclear transport. Phys. Rev. E **101**, 042404 (2020)

138. Bressloff, P.C.: Stochastic resetting and the dynamics of focal adhesions. Phys. Rev. E **102**, 022134 (2020)

139. Bressloff, P.C.: Target competition for resources under multiple search-and-capture events with stochastic resetting. Proc. Roy. Soc. A **476**, 20200475 (2020)

140. Bringuier, E.: Kinetic theory of inhomogeneous diffusion. Physica A **388**, 2588–2599 (2009)

141. Brockett, R.W.: Notes on stochastic processes on manifolds. In: Byrnes, C.I., et al. (eds.) Systems and Control in the Twenty-First Century. Birkhauser, Boston (1997)

142. Bronshtein, I., Israel, Y., Kepten, E., Mai, S., Shav-Tal, Y., Barkai, E., Garini, Y.: Transient anomalous diffusion of telomeres in the nucleus of mammalian cells. Phys. Rev. Lett. **103**, 018102 (2009)

143. Bronshtein, I., Kepten, E., Kanter, I., Berezin, S., Linder, M., Redwood, A.B., Mai, S., Gonzalo, S., Foisner, R., Shav-Tal, Y., Garini, Y.: Loss of lamin A function increases chromatin dynamics in the nuclear interior. Nat. Commun. **6**, 8044 (2015)

144. Brown, F.L.H., Leitner, D.M., McCammon, J.A., Wilson, K.R.: Lateral diffusion of membrane proteins in the presence of static and dynamics corrals: suggestions for appropriate variables. Biophys. J. **78**, 2257–2269 (2000)

145. Brown, A.: Slow axonal transport: stop and go traffic in the axon. Nat. Rev. Mol. Cell Biol. **1**, 153–156 (2001)

146. Brown, A.: Axonal transport of membranous and nonmembranous cargoes: a unified perspective. J. Cell Biol. **160**, 817–821 (2003)

147. Brown, A.: Axonal Transport. Neuroscience in the 21st Century. Springer (2013)

148. Buckwar, E., Riedler, M.G.: An exact stochastic hybrid model of excitable membranes including spatio-temporal evolution. J. Math. Biol. **63**, 1051–1093 (2011)

149. Bukauskas, F.K., Verselis, V.K.: Gap junction channel gating. Biochim. Biophys. Acta **1662**, 42–60 (2004)

150. Burada, P.S., Schmid, G., Reguera, D., Rubi, J.M., Hanggi, P.: Biased diffusion in confined media: test of the Fick-Jacobs approximation and validity criteria. Phys. Rev. E **75**, 051111 (2007)

151. Burada, P.S., Hanggi, P., Marchesoni, F., Schmid, G., Talkner, P.: Diffusion in confined geometries. Chem. Phys. Chem. **10**, 45–54 (2009)

152. Burlakov, V.M., Emptage, N., Goriely, A., Bressloff, P.C.: Phys. Rev. Lett. **108**, 028101 (2012)

153. Burov, S., Jeon, J.H., Metzler, R., Barkai, E.: Single particle tracking in systems showing anomalous diffusion: the role of weak ergodicity breaking. Phys. Chem. Chem. Phys. **13**, 1800–1812 (2011)

154. Burute, M., Kapitein, L.C.: Cellular logistics: unraveling the interplay between microtubule organization and intracellular transport. Ann. Rev. Cell Dev. Biol. **35**, 29–54 (2019)

155. Buse, O., Perez, Kuznetsov, A.: Dynamical properties of the repressilator model. Phys. Rev. E **81**, 066206 (2010)

156. Cai, L., Friedman, N., Xies, X.S.: Stochastic protein expression in individual cells at the single molecule level. Nature **440**, 358–362 (2006)

157. Camalet, S., Duke, T., Julicher, F., Prost, J.: Auditory sensitivity provided by self-tuned critical oscillations of hair cells. Proc. Natl. Acad. Sci. USA **97**, 3183–3188 (2000)

158. Cameron, D.E., Bashor, C.J., Collins, J.J.: A brief history of synthetic biology. Nat. Rev. Microbiol. **12**, 381–390 (2014)

159. Campas, O., Leduc, C., Bassereau, P., Joanny, J.- F., Prost. J.: Collective oscillations of processive molecular motors. Biophys. Rev. Lett. **4**, 163–178 (2009)

160. Campos, D., Mendez, V.: Phase transitions in optimal search times: how random walkers should combine resetting and flight scales. Phys. Rev. E **92**, 062115 (2015)

161. Cao, Y., Gillespie, D.T., Petzold, L.R.: Efficient step size selection for the tau-leaping simulation method. J. Chem. Phys. **124**, 044109 (2006)

162. Centres, P.M., Bustingorry, S.: Effective Edwards-Wilkinson equation for single-file diffusion. Phys. Rev. E **81**, 061101 (2010)

163. Chandler, D.: Introduction to Modern Statistical Mechanics. Oxford University Press, USA (1987)

164. Chechkin, A., Sokolov, I.M.: Random search with resetting: a unified renewal approach. Phys. Rev. Lett. **121**, 050601 (2018)

165. Cheng, H., Lederer, M.R., Lederer, W.J., Cannell, M.B.: Calcium sparks and waves in cardiac myocytes. Am. J. Physiol. **270**, C148–C159 (1996)

166. Cheng, H., Lederer, W.J.: Calcium sparks. Physiol. Rev. **88**, 1491–1545 (2008)

167. Cherstvy, A.G., Kolomeisky, A.B., Kornyshev, A.A.: Protein-DNA interactions: reaching and recognizing the targets. J. Phys. Chem. **112**, 4741–4750 (2008)

168. Chevalier, C., Benichou, O., Meyer, B., Voituriez, R.: First-passage quantities of Brownian motion in a bounded domain with multiple targets: a unified approach. J. Phys. A **44**, 025002 (2011)

169. Cheviakov, A.F., Ward, M.J., Straube, R.: An asymptotic analysis of the mean first passage time for narrow escape problems: Part II: The sphere. SIAM J. Multiscal Mod. Sim. **8**, 836–870 (2010)

170. Cheviakov, A.F., Ward, M.J.: Optimizing the principal eigenvalue of the Laplacian in a sphere with interior traps. Math. Comp. Modeling **53**, 042118 (2011)

171. Chipot, M., Hastings, S., Kinderlehrer, D.: Transport in a molecular motor system. Math. Model. Numer. Anal. **38**, 1011–1034 (2004)

172. Chirikjian, G.S.: Stochastic Models, Information Theory, and Lie Groups, Volume 1: Classical Results and Geometric Methods. Birkhauser, Boston (2009)

173. Choi, C.K., Vicente-Manzanares, M., Zareno, J., Whitmore, L.A., Mogilner, A., Horwitz, A.R.: Actin and α-actinin orchestrate the assembly and maturation of nascent adhesions in a myosin II motor-independent manner. Nat. Cell Biol. **10**, 1039–1050 (2008)
174. Choquet, D., Triller, A.: The role of receptor diffusion in the organization of the postsynaptic membrane. Nat. Rev. Neurosci. **4**, 251–265 (2003)
175. Choquet, D., Triller, A.: The dynamic synapse. Neuron **80**, 691–703 (2013)
176. Choquet, D.: Linking nanoscale dynamics of AMPA receptor organization to plasticity of excitatory synapses and learning. J. Neurosci. **38**, 9318–9329 (2018)
177. Chou, T.: Kinetics and thermodynamics across single-file pores: solute permeability and rectified osmosis. J. Chem. Phys. **110**, 606–615 (1999)
178. Chou, T., Lakatos, G.: Totally asymmetric exclusion processes with particles of arbitrary size. J. Phys. A **36**, 2027–2041 (2003)
179. Chou, T., Lakatos, G.: Clustered bottlenecks in mRNA translation and protein synthesis. Phys. Rev. Lett. **93**, 198101 (2004)
180. Chou, T., Mallick, K., Zia, R.K.P.: Non-equilibrium statistical mechanics: from a paradigmatic model to biological transport. Rep. Prog. Phys. **74**, 116601 (41pp) (2011)
181. Chow, C.C., White, J.A.: Spontaneous action potentials due to channel fluctuations. Biophys. J. **71**, 3013–3021 (1996)
182. Christou, C., Schadschneider, A.: Diffusion with resetting in bounded domains. J. Phys. A **48**, 285003 (2015)
183. Chu, C.L., Buczek-Thomas, J.A., Nugent, M.A.: Heparan sulfate proteoglycans modulate fibroblast growth factor-2 binding through a lipid-raft-mediated mechanism. Biochem. J. **379**, 331–341 (2004)
184. Chuang, J., Kantor, Y., Kardar, M.: Anomalous dynamics of translocation. Phys. Rev. E **65**, 011802 (2002)
185. Ciandrini, L., Stansfield, I., Romano, M.C.: Role of the particle's stepping cycle in an asymmetric exclusion process: a model of mRNA translation. Phys. Rev. E **81**, 051904 (2010)
186. Codling, E., Hill, N.: Calculating spatial statistics for velocity jump processes with experimentally observed reorientation parameters. J. Math. Biol. **51**, 527–556 (2005)
187. Collinridge, G.L., Isaac, J.T.R., Wang, Y.T.: Receptor trafficking and synaptic plasticity. Nat. Rev. Neurosci. **5**, 952–962 (2004)
188. Collins, F.C., Kimball, G.E.: Diffusion-controlled reaction rates. J. Colloid Sci. **4**, 425–439 (1949)
189. Delgado, M.I., Ward, M., Coombs, D.: Conditional mean first passage times to small traps in a 3-D domain with a sticky boundary: applications to T cell searching behavior in lymph nodes. Multiscale Model. Simul. **13**, 1224–1258 (2015)
190. Connors, B.W., Long, M.A.: Electrical synapses in the mammalian brain. Ann. Rev. Neurosci. **27**, 393–418 (2004)
191. Cookson, N.A., Mather, W.H., Danino, T., Mondragon-Palomino, O., Williams, R.J., Tsimring, L.S., Hasty, J.: Queueing up for enzymatic processing: correlated signaling through coupled degradation. Mol. Syst. Biol. **7**, 561 (2014)
192. Coombs, D., Goldstein, B.: T cell activation: kinetic proofreading, serial engagement and cell adhesion. J. Comput. Appl. Math. **184**, 121–139 (2005)
193. Coombs, D., Straube, R., Ward, M.J.: Diffusion on a sphere with localized traps: mean first passage time, eigenvalue asymptotics, and Fekete points. SIAM J. Appl. Math. **70**, 302–332 (2009)
194. Coppey, M., Benichou, O., Voituriez, R., Moreau, M.: Kinetics of target site localization of a protein DNA: a stochastic approach. Biophys. J. **87**, 1640–1649 (2004)
195. Cornell-Bell, A.H., Finkbeiner, S.M., Cooper, M.S., Smith, S.J.: Glutamate induces calcium waves in cultured astrocytes: long-range glial signaling. Science **247**, 470–473 (1990)
196. Cortini, R., Barbi, M., Care, B.R., Lavelle, C., Lesne, A., Mozziconacci, J., Victor, J.-M.: The physics of epigenetics. Rev. Mod. Phys. **88**, 025002 (2016)
197. Coulon, A., Chow, C.C., Singer, R.H., Larson, D.R.: Eukaryotic transcriptional dynamics: from single molecules to cell populations. Nature Rev. Genet. **14**, 572–584 (2013)

198. Couzin, I.D.: Collective cognition in animal groups. Trends. Cogn. Sci. **13**, 36–43 (2009)
199. Cox, D.R.: Some statistical methods connected with series of events. J. R. Stat. Soc. B **17**, 129–164 (1955)
200. Cox, D.R., Isham, V.: Point Processes. Chapman and Hall (1980)
201. Cox, J.S., Chapman, R.E., Walter, P.: The unfolded protein response coordinates the production of endoplasmic reticulum protein and endoplasmic reticulum membrane. Mol. Biol. Cell **8**, 1805–1814 (1997)
202. Craciun, G., Feinberg, M.: Multiple equilibria in complex chemical reaction networks: I. The injectivity property. SIAM J. Appl. Math. **65**, 1526–1546 (2005)
203. Craciun, G., Feinberg, M.: Multiple equilibria in complex chemical reaction networks: II. The species-reactions graph. SIAM J. Appl. Math. **66**, 1321–1338 (2006)
204. Cronshaw, J.M., Matunis, M.J.: The nuclear pore complex: disease associations and functional correlations. Trends Endocrinol. Metab. **15**, 34–39 (2004)
205. Czondora, K., Mondina, M., Garcia, M., Heine, M., Frischknecht, R., Choquet, D., Sibarita, J.B., Thoumine, O.R.: A unified quantitative model of AMPA receptor trafficking at synapses. Proc. Nat. Acad. Sci. USA **109**, 3522–3527 (2012)
206. Damm, E.M., Pelkmans, L.: Systems biology of virus entry in mammalian cells. Cell Microbiol. **8**, 1219–1227 (2006)
207. Danino, T., Mondragon-Palomino, O., Tsimring, L., Hasty, J.: A synchronized quorum of genetic clocks. Nature **463**, 326–330 (2010)
208. Danuser, G., Allard, J., Mogilner, A.: Mathematical modeling of eukaryotic cell migration: insights beyond experiments. Annu. Rev. Cell Dev. Biol. **29**, 501–528 (2013)
209. Das, R., Cairo, C.W., Coombs, D.: A hidden Markov model for single particle tracks quantifies dynamic interactions between LFA-1 and the actin cytoskeleton. PLoS Comp. Biol. **5**, e1000556 (2009)
210. Dattani, J., Barahona, M.: Stochastic models of gene transcription with upstream drives: exact solution and sample path characterization. J. R. Soc. Interface **14**, 20160833 (2017)
211. Davis, M.H.A.: Piecewise-deterministic Markov processes: a general class of non-diffusion stochastic models. J. R. Soc. Ser. B (Methodol.) **46**, 353–388 (1984)
212. de Dominicis, C.: Techniques de renormalisation de la théorie des champs et dynamique des phénomènes critiques. J. Phys. (Paris) **37**, 247–253 (1976)
213. Dembo, A., Zeitouni, O.: Large Deviations: Techniques and Applications, 2nd edn. Springer, New York (2004)
214. den Hollander, F., Majumdar, S.N., Meylahn, J.M., Touchette, H.: Properties of additive functionals of Brownian motion with resetting. J. Phys. A: Math. Theor. **52**, 175001 (2019)
215. Derrida, B.: Velocity and diffusion constant of a periodic one-dimensional hopping model. J. Stat. Phys. **31**, 433–450 (1983)
216. De Vos, K.J., Grierson, A.J., Ackerley, S., Miller, C.C.J.: Role of axonal transport in neurodegenerative diseases. Annu. Rev. Neurosci. **31**, 151–173 (2008)
217. Dibner, C., Schibler, U., Albrecht, U.: The mammalian circadian timing system: organization and coordination of central and peripheral clocks. Annu. Rev. Physiol. **72**, 517–549 (2010)
218. Dinh, A.T., Pangarkar, C., Theofanous, T., Mitragotri, S.: Understanding intracellular transport processes pertinent to synthetic gene delivery via stochastic simulations and sensitivity analysis. Biophys. J. **92**, 831–846 (2007)
219. Dix, J.A., Verkman, A.S.: Crowding effects on diffusion in solutions and cells. Annu. Rev. Biophys. **37**, 247–263 (2008)
220. Dodd, I.B., Micheelsen, M.A., Sneppen, K., Thon, G.: Theoretical analysis of epigenetic cell memory by nucleosome modification. Cell **129**, 813–822 (2007)
221. Doering, C.R., Gadoua, J.C.: Resonant activation over a fluctuating barrier. Phys. Rev. Lett. **69**, 2318–2321 (1992)
222. Doering, C.R., Sargsyan, K.V., Sander, L.M., Vanden-Eijnden, E.: Asymptotics of rare events in birth-death processes bypassing the exact solutions. J. Phys.: Condens. Matter **19**, 065145 (2007)

223. Dogterom, M., Leibler, S.: Physical aspects of the growth and regulation of microtubule structures. Phys. Rev. Lett. **70**, 1347–1350 (1993)

224. Doi, M.: Second quantization representation for classical many-particle systems. J. Phys. A. **9**, 1465–1477 (1976)

225. Doi, M.: Stochastic theory of diffusion controlled reactions. J. Phys. A. **9**, 1479–1495 (1976)

226. Doi, M., Edwards, S.F.: The Theory of Polymer Dynamics. Oxford Science Publications, Oxford (1986)

227. Dong, J., Schmittmann, B., Zia., R.K.P.: Inhomogeneous exclusion processes and protein synthesis. Phys. Rev. E **76**, 051113 (2007)

228. Driver, J.W., Rodgers, A.R., Jamison, D.K., Das, R.K., Kolomeisky, A.B., Diehl, M.R.: Coupling between motor proteins determines dynamic behavior of motor protein assemblies. Phys. Chem. Chem. Phys. **12**, 10398–10405 (2010)

229. Duan, J.: An Introduction to Stochastic Dynamics. Cambridge University Press, Cambridge (2015)

230. Duke, T.A.J., Bray, D.: Heightened sensitivity of a lattice of membrane of receptors. Proc. Natl. Acad. Sci. USA **96**, 10104–10108 (1999)

231. Dumoulin, A., Triller, A., Kneussel, M.: Cellular transport and membrane dynamics of the glycine receptor. Front. Mol. Neurosci. **2**(28), 1–11 (2010)

232. Dykman, M.I., Mori, E., Ross, J., Hunt, P.M.: Large fluctuations and optimal paths in chemical kinetics. J. Chem. Phys. A **100**, 5735–5750 (1994)

233. Dykman, M.I., Schwartz, I.B., Landsman, A.S.: Phys. Rev. Lett. **101**, 078101 (2008)

234. Dynes, J., Steward, O.: Dynamics of bidirectional transport of ARC mRNA in neuronal dendrites. J. Comp. Neurol. **500**, 433–447 (2007)

235. Earnshaw, B.A., Bressloff, P.C.: A biophysical model of AMPA receptor trafficking and its regulation during LTP/LTD. J. Neurosci. **26**, 12362–12373 (2006)

236. Earnshaw, B.A., Bressloff, P.C.: Modeling the role of lateral membrane diffusion on AMPA receptor trafficking along a spiny dendrite. J. Comput. Neurosci. **25**, 366–389 (2008)

237. Edelstein-Keshet, L., Ermentrout, G.B.: Models for the length distribution of actin filaments: simple polymerization and fragmentation. Bull. Math. Biol. **60**, 449–475 (1998)

238. Edelstein-Keshet, L., Ermentrout, G.B.: Models for spatial polymerization dynamics of rod-like polymers. J. Math. Biol. **40**, 64–96 (2000)

239. Ehlers, M.D.: Reinsertion or degradation of AMPA receptors determined by activity-dependent endocytic sorting. Neuron **28**, 511–525 (2000)

240. Ehlers, M.D., Heine, M., Groc, L., Lee, M.-C., Choquet, D.: Diffusional trapping of GluR1 AMPA receptors by input-specific synaptic activity. Neuron **54**, 447–460 (2007)

241. Eldar, A., Elowitz, M.B.: Functional roles for noise in genetic circuits. Nature **467**, 167–173 (2010)

242. Elf, J., Ehrenberg, M.: Fast evaluation of fluctuations in biochemical networks with the linear noise approximation. Genome Res. **13**, 2475–2484 (2003)

243. Elgart, V., Kamenev, A.: Rare event statistics in reaction-diffusion systems. Phys. Rev. E **70**, 041106 (2004)

244. Ellis, R.S.: Entropy, Large Deviations, and Statistical Mechanics. Springer, New York (1985)

245. Ellis, R.S.: The theory of large deviations: from Boltzmann's 1877 calculation to equilibrium macrostates in 2D turbulence. Physica D **133**, 106–136 (1999)

246. Elowitz, M.B., Leibler, S.: A synthetic oscillatory network of transcriptional regulators. Nature **403**, 335 (2000)

247. Elowitz, M.B., Levine, A.J., Siggia, E.D., Swain, P.S.: Stochastic gene expression in a single cell. Science **297**, 1183–1186 (2002)

248. Elston, T.C., Wang, H., Oster, G.: Energy transduction in ATP synthase. Nature **391**, 510–513 (1998)

249. Elston, T.C.: A macroscopic description of biomolecular transport. J. Math. Biol. **41**, 189–206 (2000)

250. Ermentrout, G.B., Terman, D.: Mathematical Foundations of Neuroscience. Springer, Berlin (2010)

251. Escudero, C., Kamanev, A.: Switching rates of multistep reactions. Phys. Rev. E **79**, 041149 (2009)
252. Evangelista, L.R., Lenzi, E.K.: Fractional Diffusion Equations and Anomalous Diffusion. Cambridge University Press (2018)
253. Evans, W.J., Martin, P.E.: Gap junctions: structure and function. Mol. Membr. Biol. **19**, 121–136 (2002)
254. Evans, M.R., Juhasz, R., Santen, L.: Shock formation in an exclusion process with creation and annihilation. Phys. Rev. E **68**, 026117 (2003)
255. Evans, M.R., Majumdar, S.N.: Diffusion with stochastic resetting. Phys. Rev. Lett. **106**, 160601 (2011)
256. Evans, M.R., Majumdar, S.N.: Diffusion with optimal resetting. J. Phys. A **44**, 435001 (2011)
257. Evans, M.R., Majumdar, S.N.: Diffusion with resetting in arbitrary spatial dimension. J. Phys. A **47**, 285001 (2014)
258. Evans, M.R., Majumdar, S.N.: Run and tumble particle under resetting: a renewal approach. J. Phys. A: Math. Theor. **51**, 475003 (2018)
259. Evans, M.R., Majumdar, S.N., Schehr, G.: Stochastic resetting and applications. J. Phys. A (2019)
260. Evans, M.R., Majumdar, S.N.: Effects of refractory period on stochastic resetting J. Phys. A: Math. Theor. **52**, 01LT01 (2019)
261. Evans, M.R., Majumdar, S.N., Schehr, G.: Stochastic resetting and applications. J. Phys. A: Math. Theor. **53**, 193001 (2020)
262. Faccioli, P., Sega, M., Pederiva, F., Orland, H.: Dominant pathways in protein folding. Phys. Rev. Lett. **97**, 108101 (2006)
263. Faggionato, A., Gabrielli, D., Crivellari, M.R.: Non-equilibrium thermodynamics of piecewise deterministic Markov Processes. J. Stat. Phys. **137**, 259–304 (2009)
264. Faggionato, A., Gabrielli, D., Crivellari, M.: Averaging and large deviation principles for fully-coupled piecewise deterministic Markov processes and applications to molecular motors. Markov Process. Relat. Fields **16**, 497–548 (2010)
265. Falcke, M., Tsimiring, L., Levine, H.: Stochastic spreading of intracellular Ca^{2+} release. Phys. Rev. E **62**, 2636–2643 (2000)
266. Falcke, M.: On the role of stochastic channel behavior in intracellular Ca^{2+} dynamics. Biophys. J. **84**, 42–56 (2003)
267. Falcke, M.: Reading the patterns in living cells - the physics of Ca^{2+} signaling. Adv. Phys. **53**, 255–440 (2004)
268. Fedotov, S., Al-Shami, H., Ivanov, A., Zubarev, A.: Anomalous transport and nonlinear reactions in spiny dendrites. Phys. Rev. E **82**, 041103 (2010)
269. Fedotov, S., Tanand, A., Zubarev, A.: Persistent random walk of cells involving anomalous effects and random death. Phys. Rev. E **91**, 042124 (2015)
270. Fedotov, S., Korabel, N.: Emergence of Levy walks in systems of interacting individuals. Phys. Rev. E **95**, 030107(R) (2017)
271. Feinberg, M.: Complex balancing in general kinetic systems. Arch. Rational Mech. Anal. **49**, 187–194 (1972)
272. Feinberg, M.: Lectures on chemical reaction networks, Delivered at the Mathematics Research Center, Univ. Wisc. Madison. Available for download at http://www.che.eng.ohio-state.edu/?feinberg/LecturesOnReactionNetworks (1979)
273. Feinberg, M.: Chemical reaction network structure and the stability of complex isothermal reactors - I. the deficiency zero and deficiency one theorems. Chem. Eng. Sci. **42**, 2229–2268 (1987)
274. Feng, J., Kurtz, T.G.: Large Deviations for Stochastic Processes. American Mathematical Society (2006)
275. Ferreira, T., Wilson, S.R., Choi, Y.G., Risso, D., Dudoit, S., Speed, T.P., Ngai, J.: Silencing of odorant receptor genes by G protein signaling ensures the expression of one odorant receptor per olfactory sensory neuron. Neuron **81**, 847–859 (2014)

276. Fisher, M., Kolomeisky, A.: Simple mechanochemistry describes the dynamics of kinesin molecules. Proc. Natl. Acad. Sci. USA **98**, 7748–7753 (2001)
277. Flyvbjerg, H., Holy, T., Leibler, S.: Stochastic dynamics of microtubules - a model for caps and catastrophes. Phys. Rev. Lett. **73**, 2372–2375 (1994)
278. Flyvbjerg, H., Holy, T., Leibler, S.: Microtubule dynamics: caps, catastrophes, and coupled hydrolysis. Phys. Rev. E **54**, 5538–5560 (1996)
279. Fogelson, B., Keener, J.P.: Enhanced nucleocytoplasmic transport due to competition for elastic binding sites. Biophys. J. **115**, 108–116 (2018)
280. Fogelson, B., Keener, J.P.: Transport facilitated by rapid binding to elastic tethers. SIAM J. Appl. Math. **79**, 1405–1422 (2019)
281. Forger, D.B., Peskin, C.S.: A detailed predictive model of the mammalian circadian clock. Proc. Natl. Acad. Sci. USA **100**, 14806–14811 (2003)
282. Forger, D.B., Peskin, C.S.: Stochastic simulation of the mammalian circadian clock. Proc. Natl. Acad. Sci. USA **102**, 321–324 (2005)
283. Forger, D.B., Kim, J.K.: A mechanism for robust circadian timekeeping via stoichiometric balance. Mol. Syst. Biol. **8**, 630 (pp. 1–12) (2012)
284. Foulaadvand, M.E., Kolomeisky, A.B., Teimouri, H.: Asymmetric exclusion processes with disorder: effect of correlations. Phys. Rev. E **78**, 061116 (2008)
285. Fox, R.F., Lu, Y.N.: Emergent collective behavior in large numbers of globally coupled independent stochastic ion channels. Phys. Rev. E **49**, 3421–3431 (1994)
286. Freche, D., Pannasch, U., Rouach, N., Holcman, D.: Synapse geometry and receptor dynamics modulate synaptic strength. PLoS One **6**, e25122 (2011)
287. Freidlin, M.I., Wentzell, A.D.: Random Perturbations of Dynamical Systems. Springer, New York (1998)
288. Friedman, A., Craciun, G.: A model of intracellular transport of particles in an axon. J. Math. Biol. **51**, 217–246 (2005)
289. Friedman, A., Craciun, G.: Approximate traveling waves in linear reaction-hyperbolic equations. SIAM J. Math. Anal. **38**, 741–758 (2006)
290. Friedman, A., Hu, B.: Uniform convergence for approximate traveling waves in linear reaction-hyperbolic systems. Indiana Univ. Math. J. **56**, 2133–2158 (2007)
291. Friedman, N., Cai, L., Xie, X.S.: Linking stochastic dynamics to population distribution: an analytical framework of gene expression. Phys. Rev. Lett. **97**, 168302 (2006)
292. Fulinski, A.: Noise-stimulated active transport in biological cell membranes. Phys. Lett. A **193**, 267–273 (1994)
293. Fulton, A.: How crowded is the cytoplasm? Cell **30**, 345–347 (1982)
294. Furriols, M., Casanova, J.: In and out of Torso RTK signalling. EMBO J. **22**, 1947–1952 (2003)
295. Gander, M.J., Mazza, C., Rummler, H.: Stochastic gene expression in switching environments. J. Math. Biol. **55**, 259–294 (2007)
296. Gang, Hu.: Lyapunov function and stationary probability distributions. Z. Phys. B. Cond. Matt. **65**, 103–106 (1986)
297. Gao, Y.Q.: A simple theoretical model explains dynein's response to load. Biophys. J. **90**, 811–821 (2006)
298. Garai, A., Chowdhury, D., Ramakrishnan, T.V.: Stochastic kinetics of ribosomes: single motor properties and collective behavior. Phys. Rev. E **80**, 011908 (2009)
299. Garcia-Ojalvo, J., Elowitz, M.B., Strogatz, S.H.: Modeling a synthetic multicellular clock: repressilators coupled by quorum sensing. Proc. Natl. Acad. Sci. U.S.A. **101**, 10955–10960 (2004)
300. Gardiner, C.W.: Handbook of Stochastic Methods, 4th edn. Springer, Berlin (2009)
301. Gardner, T.S., Cantor, C.R., Collins, J.J.: Construction of a genetic toggle switch in *E. coli*. Nature **403**, 339–342 (2000)
302. Gaspard, P.: The correlation time of mesoscopic chemical clocks. J. Chem. Phys. **117**, 8905–8916 (2002)

303. Gennerich, A., Schild, D.: Anisotropic diffusion in mitral cell dendrites. Phys. Biol. **3**, 45 (2006)

304. Gerland, U., Bundschuh, R., Hwa, T.: Translocation of structured polynucleotides through nanopores. Phys. Biol. **1**, 19–25 (2004)

305. Gerrow, K., Triller, A.: Synaptic stability and plasticity in a floating world. Curr. Opin. Neurobiol. **20**, 631–639 (2010)

306. Gerstner, J.R., Yin, J.C.P.: Circadian rhythms and memory formation. Nat. Rev. Neurosci. **11**, 577–588 (2010)

307. Ghosh, S., Gopalakrishnan, M., Forsten-Williams, K.: Self-consistent theory of reversible ligand binding to a spherical cell. Phys. Biol. **4**, 344–345 (2007)

308. Ghosh, A., Gov, N.S.: Dynamics of active semiflexible polymers. Biophys. J. **107**, 1065–1073 (2014)

309. Gibson, M., Bruck, J.: Efficient exact stochastic simulation of chemical systems with many species and many channels. J. Phys. Chem. **104**, 1876–1889 (2000)

310. Gillespie, D.T.: Exact stochastic simulation of coupled chemical reactions. J. Phys. Chem. **81**, 2340–2361 (1977)

311. Gillespie, D.T.: Approximate accelerated stochastic simulation of chemically reacting systems. J. Chem. Phys. **115**, 1716–1733 (2001)

312. Gillespie, D.T., Hellander, A., Petzold, L.R.: Perspective: stochastic algorithms for chemical kinetics. J. Chem. Phys. **138**, 170901 (2013)

313. Glansdorff, P., Prigogine, I.: Thermodynamic Theory of Structure, Stability and Fluctuations. Wiley-Interscience, Chichester (1971)

314. Glendinning, P.: Stability, Instability and Chaos: An Introduction to the Theory of Nonlinear Differential Equations. Cambridge University Press, Cambridge (1994)

315. Godec, A., Metzler, R.: First passage time statistics for two-channel diffusion. J. Phys. A **50**, 084001 (2017)

316. Goldbeter, A., Koshland, D.E.: An amplified sensitivity arising from covalent modification in biological systems. Proc. Natl. Acad. Sci. USA **78**, 6840–6844 (1981)

317. Goldbeter, A.: A model for circadian oscillations in the Drosophila period protein (PER). Proc. R. Soc. Lond. B Biol. Sci. **261**, 319–324 (1995)

318. Goldbeter, A., Gerard, C., Gonze, D., Leloup, J.C., Dupont, G.: Systems biology of cellular rhythms. FEBS Lett. **586**, 2955–2965 (2012)

319. Golding, I., Cox, E.C.: Physical nature of bacterial cytoplasm. Phys. Rev. Lett. **96**, 098102 (2006)

320. Goldstein, S.: On diffusion by discontinuous movements and the telegraph equation. Quart. J. Mech. Appl. Math. **4**, 129–156 (1951)

321. Goldstein, L., Yang, Z.: Microtubule-based transport systems in neurons: the roles of kinesins and dyneins. Ann. Rev. Neurosci. **23**, 39–71 (2000)

322. Goldstein, B., Faeder, J.R., Hlavacek, W.S.: Mathematical and computational models of immune-receptor signaling. Nat. Rev. Immun. **4**, 445–456 (2004)

323. Goldwyn, J.H., Shea-Brown, E.: The what and where of adding channel noise to the Hodgkin-Huxley equations. PLoS Comp. Biol. **7**, e1002247 (2011)

324. Gonze, D., Halloy, J., Gaspard, P.: Biochemical clocks and molecular noise: theoretical study of robustness factors. J. Chem. Phys. **116**, 10997–11010 (2002)

325. Gonze, D., Halloy, J.: Robustness of circadian rhythms with respect to molecular noise. Proc. Natl. Acad. Sci. USA **99**, 673–678 (2002)

326. Goodenough, D.A., Paul, D.L.: Gap junctions. Cold Spring Harb. Perspect. Biol. **1**, a002576 (2009)

327. Goodrich, C.P., Brenner, M.P., Ribbeck, K.: Enhanced diffusion by binding to the crosslinks of a polymer gel. Nat. Comm. **9**, 4348 (2018)

328. Gopalakrishnan, M., Forsten-Williams, K., Nugent, M.A., Taubery, U.C.: Effects of receptor clustering on ligand dissociation kinetics: theory and simulations. Biophys. J. **89**, 3686–3700 (2005)

329. Gopalakrishnan, M., Forsten-Williams, K., Cassino, T.R., Padro, L., Ryan, T.E., Tauber, U.C.: Ligand rebinding: self-consistent mean-field theory and numerical simulations applied to surface plasmon resonance studies. Eur. Biophys. J. **34**, 943–958 (2005)

330. Gopalakrishnan, M., Govindan, B.S.: A first-passage-time theory for search and capture of chromosomes by microtubules in mitosis. Bull. Math. Biol. **73**, 2483–2506 (2011)

331. Gopich, I.V., Szabo, A.: Diffusion modifies the connectivity of kinetic schemes for multisite binding and catalysis. Proc. Natl. Acad. Sci. **110**, 19784–19789 (2013)

332. Gordon, P., Sample, C., Berezhkovskii, A.M., Muratov, C.B., Shvartsman, S.Y.: Local kinetics of morphogen gradients. Proc. Natl. Acad. Sci. **108**, 6157–6162 (2011)

333. Gorman, J., Greene, E.C.: Visualizing one-dimensional diffusion of proteins along DNA. Nat. Struct. Mol. Biol. **15**, 768–774 (2008)

334. Gou, J., Li, Y.X., Nagata, W., Ward, M.J.: Synchronized oscillatory dynamics for a 1-D model of membrane kinetics coupled by linear bulk diffusion. SIAM J. Appl. Dyn. Syst. **14**, 2096–2137 (2015)

335. Gou, J., Ward, M.J.: Asymptotic analysis of a 2-D Model of dynamically active compartments coupled by bulk diffusion. J. Nonlinear Sci. **26**, 979–1029 (2016)

336. Graham, R., Tel, T.: On the weak-noise limit of Fokker-Planck models. J. Stat. Phys. **35**, 729 (1984)

337. Graham, R., Tel, T.: Weak-noise limit of Fokker-Planck models and non-differentiable potentials for dissipative dynamical systems. Phys. Rev. **31**, 1109 (1985)

338. Grandell, J.: Doubly Stochastic Process, 1st edn. Springer, New York (1976)

339. Greive, S.J., von Hippel, P.H.: Thinking quantitatively about transcriptional regulation. Nat. Rev. Mol. Cell Biol. **6**, 221–232 (2005)

340. Grewal, S.I., Moazed, D.: Heterochromatin and epigenetic control of gene expression. Science **150**, 563–576 (2003)

341. Griffith, J.S.: Mathematics of cellular control processes. I. Negative feedback to one gene. J. Theor. Biol. **20**, 202–208 (1968)

342. Griffith, J.S.: Mathematics of cellular control processes. II. Positive feedback to one gene. J. Theor. Biol. **20**, 209–216(1968)

343. Grimmett, G.R., Stirzaker, D.R.: Probability and Random Processes, 3rd edn. Oxford University Press, Oxford (2001)

344. Groff, J.R., DeRemigio, H., Smith, G.D.: *Stochastic Methods in Neuroscience* chap. 2. Markov chain models of ion channels and calcium release sites, pp. 29–64. Oxford University Press, Oxford (2009)

345. Gross, S.P.: Hither and yon: a review of bi-directional microtubule-based transport. Phys. Biol. **1**, R1-11 (2004)

346. Guerin, T., Prost, J., Martin, P., Joanny, J.-F.: Coordination and collective properties of molecular motors: theory. Curr. Opin. Cell Biol. **22**, 14–20 (2010)

347. Guerrier, C., Holcman, D.: The first one hundred nanometers inside the pre-synaptic terminal where calcium diffusion triggers vesicular release (2019)

348. Gumbel, E.J.: Statistics of Extremes. Columbia University Press (1962)

349. Gumy, L.F., Katrukha, E.A., Grigoriev, I., Jaarsma, D., Kapitein, L.C., Akhmanova, A., Hoogenraad, C.C.: MAP2 defines a pre-axonal filtering zone to regulate KIF1- versus KIF5-dependent cargo transport in sensory neurons. Neuron **94**, 347–362 (2017)

350. Gumy, L.F., Hoogenraad, C.C.: Local mechanisms regulating selective cargo entry and long-range trafficking in axons. Curr. Opin. Neurobiol. **51**, 23–28 (2018)

351. Gunawardena, J.: Chemical reaction network theory for in-silico biologists. Notes available for download at http://vcp.med.harvard.edu/papers/crnt.pdf (2003)

352. Gundelfinger, E.D., Kessels, M.M., Qualmann, B.: Temporal and spatial coordination of exocytosis and endocytosis. Nat. Rev. Mol. Cell Biol. **4**, 127–139 (2003)

353. Gupta, A., Milias-Argeitis, A., Khammash, M.: Dynamic disorder in simple enzymatic reactions induces stochastic amplification of substrate. J. R. Soc. Interface **14**, 20170311 (2017)

354. Halford, S.E., Marko, J.F.: How do site-specific DNA-binding proteins find their targets? Nucl. Acid Res. **32**, 3040–3052 (2004)

355. Halford, S.E.: An end to 40 years of mistakes in DNA-protein association kinetics? Biochem. Soc. Trans. **37**, 343–348 (2009)
356. Hanggi, P., Grabert, H., Talkner, P., Thomas, H.: Bistable systems: master equation versus Fokker-Planck modeling. Phys. Rev. A **29**, 371–378 (1984)
357. Hanggi, P., Talkner, P., Borkovec, M.: Reaction rate theory: fifty years after Kramers. Rev. Mod. Phys. **62**, 251–341 (1990)
358. Haselwandter, C.A., Calamai, M., Kardar, M., Triller, A., da Silveira, R.A.: Formation and stability of synaptic receptor domains. Phys. Rev. Lett. **106**, 238104 (2011)
359. Haselwandter, C.A., Kardar, M., Triller, A., da Silveira, R.A.: Self-assembly and plasticity of synaptic domains through a reaction-diffusion mechanism. Phys. Rev. E **92**, 032705 (2015)
360. Hasty, J., McMillen, D., Collins, J.J.: Engineered gene circuits. Nature **420**, 224–230 (2002)
361. Hasty, J., Isaacs, F., Dolnik, M., McMillen, D., Collins, J.J.: Designer gene networks: towards fundamental cellular control. Chaos **11**, 207 (2001)
362. Hasty, J., Dolnik, M., Rottschafer, V., Collins, J.J.: Synthetic gene network for entraining and amplifying cellular oscillations. Phys. Rev. Lett. **88**, 148101 (2002)
363. Havlin, S., ben-Avraham D.: Diffusion in disordered media. Adv. Phys. **51**, 187–292 (2002)
364. He, Y., Burov, S., Metzler, R., Barkai, E.: Random time-scale invariant diffusion and transport coefficients. Phys. Rev. Lett. **101**, 058101 (2008)
365. He, G., Qian, H., Qian, M.: Stochastic theory of nonequilibrium steady states and its applications. Part II. Phys. Rep. **510**, 87–118 (2012)
366. Hendricks, A.G., Perlson, E., Ross, J.L., Schroeder, H.W., Tokito, M., Holzbaur, E.L.K.: Motor coordination via a tug-of-war mechanism drives bidirectional vesicle transport. Cur. Opin. Biol. **20**, 697–702 (2010)
367. Henley, J.M., Barker, E.A., Glebov, O.O.: Routes, destinations and delays: recent advances in AMPA receptor trafficking. Trends Neurosci. **34**, 258–268 (2011)
368. Henley, J.M., Wilkinson, K.A.: Synaptic AMPA receptor composition in development, plasticity and disease. Nat. Rev. Neurosci. **17**, 337–350 (2016)
369. Herbert, K., La Porta, A., Wong, B., Mooney, R., Neuman, K., Landick, R., Block, S.: Sequence-resolved detection of pausing by single RNA polymerase molecules. Cell **125**, 1083–1094 (2006)
370. Higham, D.J.: An algorithmic introduction to numerical simulation of stochastic differential equations. SIAM Rev. **43**, 525–546 (2001)
371. Hill, T.L.: Effect of rotation on the diffusion-controlled rate of ligand-protein association. Proc. Natl. Acad. Sci. U.S.A. **72**, 4918–4922 (1975)
372. Hill, N., Hader, D.P.: A biased random walk model for the trajectories of swimming microorganisms. J. Theor. Biol. **186**, 503–526 (1997)
373. Hille, B.: Ionic Channels of Excitable Membranes, 3rd edn. Sinauer Associates, Massachusetts (2001)
374. Hillen, T., Othmer, H.: The diffusion limit of transport equations derived from velocity-jump processes. SIAM J. Appl. Math. **61**, 751–775 (2000)
375. Hillen, T., Swan, A.: The diffusion limit of transport equations in biology. In: Preziosi, L., et al. (eds.) *Mathematical Models and Methods for Living Systems*, pp. 3–129 (2016)
376. Hillen, T., Painter, K.J., Swan, A.C., Murtha, A.D.: Moments of von Mises and Fisher distributions and applications. Mathematica Biosc. Eng. **14**, 673–694 (2017)
377. Hinch, R.: A mathematical analysis of the generation and termination of calcium sparks. Biophys. J. **86**, 1293–1307 (2004)
378. Hinch, R., Chapman, S.J.: Exponentially slow transitions on a Markov chain: the frequency of calcium sparks. Eur. J. Appl. Math. **16**, 427–446 (2005)
379. Hirokawa, N., Takemura, R.: Molecular motors and mechanisms of directional transport in neurons. Nat. Rev. Neurosci. **6**, 201–214 (2005)
380. Hodgkin, A.L., Huxley, A.F.: A quantitative description of membrane and its application to conduction and excitation in nerve. J. Physiol. **117**, 500–544 (1952)

381. Hoerndli, F.J., Maxfield, D.A., Brockie, P.J., Mellem, J.E., Jensen, E., Wang, R., Madsen, D.M., Maricq, A.V.: Kinesin-1 regulates synaptic strength by mediating the delivery, removal, and redistribution of AMPA receptors. Neuron **80**, 1421–1437 (2013)

382. Hoffmann, A., Levchenko, A., Scott, M.L., Baltimore, D.: The IκB-NF-κB signaling module: temporal control and selective gene activation. Science **298**, 1241–1245 (2002)

383. Hofling, F., Franosch, T.: Anomalous transport in the crowded world of biological cells. Rep. Prog. Phys. **76**, 046602 (2013)

384. Holcman, D.: Modeling DNA and virus trafficking in the cell cytoplasm. J. Stat. Phys. **127**, 471–494 (2007)

385. Holcman, D., Schuss, Z.: Escape through a small opening: receptor trafficking in a synaptic membrane. J. Stat. Phys. **117**, 975–1014 (2004)

386. Holcman, D., Triller, A.: Modeling synaptic dynamics driven by receptor lateral diffusion. Biophys. J. **91**, 2405–2415 (2006)

387. Holcman, D., Schuss, Z.: Diffusion laws in dendritic spines. J. Math. Neurosci. **1**(10) (2011)

388. Holcman, D., Schuss, Z.: Control of flux by narrow passages and hidden targets in cellular biology. Rep. Prog. Phys. **76**, 074601 (2013)

389. Holcman, D., Schuss, Z.: The narrow escape problem. SIAM Rev. **56**, 213–257 (2014)

390. Holcman, D., Schuss, Z.: Time scale of diffusion in molecular and cellular biology. J. Phys. A **47**, 173001 (2014)

391. den Hollander, F.: Large Deviations. AMS, Providence (2000)

392. Holmes, W.R., Edelstein-Keshet, L.: A comparison of computational models for eukaryotic cell shape and motility. PLoS Comput. Biol. **8**, e1002793 (2012)

393. Hong, J.W., Hendrix, D.A., Papatsenko, D., Levine, M.S.: How the Dorsal gradient works: insights from postgenome technologies. Proc. Natl. Acad. Sci. **105**, 20072–20076 (2008)

394. Hopfield, J.J.: Kinetic proofreading: a new mechanism for reducing errors in biosynthetic processes requiring high specificity. Proc. Natl. Acad. Sci. USA **71**, 4135–4139 (1974)

395. Horn, F.J.M.: Necessary and sufficient conditions for complex balancing in chemical kinetics. Arch. Rat. Mech. Anal. **49**, 172–186 (1972)

396. Horn, F.J.M., Jackson, R.: General mass action kinetics. Arch. Rat. Mech. Anal. **47**, 81–116 (1972)

397. Howard, J.: Mechanics of Motor Proteins and the Cytoskeleton. Sinauer (2001)

398. Hsu, E.P.: Stochastic Analysis on Manifolds. Graduate Studies in Mathematics, vol. 38. American Mathematical Society, Providence, RI (2002)

399. Hu, T., Grossberg, A.Y., Shklovskii, B.I.: How proteins search for their specific sites on DNA: the role of DNA conformation. Biophys. J. **90**, 2731–2744 (2006)

400. Hu, J., Matzavinos, A., Othmer, H.G.: A theoretical approach to actin filament dynamics. J. Stat. Phys. **128**, 111–138 (2007)

401. Huet, S., Karatekin, E., Tran, V.S., Fanget, I., Cribier, S., Henry, J.P.: Analysis of transient behavior in complex trajectories: application to secretory vesicle dynamics. Biophys. J. **91**, 3542–3559 (2006)

402. Hufton, P.G., Lin, Y.T., Galla, T., McKane, A.J.: Intrinsic noise in systems with switching environments. Phys. Rev. E **93**, 052119 (2016)

403. Hughes, B.D.: Random Walks and Random Environments Vol. 1: Random Walks. Oxford University, Oxford (1995)

404. Hughey, J.J., Gutschow, M.V., Bajar, B.T., Covert, M.W.: Single-cell variation leads to population invariance in NF-κB. Mol. Biol. Cell **26**, 583–590 (2015)

405. Huxley, A.F.: Muscle structure and theories of contraction. Prog. Biophys. Biophys. Chem. **7**, 255–318 (1957)

406. Ingalls, B.P.: Mathematical Modeling in Systems Biology: An Introduction. MIT Press (2013)

407. Isaacson, S.A., McQueen, D.M., Peskin, C.S.: The influence of volume exclusion by chromatin on the time required to find specific DNA binding sites by diffusion. Proc. Natl. Acad. Sci. U.S.A. **108**, 3815–3820 (2011)

408. Israelachvili, J.N.: Intermolecular and Surface Forces, 2nd edn. Academic Press, New York (1991)
409. Jackson, M.B.: Molecular and Cellular Biophysics. Cambridge University Press, Cambridge (2006)
410. Jacob, F., Monod, J.: Genetic regulatory mechanisms in the synthesis of proteins. J. Mol. Biol. **3**, 318–356 (1961)
411. Jacobs, M.H.: Diffusion Processes. Springer, New York (1967)
412. Jacobs, K.: Stochastic Processes for Physicists. Cambridge University Press, Cambridge (2010)
413. Jacobson, K., Mouritsen, O.G., Anderson, R.G.W.: Lipid rafts: at a crossroad between cell biology and physics. Nat. Cell Biol. **9**, 7–14 (2007)
414. Janson, M.E., de Dood, M.E., Dogterom, M.: Dynamic instability of microtubules is regulated by force. J. Cell Biol. **161**, 1029–1034 (2003)
415. Janssen, H.-K.: On a Lagrangian for classical field dynamics and renormalization group calculations of dynamical critical properties. Z. Phys. B **23**, 377–380 (1976)
416. Jeon, J.-H., Metzler, R.: Inequivalence of time and ensemble averages in ergodic systems: exponential versus power-law relaxation in confinement. Phys. Rev. E **85**, 021147 (2012)
417. Jeon, J.-H., Tejedor, V., Burov, S., Barkai, E., Selhuber-Unkel, C., Berg-Sorensen, K., Oddershede, L., Metzler, R.: In vivo anomalous diffusion and weak ergodicity breaking of lipid granules. Phys. Rev. Lett. **106**, 048103 (2011)
418. Jose, J.V., Saletan, E.J.: Classical Dynamics: A Contemporary Approach. Cambridge University Press, Cambridge (2013)
419. Julicher, F., Prost, J.: Cooperative molecular motors. Phys. Rev. Lett. **75**, 2618–2621 (1995)
420. Julicher, F., Prost, J.: Spontaneous oscillations of collective molecular motors. Phys. Rev. Lett. **78**, 4510–4513 (1997)
421. Julicher, F., Ajdari, A., Prost, J.: Modeling molecular motors. Rev. Mod. Phys. **69**, 1269–1281 (1997)
422. Jung, P., Brown, A.: Modeling the slowing of neurofilament transport along the mouse sciatic nerve. Phys. Biol. **6**, 046002 (2009)
423. Kac, M.: On the distribution of certain Wiener functionals. Trans. Am. Math. Soc. **65**, 1–13 (1949)
424. Kac, M.: A stochastic model related to the telegrapher's equation. Rocky Mountain J. Math. **3**, 497–509 (1974)
425. Kaern, M., Elston, T.C., Blake, W.J., Collins, J.J.: Stochasticity in gene expression: from theories to phenotypes. Nat. Rev. Genetics **6**, 451–464 (2005)
426. Kahana, A., Kenan, G., Feingold, M., Elbaum, M., Granek, R.: Active transport on disordered microtubule networks: the generalized random velocity model. Phys. Rev. E **78**, 051912 (2008)
427. Kalinay, P., Percus, J.K.: Corrections to the Fick-Jacobs equation. Phys. Rev. E **74**, 041203 (2006)
428. Karamched, B.R., Bressloff, P.C.: Effects of cell geometry on reversible vesicular transport. J. Phys. A **50**, 055601 (2017)
429. Kardar, M.: Statistical Physics of Particles. Cambridge University Press, Cambridge (2007)
430. Karmakar, R., Bose, I.: Graded and binary responses in stochastic gene expression. Phys. Biol. **1**, 197–204 (2004)
431. Keener, J.P., Sneyd, J.: Mathematical Physiology I: Cellular Physiology, 2nd edn. Springer, New York (2009)
432. Keener, J.P., Sneyd, J.: Mathematical Physiology II: Systems Physiology, 2nd edn. Springer, New York (2009)
433. Keener, J.P., Newby, J.M.: Perturbation analysis of spontaneous action potential initiation by stochastic ion channels. Phy. Rev. E **84**, 011918 (2011)
434. Keizer, J.: Nonequilibrium statistical thermodynamics and the effect of diffusion on chemical reaction rates. J. Phys. Chem. **86**, 5052–5067 (1982)

435. Keizer, J., Smith, G.D.: Spark-to-wave transition: saltatory transmission of calcium waves in cardiac myocytes. Biophys. Chem. **72**, 87–100 (1998)

436. Keller, D., Bustamante, C.: The mechanochemistry of molecular motors. Biophys. J. **78**, 541–556 (2000)

437. Kenkre, V.M., Giuggioli, L., Kalay, Z.: Molecular motion in cell membranes: analytical study of fence-hindered random walks. Phys. Rev. E **77**, 051907 (2008)

438. Kepler, T.B., Elston, T.C.: Stochasticity in transcriptional regulation: origins, consequences, and mathematical representations. Biophys. J. **81**, 3116–3136 (2001)

439. Kerr, J.M., Blanpied, T.A.: Super-resolution imaging reveals that AMPA receptors inside synapses are dynamically organized in nanodomains regulated by PSD95. J. Neurosci. **33**, 13204–13224 (2013)

440. Kifer, Y.: Large deviations and adiabatic transitions for dynamical systems and Markov processes in fully coupled averaging. Mem. AMS **201**, issue 944 (2009)

441. Kim, H., Bressloff, P. C.: Impulsive signaling model of cytoneme-based morphogen gradient formation. Phys. Biol. **16** 056005 (2019)

442. Kim, K., Lepzelater, D., Wang, J.: Single molecule dynamics and statistical fluctuations of gene regulatory networks: a repressilator. J. Chem. Phys. **126**, 034702 (2007)

443. Kim, E., Jung, H.: Local mRNA translation in long-term maintenance of axon health and function. Curr. Opin. Neurobiol. **63**, 15–22 (2020)

444. Kimmel, M., Axelrod, D.E.: Mathematical models of gene amplification with applications to cellular drug resistance and tumorigenicity. Genetics **125**, 633–644 (1990)

445. Kimmel, M., Axelrod, D.E.: Branching Processes in Biology. Springer, New York (2002)

446. King, D.P., Takahashi, J.S.: Molecular genetics of circadian rhythms in mammals. Ann. Rev. Neurosci. **23**, 713–742 (2000)

447. Kingman, J.F.C.: Poisson Processes. Clarendon Press, Oxford (1993)

448. Klann, M., Koeppl, H., Reuss, M.: Spatial modeling of vesicle transport and the cytoskeleton: the challenge of hitting the right road. PLoS One **7**, e29645 (2012)

449. Klimontovich, Y.L.: Nonlinear Brownian motion. Physics-Uspekhi **37**, 737–767 (1994)

450. Kloeden, P.E., Platen, E.: Numerical Solution of Stochastic Differential Equations. Springer, New York (2000)

451. Klumpp, S., Lipowsky, R.: Traffic of molecular motors through tube-like compartments. J. Stat. Phys. **113**, 233–268 (2003)

452. Klumpp, S., Lipowsky, R.: Cooperative cargo transport by several molecular motors. Proc. Natl. Acad. Sci. USA **102**, 17284–17289 (2005)

453. Knessl, C., Matkowsky, B.J., Schuss, Z., Tier, C.: An asymptotic theory of large deviations for Markov jump processes. SIAM J. Appl. Math. **46**, 1006–1028 (1985)

454. Knowles, R.B., Sabry, J.H., Martone, M.E., Deerinck, T.J., Ellisman, M.H., Bassell, G.J., Kosik, K.S.: Translocation of RNA granules in living neurons. J. Neurosci. **16**, 7812–7820 (1996)

455. Kochugaeva, M.P., Shvets, A.A., Kolomeisky, A.B.: How conformational dynamics influences the protein search for targets on DNA. J. Phys. A Math. Theor. **49**, 444004 (2016)

456. Kochugaeva, M.P., Berezhkovskii, A.A., Kolomeisky, A.B.: Optimal length of conformational transitions region in the protein search for targets on DNA. J. Phys. Chem. Lett. **8**, 4049–4054 (2017)

457. Kolomeisky, A.B.: Asymmetric simple exclusion model with local inhomogeneity. J. Phys. A **31**, 1153–1164 (1998)

458. Kolomeisky, A.B., Schutz, G.M., Kolomeisky, E.B., Straley, J.P.: Phase diagram of one-dimensional driven lattice gases with open boundaries. J. Phys. A **31**, 6911–6919 (1998)

459. Kolomeisky, A.B., Fisher, M.E.: A simple kinetic model describes the processivity of myosin-V. Biophys. J. **84**, 1642–1650 (2003)

460. Kolomeisky, A.B.: Channel-facilitated molecular transport across membranes: attraction, repulsion and asymmetry. Phys. Rev. Lett. **98**, 048105 (2007)

461. Kolomeisky, A., Fisher, M.: Molecular motors: a theorist's perspective. Ann. Rev. Phys. Chem. **58**, 675–695 (2007)

462. Kolomeisky, A.B.: Physics of protein-DNA interactions: mechanisms of facilitated target search. Phys. Chem. Chem. Phys. **13**, 2088–2095 (2011)

463. Kosik, K.S., Joachim, C.L., Selkoe, D.J.: Microtubule-associated protein tau is a major antigenic component of paired helical filaments in Alzheimer disease. Proc. Natl. Acad. Sci. U.S.A. **83**, 4044–4048 (1986)

464. Koster, G., VanDuijn, M., Hofs, B., Dogterom, M.: Membrane tube formation from giant vesicles by dynamic association of motor proteins. Proc. Natl. Acad. Sci. USA **100**, 15583–15588 (2003)

465. Kramer, K.L., Yost, H.J.: Heparan sulfate core proteins in cell-cell signaling. Annu. Rev. Genet. **37**, 461–484 (2003)

466. Krapf, D.: Mechanisms underlying anomalous diffusion in the plasma membrane. Lipid Domains, Current Topics in Membranes **75**, 167–207 Elsevier (2005)

467. Krapivsky, P.L., Mallick, K.: Fluctuations in polymer translocation. J. Stat. Mech. **P07007** (2010)

468. Krapivsky, P.L., Redner, S., Ben-Naim, E.: A Kinetic View of Statistical Physics. Cambridge University Press, Cambridge (2010)

469. Krug, J.: Boundary-induced phase transitions in driven diffusive systems. Phys. Rev. E **67**, 1882–1185 (1991)

470. Kubo, R.: Fluctuation, relaxation and resonance in magnetic systems. In: TerHaar, D. (ed.) Stochastic Theory of Line Shape. Oliver and Boyd, Edinburgh (1962)

471. Kubo, R., Toda, M., Hahitsume, N.: Statistical Physics II: Nonequilibrium Statistical Mechanics. Springer (1991)

472. Kumar, N., Singh, A., Kulkarni, R.V.: Transcriptional bursting in gene expression: analytical results for general stochastic models. PLoS Comp. Biol. **11**, e1004292 (2015)

473. Kural, C., Kim, H., Syed, S., Goshima, G., Gelfand, V.I., Selvin, P.R.: Kinesin and dynein move a peroxisome in vivo: a tug-of-war or coordinated movement? Science **308**, 1469–1472 (2005)

474. Kurella, V., Tzou, J.C., Coombs, D., Ward, M.J.: Asymptotic analysis of first passage time problems inspired by ecology. Bull. Math. Biol. **77**, 83–125 (2015)

475. Kurtz, T.G.: Limit theorems and diffusion approximations for density dependent Markov chains. Math. Prog. Stud. **5**, 67–78 (1976)

476. Kurtz, T.G.: Representations of Markov processes as multiparameter changes. Ann. Prob. **8**, 682–715 (1980)

477. Kusmierz, L., Majumdar, S.N., Sabhapandit, S., Schehr, G.: First order transition for the optimal search time of Levy flights with resetting. Phys. Rev. Lett. **113**, 220602 (2014)

478. Kustanovich, T., Rabin, Y.: Metastable network model of protein transport through nuclear pores. Biophys. J. **86**, 2008–2016 (2004)

479. Kusumi, A., Sako, Y., Yamamoto, M.: Confined lateral diffusion of membrane receptors as studied by single particle tracking (nanovid microscopy): effects of calcium-induced differentiation in cultured epithelial cells. Biophys. J. **65**, 2021–2040 (1993)

480. Kusumi, A., Nakada, C., Ritchie, K., Murase, K., et al.: Paradigm shift of the plasma membrane concept from the two-dimensional continuum fluid to the partitioned fluid: high-speed single-molecule tracking of membrane molecules. Annu. Rev. Biophys. Biomol. Struct. **34**, 351–354 (2005)

481. Kusumi, A., Shirai, Y.M., Koyama-Honda, I., Suzuki, K.G.N., Fujiwara, T.K.: Hierarchical organization of the plasma membrane: investigations by single-molecule tracking vs. fluorescence correlation spectroscopy. FEBS Lett. **584**, 1814–1823 (2010)

482. Kuznetsov, A.V., Avramenko, A.A.: The method of separation of variables for solving equations describing molecular-motor-assisted transport of intracellular particles in a dendrite or axon. Proc. Roy. Soc. A **464**, 2867–2886 (2008)

483. Kuznetzov, Y.A.: Elements of Applied Bifurcation Theory, 3rd edn. Springer (2010)

484. Lagache, T., Holcman, D.: Effective motion of a virus trafficking inside a biological cell. SIAM J. Appl. Math. **68**, 1146–1167 (2009)

485. Lagache, T., Dauty, E., Holcman, D.: Physical principles and models describing intracellular virus particle dynamics. Curr. Opin. Microbiol. **12**, 439–445 (2009)

486. Lander, A.D., Nie, W., Wan, F.Y.: Do morphogen gradients arise by diffusion? Dev Cell **2**, 785–796 (2002)

487. Lander, A.D.: Pattern, growth and control. Cell **144**, 955–969 (2011)

488. Lagerholm, B.C., Thompson, N.L.: Theory for ligand rebinding at cell membrane surfaces. Biophys. J. **74**, 1215–1228 (1998)

489. Lange, M., Kochugaeva, M., Kolomeisky, A.B.: Protein search for multiple targets on DNA. J. Chem. Phys. **143**, 105102 (2015)

490. Lange, M., Kochugaeva, M., Kolomeisky, A.B.: Dynamics of the protein search for targets on DNA in the presence of traps. J. Phys. Chem. B **119**, 12410–12416 (2016)

491. Lau, A.W.C., Lubensky, T.C.: State-dependent diffusion. Thermodynamic consistency and its path integral formulation. Phys. Rev. E **76**, 011123 (2007)

492. Lauffenburger, D.A., Linderman, J.J.: Receptors: Models for Binding, Trafficking, and Signaling. Oxford University Press, Oxford (1996)

493. Lawley, S.D., Mattingly, J.C., Reed, M.C.: Stochastic switching in infinite dimensions with applications to random parabolic PDEs (2015)

494. Lawley, S.D., Tuft, M., Brooks, H.A.: Coarse-graining intermittent intracellular transport: two- and three-dimensional models. Phys. Rev. E **92**, 042709 (2015)

495. Lawley, S.D.: Boundary value problems for statistics of diffusion in a randomly switching environment: PDE and SDE perspectives. SIAM J. Appl. Dyn. Syst. **15**, 1410–1433 (2016)

496. Lawley, S.D., Best, J., Reed, M.C.: Neurotransmitter concentrations in the presence of neural switching in one dimension. Disc. Cont. Dyn. Syst. B **21**, 2255–2273 (2016)

497. Lawley, S.D., Keener, J.P.: Including rebinding reactions in well-mixed models of distributive biochemical reactions. Biophys. J. **111**, 2317–2326 (2016)

498. Lawley, S.D., Keener, J.P.: Rebinding in biochemical reactions on membranes. Phys. Biol. **14**, 056002 (2017)

499. Lawley, S.D., Madrid, J.B.: A probabilistic approach to extreme statistics of Brownian escape times in dimensions 1, 2, and 3 (2019)

500. Lebowitz, J.L., Percus, J.K.: Kinetic equations and density expansions: exactly solvable one-dimensional system. Phys. Rev. **155**, 122–138 (1967)

501. Lechleiter, J., Girard, S., Peralta, E., Clapham, D.: Spiral calcium wave propagation and annihilation in Xenopus laevis oocytes. Science **252**, 123–126 (1991)

502. Leduc, C., et al.: Cooperative extraction of membrane nanotubes by molecular motors. Proc. Natl. Acad. Sci. USA **101**, 17096–17101 (2004)

503. Leibler, S., Huse, D.A.: Porters versus rowers: a unified stochastic model of motor proteins. J. Cell Biol. **121**, 1357–1368 (1993)

504. Leitner, D.M., Brown, F.L.H., Wilson, K.R.: Regulation of protein mobility in cell membranes: a dynamic corral model. Biophys. J. **78**, 125–135 (2000)

505. Leloup, J.-C., Gonze, D., Goldbeter, A.: Limit cycle models for circadian rhythms based on transcriptional regulation in Neurospora and Drosophila. J. Biol. Rhythms **14**, 433–448 (1999)

506. Leloup, J.-C., Goldbeter, A.: Toward a detailed computational model for the mammalian circadian clock. Proc. Natl. Acad. Sci. USA **100**, 7051–7056 (2003)

507. Lemieux, M., Labrecque, S., Tardif, C., Labrie-Dion, E., LeBel, E., De Koninck, P.: Translocation of CaMKII to dendritic microtubules supports the plasticity of local synapses. J. Cell Biol. **198**, 1055–1073 (2012)

508. Lever, M., Maini, P.K., van der Merwe, P.A., Dushek, O.: Phenotypic models of T cell activation. Nat. Rev. Immun. **14**, 619–629 (2014)

509. Levin, S.A.: The problem of pattern and scale in ecology. Ecology **73**, 1943–1967 (1992)

510. Levine, J., Kueh, H.Y., Mirny, L.: Intrinsic fluctuations, robustness, and tunability in signaling cycles. Biophys. J. **92**, 4473–4481 (2007)

511. Levitt, D.G.: Dynamics of a single-file pore: non-Fickian behavior. Phys. Rev. A **8**, 3050–3054 (1973)

512. Levitt, D.G.: Modeling of ion channels. J. Gen. Physiol. **113**, 789–794 (1999)
513. Lèvy, P.: Sur certaines processus stochastiques homogenes. Compos. Math. **7**, 283 (1939)
514. Leybaert, L., Paemeleire, K., Strahonja, A., Sanderson, M.J.: Inositol-trisphosphate-dependent intercellular calcium signaling in and between astrocytes and endothelial cells. Glia **24**, 398–407 (1998)
515. Leybaert, L., Sanderson, M.J.: Intercellular Ca^{2+} waves: mechanisms and function. Physiol. Rev. **92**, 1359–1392 (2012)
516. Li, G.W., Berg, O.G., Elf, J.: Effects of macromolecular crowding and DNA looping on gene regulation kinetics. Nat. Phys. **4**, 294–297 (2009)
517. Li, Y., Rinzel, J.: Equations for InsP$_3$ receptor-mediated calcium oscillations derived from a detailed kinetic model: a Hodgkin-Huxley like formalism. J. Theor. Biol. **166**, 461–473 (1994)
518. Li, Y., Jung, P., Brown, A.: Axonal transport of neurofilaments: a single population of intermittently moving polymers. J. Neurosci. **32**, 746–758 (2012)
519. Li, P., et al.: Phase transitions in the assembly of multivalent signalling proteins. Nature **483**, 336–340 (2012)
520. Liepelt, S., Lipowsky, R.: Kinesin's network of chemomechanical motor cycles. Phys. Rev. Lett. **98**, 258102 (2007)
521. Linder, B., Garcia-Ojalvo, J., Neiman, A., Schimansky-Geier, L.: Effects of noise in excitable systems. Phys. Rep. **392**, 321–424 (2004)
522. Lindsay, A.E., Kolokolnikov, T., Tzou, J.C.: Narrow escape problem with a mixed target and the effect of orientation. Phys. Rev. E **91**, 032111 (2015)
523. Lindsay, A.E., Spoonmore, R.T., Tzou, J.C.: Hybrid asymptotic-numerical approach for estimating first-passage-time densities of the two-dimensional narrow capture problem. Phys. Rev. E **94**, 042418 (2016)
524. Lindsay, A.E., Bernoff, A.J., Ward, M.J.: First passage statistics for the capture of a Brownian particle by a structured spherical target with multiple surface targets. Multiscale Model. Simul. **15**, 74–109 (2017)
525. Lipowsky, R., Klumpp, S.: 'Life is motion': multiscale motility of molecular motors. Physica A: Stat. Mech. Appl. **352**, 53–112 (2005)
526. Lippincott-Schwartz, J., Roberts, T.H., Hirschberg, K.: Secretory protein trafficking and organelle dynamics in living cells. Ann. Rev. Cell Dev. Biol. **16**, 557–589 (2000)
527. Little, J.D.C.: A proof for the Queuing formula: $L = \lambda W$. Oper. Res. **9**, 383–387 (1961)
528. Liu, L., Kashyap, B.R.K., Templeton, J.G.C.: On the GI X /G/∞ system. J. Appl. Prob. **27**, 671–683 (1990)
529. Logan, J.D.: Applied Mathematics, 4th edn. Wiley, New Jersey (2013)
530. Loinger, A., Biham, O.: Stochastic simulations of the repressilator circuit. Phys. Rev. E **76**, 051917 (2007)
531. Loverdo, C., Benichou, O., Moreau, M., Voituriez, R.: Enhanced reaction kinetics in biological cells. Nat. Phys. **4**, 134–137 (2008)
532. Loverdo, C., Benichou, O., Moreau, M., Voituriez, R.: Robustness of optimal intermittent search strategies in one, two, and three dimensions. Phys. Rev. E **80**, 031146 (2009)
533. Low, D.A., Weyand, N.J., Mahan, M.J.: Roles of DNA adenine methylation in regulating bacterial gene expression and virulence. Infect. Immun. **69**, 7197–7204 (2001)
534. Luby-Phelps, K.: Cytoarchitecture and physical properties of cytoplasm: volume, viscosity, diffusion, intracellular surface area. Int. Rev. Cytol. **192**, 189–221 (1999)
535. Lugo, C.A., McKane, A.J.: Quasi-cycles in a spatial predator-prey model. Phys. Rev. E **78**, 051911 (2008)
536. Luo, C.H., Rudy, Y.: A dynamic model of the cardiac ventricular action potential. II. Afterdepolarizations, triggered activity, and potentiation. Circ. Res. **74**, 1097–1113 (1994)
537. Lyons, L.: Bayes and frequentism: a particle physicist's perspective. Contemp. Phys. **54**, 1–16 (2013)
538. Lyons, D.B., Allen, W.E., Goh, T., Tsai, L., Barnea, G., Lomvardas, S.: An epigenetic trap stabilizes singular olfactory receptor expression. Cell **154**, 325–336 (2013)

539. Macara, I.G.: Transport into and out of the nucleus. Microbiol. Mol. Biol. Rev. **65**, 570–594 (2001)
540. MacDonald, C.T., Gibbs, J.H., Pipkin, A.C.: Kinetics of biopolymerization on nucleic acid templates. Biopolymers **6**, 1–25 (1968)
541. MacDonald, C.T., Gibbs, J.H.: Concerning the kinetics of polypeptide synthesis on polyribosomes. Biopolymers **7**, 707–725 (1969)
542. MacDonald, J.L., Gin, C.S.Y., Roskams, A.J.: State-specific induction of DNA methyltransferases in olfactory receptor neurons development. Dev. Biol. **288**, 461–473 (2005)
543. MacGillavry, H.D., Kerr, J.M., Blanpied, T.A.: Lateral organization of the postsynaptic density. Mol. Cell. Neurosci. **48**, 321–331 (2011)
544. Mackey, M.C., Tyran-Kaminska, M.: Dynamics and density evolution in piecewise deterministic growth processes. Ann. Polon. Math. **94**, 111–129 (2008)
545. Mackey, M.C., Tyran-Kaminska, M., Yvinec, R.: Dynamic behavior of stochastic gene expression in the presence of bursting. SIAM J. Appl. Math. **73**, 1830–1852 (2013)
546. Maday, S., Twelvetrees, A.E., Moughamian, A.J., Holzbaur, E.L.F.: Axonal transport: cargo-specific mechanisms of motility and regulation. Neuron **84**, 292–309 (2014)
547. Maeder, C.I., San-Miguel, A., Wu, E.Y., Lu, H., Shen, K.: vivo neuron-wide analysis of synaptic vesicle precursor trafficking. Traffic **15**, 273–291 (2014)
548. Maeder, C.I., Shen, K., Hoogenraad, C.C.: Axon and dendritic trafficking. Curr. Opin. Neurobiol. **27**, 165–170 (2014)
549. Maguire, L., Stefferson, M., Betterton, M.D., Hough, L.E.: Design principles of selective transport through biopolymer barriers. Phys. Rev. E **100**, 042414 (2019)
550. Maheshri, N., O'Shea, E.K.: Living with noisy genes: how cells function reliably with inherent variability in gene expression. Annu. Rev. Biophys. Biomol. Struct. **36**, 413–434 (2007)
551. Maier, R.S., Stein, D.L.: Limiting exit location distribution in the stochastic exit problem. SIAM J. Appl. Math. **57**, 752–790 (1997)
552. Majumdar, S.N.: Brownian functionals in physics and computer science. Curr. Sci. **89**, 2076–2092 (2005)
553. Majumdar, S.N., Sabhapandit, S., Schehr, G.: Dynamical transition in the temporal relaxation of stochastic processes under resetting. Phys. Rev. E **91**, 052131 (2015)
554. Makhnovskii, Yu.A., Berezhkovskii, A.M., Sheu, S.Y., Yang, D.-Y., Kuo, J., Lin, S.H.: Stochastic gating influence on the kinetics of diffusion-limited reactions. J. Chem. Phys. **108**, 971–983 (1998)
555. Mandelbrot, B.B., Van Ness, J.W.: Fractional Brownian motions, fractional noise and applications. SIAM Rev. **10**, 422–437 (1968)
556. Maoileidigh, D., Tadigotla, V.R., Nudler, E., Ruckenstein, A.E.: A unified model of transcription elongation: what have we learned from single-molecule experiments? Biophys. J. **100**, 1157–1166 (2011)
557. Marban, E., Robinson, S.W., Wier, W.G.: Mechanisms of arrhythmogenic delayed and early afterdepolarizations in ferret ventricular muscle. J. Clin. Invest. **78**, 1185–1192 (1986)
558. Mardia, K.V., Jupp, P.E.: Directional Statistics, 2nd edn. Wiley Series in Probability and Statistics. Wiley, Chichester (2000)
559. Martin, P.C., Siggia, E.D., Rose, H.A.: Statistical dynamics of classical systems. Phys. Rev. A **8**, 423–437 (1973)
560. Marzen, S., Garcia, H.G., Phililps, R.: Statistical mechanics of Monod-Wyman-Changeux (MWC) models. J. Mol. Biol. **425**, 1433–1460 (2013)
561. Mather, W., Bennett, M.R., Hasty, J., Tsimring, L.S.: Delay-induced degrade-and-fire oscillations in small genetic circuits. Phys. Rev. Lett. **102**, 068105 (2009)
562. Mather, W., Hasty, J., Tsimring, L.S.: Synchronization of degrade-and-fire oscillations via a common activator. Phys. Rev. Lett. **113**, 128102 (2014)
563. Matkowsky, B.J., Schuss, Z.: The exit problem for randomly perturbed dynamical systems. SIAM J. Appl. Math. **33**, 365–382 (1977)
564. McAdams, H.H., Arkin, A.: Stochastic mechanisms in gene expression. Proc. Natl. Acad. Sci. USA **94**, 814–819 (1997)

565. McKean, H.P.: Brownian local time. Adv. Math. **15**, 91–111 (1975)
566. McKeithan, K.: Kinetic proofreading in T-cell receptor signal transduction. Proc. Natl. Acad. Sci. USA **92**, 5042–5046 (1995)
567. McMillen, D., Kopell, N., Hasty, J., Collins, J.J.: Synchronizing genetic relaxation oscillators by intercell signaling. Proc. Natl. Acad. Sci. U.S.A. **99**, 679–684 (2002)
568. Meier, J., Vannier, C., Serge, A., Triller, A., Choquet, D.: Fast and reversible trapping of surface glycine receptors by gephyrin. Nature Neurosci. **4**, 253–260 (2001)
569. Maso-Puigdellosas, A., Campos, D., Mendez, V.: Transport properties of random walks under stochastic noninstantaneous resetting. Phys. Rev. E **100**, 042104 (2019)
570. Maso-Puigdellosas, A., Campos, D., Mendez, V.: Stochastic movement subject to a reset-and-residence mechanism: transport properties and first arrival statistics. J. Stat. Mech. **033201** (2019)
571. Mehta, A.D., Rock, R.S., Rief, M., Spudich, J.A., Mooseker, M.S., Cheney, R.E.: Myosin-V is a processive actin-based motor. Nature **400**, 590–593 (2009)
572. Metzler, R., Klafter, J.: The random walk's guide to anomalous diffusion: a fractional dynamics approach. Phys. Rep. **339**, 1–77 (2000)
573. Metzler, R., Klafter, J.: The restaurant at the end of the random walk: recent developments in the description of anomalous transport by fractional dynamics. J. Phys. A **37**, R161 (2004)
574. Metzler, R., Jeon, J.-H., Cherstvy, A.G., Barkai, E.: Anomalous diffusion models and their properties: non-stationarity, non-ergodicity, and aging at the centenary of single particle tracking. Phys. Chem. Chem. Phys. **16**, 24128–24164 (2014)
575. Meyer, B., Benichou, O., Kafri, Y., Voituriez, R.: Geometry-induced bursting dynamics in gene expression. Biophys. J. **102**, 2186–2191 (2012)
576. Meylahn, J.M., Sabhapandit, S., Touchette, H.: Large deviations for Markov processes with resetting. Phys. Rev. E **92**, 062148 (2015)
577. Micheelsen, M.A., Mitarai, N., Sneppen, K., Dodd, I.B.: Theory for the stability and regulation of epigenetic landscapes. Phys. Biol. **7**, 026010 (2010)
578. Mileyko, Y., Joh, R.I., Weitz, J.S.: Small-scale copy number variation and large-scale changes in gene expression. Proc. Natl. Acad. Sci. USA **105**, 16659 (2008)
579. Millecamps, S., Julien, J.P.: Axonal transport deficits and neurodegenerative diseases. Nat. Rev. Neurosci. **14**, 161–176 (2013)
580. Mirny, L., Slutsky, M., Wunderlich, Z., Tafvizi, A., Leith, J., Kosmrlj, A.: How a protein searches for its site on DNA: the mechanism of facilitated diffusion. J. Phys. A **42**, 434013 (2009)
581. Misteli, T.: Self-organization in cell architecture. J. Cell Biol. **155**, 181–186 (2001)
582. Mitchison, T.J., Kirschner, M.W.: Dynamic instability of microtubule growth. Nature **312**, 237–242 (1984)
583. Mitchison, T.J.: Self-organization of polymer-motor systems in the cytoskeleton. Phil. Trans. R. Soc. Lond. B. Biol. Sci. **336**, 99–106 (1992)
584. Mogilner, A., Oster, G.: Cell motility driven by actin polymerization. Biophys. J. **71**, 3030–3045 (1996)
585. Mogilner, A., Fisher, A.J., Baskin, R.: Structural changes in the neck linker of kinesin explain the load dependence of the motor's mechanical cycle. J. Theor. Biol. **211**, 143–157 (2001)
586. Mogilner, A., Oster, G.: Force generation by actin polymerization II: the elastic ratchet and tethered filaments. Biophys. J. **84**, 1591–605 (2003)
587. Mogilner, A.: Mathematics of cell motility: have we got its number? J. Math. Biol. **58**, 105–134 (2009)
588. Monk, N.A.: Oscillatory expression of Hes1, p53, and NF-jB driven by transcriptional time delays. Curr. Biol. **13**, 1409 (2003)
589. Monod, J., Wyman, J., Changeux, J.-P.: On the nature of allosteric transitions: a plausible model. J. Mol. Biol. **12**, 88–118 (1965)
590. Montero, M., Maso-Puigdellosas, A., Villarroel, J.: Continuous-time random walks with reset events: historical background and new perspectives. Eur. Phys. J. B **90**, 176–186 (2017)

591. Morelli, M.J., Allen, R.J., ten Wolde, P.R.: Effects of macromolecular crowding on genetic networks. Biophys. J. **101**, 2882–2891 (2001)
592. Morris, C., Lecar, H.: Voltage oscillations in the barnacle giant muscle fiber. J. Biophys. **35**, 193–213 (1981)
593. Mugler, A., Bailey, A.G., Takahashi, K., ten Wolde, P.R.: Membrane clustering and the role of rebinding in biochemical signaling. Biophys. J. **102**, 1069–1078 (2012)
594. Mulder, B.M.: Microtubules interacting with a boundary: mean length and mean first-passage times. Phys. Rev. E **86**, 011902 (2012)
595. Muller, S., Hofbauer, J., Endler, L., Flamm, C., Widder, S., Schuster, P.: A generalized model of the repressilator. J. Math. Biol. **53**, 905–937 (2006)
596. Muller, M.J.I., Klumpp, S., Lipowsky, R.: Tug-of-war as a cooperative mechanism for bidirectional cargo transport by molecular motors. Proc. Natl. Acad. Sci. U.S.A. **105**, 4609–4614 (2008)
597. Muller, M.J.I., Klumpp, S., Lipowsky, R.: Motility states of molecular motors engaged in a stochastic tug-of-war. J. Stat. Phys. **133**, 1059–1081 (2008)
598. Murray, J.D.: Mathematical Biology, vols. I and II, 3rd edn. Springer, Berlin (2008)
599. Muthukumar, M.: Polymer translocation through a hole. J. Chem. Phys. **111**, 10371–10374 (1999)
600. Muthukumar, M.: Polymer Translocation. CRC Press, Boca Raton, Florida (2011)
601. Nadler, B., Schuss, Z., Singer, A., Eisenberg, R.S.: Ionic diffusion through confined geometries: from Langevin equations to partial differential equations. J. Phys. Cond. Matt. **16**, S2153–S2165 (2004)
602. Naeh, T., Klosek, M.M., Matkowsky, B.J., Schuss, Z.: A direct approach to the exit problem. SIAM J. Appl. Math. **50**, 595–627 (1990)
603. Nedelec, F., Surrey, T., Maggs, A.C., Leibler, S.: Self-organization of microtubules and motors. Nature **389**, 305–308 (1997)
604. Nelson, P.: Biological Physics: Energy, Information. Life W. H. Freeman, New York (2013)
605. Neubert, M.G., Caswell, H.: Alternatives to resilience for measuring the responses of ecological systems to perturbations. Ecology **78**, 653–665 (1997)
606. Neubert, M.G., Caswell, H., Murray, J.: Transient dynamics and pattern formation: reactivity is necessary for Turing instabilities. Math. Biosci. **175**, 1–11 (2002)
607. Neuman, K.C., Abbondanzieri, E.A., Landick, R., Gelles, J., Block, S.M.: Ubiquitous transcriptional pausing is independent of RNA polymerase backtracking. Cell **115**, 437–447 (2003)
608. Newby, J.M., Bressloff, P.C.: Directed intermittent search for a hidden target on a dendritic tree. Phys. Rev. E **80**, 021913 (2009)
609. Newby, J.M., Bressloff, P.C.: Quasi-steady state reduction of molecular-based models of directed intermittent search. Bull. Math. Biol. **72**, 1840–1866 (2010)
610. Newby, J.M., Bressloff, P.C.: Local synaptic signalling enhances the stochastic transport of motor-driven cargo in neurons. Phys. Biol. **7**, 036004 (2010)
611. Newby, J.M., Bressloff, P.C.: Random intermittent search and the tug-of-war model of motor-driven transport. J. Stat. Mech. **P04014** (2010)
612. Newby, J.M., Keener, J.P.: An asymptotic analysis of the spatially inhomogeneous velocity-jump process. SIAM Multiscale Model. Simul. **9**, 735–765 (2011)
613. Newby, J.M.: Isolating intrinsic noise sources in a stochastic genetic switch. Phys. Biol. **9**, 026002 (2012)
614. Newby, J.M., Bressloff, P.C., Keeener, J.P.: The effect of potassium channels on spontaneous action potential initiation by stochastic ion channels. Phys. Rev. Lett. **111**, 128101 (2013)
615. Newby, J.M., Chapman, S.J.: Metastable behavior in Markov processes with internal states: breakdown of model reduction techniques. J. Math. Biol. **69**, 941–976 (2014)
616. Newby, J.M.: Spontaneous excitability in the Morris-Lecar model with ion channel noise. SIAM J. Appl. Dyn. Syst. **13**, 1756–1791 (2014)
617. Newby, J.M.: Bistable switching asymptotics for the self regulating gene. J. Phys. A **48**, 185001 (2015)

618. Ninio, J.: Kinetic amplification of enzyme discrimination. Biochimie **57**, 587–595 (1975)
619. Novak, B., Tyson, J.J.: Design principles of biochemical oscillators. Nat. Rev. Mol. Cell Biol. **9**, 981–991 (2008)
620. Novak, I.L., Kraikivski, P., Slepchenko, B.M.: Diffusion in cytoplasm: effects of excluded volume due to internal membranes and cytoskeletal structures. Biophys. J. **97**, 758–767 (2009)
621. Nowak, S., Fok, P.W., Chou, T.: Dynamic boundaries in asymmetric exclusion processes. Phys. Rev. E **76**, 031135 (2007)
622. Nudler, E.: RNA polymerase backtracking in gene regulation and genome instability. Cell **149**, 1438–1445 (2012)
623. O'Brien, E.L., Van Itallie, E., Bennett, M.R.: Modeling synthetic gene oscillators. Math. Biosci. **236**, 1–15 (2012)
624. Oksendal, B.: Stochastic Differential Equations: An Introduction with Applications, 5th edn. Springer (1998)
625. Oshanin, O., Lindenberg, K., Wio, H.S., Burlatsky, S.: Efficient search by optimized intermittent random walks. J. Phys. A **42**, 434008 (2009)
626. Othmer, H.G., Hillen, T.: The diffusion limit of transport equations II: Chemotaxis equations. SIAM J. Appl. Math. **62**, 1222–1250 (2002)
627. Pakdaman, K., Thieullen, M., Wainrib, G.: Fluid limit theorems for stochastic hybrid systems with application to neuron models. Adv. Appl. Prob. **42**, 761–794 (2010)
628. Pakdaman, K., Thieullen, M., Wainrib, G.: Asymptotic expansion and central limit theorem for multiscale piecewise-deterministic Markov processes. Stoch. Process. Appl. **122**, 2292–2318 (2012)
629. Pal, A.: Diffusion in a potential landscape with stochastic resetting. Phys. Rev. E **91**, 012113 (2015)
630. Pal, A., Reuveni, S.: First passage under restart. Phys. Rev. Lett. **118**, 030603 (2017)
631. Pal, A., Prasad, V.V.: First passage under stochastic resetting in an interval. Phys. Rev. E **99**, 032123 (2019)
632. Pal, A., Kusmierz, L., Reuveni, S.: Invariants of motion with stochastic resetting and space-time coupled returns New. J. Phys. **21**, 113024 (2019)
633. Pal, A., Kusmierz, L., Reuveni, S.: Home-range search provides advantage under high uncertainty. arXiv:1906.06987 (2020)
634. Panda, S., Hogenesch, J.B., Kay, S.A.: Circadian rhythms from flies to human. Nature **417**, 329–335 (2002)
635. Panja, D., Barkema, G.T., Kolomeisky, A.B.: Through the eye of the needle: recent advances in understanding biopolymer translocation. J. Phys. Condens. Matter **25**, 413101 (2013)
636. Papanicolaou, G.C.: Asymptotic analysis of transport processes. Bull. Amer. Math. Soc. **81**, 330–392 (1975)
637. Parker, I., Yao, Y.: Regenerative release of calcium from functionally discrete subcellular stores by inositol triphosphate. Proc. Roy. Soc. Lond. B **246**, 269–274 (1991)
638. Parmeggiani, A., Julicher, F., Ajdari, A., Prost, J.: Energy transduction of isothermal ratchets: generic aspects and specific examples close to and far from equilibrium. Phys. Rev. E **60**, 2127–2140 (1999)
639. Parmeggiani, A., Franosch, T., Frey, E.: Phase coexistence in driven one-dimensional transport. Phys. Rev. Lett. **90**, 086601 (2003)
640. Parmeggiani, A., Franosch, T., Frey, E.: Totally asymmetric simple exclusion process with langmuir kinetics. Phys. Rev. E **70**, 046101 (2004)
641. Parker, I., Ivorra, I.: Localized all-or-none calcium liberation by inositol triphosphate. Science **250**, 977–979 (1990)
642. Parsons, J.T., Horwitz, A.R., Schwartz, M.A.: Cell adhesion: integrating cytoskeletal dynamics and cellular tension. Nat. Rev. Mol. Cell Biol. **11**, 633–643 (2010)
643. Partch, C.L., Green, C.B., Takahashi, J.S.: Molecular architecture of the mammalian circadian clock. Trends Cell Biol. **24**, 90–99 (2013)

644. Paulauskas, N., Pranevicius, M., Pranevicius, H., Bukauskas, F.F.: A stochastic four-state model of contingent gating of gap junction channels containing two "fast" gates sensitive to transjunctional voltage. Bophys. J. **96**, 3936–3948 (2009)

645. Paulsson, J.: Models of stochastic gene expression. Phys. Life Rev. **2**, 157–175 (2005)

646. Pavliotis, G.A., Stuart, A.M.: Multiscale Methods: Averaging and Homogenization. Springer, New York (2008)

647. Pavliotis, G.A.: Stochastic Processes and Applications. Springer, New York (2014)

648. Pechukas, P., Ankerhold, J.: Agmon-Hopfield kinetics in the slow diffusion regime. J. Chem. Phys. **107**, 2444 (1997)

649. Peliti, L.: Path integral approach to birth-death processes on a lattice. Journal de Physique **46**, 1469–1483 (1985)

650. Penn, A.C., Zhang, C.L., Georges, F., Royer, L., Breillat, C., Hosy, E., Petersen, J.D., Humeau, Y., Choquet, D.: Hippocampal LTP and contextual learning require surface diffusion of AMPA receptors. Nature **549**, 384–388 (2017)

651. Percus, J.K.: Anomalous self-diffusion for one-dimensional hard cores. Phys. Rev. A **9**, 557–559 (1974)

652. Pereira, P., Forni, L., et al.: Autonomous activation of B and T cells in antigen-free mice. Eur. J. Immunol. **16**, 685–688 (1986)

653. Perez-Carrasco, R., Guerrero, P., Briscoe, J., Page, K.M.: Intrinsic noise profoundly alters the dynamics and steady state of morphogen controlled bistable genetic switches. PLoS Comput. Biol. **12**, e1005154 (2016)

654. Persson, F., Linden, M., Unoson, C., Elf, J.: Extracting intracellular diffusive states and transition rates from single-molecule tracking data. Nat. Meth. **10**, 265–269 (2013)

655. Perthame, B., Souganidis, P.E.: Asymmetric potentials and motor effect: a homogenization approach. Ann. I. H. Poincare **26**, 2055–2071 (2009)

656. Peskin, C.P., Odell, G.M., Oster, G.F.: Cellular motions and thermal fluctuations: the brownian ratchet. Biophys. J. **65**, 316–324 (1993)

657. Peskin, C.S., Ermentrout, B., Oster, G.: The correlation ratchet: a novel mechanism for generating directed motion by ATP hydrolysis. In: Mow, V.C., et al. (eds.) Cell Mechanics and Cellular Engineering. Springer, New-York (1995)

658. Phillips, R., Kondev, J., Theriot, J., Garcia, H.: Physical Biology of the Cell, 2nd edn. Garland Science (2012)

659. Pierce-Shimomura, J., Morse, T., Lockery, S.: The fundamental role of pirouettes in C. elegans chemotaxis. J. Neurosci. **19**, 9557–9569 (1999)

660. Pillay, S., Ward, M.J., Peirce, A., Kolokolnikov, T.: An asymptotic analysis of the mean first passage time for narrow escape problems: Part I: two-dimensional domains. SIAM Multiscale Model. Sim. **8**, 803–835 (2010)

661. Popkov, V., Rakos, A., Williams, R.D., Kolomesisky, A.B., Schutz, G.M.: Localization of shocks in driven diffusive systems without particle number conservation. Phys. Rev. E **67**, 066117 (2003)

662. Popovic, N., Marr, C., Swain, P.S.: A geometric analysis of fast-slow models for stochastic gene expression. J. Math. Biol. **72**, 87–122 (2006)

663. Porcher, A., Dostatni, N.: The Bicoid morphogen system. Curr. Biol. **20**, R249–R254 (2010)

664. Posta, F., D'Orsogna, M.R., Chou, T.: Enhancement of cargo processivity by cooperating molecular motors. Phys. Chem. Chem. Phys. **11**, 4851–4860 (2009)

665. Potvin-Trottier, L., Lord, N.D., Vinnicombe, G., Paulsson, J.: Synchronous long-term oscillations in a synthetic gene circuit. Nature **538**, 514–517 (2016)

666. Pronina, E., Kolomeisky, A.: Spontaneous symmetry breaking in two-channel asymmetric exclusion processes with narrow entrances. J. Phys. A **40**, 2275–2287 (2007)

667. Prost, J., Chauwin, J.F., Peliti, L., Ajdari, A.: Asymmetric pumping of particles. Phys. Rev. Lett. **72**, 2652–2655 (1994)

668. Ptashne, M.: A Genetic Switch, 3rd edn. Cold Spring Harbor Laboratory Press (2004)

669. Purcell, O., Savery, N.J., Grierson, C.S., di Bernardo, M.: A comparative analysis of synthetic genetic oscillators. J. Roy. Soc. Inter. **7**, 1503 (2010)

670. Qian, H.: Cooperativity in cellular biochemical processes. Annu. Rev. Biophys. **41**, 179–204 (2012)

671. Raj, A., van Oudenaarden, A.: Nature, nurture, or chance: stochastic gene expression and its consequences. Cell **135**, 216–226 (2008)

672. Raveh, B., Karp, J.M., Cowburn, D.: Slide-and-exchange mechanism for rapid and selective transport through the nuclear pore complex. Proc. Natl. Acad. Sci. USA **113**, E2489–E2497 (2016)

673. Redner, S.: Survival probability in a random velocity field. Phys. Rev. E **56**, 4967–4972 (1997)

674. Redner, S.: A Guide to First-Passage Processes. Cambridge University Press, Cambridge, UK (2001)

675. Reed, M.C., Venakides, S., Blum, J.J.: Approximate traveling waves in linear reaction-hyperbolic equations. SIAM J. Appl. Math. **50**, 167–180 (1990)

676. Reese, L., Melbinger, A., Frey, E.: Crowding of molecular motors determines microtubule depolymerization. Biophys. J. **101**, 2190–2200 (2011)

677. Reguera, D., Rubi, J.M.: Kinetic equations for diffusion in the presence of entropic barriers. Phys. Rev. E **64**, 061106 (2001)

678. Reguera, D., Schmid, G., Burada, P.S., Rubi, J.M., Reimann, P., Hanggi, P.: Entropic transport: kinetics, scaling, and control mechanisms. Phys. Rev. Lett. **96**, 130603 (2006)

679. Reichenbach, T., Franosch, T., Frey, E.: Exclusion processes with internal states. Phys. Rev. Lett. **97**, 050603 (2006)

680. Reimann, P.: Thermally activated escape with potential fluctuations driven by an Ornstein-Uhlenbeck process. Phys. Rev. E **52**, 1579–1600 (1995)

681. Reimann, P., Hanggi, P.: Chimansky-Geier, L., Poschel, T.: Stochastic Dynamics: Lecture Notes in Physics, vol. 484, pp. 127–139. Springer, Berlin (1997)

682. Reimann, P., Bartussek, R., Hanggi, P.: Reaction rates when barriers fluctuate: a singular perturbation approach. Chem. Phys. **235**, 11–26 (1998)

683. Reimann, P.: Brownian motors: noisy transport far from equilibrium. Phys. Rep. **361**, 57–265 (2002)

684. Reimann, P., van den Broeck, C., Linke, H., Hanggi, P., Rubi, J.M., Perez-Madrid, A.: Diffusion in tilted periodic potentials: enhancement, universality and scaling. Phys. Rev. E **65**, 031104 (2002)

685. Reingruber, J., Holcman, D.: Narrow escape for a stochastically gated Brownian ligand. J. Phys. Cond. Matt. **22**, 065103 (2010)

686. Reinitz, J., Vaisnys, J.R.: Theoretical and experimental analysis of the phage lambda genetic switch implies missing levels of co-operativity. J. Theor. Biol. **145**, 295–318 (1990)

687. Reits, E.A., Neefjes, J.J.: From fixed to FRAP: measuring protein mobility and activity in living cells. Nat. Cell Biol. **3**, E145–E147 (2001)

688. Reuveni, S., Urbakh, M., Klafter, J.: Role of substrate unbinding in Michaelis-Menten enzymatic reactions. Proc. Natl. Acad. Sci. USA **111**, 4391 (2014)

689. Reuveni, S.: Optimal stochastic restart renders fluctuations in first-passage times universal. Phys. Rev. Lett. **116**, 170601 (2016)

690. Reynaud, K., Schuss, Z., Rouach, N., Holcman, D.: Why so many sperm cells? Commun. Integr. Biol. **8**, e1017156 (2015)

691. Rhee, A., Cheong, R., Levchenko, A.: Noise decomposition of intracellular biochemical signaling networks using nonequivalent reporters. Proc. Natl. Acad. Sci. **11**, 17330–17335 (2014)

692. Ribbeck, K., Gorlich, D.: Kinetic analysis of translocation through nuclear pore complexes. EMBO J. **21**, 2664–2671 (2001)

693. Ribbeck, K., Gorlich, D.: The permeability barrier of nuclear pore complexes appears to operate via hydrophobic exclusion. EMBO J. **21**, 2664–2671 (2002)

694. Ribeiro, A.S.: Dynamics and evolution of stochastic bistable gene networks with sensing in fluctuating environments. Phys. Rev. E **78**, 061902 (2008)

695. Ribiero, A.S., Smolander, O.P., Rajala, T., Hakkinen, A., Yli-Harja, O.: Delayed stochastic model of transcription at the single nucleotide level. J. Comput. Biol. **16**, 539–353 (2009)

696. Rice, S.A.: Diffusion-Limited Reactions. Elsevier, Amsterdam (1985)

697. Richter, P.H., Eigen, M.: Diffusion-controlled reaction rates in spheroidal geometry: application to repressor-operator association and membrane bound enzymes. Biophys. Chem. **2**, 255–263 (1974)

698. Riggs, A.D., Bourgeois, S., Cohn, M.: The lac repressor-operator interaction III: kinetic studies. J. Mol. Biol. **53**, 401–417 (1974)

699. Risken, H.: The Fokker-Planck Equation: Methods of Solution and Applications, 3rd edn. Springer, Berlin (1996)

700. Rodenbeck, C., Karger, J., Hahn, K.: Calculating exact propagators in single-file systems via the reflection principle. Phys. Rev. E **57**, 4382–4397 (1998)

701. Roldan, E., Lisica, A., Sanchez-Taltavull, D., Grill, S.W.: Stochacstic resetting in backtrack recovery by RNA polymerases. Phys. Rev. E **93**, 062411 (2016)

702. Roma, D.M., O'Flanagan, R.A., Ruckenstein, A.E., Sengupta, A.M.: Optimal path to epigenetic switching. Phys. Rev. E **71**, 011902 (2005)

703. Romero, J.M., Gonzalez-Gaxiola, O., Chacon-Acosta, G.: Exact solutions to Fick-Jacobs equation. Int. J. Pure Appl. Math. **82**, 41–52 (2013)

704. Rook, M.S., Lu, M., Kosik, K.S.: CamKIIα 3' untranslated regions-directed mRNA translocation in living neurons: visualization by GFP linkage. J. Neurosci. **20**, 6385–6393 (2000)

705. Roth, R.H., Zhang, Y., Huganir, R.L.: Dynamic imaging of AMPA receptor trafficking in vitro and in vivo. Curr. Opin. Neurobiol. **45**, 51–58 (2017)

706. Rothart, T., Reuveni, S., Urbakh, M.: Michaelis-Menten reaction scheme as a unified approach towards the optimal restart problem. Phys. Rev. E **92**, 060101 (2015)

707. Rout, M.P., Aitchison, J.D., Magnasco, M.O., Chait, B.T.: Virtual gating and nuclear transport: the hole picture. Trends Cell Biol. **13**, 622–628 (2003)

708. Roux, B., Allen, T., Berneche, S., Im, W.: Theoretical and computational models of biological ion channels. Q. Rev. Biophys. **37**, 15–103 (2004)

709. Rubi, J.M., Reguera, D.: Thermodynamics and stochastic dynamics of transport in confined media. Chem. Phys. **375**, 518–522 (2010)

710. Sabatini, B.L., Maravall, M., Svoboda, K.: Ca^{2+} signaling in dendritic spines. Curr. Opin. Neurobiol. **11**, 349–356 (2001)

711. Sabhapandit, S., Majumdar, S.N., Comtet, A.: Statistical properties of the paths of a particle diffusing in a one-dimensional random potential. Phys. Rev. E **73**, 051102 (2006)

712. Saez, J.C., Berthoud, V.M., Branes, M.C., Martinez, A.D., Beyer, E.C.: Plasma membrane channels formed by connexins: their regulation and functions. Physiol. Rev. **83**, 1359–1400 (2003)

713. Saffman, P., Delbruck, M.: Brownian motion in biological membranes. Proc. Natl. Acad. Sci. USA **72**, 3111–3113 (1975)

714. Sagara, H., Ohshima, D., Ichikawa, K.: Regulation of signal transduction by spatial parameters: a case in NF-κB oscillation. IET Syst. Biol. **9**, 41–51 (2015)

715. Sahoo, S., Klumpp, S.: Backtracking dynamics of RNA polymerase: pausing and error correction. J. Phys. Condens. Matter **25**, 374104 (2013)

716. Salman, H., Abu-Arish, A., Oliel, S., Loyter, A., Klafter, J., Granek, R., Elbaum, M.: Nuclear localization signal peptides induce molecular delivery along microtubules. Biophys. J. **89**, 2134–2145 (2005)

717. Salsa, S.: Partial Differential Equations in Action. Springer (2009)

718. Salvatico, C., Specht, C.G., Triller, A.: Synaptic receptor dynamics: From theoretical concepts to deep quantification and chemistry in cellulo. Neuropharm. **88**, 2–9 (2015)

719. Sanchez, A., Garcia, H.G., Jones, D., Phillips, R., Kondev, J.: Effect of promoter architecture on the cell-to-cell variability in gene expression. PLoS Comput. Biol. **7**, e1001100 (2011)

720. Sanchez, A., Choubey, S., Kondev, J.: Regulation of noise in gene expression. Annu. Rev. Biophys. **42**, 469–491 (2013)

721. Santamaria, F., Wils, S., de Schutter, E., Augustine, G.J.: Anomalous diffusion in Purkinje cell dendrites caused by spines. Neuron **52**, 635–648 (2006)
722. Santillan, M., Mackey, M.C.: Influence of catabolite repression and inducer exclusion on the bistable behavior of the lac operon. Biophys. J. **86**, 1282–1292 (2004)
723. Sasai, M., Wolynes, P.G.: Stochastic gene expression as a many-body problem. Proc. Natl. Acad. Sci. **100**, 2374–2379 (2003)
724. Saxton, M.J.: The membrane skeleton of eryrhrocytes: a percolation model. I erythrocytes: a percolation analysis. Biophys. J. **57**, 1167–1177 (1990)
725. Saxton, M.J.: Anomalous diffusion due to obstacles: a Monte Carlo study. Biophys. J. **66**, 394–401 (1994)
726. Saxton, M.J.: Single particle tracking: effects of corrals. Biophys. J. **69**, 389–398 (1995)
727. Saxton, M.J.: Anomalous diffusion due to binding: a Monte Carlo study. Biophys. J. **70**, 1250–1262 (1996)
728. Saxton, M.J., Jacobson, K.: Single-particle tracking: applications to membrane dynamics. Annu. Rev. Biophys. Biomol. Struct. **26**, 373–399 (1997)
729. Saxton, M.J.: A biological interpretation of transient anomalous subdiffusion I: qualitative model. Biophys. J. **92**, 1178–1191 (2007)
730. Schadschneider, A., Chowdhury, D., Nishinari, K.: Stochastic Transport in Complex Systems: From molecules to Vehicles. Elsevier, Amsterdam (2010)
731. Scher, H., Montroll, E.W.: Anomalous transit-time dispersion in amorphous solids. Phys. Rev. B **12**, 2455–2477 (1975)
732. Schilling, T.F., Nie, Q., Lander, A.D.: Dynamics and precision in retinoic acid morphogen gradients. Curr. Opin. Gen. Dev. **22**, 562–569 (2012)
733. Schmidt, H.B., Gorlich, D.: Transport selectivity of nuclear pores, phase separation, and membraneless organelles. Trends Biochem. Sci. **41**, 46–61 (2016)
734. Schnapp, B.J., Gelles, J., Sheetz, M.P.: Nanometer-scale measurements using video light microscopy. Cell Motil. Cytoskeleton **10**, 47–53 (1988)
735. Schnell, S., Turner, T.: Reaction kinetics in intracellular environments with macromolecular crowding: simulations and rate laws. Prog. Biophys. Molec. Biol. **85**, 235–260 (2004)
736. Schnitzer, M., Visscher, K., Block, S.: Force production by single kinesin motors. Nat. Cell Biol. **2**, 718–723 (2000)
737. Schuss, Z., Nadller, B., Eisenberg, R.S.: Derivation of PNP equations in bath and channel from a molecular model. Phys. Rev. E **64**, 036116 (2001)
738. Schuss, Z., Singer, A., Holcman, D.: The narrow escape problem for diffusion in cellular microdomains. Proc. Natl. Acad. Sci. USA **104**, 16098–16103 (2007)
739. Schuss, Z.: Theory and Applications of Stochastic Processes: An Analytical Approach. Applied Mathematical Sciences, vol. 170. Springer, New York (2010)
740. Schuss, Z., Basnayake, K., Holcman, D.: Redundancy principle and the role of extreme statistics in molecular and cellular biology. Phys. Life Rev. **28**, 52–79 (2019)
741. Schuster, M., Lipowsky, R., Assmann, M.A., Lenz, P., Steinberg, G.: Transient binding of dynein controls bidirectional long-range motility of early endosomes. Proc. Natl. Acad. Sci. USA **108**, 3618–3623 (2011)
742. Schwabe, A., Rybakova, K.N., Bruggeman, F.J.: Transcription stochasticity of complex gene regulation models. Biophys. J. **103**, 1152–1161 (2011)
743. Scott, M., Ingalls, B., Kearns, M.: Estimation of intrinsic and extrinsic noise in models of nonlinear genetic networks. Chaos **16**, 026107 (2006)
744. Segel, L.A., Edelstein-Keshet, L.: A Primer on Mathematical Models in Biology. SIAM, Philadelphia (2013)
745. Seisenberger, G., Ried, M.U., Endress, T., Buning, H., Hallek, M., Brauchle, C.: Real-time single-molecule imaging of the infection pathway of an adeno-associated virus. Science **294**, 1929–1932 (2001)
746. Sekimoto, K., Triller, A.: Compatibility between itinerant synaptic receptors and stable postsynaptic structure. Phys. Rev. E **79**, 031905 (2009)

747. Shahrezaei, V., Swain, P.S.: Analytical distributions for stochastic gene expression. Proc. Natl. Acad. Sci. USA **105**, 17256–17261 (2008)
748. Shaw, L.B., Zia, R.K.P., Lee, K.H.: Totally asymmetric exclusion process with extended objects: a model for protein synthesis. Phys. Rev. E **68**, 021910 (2003)
749. Sheinman, M., Benichou, O., Kafri, Y., Voituriez, R.: Classes of fast and specific search mechanisms for proteins on DNA. Rep. Prog. Phys. **75**, 026601 (2012)
750. Sherman, A., Rinzel, R., Keizer, J.: Emergence of organized bursting in clusters of pancreatic beta-cells by channel sharing. Biophys. J. **54**, 411–425 (1988)
751. Shomar, A., Geyrhofer, L., Ziv, N.E., Brenner, N.: Cooperative stochastic binding and un-binding explain synaptic size dynamics and statistics. PLoS Comp. Biol. **13**, e1005668 (2017)
752. Shoup, D., Szabo, A.: Role of diffusion in ligand binding to macromolecules and cell-bound receptors. Biophys. J. **40**, 33–39 (1982)
753. Shouval, H.Z.: Clusters of interacting receptors can stabilize synaptic efficacies. Proc. Natl. Acad. Sci. USA **102**, 14440–14445 (2005)
754. Shuai, J.W., Jung, P.: Stochastic properties of Ca^{2+} release of Inositol 1,4,5-Triphosphate receptor clusters. Biophys. J. **83**, 87–97 (2002)
755. Shvartsman, S.Y., Baker, R.E.: Mathematical models of morphogen gradients and their effects on gene expression. Rev. Dev. Biol. **1**, 715–730 (2012)
756. Shvets, A.A., Kolomeisky, A.B.: Sequence heterogeneity accelerates protein search for targets on DNA. J. Chem. Phys. **143**, 245101 (2015)
757. Shvets, A.A., Kolomeisky, A.B.: Crowding on DNA in protein search for targets. J. Phys. Chem. Lett. **7**, 2502–2506 (2015)
758. Shvets, A.A., Kolomeisky, A.B.: The role of DNA looping in the search for specific targets on DNA by multisite proteins. J. Phys. Chem. Lett. **7**, 5022–5027 (2016)
759. Shvets, A.A., Kochugaeva, M.P., Kolomesisky, A.B.: Mechanisms of protein search for targets on DNA: theoretical insights. Molecules **23**(2106), 1–18 (2018)
760. Simon, S.M., Peskin, C.P., Oster, G.F.: What drives the translocation of proteins? Proc. Natl. Acad. Sci. USA **89**, 3770–3774 (1992)
761. Simon, C.M., Hepburn, I., Chen, W., De Schutter, E.: The role of dendritic spine morphology in the compartmentalization and delivery of surface receptors. J. Comput. Neurosci. **36**, 483–497 (2014)
762. Singer, S.J., Nicolson, G.L.: The fluid mosaic model of the structure of the cell membrane. Science **175**, 720–731 (1972)
763. Singer, A., Schuss, Z., Holcman, D.: Narrow escape, Part II: the circular disc. J. Stat. Phys. **122**, 465–489 (2006)
764. Singer, A., Schuss, Z., Holcman, D.: Narrow escape, Part III: nonsmooth domains and Riemann surfaces. J. Stat. Phys. **122**, 491–509 (2006)
765. Slator, P.J., Cairo, C.W., Burroughs, N.J.: Detection of diffusion heterogeneity in single particle tracking trajectories using a hidden Markov model with measurement noise propagation. PLoS ONE **10**, e0140759 (2015)
766. Smiley, M.W., Proulx, S.R.: Gene expression dynamics in randomly varying environments. J. Math. Biol. **61**, 231–251 (2010)
767. Smith, G.D.: Modeling the stochastic gating of ion channels. In: Fall, C., Marland, E.S., Wagner, J.M., Tyson, J.J. (eds.) Computational Cell Biology. chap. 11. Springer, New York (2002)
768. Smith, D.A., Simmons, R.M.: Models of motor-assisted transport of intracellular particles. Biophys. J. **80**, 45–68 (2001)
769. Smith, H.: An Introduction to Delay Differential Equations with Applications to the Life Sciences. Springer, New York (2010)
770. Smolen, P., Baxter, D.A., Byrne, J.H.: Modeling circadian oscillations with interlocking positive and negative feedback loops. J. Neurosci. **21**, 6644–6656 (2001)
771. Smoluchowski, M.V.: Z. Phys. Chem. **92**, 129–168 (1917)

772. Sneppen, K.: Models of Life: Dynamics and Regulation in Biological Systems. Cambridge University Press, Cambridge (2014)

773. Sneppen, K.: Models of life: epigenetics, diversity and cycles. Rep. Prog. Phys. **80**, 042601 (2017)

774. Sneyd, J., Wetton, B.T., Charles, A.C., Sanderson, M.J.: Intercellular calcium waves mediated by diffusion of inositol trisphosphate: a two-dimensional model. Am. J. Physiol. Cell Physiol. **268**, C1537–C1545 (1995)

775. Sokolov, I.M.: Cyclization of a polymer: first-passage problem for a non-Markovian process. Phys. Rev. Lett. **90**, 080601 (2003)

776. Sokolov, I.M., Metzler, R., Pant, K., Williams, M.C.: First passage time of n excluded-volume particles on a line. Phys. Rev. E **72**, 041102 (2005)

777. Song, C., Phenix, H., Abed, V., Scott, M., Ingalls, B.P., Kaern, M., Perkins, T.J.: Estimating the stochastic bifurcation structure of cellular networks. PLoS Comput. Biol. **6**, e1000699 (2010)

778. Soppina, V., Rai, A.K., Ramaiya, A.J., Barak, P., Mallik, R.: Tug-of-war between dissimilar teams of microtubule motors regulates transport and fission of endosomes. Proc. Natl. Acad. Sci. U.S.A. **106**, 19381–19386 (2009)

779. Sorra, K.E., Harris, K.M.: Overview on the structure, composition, function, development, and plasticity of hippocampal dendritic spines. Hippocampus **10**, 501–511 (2000)

780. Spouge, J.L., Szabo, A., Weiss, G.H.: Single-particle survival in gated trapping. Phys. Rev. E **54**, 2248–2255 (1996)

781. Stauffer, D., Aharony, A.: Introduction to Percolation Theory. Taylor and Francis (1994)

782. Stratonovich, R.L.: Radiotekhnika i elektronika **3**, 497–511 (1958)

783. Straube, R., Ward, M.J., Falcke, M.: Reaction rate of small diffusing molecules on a cylindrical membrane. J. Stat. Phys. **129**, 377–405 (2007)

784. Strelkowa, N., Barahona, M.: Transient dynamics around unstable periodic orbits in the generalized repressilator model. Chaos **21**, 023104 (2011)

785. Stricker, J., Cookson, S., Bennett, M.R., Mather, W.H., Tsimring, L.S., Hasty, J.: A fast, robust and tunable synthetic gene oscillator. Nature **456**, 516 (2008)

786. Stukalin, E.B., Kolomeisky, A.B.: Simple growth models of rigid multifilament biopolymers. J. Chem. Phys. **121**, 1097–1104 (2004)

787. Suel, G.M., Garcia-Ojalvo, J., Liberman, L.M., Elowitz, M.B.: An excitable gene regulatory circuit induces transient cellular differentiation. Nature **440**, 545–550 (2006)

788. Sung, B.J., Yethiraj, A.: Lateral diffusion of proteins in the plasma membrane: spatial tessellation and percolation theory. J. Phys. Chem. B **112**, 143–149 (2008)

789. Sung, W., Park, P.J.: Polymer translocation through a pore in a membrane. Phys. Rev. Lett. **77**, 783–786 (1996)

790. Swillens, S., Dupont, G., Combettes, L., Champeil, P.: From calcium blips to calcium puffs: theoretical analysis of the requirement for interchannel communication. Proc. Natl. Acad. Sci. (USA) **96**, 13750–13755 (1999)

791. Szabo, A., Schulten, K., Schulten, Z.: First passage time approach to diffusion controlled reactions. J. Chem. Phys. **72**, 4350–4357 (1980)

792. Szabo, A., Shoup, D., Northrup, S.H., McCammon, J.A.: Stochastically-gated diffusion-influenced reactions. J. Chem. Phys. **77**, 4484–4493 (1982)

793. Szymanski, J., Weiss, M.: Elucidating the origin of anomalous subdiffusion in crowded fluids. Phys. Rev. Lett. **103**, 038102 (2009)

794. Tabareau, N., Slotine, J.J., Pham, Q.C.: How synchronization protects from noise. PLoS Comput. Biol. **6**, e1000637 (2010)

795. Takacs, L.: Introduction to the Theory of Queues. Oxford University Press (1962)

796. Takahashi, K., Tanase-Nicola, S., ten Wolde, P.R.: Spatio-temporal correlations can drastically change the response of a MAPK pathway. Proc. Natl. Acad. Sci. U.S.A. **107**, 2473–2478 (2010)

797. Taloni, A., Marchesoni, F.: Single-file diffusion on a periodic substrate. Phys. Rev. Lett. **96**, 020601 (2006)

798. Taloni, A., Lomholt, M.A.: Langevin formulation for single-file diffusion. Phys. Rev. E **78**, 051116 (2008)
799. Teimouri, H., Kolomeisky, A.B.: Mechanisms of the formation of biological signaling profiles. J. Phys. A: Math. Theor. **49**, 483001 (2016)
800. Telley, I.A., Bieling, P., Surrey, T.: Obstacles on the microtubule reduce the processivity of Kinesin-1 in a minimal in vitro system and in cell extract. Biophys. J. **96**, 3341–3353 (2009)
801. Tetenbaum-Novatt, J., Hough, L.E., Mironska, R., McKenney, A.S., Rout, M.P.: Nucleocytoplasmic transport: a role for nonspecific competition in karyopherin-nucleoporin interactions. Mol. Cell. Proteomics **11**, 31–46 (2012)
802. Thattai, M., van Oudenaarden, A.: Intrinsic noise in gene regulatory networks. Proc. Natl. Acad. Sci. USA **98**, 8614–8619 (2001)
803. Thattai, M., van Oudenaarden, A.: Attenuation of noise in ultrasensitive signaling cascades. Biophys. J. **82**, 2943–2950 (2001)
804. Thattai, M., van Oudenaarden, A.: Stochastic gene expression in fluctuating environments. Genetics **167**, 523–530 (2004)
805. Thomas, P., Popovic, N., Grima, R.: Phenotypic switching in gene regulatory networks. Proc. Natl. Acad. Sci. USA **111**, 6994–6999 (2014)
806. Thon, G., Friis, T.: Epigenetic inheritance of transcriptional silencing and switching competence in fission yeast. Genetics **145**, 685–696 (1997)
807. Tijana, J.-T., Zilman, A.: Protein transport by the nuclear pore complex: simple biophysics of a complex biomachine. Biophys. J. **113**, 6–14 (2017)
808. Torquato, S.: Random Heterogeneous Materials. Springer, New York (2002)
809. Touchette, H.: The large deviation approach to statistical mechanics. Phys. Rep. **478**, 1–69 (2009)
810. Touchette, H.: A basic introduction to large deviations: theory, applications, simulations. arXiv:1106.4146v3 (2012)
811. Tran, E.J., Wente, S.R.: Dynamic nuclear pore complexes: life on the edge. Cell **125**, 1041–1053 (2006)
812. Triesch, J., Vo. A.D., Hafner, A.-S.: Competition for synaptic building blocks shapes synaptic plasticity. eLIFE **7**, e37836 (2018)
813. Triller, A., Choquet, D.: Surface trafficking of receptors between synaptic and extrasynaptic membranes: and yet they do move! Trends Neurosci. **28**, 133–139 (2005)
814. Triller, A., Choquet, D.: New concepts in synaptic biology derived from single-molecule imaging. Neuron **59**, 359–374 (2008)
815. Tsimiring, L.S., Volfson, D., Hasty, J.: Stochastically driven genetic circuits. Chaos **16**, 026103 (2006)
816. Tsimiring, L.S.: Noise in biology. Rep. Prog. Phys. **77**, 026601 (2014)
817. Tu, L.-C., Fu, G., Zilman, A., Musser, S.M.: Large cargo transport by nuclear pores: implications for the spatial organization of FG-nucleoporins. EMBO J. **32**, 3220–3230 (2013)
818. Tyson, J.J., Hong, C.I., Thron, C.D., Novak, B.: A simple model of circadian rhythms based on dimerization and proteolysis of PER and TIM. Biophys. J. **77**, 2411–2417 (1999)
819. Vale, R.D., Funatsu, T., Pierce, D.W., Romberg, L., Harad, Y., Yanagida, T.: Direct observation of single kinesin molecules moving along microtubules. Nature **380**, 451–453 (1996)
820. Vance, W., Ross, J.: Fluctuations near limit cycles in chemical reaction systems. J. Chem. Phys. **105**, 479–487 (1996)
821. van den Engh, G., Sachs, R., Trask, B.J.: Estimating genomic distance from DNA sequence location in cell nuclei by a random walk model. Science **257**, 1410–1412 (1992)
822. van Kampen, N.G.: Stochastic Processes in Physics and Chemistry. North-Holland, Amsterdam (1992)
823. Varadhan, S.R.S.: Asymptotic probabilities and differential equations. Comm. Pure Appl. Math. **19**, 261–286 (1966)
824. Vastola, J.J., Holmes, W.R.: Stochastic path integrals can be derived like quantum mechanical path integrals arXiv:1909.12990 (2020)

825. Veksler, A., Kolomeisky, A.B.: Speed-selectivity paradox in the protein search for targets on DNA: Is it real or not? J. Phys. Chem. B **117**, 12695–12701 (2013)

826. Vellela, M., Qian, H.: Stochastic dynamics and non-equilibrium thermodynamics of a bistable chemical system: the Schlögl model revisited. J. R. Soc. Interface **6**, 925–940 (2009)

827. Vereb, G., Szollosi, J., Nagy, J.M.P., Farkas, T.: Dynamic, yet structured: the cell membrane three decades after the Singer-Nicolson model. Proc. Natl. Acad. Sci. USA **100**, 8053–8058 (2003)

828. Vershinin, M., Carter, B.C., Razafsky, D.S., King, S.J., Gross, S.P.: Multiple-motor based transport and its regulation by Tau. Proc. Natl. Acad. Sci. U.S.A. **104**, 87–92 (2007)

829. Visscher, K., Schnitzer, M., Block, S.: Single kinesin molecules studied with a molecular force clamp. Nature **400**, 184–189 (1999)

830. Viswanathan, G., Buldyrev, S., Havlin, S., da Luz, M., Raposo, E., Stanley, H.: Optimizing the success of random searches. Nature **401**, 911–914 (1999)

831. Viswanathan, G., Bartumeus, F., Buldyrev, S., Catalan, J., Fulco, U., Havlin, S., da Luz, M., Lyra, M., Raposo, E., Stanley, H.: Levy flight random searches in biological phenomena. Physica A **314**, 208–213 (2002)

832. Voliotis, M., Cohen, N., Molina-Paris, C., Liverpool, T.: Fluctuations, pauses, and backtracking in DNA transcription. Biophys. J. **94**, 334–348 (2008)

833. Vorotnikov, D.: Analytical aspects of the Brownian motor effect in randomly flashing ratchets. J. Math. Biol. (2013)

834. Vousden, K.H., Lane, D.P.: p53 in health and disease. Nat. Rev. Mol. Cell Biol. **8**, 275–283 (2007)

835. Wagner, C.: Theorie der Alterung von Niederschlägen durch Umlösen. Z. Elektrochemie **65**, 581–594 (1961)

836. Wainrib, G., Thieullen, M., Pakdaman, K.: Reduction of stochastic conductance-based neuron models with time-scales separation. J. Comput. Neurosci. **32**, 327–46 (2012)

837. Walczak, A.M., Sasai, M., Wolynes, P.G.: Self-consistent proteomic field theory of stochastic gene switches. Biophys. J. **88**, 828–850 (2005)

838. Walter, W.: Nonlinear parabolic differential equations and inequalities. Disc. Cont. Dyn. Syst. **8**, 451–468 (2002)

839. Wang, K.G.: Long-time correlation effects and biased anomalous diffusion. Phys. Rev. A **45**, 833–837 (1992)

840. Wang, J., Wolynes, P.: Survival paths for reaction dynamics in fluctuating environments. Chem. Phys. **180**, 141–156 (1994)

841. Wang, H., Elston, T.C., Mogilner, A., Oster, G.: Force generation in RNA polymerase. Biophys. J. **74**, 1186–1202 (1998)

842. Wang, L., Ho, C., Sun, D., Liem, R.K.H., Brown, A.: Rapid movement of axonal neurofilaments interrupted by prolonged pauses. Nat. Cell. Biol. **2**, 137–141 (2000)

843. Wang, S.Q., Song, L.S., Xu, L., Meissner, G., Lakatta, E.G., Rios, E., Stern, M.D., Cheng, H.: Thermodynamically irreversible gating of ryanodine receptors in situ revealed by stereotyped duration of release of Ca21 sparks. Biophys. J. **83**, 242–251 (2002)

844. Wang, Q., Holmes, W.R., Sosnik, J., Schilling, T., Nie, Q.: Cell sorting and noise-induced cell plasticity coordinate to sharpen boundaries between gene expression domains. PLoS Comp. Biol. **13**, e1005307 (2017)

845. Ward, M.J., Henshaw, W.D., Keller, J.B.: Summing logarithmic expansions for singularly perturbed eigenvalue problems. SIAM J. Appl. Math. **53**, 799–828 (1993)

846. Wartlick, O., Kicheva, A., Gonzalez-Gaitan, M.: Morphogen gradient formation. Cold Spring Harb. Perspect. Biol. **1**, a001255 (2009)

847. Weake, V.M., Worlkman, J.L.: Inducible gene expression: diverse regulatory mechanisms. Nat. Rev. Genet. **11**, 426–437 (2010)

848. Weber, S.C., Spakowitz, A.J., Theriot, J.A.: Bacterial chromosomal loci move subdiffusively through a viscoelastic cytoplasm. Phys. Rev. Lett. **104**, 238102 (2010)

849. Weber, M.F., Frey, E.: Master equations and the theory of stochastic path integrals. Rep. Prog. Phys. **80**, 046601 (2017)

850. Weigel, A.V., Simon, B., Tamkun, M.M., Krapf, D.: Ergodic and nonergodic processes coexist in the plasma membrane as observed by single-molecule tracking. Proc. Nat. Acad. Sci. USA **108**, 6438–6443 (2011)

851. Weigel, A.V., Tamkun, M.M., Krapf, D.: Quantifying the dynamic interactions between a clathrin-coated pit and cargo molecules. Proc. Nat. Acad. Sci. USA **111**, 4591–4600 (2013)

852. Weiss, G.H., Shuler, K.E., Lindenberg, K.: Order statistics for first passage times in diffusion processes. J. Stat. Phys. **31**, 255–278 (1983)

853. Weiss, M., Elsner, M., Kartberg, F., Nilsson, T.: Anomalous subdiffusion Is a measure for cytoplasmic crowding in living cells. Biophys. J. **87**, 3518–3524 (2004)

854. Welte, M.A.: Bidirectional transport along microtubules. Curr. Biol. **14**, R525-537 (2004)

855. White, J.A., Budde, T., Kay, A.R.: A bifurcation analysis of neuronal subthreshold oscillations. Biophys. J. **69**, 1203–1217 (1995)

856. White, J.A., Rubinstein, J.T., Kay, A.R.: Channel noise in neurons. Trends. Neurosci. **23**, 131–137 (2000)

857. Whittaker, G.R., Kann, M., Helenius, A.: Viral entry into the nucleus. Annu. Rev. Cell Dev. Biol. **16**, 627–651 (2000)

858. Wiley, H.S., Shvartsman, S.Y., Lauffenburger, D.A.: Computational modeling of the EGF-receptor system: a paradigm for systems biology. Trends Cell Biol. **13**, 43–50 (2003)

859. Williams, D.: Probability with Martingales. Cambridge University Press, Cambridge (1991)

860. Williams, G.S.B., Huertas, M.A., Sobie, E.A., Jafri, M.S., Smith, G.D.: A probability density approach to modeling local control of calcium-induced calcium release in cardiac myocytes. Biophys. J. **92**, 2311–2328 (2007)

861. Williams, G.S.B., Huertas, M.A., Sobie, E.A., Jafri, M.S., Smith, G.D.: Moment closure for local control models of calcium-induced calcium release in cardiac myocytes. Biophys. J. **95**, 1689–1703 (2008)

862. Winter, R.B., von Hippel, P.H.: Diffusion-driven mechanisms of protein translocation on nucleic acids. 1. Models and theory. Biochem. **20**, 6929–6948 (1981)

863. Wolpert, L.: Positional information and the spatial pattern of cellular differentiation. J. Theor. Biol. **25**, 1–47 (1969)

864. Wolpert, L.: Principles of Development. Oxford University Press, Oxford, UK (2006)

865. Wong, P., Gladney, S., Keasling, J.D.: Mathematical model of the lac operon: inducer exclusion, catabolite repression, and diauxic growth on glucose and lactose. Biotech. Prog. **13**, 132–143 (1997)

866. Wong, M.Y., Zhou, C., Shakiryanova, D., Lloyd, T.E., Deitcher, D.L., Levitan, E.S.Z.: Neuropeptide delivery to synapses by long-range vesicle circulation and sporadic capture. Cell **148**, 1029–1038 (2012)

867. Wu, Y., Han, B., Li, Y., Munro, E., Odde, D.J., Griffin, E.E.: Rapid diffusion-state switching underlies stable cytoplasmic gradients in the Caenorhabditis elegans zygote. Proc. Natl. Acad. Sci. USA **115**, E8440 (2018)

868. Xu, Y.Y., Zawadzki, K.A., Broach, J.R.: Single-cell observations reveal intermediate transcriptional silencing states. Mol. Cell **23**, 219–229 (2006)

869. Xue, C., Jameson, G.: Recent mathematical models of axonal transport. In: Holcman, D. (ed.) Stochastic Processes, Multiscale Modeling, and Numerical Methods for Computational Cellular Biology, pp. 265–285 (2017)

870. Yamada, Y., Peskin, C.: A look-ahead model for the elongation dynamics of transcription. Biophys. J. **96**, 3015–3031 (2009)

871. Yamamoto, E., Akimoto, T., Kalli, A.C., Yasuoka, K., Sansom, M.S.P.: Dynamic interactions between a membrane binding protein and lipids induce fluctuating diffusivity. Sci. Adv. **3**, e1601871 (2017)

872. Yang, Y.M., Austin, R.H., Cox, E.C.: Single molecule measurements of repressor protein 1D diffusion on DNA. Phys. Rev. Lett. **97**, 048302 (2006)

873. Yang, Y.J., Mai, D.J., Dursch, T.J., Olsen, B.D.: Nucleopore-inspired polymer hydrogels for selective biomolecular transport. Biomacromolecules **19**, 3905–3916 (2018)

874. Yao, Y., Choi, J., Parker, I.: Quantal puff of intracellular Ca^{2+} evoked by inositol triphosphate in Xenopus oocytes. J. Physiol. **482**, 533–553 (1995)

875. Yildirim, N., Mackey, M.C.: Feedback regulation in the lactose operon: a mathematical modeling study and comparison with experimental data. Biophys. J. **84**, 2841–2851 (2003)

876. Yildirim, N., Santillan, M., Horike, D., Mackey, M.C.: Dynamics and bistability in a reduced model of the lac operon. Chaos **14**, 279–292 (2004)

877. Yildiz, A., Tomishige, M., Vale, R.D., Selvin, P.R.: Kinesin walks hand-over-hand. Science **303**, 676–678 (2004)

878. Young, G.W.D., Keizer, J.: A single pool IP_3-receptor model for agonist stimulated Ca^{2+} oscillations. Proc. Natl. Acad. Sci. (USA) **89**, 9895–9899 (1992)

879. Yu, J., Moffitt, J., Hetherington, C.L., Bustamante, C., Oster, G.: Mechanochemistry of a viral DNA packaging motor. J. Mol. Biol. **400**, 186–203 (2010)

880. Yuste, S.B., Lindenberg, K.: Order statistics for first passage times in one-dimensional diffusion processes. J. Stat. Phys. **85**, 501–512 (1996)

881. Yuste, S.B., Acedo, L., Lindenberg, K.: Order statistics for d-dimensional diffusion processes. Phys. Rev. E **64**, 052102 (2001)

882. Yuste, R., Majewska, A., Holtho. K.: From form to function: calcium compartmentalization in dendritic spines. Nat. Neurosci. **3**, 653–659 (2000)

883. Yvert, G.: "Particle genetics?": treating every cell as unique. Trends Genet. **30**, 49–56 (2014)

884. Yvinec, R., Zhuge, C., Lei, J., Mackey, M.C.: Adiabatic reduction of a model of stochastic gene expression with jump Markov process. J. Math. Biol. **68**, 1051–1070 (2014)

885. Zahradnikova, A., Zahradnik, I.: A minimal gating model for the cardiac Ca^{2+} release channel. Biophys. J. **71**, 2996–3012 (1996)

886. Zeiser, S., Franz, U., Wittich, O., Liebscher, V.: Simulation of genetic networks modelled by piecewise deterministic Markov processes. IET Syst. Biol. **2**, 113–135 (2008)

887. Zeitz, M., Kierfeld, J.: Feedback mechanism for microtubule length regulation by stathmin gradients. Biophsy. J. **107**, 2860–2871 (2014)

888. Zelinski, B., Muller, N., Kierfeld, J.: Dynamics and length distribution of microtubules under force and confinement. Phys. Rev. E **86**, 041918 (2012)

889. Zhang, E.E., Kay, S.A.: Clocks not winding down: unravelling circadian networks. Nat. Rev. Mol. Cell Biol. **11**, 764–776 (2010)

890. Zhang, X.-J., Qian, H., Qian, M.: Stochastic theory of nonequilibrium steady states and its applications. Part I. Phys. Rep. **510**, 1–89 (2012)

891. Zhang, L., Radtke, K., Zheng, L., Cai, A.Q., Schilling, T.F., Nie, Q.: Noise drives sharpening of gene expression boundaries in the zebrafish hindbrain. Mol. Syst. Biol. **8**, 613 (2012)

892. Zhou, H.-X., Szabo, A.: Theory and simulation of stochastically-gated diffusion-influenced reactions. J. Phys. Chem. **100**, 2597–2604 (1996)

893. Zhou, H.X.: A model for the mediation of processivity of DNA-targeting proteins by nonspecific binding: dependence on DNA length and presence of obstacles. Biophys. J. **88**, 1608–1615 (2005)

894. Zhou, H.-X., Rivas, G., Minton, A.P.: Macromolecular crowding and confinement: biochemical, biophysical, and potential physiological consequences. Annu. Rev. Biophys. **37**, 375–397 (2008)

895. Zhou, T., Zhang, J., Yuan, Z., Chen, L.: Synchronization of genetic oscillators. Chaos **18**, 037126 (2008)

896. Zia, R.K.P., Dong, J.J., Schmittmann, B.: Modeling translation in protein synthesis with TASEP: a tutorial and recent developments. J. Stat. Phys. **144**, 405–428 (2011)

897. Zilman, A., Talia, S.D., Chait, B.T., Rout, M.P., Magnasco, M.O.: Efficiency, selectivity and robustness of nucleocytoplasmic transport. PLoS Comp. Biol. **3**, e125 (2007)

898. Ziman, J.M.: Models of Disorder. Cambridge University Press, Cambridge (1979)

899. Ziv, N.E., Brenner, N.: Synaptic tenacity or lack thereof: spontaneous remodeling of synapses. Trends Neurosci. **41**, 89–99 (2017)

900. Zmurchok, C., Small, T., Ward, M., Edelstein-Keshet, L.: Application of quasi-steady state methods to nonlinear models of intracellular transport by molecular motors. Bull. Math. Biol. **79**, 1923–1978 (2017)
901. Zuckerman, D.M., Woolf, T.B.: Efficient dynamic importance sampling of rare events in one dimension. Phys. Rev. E **63**, 016702 (2000)
902. Zumofen, G., Klafter, J., Blumen, A.: Enhanced diffusion in random velocity fields. Phys. Rev. A **42**, 4601–4608 (1990)
903. Zwanzig, R.: Rate processes with dynamical disorder. Acc. Chem. Res. **23**, 148–152 (1990)
904. Zwanzig, R.: Diffusion past an entropy barrier. J. Phys. Chem. **96**, 3926–3930 (1992)

Index

A

Abel's theorem, 157
Accumulation time, 389–391
Actin
  polymerization, 178–182
Active intracellular transport
  axonal transport, 512–519
  microtubular networks, 526–535
  quasi-steady-state analysis, 522–526, 528–529
  random intermittent search, 535–547
  slow axonal transport, 514
  velocity jump process, 519–535
  virus trafficking, 531–535
Active-repressor gene network, 261–264
Adenosine triphosphate (ATP), 208, 220
Anomalous diffusion, 361–380
  continuous-time random walks, 364–368
  homogenization theory, 369–374
  molecular crowding, 369–374
  percolation theory, 374–378
  random velocity field model, 531
  single-file diffusion, 479
  types of, 361–364
Antenna effect, 398
Arrhenius formula, 88
Asymptotic analysis, 24, 413–434, 437
Autocorrelation function, 53
Axon, 133
Axonal transport, 512–519

B

Backward Fokker–Planck equation, 63, 79
Backward master equation, 131
Bayes theorem, 10

Berg, Winter and von Hippel (BMW) model, 396–399
Bernoulli distribution, 34
Bicoid gradient, 333
Binomial distribution, 10, 34, 127
Birth-death processes
  autoregulatory gene network, 283
  cooperative molecular motors, 213
  first passage time, 131–132
  ion channels, 124–128
  linear noise approximation, 129
  master equation, 124–128
  path integral representation, 624–630
  system-size expansion, 128–131
  WKB approximation, 621–624
Bistability, 86–91, 140–142, 150, 243
Boltzmann constant, 3
Boltzmann–Gibbs distribution, 14–19, 88, 120
Boundary value problems, 68–72
Branching processes, 688–695
Brownian functionals, 655–662, 669–671
Brownian motion, 49–52
  extreme statistics, 434–439
  periodic potentials, 199–202
  stochastically-gated, 450–453
  stochastic resetting, 564–568
  switching diffusivities, 453–456
Brusselator, 285

C

Calcium
  calcium-induced calcium release, 143, 147–152, 435
  cardiac myocytes, 147

© Springer Nature Switzerland AG 2021

P. C. Bressloff, *Stochastic Processes in Cell Biology*, Interdisciplinary
Applied Mathematics 41, https://doi.org/10.1007/978-3-030-72515-0

Printed in the United States
by Baker & Taylor Publisher Services